ASTEROIDS, COMETS, METEORS 1993

INTERNATIONAL ASTRONOMICAL UNION

UNION ASTRONOMIQUE INTERNATIONALE

ASTEROIDS, COMETS, METEORS 1993

PROCEEDINGS OF THE 160TH SYMPOSIUM OF THE INTERNATIONAL ASTRONOMICAL UNION, HELD IN BELGIRATE, ITALY, JUNE 14–18, 1993

EDITED BY

A. MILANI

Dipartimento di Matematica,
Università di Pisa, Italy

and

M. DI MARTINO, A. CELLINO

Osservatorio Astronomico di Torino,
Pino Torinese, Italy

SPRINGER SCIENCE+BUSINESS MEDIA, B.V.

IAU

Library of Congress Cataloging-in-Publication Data

International Astronomical Union. Symposium (160th 1993
Belgirate, Italy)
Asteroids, comets, meteors 1993 proceedings of the 160th
International Astronomical Union, held in Belgirate., Italy, June
14-18, 1993 / edited by A. Milani, M. Di Martino, A. Cellino.
p. cm.
Includes index.
ISBN 978-0-7923-2881-0 ISBN 978-94-011-1148-5 (eBook)
DOI 10.1007/978-94-011-1148-5
1. Asteroids--Congresses. 2. Comets--Congresses. 3. Meteors--Congresses. I. Milani, Andrea. II. Di Martino, Michel.
III. Cellino, A. IV. Title.
QB651.I57 1993
523.4'4--dc20 94-16936

ISBN 978-0-7923-2881-0

Printed on acid-free paper

TABLE OF CONTENTS

DYNAMICS

PHYSICAL OBSERVATIONS AND MODELING

OBSERVATIONS FROM SPACE

ORIGIN AND EVOLUTION

DATA BASES

INTRODUCTION

THE MEETING

The IAU Symposium 160 *ASTEROIDS COMETS METEORS 1993* has been held at Villa Carlotta in Belgirate, on the shore of Lago Maggiore (Italy), from June 14 to June 18, 1993. It has been organized by the Astronomical Observatory of Torino and by the Lunar and Planetary Institute of Houston. It has been a very large meeting, with 323 registered participants from 38 countries. The scientific program included 29 invited reviews, 106 oral communications, and 215 posters. The subjects covered included all the aspects of the studies of the minor bodies of the solar system, including asteroids, comets, meteors, meteorites, interplanetary dust, with special focus on the interrelationships between these.

The meeting was structured as follows. 5 morning plenary sessions have been devoted to invited reviews on: (1) search programs (2) populations of small bodies (3) dynamics (4) physical observations and modelling (5) origin and evolution. Two afternoon plenary sessions have been devoted to space missions to small bodies and to interrelationships between the different populations. The afternoon parallel sessions have been devoted to: dynamics of comets; Toutatis, Ida, Gaspra; physical processes in cometary comae and tails; meteorites; the cosmogonic message from cometary nuclei; physics of asteroids; the interplanetary dust complex; comet nuclei; meteors; composition and material properties of comets; dynamics of asteroids.

Judging from the opinion expressed by many participants the meeting has been superb from the organization point of view, and very successful scientifically. We would like to take this opportunity to thank the members of the Scientific and Local Organizing Committees, but especially the man who was responsible for the overall organization of the meeting, Vincenzo Zappalà. Without him, all this would not have been possible.

HIGHLIGHTS

This meeting has been the last of a series of *Asteroids Comets Meteors* conferences held in Uppsala (1983, 1985, 1989) and in Flagstaff (1991); it also follows a number of important meetings devoted to more specific topics, like *Asteroids II* (Tucson, 1988), *Comets in the Post–Halley Era* (Bamberg, 1989), *Origin and Evolution of Interplanetary Dust* (Kyoto, 1990) and *Meteoroids and their Parent Bodies* (Bratislava, 1992). There is no doubt on the existence of a large community of dedicated scientists producing results on the different aspects of the small bodies; the ACM93 meeting has been one of the largest for attendance among the recent IAU Symposia. The question is, is this attendance and production of papers a manifestation of normal science (in the sense of Kuhn)? There have been, in the short time elapsed from all these other meetings, enough new discoveries and new ideas to justify such a large attendance and such a large number of papers?

Our opinion is that this meeting, apart from being a successful social event, was really necessary to spread a number of new ideas and information. The last few years have been especially interesting for the study of the minor bodies of the solar system. We cannot quote here all the new discoveries and conceptual achievements, which are recorded in the proceedings volumes, but we would like to stress two major changes occurred in our field very recently.

Until few years ago, the perception of this subject was that there are many different populations of small bodies, with such different properties and histories that their study required different technical and conceptual tools; as a result, the scientific communities studying asteroids, comets, meteors, dust, meteorites, and collisional phenomena were largely separated. We now understand that the exchanges and transitions between the objects responsible for such a variety of phenomena are the rule rather than the exception, e.g. an asteroid orbiting in the outer solar system can exhibit cometary activity if its orbit changes, a comet can become inactive and be called an asteroid, dust and meteoroids can have both asteroidal and cometary origin, craters of different sizes are excavated by any one of these populations, meteorites in the museums are samples from the asteroid belt, possibly after having been observed as meteors. All these interrelationships had been proposed long time ago, but now we have the theoretical and experimental evidence to prove their occurrence and to give quantitative estimates of their relevance. The discussion of these interrelationships and exchanges has been one of the focal points of ACM93, and this has also been the result of the choice done by the Scientific Organizing Committee, by which all the plenary sessions have included presentations on all the different populations of small bodies. This has been a significant change from what was customary before, and we have the impression it has been well received by the participants, who have had the possibility to appreciate the general scenario.

Another fundamental improvement occurred in this field in the last few years has been the enormous increase in the availability, accessibility and reliability of the relevant data. Until recently, not only the observational data were scanty, because the small bodies are dim and the necessarily powerful instruments are not easily available, but also the existing data were either not accessible or very difficult to use. This limitation is very serious in our field, since the main properties of the small bodies populations are statistical ones, and the detailed study of a few examples cannot give all the answers. It is therefore essential to have enough data collected in a systematic and unbiased way, to have them processed, to check their reliability by comparison with other data and theoretical models, and to archive all the available information where it can be accessed by the entire scientific community. In the last few years a major effort has been done to generate, maintain and make available much larger data sets. This is the result of many factors, including the space missions (Giotto, Vega, Galileo), the availability of new detector technology, and the progress in computer and networks, but also of the dedicated work of many scientists who have recognized the strategic value of such endeavour. As a result, the proceedings of this meeting contain many papers discussing the global, statistical properties of entire populations. In this volume we have done a special effort to contribute to the availability of homogeneous and reliable data sets on small bodies. The last section of this book is specifically devoted to the presentation of data bases, some of which have been made available to the entire community as a result of our specific request. The easy availability of these data sets will make research in our field a fully international enterprise, to which even scientists not belonging to the best equipped institutions can take part; it will make easier to fulfill the fundamental requirement of reproducibility of the results, by making possible the comparison of different data sets and the testing of the theories with homogeneous observational data.

We would like to quote some of the highlights of the meeting, of course without pretending to be exhaustive, but just providing some examples of the most important discoveries of the last 2–3 years. We have seen at ACM93 the images of the asteroid *951 Gaspra*, as observed by the Galileo spacecraft (after the meeting we have also received the images of *243 Ida*; the review paper by Chapman gives some preliminary results also on this second encounter). We have seen the images of the comet *Shoemaker–Levy 9* split into more than 20 pieces, and the radar images of *4179 Toutatis*. We have heard many conceptual achievements about the problem of the influx of cosmic material on the Earth, as well as data about the spectacular fireball seen over northern Italy on January 19, 1993 and about material recovered from the Tunguska impactor. We have finally begun to understand the entire process leading from the astrometric observations of asteroids, through the computation of proper elements and the identification of asteroid families, to the transport of meteoroids and the identification of a recovered meteorite with its parent body. We have heard about the surveys, which have been very successful in discovering new bodies, including some belonging to the formerly hypothetical Kuiper belt and some with the highest probability of interacting with the Earth. The subject of interplanetary dust, of its origin, dynamics, and optical properties has been widely discussed. New progress has been reported on the identification of the chemical and mineralogical composition for both asteroids and comets. We have even learned about the evidence for the existence of populations of small bodies (dust, asteroids and comets) around other stars.

We believe the ACM93 meeting has shown that the small bodies are a very active and rapidly changing field of research in planetary science, and that such large and interdisciplinary meetings are very productive.

THE PROCEEDINGS

For a large meeting like ACM93, the publication of the proceedings is an important but difficult task. To keep the size and complexity of the publications within acceptable limits, it was decided to split the proceedings into several separately edited volumes. The abstracts received before the meeting have been published in a volume edited by the Lunar and Planetary Institute (LPI Contribution no. 810), which was distributed to all participants at the meeting; it contains 329 abstracts. The invited reviews and the presentations of the data bases are collected in this volume, edited by Kluwer under the sponsorship of the IAU. The authors of contributed papers have been invited to submit their manuscripts to the journal *Planetary and Space Science*; they will be undergo the usual refereeing procedure of the journal, and will be published in a number of special issues (we understand from the editor in chief, M. Coradini, and from the subject editor on small bodies, M. Fulchignoni, that more than 100 papers have been submitted). These special issues of *PPS* will appear by the end of the year 1994.

As for this volume, we had originally planned to include all the invited review papers presented in the 5 morning sessions of the meeting, plus the data base presentations. This implies that originally the list of authors was chosen by the Scientific Organizing Committee. With respect to this original list, we have been forced to introduce some changes. Some of the invited speakers could not attend the meeting; some others presented a talk, but did not submit a paper. We tried to replace at least some of these missing papers with review articles written by other recognized specialists who had presented contributed papers on the same topics. In some cases, we have asked the authors of related reviews to slightly extend the scope of their paper to cover the missing topics. We have also added a paper on the space mission *Rosetta*, which was presented at the meeting by several speakers, and was later (November 1993) formally accepted by the European Space Agency as the next *cornerstone* mission.

We are aware that it is not possible to cover in a single volume all the aspects of the small bodies science; we understand that some omissions could be criticized, but we have done our best to give the most comprehensive possible survey of the state of the art. Each paper published in this book has been reviewed by at least two referees, and the content as published is the result of a debate between at least four people, including also some of the editors. However, we acknowledge that in many cases the points of view presented in the papers of this book reflect the opinion of the authors, and are not always universally accepted. We could have insisted more, and put more pressure upon the authors, to force them to give a more balanced account of the current debate on the many controversial issues. We have decided on the contrary to allow the authors to express themselves, and we have accepted in some cases the presentation of theories and interpretations which are not well established. On the other hand, the goal of this book –as well as the goal of the review sessions at the meeting– was to give an overview of the research going on at present, and to open a debate on the current ideas, rather than obtaining a perfectly "objective", and frozen, view of perfectly established results. The invited review speakers have been chosen among the most active in their field in the last 2–3 years; this guarantees that the work presented is up to date, but necessarily implies that the authors may emphasize their own point of view and their own work.

Andrea Milani, Mario Di Martino, Alberto Cellino

Scientific Organizing Committee

V. Zappalà, Chairman (Italy)
E. Bowell (USA)
A. Brahic (France)
A. Carusi (Italy)
Z. Ceplecha (Czech Republic)
M. Coradini (Italy)
D.R. Davis (USA)
P. Farinella (Italy)
M. Festou (France)
Cl. Froeschlé (France)
A.W. Harris (USA)
J. Henrard (Belgium)
Y. Kozai (Japan)
C.-I. Lagerkvist (Sweden)
A.-C. Levasseur Regourd (France)
B.A. Lindblad (Sweden)
D.F. Lupishko (Ukraine)
B.G. Marsden (USA)
A. Milani (Italy)
J. Rahe (USA)
H. Rickman (Sweden)
E. Shoemaker (USA)
A. Sokolsky (Russia)
G. Valsecchi (Italy)
I.P. Williams (UK)
P.R. Weissman (USA)

Local Organizing Committee

M. Di Martino (Chairman)
C. Blanco
C. Casacci
A. Cellino
G. De Sanctis
P. Jones
A. Manara

ACKNOWLEDGEMENTS

The symposium was sponsored by IAU Commission 15 (Physical Study of Comets, Minor Planets and Meteorites) and cosponsored by Commissions 7 (Celestial Mechanics), 20 (Positions and Motions of Minor Planets, Comets and Satellites), 21 (Light of the Night Sky), 22 (Meteors and Interplanetary Dust).

We thank the President of the Italian Republic for having given his High Patronage to the symposium.

We are most indebted to several organizations and agencies for financial and other support and/or sponsorship. These are the Regione Piemonte, the Provincia di Novara, the European Space Agency, the Alenia Spazio S.p.A., the Consiglio Nazionale delle Ricerche (C.N.R.) and the Gruppo Nazionale di Astronomia of the C.N.R., the International Science Foundation, and the Banca Popolare di Novara. We also wish to express our thanks to the Azienda Promozione Turistica del Lago Maggiore, to the Camera di Commercio di Novara and Martini & Rossi IVLAS S.p.A.

We especially thank Pam Jones and .the members of the Program Services Department of the Lunar and Planetary Institute (LPI) for their skilled help in organizing the meeting and to the people of Publications Services Department of LPI for the production of the abstract volume.

It is also a pleasure to thank the Local Organizing Committee as well as many individuals who worked hard to make this meeting a success. In particular Patrizia Santoro, Mara Marini, Giuseppe Minichino of the Osservatorio Astronomico di Torino and Maddalena Armando, Jeannine Casarini, Carla and Paola Mandelli, Barbara Zappalà, for their work at the reception desk and for conducting tours in the Lago Maggiore area, and E. Cozzi, S. Foglia and E. Radice for technical support.

Finally, we appreciated very much the work of the staff of the hotels Villa Carlotta and Milano in Belgirate.

LIST OF PARTICIPANTS

ADOLFSSON L. G.
University of Florida
Dept. of Astronomy 211 SSRB
Gainesville, FL 32611, U.S.A.
phone : 904 392 7677; fax : 904 392 5089
e-mail : adolfsso@astro.ufl.edu

AIKMAN C.
National Research Council
5071, West Saanich Road
Victoria, BC V8X 4M6, CANADA
phone : 604 388 0008; fax : 604 363 0045
e-mail : aikman@dao.nrc.ca

AKISAWA H.
Himeji City Astron. Observatory
Hoshinoko Yakata, 1470-24 Aoyama
Himeji City 671-22, JAPAN
phone : 792 67 3050; fax : 792 67 3055

AKSNES K.
Institute of Theoretical Astrophysics
P.O. Box 1029
Blindern, 0315 Oslo, NORWAY
phone : 22 856515; fax : 22 856505
e-mail : rolf::aksnes

ALLEN M. S.
National Research Council, HA 584
2101, Constitution Avenue
Washington, DC 20418, U.S.A.
phone : 202 334 3477

ALMÁR I.
Konkoly Observatory, P.O. Box 67
1525 Budapest, HUNGARY
fax : 1 1569 640
e-mail : pds@astro.rmki.kfki.hu

ANDRE C.
Observatoire Midi Pyrenées
14, Avenue E. Belin
31400 Toulouse, FRANCE
e-mail : omptob::prieur

ANGELI C.
EUROPA, Observatoire de Paris
5, Place Janssen
92195 Meudon, Principal Cedex, FRANCE
e-mail : 17733::angeli

ARPIGNY C.
Université de Liege
Institute d'Astrophysique
4000 Liege Ougree, BELGIUM
phone : 41 529980; fax : 41 527474
e-mail : u2182ca@bliulg11

ASPHAUG E.
University of Arizona
Lunar & Planetary Laboratory
Tucson, AZ 85721, U.S.A.
phone : 602 621 2773
e-mail : asphaug@hindmost.lpl.arizona.edu

BABADZHANOV P.B.
Institute of Astrophysics
Tajik SSR Academy of Science
22, Sviridenko Str.
Dushanbe, 734670, TAJIKISTAN
phone : 3772 23 14 32

BAGGALEY J.
University of Canterbury
Dept. of Physics and Astronomy
Christchurch, NEW ZELAND
fax : 0742 72 8079
e-mail : phys@csc.canterbury.ac.nz

BAGUHL M.
Max Planck Institute für Kernphysik
Saupfercheckwega post. 103980
6900 Heidelberg, GERMANY
phone : 6221 516543; fax : 6221 516540
e-mail : dusty::baguhl

BAILEY M.
Liverpool John Moores University
School of Comp. & Math. Science, Byrom Street
Liverpool L3 3AF, U.K.
fax : 051 207 4596
e-mail : istmbailey@livjm.uk.ac

BAKATSELOS T.
5, Argonafton Str.
Argiroupoli, 16452, Athens, GREECE

BARONI S.
via Ciconi, 8
20100 Milano, ITALY
phone : 2 4121429

BARUCCI A.
EUROPA, Observatoire de Paris
5, Place Janssen
92195 Meudon, Principal Cedex, FRANCE
phone : 1 4507 7775; fax : 1 4507 7469
e-mail : barucci@frmeu51.bitnet

BATALLI COSMOVICI C.
IFSI, CNR, Via G. Galilei, C.P. 27
00044 Frascati, (Roma) ITALY
phone : 6 9426814; fax : 6 94186230
e-mail : cosmo@hp.ifsi.fra.cnr.it

BEC-BORSENBERGER A.
Bureau des Longitudes
77, Avenue Denfert Rocherau
75014 Paris, FRANCE
phone : 1 4052 2273; fax : 1 4633 2834
e-mail : bdlmca::borsenberger

BECK P.
Max Planck Institute für Kernphysik
postfach 103980
6900 Heidelberg, GERMANY
phone : 6221 380651
e-mail : kosmo::kissel

BELLUCCI G.
IFSI, CNR, via G. Galilei, C.P. 27
00044 Frascati, (Roma), ITALY
phone : 6 94186217

BELSKAYA I.
Astronomical Observatory
Sumskaya Str., 35
Kharkov 310022, UKRAINE
phone : 43 24 28

BELTON M. J.
Kitt Peak Natl. Obs., Cherry Avenue
Tucson, AZ 85719, U.S.A.
phone : 602 326 5511; fax : 602 325 9360
e-mail : belton@noao.edu

BEVILACQUA F.
Alenia Spazio S.p.A.
Corso Marche, 41
10146 Torino, ITALY
phone : 11 725089; fax : 11 723307

BINZEL R. P.
Massachusetts Institute of Technology
Dept. of Earth, Atmospheric and Plan. Sciences
Cambridge, MA 02139, U.S.A.
phone : 617 253 6486; fax : 617 253 2886
e-mail : rpb@astron.mit.edu

BIRLAN M.
Astr. Institute of the Romanian Academy
Str. Cutitul de Argint, 5
75212 Bucaresti, Sect 4, ROMANIA
phone : 401 31 23391; fax : 401 31 23391

BLANCO C.
Università di Catania
Istituto di Astronomia, viale A. Doria, 6
95125 Catania, ITALY
phone : 95 7332245; fax : 95 330592
e-mail : cblanco@astrct.cineca.it

BOCKELÉE-MORVAN D.
Observatoire de Paris
5, Place Janssen
92195 Meudon, Principal Cedex, FRANCE
phone : 1 4507 7605; fax : 1 4507 7939
e-mail : domi@frmeu51.bitnet

BONEV T.
Bulgarian Academy of Sciences
Dept. of Astronomy
Sofia, BULGARIA
phone : 5556 401 293; fax : 5556 401 240
e-mail : jockers@linmpi.dnet.gwdg

BOROVIČKA J.
Astronomical Institute
251 65 Ondřejov, CZCH REPUBLIC
phone : 204 857153; fax : 204 85314
e-mail : astmph@csearn.bitnet

BOSWELL J.
Sheffield University, Dept. of Physics
Sheffield S3 7RH, U.K.

BOWELL E.
Lowell Observatory
1400 West Mars Hill Road
Flagstaff, AZ 86001, U.S.A.
phone : 602 774 3358; fax : 602 774 6296
e-mail : elgb@lowell.edu

BRANDT J. C.
University of Colorado
LASP, Campus Box 392
Boulder, CO 80309-0392, U.S.A.
phone : 303 492 3215; fax : 303 492 6946

BURBINE T. H.
Smithsonian Astrophysical Observatory
60, Garden Street
Cambridge, MA 02138, U.S.A.
phone : 617 495 7041
e-mail : cfa::burbine

BURCHI R.
Osservatorio Astronomico di Collurania
64100 Teramo, ITALY
phone : 861 210490; fax : 861 210492
e-mail : burchi@astrte.astro.it

BUSSOLETTI E.
Istituto Universitario Navale
Istituto di Fisica
via Acton, 38
80133 Napoli, ITALY
phone : 81 5475601; fax : 81 5519957
e-mail : 40771::bussoletti

CAMPINS U.
University of Florida
Dept. of Astronomy 211 SSRB
Gainesville, FL 32611, U.S.A.
phone : 904 392 3066; fax : 904 392 5089
e-mail : campins@astro.ufl.edu

CAPACCIONI F.
IAS, Reparto di Planetologia
viale Università, 11
00185 Roma, ITALY
phone : 6 4456951; fax : 6 4454969

CAPRIA M. T.
IAS, CNR, Via G. Galilei, C.P. 67
00022 Frascati, (Roma) ITALY
fax : 6 2252969

CARPINO M.
Osservatorio Astronomico di Brera
via Brera, 28
20121 Milano, ITALY
phone : 2 874444; fax : 2 72001600
e-mail : carpino@bach.mi.astro.it

CARUSI A.
IAS, Reparto di Planetologia
viale Università, 11
00185 Roma, ITALY
phone : 6 4456951; fax : 6 4454969
e-mail : carusi@itcaspur.bitnet

CARUSO A.
via T. Nuvolari, 219
00142 Roma, ITALY
phone : 6 5037406

CASACCI C.
Alenia Spazio S.p.A.
Corso Marche, 41
10146 Torino, ITALY
phone : 11 7180385; fax : 11 723307

CELLINO A.
Osservatorio Astronomico di Torino
10025 Pino Torinese, (Torino), ITALY
phone : 11 4619033; fax : 11 4619030
e-mail : cellino@to.astro.it

CEPLECHA Z.
Ondřejov Observatory
251 65 Ondřejov, CZECH REPUBLIC
phone : 204 857153; fax : 204 85314
e-mail : ceplecha@asu.cas.cs

CEVOLANI G.
FISBAT, CNR, via dé Castagnoli, 1
40126 Bologna, ITALY
phone : 51 6399604

CHAMBERLIN A. B.
University of California, San Diego
California Space Institute
La Jolla, CA 92093-0216, U.S.A.
phone : 619 534 3915; fax : 619 534 2674

CHAPMAN C. R.
Planetary Science Institute
620, North 6th Avenue
Tucson, AZ 85705 8331, U.S.A.
phone : 602 622 6300; fax : 602 622 8060
e-mail : chapman@psi.edu

CHEN D.-h.
Purple Mountain Observatory
Academia Sinica
Najing, CHINA
fax : 86 25 301459

CHENG A.
Applied Physics Laboratory
11100, John Hopkins road
Lauren, MD 20723-6099, U.S.A.

CHERNOVA G. P.
Institute of Astrophysics
Dushanbe 734670, TAJIKISTAN
phone : 27 37 93

CHURYMOV K. I.
Kiev University
Observatornaya 3
Kiev 254053, UKRAINE
phone : 044 266 4517

CLAIREMIDI J.
Observatoire de Besançon, BP 1615
25000 Besançon, Cedex, FRANCE
phone : 81 666915; fax : 81 666944
e-mail : earn:frobes51

CLARK B.
Planetary Geosciences Division
2525, Correa Road
Honolulu, Hi 96822, U.S.A.
phone : 808 956 3160; fax : 808 956 6322
e-mail : beth@kanana.pgd.hawaii.edu

COCHRAN A. L.
McDonald Observatory
Austin, TX 78712-1083, U.S.A.
phone : 512 471 1471; fax : 512 471 6016
e-mail : anita@zinfandel.as.utexas.edu

COCHRAN W.
McDonald Observatory
Austin, TX 78712, U.S.A.
phone : 512 471 6474; fax : 512 471 6016
e-mail : wdc@shiraz.as.utexas.edu

COLANGELI L.
Università di Cassino
Dip. di Ingegneria, via Zamosch, 43
03043 Cassino, (Frosinone), ITALY
fax : 776 310609
e-mail : colangeli@oaclan.astro.it

COLOM P.
Observatoire de Paris
5, Place Janssen
92195 Meudon, Principal Cedex, FRANCE
e-mail : colom@frmeu51.bitnet

COLWELL J. E.
University of Colorado, LASP, Campus Box 392
Boulder, CO 80309-0392, U.S.A.
phone : 303 492 6805; fax : 303 492 6946
e-mail : colwell@sargon.colorado.edu

CORADINI A.
IAS, Reparto di Planetologia
viale Università, 11
00185 Roma, ITALY
phone : 6 4456951; fax : 6 4454969
e-mail : angio@itcaspur.bitnet

CORADINI M.
European Space Agency
8-10, rue Mario Nikis
75738 Paris, Cedex 15, FRANCE
phone : 1 42737555; fax : 1 42737236
e-mail : mcoradin@esa

COZZI E.
via Borghi, 14
22076 Mozzate, (Como), ITALY
phone : 331 830704

CREMONESE G.
Università di Padova, Dip. di Astronomia
vicolo Osservatorio, 5
35122 Padova, ITALY
phone : 49 8293458; fax : 49 8799840
e-mail : cremonese@astrpd.astro.it

CRIFO J. F.
Service d'Aéronomie du CNRS, BP3
91371 Verrières le Buisson, Cedex, FRANCE
phone : 1 164474345; fax : 1 69202999
e-mail : savera::crifo

CROVISIER J.
Observatoire de Paris, 5, Place Janssen
92195 Meudon, Principal Cedex, FRANCE
phone : 1 4507 7599
e-mail : crovisie@mesioa.obspm.circe.fr

DAHLGREN M.
Astronomiska observatoriet, Box 515
Uppsala 75120, SWEDEN
phone : 18 525724; fax : 18 527583
e-mail : md@laban.astro.uu.se

DAVIS D. R.
Planetary Science Institute
620, North 6th Avenue
Tucson, AZ 85705 8331, U.S.A.
phone : 602 622 6300; fax : 602 622 8060
e-mail : drd@psi.edu

DE ANGELIS G.
ESTEC, ESA
Space Science Division, Kepleraan 1
2200 Ah Nordwijk, THE NETHERLANDS
phone : 1719 85265; fax : 1719 84697
e-mail : estsao::gdeangel

DEBEHOGNE H.
Observatoire Royal de Belgique
Avenue Circulaire, 3
1180 Bruxelles, BELGIUM
phone : 2 3730227; fax : 2 3749822
e-mail : henri@astro.oma.be

DELSEMME A. H.
University of Toledo
2509, Meadow-Wood Drive
Toledo, OH 43606, U.S.A.
fax : 419 536 2133

DE MEIS S.
via Tortona, 76
20144 Milano, ITALY
phone : 2 4233946

DERMOTT S. F.
University of Florida
Dept. of Astronomy 211 SSRB
Gainesville, FL 32611, U.S.A.
phone : 904 3935089; fax : 904 3925089
e-mail : dermott@astro.ufl.edu

DE SANCTIS G.
Osservatorio Astronomico di Torino
10025 Pino Torinese, (Torino), ITALY
phone : 11 4619033; fax : 11 4619030
e-mail : desanctis@to.astro.it

DE SANCTIS M. C.
Istituto Astronomico
via Lancisi, 29
00161 Roma, ITALY
phone : 6 4406327; fax : 6 4403673
e-mail : 40058::desanctis

DESPOIS D.
Observatoire de Bordeaux, BP 89
33270 Floirac, FRANCE
phone : 56 864330; fax : 56 404251
e-mail : despois@frobor51.bitnet

DIKOVA S.
Bulgarian Academy of Sciences
Dept. of Astronomy
Sofia, BULGARIA
e-mail : planet@bgearn.bitnet

DI MARTINO M.
Osservatorio Astronomico di Torino
10025 Pino Torinese, (Torino), ITALY
phone : 11 4619035; fax : 11 4619030
e-mail : dimartino@to.astro.it

DINEV C.
Bulgarian Academy of Sciences
Dept. of Astronomy
Sofia, BULGARIA
e-mail : planet@bgearn.bitnet

DONNISON J. R.
University of London, Goldsmith's College
Dept. of Math., New Cross
London SE14 6NW, U.K.
e-mail : maa01jrd@uk.ac.gold.scorpio

DOTTO E.
Istituto Astronomico
via Lancisi, 29
00161 Roma, ITALY
phone : 6 4406327; fax : 6 4403673
e-mail : 40058::dotto

DROBYSHEVSKI E. M.
Academy of Sciences of the USSR
A.F. Ioffe Physico-Tech. Institute
194021 St. Petersburg, RUSSIA
phone : 812 2479338; fax : 812 2472135
e-mail : emdrob@drob.pti.spb.su

DUNHAM D.
Applied Physics Laboratory
11100, John Hopkins road
Lauren, MD 20723-6099, U.S.A.

DURDA D. D.
University of Florida
Dept. of Astronomy 211 SSRB
Gainesville, FL 32611, U.S.A.
phone : 904 392 2052; fax : 904 392 5089
e-mail : durda@astro.ufl.edu

EBERHARDT P.
University of Bern, Physics Institute
Sidlerstrasse, 5
3012 Bern, SWITZERLAND
phone : 31 654404; fax : 31 654405
e-mail : eberhardt@phim.unibe.ch

ELST E. W.
Observatoire Royal de Uccle
Ringlaan, 3
1180 Uccle, BELGIUM
phone : 02 3730309

EMEL'YANENKO V. V.
Technical University, Dept. of Math.
2, Lenina 76
Chelyabinsk 454080, RUSSIA
phone : 351 2 399465; fax : 351 2 347408

ERIKSON A.
Astronomiska observatoriet, Box 515
Uppsala 75120, SWEDEN
phone : 18 530816; fax : 18 527583
e-mail : ae@pluto.astro.uu.se

FARINELLA P.
Università di Pisa, Dip. di Matematica
via Buonarroti, 2
56127 Pisa, ITALY
phone : 50 599554; fax : 50 599524
e-mail : twin2@vm.cnuce.cnr.it

FEDERICO C.
IAS, Reparto di Planetologia
viale Università, 11
00185 Roma, ITALY
phone : 6 4456951; fax : 6 4454969

FELDMAN P. D.
John Hopkins University
Dept. of Physics and Astronomy
Baltimore, MD 21218, U.S.A.
phone : 410 5167339; fax : 410 5165494
e-mail : pdf@casa.pha.jhu.edu

FERRERI W.
Osservatorio Astronomico di Torino
10025 Pino Torinese, (Torino), ITALY
phone : 11 4619025; fax : 11 4619030
e-mail : ferreri@to.astro.it

FERNÁNDEZ J. A.
Facultad de Ciencias
Dept. de Astronomia
Tristán Narvaja, 1674
11200 Montevideo, URUGUAY
phone : 2 418004; fax : 2 409973

FERRAZ-MELLO S.
Univ. de Sao Paulo
Dept. de Astronomia
Caixa Post. 30627
01051 Sao Paulo, BRASIL
e-mail : sylvio@vax.iagusp.usp.br

FLYNN G. J.
SUNY at Plattsburgh
Dept. of Physics
Plattsburgh, NY 12901, U.S.A.
phone : 518 564 3156; fax : 518 564 3152
e-mail : oneilna@snyplava.bitnet

FOGLIA S.
via F. Bisleri, 11
20148 Milano, ITALY
phone : 2 4076001

FOMENKOVA M. N.
University of California, San Diego
California Space Institute
La Jolla, CA 92093-0216, U.S.A.
phone : 619 534 0921; fax : 619 534 5306
e-mail : marifo@macvax.ucsd.edu

FORTI G.
Osservatorio di Arcetri
Largo E. Fermi, 5
50125 Firenze, ITALY
phone : 55 2752236; fax : 55 220039
e-mail : forti@arcetri.astro.it

FORZONI-ACCOLTI E.
via degli Alpini, 54A
17025 Loano, (Savona), ITALY

FROESCHLÉ Ch.
OCA, Observatoire de Nice, BP 229
06304 Nice, Cedex 4, FRANCE
phone : 92003036; fax : 92003033
e-mail : froesch@rameau.obs-nice.fr

FROESCHLÉ Cl.
OCA, Observatoire de Nice, BP 229
06304 Nice, Cedex 4, FRANCE
phone : 92003036; fax : 92003033
e-mail : claude@rameau.obs-nice.fr

FULCHIGNONI M.
Université de Paris VII
EUROPA, Observatoire de Paris
5, Place Janssen
92195 Meudon, Principal Cedex, FRANCE
phone : 1 45077539; fax : 1 45077469
e-mail : fulchignoni@frmeu51.bitnet

FULLE M.
Osservatorio Astronomico di Trieste
via Tiepolo, 11
34131 Trieste, ITALY
phone : 40 3199221; fax : 40 309418
e-mail : fulle@astrts.astro.it

GAJDOŠ Š.
Comenius University
Dept. of Astronomy
Mlynská Dolina
842-15 Bratislava, SLOVAK REPUBLIC
phone : 7 720003; fax : 7 725882
e-mail : gajdos@mff.uniba.cs

GALLI M.
Università di Bologna
Dipartimento di Fisica
via Irnerio, 46
40126 Bologna, ITALY
phone : 51 351001; fax : 51 247244

GALLINO R.
Università di Torino
Istituto di Fisica Generale
via P. Giuria, 1
10125 Torino, *ITALY*
phone : 11 6707448; fax : 11 6699579
e-mail : gallino@to.astro.it

GAMMELGAARD P.
Institute of Astronomy
8000 Aarhus C, *DENMARK*
phone : 8 6128899; fax : 8 6120740
e-mail : pg@obs.aau.dk

GEBHARD J.
Max Planck Institute für Kernphysik
Saupfercheckwega post. 103980
6900 Heidelberg, *GERMANY*
phone : 6221 516543; fax : 6221 516540
GIBLIN I.
University of Sussex, School of MaPS
Brighton BN1 9QH, *U.K.*
phone : 273 686391; fax : 273 678097
e-mail : mpfb7@central.sussex.ac.uk
GIL HUTTON R.
Obs. Astr. Felix Aguilar, Av. Benavidez 8175
5413 Chimbas, San Juan, *ARGENTINA*
phone : 64 231494; fax : 64 214586
e-mail : rgh@unsjfa.edu.ar
GOFFIN E.
Aartselaarstraat, 14
2660 Hoboken, *BELGIUM*
GOIDET B.
Service d' Aéronomie
Avenue de l'Observatoire, BP 1615
25010 Besançon, Cedex, *FRANCE*
e-mail : goidet@frobes51
GREEN S.
University of Kent
Unit for Space Sciences
Canterbury CT2 7NR, *U.K.*
phone : 227 764000; fax : 227 762616
GREENBERG J. M.
Huygens Laboratory
Nielsbohrweg 2, P.O. Box 9504
2300, Leiden, *THE NETHERLANDS*
phone : 71 275894; fax : 71 275819
e-mail : mayo@hlerul51
GRUDZIŃSKA S.
Astronomical Observatory
Ul. Chopina 12/18
87-100 Toruń, *POLAND*
e-mail : iastlibr@pltumkii
GRÜN E.
Max Planck Institute für Kernphysik
Saupfercheckwega post. 103980
6900 Heidelberg, *GERMANY*
phone : 6221 515478; fax : 6221 516324
e-mail : gruen@dusty.mpi-hd.mpg.de
GUAITA C.
via Francesco Baracca, 6
20100 Milano, *ITALY*
GUSTAFSON B.A.S.
University of Florida
Dept. of Astronomy 211 SSRB
Gainesville, FL 32611, *U.S.A.*
phone : 904 392 2052; fax : 904 392 5089
e-mail : gustaf@astro.ufl.edu
HADAMCIK E.
23, rue C. Fourier
75013 Paris, *FRANCE*
HAHN G.
Astronomiska observatoriet, Box 515
Uppsala 75120, *SWEDEN*
phone : 18 530816; fax : 18 527583
e-mail : laban::gerhard
HAJDUK A.
Astronomical Institute
Slovak Academy of Sciences
842-28 Bratislava, *SLOVAK REPUBLIC*
HAJDUKOVÁ M.
Comenius University, Astronomical Institute
842-28 Bratislava, *SLOVAK REPUBLIC*
HANNER M. S.
Jet Propulsion Laboratory, MS 183-601
Pasadena, CA 91109, *U.S.A.*
phone : 818 354 4100; fax : 818 393 4605
e-mail : msh@jplsc8.dnet.nasa.gov
HARRIS A. W.
Jet Propulsion Laboratory, MS 183-601
Pasadena, CA 91109, *U.S.A.*
phone : 818 956 8429; fax : 818 354 0966
e-mail : awharris@128.149.121.2
HARRIS N.
Sheffield University, Dept. of Physics
Sheffield S3 7RH, *U.K.*
phone : 742 768555; fax : 742 728079
e-mail : phznwh@uk.ac.sheffield.sunc
HARUYAMA J.-i
Institute of Space and Astron. Sciences
Mizutani Lab., 3-1-1 Yoshinodai
Sagamihara, Kanagawa 229, *JAPAN*
phone : 0427 59 4237
e-mail : haruyama@planetb.sci.isas.ac.jp
HE M.-f
Shanghai Observatory
Chinese Academy of Sciences
80, Nan Dan road
Shanghai, *CHINA*
phone : 021 4386191; fax : 021 4384618

HECHLER M.
ESA, ESOC, MAO
Robert Bosch Str., 5
Darmstadt, *GERMANY*
fax : 6151 902625
e-mail : mhechler@esoc

HELIN E. F.
Jet Propulsion Laboratory, MS 183-601
Pasadena, CA 91109, *U.S.A.*
phone : 818 354 4606; fax : 818 354 0966
e-mail : mipl3::efho51

HENRARD J.
Dept. de Mathematique, FUNDP
8, Rempart de la Vierge
5000 Namur, *BELGIUM*
phone : 81 724903; fax : 81 724914
e-mail : jhenrard@math.fundp.ac.be

HESTROFFER D.
Bureau des Longitudes
77, Avenue Denfert Rocherau
75014 Paris, *FRANCE*

HIRAYAMA T.
National Astronomical Observatory
Mitaka Tokyo 181, *JAPAN*
e-mail : ohiraya@c1.mtk.nao.ac.jp

HOBAN S.
NASA Goddard Space Flight Center, Code 693
Greenbelt, MD 20771, *U.S.A.*
phone : 301 286 3840; fax : 301 286 1629
e-mail : hoban@lepb93.gsfc.nasa.gov

HOFFMANN M.
University of Arizona
Lunar & Planetary Laboratory
Tucson, AZ 85721, *U.S.A.*
phone : 602 621 2692; fax : 602 621 4933
e-mail : hoffmann@panda.lpl.arizona.edu

HOWELL E.
University of Arizona
Lunar & Planetary Laboratory
Tucson, AZ 85721, *U.S.A.*
phone : 602 621 6950; fax : 602 621 4933
e-mail : ebus@lpl.arizona.edu

HUEBNER W.
Southwest Research Institute
Space Science Division, 6220, Culebra Road
San Antonio, TX 78228-0510, *U.S.A.*
phone : 512 522 2730; fax : 512 647 4325
e-mail : walter@swri.space.swri.edu

HUGHES D. W.
Sheffield University, Dept. of Physics
Sheffield S3 7RH, *U.K.*
phone : 742 768555; fax : 742 728079
e-mail : phznwh@uk.ac.sheffield.sunc

IBADOV S.
Institute of Astrophysics
Tajik SSR Acad. Science, 22 Sviridenko Str.
Dushanbe 734670, *TAJIKISTAN*
phone : 83772 225466

IPATOV S. I.
M. V. Keldysh Institute of Appl. Math.
Russian Acad. of Science, Miusskaya Sq., 4
Moscow 125047, *RUSSIA*
phone : 095 250 7833; fax : 095 972 0737
e-mail : user@applmat.msk.su

ISHIMOTO H.
Kobe University, Dept. of Earth Science
Kobe 657, *JAPAN*
phone : 78 881 1212; fax : 78 882 1549
e-mail : isimoto@icluna.kobe-u.ac.jp

IVANOVA V.
Bulgarian Academy of Sciences
Dept. of Astronomy
Sofia, *BULGARIA*
phone : 75 88 27; fax : 75 89 27
e-mail : planet@bgearn.bitnet

JACKSON W. M.
University of California, Dept. of Chemistry
Davis, CA 95616, *U.S.A.*
phone : 916 752 6310; fax : 916 752 8995
e-mail : wmjackson@ucdavis.edu

JÄMSÄ S.
University of Helsinki, Obs. PL 14
00014 Helsinki, *FINLAND*
phone : 0 1912943; fax : 0 1912952
e-mail : sauli.jamsa@esc.fi

JAYARAMAN S.
University of Florida
Dept. of Astronomy, 211 SSRB
Gainesville, FL 32611, *U.S.A.*
phone : 904 392 7678; fax : 904 392 5089
e-mail : sumita@astro.ufl.edu

JESSBERGER E. K.
Max Planck Institute für Kernphysik
Saupfercheckwega post. 103980
6900 Heidelberg, *GERMANY*
phone : 6221 516240; fax : 6221 516540

JOCKERS K.
Max Planck Institute für Aeronomie
post. 20, Max Planck Str., 2
3411 Katlenburg-Lindau, *GERMANY*
phone : 5556 401293
e-mail : jockers@linmpi.dnet.gwdg.de

JONES J.
University of Western Ontario, Physics Dept.
London, Ontario N6A 3K7, *CANADA*
phone : 519 661 3283; fax : 519 661 2033
e-mail : jimjones@uwo.ca

JONES W.
Sheffield University, Dept. of Physics
Sheffield S3 7RH, *U.K.*
phone : 742 824279; fax : 742 728079

JORDA L.
Observatoire de Paris, 5, Place Janssen
92195 Meudon, Principal Cedex, *FRANCE*
phone : 1 4507 7579; fax : 1 4507 7939
e-mail : jorda@frmeu51.bitnet

KELEMEN J.
Konkoly Observatory
1525 Budapest POB 67, *HUNGARY*
fax : 1 1569 640
e-mail : pds@astro.rmki.kfki.hu

KELLER H. U.
Max Planck Institute für Aeronomie
post. 20, Max Planck Str., 2
3411 Katlenburg-Lindau, *GERMANY*
phone : 5556 401419; fax : 5556 401141
e-mail : linmpi::keller

KIM S. J.
University of Maryland, Astron. Program
College Park, MD 20742, *U.S.A.*
phone : 301 405 1546; fax : 301 513 1608
e-mail : ndadsa::sang

KISELEV N. N.
Max Planck Institute für Aeronomie
post. 20, Max Planck Str., 2
3411 Katlenburg-Lindau, *GERMANY*
phone : 5556 401394; fax : 5556 401240
e-mail : kiselev@linmpi.dnet.gwdg.de

KISSEL J.
Max Planck Institute für Kernphysik
Saupfercheckwega post. 103980
6900 Heidelberg, *GERMANY*
phone : 6221 516479; fax : 6221 380651
e-mail : kosmo::kissel

KNEŽEVIĆ Z.
Astronomical Observatory, Volgina, 7
11050 Beograd, *YUGOSLAVIA*
phone : 11 404 513
e-mail : eaopo02@yubgss21.bitnet

KOLOKOLOVA L. O.
Astron. Obs. of Acad. of Science of Ukraine
Goloseevo, Kiev 252127, *UKRAINE*
phone : 044 266 4761
e-mail : kiev.na@ussr.eu.net

KÖLZER G.
Universität zu Köln
Abteilung Nuklearchemie
Zülpicher Str. 47
5000 Köln 1, *GERMANY*
phone : 221 4702682; fax : 221 4705115

KORLEVIĆ K.
Visnjan Observatory, Istarska 5
51463 Visnjan, *CROATIA*
phone : 531 49212; fax : 531 49205

KOTSARENKO N. Ya.
Kiev University, Physical Faculty
Vladimirskaya, 64
Kiev 252017, *UKRAINE*
phone : 044 266 4507; fax : 044 266 4517

KRESÁK Ľ.
Astron. Institute, Slovak Acad. of Sciences
842-28 Bratislava, *SLOVAK REPUBLIC*
phone : 0427 375157
e-mail : astropor@sawba.cs

KRISTENSEN L. K.
University of Aarhus, Institute of Physics
8000 Aarhus C, *DENMARK*
fax : 8 6120740

KUEPPERS M.
Germany

KUNZ J.
Max Planck Institute für Kernphysik
Saupfercheckwega post. 103980
6900 Heidelberg, *GERMANY*
phone : 6221 516453; fax : 6221 516540
e-mail : cujo@kosmo.mpi-hd.mpg.de

KWIATKOWSKI T.
A. Mickiweicz University, Astron. Obs.
Ul. Stoneczna, 36
60286 Poznań, *POLAND*
phone : 61 679670; fax : 61 536536
e-mail : tkastro@plpuam11.bitnet

LAGERKVIST C. I.
Astronomiska observatoriet, Box 515
Uppsala 75120, *SWEDEN*
fax : 18 527583
e-mail : md@laban.uu.se
LAGERROS J.
Astronomiska observatoriet, Box 515
Uppsala 75120, *SWEDEN*
phone : 18 530816; fax : 18 527583
e-mail : johan.lagerros@astro.uu.se
LARSON S.
University of Arizona
Lunar & Planetary Laboratory
Tucson, AZ 85721, *U.S.A.*
phone : 602 621 4973; fax : 602 621 4933
e-mail : slarson@lpl.arizona.edu
LAURANCE R. J.
U.K.
LAZZARIN M.
Università di Padova
Dipartimento di Astronomia
vicolo Osservatorio, 5
35122 Padova, *ITALY*
phone : 49 8293668
e-mail : lazzarin@padova.infn.it
LELIWA-KOPYSTYNSKI J.
Warsaw University
Institute of Geophysics
Ul. Pasteura, 7
02-093 Warsawa, *POLAND*
phone : 22 235281; fax : 22 235281
LEMAITRE A.
Dept. de Mathematique, FUNDP
Rempart de la Vierge, 8
5000 Namur, *BELGIUM*
phone : 81 7249038
e-mail : alemaitre@rnandp51
LEVASSEUR-REGOURD A. C.
Service d'Aéronomie du CNRS, BP3
91371 Verrières le Buisson, Cedex, *FRANCE*
phone : 1 64474293; fax : 1 69202999
LEVISON H.
Southwest Research Institute
Space Science Division
6220, Culebra Road
San Antonio, TX 78228-0510, *U.S.A.*
phone : 512 684 5111; fax : 512 647 4325
e-mail : hal@gort.space.swri.edu
LICANDRO J.
Facultad de Ciencias
Dept. de Astronomia
Tristan Narvaja, 1674
11200 Montevideo, *URUGUAY*
phone : 2 418004; fax : 2 409973
e-mail : licandro@fisica.edu.uy
LINDBLAD B. A.
Lund Observatory, Box 43
22100 Lund, *SWEDEN*
phone : 46 107301; fax : 46 104614
LINDGREN M.
Astronomiska observatoriet, Box 515
Uppsala 75120, *SWEDEN*
phone : 18 513522; fax : 18 527583
e-mail : mats.lindgreen@astro.uu.se
LIOU J. C.
University of Florida
Dept. of Astronomy 211 SSRB
Gainesville, FL 32611, *U.S.A.*
phone : 904 392 2052; fax : 904 392 5089
e-mail : liou@astro.ufl.edu
LONGO G.
Università di Bologna, Dipartimento di Fisica
via Irnerio, 46
40126 Bologna, *ITALY*
phone : 51 351087; fax : 51 247244
e-mail : longo@bologna.infn.it
LOPEZ GARCIA A.
Astronomical Observatory
Adva. Blasco Ibanez 13
46010, Valencia, *SPAIN*
phone : 6 3864773; fax : 6 3864773
e-mail : obsast@evalun11.bitnet
LUMME K.
University of Helsinki, Obs. PL 14
00014 Helsinki, *FINLAND*
phone : 0 1912910; fax : 0 1912952
e-mail : lumme@finuh
LUPISHKO D. F.
Astronomical Observatory
Sumskaya Str., 35
Kharkov 310022, *UKRAINE*
phone : 43 24 28
LUU J.
Stanford University, Physics Dept.
Stanford, CA 94305, *U.S.A.*
e-mail : luu@bigz.berkeley.edu

MAGNUSSON P.
School of Math. Sciences
Queen Mary and Westfield College
Mile and Road
London E1 4NS, *U.K.*
fax : 81 981 9587; e-mail : 19764::prm
MAKHLOUF A.
14, rue Aisset Idir
Drean Eltarf (36), *ALGERIA*
MANARA A.
Osservatorio Astronomico di Brera
via Brera, 28
20121 Milano, *ITALY*
phone : 2 874444; fax : 2 72001600
e-mail : manara@astmib.astro.it
MANN I.
Max Planck Institute für Astronomie
post. 20, Max Planck Str., 2
3411 Katlenburg-Lindau, *GERMANY*
phone : 5556 401291; fax : 5556 401240
e-mail : linmpi::mann
MAROV M. Ya.
Institute of Applied Mathematics
Academy of Sciences, Miusskaya SQ 4
125047 Moscow, *RUSSIA*
phone : 095 250 0485; fax : 095 972 0737
MARSDEN B. G.
Smithsonian Astrophysical Observatory
60, Garden Street
Cambridge, MA 02138, *U.S.A.*
phone : 617 495 7244
e-mail : 6700::brian
MARTELLI G.
University of Sussex, School of MaPS
Brighton BN1 9QH, *U.K.*
phone : 273 686391; fax : 273 678097
MARZARI F.
Università di Padova, Dip. di Fisica
via Marzolo, 8
35131 Padova, *ITALY*
phone : 49 831749; fax : 49 844245
e-mail : marzari@padova.infn.it
MATESE J. J.
University of S. W. Louisiana
Physics Dept., Box 44210
Lafayette, LA 70504-4210, *U.S.A.*
phone : 318 231 6691; fax : 318 231 6695
e-mail : matese@usl.edu
MAURY A.
Telescope Schmidt
06460 Caussols, *FRANCE*
phone : 93 626970; fax : 93 092613
Mc BRIDE N.
University of Kent
Unit for Space Science
Canterbury CT2 7NR, *U.K.*
phone : 227 764000; fax : 227 762616
e-mail : nm1@ukc.ac.uk
Mc FADDEN L. A.
University of Maryland, Astron. Dept.
College Park, MD 20742-2421, *U.S.A.*
phone : 301 405 2081; fax : 301 314 9067
e-mail : llmcfadden@ucsd.edu
MEIER R.
Physikalisches Institute UNI, Sidlerstr. 5
3012 Bern, *SWITZERLAND*
phone : 65 4424; fax : 65 4405
e-mail : rmeier@phim.unibe.ch
MENNELLA V.
Osservatorio Astronomico di Capodimonte
via Moiariello, 16
80131 Napoli, *ITALY*
phone : 81 298384; fax : 81 456710
e-mail : mennella@oaclan.astro.it
MICHAUX C.
FRANCE
MICHALOWSKI T.
A. Mickiweicz University, Astron. Obs.
Ul. Stoneczna, 36
60286 Poznań, *POLAND*
fax : 61 535535
e-mail : michastr@plpuam11.amu.edu.pl
MILANI A.
Università di Pisa
Dipartimento di Matematica
via Buonarroti, 2
56127 Pisa, *ITALY*
phone : 50 599554; fax : 50 599524
e-mail : milani@adams.dm.unipi.it
MISTRETTA R. M.
Via Cherso, 1
10100 Torino, *ITALY*
MOHAMED R. A.
Astronomical Observatory
Sumskaya Str., 35
Kharkov 310022, *UKRAINE*

MOONS M.
Dept. de Mathematique, FUNDP
Rempart de la Vierge, 8
5000 Namur, *BELGIUM*
MORANDO B.
Bureau des Longitudes
77, Avenue Denfert Rocherau
75014 Paris, *FRANCE*
phone : 1 4051 2276; fax : 1 4633 2834
e-mail : morando@friap51.bitnet
MORBIDELLI A.
OCA, Observatoire de Nice, BP 229
06304 Nice, Cedex 4, *FRANCE*
phone : 92003126; fax : 92003033
e-mail : amorby@bnandp51.bitnet
MOREELS G.
Observatoire de Besançon, BP 1615
25000 Besançon, Cedex, *FRANCE*
phone : 81 666900; fax : 81 666944
e-mail : earn:frobes51
MORRISON D.
NASA Ames Research Center
Space Science Division, M 245-1
Moffet Field, CA 90035-1000, *U.S.A.*
phone : 415 604 5028; fax : 415 604 6779
e-mail : dmorrison@nasa.mail.nasa.gov
MOTTOLA S.
DLR, German Aerospace Research Establishment
Ruddwer Chaussee
Berlin Adlershof, *GERMANY*
phone : 8153 28595; fax : 8153 2476
e-mail : mottola@terra.pe.ba.dlr.de
MUELLER B.
Kitt Peak Natl. Observatory
Cherry Avenue
Tucson, AZ 85719, *U.S.A.*
fax : 602 325 9360
e-mail : muller@noao.edu
MUINONEN K.
University of Helsinki, Obs. PL 14
00014 Helsinki, *FINLAND*
phone : 0 1912907
MUKAI T.
Kobe University
Dept. of Earth Sciences
Kobe 657, *JAPAN*
fax : 78 882 1549
e-mail : mukai@icluna.kobe-u.ac.jp
MUMMA M.
NASA Goddard Space Flight Center
Code 693
Greenbelt, MD 20771, *U.S.A.*
phone : 301 286 6994; fax : 301 286 3271
e-mail : mumma@leqvax.gsfc.nasa.gov
NAKAMURA A.
Institute of Space and Astron. Sciences
Mizutani Laboratory
3-1-1 Yoshinodai
Sagamihara, Kanagawa 229, *JAPAN*
phone : 427 513911; fax : 427 594237
e-mail : nakamura@cr.scphys.kyoto-u.ac.jp
NAKAMURA R.
Kobe University, Dept. of Earth Sciences
Kobe 657, *JAPAN*
fax : 78 882 1549
e-mail : ryosuke@icluna.kobe-u.ac.jp
NAKAMURA T.
National Astronomical Observatory
Mitaka Tokyo 181, *JAPAN*
phone : 422 34 3627; fax : 422 34 3698
e-mail : nakamura@c1.mtk.nao.ac.jp
NARANJO O. A.
Apartado 690
Merida 5101A, *VENEZUELA*
NAZARCHUK G. K.
Astronomical Observatory
Academy of Science of Ukraine
Goloseevo Kiev 252127, *UKRAINE*
NESLUŠAN L.
Astronomical Institute
Slovak Academie of Sciences
059-60 Tatranská Lomnica, *SLOVAK REPUBLIC*
phone : 969 967866; fax : 969 967656
e-mail : neslusan@asu.savba.cs
NOLAN M.
University of Arizona
Lunar & Planetary Laboratory
Tucson, AZ 85721, *U.S.A.*
phone : 602 621 4555; fax : 602 621 4933
e-mail : nolan@lpl.arizona.edu
NOLL K.
Space Telescope Science Institute
3700, San Martin Drive
Baltimore, MD 21218, *U.S.A.*
phone : 410 3381828; fax : 410 3384767
e-mail : noll@stsci.edu

OKAMOTO H.
Kobe University, Dept. of Earth Sciences
Nada 657, *JAPAN*
fax : 78 8821549
e-mail : okamoto@jasmin.kobe-u.ac.jp

OROFINO V.
Università di Lecce
Dipartimento di Fisica, C.P. 193
73100 Lecce, *ITALY*

OROSEI R.
ITALY

OSIP D. J.
University of Florida
Dept. of Astronomy, 211 SSRB
Gainesville, FL 32611, *U.S.A.*
phone : 904 392 3748

OTTACHER H.
Austria

OTTINK M.
Dutch Meteor Society
Bongweg 296
3192, Hoogvliet, Rotterdam,
THE NETHERLANDS
phone : 10 4166224; fax : 10 4166224

OWEN T.
University of Hawaii
Institute of Astronomy
2680, Woodlawn Drive
Honolulu, HI 96822, *U.S.A.*
phone : 808 956 8007

PALUMBO M. E.
Università di Catania
Istituto di Astronomia
viale A. Doria, 6
95125 Catania, *ITALY*
phone : 95 330533; fax : 95 330592

PAUWELS T.
Observatoire Royal de Belgique
Avenue Circulaire, 3
1180 Bruxelles, *BELGIUM*
phone : 2 3730225; fax : 2 3749822
e-mail : thirry@astro.oma.be

PEALE S. J.
University of California
Dept. of Physics
Santa Barbara, CA 93106, *U.S.A.*
phone : 805 893 2977; fax : 805 893 4170
e-mail : peale@io.ucsb.edu

PECHERNIKOVA G.
Otto Schmidt Institute of Physics
B. Gruzinskaya, 10
123810 Moscow, *RUSSIA*
phone : 095 254 9006; fax : 095 254 9088

PECINA P.
Astronomical Institute
251 65 Ondřejov, *CZCH REPUBLIC*
phone : 0204 85201; fax : 0204 85314
e-mail : astmph@csearn.bitnet

PITTICH E. M.
Astron. Institute, Slovak Acad. of Sciences
842-28 Bratislava, *SLOVAK REPUBLIC*
phone : 7 375157; fax : 7 376085
e-mail : pittich@savba.cs

PORUBČAN V.
Astron. Institute, Slovak Acad. of Sciences
842-28 Bratislava, *SLOVAK REPUBLIC*
phone : 7 375157; fax : 7 376085
e-mail : astropor@savba.cs

PRAVEC P.
Ondřejov Observatory
251 65 Ondřejov, *CZECH REPUBLIC*
phone : 204 85201; fax : 204 85314
e-mail : astmph@csearn.bitnet

PRIALNIK-KOVETZ D.
Tel Aviv University, Dept. of Geophysics
Ramat Aviv 19978, *ISRAEL*
phone : 3 5450528; fax : 3 6409282
e-mail : b13@taunos

PRIEUR J. L.
Obs. Midi Pyrenees, 14, Av. E. Belin
31400 Toulouse, *FRANCE*
e-mail : omptob::prieur

QUESADA V.
Osservatorio Astronomico di Cagliari
via Ospedale, 72
09124 Cagliari, *ITALY*
e-mail : quesada@astrca.astro.it

RADICE E.
via Inglesina, 24
21040 Gerenzano, Varese, *ITALY*
phone : 02 96480269

RAJCHL J.
Ondřejov Observatory
251 65 Ondřejov, *CZECH REPUBLIC*
fax : 204 85314
e-mail : astmph@csearn.bitnet

RANDALL C. E.
University of Colorado
LASP, Campus Box 392
Boulder, CO 80309-0392, *U.S.A.*
phone : 303 492 8208; fax : 303 492 6946
e-mail : 34075::randall

REBER M.
University of Bern, Physics Institute
Sidlerstrasse, 5
3012 Bern, *SWITZERLAND*
fax : 41 31 654405
e-mail : reber@phim.unibe.ch

RENARD J. B.
Service d'Aéronomie du CNRS, BP3
91371 Verrières le Buisson, Cedex, *FRANCE*
phone : 1 64474335; fax : 1 69202999

REVELLE D. O.
Nord Illinois University
Dekalb, IL 60115-2854, *U.S.A.*
phone : 815 7536837; fax : 815 7536872
e-mail : t50dor1@niv.bitnet

RICCIOLI D.
Università di Catania, Istituto di Astronomia
viale A. Doria, 6
95125 Catania, *ITALY*
phone : 95 7332245; fax : 95 330592
e-mail : driccioli@astrct.cineca.it

RICKMAN H.
Astronomiska observatoriet, Box 515
Uppsala 75120, *SWEDEN*
phone : 18 113522; fax : 18 527583
e-mail : hans@laban.uu.se

ROESSLER K.
Institute für Nuklearchemie
Forschungszentrum Jülich GmbH, P.O. Box 1913
5170 Jülich, *GERMANY*
phone : 2461 613109; fax : 2461 612535

ROTUNDI A.
Osservatorio Astronomico di Capodimonte
via Moiariello, 16
80131 Napoli, *ITALY*
fax : 81 5519957
e-mail : 40060::rotundi

SAFRONOV V. S.
Otto Schimot Institute of Physics
B. Gruzinskaya, 10
123810 Moscow, *RUSSIA*
phone : 095 254 9006; fax : 095 254 9088

SCHLEICHER D. G.
Lowell Obs., 1400 West Mars Hill Road
Flagstaff, AZ 86001, *U.S.A.*
phone : 602 774 3358; fax : 602 774 6296
e-mail : schleicher@lowell.edu

SCHLOERB F. P.
University of Massachussets
Astronomy Lederle GSWRB
Graduate Residence, Tower B
Amherst, MA 01003, *U.S.A.*
phone : 413 545 4303; fax : 413 545 4223
e-mail : schloerb@ferao1.phast.uman.edu

SCHOBER H. J.
University of Graz
Institute of Astronomy
Universitatsplatz, 5
8010 Graz, *AUSTRIA*
phone : 0316 3805273; fax : 0316 3804091
e-mail : schober@edvz.uni-graz.ada.at

SCHOLL H.
OCA, Observatoire de Nice, BP 229
06304 Nice, Cedex 4, *FRANCE*
phone : 92003041; fax : 92003058
e-mail : scholl@haendel.obs-nice.fr

SCHULZ R.
University of Maryland
Dept. of Astronomy
College Park, MD 20742, *U.S.A.*
phone : 301 405 1513; fax : 301 314 9067
e-mail : rita@astro.umd.edu

SCHWEHM G. H.
ESTEC, ESA
Space Science Division, Postbus 299
2200 Ah Noordwijk, *THE NETHERLANDS*
phone : 171983539; fax : 171984697

SCOTTI J. V.
University of Arizona
Lunar & Planetary Laboratory
Tucson, AZ 85721, *U.S.A.*
phone : 602 621 4933; fax : 602 621 4933
e-mail : jscotti@lpl.arizona.edu

SEIDELMANN P. K.
U. S. Naval Observatory
Washington, DC 20392, *U.S.A.*
phone : 202 653 0529; fax : 202 653 1744
e-mail : omd@ariel.usno.navy.mil

SEU R.
ITALY

SHIMIZU M.
Institute of Space and Astron. Sciences
Mizutani Lab., 3-1-1 Yoshinodai
Sagamihara, Kanagawa 229, *JAPAN*
phone : 0427 51 3911; fax : 0427 59 4237

SHKODROV V.
Bulgarian Academy of Sciences
Dept. of Astronomy
Sofia, *BULGARIA*
phone : 75 8827; fax : 75 8927
e-mail : planet@bgearn.bitnet

SHOEMAKER E. M.
U. S. Geological Survey
2255, North Gemini Drive
Flagstaff, AZ 86001, *U.S.A.*
phone : 602 556 7003
e-mail : astrog::gshoemaker

SHOR V. A.
Institute of Theoretical Astronomy
Naberezhnaya Kutuzova, 10
191187 St. Petersburg, *RUSSIA*
phone : 812 278 8824; fax : 812 272 7968

SHULMAN L. M.
Astron. Obs. of Acad. of Science of Ukraine
Goloseevo Kiev 252127, *UKRAINE*
phone : 266 4769
e-mail : sky@gao.kiev.ua

SICARDY B.
Observatoire de Paris, LAM
92195 Meudon, Principal Cedex, *FRANCE*
phone : 1 650771609
e-mail : mesioa::sicardy

ŠIMEK M.
Ondřejov Observatory
251 65 Ondřejov, *CZECH REPUBLIC*
fax : 204 85314
e-mail : astmph@csearn.bitnet

SITARSKI G.
Space Research Center
Bartycka, 18
00-716 Warszawa, *POLAND*
Phone : 4822 410041

SOKOLSKY A.
Institute of Theoretical Astronomy
Naberezhnaya Kutuzova, 10
191187 St. Petersburg, *RUSSIA*
fax : 812 272 7968
e-mail : sokolsky@iiii.spb.su

SOLOVAYA N. A.
Sternberg State Astron. Institute
University Prospeckt, 13
119899 Moscow V 234, *RUSSIA*
phone : 7 095 939 3764
e-mail : snn@sai.msk.su

SPENCER J. R.
Lowell Obs., 1400, West Mars Hill Road
Flagstaff, AZ 86001, *U.S.A.*
phone : 602 774 3358; fax : 602 774 6296
e-mail : spencer@lowell.edu

SPURNÝ P.
Astronomical Institute
251 65 Ondřejov, *CZECH REPUBLIC*
fax : 204 85814
e-mail : astmph@csearn.bitnet

STEEL D.
Anglo Australian Observatory
Coonabarabran, NSW 2357, *AUSTRALIA*
phone : 68 426 314; fax : 68 842 298
e-mail : dis@aaocbn3.aao.gov.au

STERN A.
Southwest Research Institute
Space Science Division, 6220, Culebra Road
San Antonio, TX 78228-0510, *U.S.A.*
phone : 512 522 5127; fax : 512 647 4325
e-mail : alan@swri.space.swri.edu

STEYAERT C.
V.V.S. Astron. Assoc., Belgium
Dr. Van de Perrestr., 83
2440 Geel, *BELGIUM*
fax : 14 221373
e-mail : chri-steyaert@f852

STRAZZULLA G.
Università di Catania
Istituto di Astronomia, viale A. Doria, 6
95125 Catania, *ITALY*
phone : 95 330533; fax : 95 330592
e-mail : 40297::gianni

SVESTKA J.
Praga Observatory
11846 Prague 1, *CZECH REPUBLIC*
phone : 2 377576; fax : 2 375970

SVOREŇ J.
Astron. Institute, Slovak Acad. of Sciences
059-60 Tatranská Lomnica, *SLOVAK REPUBLIC*
phone : 969 967866; fax : 969 967656
e-mail : astrsven@asu.savba.cs

TAKAGI Y.
Dept. of Earth Science
Nagoya University
Chikusa-ku, Nagoya 464-01, *JAPAN*
phone : 52 783 0659; fax : 52 782 7091
TAYLOR A.
University of Kent
Unit for Space Sciences
Canterbury, CT2 7NR, *U.K.*
phone : 227 764000; fax : 227 762616
TEDESCO E. F.
Mission Research Corporation
One Tara Blvd., Suite 302
Nashua, NH 03062-2801, *U.S.A.*
phone : 603 891 0070; fax : 603 891 0088
e-mail : et@ipac.caltech.edu
TERENTJEVA A. K.
Academi of Sciences, Astronomical Council
Pjatnitskaya, 48
109017 Moscow, *RUSSIA*
phone : 095 231 5461; fax : 095 230 2081
e-mail : iaas@node.ias.msk.su
THOLEN D.
University of Hawaii
Institute of Astronomy
2680, Wood Drive
Honolulu, HI 96822, *U.S.A.*
phone : 808 948 6930; fax : 808 988 2790
e-mail : tholen@uhifa.ifa.hawaii.edu
THOMAS P.
Cornell University, 322 Space Sciences
Ithaca, NY 14850, *U.S.A.*
phone : 607 255 9581; fax : 607 255 9002
e-mail : cuspif::thomas
TICHA J.
Klet Obs., Zatkovo Nabrezi, 4
370 01 Ceske Budejovice, *CZECH REPUBLIC*
TICHY M.
Klet Obs., Zatkovo Nabrezi, 4
370 01 Ceske Budejovice, *CZECH REPUBLIC*
TIMOSHKOVA E.
Institute of Theoretical Astronomy
Nabereshnaya Kutusova, 10
191187 St. Petersburg, *RUSSIA*
phone : 812 278 8832; fax : 812 272 7968
TORNATORE V.
via Madonna di Re, 2A
28037 Domodossola, Novara, *ITALY*
TOZZI G. P.
Osservatorio di Arcetri, Largo E. Fermi, 5
50125 Firenze, *ITALY*
phone : 55 2752250; fax : 55 220039
e-mail : tozzi@arcetri.astro.it
URAS S.
Osservatorio Astronomico di Cagliari
via Ospedale, 72
09124 Cagliari, *ITALY*
e-mail : uras@astrca.astro.it
VALLERANI E.
Alenia Spazio S.p.A., Corso Marche, 41
10146 Torino, *ITALY*
phone : 11 7180701; fax : 11 723307
VALSECCHI G. B.
IAS, Reparto di Planetologia
viale Università, 11
00185 Roma, *ITALY*
phone : 6 4456951; fax : 6 4454969
e-mail : giovanni@irmias.bitnet
VALTONEN M.
Tuorla Observatory
21500 Piikkio, *FINLAND*
phone : 21 2435822; fax : 21 2433767
VAN FLANDERN T.
META Research, 6327, Western Avenue
Washington, DC 20015, *U.S.A.*
phone : 202 362 9176
VANZANI V.
Università di Padova, Dip. di Fisica
via Marzolo, 8
35131 Padova, *ITALY*
phone : 49 831746; fax : 49 844245
e-mail : vanzani@padova.infn.it
VAUTERIN P.
Sterrenkundig Observatory
Krijgslaan 281 (S9)
9000 Gent, *BELGIUM*
VELICHKO F. P.
Astronomical Observatory
Sumskaya Str., 35
Kharkov 310022, *UKRAINE*
phone : 2 432438; fax : 2 457326
VERKHOGLYADOVA O. P.
Kiev University, Physical Faculty
Dept. of Astronomy, Vladimirskaya, 64
Kiev 252017, *UKRAINE*
phone : 044 266 4507; fax : 044 266 4517

VON HOERNER H.
Von Hoerner & Sulger GmbH, Schlossplatz, 8
6830 Schwetzingen, *GERMANY*
phone : 6202 24303

WALKER R.
Jamieson Science & Engineering
5321, Scotts Valley Drive, 204
Scotts Valley, CA 95066, *U.S.A.*
fax : 408 438 8930
e-mail : jamieson@bkyast.berkeley.ei

WATANABE J.-i
National Astronomical Observatory Osawa
Mitaka, 181 Tokyo, *JAPAN*
phone : 422 41 3614; fax : 422 34 3608
e-mail : owatana@c1.mtk.nao.ac.jp

WEAVER H.
Space Telescope Science Institute
3700, San Martin Drive
Baltimore, MD 21218, *U.S.A.*
phone : 301 3384765; fax : 301 3384767
e-mail : weaver@stsci.edu

WEIDENSCHILLING S.
Planetary Science Institute
620, North 6th Avenue
Tucson, AZ 85705 8331, *U.S.A.*
phone : 602 622 6300; fax : 602 622 8060
e-mail : psikey::psikey

WEISSMAN P.
Jet Propulsion Laboratory, MS 183-601
Pasadena, CA 91109, *U.S.A.*
phone : 818 354 2636; fax : 818 393 4605
e-mail : issac::weissman

WETHERILL G. W.
DTM Carnegie Institute
5241, Broad Branch Road NW
Washington, DC 20015, *U.S.A.*
phone : 202 686 4375; fax : 202 364 8726

WHIPPLE F. L.
Smithsonian Center for Astrophysics
60, Garden Street
Cambridge, MA 02138, *U.S.A.*
phone : 617 8647383; fax : 617 4957356

WILLIAMS I. P.
Queen Mary and Westfield College
Mile and Road
London E1 4NS, *U.K.*
phone : 1 9755452; fax : 1 9819587
e-mail : cpco@uk.ac.qmw.maths

WISNIEWSKI W.
University of Arizona
Lunar & Planetary Laboratory
Tucson, AZ 85721, *U.S.A.*
phone : 602 621 2773
e-mail : wieslaw@pirl.lpl.arizona.edu

WOOTTEN A.
National Radio Astronomy Observatory
Headquarters, Edgemont Road
Charlottesville, VA 22901, *U.S.A.*
phone : 804 2960 329; fax : 804 2960 278
e-mail : awootten@nrao.edu

WOSZCZYK A.
Institute of Astronomy
Ul. Chopina 12/18
87-100 Toruń, *POLAND*
phone : 48 56 26037; fax : 48 56 24602

WU C. Y. R.
University of Southern California
Space Sciences Center
Department of Physics and Astronomy
Los Angeles, CA 90089-1341, *U.S.A.*
phone : 213 740 6332; fax : 213 740 6342
e-mail : robertwu@lism.esc.edu

WU Z.
University of London
Queen Mary and Westfield College
Astronomy Unit, Mile End Road
London E1 4NS, *U.K.*
phone : 081 983 1528
e-mail : zidian@math.gmn.ac.uk

XIE X.
University of Maryland
College Park, MD 20740, *U.S.A.*
phone : 301 935 5324; fax : 301 286 3271
e-mail : xingfa@lep693.gsfc.nasa.gov

XU Y.-l
University of Florida
Dept. of Astronomy 211 SSRB
Gainesville, FL 32611, *U.S.A.*
phone : 904 392 7678; fax : 904 392 5089
e-mail : shu@astro.ufl.edu

YAGUDINA E.
Institute of Theoretical Astronomy
Naberezhnaya Kutuzova, 10
191187 St. Petersburg, *RUSSIA*
fax : 812 272 7968
e-mail : shot@iipah.spb.su

YAMAMOTO T.
Institute of Space and Astron. Sciences
Mizutani Lab., 3-1-1 Yoshinodai
Sagamihara, Kanagawa 229, *JAPAN*
phone : 427 51 3911x2533; fax : 427 59 4237
e-mail : d52321@jpnfudpc

YANO H.
University of Kent, Unit for Space Sciences
Canterbury CT2 7NR, *U.K.*
fax : 227 762616

YEOMANS D.
Jet Propulsion Laboratory, MS 183-601
Pasadena, CA 91109, *U.S.A.*
phone : 818 354 2127; fax : 818 393 1159
e-mail : dky@naif.span.nasa.gov

YOSHIKAWA M.
Communications Research Laboratory
893-1 Hirai, Kashima
Ibaraki 314, *JAPAN*
fax : 299 84 4129
e-mail : makoto@moon.cc.clr.go.jp

YUASA M.
Kinki University, Dept. of Mathematics
Higashi-Osaka 577, *JAPAN*
phone : 721 23324709; fax : 721 2353
e-mail : liyuasa@jpnknu01.bitnet

ZADUNAISKY P. E.
University of Buenos Aires
Dept. de Matematica, Ciudad Universitaria
Buenos Aires, *ARGENTINA*
phone : 1 7762913; fax : 1 7745703

ZAJTSEV A. L.
Institute of Radio Electronics
Moscow, *RUSSIA*

ZAMBRA A.
Laben S.p.A.
Strada Padana Inferiore, 290
20090 Vimodrone, (Milano), *ITALY*

ZANOTTA M.
Via Demonte, 1
20162 Milano, *ITALY*

ZAPPALÀ V.
Osservatorio Astronomico di Torino
10025 Pino Torinese, (Torino), *ITALY*
phone : 11 4619035; fax : 11 4619030
e-mail : zappala@to.astro.it

ZHENG J.-z
Tuorla Observatory
21500 Piikkio, *FINLAND*
phone : 21 2435822; fax : 21 2433767
e-mail : zheng@sara.cc.utu.fi

ZHOU H.-n
Nanjing University, Dept. of Astronomy
Najing 210008, *CHINA*
phone : 25 637551-3138; fax : 25 307965
e-mail : bmnju@ica.beijing.canet.cn

SEARCH PROGRAMS FOR COMETS

B. G. MARSDEN
Harvard-Smithsonian Center for Astrophysics,
60 Garden Street, Cambridge, MA 02138, U.S.A.
E-mail MARSDEN@CFA.HARVARD.EDU

Abstract. Although there are currently no professional search programs specifically designed for comets, comet discoveries are a significant by-product of the current search programs for near-earth asteroids. With their emphasis on the opposition point and the ecliptic, asteroid patrols are clearly biased toward discovering short-period comets. A comparison of the comet discoveries made at elongations of less than 120° from the sun during the ten-year intervals ending in 1892 and 1992 indicates that the current opposition searches are not able to displace visual searches in the vicinity of the sun. Modification of the near-opposition search region to that proposed for the Spaceguard Project would be helpful, but—as also proposed for Spaceguard—supplementation is still necessary for high-inclination objects. CCD experimentation in the regions traditionally searched visually by amateurs would be useful. Searches for particular types of object like Kreutz sungrazers and comets near Jupiter in space are briefly discussed.

1. Introduction

Ever since it became fashionable, toward the end of the nineteeth century, to draw a distinction between professional and amateur astronomers, to make searches exclusively for comets has been generally regarded as an activity for the latter. Indeed, there must be several hundred amateur astronomers around the world who, on a regular basis, scan the skies visually with small to moderate-sized telescopes in the hope of finding a new comet. The technique, and even the rate of success, has not greatly changed since the 1750s, when Charles Messier attempted to be the first person to reobserve Halley's Comet—and instead effectively inaugurated comet hunting as a serious sport. Until then, the great comet of 1680 was the only one to have been discovered telescopically.

In recent times, the only case where a substantial fraction of the resources of a professional observatory has been expended on comet hunting was the highly successful program conducted using pairs of 25 × 105 binoculars at the Skalnaté Pleso Observatory in Slovakia during the decade following World War II. About a dozen years ago Ľubor Kresák, one of the best known participants in that program, gave an extensive review of comet discoveries, including a discussion of observational selection (Kresák 1982). In that review he discussed the seminal papers on comet discoveries and observational selection (Everhart 1967a, 1967b) by Edgar Everhart, another successful visual comet hunter who at the time he wrote these papers was in the process of leaving professional physics for professional astronomy.

Even before the time of Messier it was clear that bright comets were most likely to be discovered at relatively small angular distances from the sun. During the preceding century, for example, only four of the 37 comets (or 11 percent) for which orbits have been computed were discovered at elongations $\epsilon > 120°$; these four, all naked-eye objects at discovery, were the comets of 1729 and 1747 (which, with

A. Milani et al. (eds.), Asteroids, Comets, Meteors 1993, 1–16.

perihelion distances q of 4.05 and 2.20 AU, were intrinsically stupendous objects that remained the two comets of largest known q until 1885), the comet of 1744 (which was by far the most spectacular of the eighteenth century as seen from the earth) and the comet of 1678 (which has recently been conclusively demonstrated as an early appearance of periodic comet d'Arrest on a close passage by the earth; Carusi *et al.* 1991). Fully 73 percent of the remainder were discovered west of the sun, in generally the eastern part of the sky before dawn, as the earth's orbital motion allowed them to emerge further from the sun's rays as the days wore on. Most of them were also then well inside the earth's orbit and were brightened significantly by their proximity to the sun. The smaller fraction discovered east of the sun, in the western sky after dusk, were almost exclusively comets in the latter category, frequently much closer to the earth than the sun was, their orbital motions being more important than the earth's in defining their apparent motions.

This morning-evening disparity was less pronounced in the comet searches by Messier and several of the other early comet hunters, perhaps in part for social reasons. More recent data do show a 3-to-2 bias toward the morning sky, however, and the effect is clearly enhanced if one considers only comets in retrograde orbits, i.e., with their own contributions to the prevailing westerly motion. Everhart's (1967a) study of the discovery circumstances of long-period comets since 1840 noted that 53 percent of those in direct orbits and 69 percent of those in retrograde orbits were found in the morning sky. Furthermore, on attempting to adjust the actual discovery dates to those when the comets *ought* to have been discoverable with the available instrumentation, he suggested that these fractions should be increased to 70 percent and 81 percent, respectively. These adjusted numbers are undoubtedly over-optimistic, mainly because many comets do not vary in brightness as the inverse-fourth power of heliocentric distance r—or, indeed, in any uniform manner.

Bright naked-eye comets have, of course, been discovered at very small elongations from the sun. When a telescope has been necessary for discovery, only a few comets have been found at $\epsilon < 30^\circ$. While observers might disagree whether a telescope is actually needed in specific cases, E. E. Barnard's discovery—his first—of comet 1881 VI at $\epsilon = 24^\circ.5$ must be a serious contender for the record; he was using a 0.13-m telescope, and the comet was probably of magnitude $m \sim$ 6-7. Perhaps equally noteworthy was Jérôme Coggia's discovery of comet 1890 III at $\epsilon = 26^\circ.4$ and $m \sim 8$.

Another of Everhart's conclusions (Everhart 1967b) was that more than 80 percent of the long-period comets that should be observable were being missed. This conclusion was based on both observations and observability in the northern hemisphere alone. There is an obvious north-south asymmetry in the distribution of comet observers, and Everhart noted that, in comparison to the inverse, there were six times as many comets discovered in the northern hemisphere that would never have been observable in the southern. The failure of comets to conform to the r^{-4} brightness-variation law again undoubtedly means that the 80-percent discovery failure is far too extreme. It is also unfair to expect a faint comet to be discovered visually in the Coma-Virgo region of galaxies or photographically among the crowded stars close to the Milky Way. Everhart's figure was also disputed by Kresák

(1982), principally on the grounds that many comets have accidental co-discoverers. In my capacity as director of the IAU Central Bureau for Astronomical Telegrams, I am aware of numerous additional cases of independent discoveries that are never published.

Everhart's analysis principally involved visual discoveries of comets, but it did include a number of photographic discoveries. Although there are photographic records of what was evidently a sungrazing comet during the total solar eclipse in 1882, the earliest recognized photographic discovery is that by Barnard of comet 1892 V. Two more photographic discoveries followed in 1898, and by the end of World War I the number of photographic discoveries had risen to a dozen. By the end of World War II photographic discoveries were in the majority. Since most of the photographic discoveries have been by professional astronomers as a by-product of other observations, they have tended to concentrate near opposition. Only 20 percent of them have been made at $\epsilon < 120^\circ$ and less than 7 percent at $\epsilon < 90^\circ$. Except for comets recorded during total solar eclipses, the smallest elongations for photographic discoveries seem to be $\epsilon = 30^\circ\!.8$ and $46^\circ\!.7$ for comets 1911 II and 1945 VII, but both of these objects were bright enough to have been discovered visually. More of a challenge was the rediscovery by Jeff Johnston in 1979 of periodic comet Schwassmann-Wachmann 3 some 25° from its predicted position; at $\epsilon = 49^\circ\!.5$ and $m \sim 13$, such a detection would be borderline for a visual discovery. As for comets that were definitely too faint for visual discovery, mention should be made of Richard West's discovery with the European Southern Observatory's 1.0-m Schmidt of comet 1977 IX at $\epsilon = 71^\circ\!.8$ and $m \sim 17.5$, as well as comet 1955 VI, found during the Palomar Sky Survey at $\epsilon = 73^\circ\!.9$ and $m \sim 15$.

Most of the photographic discoveries during the early years of the twentieth century were made in the course of search programs for asteroids. Although general sky surveys and search programs for extragalactic supernovae have contributed a substantial number of comets, discoveries during patrols for asteroids have continued to dominate, particularly as several such patrols have in recent times concentrated on finding asteroids that come close to the earth. The association with asteroid programs has tended to concentrate photographic comet discoveries near the ecliptic. This in turn has resulted in an excess of discoveries of comets with small orbital inclinations, and the majority of them have therefore been short-period comets of the so-called Jupiter family.

2. Cometary Discoveries, 1818-1992

Fig. 1 shows the number of comets discovered in five-year intervals from 1818 through 1992. During the past ten years the rate of discoveries has been quite extraordinary. The tallies for 1983-1987 and 1988-1992 should be further increased by four and eight, respectively, to include the discoveries of sungrazing comets in coronagraphic observations from the SOLWIND and SMM satellites. The count for 1983 does include the six comets discovered (in two cases involving also ground-based co-discoveries) by IRAS, the American-Dutch-British Infrared Astronomical Satellite. No fewer than 28 comet discoveries—a record for specific discoverers—are named for the observing team led by Carolyn and Eugene Shoemaker in a

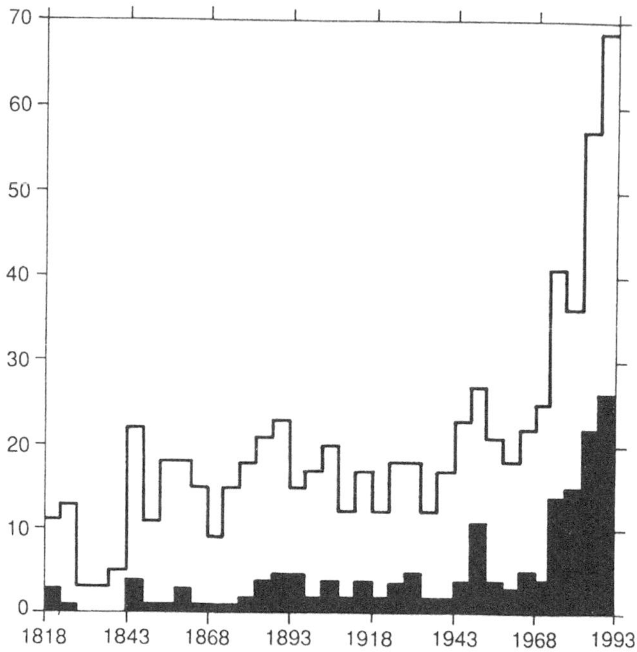

Fig. 1. Comet discoveries in five-year intervals, 1818-1992. The shaded portion refers to short-period comets.

near-earth asteroid search using the 0.46-m Schmidt at Palomar; the team led by Eleanor Helin, in a similar program with the same telescope, produced 11 more comets during the same decade. The years 1991 and 1992 yielded the first CCD discoveries of comets, specifically by using the scanning Spacewatch camera (also designed largely for a near-earth-asteroid search) on a 0.9-m telescope on Kitt Peak; at magnitude 21, the 1991 object is the faintest cometary discovery hitherto made.

The shaded area in Fig. 1 shows the comets that are of short period, an upper limit of 20 years (for Jupiter-family membership), rather than the conventional 200 years, being adopted here for this definition. The year 1818 is significant for producing the observations that led to the final recognition of Encke's Comet, and the following year brought P/Blanpain and the first observations of what later became known as P/Pons-Winnecke. Although there was already some suspicion that what came to be known as P/Biela had already been observed in both 1772 and 1805, this was not proven until Biela found the comet in 1826. The discovery of P/Faye in 1843 started the sequence of frequent discoveries of short-period comets that continues to this day. Although the fraction of comets discovered that are of short period remained fairly consistently at one out of four or five until around 1970, this has now increased to more than one out of three, mainly as a result of the near-earth-asteroid search programs mentioned above.

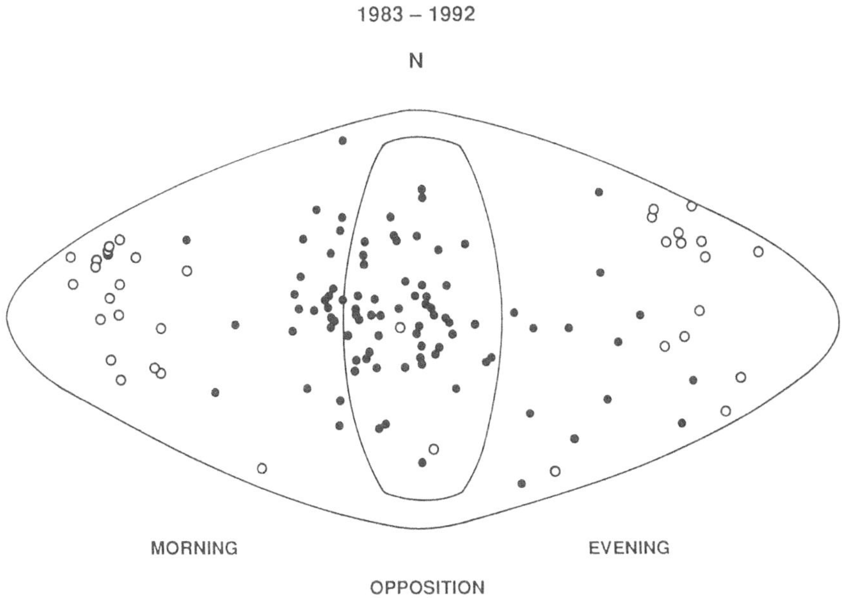

Fig. 2. Equal-area plot of discovery positions of comets during 1983-1992, centered on opposition, around which the prime "Spaceguard" search region is marked. The ecliptic north pole is at the top. Open circles denote visual discoveries, closed circles non-visual discoveries.

Fig. 2 is an all-sky, equal-area plot showing the discovery positions for the comets found during 1983-1992. Ecliptic coordinates are used, and the plot is centered on opposition. The open circles show the visual discoveries (or, at least, cases where the first of independent discoveries were visual). As expected, the non-visual (i.e., essentially photographic) discoveries are concentrated near opposition, while the visual discoveries are generally far from opposition. It is useful to term the region of the sky with $\epsilon > 120°$ the "opposition region", and during the decade in question 63 percent of the comet discoveries were in this region. On the other hand, only eight discoveries, or 6 percent, were made when $90° < \epsilon < 120°$, which I shall term the "transition region". Actually, the transition region is dominated by the six comets (counting independent discoveries) found by IRAS, which was designed to survey the sky near $\epsilon = 90°$; if all the comets of 1983—the year IRAS was operating—are removed from the sample, the fraction in the transition region drops to less than 3 percent. The minimum value of ϵ for the comets during this decade was 29°.1 in the case of comet 1983 V.

The morning-evening asymmetry is evident in Fig. 2, but *only in the opposition region*, where there are precisely *twice* as many on the morning side as on the evening. Given the concentration of professional programs near opposition and the fact that, unless they are close to the earth, all comets are moving from morning to evening and are essentially equally bright at the same elongation before and after opposition, this bias is really not at all surprising.

Fig. 2 also shows the north-south bias, with 57 percent of the discoveries at positive ecliptic latitudes β. This separates into 56 percent for the opposition region and 60 percent for the remainder. More significant is the fact that only 24 percent of the discoveries are in the half of the sky with $|\beta| > 30°$, and this fraction drops to 14 percent in the opposition region, where the extreme latitude values are $\beta = +38^{\circ}\!.4$ and $-42^{\circ}\!.1$, the maximum possible latitude being 60°. Each hemisphere has five discoveries at higher latitudes outside the opposition region, the extremes being $\beta = +70^{\circ}\!.2$ and $-62^{\circ}\!.9$.

It would appear that analysis of cometary discoveries can benefit by separating the discoveries into those that were made in the opposition region and those that were not. The opposition region has discoveries that are mainly photographic and nowadays faint and frequently short-period comets. The non-opposition region has mainly visual discoveries of bright, long-period comets. The boundary point in ϵ is not particularly critical, and the outcome would not be greatly different if the transition region were included with the opposition region.

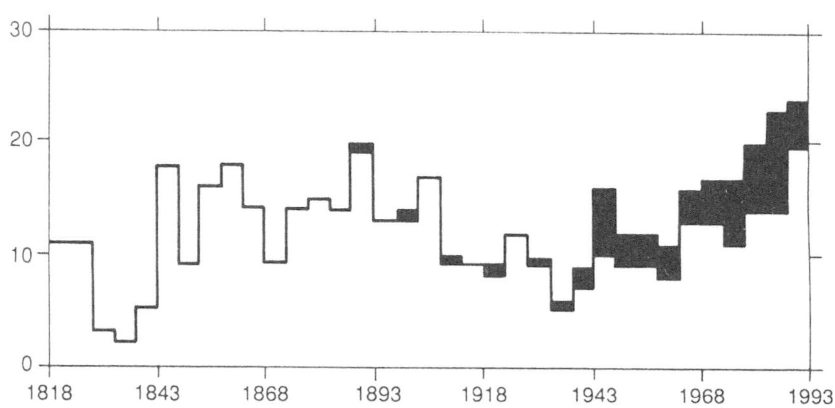

Fig. 3. Non-opposition comet discoveries in five-year intervals, 1818-1992. The shaded portion refers to non-visual discoveries.

Fig. 3 is the subset of Fig. 1 that contains the comets discovered in the non-opposition region. Short-period comets are not distinguished, the shaded area this time showing comets not discovered visually. For the entries before 1982 this plot utilized Rudenko's (1986) convenient compilation of cometary discovery data. The distribution with time in Fig. 3 is, understandably, much more uniform than in Fig. 1. After photography came widely into use the number of comet discoveries at small elongations diminished, and it is only recently that the level of non-opposition discoveries has returned to that of the late-nineteenth century.

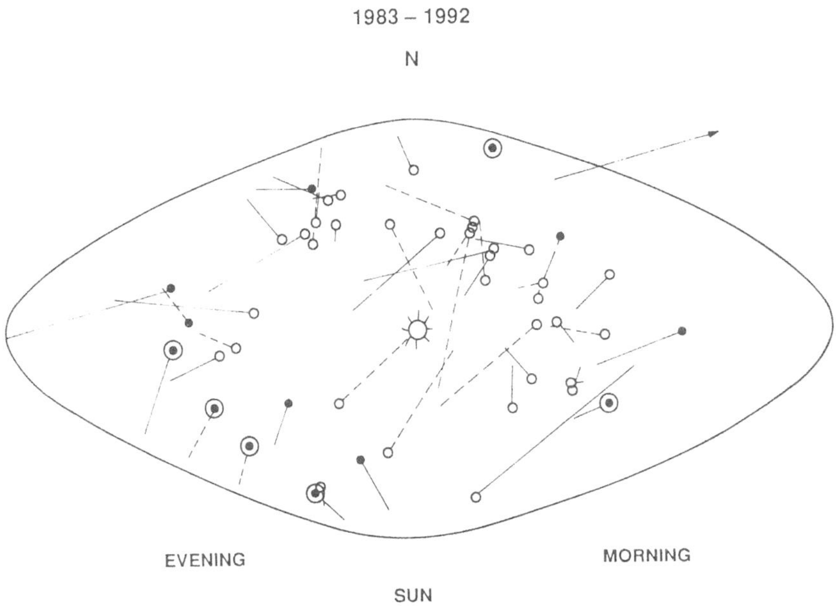

Fig. 4. Equal-area plot of discovery positions of comets during 1983-1992, centered on the sun. See also the caption to Fig. 2 and the text.

3. Comparison of Discoveries during 1883-1892 and 1983-1992

The near equality of the numbers suggests that it might be useful to compare the non-opposition discoveries during 1983-1992 with those during 1883-1892, the end of which decade saw the first photographic discovery of a comet.

Fig. 4 is a plot of the 47 non-opposition 1983-1992 discoveries, much in the form of Fig. 2 and as already discussed, except that the center is now the sun, rather than the opposition point. An outer open circle identifies each IRAS discovery (or co-discovery), the aforementioned extreme case of $\beta = +70^{\circ}\!.2$ being comet 1983 VII (IRAS-Araki-Alcock). The straight lines attached to the comet points are drawn from the positions (again with respect to the sun) the comets would have occupied 30 days earlier. A solid line indicates that, according to an r^{-4} law, the comet should have been brightening during that month, a broken line that it should have been fading. One comet, 1989 IX, passed through opposition during that month, so the line, extending outside the diagram, is a particularly bad representation of the comet's motion in this instance. The comets should have been brightening in 70 percent and by more than three magnitudes in 11 percent of the cases. In six cases, or 43 percent of the comets that were fading, the elongation the previous month would clearly have been too small to permit discovery; all of them would in

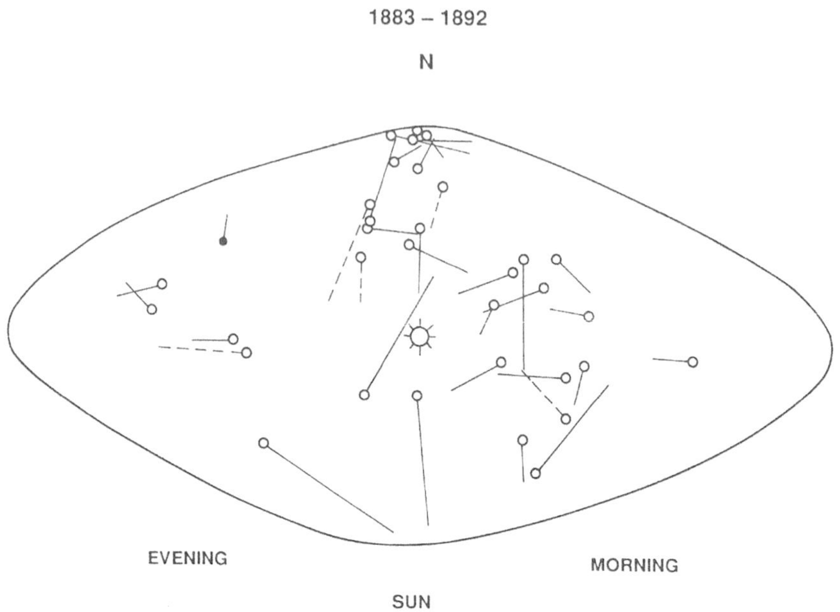

Fig. 5. Equal-area plot of discovery positions of comets during 1883-1892, centered on the sun. See also the caption to Fig. 2 and the text.

fact have had $\epsilon < 24^\circ\!.5$, the value for the record case of comet 1881 V. In no case was it particularly surprising that the discovery was not made one month earlier.

Fig. 5 is the corresponding plot for the 34 comets discovered in the non-opposition region during 1883-1892, only one of them, of course, being not a visual discovery. As for the 1983-1992 non-opposition discoveries, there is *no excess* on the morning side; in fact, as long as one includes the transition region, which accounts for 15 percent of the discoveries, there is a 19-to-15 bias in favor of *evening* discoveries. Almost 65 percent of the values of β are positive, which is not unreasonable, considering that 85 percent of the comets were discovered from sites at north-temperate latitudes; the maximum southern value was $\beta = -48^\circ\!.2$ for comet 1891 IV, which was actually discovered by Barnard from northern California. The β distribution north of the ecliptic shows a sharp difference from 1983-1992, 65 percent of the latitudes being north of $+30^\circ$; four comets were found further north than any of those of the past decade, the maximum latitude being $+84^\circ\!.7$ for comet 1887 II. Comparison with the situation 30 days before discovery shows that brightening should have occurred for 85 percent of the comets, by more than three magnitudes in the case of the sungrazing comet 1887 I, which was of first magnitude and only $16^\circ\!.2$ from the sun at discovery. At least seven of the brighteners and one of the faders would have been too close to the sun for discovery a month earlier.

In terms of searches with modern professional instrumentation, however, the real test for possible earlier discovery involves examination of the circumstances when a comet was previously near opposition. The proposed "Spaceguard Project" (Morrison *et al.* 1992) for searching for near-earth objects—principally asteroids—specified the need for wide-field CCD instrumentation with a limiting magnitude of 22 and concentration on the "standard Spaceguard region" (SSR) extending ±30° in ecliptic longitude and ±60° in ecliptic latitude about the opposition point. The SSR is delineated in Fig. 2. At least for objects near the ecliptic, the peak longitude for the discovery of comets in the opposition region during the past decade, ~ 20° on the morning side of opposition, is included in the SSR. The discovery statistics for 1983-1992 suggest that these searches have fallen far short of the SSR limit in latitude. In any case, Marsden and Steel (1993) have shown that strict adherence to the SSR, particularly in latitude, would result in the failure to detect some 25 percent of the long-period comets that may be on collision courses with the earth, and the same logic applies to the discovery of all long-period comets with perihelia near and inside the earth's orbit.

TABLE I
1983-1992 comets at previous opposition.

Comet	β (deg)	t (day)	r (AU)	m	
1983 I	+25.4	263	2.47	17*	
1983 XII	+37.2	237	3.99	13	MW
1984 XXIII	−25.6	89	2.12	12	
1986 XVIII	+36.2	352	4.77	18*	
1987 III	+15.2	47	1.93	10	MW
1987 XXIX	−31.4	88	2.83	13	
1987 XXX	−34.4	421	4.18	15	MW
1988 III	−27.0	480	4.95	17*	MW
1988 XV	− 7.1	324	5.47	18*	
1990 III	− 6.1	189	2.95	14	
1991 I	+16.6	141	2.12	10	SP
1991 XI	−23.9	391	5.01	17*	MW
1992 XIV	−22.4	423	3.57	13	SP

On computing the previous-opposition circumstances for the 47 comets discovered outside the opposition region during the last decade, we find that 26 percent of them would escape Spaceguard detection because $|\beta| > 60°$. As noted, however, the latitude range actually scanned is much smaller, and 57 percent of the comets would have escaped prior opposition detection because $|\beta| > 40°$. Some 30 percent of the comets would have escaped detection because the r^{-4} law predicts them to have been fainter than magnitude 19. Combination of the faintness and SSR limit would prevent the detection of 53 percent of the comets, rising to 72 percent for the 40° latitude limit. The 13 comets that ought to have been detectable near their previous opposition are shown in Table I. In terms of the opposition-region comets that *were* discovered, this represents a failure of 16 percent—considerably less than Everhart's 80 percent. The 16 percent is also an overestimate, in the sense that comet 1987 III, for example, expected at $m = 10$ at opposition $t = 47$ days before its actual discovery, would probably not have been recorded at opposition because it was then near the Milky Way; it is one of five comets probably rendered undiscoverable for this reason and noted by "MW" in the table). Comets 1991 I and 1992 XIV, which are of short period ("SP" in the table), must surely have brightened more rapidly than by an r^{-4} law; the former, an accidental rediscovery of P/Metcalf, previously observed only in 1906, undoubtedly experienced an anomalous brightening before discovery. Comets 1987 XXX and 1988 III are known to have separated from each other at their previous perihelion passage and may also therefore have brightened more rapidly than indicated. Actually, the r^{-4} law is always suspect when one tries to relate the total visual magnitude of a bright, nearby comet and the magnitude that would be recorded with a large telescope when the comet is much farther away. The comets computed here to be no brighter than magnitude 17 (shown with asterisks) would undoubtedly have been undetected in normal photographic searches. Perhaps the most surprising prior failure is 1984 XXIII, predicted to be at $m = 12$ and at opposition $t = 89$ days before discovery (but at a declination of $-37°$), and one wonders to some extent about the failure to record 1987 XXIX and 1990 III.

By way of stressing the problem of cometary brightness variability, it should be noted that two of the comets shown in Fig. 2 in the opposition region were discovered visually at $m = 9$ or 10 but that they clearly experienced significant brightness variations. Comet 1984 VII, a new short-period comet, had in fact been noticed by a photographic observer some three weeks earlier, when it was much fainter than would have been expected from geometry alone, and this observer had not reported the discovery because he could not confirm the object on an exposure two days before that; he did subsequently find the comet on that exposure, when it was four magnitudes fainter still. Comet 1988 I, only 0.4 AU from the earth at discovery, was expected to brighten further as it approached the earth and the sun, but instead, it quickly fizzled out.

Although photographic opposition searches were essentially unknown for most of 1883-1892—they came into play for asteroid work at the end of 1891—it is instructive to carry out the same exercise on the comets discovered in the non-opposition region during that decade. Of these comets the fraction that would have escaped opposition detection is similar to that during the latest decade—24 percent because $|\beta| > 60°$ and 56 percent because $|\beta| > 40°$. I adopt a limiting magnitude of 12.0, thereby clearly favoring our nineteenth-century predecessors, but this is a good practical bound for visual discoveries. More than 82 percent would therefore have escaped opposition detection on faintness grounds alone, increasing to 85 percent because of both faintness and high latitude. Only five comets would therefore have been potentially observable, and these are listed in Table II. The brightest one predicted, 1886 IV, was a short-period comet that has not been subsequently

TABLE II
1883-1892 comets at previous opposition.

Comet	β (deg)	t (day)	r (AU)	m	
1886 IV	+27.9	76	1.65	10	SP
1889 I	−14.6	227	4.61	12*	
1889 II	− 2.8	67	2.74	12*	
1889 VI	+ 3.3	77	1.73	12*	SP
1890 II	−23.5	209	3.76	10	

observed; it was probably therefore erratic in brightness, and in any case, it was located in the Coma-Virgo region, a booby-trap that visual comet hunters usually avoid. Comet 1889 VI was also of short period (SP) and undoubtedly too faint, and the other magnitude 12 entries (shown with asterisks) would be very borderline. The principal failure was comet 1890 II, although even in this case, extrapolation of the magnitude formula to the distance involved is probably questionable. One is tempted to claim that the comet searches actually made during the decade were almost completely successful.

4. The Comets of W. A. Bradfield

The champion visual discoverer of the twentieth century is William Bradfield, of Dernancourt, South Australia. He has found 16 comets since 1972, generally using a 0.15-m refractor at 26×, and it is quite remarkable that in no case has he had to share a comet with a co-discoverer. This is probably indicative of an absence of competition in the southern hemisphere, but it is also the case that, after Bradfield's first half-dozen or so discoveries, the Central Bureau made a point of announcing his comets very quickly—often before they had been confirmed by other observers, although that confirmation has always been forthcoming. Only two individuals, Jean-Louis Pons and William R. Brooks, have made more visual discoveries of comets than he has, but their actual totals are unclear, because the published records do contain objects that were not confirmed by others.

Bradfield's comets therefore make a convenient set for our continuing discussion, although it should be noted that 11 of the 16 were discovered during the three-month summer season beginning in mid-December. The relevant discovery information is listed in Table III; not surprisingly, all the comets were south of the ecliptic, and the fact that 10 of the 16 were in the morning sky (an "M" in the ϵ column, as opposed to "E" for evening sky) is surely the discoverer's personal preference. Two of the comets were particularly bright and at significantly less than 30° from the sun. The remainder were in the magnitude range 8-12 (all but three of them magnitude 9-10), and ten were confined to the surprisingly narrow range of $30° < \epsilon < 48°$ and $-38° < \beta < -16°$. The comets of $m = 11$ or 12 had $\epsilon > 40°$.

TABLE III
The comets of Bradfield.

Comet	β (deg)	ϵ (deg)	Δ (AU)	r (AU)	m	β (deg)	ϵ (deg)	m
1972 III	−23.1	M 30.7	1.67	0.96	10	− 9.0	E 9.6**	12*
1974 III	−28.5	E 32.9	1.57	0.92	9	−26.1	E 39.1	12*
1975 V	−26.7	E 30.4	2.02	1.27	9	−18.0	E 28.2*	10
1975 XI	−40.3	M 57.5	1.25	1.10	10	−32.2	M 60.2	13**
1976 IV	−41.6	E 56.2	0.78	0.85	9	−39.0	E 56.8	11
1976 V	−30.4	M 44.4	0.64	0.70	9	+ 1.9	M 25.8*	12*
1978 VII	−26.5	M 47.8	1.41	1.04	8	−35.6	M 53.7	11
1978 XVIII	−23.3	M 31.8	0.83	0.53	9	+ 2.0	E 6.6**	11
1979 VII	−19.2	E 43.9	1.16	0.82	10	−65.5	E 88.5	12*
1979 X	−13.7	M 26.5	1.21	0.55	5	− 1.9	E 15.5**	7
1980 XV	−14.6	M 21.7	1.19	0.45	6	−17.8	E 18.6**	11
1983 XIX	−25.9	M 46.4	1.84	1.37	11	−12.2	M 39.3	11
1987 XXIX	− 9.6	E 80.7	1.56	1.72	10	−19.5	E106.7	11
1988 XXIII	−37.9	E 40.5	1.35	0.88	12	− 8.2	M 14.4**	9
1992 VII	−20.2	M 65.2	1.14	1.16	10	−19.2	M 69.0	13**
1992 XIII	−16.6	M 47.2	0.87	0.76	10	− 6.3	M 34.0	14**

The last three columns of Table III also show β, ϵ and m 30 days before discovery. All of the comets should then have been fainter, except for comet 1988 XXIII, predicted at $m = 9$ but at an impossibly small elongation from the sun. Predicted magnitudes of 12 are asterisked, and fainter objects are double-asterisked. Elongations less than 30° are asterisked, and those less than 20° are double-asterisked. Obviously, the nine comets with a total of two or more asterisks could not then have been discovered. The three with single asterisks would have been very borderline for a visual discovery, and although comet 1979 VII was almost in the transition region, its high southern ecliptic latitude would count against discovery (particularly by Bradfield, for this was a discovery in June, not his favored time of the year). This leaves the four cases of 1976 IV, 1978 VII, 1983 XIX and 1987 XXIX, all calculated at $m = 11$. Since 1983 XIX was the first comet as faint as magnitude 11 that Bradfield has discovered, the 1976 and 1978 cases should probably also be classified as borderline. Perhaps he could have found 1983 XIX a month earlier. As for 1987 XXIX, the time of year (August) was again not favored, and it is not obvious that Bradfield's searches extend into the transition region anyway; but this comet might have been found then by some other observer.

Table IV shows the situation with regard to the Bradfield comets at their preceding opposition. Four of the comets (with double-asterisked magnitudes) would obviously then have been impossibly faint—two of them were also in the Milky Way—and another faint case (single asterisks denoting m = 17 or 18) was also in the Milky Way. Two more cases (with β double asterisked) were outside the SSR, and the five comets with a single asterisk had $|\beta| > 40°$. Six comets (1974 III, 1975 V, 1976 IV, 1979 X, 1980 XV and 1987 XXIX) would have been viable candidates for a prior discovery at opposition, and one of them (1980 XV) was in fact subsequently identified on a U.K. Schmidt plate taken about a month after opposition. The brightest comet, 1979 X, should have been a straightforward opposition discovery in the northern hemisphere, as should 1975 V, and the failure to detect 1987 XXIX has already been noted.

TABLE IV
Bradfield comets at previous opposition.

Comet	β (deg)	t (day)	r (AU)	m	
1972 III	+40.5*	171	2.93	15	
1974 III	−22.7	171	3.39	16	
1975 V	+24.9	207	3.34	14	
1975 XI	−24.6	317	5.32	20**	MW
1976 IV	−25.4	153	2.62	16	
1976 V	+45.6*	193	3.04	18*	
1978 VII	−59.4*	345	5.43	18*	MW
1978 XVIII	+59.5*	141	2.45	17*	
1979 VII	−78.6**	45	1.62	13	
1979 X	+36.8	130	2.33	12	
1980 XV	−27.4	179	3.40	16	
1983 XIX	+60.8**	363	4.39	18*	
1987 XXIX	−31.4	88	2.83	13	
1988 XXIII	+52.7*	436	5.42	23**	MW
1992 VII	−18.7	353	5.52	20**	
1992 XIII	+ 7.0	231	3.90	20**	

5. Concluding Remarks

Photographic searches for comets near opposition will continue to be effective, particularly for discovering short-period comets, but it would clearly be useful to extend the latitude range to $|\beta| = 60°$, as recommended in the Spaceguard Survey, even if this is accomplished at the expense of the present more extended range in longitude. Progressive replacement of photography by wide-field CCD operation would be effective in increasing the magnitude limit, and at some point in the not-too-distant future a CCD survey will become more efficient than a photographic survey in terms of sky coverage.

While it will continue to be the case that some comets, subsequently observable far from opposition, are always too faint for prior opposition detection, the Spaceguard search region has the principal disadvantage that some of the comets of high orbital inclination will be missed. The loss is not so great for comets of large q, but it increases to $\sim$ 25 percent for $q \lesssim 1$ AU. If one wishes to accommodate this problem by increasing the latitude coverage still further, it does become necessary to extend the longitude coverage—into the transition region and perhaps even beyond.

Extension of "professional" comet searches into the realm of the amateur is probably inevitable. It is rather curious, however, that photographic forays into this field have not been particularly productive. Non-apocryphal information on negative searches is difficult to assimilate, but it is not impossible that intrinsically faint long-period comets simply do not exist. Drummond *et al.* (1993) have recently mentioned a promising but unproductive survey covering $28^\circ < \epsilon < 40^\circ$ in 1989 by Martin Hoffmann and Edward Geyer with the 1.0-m Schmidt near Mérida, Venezuela. Of course, plate fogging in a twilit sky at the solar elongations favored by the most successful amateurs has obviously been a problem, there is magnitude loss at the inevitable low altitudes concerned, and many professional telescopes are not designed to operate at the extreme hour angles that would be necessary. As already remarked, the principal professional accomplishment at a moderately small ϵ was the discovery of comet 1977 IX. From the point of view of a discovery that could not otherwise be made, however, this was a wasted effort, for this particular comet has $q = 5.6$ AU and was later found on plates taken almost two years earlier! CCD searches at small elongations from the sun will have several advantages over photographic searches, and the experience of the visual searches suggests that it will be essential to experiment at small elongations before simply expanding the SSR about opposition.

Some will read these comments with sadness. The point about Spaceguard is that it is to be an *automated* survey. For more than two centuries comet hunting has been a sport, often highly competitive, and this competition has also been clear in many of the professional discoveries : this is one of the reasons that the professional discoveries encroach more on the morning side of opposition. Perhaps the time has already come for competition to be replaced by cooperation. The aim should not simply be to beat one's competition by a matter of a day or two, or even hours, but to ensure that, at a given time, the appropriate parts of the sky are properly covered to a useful magnitude limit. Automation implies that each comet found would be confirmed on a second night, with the necessary astrometry provided as a matter of course. Cooperation would undoubtedly also mean a departure from the custom of naming comets for their discoverers. I know of at least one successful comet hunter who would give up searching if this were done. To some extent this attitude is understandable, but it is rather pathetic that someone would hunt for comets *solely* because discoveries would be named for him. This is not science!

The emphasis of this paper has been on the general discovery of comets in the course of observing programs—professional or amateur—that may or may not have comet hunting as their principal aim. There are also searches for a specific comet or type of comet. These generally involve a one-dimensional sky search using a particular cometary orbit, generally treating the perihelion time as the only variable. Such a search led to the recovery of P/Swift-Tuttle in 1992, and the technique has also been used to search for members of the Kreutz sungrazing group. Kreutz searches were undertaken by a few observers in 1988, following the SMM discovery of 1987 XXII and XXV, only 12 days apart in the same orbit. It is evidently the case that the Kreutz sungrazers found near the sun by SMM later in 1988 and 1989 were all too faint for earlier detection in a dark sky. To the best of my knowledge, the only object actually discovered on the basis of the Kreutz ephemeris was the great comet 1970 II (Bennett), which was not a sungrazer at all! Bradfield's comet 1975 XI was found on the Kreutz track, as the discoverer quickly realized, but he does not specifically isolate this track in his comet hunting.

Another type of specific comet-search program is for comets near Jupiter in space, and thus within a few degrees of Jupiter in the sky. Such a search was discussed by Tancredi and Lindgren (1992), and these authors obtained the plates appropriate for their first search with the 1.0-m Schmidt at the European Southern Observatory around Jupiter's opposition in 1992. No cometary candidate was then found to a limiting magnitude—for a stationary object—of $B = 22.1$ (Tancredi and Lindgren 1993). As is now well known, such a comet, 1993e, was found a few days before Jupiter's opposition in 1993, but it was as bright as m = 12-13. First reported near opposition by the Shoemakers' near-earth-asteroid program, comet 1993e was independently discovered by at least three other groups, including Tancredi and Lindgren in their 1993 search. This is encouraging from the point of view of the completeness of general searches for comets near opposition. The discouraging part is that the comet must already have substantially brightened more than eight months earlier, around the time it passed inside Jupiter's Roche limit and broke up into at least eleven components. Jupiter was approaching conjunction at that time, and comet 1993e was evidently not bright enough for visual comet searches then or during the months following conjunction. Nevertheless, this is one case where the professional opposition-region searches might have been usefully extended into the morning sky!

References

Carusi, A., Kresák, Ľ, Kresáková, M. and Valsecchi, G. B. : 1991, "Observations of Periodic Comet d'Arrest in 1678 and implications for its evolutionary history." ***Astron. Astrophys.***, **252**,377–384.

Drummond, J., Rabinowitz, D. and Hoffmann, M. : 1993, "On the search for near-earth asteroids." In ***Resources of Near-Earth Space***, (J.S.Lewis, Ed.), in press, Univ. of Arizona Press.

Everhart, E. : 1967a, "Comet discoveries and observational selection." ***Astron. J.***, **72**,716–726.

Everhart, E. : 1967b, "Intrinsic distributions of cometary perihelia and magnitudes." ***Astron. J.***, **72**,1002–1011.

Kresák, Ľ. : 1982, "Comet discoveries, statistics and observational selection." In ***Comets***, (L.Wilkening, Ed.), 56–82, Univ. of Arizona Press.

Marsden, B. G. and Steel, D. I. : 1993, "Warning times and impact probabilities for long-period comets." In ***Hazards due to Comets and Asteroids***, (M.S.Matthews, Ed.), in press, Univ. of Arizona Press.

Morrison, D., *et al.* : 1992, ***The Spaceguard Survey***, NASA, Washington.

Rudenko, M. : 1986, "Catalogue of cometary discovery positions." ***Internat. Comet Q.***, **8**,117–129.

Tancredi, G. and Lindgren, M. : 1992, "The vicinity of Jupiter : a region to look for comets." In ***Asteroids, Comets and Meteors 1991***, (A.W.Harris and E.Bowell, Eds.), 601–604, Lunar and Planetary Inst., Houston.

Tancredi, G. and Lindgren, M. : 1993, "Searching for comets encountering Jupiter." ***Icarus***, submitted.

COMPUTER AIDED NEAR EARTH OBJECT DETECTION

JAMES V. SCOTTI

Lunar and Planetary Laboratory
The University of Arizona
Tucson, AZ, 85721, USA
E–mail jscotti@lpl.arizona.edu.INTERNET

Abstract.
The Spacewatch program at the University of Arizona has pioneered automatic methods of detecting Near Earth Objects. Our software presently includes three modes of object detection : automatic motion identification; automatic streak identification; and visual streak identification. For automatic motion detection at sidereal drift rates, the 4σ detection threshold is near magnitude $V = 20.9$ for nearly stellar asteroid images. The automatic streak detection is able to locate streaks whose peak signal is above $\sim 4\sigma$ and whose length is longer than about 10 pixels. Some visually detected streaks have had peak signals near $\sim 1\sigma$.

Between 1990 September 25 and 1993 June 30, 45 new Near Earth asteroids, two comets and two Centaur's have been discovered with the system. An additional six comets, five Near Earth asteroids, and one Centaur were also "re–discovered". The system has directly detected for the first time Near Earth Objects in the complete size range from about 5 kilometers to about 5 meters. Each month ~2,000 main belt asteroids are also detected.

Future upgrades in both hardware, software, and telescope aperture may allow an order of magnitude increase in the rate of discovery of Near Earth Objects in the next several years. Several of the techniques proposed for the Spaceguard Survey have already been tested by Spacewatch, and others will need to be tested in the near future before such a survey can be implemented.

1. Introduction

The University of Arizona's 0.91-meter Spacewatch Telescope on Kitt Peak is being used during 18 nights centered on the new moon each lunation to survey for Near Earth Objects (NEOs). A Tektronix TK2048E thinned backside illuminated Charge Coupled Device (CCD) with 2048 × 2048 pixels of 24μm size is placed at the f/5 Newtonian focus. The resulting image scale is 1.076 arcseconds per pixel. The CCD is normally operated in slow scanning mode with the accumulating electronic charge being transferred along the CCD rows in sync with the drift of the sky across the CCD. Scans are then 2048 pixels high in declination and an arbitrary number of rows long in right ascension. A survey region consists of three scans ("passes") of the same length at the same location on the sky. Survey regions are usually selected near the opposition point and are normally done sequentially or by alternating passes from different survey regions. The resulting interval between passes is typically between thirty minutes and one hour. The field of view is ~32 arcminutes tall in declination and the sidereal drift time across the CCD is 147 seconds at the equator and varies with the secant of the declination. Survey regions of 10 frames in length result in 3.3 square degrees of area coverage every 1.5 hours. The limiting magnitude is $V_{lim} \sim 20.9$ at about 4σ for stellar images.

A. Milani et al. (eds.), Asteroids, Comets, Meteors 1993, 17–30.

Prior to the 1992 September observing run, a Tektronix TK2048 thick frontside illuminated CCD with 2048 × 2048 pixels of 27μm size was used. The lower quantum efficiency but larger pixels resulted in $V_{lim} \sim 20.5$, an image scale of 1.211 arcseconds per pixel, an integration time of 165 seconds at the equator and a field of view ~40 arcminutes tall.

2. Detection methods

The pedigree of our automated asteroid detection software includes analogs of the methods used in detection moving objects on photographic plates. For example, fast electronic blinking and registration and subtraction of pairs of frames were both used (for example, see Taff, 1981), but each of these methods rely on humans for the actual detection of the moving asteroid image.

The Moving Object Detection Program (MODP) was designed to combine automated detection methods with a real-time user interface (Rabinowitz, 1991). MODP includes three modes of object detection.

The automated motion detection method is used for detecting moving objects whose appearances are nearly stellar and whose rates of motion do not exceed about 1 degree per day. The first implementation of this method is described by McMillan, *et al.* (1986). Objects whose peak signals are above $\sim 3.5\sigma$ and whose rates of motion are below about 1.0 degrees per day are within the detection limits of automated motion detection. Near opposition, the orbital characteristics of a moving object correlate well with the resulting ecliptic angular rates of motion (Bowell, *et al.* 1990). The observer compares the observed rates of motion of a newly detected object with the results of a simulated sample of orbits at the observed opposition geometry (Rabinowitz, 1991; Scotti, *et al.* 1992).

Figure 1 shows an example of a simulated rate plot. Near Earth Asteroids (NEAs), even at the distance of the Main Belt can usually be distinguished by their ecliptic rates of motion alone. Figure 2 is a sample screen from MODP showing the third pass detection of 1991 VG, whose brightness was about $V = 20.8$ and whose rate of motion at discovery was $\sim 0.8°/day$. Three typical main belt asteroids are also shown.

The automated streak detection method is used to detect relatively bright trailed images of fast moving nearby objects. This method is effective in identifying trails whose length is longer than about 10 pixels and whose peak signal is above $\sim 4\sigma$. Figure 3 is a sample screen from MODP showing the automated detection of 1992 JD which was magnitude $V = 16.7$ and moving $\sim 13.8°/day$ at the time of discovery.

The visual streak detection method allows the observer to detect the images of faint fast moving objects whose images are not bright enough to be detected automatically. MODP allows the observer to identify faint trailed images and to compute rates and predicted future positions for trails. The real-time response of the observer has allowed faint Very Fast Moving Objects (VFMO's) whose rates of motion have approached 100 degrees per day and whose peak signals have been $\leq 1\sigma$ to be followed (Scotti, *et al.* 1991; Rabinowitz, *et al.* 1993). Figure 4 shows eight trailed images of 1991 BA, a magnitude V=18.2 object moving $\sim 27.5°/day$ at the time of discovery. It was observed over about 4.6 hours on 1991 January 18.

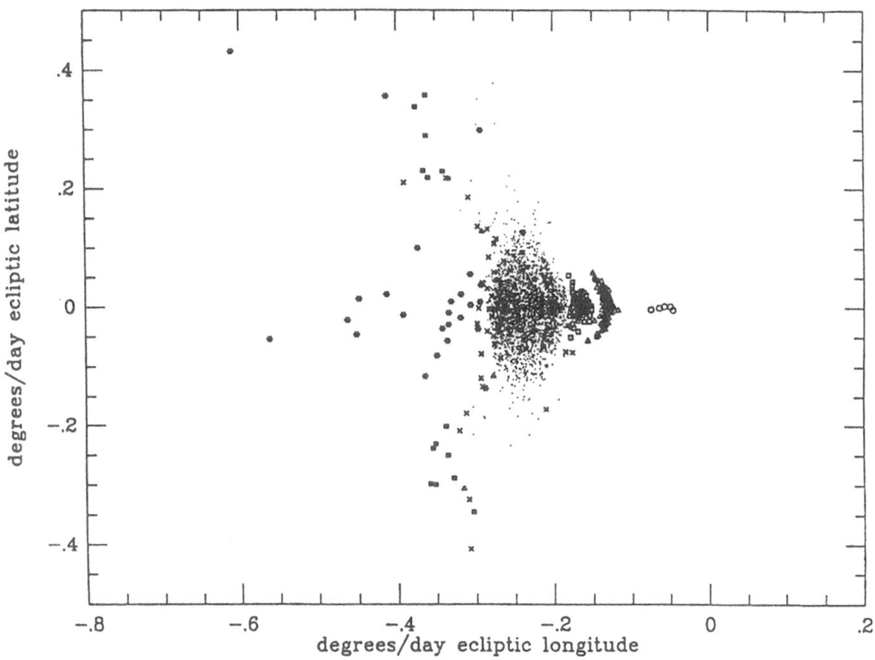

Fig. 1. A simulated rate plot of asteroids near opposition. Dots represent Main Belt asteroid rates, and (*) represent the rates of near Earth asteroids. (.) are main belt asteroids, (×) are Mars crossing asteroids, (□) are Hungaria type asteroids, (△) are Phocaea type asteroids, (□) are Hilda type asteroids, (△) are Trojan type asteroids and (o) are Centaur type asteroids.

3. Astrometry

During the first pass of a survey region, MODP identifies the images of Hubble Space Telescope Guide Star Catalog Stars ("GSC" stars hereafter, see Russell, *et al.* 1990) that are present within the scan and measures their locations. A linear least-squares solution of the measured pixel coordinates with respect to the apparent "of date" right ascension and declination are made for the set of identified GSC stars. Later, when measurements of an objects pixel coordinates are made, those coordinates can then be transformed into "astrometric" J2000 right ascension and declination. Figure 5 shows plots of a typical set of GSC star residuals with respect to the solution for that scan. One can often identify the original GSC plate boundaries in the residual plots. The scan to scan consistency of the GSC star measurements is ~ 0.1 arcseconds. The systematic trends in the residuals are reproducible from scan to scan.

The astrometric measurements of moderately bright asteroids produces residuals

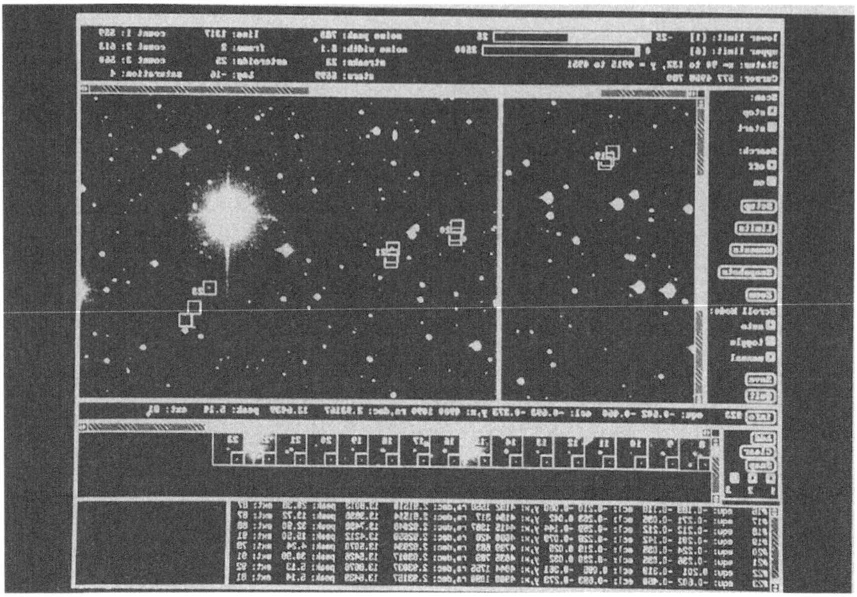

Fig. 2. The MODP display during the automated motion detection of asteroid 1991 VG. Three other main belt asteroids are also marked by MODP in the same display. The image of 1991 VG is indicated by the the small boxes widely spaced on the right. The narrower spaced boxes mark the 3 anonymous main belt asteroids.

on the order of a few tenths of an arcsecond. Table I shows parts of two typical pages from the Minor Planet Circulars showing the orbit solution and resulting residuals for two asteroids detected by Spacewatch. 1993 DT_1 is a case where all of the astrometry was produced completely hands-off with the observer only indicating that the object was real. All the measurements of the GSC stars and of the asteroid were made automatically. 1993 DT_1 was approximately magnitude V=19. (5693) 1993 EA is an NEA discovered and specifically followed up by Spacewatch. (5693) 1993 EA was about magnitude V=19 at the time of discovery and faded past V=20 during the last measurements reported in this sample. For bright objects ($V < 20$), the consistency of residuals during any night are within $\sim$ 0.2 arcseconds. For fainter objects ($V > 20$), the nightly consistency is within $\sim$ 0.6 arcseconds. The night-to-night consistency of residuals reflects the global precision of the GSC and is normally better than $\sim$ 1.0 arcseconds.

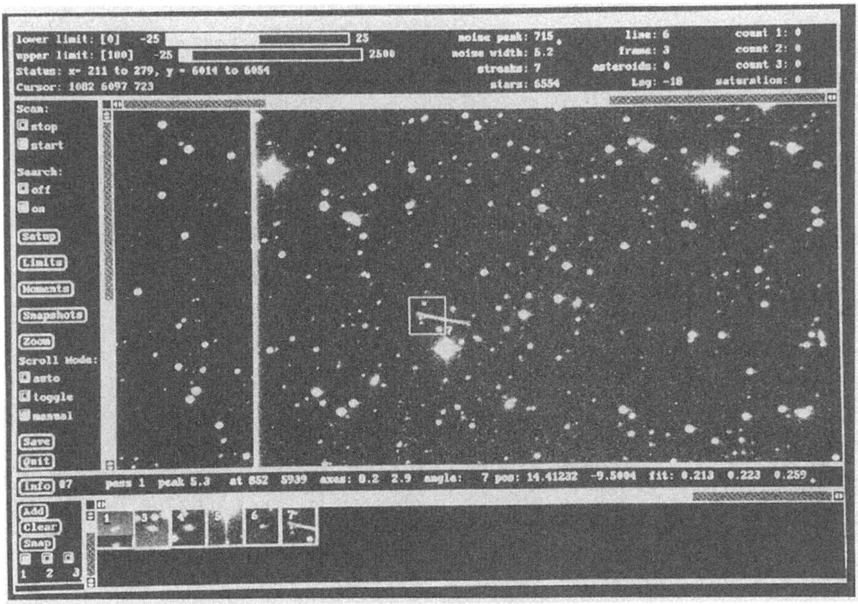

Fig. 3. The MODP display during the automated streak detection of asteroid 1992 JD.

4. Survey results

4.1 NEAR EARTH ASTEROIDS

Spacewatch discovered 1990 SS on 1990 September 25, making this the first fully automatic discovery of a NEA. During 29 lunations through 1993 June 30, a total of 45 new NEAs have been discovered with Spacewatch. Amongst the objects discovered by Spacewatch are the smallest asteroids ever detected and the objects making the closest observed approaches to the Earth. Five previously known NEAs were "re-discovered" during this time.

Figure 6 shows a histogram of the number of NEAs discovered per absolute magnitude, H. Spacewatch discoveries show a bimodal distribution of larger distant objects with a peak near $H = 21$ and a second peak of smaller nearby objects with a peak near $H = 27$. Spacewatch finds large NEAs far from the Earth and small NEAs close to the Earth. The first direct sampling of NEAs over the entire range $15 < H < 29$ has been made.

Table II shows the number of NEAs discovered during each of the first three years, scaled to a 10 month observing season (normally September through June), distributed according to the size of the object. The first VFMO was found in

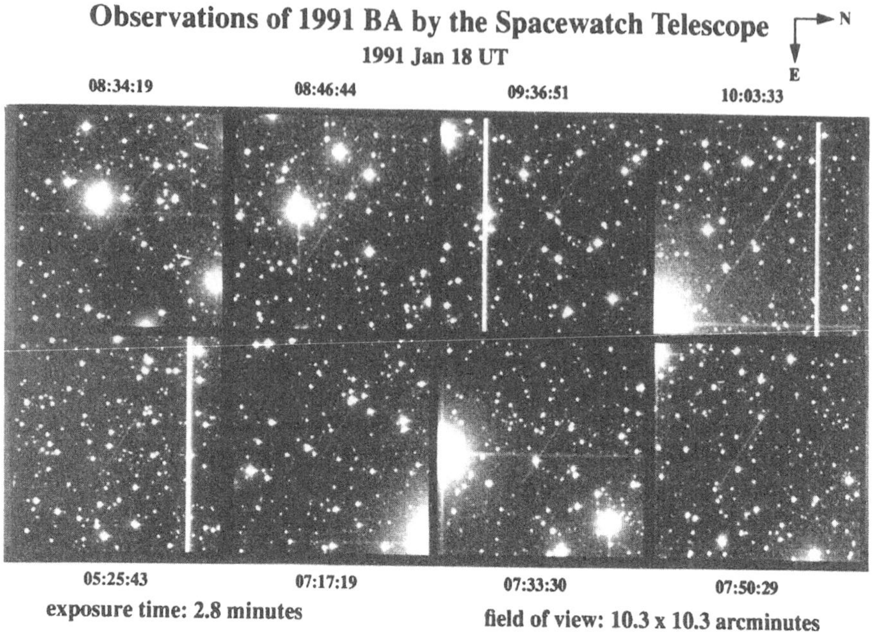

Fig. 4. The trailed image of 1991 BA from Scotti, *et al.* 1991.

January 1991, during the fifth month of the first season. More VFMOs may have been recognized during the first year had the significance of their long trailed images been recognized earlier. Also reflected in the table is a decreased discovery rate due to poorer weather conditions during the second season. The increased detection rate during the third season is due to the installation of a new thinned, backside illuminated Tektronix CCD with significantly higher quantum efficiency than that of our earlier detector.

The discovery of objects smaller than 100 meters diameter has led to the identification of an enhancement in the flux of small objects compared to an extrapolation of the magnitude/frequency of the larger NEAs down to the sizes of the small objects. These discoveries have allowed the first direct measurements of the magnitude/frequency distribution from the size of fireballs, to the smallest photographically discovered NEAs whose smallest object is estimated to be a few hundred meters in diameter and to the larger NEAs (Rabinowitz, 1992; 1993).

The first analysis of the orbital characteristics of these small objects was done by modeling the detection biases of Spacewatch and comparing the orbits of the detected small objects with the orbits of objects with a simulated initial orbit distribution of the larger NEAs. The magnitude/frequency distribution of the small objects has suggested that their orbits are different than those of large NEAs. The

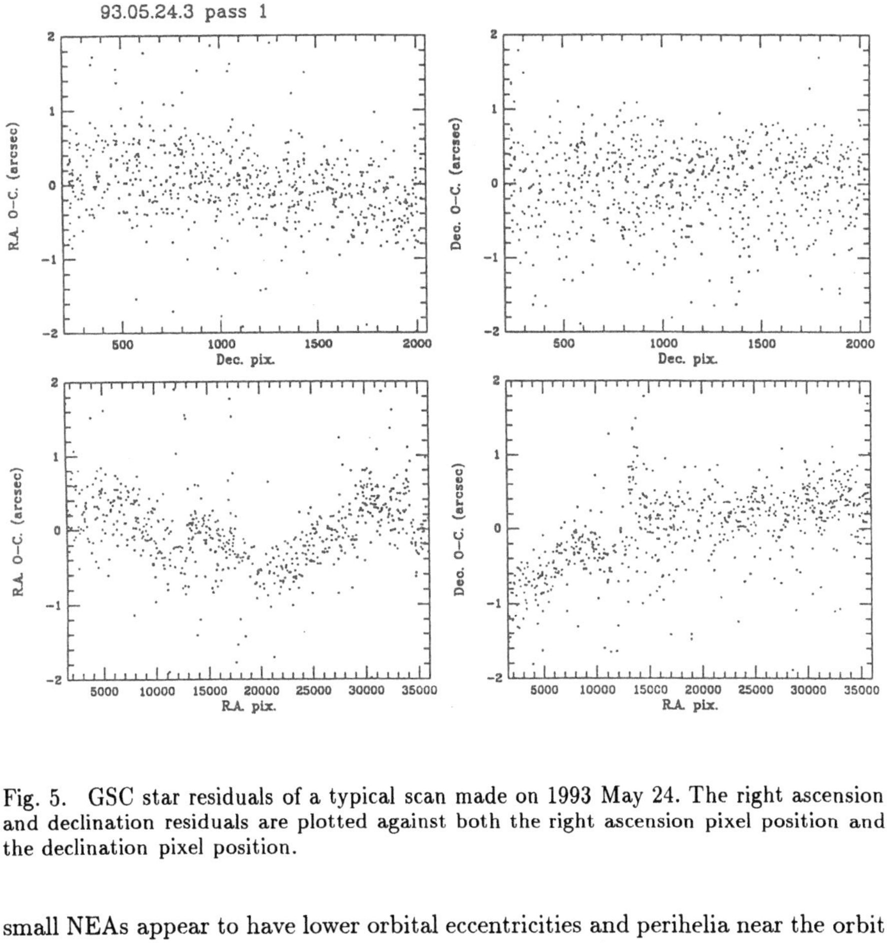

Fig. 5. GSC star residuals of a typical scan made on 1993 May 24. The right ascension and declination residuals are plotted against both the right ascension pixel position and the declination pixel position.

small NEAs appear to have lower orbital eccentricities and perihelia near the orbit of the Earth (Rabinowitz, 1993; Rabinowitz, *et al.* 1993).

Preliminary physical studies of some of the small NEAs also indicates that they have significantly different colors than the larger NEAs and main belt asteroids (Rabinowitz, *et al.* 1993).

4. 2. CENTAURS

Two new trans-Saturnian "Centaur" asteroids have been discovered by Spacewatch. (5145) Pholus (=1992 AD) was discovered on 1992 January 9 near its perihelion at 8.7 AU, just inside the distance of Saturn. On 1993 April 26, asteroid 1993 HA_2 was discovered just outside of its perihelion distance of 11.2 AU. Three weeks after the discovery of Pholus, (2060) Chiron was accidentally "re-discovered" and its subtle

TABLE I

Comparison of automated astrometry with orbit calculations. The residuals for each published observation of the object are shown with the date (YYMMDD), the observatory code (Spacewatch is identified by observatory code 691), and the right ascension and declination residuals in arcseconds.

M. P. C. 22 060 — 1993 MAY 6

1993 DT1 = 1981 JA6

Epoch 1993 Aug. 1.0 TT = JDT 2449200.5					Williams
M 315.98082			(2000.0)	P	Q
n	0.23556870	Peri.	55.00602	-0.25802517	+0.96594907
a	2.5965159	Node	200.06695	-0.90357903	-0.24827420
e	0.2413425	Incl.	3.19364	-0.34200578	-0.07281694
P	4.18	H	15.0	G 0.15	

Residuals in seconds of arc

810508	675	1.0-	0.1+	930303	691	0.4+	0.2-	930303	691	0.2-	0.4-
810509	675	1.0+	0.1-	930303	691	0.1-	0.1-	930319	691	0.2+	0.3+
930226	691	0.2+	0.3+	930303	691	0.2-	0.3-	930319	691	0.2-	0.0
930226	691	0.3-	0.2+	930303	691	0.1+	0.1-	930319	691	0.1+	0.2-
930226	691	0.1-	0.1+	930303	691	0.2+	0.3+				

M. P. C. 22 585 — 1993 SEPT. 30

(5693)* 1993 EA = 1984 AJ

Discovered 1993 Mar. 3 by Spacewatch at Kitt Peak.

Id. B. G. Marsden (MPC 22060)

Epoch 1993 Aug. 1.0 TT = JDT 2449200.5					Marsden
M 284.81470			(2000.0)	P	Q
n	0.68703159	Peri.	258.60945	+0.99360789	+0.07144495
a	1.2719936	Node	97.24980	-0.03233376	+0.92191972
e	0.5854749	Incl.	5.05463	-0.10815676	+0.38073566
P	1.43	H	17.0	G 0.15	

Residuals in seconds of arc

840110	675	0.8-	1.4+	890513	413	(1.7+	3.8-)	930413	691	0.1-	0.6-
840110	675	1.1+	0.1-	890513	413	0.8-	0.1+	930413	691	0.4-	0.1-
860408	413	1.3-	0.2+	930303	691	0.1+	0.4-	930427	691	0.3-	0.5-
860413	413	1.7+	1.1-	930303	691	0.1+	0.6-	930427	691	0.4-	0.4-
860413	413	0.3+	0.9+	930303	691	0.2+	0.5-	930511	413	1.3+	0.6+
860413	413	0.1+	1.0-	930303	691	0.2+	0.6-	930514	691	0.6-	0.2-
860413	413	0.5+	0.4-	930303	691	0.2+	0.5-	930514	691	0.6-	0.3-
890414	413	0.5+	0.9-	930303	691	0.0	0.3-	930514	691	0.1+	0.4+
890414	413	0.5-	0.9+	930304	691	0.0	0.2-	930526	691	0.8-	0.4+
890502	675	(3.7+	3.6-)	930304	691	0.1+	0.3-	930526	691	0.7-	0.5+
890502	675	(1.8+	3.6-)	930304	691	0.1-	0.3-	930526	691	0.4-	0.0
890503	675	(0.2-	3.6-)	930319	691	0.2+	0.4+	930617	691	0.7-	0.6-
890503	675	(8.8+	9.3-)	930319	691	0.0	0.2+	930617	691	1.1-	0.9-
890504	675	1.4+	1.7+	930319	691	0.0	0.3+	930617	691	1.3-	0.6-
890504	675	0.0	1.4+	930413	691	0.2-	0.7-				

cometary appearance was recognized at the time of observation.

Pholus has been found to be extraordinarily red, possibly indicating that this object is a primordial object or an object unmodified by close approaches to the sun

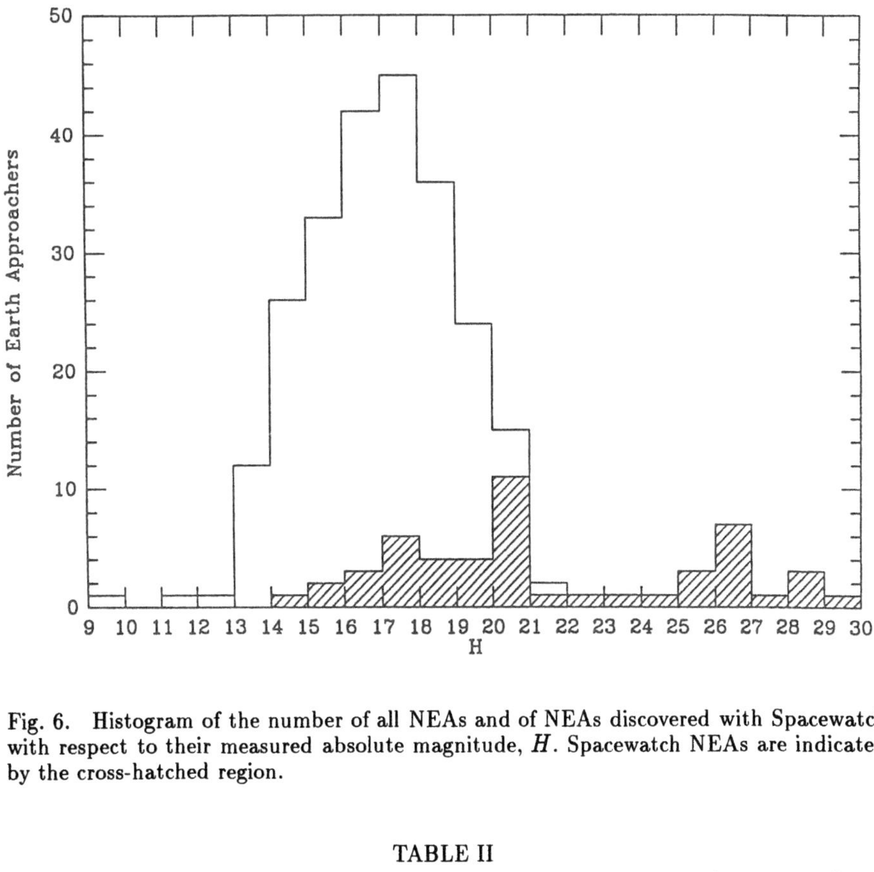

Fig. 6. Histogram of the number of all NEAs and of NEAs discovered with Spacewatch with respect to their measured absolute magnitude, H. Spacewatch NEAs are indicated by the cross-hatched region.

TABLE II

Table of the number of NEAs discovered per size per ten month observing season. Large refers to objects with diameter $D > 1.0$ km ($H < 18.3$), medium refers to 1.0 km $> D > 0.1$ km ($18.3 > H > 23.3$), and small refers to $D < 0.1$ km ($H > 23.3$).

	#	large	medium	small
1990–1991	16	8	6	2
1991–1992	12	3	2	7
1992–1993	24	6	9	9

(Mueller, *et al.* 1992; Fink, *et al.* 1992; Buie and Bus, 1992). No evidence has yet been found for cometary outgassing by Pholus (Zellner, *et al.* 1992). 1993 HA_2 has also recently been found to be extremely red, perhaps redder than Pholus (Tholen, 1993).

The dynamical evolution of Pholus is chaotic and presently under the control of

Saturn. It is presently in the 1 : 3 mean-motion resonance with Saturn. The half-life to ejection from the solar system is about 2 Myrs. The probability of Pholus-like orbits to evolve to (or from) a Jupiter-family short period comet orbit is about 40% and about 1–2% per million years evolve to Earth crossing orbits (Asher and Steel, 1993; Bailey, *et al.* 1993; Bailey, *et al.* 1994a;).

These Centaurs may be the largest members of a population of objects in intermediate orbits between the Jupiter Family comets and the Kuiper disk or Oort cloud. The possible significance of one of these large objects becoming a Jupiter Family comet or even a NEA has only begun to be considered (Bailey, *et al.* 1994b)

4. 3. COMETS

Two new comets have been discovered during the course of the Spacewatch Survey and six other comets (not including the re-discovery of Chiron) were also rediscovered. The long faint tail of comet P/Spacewatch (1991x) was recognized by the observer on 1991 September 8. The comet was nearly a year past perihelion and its total coma magnitude was near V=21, making it the faintest comet yet at the time of discovery. Comet Spacewatch (1992h) was discovered on 1992 May 1. This comet was the first new comet discovered automatically by Spacewatch. It was detected by its motion and the extremely subtle coma was only recognized by the observer. Spacewatch software is not yet able to discriminate comets by their appearance.

A preliminary estimate of the number of comets that Spacewatch should have detected suggests that there should have been at least twice as many comets discovered than have actually been found. This discrepancy is still not understood.

4. 4. MAIN BELT ASTEROIDS

The automatic motion detection software finds between 15 and 30 asteroids per square degree when surveying close to the opposition point and an average of about 10 to 15 asteroids per square degree during a month of surveying at a large range of opposition geometries. The automated astrometry discussed in section 3 has allowed completely automatic, hands off astrometry of every asteroid detected. Normally, astrometric measurements are made for each of the 3 images of each asteroid detection. Approximately 2000 asteroid detections are thus made available each month to the Minor Planet Center. Approximately 20% of these objects have been identified either with known asteroids or with other objects detected by Spacewatch (Williams, 1992). These observations can also be made available to other researchers.

5. Future Directions

5. 1. PHYSICAL AND ASTROMETRIC FOLLOW-UP

The rate at which NEA's are being detected has been increasing over the past 10 years from 7 in 1983 up to nearly 40 in 1993. That rate should increase to about 300 per year within 5–10 years as planned survey programs and improvements to existing survey programs are implemented. There will be a corresponding increase in the need for astrometric and physical follow-up on these new objects in order to characterize their dynamical evolution and hazard to the Earth, and to study their composition and origin.

The projected area coverage of these near term surveys will not allow complete astrometric follow-up to be built into the survey strategy. An astrometric follow-up strategy needs to consider the brightness distribution of the discovered NEA's, the requirements for adequate orbit determination so that at least the largest detected objects can be recovered at some time in the future, and the limited time before the object becomes unavailable due to its faintness or its viewing geometry. In order to maximize the discovery rate, the survey telescopes should not contribute significant follow-up observations. Dedicated astrometric follow-up telescopes with apertures comparable to the survey telescopes should be used.

In order to estimate the magnitude of astrometric follow-up required in the near term, we assume that new NEA's are identified by their rates of motion near opposition as they are presently found by Spacewatch and that about half of the objects to be followed are objects with rates at the margins of the range expected for NEA's, and will turn out not to be NEA's. The objects which are not found to be NEA's are assumed to require only 2 observations in addition to the discovery observation in the discovery lunation and 2 additional observations the following lunation. NEA's are assumed to be observed twice in the discovery lunation, twice in the following lunation, and once in each of two subsequent lunations. Using reasonable integration times with an unfiltered Tektronix 2048 $\times$ 2048 pixel CCD array and accounting for its long readout time, this astrometric follow-up program would keep a 0.9 meter telescope occupied full time for about 10 nights per lunation. A 1.8 meter telescope would be occupied full time for about 7 nights per lunation.

Follow-up of objects in order to determine color and light curves requires much more telescope time than that required for astrometry. Filtered photometry could be combined with astrometric follow-up, but the increased exposure times required will frequently result in trailed reference stars, degrading the astrometry. A larger aperture would be preferable and the additional load would saturate the monthly load on a single 1.8 meter telescope dedicated to doing the astrometric follow-up and a minimal program of photometric follow-up. Ideally, several larger aperture telescopes outfitted with CCDs would share the load and be distributed globally for better coverage and weather insurance.

5. 2. THE SPACEGUARD SURVEY

Many of the necessary methods that will be required to implement a survey to locate the most dangerous NEOs have been demonstrated by Spacewatch. In the process, a number of challenges have also been identified.

As proposed by the Spaceguard Report (Morrison, 1992; Morrison and Chapman, 1993), the Spaceguard survey would be comprised of six 2-3 meter class telescopes outfitted with CCD detectors. The goal of the Spaceguard survey would be to detect the majority of NEAs larger than $\sim$ 1 km in a reasonable interval of time ($\sim$ 20 years). The Spaceguard survey would detect about 2 orders of magnitude more NEAs each month than are presently being found. Although the Spaceguard survey has not been funded and is not likely to be implemented exactly as described by Morrison (1992), it serves as a useful model for future NEO survey programs.

In order to be most effective, these telescopes will have to make the best use of the focal plane and field of view by incorporating large acreage CCD's and/or CCD mosaics with the highest detective quantum efficiencies and an optimal image scale (Morrison, 1992). Presently, scanning CCDs may produce the most efficient use of observing time due to the relatively long readout times of currently available astronomical quality large format CCDs. Faster readout times might make staring efficient as long as the problem of flat fielding the CCD is properly addressed. Faster scanning readout with the telescope moving along a great circle will require processing of tens of Gigabytes of raw data per night at each telescope. About 2000 main belt asteroids and 4 NEAs would be found each night with each telescope, resulting in the need for extensive follow-up support either by the survey telescope or by dedicated follow-up telescopes. With the full complement of six telescopes, extensive coordination will be required so that the survey and follow-up proceeds most efficiently with the least duplication of effort. If all six telescopes are available, there will be enough sky coverage to build most of the follow-up into the survey strategy by multiple coverage. If, however, there is not enough sky coverage, then the follow-up will greatly tax the system (see section 5. 1.).

Surveying should cover the area of opposition at least to ± 30 degrees in ecliptic longitude and ± 60 degrees in ecliptic latitude. In order to locate the Athens and to locate any previously unknown classes of objects, surveying far from opposition should also be done (Morrison, 1992). Dangerous comets, for example, will require maximal sky coverage and their danger will persist even after all of the dangerous NEAs have been identified (Marsden and Steel, 1994).

During a Spaceguard Survey, the majority of NEA detections will be of objects smaller than about 1 km diameter. A significant number of detectable NEAs will be VFMOs whose diameters are under 100 meters and which are passing through the neighborhood of the Earth-Moon system. The study of these objects will be very attractive to the researcher who is interested in studying the entire solar system, but the follow-up of such objects will drain the resources of the Spaceguard network if the best orbital data are desired. A decision will have to be made early on for each object. In general, perhaps objects whose estimated diameters are $\ll 1$ km should not be followed, or only a random subset should be followed if such a survey is to be successful in finding objects with diameters larger than 1 kilometer.

6. Summary

The Spacewatch project has demonstrated the feasibility of automated NEO detection. Fainter detection limits and inherently higher detective quantum efficiencies of CCDs with respect to traditional photographic detection methods for NEAs has more than compensated for the limited field of view of the physically much smaller detectors.

The foreseeable improvements in both telescopes and detectors can produce an order of magnitude increase in the rate of discovery of NEOs in the next few years. An additional order of magnitude is possible with a modest additional improvement in telescopes and detectors as recommended by the Spaceguard Report. The improvements are within our present technological capabilities.

Continued discovery of NEOs with Spacewatch will allow not only the detection of potentially hazardous large NEOs, but also may unlock the new puzzle of the origins of the small NEOs that thus far have only been detectable by Spacewatch, and may also produce discoveries of other classes of solar system objects. Other automated NEO detection systems have been started and may come on line in the near future, in particular project EUNEASO (Hahn and Maury, 1993). Others are in their initial design stages or are awaiting funding (For example, see Steel, *et al.* 1993).

With some modification of the survey technique, the large members of the population of the outer solar system can also be studied as the discovery of the Centaurs have indicated. Understanding of the characteristics of this population will aid in the understanding of the evolution of objects from the outer solar system into the observed comet population and perhaps also the observed NEA population.

Acknowledgements

The author would like to thank T. Gehrels, R. S. McMillan, D. L. Rabinowitz and an anonymous referee for their comments and suggestions during the preparation of this manuscript. The author would also like to thank A. Maury for some helpful discussions. This work was supported by the U.S. Air Force Office of Scientific Research; the NASA Planetary Astronomy Program; the National Geographic Society; the NSF Astronomical Instrumentation Program; private donations; and the director of Steward Observatory by his generous dedication of the 0.91-meter Spacewatch Telescope to our research programs.

References

Asher, D.J., and Steel, D.I. : 1993, "Future orbital evolution of giant Comet/Asteroid (5145) Pholus = 1992 AD." In ***Observations and Physical Properties of Small Solar System Bodies*** (J. Surdej and J.C. Gerard, Eds.) ***Proc. 30th Liège Int. Astrophys. Coll.***, Univ of Liège, Belgium, 263–266.

Bailey, M.E., Chambers, J.E., Hahn, G., Scotti, J.V., and Tancredi, G. : 1993, "Transfer probabilities between Jupiter and Saturn-family orbits : Applications to 1992 AD = 5145." In ***Observations and Physical Properties of Small Solar System Bodies*** (J. Surdej and J.C. Gerard, Eds.) ***Proc. 30th Liège Int. Astrophys. Coll.***, Univ of Liège, Belgium, 285–287.

Bailey, M.E., Chambers, J.E., Hahn, G., Scotti, J.V., and Tancredi, G. : 1994a, "Chaotic orbital evolution of 5145 Pholus." In preparation.

Bailey, M.E., Clube, S.V.M., Hahn, G., Napier, W.M., and Valsecchi, G.B. : 1994b, "Hazards due to Chaotic Chirons : Climate and short-term Catastrophism." In ***Hazards due to Comets and Asteroids*** (T. Gehrels, Ed.) in the press.

Bowell, E., Skiff, B.A., Wasserman, L.H., and Russell, K.S. : 1990, "Orbital information from Asteroid motion vectors." In ***Asteroids, Comets, Meteors III*** (C.-I. Lagerkvist, H. Rickman, B.A. Lindblad, and M. Lindgren, Eds.) 19–24.

Buie, M.W., and Bus, S.J. : 1992, "Physical observations of (5145) Pholus." ***DPS Meeting abstracts, Munich, 1992.***

Fink, U., Hoffmann, M. Grundy, W., Hicks, M., and Sears, W. : 1992, "The steep red spectrum of 1992 AD : an Asteroid covered with organic material?" *Icarus*, **97**, 145–149.
Gehrels, T. : 1991, "Scanning with Charge–Coupled Devices." *Space Sci. Rev.*, **58**, 347–375.
Hahn, G., and Maury, A. : 1993, "Project EUNEASO An initiative for a European Near Earth Asteroid Search Observatory." In *Observations and Physical Properties of Small Solar System Bodies* (J. Surdej and J.C. Gerard, Eds.) *Proc. 30th Liège Int. Astrophys. Coll.*, Univ of Liège, Belgium, 239–241.
Marsden, B.G., and Steel, D.I. : 1994, "Detection, warning times, impact probabilities and Impact Velocities for Long-period Comets." In *Hazards due to Comets and Asteroids* (T. Gehrels, Ed.) in the press.
McMillan, R.S., Scotti, J.V., Frecker, J.E., Gehrels, T., and Perry, M.L. : 1986, "Use of a scanning CCD to discriminate Asteroid images moving in a field of stars." In *Instrumentation in Astronomy - VI* (D.L. Crawford, Ed.) *Proceedings of the SPIE*, **627**, 141-154.
Morrison, D. (Ed.) : 1992, "The Spaceguard Survey : Report of the NASA International Near-Earth-Object Detection Workshop." Jet Propulsion Laboratory/California Institute of Technology, Pasadena, California.
Morrison, D., and Chapman, C.R. : 1993, "Impact hazard and the International Spaceguard Survey." In *Observations and Physical Properties of Small Solar System Bodies* (J. Surdej and J.C. Gerard, Eds.) *Proc. 30th Liège Int. Astrophys. Coll.*, Univ of Liège, Belgium, 223–230.
Mueller, B.E.A., Tholen, D.J., Hartmann, W.K., and Cruikshank, D.P. : 1992, "Extraordinary colors of Asteroidal object (5145) 1992 AD." *Icarus*, **97**, 150-154.
Rabinowitz, D.L. : 1991, "Detection of Earth-approaching Asteroids in near real time." *Astron. J.* **101**, 1518–1529.
Rabinowitz, D.L. : 1992, "The flux of small Asteroids near the Earth." In *Asteroids, Comets, Meteors 1991* (A.W. Harris and E. Bowell, Eds.), 481–485.
Rabinowitz, D.L. : 1993, "The size distribution of the Earth-approaching Asteroids." *Astrophys. J.* **407**, 412–427.
Rabinowitz, D.L., Gehrels, T., Scotti, J.V., McMillan, R.S., Perry, M.L., Wisniewski, W., Larson, S.M., Howell, E., and Mueller, B.E.A. : 1993, "Evidence for a near-Earth asteroid belt." *Nature* **363**, 704–706.
Russell, J.L., Lasker, B.M., McLean, B.J., Sturch, C.R., and Jenkner, H. : 1990, "The Guide Star Catalog. II. Photometric and astrometric models and solutions." *Astron. J.* **99**, 2059–2081.
Scotti, J.V., Gehrels, T., and Rabinowitz, D.L. : 1992, "Automated detection of asteroids in real-time with the Spacewatch Telescope." In *Asteroids, Comets, Meteors 1991* (A.W. Harris and E. Bowell, Eds.), 541–544.
Scotti, J.V., Rabinowitz, D.L., and Marsden, B.G. : 1991, "Near miss of the Earth by a small Asteroid." *Nature* **354**, 287–289.
Steel, D.I., McNaught, R.H., and Russell, K.S. : 1993, "The Anglo-Australian Near Earth Asteroid Survey." In *Observations and Physical Properties of Small Solar System Bodies* (J. Surdej and J.C. Gerard, Eds.) *Proc. 30th Liège Int. Astrophys. Coll.*, Univ of Liège, Belgium, 219–221.
Taff, L.G. : 1981, "A new asteroid observation and search technique." *Publ. Astron. Soc. Pacific* **93**, 658–660.
Tholen, D.J. : 1993, personal communication.
Williams, G.V. : 1992, personal communication.
Zellner, B., Meech, K.J., Weaver, H., and Noll, K. : 1992, "Imaging of 1992 AD with the Hubble Space Telescope" *DPS Meeting abstracts, Munich, 1992.*

THE KUIPER BELT

JANE LUU[1]

Physics Department,
Stanford University
Stanford, CA 94305, USA

Abstract. The existence of a belt of comets in the outer solar system (the "Kuiper belt") has been postulated for a variety of reasons, including the need for a source for the short-period comets. The existence of the belt seems supported by the discoveries of the trans-Neptunian objects 1992 QB_1, 1993 FW, 1993 RO, 1993 RP, 1993 SB, and 1993 SC. If these objects are members of the Kuiper belt, crude lower limits on the belt population can be established from the discoveries. The Kuiper belt comets are likely to be primordial remnants of the disk from which the solar system accreted. According to the current theories of cometary nucleus evolution, these objects are expected to possess mantles ("irradiation mantles") which are different from mantles of comets which have been heated to the point of sublimation ("rubble mantles"). Kuiper belt comets on their way to short-period comet orbits may exist among the Centaur objects.

1. Introduction

Nearly 200 yrs after Edmond Halley discovered that comets were permanent members of the solar system, Jan Oort wrote his classic paper on the origin of comets (Oort 1950). He proposed the existence of the "Oort cloud", a vast swarm of 10^{12} comets which lies at the edge of the solar system and forms the source of the long-period (LP) comets (orbital periods > 200 yrs). Oort's theory is still accepted today. However, the notion of the Oort cloud also supplying the short-period (SP) comets (orbital periods < 200 yrs) has come under attack in the last two decades. Joss (1973) pointed out that the Oort cloud could not adequately explain the observed flux of the SP comets. This problem might be solved if a very massive ($\sim$ 700 M_{Earth}, where M_{Earth} is one Earth mass) inner Oort cloud existed (Bailey 1986). Alternatively, it was suggested that the SP comets were derived from a belt of low-inclination comets beyond the orbit of Neptune (e.g., Fernández 1980). The latest serious challenge of the Oort cloud as the major source of the SP comets came from Duncan *et al.* (1988), who demonstrated that the inclination distribution of comets with large perihelia (perihelion distance $q \leq 30$ AU) is conserved during capture. As a consequence, the prograde, low-inclination orbits of the SP comets were incompatible with a spherical source such as the Oort cloud. Duncan *et al.* then reiterated the hypothesis that a better alternative source may be found in a comet belt beyond the outer planets.

The concept of a population of trans-Neptunian objects is not a new one. It was first championed by Gerard Kuiper, who did not like the apparent emptiness of the outer solar system (Kuiper 1951). Kuiper saw no reason for accretion to end

[1] Hubble Fellow

A. Milani et al. (eds.), Asteroids, Comets, Meteors 1993, 31–44.

abruptly at Neptune. As he saw it, the planetary system formed by coagulation in a disk of gas and dust orbiting the Sun. Near the outer edge of the disk, the icy-dusty material condensed into comet-size planetesimals which later agglomerated to form the cores of the outer planets. Planet formation stopped at Neptune because, beyond this distance, the low disk density combined with the progressively longer orbital periods to prohibit the formation of planet-size bodies on a reasonable time scale ($\sim$ a few 10^7 yrs). However, a disk of icy planetesimals could exist, and lying beyond the outer planets, may be able to survive until today. This disk of partially agglomerated icy planetesimals, christened the Kuiper belt, was presumed to lie in the 40–100 AU region, just beyond Neptune. From the massive cores of Uranus and Neptune (several tens of Earth masses), Kuiper estimated that his disk of icy planetesimals was comparably massive.

Following Kuiper, the concept of a residual comet belt in the outer solar system was investigated by numerous researchers (e.g., Whipple 1964, Cameron 1972, Safronov 1972, Fernández 1980). Fernández and Ip (1981) showed that such a belt could be stable over the age of the solar system, and Duncan *et al.* (1989) discovered stable bands between Saturn and Uranus and between Uranus and Neptune, which may contain members from the Belt. In this paper I will concentrate on the observational aspects of the Kuiper belt. The interested reader who wishes to learn more about the dynamical aspects of the Kuiper belt is referred to reviews by Fernández and Ip (1991) and Duncan and Quinn (1993).

2. Observations of the Kuiper Belt

The Oort cloud is completely unobservable with current technology and thus escapes the scrutiny of observers (a Halley-size comet at distance 10^4 AU would have magnitude ~ 61). But the Kuiper belt is not so fortunate. At heliocentric distance $R \sim 40$ AU (roughly the inner edge of the belt), its largest members are within the reach of available optical telescopes.

The earliest surveys for outer solar system objects employed photographic plates and yielded the planet Pluto (Tombaugh 1961) and the distant comet Chiron (Kowal 1989). The first survey based on modern technology, (i. e., CCDs) was performed by Luu and Jewitt (1988), who covered 0.34 square degrees to the limiting magnitude $m_R = 24$ mag with the Michigan-Dartmouth-MIT 1.3-m telescope on Kitt Peak, Arizona. Their strategy was as follows :

1) Consecutive images were taken of fields centered on the ecliptic. The ecliptic plane (or near it) is where most solar system objects, including the SP comets, are likely to be concentrated.

2) The fields have to be located near opposition, in order to allow the estimation of the distance of any object found during the survey. At opposition, most of an object's motion is due to parallax. Assuming concentric, coplanar circles for the orbits of the Earth and the Kuiper belt candidate, the apparent motion, $d\theta/dt$ [arcsec/hr] of the candidate is

$$\frac{d\theta}{dt} = 148\left(\frac{1 - -R^{-0.5}}{R - -1}\right)$$

where $R > 1$ is the heliocentric distance in AU.

3) Observations for the survey were carried out at the time of the equinoxes, when the Milky Way is at high ecliptic latitude. This minimizes the density of field stars and facilitates the identification of faint, slow moving objects.

4) Three consecutive images were taken of each field, followed by a fourth image separated by at least 2 hrs in time. The images were then displayed rapidly ("blinked") to detect moving objects. Although 2 images are sufficient to identify motion, the redundancy eliminates confusion with cosmic ray hits in the field.

More detail on the search strategy can be found in Luu and Jewitt (1988). The survey found no new objects beyond Saturn.

The Luu and Jewitt survey was followed by searches by several other groups, the largest of which was carried out by Levison and Duncan (1990), who used the U. S. Naval Observatory 40-in. telescope to cover nearly 5 sq. degrees to $m_V = 22.5$ ($m_R \sim 22$). No evidence for the Kuiper belt was found until August 1992, when Jewitt and Luu (1993a) discovered the faint, slow-moving object 1992 QB_1 with the University of Hawaii 2.2m. Other discoveries followed quickly : the second Kuiper belt candidate, 1993 FW, was found 6 months later, followed by 4 more objects (1993 RO, 1993 RP, 1993 SB, 1993 SC) 6 months after that. The discoveries are summarized in Table I.

TABLE I

Trans-Neptunian Objects

Object	a [AU]	e	i [deg]	Mag	Discovery Date	Discoverers
1992 QB_1	43.8	0.07	2.20	22.8	Aug 1992	Jewitt and Luu (1993a)
1993 FW	42.4	0.04	7.75	22.8	Mar 1993	Luu and Jewitt (1993a)
1993 RO	32.3	0.00 [a]	2.53	23.2	Sep 1993	Jewitt and Luu (1993b)
1993 RP	35.4	0.00 [a]	2.80	24.5	Sep 1993	Luu and Jewitt (1993b)
1993 SB	33.2	0.00 [a]	2.28	22.7	Sep 1993	Williams *et al.* (1993)
1993 SC	34.5	0.00 [a]	5.57	21.7	Sep 1993	Williams *et al.* (1993)

[a] assumed

The problem of determining the orbits of the newly discovered trans-Neptunian objects is much more difficult than that involving asteroids and comets in the inner planetary region, due to the large distances and long orbital periods of the trans-Neptunian objects. Continuing astrometry is essential, yet the task is complicated by these objects' faint appearances. 1992 QB_1 has been confirmed to have a nearly circular orbit beyond Neptune (Marsden 1993) but it is yet too early to tell for the other objects.

A graphical summary of recent outer solar system surveys can be found in Fig. 1, which plots the surface densities of the searches versus their limiting magnitudes. The Figure seems to suggest that flux-limited surveys (i.e., all the mentioned searches) are likely to succeed by covering at least ~ 1 sq. deg. to $m_R \sim 23$.

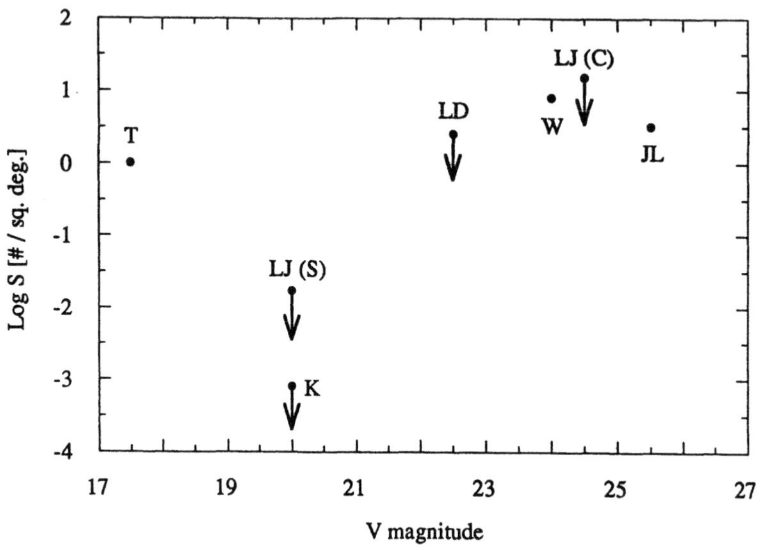

Fig. 1. Surface density of objects beyond Neptune as a function of limiting magnitudes for various surveys. Downward arrows indicate 99%-confidence upper limits. The symbols are : T = Tombaugh (1961), K = Kowal (1989), LJ (S) = Luu and Jewitt (1988, with Schmidt), LJ (C) = Luu and Jewitt (1988, with CCD), LD = Levison and Duncan 1990, JL = Jewitt and Luu (1993), W = Williams *et al.* (1993).

3. The Kuiper Belt

A sample of six objects hardly allows meaningful statistical analysis, and furthermore, it is not clear that all six objects belong to the postulated Kuiper belt. In this section, for the sake of argument, we assume that they are members of the belt, and we explore the implications of the discoveries upon the Kuiper belt population.

3.1 MASS AND POPULATION

The six objects were detected in ~ 1.5 sq. degree of the sky. 1993 FW currently possesses the most inclined orbit at ~ 8 deg. The projected area of the belt as seen from the Earth is thus 2 x 8 deg x 360 deg, or 5760 sq. degrees. The extrapolated minimum number of objects brighter than $m_R \sim 25$ is then $N_{min} \sim 2 \times 10^4$ objects.

Assuming a density of 1000 kg m^{-3}, the minimum mass of the belt (counting only objects with diameter larger than $\sim$ 100-km) is $M \geq 0.03 M_{Earth}$.

It should be emphasized that this estimated mass can be no more than a lower limit, as all the surveys up to date are necessarily biased toward the largest, closest members of the belt. Smaller, as well as more distant, objects must exist which are currently beyond the practical limit of ground-based telescopes. Furthermore, the Kuiper belt is likely to be thicker than 16 deg (the inclination distribution of the SP comets is $\sim$ 30 deg). We note that this minimum mass is consistent with previous mass estimates of the Kuiper belt. The absence of perturbations on the orbit of comet Halley placed an upper limit of $M_{belt} \leq 1 M_{Earth}$ (Hamid *et al.* 1968, Yeomans 1986, Hogg *et al.* 1991), while Duncan *et al.* (1988) estimated at least $M_{belt} \sim 0.02\ M_{Earth}$ was needed to produce the observed flux of SP comets. Stronger constraints on the mass of the Kuiper belt can only be reached with an understanding of the size and spatial distribution of the belt.

We know very little about the size distribution of the Kuiper belt. If it resembles the asteroid belt, the familiar asteroidal power-law size distribution predicts that large objects should be more abundant than smaller objects. So far, the surveys have yielded fewer faint objects near the detection limit ($m_R \sim 25$) than the relatively bright objects such as 1992 QB_1. The apparent "scarcity" of smaller objects may be understood if we keep in mind that the Jewitt and Luu survey is flux-limited, not volume-limited. In a flux-limited survey, the limiting magnitude determines the volume of space, as well as the size range, that the survey can sample. Because of the dependence on geometry, the apparent size distribution is most likely not the true size distribution. Furthermore, the Kuiper belt size distribution may also be very different from an asteroidal size distribution. Again, the character of the belt population can only be revealed with a larger sample of possible belt members.

3. 2. DYNAMICS

The dynamical properties of the belt have not been clearly established. Although the belt has been identified as an alternative source for SP comets, dynamical processes operating within the belt are poorly understood. The various mechanisms suggested to transfer comets into the inner solar system include perturbations by the largest members of the belt (Duncan *et al.* 1988, Ip and Fernández 1991), and chaos. Torbett and Smoluchowski (1990) discovered a chaotic zone beyond Neptune for objects with perihelia between 30 and 45 AU. Holman and Wisdom (1993) confirmed the important role of chaos in delivering SP comets from the Kuiper belt, and pointed out that the delivery process may be analogous to the delivery of meteorites from resonances in the asteroid belt (Wisdom 1985, Froeschlé and Scholl 1986). Numerical simulations also revealed that the belt was likely to have an inner edge near 40 AU (Holman and Wisdom 1993), beyond which existed stable regions interspersed with unstable regions. Although its inner edge may be slowly eroded, the Kuiper belt thus could survive for the age of the solar system and can be the source of the SP comets (Gladman and Duncan 1990, Levison 1991).

The orbit of 1992 QB_1 appears stable against perturbations by Neptune on timescales of 10^9 yrs (Levison and Duncan 1993). At the present time, 1993 RO,

1993 RP, 1993 SB and 1993 SC have semimajor axes estimated at 32–35 AU. If these estimates are correct, some of these objects are likely to have future close encounters with Neptune. Once on Neptune-crossing orbits, the comets may be on their way to the inner planetary region. A fraction of the Neptune-crossing comets will be gravitationally scattered by the giant planets into Jupiter-crossing orbits, from which they may become observable SP comets (Everhart 1977). The timescale in which the transfer process takes place is short compared to the age of the solar system ($10^5 - -10^7$ yrs, Gladman and Duncan 1990). Is it then possible to observe Kuiper belt comets during their journey to the inner planetary region?

We may witness some of these transition comets in the Centaur population. Centaurs are objects on elliptical orbits which cross the orbits of one or more giant planets. Planet-crossing orbits are very unstable, evolving on the timescale of $10^5 - -10^6$ yrs (Oikawa and Everhart 1979), implying that the Centaur must have originated somewhere else, quite possibly the Kuiper belt. The known Centaurs currently include 2060 Chiron, 5145 Pholus and 1993 HA_2, all with semimajor axes in the Saturn-Uranus region. These objects also possess unusual physical properties that suggest a non-asteroidal origin, and they will be discussed in more detail in section 5.2.

3. 3. FUTURE WORK

Several surveys of the outer solar system are still in progress, fueled by the evidence that the Kuiper belt may exist and is within the grasp of ground-based telescopes. Most surveys employ medium to large-size telescopes ($\sim 2-4$ m range) and are aimed toward optical detection of the belt comets via reflected sunlight. An alternative observational strategy with great promise relies on the detection of Kuiper belt comets via their occultation of background stars (Bailey 1976, Dyson 1992). Axelrod *et al.* (1992) outline a survey which would make use of a row of small telescopes equipped with wide field CCDs. The telescopes would monitor a large number of stars of small angular diameter to search for very short duration occultations, during which the stars would appear to blink out because a comet blocks their light. This strategy has the advantage of being more sensitive to more numerous smaller (as opposed to the largest and rarest) Kuiper belt objects, as well as relying only on dedicated small, inexpensive telescopes and is thus impervious to the ubiquitous problem of oversubscription on medium-large telescopes.

4. Surface Composition : the Irradiation Mantle

Whether in the Oort cloud or the Kuiper belt, comets are continually exposed to galactic cosmic rays, consisting of $\sim$ 93% protons, 6% helium nuclei, and 1% heavier nuclei (Meyer *et al.* 1974). The mean energy deposited in a cometary nucleus over 4.5 billion years was independently estimated by Donn (1976) and Whipple (1977) to be $\sim 3 \times 10^{19}$ MeV cm^{-2}, enough to drive significant chemical modification of the surface layers. For example, irradiation of mixtures of astrophysically relevant ices such as H_2O, NH_3, CH_4, N_2, CO and CO_2 with a dose of $\sim$ 100 eV/molecule produced C_2H_6, CO_2, CO, N_2O, NO and CH_4 (see TABLE II, also review by Strazzulla and Johnson 1991). The typical dose estimated to be deposited in the top 20-cm layer of a comet stored in the Oort cloud is ~ 40 eV/molecule (Moore *et al.* 1983). It is not clear what fraction of the newly formed radicals and molecules will survive until the comet's first entry into the planetary region. What is postulated to survive the residence time in the Oort cloud/Kuiper belt, however, is a crust of nonvolatile organic residue (Moore *et al.* 1983, Johnson *et al.* 1987, Khare *et al.* 1989). This residue has been observed in laboratory experiments with irradiated ices, and is formed by bombardment of carbon and sulfur-containing species. Evidence for such organic refractory residues in space exists in the 3.4 μm feature seen in the galactic center sources IRS7 and Sgr AW, and identified with residues created in the laboratory (Greenberg 1987). The formation of the residue appears independent of the particle energy in the range ~ 100 eV to 1 MeV.

TABLE II

Identified products of irradiated ices[a]

Irradiated Ices	Irradiation type	Identified products
$H_2O + NH_3 + CH_4$	proton	C_2H_6, CO_2, $^{13}CO_2$, CO, N-N-N species
$H_2O + N_2 + CO_2$	proton	^{12}CO, ^{13}CO, NO, CH_4, CO_2
$H_2O + N_2 + CO$	proton	$^{12}CO_2$, $^{13}CO_2$, CH_4, CO_2
$NH_3 + CO_2$	UV	CN, HCO, NH_2, NH_2CO, CO_2, H_2CO, NH_4^+
$H_2O + NH_3 + CO + CO_2$	UV	NO, HCO, CO_2, NO_2^-, NH_4^+, OCN^-, $HCONH_2$

[a] The products of proton-irradiated ices were adopted from Moore *et al.* (1983), while those of UV-irradiated ices were adopted from Grim *et al.* (1989).

vertically

The typical chemical reaction when ices are irradiated is the preferential liberation of hydrogen, encouraging the formation of carbon bonds such as C-C, C-N, C-O, C-S. The irradiated ices become increasingly darker as part of the volatile ice mass is converted into the refractory residue (Moore *et al.* 1983, Khare *et al.* 1989). The residue thickness is a function of the initial porosity and particle fluence (Strazzulla and Johnson 1991). In general, it is observed to increase proportionally with the received irradiation, and is predicted to reach a few meters over the age of the solar system. The crust of thickened residue is usually referred to as the "irradiation mantle," to be distinguished from the "rubble mantle" which is a direct consequence of sublimation (e.g., Jewitt 1992). The irradiation mantle should cover most of the surface due to the isotropy of the galactic cosmic ray field and the rotation of the nucleus. The irradiated crust is also likely to be porous, with cracks and fissures in the mantle which allow gas diffusion upon heating. The thickness of the crust, the gas diffusion capability, together with the slow rate of heating by a comet as it drifts toward the Sun, have led Strazzulla *et al.* (1991) to believe that a significant part of the crust should be able to survive the first few orbits in the planetary region. A rough illustration of the effects of irradiation upon the nucleus surface is given in Fig. 2.

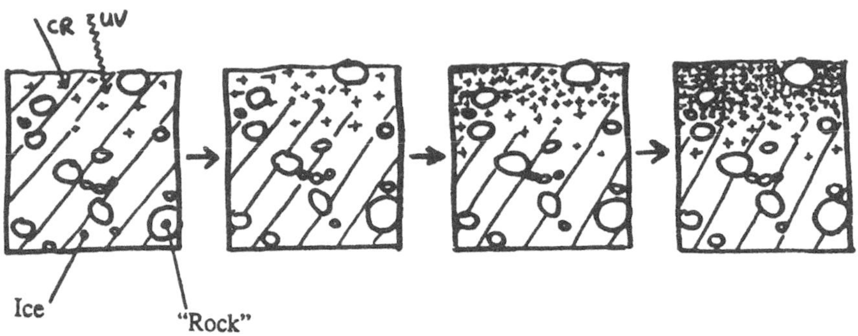

Fig. 2. The creation of an irradiation mantle. The nucleus surface, consisting of ice and rocky material, is irradiated by high energy particles and produces an irradiation-induced crust. Figure adapted from Jewitt : 1992, "

It is not known how or whether comets in the Oort cloud physically differ from those in the Kuiper belt. Both suffer from long-term exposure to cosmic ray irradiation, but where their evolutionary paths diverge is not known. Available observations indicate that the interplanetary cosmic ray intensity should reach the average large scale galactic value at roughly the distance of the belt ($R \sim$ 50-60 AU, Webber 1987) so the irradiation damage may not be much different between the Oort cloud and Kuiper belt comets. On the other hand, the Kuiper belt straddles the termination shock ($R \sim$ 50 AU, Suess 1990) and the heliopause ($R \sim$ 80 AU, Webber 1987) and may experience additional processing particular to these areas (e.g., cosmic ray production from the interaction of interstellar neutral gas with the solar wind, Suess 1990). It has also been suggested that the Oort cloud comets are heated by passing stars (Stern and Shull 1988), whereas the Kuiper belt comets would be much less susceptible to their effects to their smaller distances. The main effect of UV photons on cometary ices such as H_2O, CH_4 and NH_3 is to break the molecular bonds, leaving behind radicals (e.g., TABLE II, also Greenberg 1983). These radicals may remain as is or recombine to form new molecules. Thus, if the heating effect of passing stars is significant, it may result in somewhat different surface compositions between the Oort cloud and Kuiper belt comets. These hypotheses remain untested until observations of the nuclei of long-period comets are available.

5. Properties of Kuiper Belt Candidates

5. 1. THE TRANS-NEPTUNIAN OBJECTS

At the time of discovery, the apparent magnitudes of the trans-Neptunian objects range from $m_R \sim 22.8$ to 24.5. What little is known about their physical properties is summarized in TABLE III. Assuming an albedo of 0.04 (typical of known comet nuclei), they have diameters in the range 100–250 km. The few colors available indicate that they are red, although possibly with different degrees of redness. 1992 QB_1 and 1993 RO seem to be substantially redder than the Sun, while 1992 FW possesses a more moderate red color.

TABLE III
Properties of Trans-Neptunian Objects

	Diameter [km]	V - R	R - I	Cometary Activity?
1992 QB_1	250 [a]	0.6 ± 0.1 [b]	1.0 ± 0.2 [b]	No
1993 FW	250 [a]	0.4 ± 0.1 [c]	---	No
1993 RO	120 [a]	---	---	No
1993 RP	100 [a]	---	---	No
1993 SB	160 [a]	---	---	No
1993 SC	280 [a]	---	---	No
Solar colors [d]		0.32	0.4	

[a] approximate, assuming an albedo of 0.04
[b] Jewitt and Luu (1993a)
[c] Luu and Jewitt (1993a)
[d] Fernie (1983) and Hartmann *et al.* (1990)

We have few constraints on the origin of these objects, since the size distribution is not known. The collision time between bodies with diameters ≥ 100 km is comparable to or greater than the age of the solar system, but collisions with smaller, more abundant objects are likely. However, collisions in the Kuiper belt occur at lower velocities than in the asteroid belt (a few x 0.1 km s^{-1} as opposed to a few x km s^{-1}) and therefore should be less disruptive.

5. 2. THE CENTAURS

A summary of the known physical properties of the Centaurs is provided in TABLE IV.

TABLE IV
Properties of Centaur Objects

	Diameter [km]	V - R	R - I	S' [a] [%/10^3 Å]	Cometary Activity?
2060 Chiron	≤ 300 [b]	0.34 ± 0.03 [c]	0.28 ± 0.03 [c]	-3.9 ± 0.4 [d]	Yes
5145 Pholus	189 ± 26 [e]	0.7 [f]	0.72 ± 0.06 [f]	46.1 ± 0.3 [d]	No
1993 HA_2	60 [g]	~ 0.7 [h]	----	----	No
Solar colors [i]			0.32	0.4	

[a] Normalized reflectivity gradient
[b] 3σ upper limit (Jewitt and Luu 1992)
[c] Hartmann *et al.* (1990)
[d] Luu (1993)
[e] minimum diameter (Davies *et al.* 1993)
[f] Mueller *et al.* (1992)
[g] assuming an albedo of 0.04
[h] Jewitt and Luu 1994, unpublished data
[i] Fernie (1983) and Hartmann *et al.* (1990)

Classified as an asteroid upon discovery, Chiron is now known to be a comet (Tholen *et al.* 1988, Meech and Belton 1989). No sign of cometary activity has been detected in Pholus or 1993 HA_2, but Pholus possesses a very red color (V-R = 0.7; e.g., Mueller *et al.* 1992) which may be rivaled only by 1993 HA_2 and the trans-Neptunian objects (see TABLE III). A good (but non-unique) match for the spectra of Pholus has been found in tholins, organic residues formed by irradiation of CH_4 and other organic molecules (Davies and Sykes 1992, Wilson *et al.* 1994). A tholin-rich surface is consistent with an irradiation mantle and an origin in the Kuiper belt. From thermal measurements, Davies *et al.* (1993) reported that the albedo of Pholus could not exceed 0.044 ± 0.013, corresponding to a minimum diameter of 189 ± 26 km. The low albedo is similar to that of known cometary nuclei (Jewitt 1991). Recent work by Yabushita (1993) lends emphasis to the irradiation mantle hypothesis by showing that large (radius ≥ 40 km) nuclei are especially susceptible to irradiated ices, since they have a layered structure where highly volatile materials are pushed toward the surface.

The red color of Pholus stands in sharp contrast with the neutral-blue color of Chiron (Luu and Jewitt 1990, Luu 1993). The color difference has been attributed to the presence of cometary activity and the lack thereof on Pholus. In this scenario, sublimation has blown off or covered with debris the original irradiation mantle of Chiron, while that of Pholus remains intact. We caution that this theory is only speculative and can be tested if cometary activity is detected in Pholus or 1993 HA_2.

6. Conclusions

Our understanding of the putative Kuiper belt and its members may be summarized as follows :
1) The existence of the Kuiper belt is supported by the discovery of the trans-Neptunian objects 1992 QB_1, 1993 FW, 1993 RO, 1993 RP, 1993 SB and 1993 SC. The orbit of 1992 QB_1 is consistent with a residence in the Kuiper belt. The orbits of the other trans-Neptunian objects are still indeterminate at the time of writing.
2) The Kuiper belt is accessible to ground-based studies and, as a fossil from the formation of the solar system, is an important reservoir of primordial material. This reservoir awaits exploration and may yield important information on early solar system conditions.
3) The dynamics of the Kuiper belt remain largely unexplored. Recent numerical simulations suggest that chaos and gravitational scattering by large belt members play a role in the delivery of comets from the belt to the inner planetary region.
4) Kuiper belt objects are expected to retain primordial irradiation mantles, produced by prolonged irradiation of ices by cosmic rays. The irradiation mantles may exhibit diverse colors.
5) The Centaur objects (2060 Chiron, 5145 Pholus, and 1993 HA_2) may be Kuiper belt comets in transition to short-period comet orbits.

Acknowledgements

I thank MDM Observatories and the University of Hawaii for allocation of telescope time, and Brian Marsden for his invaluable help with orbit determination. Support for this work was provided by NASA through grant # HF-1035.01-92A awarded by the Space Telescope Science Institute which is operated by the Association of Universities for Research in Astronomy, Inc., for NASA under contract NAS5-26555.

References

Axelrod, T. S., Alcock, C., Cook, K. H. and Park, H.-S. : 1992, "A direct census of the Oort Cloud with a robotic telescope." In ***Robotic Telescopes in the 1990's,*** (A. V. Filippenko, Ed.), ASP Conference Series, vol. 34, 171–181, Astronomical Society of the Pacific, San Francisco.

Bailey, M. : 1976, "Can "invisible" bodies be observed in the Solar System?" ***Nature,*** **259**, 290–291.

Bailey, M. : 1986, "The mean energy transfer rate to comets in the Oort cloud and implications for cometary origins." ***Mon. Not. Roy. Astron. Soc.***, **218**, 1–30.

Cameron, A. G. W. : 1972, "The formation of the sun and planets." ***Icarus***, **1**, 13–69.

Davies, J. and Sykes, M. : 1992, "IAU Circular 5480".

Davies, J., Spencer, J., Sykes, M., Tholen, D. and Green, S. : 1993, "IAU Circular 5698."

Donn, B. : 1976, "The nucleus : panel discussion." In ***The Study of Comets*** (B. Donn *et al.*, Eds.), 611–621, NASA SP-393, Government Printing Office, Washington, DC.

Duncan, M., Quinn, T. and Tremaine, S. : 1988, "The origin of short-period comets." ***Astrophys. J. Lett.***, **328**, L69–L73.

Duncan, M., Quinn, T. and Tremaine, S. : 1989, "The long-term evolution of orbits in the solar system : a mapping approach." ***Icarus***, **82**, 402–418.

Duncan, M. J. and Quinn, T. : 1993, "The long-term dynamical evolution and stability of the solar system." In ***Protostars and Planets III*** (E.H. Levy, J.I. Lunine, M.S. Matthews, Eds.), 1371–1394, University of Arizona Press, Tucson.

Dyson, F. J. : 1992, "Hunting for comets and planets." ***Quart. J. Roy. Astron. Soc.***, **33**, 45–57.

Everhart, E. : 1977, "The evolution of comet orbits as perturbed by Uranus and Neptune." In ***Comets-Asteroids-Meteorites*** (A. H. Delsemme, Ed.), 99–107, University of Toledo Press, Ohio.

Fernández, J. A. : 1980, "On the existence of a comet belt beyond Neptune." ***Mon. Not. Roy. Astron. Soc.***, **192**, 481–491.

Fernández, J. A. and Ip, W. H. : 1981, "Dynamical evolution of a cometary swarm in the outer planetary region." ***Icarus***, **47**, 470–479.

Fernández, J. A. and Ip, W. H. : 1991, "Statistical and evolutionary aspects of cometary orbits." In ***Comets in the Post-Halley Era*** (R. L. Newburn, Jr., M. Neugebauer, J. Rahe, Eds.), 487–535, Kluwer, Dordrecht.

Fernie, J. D. : 1983, "Relationships between the Johnson and Kron-Cousins VRI photometric systems." ***P. A. S. P.***, **95**, 782–785.

Froeschlé, Ch. and Scholl, H. : 1986, "The secular resonance ν_6 in the asteroidal belt." ***Astron. Astrophys.***, **166**, 326–336.

Gladman, B., and Duncan, M. : 1990, "On the fate of minor bodies in the outer solar system." ***Astron. J.***, **100**, 1680–1693.

Greenberg, J.M. : 1983, "The largest molecules in space : interstellar dust." In ***Cosmochemistry and the Origin of Life*** (C. Ponnamperuma, Ed.), 71–1, Reidel, Dordrecht.

Greenberg, J. M. : 1987, "Basic laboratory studies of grains." In ***Astrochemistry*** (M. S. Vardya, S. P. Tarafdar, Eds.), 501–523, Reidel, Dordrecht.

Grim, R. J. A., Greenberg, J. M., de Groot, M. S., Baas, F., Schutte, W. A. and Schmitt, B. : 1989, "Infrared spectroscopy of astrophysical ices : new insight in the photochemistry." ***Astron. Astrophys. Suppl. Ser.***, **78**, 161–186.

Hamid, S. E., Marsden, B. and Whipple, F. : 1968, "Influence of a comet belt beyond Neptune on the motions of periodic comets." ***Astron. J.***, **73**, 727.

Hartmann, W. K., Tholen, D. J., Meech, K. and Cruikshank, D. P. : 1990, "2060 Chiron : Colorimetry and cometary behavior." ***Icarus*** **83**, 1–15.

Hogg, D. W., Quinlan, G. D. and Tremaine, S. : 1991, "Dynamical limits on dark mass in the outer solar system." ***Astron. J.***, **101**, 2274–2286.

Holman, M. and Wisdom, J. : 1993, "Dynamical stability in the outer solar system and the delivery of short-period comets." ***Astron. J.***, **105**, 1987–1999.

Ip, W.-H. and Fernández, J. A. : 1991, "Steady-state injection of short- period comets from the trans-Neptunian cometary belt." ***Icarus***, **92**, 185–193.

Jewitt, D. C. : 1991, "Cometary Photometry." In ***Comets in the Post-Halley Era*** (R. L. Newburn, Jr., M. Neugebauer, J. Rahe, Eds.), 19–66, Kluwer, Dordrecht.

Jewitt, D. C. : 1992, " Physical Properties of Cometary Nuclei." In ***Proceedings of the 30th Liege International Astrophysical Colloquium*** (A. Brahic, J.-C. Gerard, J. Surdej, Eds.), 85–112, Univ. Liege Press.

Jewitt, D. C. and Luu, J. X. : 1992, "Submillimeter continuum observations of 2060 Chiron." *Astron. J.*, **104**, 398–404.
Jewitt, D. C. and Luu, J. X. : 1993a, "The discovery of the slow moving object 1992 QB_1." *Nature*, **362**, 730–732.
Jewitt, D. C. and Luu, J. X. : 1993b, "IAU Circular 5865."
Johnson, R. E., Cooper, J. F., Lanzerotti, L. J. and Strazzulla, G. : 1987, "Radiation formation of a non-volatile comet crust." *Astron. Astrophys.*, **187**, 889–892.
Joss, P. C. : 1973, "On the origin of short-period comets." *Astron. Astrophys.*, **25**, 271–273.
Khare, B. N., Thompson, W. R., Murray, B. G. J. P. T., Chyba, C. F. and Sagan, C. : 1989, "Solid organic residues produced by irradiation of hydrocarbon-containing H_2O and H_2O/NH_3 ices : infrared spectroscopy and astronomical implications." *Icarus*, **79**, 350.
Kowal, C. : 1989, "A solar system survey." *Icarus*, **77**, 118.
Kuiper, G. P. : 1951, "On the origin of the solar system." In *Astrophysics* (J. A. Hynek, Ed.), 357–427, McGraw-Hill, New York.
Levison, H. F. : 1991, "The long-term dynamical behavior of small bodies in the Kuiper belt." *Astron. J.*, **102**, 787–794.
Levison, H. and Duncan, M. : 1990, "A search for proto-comets in the outer regions of the solar system." *Astron. J.*, **100**, 1669–1675.
Levison, H. and Duncan, M. : 1993, "The gravitational sculpting of the Kuiper belt." *Astrophys. J. Lett.*, **406**, L35–L38.
Luu, J. X. : 1993, "Spectral diversity in cometary nuclei." *Icarus*, **104**, 138–148.
Luu, J. X. and Jewitt, D. C. : 1988, "A two-part search for slow moving objects." *Astron. J.*, **95**, 1256.
Luu, J. X. and Jewitt, D. C. : 1990, "Cometary activity in 2060 Chiron." *Astron. J.*, **100**, 913–923.
Luu, J. X. and Jewitt, D. C. : 1993a, "IAU Circular 5730."
Luu, J. X. and Jewitt, D. C. : 1993b, "IAU Circular 5867."
Marsden, B. : 1993, "Minor Planet Circular 22594."
Meech, K. and Belton, M. : 1989, "IAU Circular 4770."
Meyer, P., Ramaty, R. and Webber, W. R. : 1974, Cosmic ray astronomy with energetic particles." *Physics Today*, **27**, 23–32.
Moore, M. H., Donn, B., Khanna, R. and A'Hearn, M. F. : 1983, "Studies of proton-irradiated cometary-type ice mixtures." *Icarus*, **54**, 388–405.
Mueller, B. E. A., Tholen, D. J., Hartmann, W. K. and Cruikshank, D.P. : 1992, "Extraordinary colors of asteroidal object 1992 AD." *Icarus*, **97**, 150–154.
Oikawa, S. and Everhart, E. : 1979, "Past and future orbit of 1977 UB, object Chiron." *Astron. J.*, **84**, 134–139.
Oort, J. H. : 1950, "The structure of the cloud of comets surrounding the solar system and a hypothesis concerning its origin." *Bull. Astr. Inst. Neth.*, **11**, 91.
Safronov, V. S. : 1972, "Ejection of bodies from the solar system in the course of the accumulation of the giant planets and the formation of the cometary cloud." In *The Motion, Evolution of Orbits and Origin of Comets* (G. A. Chebotarev, E. I. Kazimir, B. G. Marsden, Eds.), 329–334, Reidel, Dordrecht.
Stern, S. A. and Shull, J. M. : 1988, "The thermal evolution of comets in the Oort cloud by stars and supernovae." *Nature*, **332**, 407–411.
Strazzulla, G., Baratta, G. A., Johnson, R. E. and Donn, B. : 1991, "Primordial comet mantle : irradiation production of a stable, organic crust." *Icarus*, **91**, 101–104.
Strazzulla, G. and Johnson, R. E. : 1991, "Irradiation effects on comets and cometary debris." In *Comets in the Post-Halley Era* (R. L. Newburn, Jr., M. Neugebauer, J. Rahe, Eds.), 243–276, Kluwer, Dordrecht.
Suess, S. T. : 1990, "The heliopause." *Reviews of Geophysics*, **28**, 97–115.
Tholen, D. J., Hartmann, W. K. and Cruikshank, D. P. : 1988, "IAU Circular 4554."

Tombaugh, C. W. : 1961, In ***Planets and Satellites*** (G. P. Kuiper, B. M. Middlehurst, Eds.), 12–30, University of Chicago, Chicago.
Torbett, M. V. and Smoluchowski, R. : 1990, "Chaotic motion in a primordial comet disk beyond Neptune and comet influx." ***Nature***, **345**, 49–51.
Webber, W. R. : 1987, "The interstellar cosmic ray spectrum and energy density. Interplanetary cosmic ray gradients and a new estimate of the boundary of the heliosphere." ***Astron. Astrophys.***, **179**, 277–284.
Whipple, F. L. : 1964, "The history of the solar system." ***Proc. Natl. Acad. Sci. USA***, **51**, 711–718.
Whipple, F. L. : 1977, "The constitution of cometary nuclei." In ***Comets, Asteroids, Meteorites*** (A. H. Delsemme, Ed.), 25–35, University of Toledo Press, Toledo.
Williams, I., Fitzsimmons, A., and O'Ceallaigh, D. : 1993, "IAU Circular 5869."
Wilson, P. D., Sagan, C., and Thompson, W. R. : 1994, "The organic surface of 5145 Pholus : constraints set by scattering theory." Preprint.
Wisdom, J. : 1985, "Meteorites may follow a chaotic route to Earth. ***Nature***, **315**, 731–733.
Yabushita, S. : 1993, "Thermal evolution of cometary nuclei by radioactive heating and possible formation of organic chemicals." ***Mon. Not. Roy. Astron. Soc.***, **260**, 819–825.
Yeomans, D. K. : 1986, In ***Proc. 20th ESLAB Symp.***, p. 419, ESA SP-250, Heidelberg.

DENSITY OF METEOROIDS AND THEIR MASS INFLUX ON THE EARTH

PULAT B. BABADZHANOV

Institute of Astrophysics, Dushanbe 734042, Tajikistan

Abstract. According to the investigation technique it is customary to divide meteoroids into three groups : 1) micrometeorites, with masses between 10^{-13} g and 10^{-6} g, and densities between 0.4 and 4 g/cm^3; 2) meteorites, for which the chemical composition is studied in detail, and having densities from 2 to 8 g/cm^3; 3) meteoroids proper, with masses larger than 10^{-6} g, which produce meteoric phenomena in the Earth's atmosphere detected by optical and radar means.

On the basis of available photographic and radar observations in Dushanbe the influx $M(m)$ of meteoroids with mass equal or greater than m is determined as $\log M(m) = 3.7 - 0.1 \cdot \log m$ in kg to the Earth per day. This formula is applicable to a mass range from 10^{-6} g to 10^2 g.

The phenomenon of meteoroid fragmentation in the atmosphere was observed repeatedly by means of different methods and especially using the photographic technique of instantaneous exposure. Among four principal forms of fragmentation, the quasi-continuous fragmentation, *i.e.* a gradual release of smallest fragments from the surface of a parent meteoroid and their subsequent evaporation, is most common. The analysis of photographic observations shows that about 30% of meteoroids display this type of fragmentation. According to the theory of quasi-continuous fragmentation and on the basis of lightcurves of meteors, the densities of 85 meteoroids have been determined, which vary in the range from 0.1 to 8 g/cm^3. Not only porous and crumbly meteoroids but more dense stony and stony-iron meteoroids are also the subjects of fragmentation as well.

1. Introduction

The smallest bodies moving in the interplanetary space around the Sun and having sizes larger than molecules and smaller than asteroids are called meteoroids. Meteoroids are the products of disintegration of larger bodies : comets and asteroids. Hence, the investigation of the orbits and physical characteristics of meteoroids is important not only for meteor astronomy but also for understanding the physical nature of cometary nuclei and asteroids and for the cosmogony of the Solar system.

Compared with other celestial bodies the range of meteoroidal masses is large - from 10^{-13} g to hundreds tons, *i.e.* more than 20 orders of magnitude. This creates certain difficulties in the interpretation of meteor observations carried out by different methods, because different techniques (for example photographic or radar) may detect meteors produced by particles whose masses differ no more than three orders of magnitude, only a small part of the whole range.

According to the detection technique it is customary to divide the meteoroids into three groups : micrometeorites, meteorites, and meteoroids. In this paper we give brief information on micrometeorites and meteorites and dwell on the results

A. Milani et al. (eds.), Asteroids, Comets, Meteors 1993, 45–54.

of the investigation of meteoroid densities according to photographic observations of bright meteors.

1. 1. MICROMETEORITES

Micrometeorites or micrometeoroids are the smallest particles under consideration here. Because of their large values of the surface to mass ratio, entering the Earth's atmosphere they radiate away the heat generated by collisions with air atoms and molecules. Therefore such particles do not reach the fusion point and they decelerate and drift down to the surface. The sizes of these particles are about tens of microns and smaller. The upper limit of their mass is equal to about 10^{-6} g which corresponds to the lower mass limit of the particles producing in the atmosphere the meteors of $+12^m - +13^m$ registered by high sensitive meteor radar systems (Kascheev and Tkachuk 1980; Baggaley *et al.*, 1992).

The main tools for micrometeoroid detection are special detectors aboard space vehicles, meteor traps on high-altitude rockets and balloons, analyses of micro-craters on lunar specimens, particle collection from high-flying aircrafts and from glacial accumulations, observations of the zodiacal light and gegenschein. These methods allow us to estimate the flux density of meteoroids of different mass ranges and to investigate their physical and chemical properties (structure, density and composition).

An analysis of micrometeorites collected in the high atmosphere shows that according to their physical properties and chemical composition they are similar to carbonaceous chondrites. Thus, Sutton and Flynn (1989) estimated the density of 11 cosmic dust particles to be in the range from 0.4 to 2 g/cm^3. According to studies of micrometeorite craters on the lunar specimens it was concluded that the mean density of micrometeorites lies in the range from 2 to 4 g/cm^3, i.e. close to the density of stone meteorites (Lebedinets 1981).

1. 2. METEORITES

The chemical and mineralogical composition of meteorites has been studied in detail. According to chemical composition the meteorites are divided into three groups : a) iron, having an average density of 7.7 g/cm^3; b) stony-iron, with a mean density of 5.5 g/cm^3; c) stony, mainly chondrites, with a mean density of 3.6 g/cm^3.

Of particular interest are carbonaceous chondrites, which differ from other meteorites by high contents of volatiles, are enriched in carbon (in the form of organic compounds), and contain water (up to 20 %), sulphur, and a number of other minerals. Many authors consider that, according to their composition, carbonaceous chondrites are very close to the original matter of the Solar system. The density of carbonaceous chondrites varies in the range from 2 to 3.5 g/cm^3.

1. 3. METEOROIDS

Meteoroids are the particles entering the atmosphere from interplanetary space and producing meteoric phenomena, *i.e.*, creating sufficient luminescence and ionization to be detected by optical and radar methods. Their mass ranges between 10^{-6} g and

10^3 g and beyond. The flux density of meteoroids, *i.e.* the number of meteoroids colliding with the unit surface per unit of time, is one of the main parameters for the studies of the interplanetary meteor matter and may be determined from photographic and radar observations of meteors.

The integrated meteoroid mass distribution may be expressed as :

$$N(m) = bm^{1-s}, \tag{1}$$

where $N(m)$ is the number of meteoroids with mass equal to or greater than m, which intersect a fixed area within a fixed time t, and b and s are parameters to be determined.

To estimate the incident flux density using radar observations, we selected data on sporadic meteors producing overdense trails of duration $T > 0.1$ s corresponding to masses greater than $m = 1.4 \times 10^{-3}$ g. Some 226,000 meteors of that type were recorded in Dushanbe during 6800 hours of observation.

From double-station photographic observations of Super-Schmidt faint meteors and of small-camera bright meteors, and from the above radar observations, the mass index is found to be $s = 2.1$ for meteoroids with mass from 10^{-3} g to 10^2 g. The incident meteoroid flux density turns out to be (Babadzhanov et al. 1990)

$$\log \Phi(m) = -13.7 - 1.1 \log m, \tag{2}$$

where m is in g, and $\Phi(m)$ in $m^{-2}s^{-1}(2\pi ster)^{-1}$. The logarithm of the cumulative number of meteoroids with mass equal or greater than m coming to the whole Earth per day may be determined as

$$\log N(m) = 2.6 - 1.1 \log m, \tag{3}$$

where m is in kg.

If the incident flux density of meteoroids with mass equal or greater than m is known, then the influx of meteor matter to the terrestrial surface per day is given by

$$\log M(m) = 3.7 - 0.1 \log m, \tag{4}$$

where $M(m)$ and m are in kg. This formula is applicable in an interval of mass from 10^{-9} to 10^{-1} kg (Babadzhanov et al., 1990).

According to (4) at $m \geq 10^{-9}$kg the influx of meteoric matter to the Earth is estimated to be 40 tons per day or about 13,000 tons per year.

Present estimates of the meteoroidal mass influx on Earth, including meteoroids with masses between 10^{-17} kg and 10^{-1} kg, meteorites and crater-forming bodies (asteroids and comets) are given in the paper by Kyte and Wasson (1986), which shows that in the total influx of all interplanetary bodies to Earth, the portion of large bodies with masses of 10^{10} kg and more is substantial.

Ceplecha (1992) estimated the total influx to the Earth of all interplanetary bodies in the mass range between 10^{-21} kg and 10^{15} kg to be 1.7×10^8 kg per year for the entire Earth surface. Recently Hughes (1993) estimated the total annual influx of meteoroids to the Earth around 2.2×10^8 kg.

Lebedinets (1991) came to the conclusion that Draconid type meteors having anomalous heights compared with the meteor luminosity theory are produced by organic meteoroids similar to CHON-particles. He supposed that in the interplanetary space at the distance $R = 1$ AU from the Sun about half of all meteoroids and 90 % of Draconids must be organic. If this is the case the influx of meteoroids into the Earth must be at least twice as larger as estimated. This conclusion needs further experimental examination.

It should be noted that meteoric phenomena at high altitude do not necessarily require an organic composition of meteoroids, but may be caused by the existence of other volatiles in meteoroids of usual composition. This is a subject of extensive research.

A new method of determination of the sporadic meteoroid flux density, which is based on the meteor ionization of the E-layer of the Earth's atmosphere at night time, was developed in Dushanbe (Kolmakov 1990). It was shown that the maximum electron concentration of the E-layer before sunrise equal to 2×10^3 cm^{-3}, corresponding to a flux density of

$$\lg \Phi' = -13.4 - 1.1 \lg M \quad [m^{-2} s^{-1} (2\pi ster)^{-1}], \tag{5}$$

in a good agreement with the relation (2) determined directly according to meteor observations.

This method was applied to estimate the meteoroidal flux density on Venus and Mars using space probes data on the ionospheres of these planets. The obtained ratio of flux densities of sporadic meteoroids of masses larger than 10^{-7} g on Venus, Earth and Mars is 4 : 1 : 10. These results may be explained by the fact that Venus is located closer to the Sun, where intensive disintegration of cometary nuclei take place and the perihelia of meteoroid orbits are concentrated. Mars is close to the asteroid belt and moreover has two small satellites, which pollute intensively the circum-martian space due to their own impacts with meteoroids.

Using observational data on the heights of meteors and fireballs produced by the meteoroids in a wide mass range from 10^{-4} g to 10^8 g Ceplecha (1977) concluded that according to their structure and composition the meteoroids form three main groups : 1) ordinary chondrites, with a mean density of 3.7 g/cm^3; 2) carbonaceous chondrites, with a mean density of 2.1 g/cm^3; 3) "cometary" material with a density from 0.2 to 1 g/cm^3. These conclusions are confirmed by the results of our investigations on meteoroid densities taking into account quasi-continuous fragmentation (see later).

2. Fragmentation and density of meteoroids

The composition and density of meteoroids has been the subject of a number investigations in past years (Jacchia et al. 1967; Verniani 1967, 1969). The opinion that meteoroids have low densities was widely spread till recently. For example, analyzing the heights and decelerations of meteors photographed by Super-Schmidt cameras Jacchia et al. (1967) came to the conclusion that they have a crumbly structure and low densities (equal on the average to 0.26 g/cm^3). The main shortcoming of this and other papers is that all their results are based on a theory and methods which are applicable only to single non-fragmenting meteoroids, what cannot be considered as satisfactory.

Fragmentation of meteoroids in the Earth's atmosphere has been recorded repeatedly by both visual and photographic observations of meteors. The results of photographic observations carried out using the method of instantaneous exposure (5.6×10^{-4} sec each) show this phenomenon the most strikingly and completely (Babadzhanov 1983; Babadzhanov and Kramer 1967). At present one can consider as established that fragmentation is a normal type of ablation for the majority of meteoroids producing meteors registered by photographic and radar methods. Levin (1961) showed that neglecting fragmentation in the interpretation of meteor observations may lead to erroneous results.

Analyzing the data on photographic observations of meteors Levin (1961) distinguished four main forms of meteoroidal fragmentation : 1) the decay of a meteoroid into comparably large non-fragmenting debris; 2) the progressive disintegration of the original meteoroid into fragments, which continue to crumble into smaller debris; 3) the instantaneous spray of a large number of small particles giving rise to meteor flares; 4) the quasi-continuous fragmentation : a gradual release of the smallest fragments from the surface of a parent meteoroid and their subsequent evaporation.

Photographic observations of meteors show that along the meteor trajectory the fragmentations of types 1) and 3) may occur more than once. It is quite probable that a meteoroid may undergo different combinations of these fragmentation forms in the atmosphere (Babadzhanov *et al.*, 1989).

Among the different forms of meteoroidal disintegration in the atmosphere described above, the quasi-continuous fragmentation is of the greatest interest. Both meteor wakes and terminal blendings, as well as the results of laboratory experiments provide evidence in favour of this mechanism of meteoroid disintegration.

The most plausible mechanisms leading to this type of fragmentation may be (Lebedinets 1980) : 1) the blowing off of the melted meteoroid layer in the case of iron meteoroids; 2) the ejection of the heated surface owing to fast evaporation of more volatile admixtures; 3) the thermodestruction and blowing off of the surface layer; 4) the separation of solid grains, assuming a grainy structure of the body, under the aerodynamic pressure or due to the melting of glue between grains (Hawkes and Jones 1975).

Thus the mere fact of meteoroidal fragmentation in the Earth's atmosphere is not yet an indication of a crumbly structure of meteoroids. Comparatively large and dense stony and iron meteoroids undergo fragmentation, too.

Proceeding from the results of an analysis of possible physical mechanisms of quasi-continuous fragmentation of different types of meteoroids Lebedinets (1987) *a priori* distinguished the following types of meteoroids according to their densities and fragmentation energies (Table I). These data are generally used by the different methods of determination of meteoroid densities which account for quasi-continuous fragmentation.

From the data of Table I the dependence of fragmentation energy on density may be approximated as

$$Q_f \times 10^{-10} = 0.10 + 0.18\,\delta_0 \tag{6}$$

TABLE I
Densities δ_0 and fragmentation energy Q_f for different types of meteoroids

Meteoroid type	δ_0 (g/cm^3)	$Q_f \times 10^{-10}$ erg/g
Iron	7.7	1.3
Stony-iron	5.5	1.0 - 1.3
Ordinary chondrite	3.5	0.7 - 1.3
Carbonaceous chondrite CII	2.5	0.5 - 0.7
Carbonaceous chondrite CI	2.0	0.3 - 0.5
Carbonaceous chondrite C0	2.0	0.15- 0.25
Dust ball	1.0	0.2 - 0.6
Draconids type	1.0	0.1 - 0.2

First estimations of meteoroidal densities taking into account quasi-continuous fragmentation were obtained by Novikov *et al.* (1984), but a shortcoming of their work is that the theoretical heights of fragmentation beginning and the termination of meteoric phenomena were identified by means of the photographed heights of meteor appearance and disappearance, and this cannot be considered as satisfactory.

Basic equations of the theory of meteoroid quasi-continuous fragmentation were used by Lebedinets (1987a, 1987b) for the analysis of observed decelerations of faint photographic meteors and estimation of meteoroid densities. Using a mathematical simulation of the observed deceleration of Super-schmidt meteors, Lebedinets (1987a, 1987b) estimated density and fragmentation energy of 190 meteoroids (92 sporadic and 98 shower) of mass from 0.1 to 1 g (Table II).

It was found that out of 92 sporadic meteors 57 (or 62 %) were produced by meteoroids similar to carbonaceous chondrites, 22 (or 24 %) by meteoroids similar to ordinary chondrites, 8 (or 9 %) by iron meteoroids, 4 by stony-iron meteoroids, and 3 by very crumbly bodies of "dust ball" type. In the case of shower meteors, meteoroids of carbonaceous chondrite type and ordinary chondrite type were found more often, namely 48 % and 29 %, but the meteoroids of iron and of stony-iron types, and very crumbly bodies interspersed with free ice were found rarely, only 5 %, 6 % and 12 %, respectively (Table II).

The main shortcoming of Lebedinets (1987a, 1987b) work is that when determining meteoroid densities the meteor light curves are not taken into consideration. The value and reliability of the obtained results would increase considerably if the same meteors were analyzed by different methods.

TABLE II
Distribution of sporadic and shower meteoroids according to Lebedinets (1987)

Group	Types of meteoroids in %						
	Iron	Iron-stone	Ordin.chondr.	CII	CI	C0-DB	Draconids
Sporadic meteors (92)	9	4	22	21	41	3	0
Shower meteors (98)	5	6	29	15	33	10	2

3. Determination of meteoroid densities

The light curves of meteors are indicative of meteoroid ablation processes during their flight in the atmosphere. Since 1964 photographic observations of meteors were carried out in Dushanbe using both usual and instantaneous methods. Some unique meteor photographs obtained in Dushanbe by the method of instantaneous exposure with exposure times of 5.6×10^{-4} sec show clearly the effects of different types of meteoroid fragmentation. The most common type is the quasi-continuous fragmentation forming short-lived meteor wakes of up to several tens of metres in length. Instantaneous images of such meteors have a drop-like shape. (Babadzhanov 1983; Babadzhanov and Kramer 1967).

A theory of meteoroid quasi-continuous fragmentation is well developed at present. The formulas describing the variation of meteor luminosity along its visible trajectory depending on atmospheric density and parameters of quasi-continuous fragmentation have been obtained. According to this theory (Novikov *et al.*., 1984; Babadzhanov *et al.*, 1984, 1988) the common expression describing the meteor luminosity along its trajectory is :

$$I(\rho) = \frac{9\tau_\nu M_0^3 V_0 \cos Z_R}{2H(R_0 R_1)^3} \times \tag{7}$$

$$\times \{F_1(\rho)\Theta(a-\rho) + R_1^3 F_2(\rho)\Theta(\rho-a)\Theta(b-\rho) + \frac{1}{30}F_3(\rho)\Theta(\rho-b)\Theta(\rho_e-\rho)\}$$

where

$$F_1(\rho) = \rho\{\frac{1}{3}(\rho_e-\rho)^2[R_1^3-(a-\rho)^3] - \frac{1}{2}(\rho_e-\rho)[R_1^4-(a-\rho)^4] + \frac{1}{5}[R_1^5-(a-\rho)^5]\} \tag{8}$$

$$F_2(\rho) = \rho\{\frac{1}{3}(\rho_e-\rho)^2 - \frac{1}{2}R_1(\rho_e-\rho) + \frac{1}{5}R_1^2\} \tag{9}$$

$$F_3(\rho) = \rho(\rho_e - \rho)^5 \tag{10}$$

$$a = \rho_e - R_0, \quad b = \rho_e - R_1 \tag{11}$$

$I(\rho)$ is the meteor luminosity at the height where the atmospheric density is equal to $\rho; \tau_\nu, M_0, V_0$ are the luminous efficiency, pre-atmospheric meteoroid mass and velocity, respectively; H is the scale height; Z_R the zenith distance of the meteor radiant; ρ_e the atmospheric density at the end height of meteor phenomena :

$$\rho_e = 2\rho_m + 0,75R_1 + \sqrt{\rho_m^2 - 0,0375R_1^2} \tag{12}$$

deduced from a condition that the maximum of light curve is in the range $a \leq \rho_m \leq b$; ρ_m is the atmospheric density at the height of maximum luminosity, and ρ the atmospheric density in the arbitrary point of the meteor trajectory; a is the atmospheric density at the height of complete evaporation of fragments released at the moment of fragmentation beginning, while b is the atmospheric density at the height of the end of fragmentation; R_0 and R_1 are the parameters determining the quasi-continuous fragmentation :

$$R_0 = \frac{6Q_f M_0^{1/3} \delta_0^{2/3} \cos Z_R}{\Lambda A H V_0^2}, \quad R_1 = \frac{9.67(Q - Q_f)(r\delta) \cos Z_R}{\Lambda' A' H V_0^2} \tag{13}$$

Q_f being the specific energy of meteoroid fragmentation, while $Q = 8 \times 10^{10}$ erg/g is the specific energy of heating and evaporation of meteoroidal matter; Λ, A, δ_0 are the heat transfer coefficient, the shape factor and the meteoroid density, respectively; Λ', A', δ - the same values for the fragments; r is the fragment radius; $\Theta(x)$ is the Heaviside function : $\Theta(x) = 1$ at $x > 0$ and $\Theta(x) = 0$ at $x \leq 0$.

In the case $R_1 \to 0$, that corresponds to very small fragments and to a condition $Q \simeq Q_f$, the equation (7) coincides with the conventional equation of meteor light curves :

$$I = \frac{9I_{\max}}{4} \frac{\rho}{\rho_{\max}} [1 - \frac{\rho}{3\rho_{\max}}]^2 \tag{14}$$

where

$$I_{\max} = \frac{2\tau_\nu M_0 V_0^3 \cos Z_R}{9H}, \quad \rho_{\max} = \frac{2Q M_0^{1/3} \delta_0^{2/3} \cos Z_R}{\Lambda A H V_0^2} \tag{15}$$

Using the observed values of M_0, V_0, Z_R and the light curves of meteors, by the method of successive approximations we determined the values of the parameters R_0 and R_1 of equation (7) for which the theoretical and observed meteor lightcurves are in the best agreement.

The meteoroid density and mass of fragments may be estimated from the derived values of R_0 and R_1 if the values of $\Lambda, \Lambda', A, A', Q_f$ and product $r\delta$ are known *a priori*. Meteors detected in the Dushanbe photographs are bright and produced by meteoroids of mass greater than 0.01 g. According to Lebedinets (1991), for large meteoroids of mass between 0.01 g and 10 g the heat transfer coefficient depends on the mass M_0 as

$$\Lambda = \Lambda_0 + (1 - \Lambda_0)\exp(-kM_0) \tag{16}$$

where $\Lambda_0 = 0.03, k = 0.25g^{-1}$. It is assumed that $A = 1.5$, $A' = 1.21$, $H = 6$ km, $\Lambda' = 1$. Simulations of the meteor lightcurves were carried out for meteoroid densities between 10^{-2} and 10 g/cm^3 and products $r\delta$ of fragments between 10^{-4} and 10^{-1} g/cm^2. We adopted as reliable the values of meteoroid densities for which the mean deviation of the theoretical lightcurve from the observed one was minimum.

The results of the determination of densities of meteoroids belonging to different streams are given in Table III. The analysis of results shows that meteor matter contains at least three types of meteoroids : 1) dense stony bodies of ordinary chondrites or iron meteorite types (49 meteoroids); 2) meteoroids of carbonaceous chondrites of C II and C I types (11 meteoroids) 3) meteoroids of the carbonaceous chondrites of C 0 or "dust ball" types (25 meteoroids).

TABLE III

Meteoroid densities

Shower	Parent	Number of meteoroids having different densities (g/cm^3)						All	Mean density
		< 1	1-2	2-3	3-4	4-5	5-8		g/cm^3
Perseids	P/Swift-Tuttle	3	5	5	13	8	3	37	3.4
Leonids	P/Tempel-Tuttle	-	4	2	1	-	-	7	2.0
Geminids	3200 Phaeton	1	-	1	-	1	2	5	4.2
κ Cygnids	?	-	-	1	-	-	1	2	5.0
Quadrantids	P/Machholz	1	1	-	1	-	1	4	3.3
δ Aquarids	P/Machholz	-	-	-	2	2	3	7	5.4
Taurids	P/Encke	1	-	-	-	3	-	4	3.7
Sporadic		4	5	2	4	2	2	19	2.4
All		10	15	11	21	16	12	85	3.3

Small differences of the results of the determination of meteoroid densities presented in Table III with respect to previous results (Babadzhanov 1992) are due to the fact that here we took into account two different kinds of quasi-continuous fragmentation (QCF), namely slow and fast QCF (Babadzhanov *et al.*, 1984), whereas before only the first form of QCF was considered.

Slow fragmentation means that the fragments released at the beginning of fragmentation evaporate before the parent meteoroid body breaks completely into fragments. Fast fragmentation means that the complete destruction of the parent meteoroid body takes place before the disappearance of the fragments released at the moment of the fragmentation onset.

From Table III it follows that the composition of meteoroids in each stream may be quite different, from carbonaceous chondrites of C 0 type or "dust balls" to iron type. The small amount of observational evidence does not permit us to carry out a more detailed analysis at present. The main conclusions of the present research can be summarized here : 1) not only friable meteoroids but also dense ones are subjects of fragmentation; 2) in the majority of the cases the meteoroid densities are similar to the densities of meteorites of different types; 3) according to these results all three groups of meteoroids (namely micrometeorites, meteoroids and meteorites) on the whole are similar.

References

Babadzhanov, P.B. : 1983, in *Asteroids, Comets, Meteors* (C.-I.Lagerkvist, H. Rickman, Eds.), 439–445, Uppsala universitet, Reprocetralen HSC, Uppsala, Sweden.

Babadzhanov, P.B. : 1992, in *Asteroids, Comets, Meteors* (A.W. Harris, E. Bowell, Eds.), 23–26, L.P.I., Houston.

Babadzhanov, P.B., Bibarsov, R.Sh. and Kolmakov, V.M. : 1990, *Mass distribution and the flux of sporadic meteoroids*, 56, Sov. Geophys. Committee, Moscow.

Babadzhanov, P.B. and Kramer, E.N. : 1967, in *Physics and Dynamics of Meteors* (Kresák L., Millman P.M., Eds.), 128–142, Reidel-Dordrecht.

Babadzhanov, P.B., Novikov, G.G. and Konovalova, N.A. : 1989, *Astron. Vestnik* **XXIII**, 277–281.

Babadzhanov, P.B., Novikov, G.G., Lebedinets, V.N. and Blokhin, A.V. : 1984, *Preprint of Physical Technology Institute*, **919**, 55, Leningrad.

Babadzhanov, P.B., Novikov, G.G., Lebedinets, V.N. and Blokhin, A.V. : 1988, *Astronomicheskij Vestnik*, **XXVII**, 71–78.

Baggaley, W.J., Steel, D.I. and Taylor, A.D. : 1992, in *Asteroids, Comets, Meteors* (A.W. Harris, E. Bowell, Eds.), 37–40, L.P.I., Houston.

Ceplecha, Z. : 1977, in *Comets, Asteroids, Meteoroids. Interrelation, evolution and origin* (A.H. Delsemme, Ed.), 143–152, The University of Toledo.

Ceplecha, Z. : 1992, *Bull. of the Amer. Astron. Soc.*, **24**, 952.

Hawkes, R.L. and Jones, J. : 1975, *Mon. Not. Royal Astron. Soc.*, **173**, 339–356.

Hughes, D.W. : 1993, in *Workshop on the Activity of Distant Comets* (W.F. Huebner, H.U. Keller, D. Jewitt, J. Klinger, R. West, Eds.), 83-99, Southwest Research Institute, San Antonio, Texas.

Jacchia, L.G., Verniani, F. and Briggs, R.E. : 1967, *Smithson. Contr. Astrophys.*, **10**, No 1, 1–139.

Kascheev, B.L. and Tkachuk, A.A. : 1980, *Result of Radar Observations of Meteor Orbits to* $+12^m$, 232, Sov. Geophys. Committee, Moscow.

Kete, F.T. and Wasson, J.T. : 1986, *Science*, **232**, 1225–1229.

Lebedinets, V.N. : 1980, *Dust in the high atmosphere and space. Meteors.*, 247, Gydrometeoizdat, Leningrad.

Lebedinets, V.N. : 1981, *Aerosol in the upper atmosphere and interplanetary dust*, 280, Gydrometeoizdat, Leningrad.

Lebedinets, V.N. : 1987a, *Astron. Vestnik*, **XXI**, 65–74.

Lebedinets, V.N. : 1987b, *Astron. Vestnik*, **XXI**, 262–271.

Lebedinets, V.N. : 1991, *Astron. Vestnik*, **XXV**, 200–207.

Levin, B.Yu. : 1961, *Bull. Komissii po kometam i meteoram*, No 6, 3–10.

Lindblad, B.A. : 1976, In *Interplanetary Dust and Zodiacal Light* (H.Elsasser, H.Fechtig, Eds.), 373-378, Springer Verlag, Berlin.

Novikov, G.G., Lebedinets, V.N. and Blokhin, A.V. : 1984, *Pisma v Astron. Journal.*, **X**, 71–76.

Sutton, S.R. and Flynn, C.J. : 1989, *Meteoritics*, **24**, 329–330.

Verniani, F. : 1967, *Smithson. Contr. Astrophys.*, **10**, 181–195.

Verniani, F. : 1969, *Space Sci. Rev.*, **10**, 230–261.

ASTEROID ALBEDOS AND DIAMETERS

EDWARD F. TEDESCO
Mission Research Corporation, One Tara Boulevard, Nashua, NH 03062 USA
E–mail etedesco@pldac.plh.af.mil

Abstract. This paper discusses the status of our knowledge regarding asteroid sizes and albedos. It presents a brief discussion of the methods used to determine sizes and albedos beginning with attempts using filar micrometers as early as 1802 and culminating with the 1993 publication of the IRAS Minor Planet Survey. It addresses the completeness and reliability of the IRAS Minor Planet Survey, summarizes the statistical adjustments made to the derived results, and considers the implications of those results.

1. Techniques

The first attempts to measure diameters of asteroids employed direct measurements of their seeing disks using filar micrometers. The earliest such measurements were made by Herschel in 1802 and Schroeter in 1805 (*cf.*, Barnard, 1894). By 1881 thirteen determinations of the diameters of the four brightest asteroids had been published. The most notable feature of these measurements was their large scatter, *e.g.*, up to a factor of ten in the diameter of Ceres. This led Barnard (1894, p. 342) to write that : "... the previous measures of these bodies are very discordant, and were evidently made with instruments too small to deal with the subject." Thus between 1894 and 1900, using a filar micrometer on the Lick Observatory 36-inch, and Yerkes 40-inch, refractors Barnard undertook to measure the diameters of the four brightest asteroids (Barnard, 1894, 1895, 1900a,b). His results are compared with those from current techniques in the section on the largest asteroids, below. See Dollfus (1971) for a summary of direct visual diameter determinations made through 1969.

Modern techniques used to determine asteroid diameters include polarimetry, radar, speckle imaging, stellar occultations, and thermal radiometry. Soon imaging at optical and near-infrared wavelengths, using the Hubble Space Telescope and large ground-based telescopes equipped with adaptive optics systems, will permit direct diameter determinations for many of the larger asteroids. Of these, only the radiometric method is capable of providing results for significant numbers of asteroids.

Measurements of the linear polarization of the reflected light from asteroids as a function of phase angle has been used to determine their diameters. The polarimetric method yields the albedo directly, through the empirically-derived so-called "slope-albedo relation" (KenKnight, *et al.*, 1967). To date "polarimetric diameters" have been published for about 30 asteroids (*cf.*, Morrison and Zellner, 1979). Polarimetric observations to derive diameters have not been used since the late 1970's for two reasons. First, unlike the other methods discussed here which, in principle, require observations from only a single night, the polarimetric method requires measurements from several nights spaced over a number of weeks in order to define the slope of the polarization – phase angle curve. Second, it was found

A. Milani et al. (eds.), Asteroids, Comets, Meteors 1993, 55–74.

that the slope-albedo relation saturates for low-albedo materials (Zellner *et al.*, 1977a,b). This means that one cannot distinguish among asteroids with albedos less than about 0.1 on the current H, G magnitude system albedo scale. See Dollfus *et al.* (1989) for a review of the current state of the art in asteroid polarimetry.

Radar observations in which the target is not actually resolved, *i.e.*, those in which one-dimensional images are obtained, require knowledge of the asteroid's spin rate and pole orientation for an accurate diameter determination. In the absence of these parameters, radar observations set a lower bound on the asteroid's dimensions and pole-on projected area. In the ten years following its proposed upgrade the Arecibo radar telescope will be capable of providing two-dimensional images ($\gg$10 pixels on target) for two dozen main belt asteroids and a similar number of near-Earth objects, plus accurate lower bounds on the dimensions of over 100 objects (Ostro, 1989).

Speckle imaging has been used to derive shapes, pole orientations, and sizes for six asteroids (Drummond and Hege, 1989). Due to the relatively bright limiting magnitude of this technique its future application is probably limited to a few score asteroids.

Stellar occultations, the passage of an asteroid in front of a star, provide the most accurate asteroid diameters. Measurement of the time interval during which the star is occulted gives the length of the chord across the asteroid. Given a number of well-placed chords one can map the apparent limb profile of the asteroid with great accuracy. If asteroids were spherical, a single well-observed occultation would result in an accurate diameter. In fact, however, most asteroids are significantly non-spherical as indicated by their lightcurves. Hence, a single occultation observation can yield a highly *precise* result which turns out to be of low *accuracy*.

Occultation observers are well aware of this situation. Where possible, lightcurves contemporaneous with the occultation event are obtained and used to estimate the three-dimensional shape and thus the mean diameter $\equiv (a * b * c)^{\frac{1}{3}}$ and to adjust the uncertainty.

While occultations of stars by asteroids is a common occurrence, observations of these events remain rare – perhaps one or two per year on average. To date accurate diameters based upon occultation data are available for fewer than 20 asteroids. Even at the rate of two additional asteroids per year it will be many decades before this number reaches 100. Thus, in the context of determining diameters for statistically significant numbers of asteroids, the greatest value in occultation diameters lies in the calibration of the indirect techniques, primarily the polarimetric and radiometric methods. See Millis and Dunham (1989) for a review of the asteroid stellar occultation method.

The radiometric method makes use of the fact that the thermal radiation emitted by an asteroid must equal the solar radiation it absorbs. For an asteroid at a given heliocentric distance the amount of this energy depends upon the asteroid's size and albedo. By measuring the visual (reflected) brightness and the thermal (emitted) flux a size and albedo can be deduced. The accuracy of the derived size and albedo is a function of the assumptions made in the calculation of the emitted thermal energy, *i.e.*, in the adequacy of the "thermal model". An explication of the thermal model is beyond the scope of this paper. See Lebofsky and Spencer (1989) for a recent review. The thermal model used to reduce the IRAS asteroid data is that of Lebofsky *et al.* (1986a,b).

The accuracy of a radiometric diameter is also limited by the infrared flux calibration and knowledge of the visual brightness at the time of the infrared observation and its extrapolation to zero phase angle. (The visual flux at the time of a thermal infrared observation is almost always calculated from the asteroid's absolute magnitude and an assumed phase function.) Finally, flux variations due to rotation and aspect changes further compromise the accuracy. Thus, a diameter based upon a single infrared observation is unlikely to have an uncertainty less than 10% regardless of the precision of the infrared measurement.

2. The Largest Asteroids

By 1900 Barnard had established that Ceres was the largest of the "big four" (i.e., the four brightest asteroids) followed by Pallas and Vesta (which, to within one standard deviation, he found to be the same size). He found that Juno, at about one-fourth the size of Ceres, was the smallest of the big four. Compared with the best current values for the diameters of these four asteroids Barnard's sizes are systematically small by about 20%; the relative sizes he determined are correct to within 12%.

Using published photometry together with his diameter determinations Barnard derived albedos and noted that the albedo of Vesta was about four times greater than that of Ceres. (The best current estimate of the Vesta/Ceres albedo ratio is 3.7.)

Table I summarizes diameter determinations for the four brightest asteroids published between 1900 and 1993. Note that the radiometric diameters published by Allen (1971) and TRIAD (Morrison and Zellner, 1979) are not directly comparable to one another or to the those from the IRAS Minor Planet Survey (IMPS) due to changes in thermal models and calibrations of infrared standard stars. TRIAD diameters are about 17 $\pm$ 3% larger with respect to IMPS results for the three largest asteroids. If the TRIAD polarimetric albedos are used together with the absolute magnitudes on the H,G system (rather than on the V(1,0) system) the polarimetric diameters increase by about 15% compared to the TRIAD values given below.

The diameters given in Table I are formally "mean diameters" only in the case of the Speckle and Occultation results. The others are approximations to mean diameters. The more measurements there are, averaged over both rotational phase and aspect angle, the closer to the mean diameter they will be. Because the four asteroids given in Table I are only slightly nonspherical the uncertainties in their diameters are dominated by random and systematic effects specific to the different techniques used to measure them. For example, before 1979 it is clear that the results are greatly influenced by systematic effects and the quoted uncertainties are, in most cases, seriously underestimated. With the exception of Vesta (observed only once), the IMPS results, although all obtained at essentially a single aspect angle, sample between six and eight different rotational phases. IMPS diameters reported here and for Ceres, Pallas, and Juno and actual mean diameters, had sufficient observations been made to sample them, would be well within the quoted uncertainties. Note, however, that even had such observations been available the systematic differences, discussed in section 8, below, would remain.

Diameter estimates published over the past 90 years for the three largest asteroids have been within about 1.5σ of the actual mean diameters determined from observations of stellar occultations by these asteroids. The real improvement has been in the confidence in the quoted diameters. Prior to the mid-1980's the absolute uncertainties were about ten percent whereas today, thanks to the application of the stellar occultation technique, the uncertainty in the mean diameters of these asteroids is about 1%.

TABLE I
A Summary of Diameter Determinations for the Four Brightest Asteroids (km).

Minor Planet	Barnard (1900) Filar Microm.[1]	Allen (1971) Radiometry	TRIAD (1979) Polarimetry	TRIAD (1979) Radiometry	Drummond and Hege (1989) Speckle	Occultation[2]	IMPS (1993) Radiometry[3]
Ceres	770 ± 90	1160 ± 80	1016±100	1014±100		933 ± 5	848 ± 40
Pallas	490 ±120		692 ± 70	589 ± 60	503 ± 33	533 ± 6	498 ± 38
Juno	195 ± 25	290 ± 20	252 ± 25	247 ± 25		267 ± 5	234 ± 22
Vesta	385 ± 50	570 ± 10	579 ± 60	530 ± 50	520 ± 25	520 ± 10	468 ± 54

Notes to Table I.

1. Uncertainties are one standard deviation of Barnard's mean disk diameters at 1 AU converted to km. (All of Barnard's published diameters were in units of "english miles".)
2. Ceres – Millis *et al.* (1987); Pallas – Dunham *et al.* (1990); Juno – Millis *et al.* (1981); Vesta – Dunham (1991) [Uncertainty is from Dunham, personal communication.]
3. Uncertainties given here are twice the uncertainty in the precision tabulated in *The IRAS Minor Planet Survey* (1992).

The IRAS-derived radiometric diameters, although calibrated using the occultation diameters for Ceres and Pallas are systematically low relative to the occultation diameters for reasons discussed in the Summary section below.

The vast majority of measured asteroid diameters are from radiometric observations. This has been the case since the late 1970's; TRIAD (Morrison and Zellner, 1979) lists polarimetric diameters for 28 asteroids and radiometric diameters for 192. Since then diameters from the other techniques discussed above have become available for about 25 asteroids. When available, quality A occultation diameters (as defined by Millis and Dunham, 1989) are generally the most reliable. Radiometric diameters, however, remain the most common due to the results from the IRAS survey which increased the number of asteroids with radiometric diameters to over 2,000.

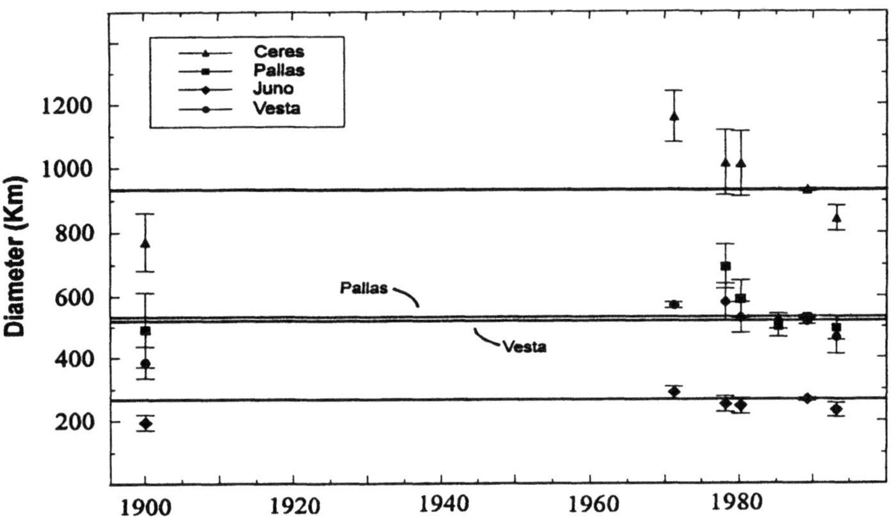

Fig. 1. Published diameters for the "big four" as a function of time. Horizontal lines indicate the best diameter estimates – all from stellar occultations. The techniques from which the diameters shown were obtained are, in chronological order, Filar Micrometry (Barnard, 1900), Infrared Radiometry (Allen, 1971), Polorimetry (Morrison and Zellner, 1979), Infrared Radiometry (Morrison and Zellner, 1979), Speckle (Drummond and Hege, 1989), Occultation (see note to Table I for references), and Infrared Radiometry (the IRAS Minor Planet Survey)

The first release of IRAS results (*The IRAS Asteroid and Comet Survey, 1986*) contained diameters for just under 1,800 asteroids. Many useful scientific results were obtained with this version. For example, it led to new estimates of the size-frequency distributions of asteroids, an improvement in the standard asteroid thermal model, and the discovery of new taxonomic classes, to mention but a few. See Cellino *et al.* (1991), Matson *et al.* (1989), Tedesco *et al.* (1989a,b), and Veeder *et al.* (1989b) for further details.

As good as the 1986 data products were, it was clear that (just as in the case of the *IRAS Point Source Catalog*) a better product could be produced using the experience gained during production of the first data set. In the case of the asteroids this meant searching the data stream using significantly more sets of (and more reliable) orbital elements, incorporating improved visual absolute magnitudes, accounting for the effects of flux overestimation (which affected about 40% of the observed asteroids), and better understanding the completeness versus reliability issue.

Thus between 1988 and 1992, under support provided by the Air Force Geophysics Laboratory, improved extraction and reduction routines were devised and implemented. Results for the approximately 2,000 asteroids with reliable IRAS observations are presented in an explanatory supplement entitled : *The IRAS Minor Planet Survey* (1992). There the history, processing, and analysis of these data are described. This document, together with machine-readable files of the final data products, have been deposited at the NSSDC under the name *The IRAS Minor Planet Survey Database and Catalog* supplanting the 1986 version *IRAS Asteroid and Comet Database and Catalog.* See the chapter in this book entitled : *The IRAS Minor Planet Survey Data Base.*

Due to the quantity and homogeneous nature of the IRAS Minor Planet Survey (IMPS) asteroid diameters and albedos the remainder of this paper will be devoted to an overview of the properties of that data set.

3. IMPS Completeness

The question of geometrical completeness of IRAS asteroid coverage is a complicated one. Because the survey was primarily concerned with inertially fixed sources, different parts of the sky were covered at different times, with the result that coverage was highly fragmented for any class of orbital elements (e.g., objects with the same elements except for the longitude of the ascending node). Here we use the expression "geometrical coverage" to mean that a given asteroid was swept by the IRAS field of view; a detection may or may not have occurred.

There are at least three geometrical factors that make the completeness of the IRAS survey different for the asteroids than for the inertially-fixed sources. The first two factors are independent of asteroid motion :

A. The observed brightness of an asteroid depends on the IRAS viewing geometry.

As the solar elongation angle of the IRAS scan changed, both the distance and phase angle (and hence the observed brightness) of an asteroid at a given ecliptic longitude varied. Therefore, the asteroid brightness distribution for a given ecliptic longitude was a function of the solar elongation angle(s) at which the observation scans were obtained.

The maximum variation occurred for the closest asteroids. For a ring of asteroids at 2 AU from the sun, the geometrical distance effect causes a maximum variation in the flux of a factor of 3.1. The phase angle variation is too small to significantly affect the infrared flux. (The minimum phase angle is 25.7° at 60° or 120° elongation, while the maximum phase angle is 30° at 90° elongation).

In addition, the essentially Earth-centered IRAS viewing geometry caused an apparent variation in the density of asteroids with ecliptic longitude. The maximum observed density enhancement, caused by a variation of solar elongation scan angle from 60° to 120°, is a factor of 1.8.

B. The rate of coverage of the asteroid belt is significantly lower than the coverage of the inertial sky due to the Earth-centered observation frame

Although IRAS covered the inertially-fixed sky completely after six months, a gap remained in the asteroid coverage due to the motion of the Earth (see Fig. 2). Specifically, IRAS began the survey at ecliptic longitudes of approximately 60° and 240°. Six months later, IRAS finished the first survey of the sky at those same longitudes. However, since the Earth was on the opposite side of the Sun from its original position, the part of the asteroid belt currently at ecliptic longitudes from approximately 62° to 108° (for asteroids at 2 AU from the sun) remained unsurveyed. (Of course, since these scans went somewhere, this implies that other parts of the asteroid belt had a higher density of scans than the inertially-fixed sky.)

The third hours-confirming survey of the sky began by backing up to almost the maximum scan angle and resurveying the area of the sky just previously surveyed. When 60° longitude was reached again, those asteroids were being seen for the first time, even though inertial sources were being resurveyed. Thus, ecliptic longitudes from ~ 62° to ~ 108° contain two, independent, surveys of different asteroids.

C. Asteroidal motion

Asteroidal motion causes several different effects. The major effect for asteroids with prograde orbits is to slightly lower the rate of coverage for those objects. For example, as mentioned above, without asteroid motion, for asteroids at 2 AU from the sun, ecliptic longitudes from ~ 62° to ~ 108° were unsurveyed at the end of the first survey of the sky. Asteroidal motion amounts to about 20 degrees during the time it takes IRAS to observe that area, and hence IRAS must observe until longitude 130° to entirely survey those asteroids.

In fact, ignoring the two major five degree gaps of the inertial coverage, IRAS was able to completely survey the asteroid belt except for prograde-orbiting asteroids having circular orbits with radii less than about 2.7 AU. Asteroids in such orbits moved too fast for IRAS to completely catch up to them.

A second effect due to asteroidal motion is caused by the short-term variations in the IRAS coverage. The IRAS survey used only about 60% of the IRAS observation time, with most of the rest of the time devoted to pointed observations. In addition, the survey was suspended due to the presence of the moon in a particular hemisphere for about 3 days per month. Because the IRAS survey was designed to ensure complete coverage of the inertial sky, these effects were compensated in carrying out the survey. However, because of Earth motion and asteroidal motion during those short-term hiatuses, there are significant local variations in coverage of the asteroids.

In order to obtain some feeling for the latter two effects, we ran a simulation using the actual survey observations and a set of hypothetical tracer asteroids.

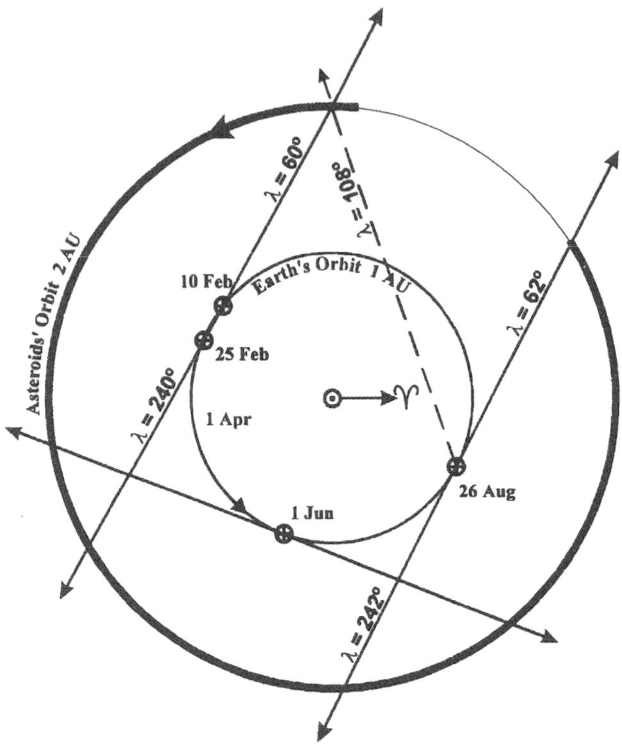

Fig. 2. Geometrical coverage of a ring at 2 AU by the first six months of the IRAS mission. IRAS began the survey at ecliptic longitudes of ~ 60° and 240°. Approximately six months later, IRAS finished the first survey of the sky at about those same longitudes. However, since the Earth was then on the opposite side of the Sun from its position at the start of the mission, the part of the asteroid belt now at ecliptic longitudes from approximately 62° to 108° remained unsurveyed.

These tracer asteroids had circular orbits with zero inclinations. Ten evenly-spaced rings with semimajor axes from 2.0 to 5.5 AU were populated with a total of 3,000 asteroids. These "asteroids" were assigned longitudes of the ascending node such that the linear density along the rings was constant. For each IRAS scan, all asteroids within 15 arcminutes of the IRAS boresight were counted as observed and the number of coverages was computed for each asteroid.

For this class of orbital elements, more than 94% of the tracers were observed at least once. The inner ring is the least complete because the higher rate of motion allows tracers which lagged the coverage early to avoid being "lapped". The completeness varied over semimajor axis as shown in Table II.

Note that, except for the innermost orbits where the asteroidal motion prevented a full survey, the completeness for this set of asteroids is actually higher than that of the inertial survey! This is a direct result of the "slower" asteroidal-survey; the five degree gaps in the inertial sky are not quite as big for the asteroids. However, note that retrograding asteroids can (and a few near-Earth asteroids actually did) easily skip between IRAS scans, and hence the values given in Table II apply only to asteroids with elements similar to those assumed in this simulation.

TABLE II
Completeness Versus Semimajor Axis Ring for the Simulated Asteroids

Semimajor Axis (AU)	Completeness
2.0 - 3.0	0.944
3.0 - 4.0	0.988
4.0 - 5.0	0.983
5.0 - 5.5	0.979
Infinity	0.973

Finally, Table III shows detailed results of the number of times a simulated asteroid from a 2,000 asteroid per ring simulation was seen. Note that the mean coverage must always be the same within each ring. The "slower" coverage of the asteroids causes there to be many more asteroids observed a large number of times than for inertial sources.

TABLE III
Number of Times Simulated Asteroids Were Observed (2,000 in each ring)

Semimajor axis range	0	1	2	3	4	5	6	7	8	9	10	>10	mean	sigma
2.0-3.0	112	7	237	183	201	223	221	248	176	134	95	163	5.82	3.33
3.0-4.0	24	7	246	106	304	266	304	286	173	107	68	109	5.82	2.86
4.0-5.0	34	7	214	56	377	220	360	325	175	75	60	97	5.82	2.78
5.0-5.5	43	10	182	36	344	251	434	320	165	76	55	85	5.82	2.70
Infinity	55	3	2	2	441	174	726	384	122	35	16	40	5.82	1.99

4. Reliability

There were 38,281 predicted asteroid crossings of the IRAS focal plane while it was in survey mode. Of these predicted sightings, 1,123 (a total of 232 different asteroids) had predicted 25 μm flux densities that never exceeded 0.140 Jy (the 1σ limit).

Table IV summarizes the results of a test to assess the reliability of sources accepted as asteroids : 155 of the 1,123 potential sightings, for 31 of the 232 different asteroids, generated an association, *i.e.*, a source with a 25 μm flux density having an SNR > 3 was found within the association ellipse for the putative asteroid. This implies that the false association rate is about 14%. Note that all 155 of these false sightings were rejected by the IMPS data processing system; none were used in producing an accepted IMPS albedo and diameter. This further validates the reliability of the IMPS acceptance criteria. Furthermore, there is no significant difference in the false association rate between numbered and two or more opposition unnumbered asteroids. This is to be expected since these 232 asteroids are essentially a set of random positions near the ecliptic.

TABLE IV
Summary of $< 1\sigma$ Flux Limit Association Test

ID Type	Number of Predicted Sightings	Number of Predicted Asteroids	Number (%) of Predictions Associated	Number (%) of Asteroids Associated
1	140	26	24 (17%)	6 (23%)
2	983	206	131 (13%)	25 (11%)
Totals:	1,123	232	155 (14%)	31 (13%)

5. IMPS Vs. ADAS

This section summarizes the major differences between ADAS (documented in *IRAS Asteroid and Comet Survey, 1986*) and the IMPS processing described in *The IRAS Minor Planet Survey, 1992*. There was one major difference in the association algorithm and two in the algorithms that computed the final albedos and diameters. These are discussed in detail below. The ground-based data sets used differed as well. ADAS processed asteroids through number 3318 plus 135 asteroids with two-or-more-opposition orbits while for IMPS these numbers were 4679 and 2,632, respectively. ADAS used asteroid absolute magnitudes and slope parameters on the 1985 IAU system, while IMPS used the 1991 system.

5. 1. THE ASSOCIATION ALGORITHM

The ADAS association algorithm used a one-sigma positional uncertainty attributed to asteroids that was a function of the reliability of the orbital elements. In the course of the IMPS processing it was realized that the larger acceptance area used for asteroids with less-reliable orbital elements led to many spurious associations. This is especially true for faint asteroids. In IMPS a single one-sigma asteroid positional uncertainty was adopted. This single value (ten arcseconds) was, in effect, root-sum-squared with the IRAS Convolved Gaussian-Uniform positional uncertainty. This change virtually eliminated spurious associations for faint asteroids.

5. 2. THE FLUX OVERESTIMATION CORRECTION

There is a systematic error associated with the measured flux densities near the noise limit of the detectors which can increase the reported value by as much as a factor of two compared with that of the true value. Weak sources were often detected when positive noise excursions pushed the source signal above the 3σ SNR cutoff imposed by the IRAS processing. Negative excursions dropped the signal below the SNR cutoff and so were not detected. Thus, the flux densities of weak sources were systematically overestimated. This overestimation of the flux for low-SNR asteroid detections results in an overestimation of their diameters and an underestimation of their albedos.

Two correction methods have been used to correct this effect for inertial sources. Each of these methods assumes that the source is non-variable and the noise Gaussian. Unfortunately, (as shown by J. Fowler in Tedesco *et al.*, 1992) each of the methods have an error in their derivations. Even were this not the case, without at least *a priori* knowledge of the luminosity function, these methods cannot be used to derive flux overestimation corrections.

Besides the above, the following are true of asteroidal sources but not of inertial sources.

1) Asteroids are intrinsically variable.
2) Asteroids move, thus increasing their "variability" through confusion.
3) About one-third of the accepted asteroids have fewer than three sightings; all inertial sources accepted into the PSC were sighted at least three times.
4) Over 95% of the accepted asteroid sightings are within 20° of the ecliptic plane and hence were observed through the emission from the zodiacal cloud; the vast majority of inertial sources are outside this band and therefore were observed against a less noisy background.

The SNR correction method is rendered less certain by points 1, 2, and 3, while point 4 implies that the SNR-based correction factors, as published by Cohen *et al.* (1987), are not applicable to asteroidal sources. For these reasons neither of these methods, even if they were valid, could be applied to asteroidal sources.

Thirty-nine percent of the accepted IRAS 25 μm asteroid sightings, and 47% of the 12 μm and 60 μm sightings, are "weak sources", *i.e.*, have SNR < 10. IRAS asteroid diameters derived from "weak sources" are systematically large by an average of ~33% with respect to ground-based observations. In our judgment, this effect was too large to ignore. We therefore decided to use the method originally proposed by Tedesco and Gradie (1988), *i.e.*, to use the results from ground-based IRTF observations to derive a statistical correction for those IRAS asteroids affected by the flux-overestimation problem.

The upper portion of Fig. 3 shows a plot of the IRAS SNR versus the ratio of the 25 μm flux density derived from ground-based (IRTF) observations to that reported by IRAS. The sample used consists of 801 accepted IRAS sightings for which high-quality albedos and diameters are available from IRTF observations by Gradie and Tedesco (1988). The albedo and diameter derived from each IRTF observation were used to predict the 25 μm flux density at the time of each IRAS sighting of that asteroid. The data were grouped into logarithmic bins 0.1 unit wide and the mean IRTF/IRAS flux density ratio and its standard deviation were then computed for each bin.

The quantities plotted were chosen to facilitate comparison between the flux overestimation correction method derived here and that used in correcting the IRAS Point Source Catalog Ver. 1 flux densities (*cf.*, ***IRAS Explanatory Supplement***, 1988 : Fig. XII.A.2, p. XII-6). Note that for SNR > 10 the agreement is good but that below this value the IRAS fluxes are systematically higher than those predicted on the basis of the IRTF observations.

Because the IRTF data do not have an SNR-imposed cutoff the fluxes derived from them do not have an SNR-related systematic error. We therefore derived a linear correction factor as a function of SNR based upon the departure between the IRTF and IRAS fluxes. The correction factor was 0.725 at SNR = 3.0 and 1.0 at SNR = 10.0.

The correction was applied to all detected fluxes with signal-to-noise ratios (SNR) between 3.0 and 10.0. The results of applying this correction factor to the IRAS measurements in the sample used to derive the correction factors is shown in the lower portion of Fig. 3.

To within the limits of measurement the correction factor at 12 μm is the same as that for 25 μm. Because no ground-based measurements are available at 60 μm we are unable to derive a correction for observations made in this band and simply assume that it is the same as that for the shorter bands. Thus, we subsequently applied this single correction factor to all IRAS asteroid detections with SNR < 10 before using them to derive IRAS albedos and diameters. This was done before iteratively computing the albedo and diameter. The flux uncertainty for corrected bands was increased by root-sum-squaring it with the flux correction. The flux overestimation correction factor, if any, applied to each accepted sighting is part of the IMPS data set.

6. Band-to-Band Albedo Discrepancies

Albedos derived from 25 or 60 μm IRAS fluxes are systematically about 10% higher than those derived from 12 μm IRAS fluxes. This could, for example, be caused by an error in the IRAS flux calibration or an invalid assumption in the asteroid thermal model (perhaps the thermal emissivity and/or beaming parameter are wavelength-dependent). Regardless of the cause, the effect is to introduce erroneous differences in the albedos derived from 12 μm detections and those derived from 25 and 60 μm detections.

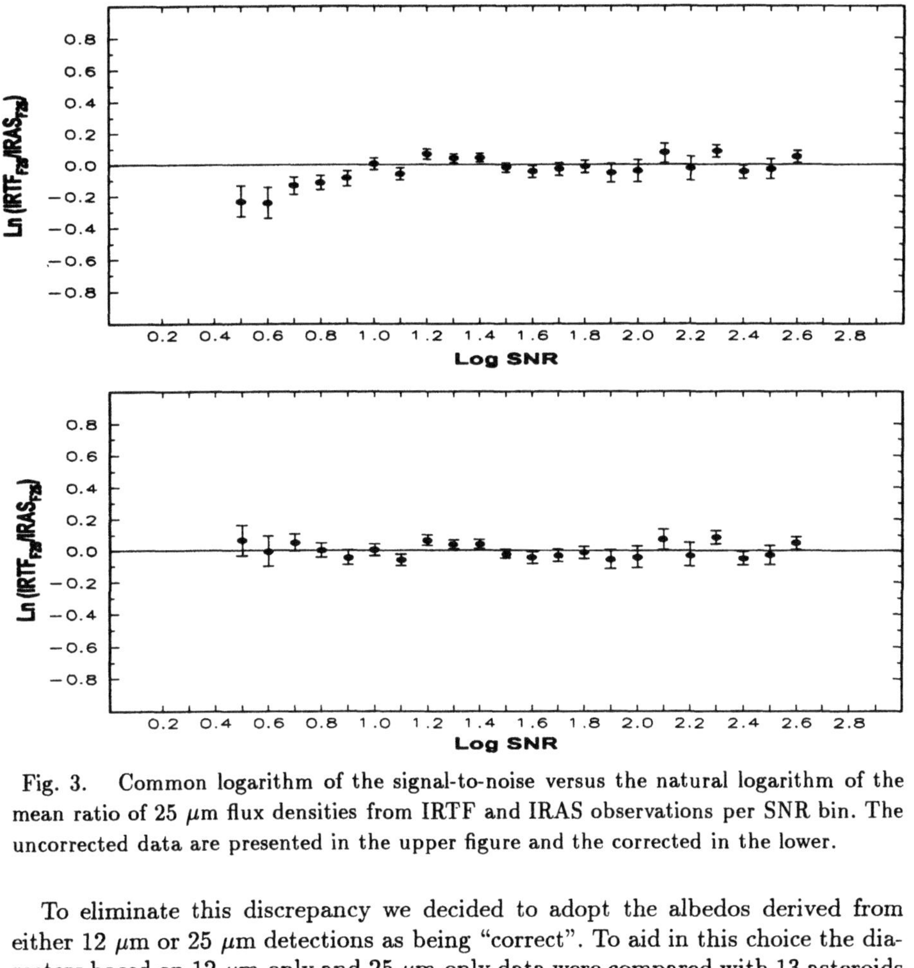

Fig. 3. Common logarithm of the signal-to-noise versus the natural logarithm of the mean ratio of 25 μm flux densities from IRTF and IRAS observations per SNR bin. The uncorrected data are presented in the upper figure and the corrected in the lower.

To eliminate this discrepancy we decided to adopt the albedos derived from either 12 μm or 25 μm detections as being "correct". To aid in this choice the diameters based on 12 μm-only and 25 μm-only data were compared with 13 asteroids with high-quality diameters obtained from stellar occultations. The mean difference between the occultation-derived diameters and the 12 μm-derived diameters was 6.5% while that for the 25 μm-derived diameters was 7.0%. We therefore adopted the 12 μm-derived diameters as being "correct". Subsequently, we learned that this choice is consistent with the IRAS calibration error discussed in Cohen *et al.*, (1992). Hence, a correction factor of 1.12 was applied to the albedos (and therefore, indirectly, to the diameters computed from them) to all albedos derived from 25 μm and 60 μm fluxes. Following this correction the mean difference between the occultation-derived diameters and the 25 μm-derived diameters was reduced to 6.7%

Figure 4 shows plots of the differences between albedos derived from 12 μm fluxes and those determined from 25 and 60 μm fluxes.

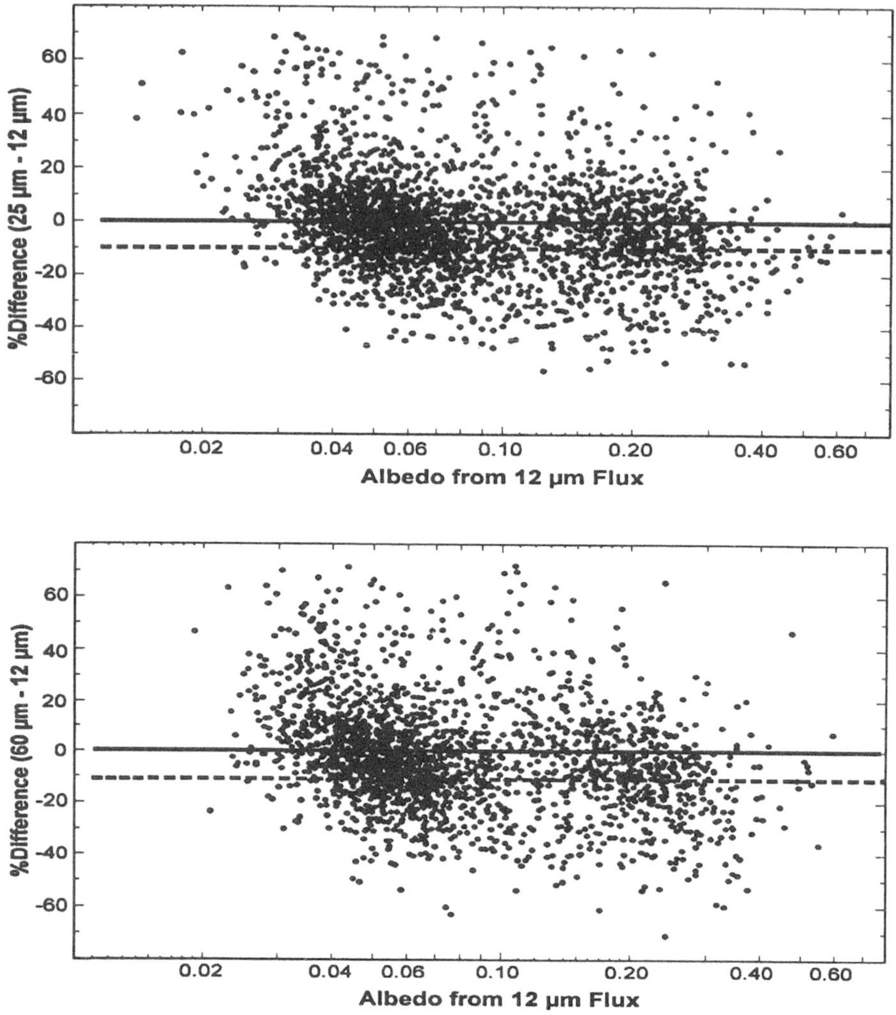

Fig. 4. Comparison of model albedos obtained from data at 12, 25, and 60 μm. The 12 vs. 25 μm results are displayed at the top and the 12 vs. 60 μm results at the bottom. In each case the uncorrected mean difference is indicated by the dashed line and the corrected difference by the solid line. Due to the large size of the graphics file ($\sim$3 Mb) only every other data point is plotted in these figures.

Table V presents statistics on the 12/25 μm and 12/60 μm distributions before and after application of the band-to-band correction.

TABLE V
Sample Statistics for Corrected Vs. Uncorrected 12 μm/25 μm and 12 μm/60 μm Albedos

Bands	N	Mean % Diff		Sample σ		% > 2σ	
		Before	After	Before	After	Before	After
12/25	5,763	-9.9	+0.9	20.7	23.2	5.9	4.9
12/60	4,884	-11.1	-0.5	21.7	24.3	2.9	4.6

The albedos and diameters given in the IMPS final data products have had both the flux-overestimation and band-to-band corrections applied. After applying these corrections the data were again examined for systematic band-to-band effects. It was found that the mean difference in albedos derived from 12 μm flux densities and those derived from 25 μm or 60 μm flux densities is less than 1% for the entire sample.

6. 1. ADAS VERSUS IMPS SIGHTINGS OF NUMBERED ASTEROIDS

Using orbital elements for asteroids numbered through 3318 ADAS processing found 1,790 asteroids with one or more accepted sightings whereas the IMPS processing produced 1,678 such asteroids. As discussed above, this difference is due to the more-stringent positional match requirement used in the IMPS processing. The total number of asteroids with accepted IMPS sightings is 1,890 due to the greater number of available orbital elements. IMPS also produced fewer rejected asteroid sightings (and more missed sightings) than ADAS, implying that the initial tagging by IMPS of an IRAS detection as a "potential sighting" is more reliable than the association method used in the ADAS processing. See Fig. 5.

7. Apparent Albedo Distribution

Figure 6 shows the apparent albedo distribution of the asteroids observed by IRAS. The data are shown separately for the larger and smaller asteroids with 44 km being chosen as the dividing diameter. Note that the distribution for the larger asteroids is clearly bimodal with the dividing minimum near an albedo of 0.1. The distribution for the smaller asteroids is certainly different. Part of this difference can be attributed to differences in the selection effects for small and large numbered asteroids. The larger numbered asteroids contain a disproportionate number of higher albedo members. Not so easily explained is the difference in the proportion of intermediate albedo asteroids in the two samples : Only 5% of asteroids with $D > 44$ km have IMPS albedos between 0.09 and 0.14 whereas 20% of the smaller asteroids have IMPS albedos in this range.

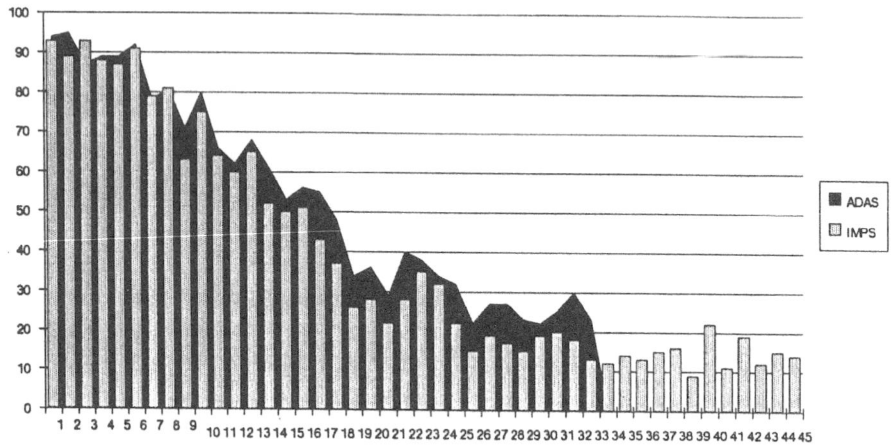

Fig. 5. Percent of numbered asteroids with accepted ADAS and IMPS observations. The area and bar chart peaks represent the actual number of asteroids, per 100 asteroid wide bin, with accepted ADAS and IMPS observations, respectively. Note that ADAS has more accepted asteroid observations in most bins, a fact attributable to the less-stringent positional match requirement compared with that used in the IMPS processing. ADAS processed 3,318 numbered asteroids and IMPS 4,679. Note that more than 10% of asteroids with numbers greater than 2500 yield reliable data.

8. Summary

IMPS diameters for the largest asteroids are systematically low with respect to occultation diameters (*cf.*, Fig. 1). This difference is about 10% or 1σ except for Ceres which differs by 9% or 2σ. The reason for this is that although, on average, the band-to-band correction improved the agreement in the albedos among results derived from 12 μm fluxes and those derived from 25 and 60 μm fluxes (from 10% to <1%) the agreement actually worsened for a few of the largest asteroids. In particular, the diameter obtained for Ceres from 12 μm fluxes alone differs, following the band-to-band correction, by about 8% (which corresponds to 10σ!) with respect to that derived from 25 μm fluxes.

Apparently, the need for the band-to-band correction is due to the fact that Ceres was used to "calibrate" the standard thermal model by requiring results based upon observations at 10 and 20 μm to simultaneously match the diameter obtained from the stellar occultation (Lebofsky *et al.*, 1986a,b), a not unreasonable restriction. However, this requirement led to the difference seen in the IMPS-derived albedos and diameters for virtually all other asteroids. Thus the parameters in the standard

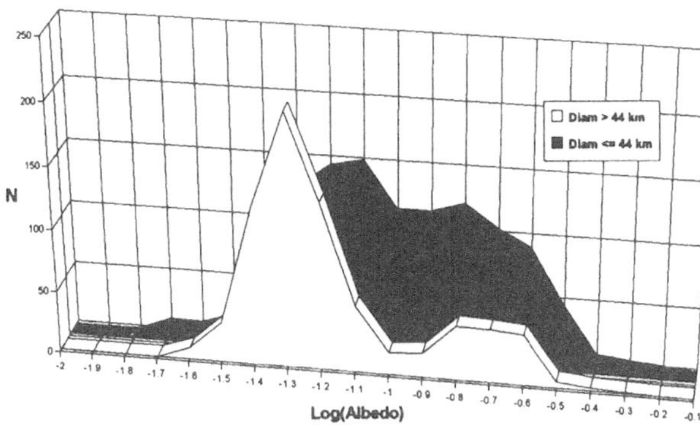

Fig. 6. IMPS albedo distribution for two asteroid size-groups.

thermal model for wavelengths > 10 μm are correct for Ceres but not for the vast majority of smaller asteroids. Hence, the thermal properties of Ceres, and perhaps those of other asteroids with diameters $\gg$ 200 km, differ significantly from those of smaller asteroids. For example, the thickness, mineralogy, and/or particle-size distribution in the regoliths of these largest asteroids may differ appreciably from those of smaller asteroids.

For intermediate-size asteroids, *i.e.*, those with diameters greater than 100 km but less than 350 km, the IMPS results appear accurate to better than 10%. This is based upon a comparison of 13 IMPS asteroids for which there exist high-quality results from stellar occultations. The mean difference in the diameters from these two techniques is 7%.

The albedo size-dependence seen in the IMPS data may be real and due to the lack of a mature dusty regolith on many small asteroids (Veeder,1991). This view is supported by results from ground-based observations as well. For example, Tedesco *et al.* (1990) computed the "10 μm albedo"/"20 μm albedo" ratio (p_{10}/p_{20}) for 352 dual-wavelength observations obtained in the Gradie and Tedesco IRTF Asteroid Radiometry Survey (Gradie and Tedesco, 1988) and then performed linear least-squares fits to this ratio versus the diameter, heliocentric distance at the time of observation, and the "10 μm albedo". They found that there is no correlation between the albedo ratio and the heliocentric distance or albedo but a strong correlation between the albedo ratio and the diameter. They concluded that the thermal properties of asteroids vary in a statistically significant way with size but not with albedo or distance. The sense of this variation is consistent with many smaller asteroids having surfaces with a larger rock/dust component (larger mean particle size) than the typical large asteroid.

If this is indeed the case, then radiometric observations of asteroids with diameters less than about 40 km cannot be used to derive accurate albedos or diameters. For asteroids below this size the standard thermal model can produce albedos which are too high by as much as a factor of two (*cf.*, Veeder *et al.*, 1989a). Thus applications such as derivations of size-frequency distributions or albedo-dependent taxonomic classifications may produce unreliable results if they include asteroids below this size.

To the extent that "the science is in the details" the IMPS data base is a significant improvement over the earlier ADAS product. Due to space limitations, and to avoid being overly pedantic, I have not attempted to contrast every difference between the two data bases. Suffice it to say that the major conclusions drawn from the ADAS data base are supported by IMPS but significant differences in the details do exist.

Acknowledgements

The work described herein was made possible through support provided by Dr. Stephan D. Price, Phillips Laboratory, Geophysical Directorate. I also thank Paul Noah of MRC for creating Fig. 2 and for sharing his knowledge on how IRAS surveyed the non-inertial sky. Finally, I hereby acknowledge the contributions by Tom Chester and John Fowler of IPAC to Chapter 8 of The *IRAS Minor Planet Survey, 1992* from which portions of the present paper have been abstracted, and to Drs. R. Binzel and V. Zappalá for constructive criticisms on the penultimate version of this paper.

References

Allen, D.A. : 1971, "The method of determining infrared diameters". In ***Physical Studies of Minor Planets*** (T. Gehrels, Ed.), NASA SP-267, pp. 41-44.

Barnard, E.E. : 1894, "On the diameters of Ceres, Pallas, and Vesta". ***Astron. & Astrophys.* 13**, 341-344.

Barnard, E.E. : 1895, "Micrometrical determinations of the diameters of the minor plan ets Ceres (1), Pallas (2), Juno (3), and Vesta (4), made with the filar micrometer of the 36-inch equatorial of the Lick Observatory, and on the albedos of those planets". ***Monthly Not. Roy. Astron. Soc.* 56**, 55-63.

Barnard, E.E. : 1900a, "On the diameter of Ceres and Vesta". ***Monthly Not. Roy. Astron. Soc.* 60**, 261-263.

Barnard, E.E. : 1900b, "The diameter of the asteroid Juno (3) determined with the micrometer of the 40-inch refractor of the Yerkes Observatory, with remarks on some of the other asteroids". ***Monthly Not. Roy. Astron. Soc.* 61**, 68-69.

Cellino, A., Zappala, V., and Farinella, P. : 1991, "The size distribution of main-belt asteroids from IRAS data". ***Monthly Not. Roy. Astron. Soc.* 253**, 561-574.

Cohen, M., Schwartz, D.E., Chokshi, A., and Walker, R.G. : 1987, "IRAS colors of normal stars". ***Astron. J.* 93**, 1199-1219.

Dollfus, A. : 1971, "Diameter measurements of asteroids". In ***Physical Studies of Minor Planets*** (T. Gehrels, Ed.), NASA SP-267, pp. 25-32.

Dollfus, A., Wolff, M., Geake, J.E., Lupishko, D.F., and Dougherty, L.M. : 1989, "Photopolarimetry of asteroids". In ***Asteroids II*** (R.P. Binzel, T. Gehrels, and M.S. Matthews, Eds.), University of Arizona Press, Tucson, pp. 594-616.

Drummond, J.D. and Hege, E.K. : 1989, "Speckle interferometry of asteroids". ***Ibid***, pp. 171-191.

Dunham, D.W. : 1991, "Asteroid news". ***Occultation Newsletter* 5**, 93.

Dunham, D.W. and 45 others : 1990, "The size and shape, of (2) Pallas from the 1983 occultation of 1 Vulpeculae". ***Astron. J.* 99**, 1636-1662.

Gradie, J. and Tedesco, E.F. : 1988, "Albedos and diameters for 350 asteroids from the IRTF 10 μm and 20 μm radiometry survey". ***Bull. Amer. Astron. Soc.* 20**, 866.

The IRAS Minor Planet Survey, 1992, edited by E.F. Tedesco. Phillips Laboratory Technical Report No. PL-TR-92-2049. Hanscom Air Force Base, MA.

Infrared Astronomical Satellite Asteroid and Comet Survey : Preprint Version No. 1, 1986, edited by Matson, D.L., JPL Internal Document No. D-3698. (Jet Propulsion Laboratory, Pasadena)

IRAS Catalogs and Atlases, Volume 1, Explanatory Supplement, 1988, edited by Beichman, C.A., Neugebauer, G., Habing, H.J., Clegg, P.E. and Chester, T.J. (NASA Publication No. RP-1190, Washington D.C. : U.S. Government Printing Office).

KenKnight, C.E., Rosenberg, D.L., and Wehner, G.K. : 1967. "Parameters of the optical properties of the lunar surface powder in relation to solar wind bombardment". ***J. Geophys. Res.* 72**, 3105-3117.

Lebofsky, L.A. and Spencer, J.R. : 1989. "Radiometry and thermal modeling of asteroids". In *Asteroids II* (R.P. Binzel, T. Gehrels, and M.S. Matthews, Eds.), University of Arizona Press, Tucson, pp. 128-147.

Lebofsky, L.A., Matson, D.L., Veeder, G.J., and Tedesco, E.F. : 1986a. "The IRAS asteroid thermal model". In *Infrared Astronomical Satellite Asteroid and Comet Survey,* D.L. Matson, ed. (JPL D-3698, Jet Propulsion Laboratory, Pasadena), pp. 7-1 to 7-18.

Lebofsky, L.A., Sykes, M.V., Tedesco, E.F., Veeder, G.J., Matson, D.L., Brown, R.H., Gradie, J.C., Feierberg, M.A. and Rudy, R.J. : 1986b. "A refined "standard" thermal model for asteroids based on observations of 1 Ceres and 2 Pallas". ***Icarus* 68**, 239-251.

Matson, D.L., Veeder, G.J., Tedesco, E.F., and Lebofsky, L.A. : 1989. "The IRAS Asteroid and Comet Survey". In *Asteroids II,* R. Binzel, T. Gehrels and M. Matthews, eds. (University of Arizona Press, Tucson), pp. 269-281.

Millis, R.L. and Dunham, D.W. : 1989. "Precise measurement of asteroid sizes and shapes from occultations". In *Asteroids II* (R.P. Binzel, T. Gehrels, and M.S. Matthews, Eds.), University of Arizona Press, Tucson, pp. 148-170.

Millis, R.L. and 37 others : 1981. "The diameter of Juno from its occultation of AG +0° 1022". ***Astron. J.* 86**, 306-313.

Millis, R.L. and 41 others : 1987, "The size, shape, density, and albedo of Ceres from its occultation of BD +8° 471". ***Icarus* 72**, 507-518.

Morrison, D. and Zellner, B. : 1979, "Polarimetry and radiometry of the asteroids". In *Asteroids II* (R.P. Binzel, T. Gehrels, and M.S. Matthews, Eds.), University of Arizona Press, Tucson, pp. 1090-1097.

Ostro, S.J. : 1989, "Radar observations of asteroids". In *Asteroids II* (R.P. Binzel, T. Gehrels, and M.S. Matthews, Eds.), University of Arizona Press, Tucson, pp. 192-212.

Tedesco, E.F. : 1989a, "Introduction to the Asteroids II data base". In *Asteroids II,* (R. Binzel, T. Gehrels and M. Matthews, Eds.) pp. 997-1001, University of Arizona Press, Tucson.

Tedesco, E.F. : 1989b, "Asteroid magnitudes, UBV colors and IRAS albedos and diameters". In *Asteroids II,* (R. Binzel, T. Gehrels and M. Matthews, Eds.) pp. 1090-1138, University of Arizona Press, Tucson.

Tedesco, E.F. and Gradie, J.C. : 1988, "IRAS Vs. IRTF asteroid radiometric albedos". ***Bull. Amer. Astron. Soc.* 20**, 865.

Tedesco, E.F., Gradie, J.C., and Lebofsky, L.A. : 1990, "Size dependence of radiometric diameters and albedos". ***Bull. Amer. Astron. Soc.* 22**, 1113.

Tedesco, E.F., Fowler, J.W., and Chester, T.J. : 1992, "Summary". In *Infrared Astronomical Satellite Minor Planet Survey Catalog, 1992* (E.F. Tedesco, Ed.), pp. 127-148. Phillips Laboratory Technical Report No. PL-TR-92- 2049. Hanscom Air Force Base, MA.

Tedesco, E.F., Matson, D.L. and Veeder, G.J. : 1989a, "Classification of IRAS asteroids". In ***Asteroids II,*** R. Binzel, T. Gehrels and M. Matthews, eds. (University of Arizona Press, Tucson), pp. 290-297.

Tedesco, E.F., Williams, J.G., Matson, D.L., Veeder, G.J., Gradie, J.C. and Lebofsky, L.A. : 1989b, "A three-parameter asteroid taxonomy". ***Astron. J.*** **97**, 580-606.

Veeder, G.J. : 1991, "Evidence against dusty regoliths on small main belt asteroids". ***Bull. Amer. Astron. Soc.*** **23**, 1155.

Veeder, G.J., Hanner, M.S., Matson, D.L., Tedesco, E.F., Lebofsky, L.A. and Tokunaga, A.T. : 1989a, "Radiometry of near-Earth asteroids". ***Astron. J.*** **97**, 1211-1219.

Veeder, G.J., Tedesco, E.F., and Matson, D.L. : 1989b, "Asteroid results from the IRAS survey". In ***Asteroids II,*** R. Binzel, T. Gehrels and M. Matthews, eds. (University of Arizona Press, Tucson), pp. 282-289.

Zellner, B. Leake, M., Le Bertre, T., Duseaux, M., and Dollfus, A. : 1977a, "The asteroid albedo scale. I – Laboratory polarimetry of meteorites". ***Proc. Lunar Sci. Conf.*** **8**, 1091-1100.

Zellner, B. Le Bertre, T., and Day, K. : 1977b, "The asteroid albedo scale. II – Laboratory polarimetry of dark carbon-bearing silicates". ***Proc. Lunar Sci. Conf.*** **8**, 1111-1117.

IN MEMORIAM - ĽUBOR KRESÁK

After Jan Štohl, another dear friend from Slovakia has left us : Ľubor Kresák died on 20 January 1994, in Bratislava, Slovakia, not long after the end of ACM 93 meeting; only one and a half years before, many of the participants of the Belgirate meeting had attended the International Symposium on Meteors and their Parent Bodies, held in Smolenice, Slovakia, to celebrate Ľubor's 65th anniversary.

He was born on 23 August 1927, in Topoľčany, Slovakia, and studied at Charles University, in Prague, becoming RNDr. in 1951, with the thesis "Structure, mass and age of the comet Halley meteoroid stream". He started as External Lecturer of Astronomy at Comenius University in Bratislava in 1956, and was then affiliated to the Czechoslovak Academy of Sciences in 1957, with the CSc. title; afterwards, he became Docent at Comenius University in 1962, DrSc. in 1967, Corresponding

A. Milani et al. (eds), Asteroids, Comets, Meteors 1993, 75–76

Member of the Slovak Academy of Sciences in 1968 and of the Czechoslovak Academy of Sciences in 1989, Associated Member of the Royal Astronomical Society in 1987, and finally Professor at Charles University in Prague in 1992. Asteroid 1849 Kresák is named after him, in recognition of his achievements. He married Margita, also astronomer, in 1954, and leaves a daughter, Katka.

Very deep was his involvement in the international activities : he was Acting President of IAU Commission 22 (Meteors and Meteorites) in 1961; became Vice-President, and then President, of IAU Commission 20 (Position and Motion of Minor Planets, Comets and Satellites) in 1970-1973 and 1973-1976; was Vice-President of the IAU in 1979, until 1985; was Vice-President, and then President, of the IAU Commission 15 (Physical Studies of Comets, Minor Planets and Meteorites) in 1982-1985 and 1985-1988.

Ľubor Kresák has worked mainly in the field of solar system minor bodies; he was also a very skilled observer. In 1946, at the Skalnaté Pleso Observatory, while he was still a student, he observed an exceptional outburst in activity of the Giacobinids. In 1951, at the start of his scientific career, he returned at that observatory and stayed there until 1955, hunting for comets. The search was successful, since in 1954 he discovered P/Tuttle-Giacobini-Kresák and Kresák-Peltier. After that period he moved to the Astronomical Institute of the Slovak Academy of Sciences in Bratislava.

The work done by him on small solar system bodies, especially comets and meteoroids, has been very influential. Among the many fields, related to minor bodies, to which he gave outstanding contributions let us recall just a few, such as : dynamics of short-period comets, interrelations between comets and asteroids, observational biases, ensemble of available observations and general conclusions on their statistical significance. He had the special ability to treat equally successfully both the observational and the dynamical aspects of the problems, so as to get the most from the observations and, at the same time, make clear to himself and the rest of the astronomical community what were the limits beyond which the conclusions could not be pushed. He worked at several important catalogues concerning short-period comets, namely the one in cooperation with the Institute for Theoretical Astronomy in Leningrad, on orbital evolution and observational circumstances, that in cooperation with the Istituto di Astrofisica Spaziale in Roma, on orbital evolution over a longer time-span, and that in cooperation with his wife Margita, on absolute magnitudes.

But as important as his scientific contributions to astronomy have been his personal qualities : he was a friendly person and an excellent teacher, and educated practically all astronomers in Slovakia. He was always ready to help and advice younger astronomers, and we know this well, since in the last fifteen years we have had the unique opportunity to work in very close cooperation with him : we owe to Ľubor most, if not all, we know about cometary dynamics, and will always, as for sure many others, miss our dear friend.

Andrea Carusi and Giovanni Valsecchi

COMETS (EXISTING POPULATIONS)

Ľ. KRESÁK

Astronomical Institute of the Slovak Academy of Sciences
842 28 Bratislava, Slovakia

Abstract. The definition, population, extent, origin and evolution of the individual subsystems of comets and transitions between them are discussed, together with presentation of the relevant statistical data and their changes with time. The largest outer subsystems are unobservable, but their existence is documented by the necessity of progressive replenishment of the observable populations, with limited survival times. There is persuasive evidence for two different evolutionary paths, one from the Oort cloud and another from the Kuiper belt. While the extent and accuracy of the data available is increasing rapidly, the Jupiter family of comets is the only one for which the evolutionary time scales do not exceed by many orders of magnitude the history of astronomical observations. The individual comet populations differ from one another not only by the distribution of orbits, but also by the size distribution and aging rate of their members. Their dynamical evolution is coupled with disintegration processes, which make it questionable whether the present state can be interpreted as a long-term average.

Keywords : Comets – populations, dynamics, evolution

1. Introduction

The huge system of comets consists of several different populations. As to the appearance, volatility and activity, there are no systematic differences between their members (Whipple, 1992a,b) which can be identified only by the orbit determination. However, there are enormous differences in their populations, in the space occupied by them, and in their evolutionary time scales. The volume of the Oort cloud is about 10^{12}-times larger than that of the Jupiter family of comets. The revolution period of a new comet is only about 1/1000 of its age, and the whole active lifetime of a comet of the Jupiter family is still 1000-times shorter. This, and the long-term dynamical integrations, require progressive transitions between the individual populations, replenishment of those evolving on shorter time scales, and transformation into inactive asteroid-like objects and interplanetary dust.

The main restriction of our knowledge is that the comets become detectable only when they approach the Sun within less than 1/10000 of the radius of the Oort cloud. Comets with perihelia beyond this limit remain unobservable all the time. New comets in Oort's sense, passing inside for the first time - and most of them also for the last time - can only be traced for 10^{-7} to 10^{-6} of their revolution. Another complication comes from the nongravitational accelerations, which do not allow an extrapolation of the motion of the individual objects over sufficiently long time spans. And finally, one cannot be sure that the available observations, covering a few centuries, are representative for a long-term average or a state of equilibrium.

A. Milani et al. (eds.), Asteroids, Comets, Meteors 1993, 77–94.

Stellar passages through the Oort cloud, encounters with giant molecular clouds, or disruptions of large objects into a number of smaller ones can introduce substantial temporary changes. Thus all that we know about the cometary populations reflects only the present state.

2. Classification

The main classification of cometary populations is based on the revolution periods. The boundary between the short- and long-period comets, at $P = 200$ years, is rather conventional. It was introduced as the upper limit for those which already could have been reliably observed at more than one apparition, and because at that time there was no well-determined comet orbit with P between 160 and 250 years. Much better defined is the upper limit of the Jupiter family, $P = 20$ years. We currently know about 150 of its members, all moving in direct orbits, while only four comets have P between 20 and 50 years, and the orbits of two of them are retrograde. Comets with P between 20 and 200 years are called the Halley type. The third limit, at $P = 1\ 000\ 000$ years, separates the old long-period comets from the new ones, in Oort's sense (1950). Due to the third Kepler's law, the classification according to the revolution period is equivalent to that according to the semimajor axis a $(=P^{2/3})$, or binding energy $1/a$ $(= P^{-2/3})$.

Another classification with good dynamical reasoning is that by the Tisserand invariant with respect to Jupiter

$$J = \frac{a_J}{a} + 2\sqrt{\frac{a}{a_J}(1-e^2)}\cos i = \frac{2a_J}{q+Q} + 2\sqrt{\frac{2qQ}{a_J(q+Q)}}\cos i$$

where a, e, i, q and Q are the semi-major axis, eccentricity, inclination, perihelion distance and aphelion distance of the comet's orbit, and a_J the semi-major axis of Jupiter. Since J is a function of the unperturbed encounter velocity with Jupiter, proportional to $\sqrt{3-J}$, this parameter measures also the degree of stability of the comet's orbit. Due to the substantial impact of orbital inclination, J does not present any separation between Halley type, old and new comets. On the other hand, for the Jupiter family it allows a meaningful subdivision by $J = 2$, $2^{3/2}$ and 3. In the restricted three-body problem Sun/Jupiter/comet, the subgroup of $2 < J < 2^{3/2}$ may be subject to the ejection from the solar system; that of $2^{3/2} < J < 3$ to a temporary satellite capture; and that of $J > 3$ to the transfer of the orbit from completely outside to completely inside the orbit of Jupiter, and inversely (Kresák 1982, Carusi and Valsecchi, 1985). The limit of $J = 2$ separates the Jupiter family from all the other cometary populations. It coincides with the classification by the revolution periods for more than 99% of the known comets. $P < 20$ years and $J < 2$ applies only to P/Tuttle, P/Machholz and P/IRAS - the three Jupiter family comets with highest inclinations; and $P > 20$ years and $J > 2$ to 1905 IV Kopff, 1957 VI Wirtanen and 1982 I Bowell - all with low inclinations and abnormally large perihelion distances.

A third acceptable classification is that by the aphelion distance. For the long-period comets it is essentially identical with that by the revolution period, because $Q \simeq 2a$, and for the Halley type it is rather close to it. For the Jupiter family it presents a better view on the degree of dynamical stability, including also information on the potential perturbations by other planets than Jupiter. All of the classification parameters are obviously varying with the planetary perturbations; and this is just an indication of the dynamical evolution and transfer between the individual comet populations.

3. Observed populations

Figure 1 shows the distribution of comets according to the aphelion distance Q (above) and Tisserand invariant J (below). Only the best determined orbits from Marsden's catalogue (1992) are included : those of short-period comets of more than one apparition, long-period comets of class 1A or 1B, and comets of Halley type with comparable orbital accuracy. The data for the individual populations are approximately normalized to their observed proportion, and the steps are twice as long for the long-period comets, where the data are also smoothed by taking the averages from pairs of neighbouring intervals. Even on the logarithmic scale of Q, the required accuracy of astrometric observations increases steeply towards right, and thus the peak corresponding to the new comets may be in fact much narrower and higher.

There is a clear separation of the four main comet populations in the lower, complete part of the diagram, based on the original values of $\log Q$ before entering the inner solar system. In the upper part of the diagram the future values, after leaving it, are plotted. The immediate mixing of the populations of new and old comets is evident, as well as the ejection of nearly one half of the new comets from the solar system - outside the right margin of the diagram.

In the upper left corner the positions corresponding to the mean heliocentric distances of the four giant planets are indicated. The highly prevailing binding effect of Jupiter is evident. The perturbations by Jupiter are statistically 10-times stronger than those by Saturn, and 100-times stronger than those by any other planet (Everhart and Raghavan, 1970; Rickman and Huebner, 1990). Therefore, it has little sense to speak about observed populations associated with other planets.

The same concentration of short-period comets into the Jupiter family is reflected by the lower diagram, with the relative number of observed comets plotted against the Tisserand invariant J (lower horizontal scale) and the corresponding unperturbed encounter velocity with Jupiter V (upper horizontal scale, in km/s). This diagram shows a close similarity between the distribution of new and old comets, a broad range for the Halley type, and a rather sharp separation of the Jupiter family.

A general view on the whole system of comets is presented by Fig. 2. Although this is meant as a scheme, its co-ordinates are scaled. The vertical scale of $\log Q$,

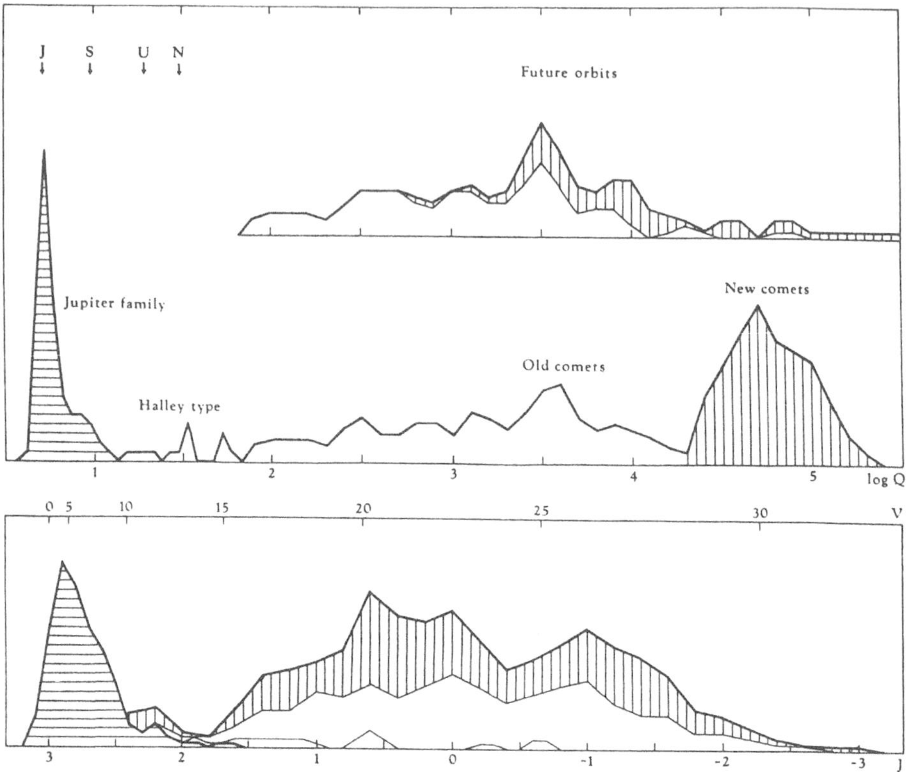

Fig. 1. The distribution of comets with well determined orbits according to the aphelion distance Q (on a logarithmic scale, above), and according to the Tisserand invariant J with respect to Jupiter (below).

identical with the horizontal scale of Fig. 1, delimits the individual populations; each rectangle includes more than 90% of their members, and the medians are marked by dots. The position on the horizontal scale of $\log q$ governs the detectability of each comet. Only those situated on the right, i.e. passing close enough to the Sun, can be observed.

The void space on the left can be called "terra incognita" or "hic sunt leones". The high occurrence rate of new comets demonstrates the presence of the Oort cloud, in the upper left corner. This is subject to the perturbations by our galactic environment (Biermann et al., 1983; Bailey, 1986; Delsemme, 1987; Matese and Whitman, 1992). Stars passing through it remove one half of the objects encountered from the solar system (upwards), and transfer the other half into smaller orbits. For original low-eccentricity orbits in the outer Oort cloud the heliocentric

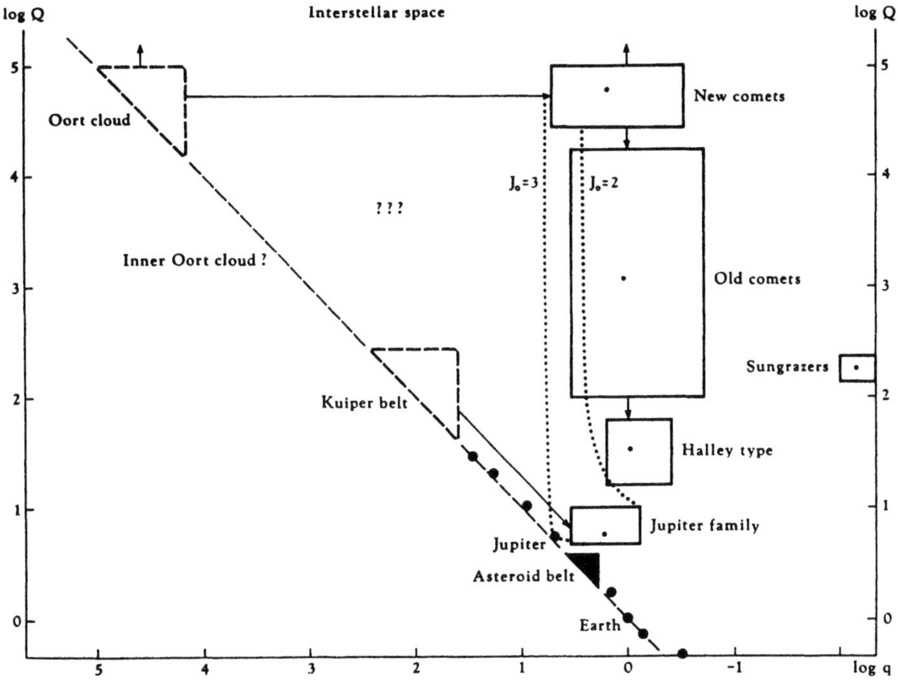

Fig. 2. A schematic view on the distribution and evolution of the cometary populations in the $\log Q / \log q$ reference frame.

velocities are typically 100 - 150 m/s, but the comet must be decelerated to less than 2 - 3 m/s to move towards the Sun, start its activity, and become observable 2 - 3 $\times 10^6$ years later. This strict condition is the main proof of the enormously large population, where 10^{12} of individual objects is considered as the lower limit (Weissman, 1985).

According to Yabushita (1991) there is statistical evidence that some of the new comets - those with original values of $1/a < -4 \times 10^{-5} \mathrm{AU}^{-1}$ - did not come from the Oort cloud but from the interstellar space. A more detailed examination has shown, however, that there is not yet a single case where this hyperbolic excess could not be attributed to the nongravitational accelerations and computing uncertainties (Kresák, 1992).

4. Origin and evolution

After the first passage near the Sun, nearly one half of the new comets escape from the solar system (upwards), and the other half joins the population of old comets.

Only for 5% of them planetary perturbations on the path inwards and outwards balance so that a and Q remain within the limits acceptable for new comets. The high proportion of new comets among those of long period - more than one third - is indicative of their total extinction, substantial reduction of their gas and dust production (Oort, 1990), or some so far unidentified long-term rejuvenation process within the Oort cloud (Kresák, 1977). However, as already mentioned, the present state cannot be interpreted with certainty as a long-term average.

The probability of a direct conversion into a comet of Halley type is about 1/1000, and no such transfer has been recorded as yet. Thus this rough estimate is only based on the extrapolation of the distribution of the absolute energy perturbations taken from Marsden's catalogue (1992) and reproduced in Fig. 3. A direct conversion would require $\Delta E < 29 \times 10^{-4}$ AU^{-1}. Since all comets cannot be considered as objects of equal active survival times, and the number of comets of Halley type is relatively low, there is no contradiction to the assumption that they have evolved via new - old comets. This interpretation is supported by the low upper limit of their perihelion distances ($q < 2$), compatible with the blow–off effect (Whipple, 1978; Rickman et al., 1991). Both the distribution of their perihelion distances and their inclinations (3 : 1 in favour of direct orbits) agrees very well with the dependence of the energy perturbations on these two elements, as shown in Fig. 3, and with the expected role of the Tisserand invariant.

Much more difficult is the explanation of the origin of the Jupiter family. The nongravitational effects do not allow an extrapolation of their motion back over the requisite time spans. The only feasible approach is the use of modelling experiments, the results of which depend on the adopted starting conditions and are still contradictory (Duncan et al., 1988; Quinn et al., 1990; Rickman and Huebner, 1990; Valsecchi, 1992). For a long-period comet of very low inclination and perihelion near the orbit of Jupiter, an extremely close encounter with this planet can, in principle, result in such a transformation. The potential way is limited by the two dotted curves corresponding to $J = 2$ and $J = 3$ for $i = 0^o$; with increasing inclination these limits shift to the left until they disappear at $i = 45^o$. However, the occurrence rate of such phenomena must be very low, definitely incompatible with the current population of the Jupiter family and active survival times of its members, estimated at 3,000 - 10,000 years (Kresák, 1985; Rickman, 1992). Even the extension of the total lifetimes by dormant phases (Kresák, 1987) does not appear sufficient to solve this problem.

This was the main reason for suggesting the presence of the Kuiper belt (Kuiper, 1951; Bailey, 1992). Its position is indicated by the triangle in the middle of Fig. 2. Transfer of its invisible members would require a longer interaction with the outer planets, because the effect of the galactic environment is too weak and rare. Objects on this way may be 2060 Chiron, 5145 Pholus, 1992 QB1, 1993 FW and 1993 HA2; all of them are formally classified as asteroids, but for Chiron an outburst of cometary activity was recently recorded (Meech and Belton, 1989; Meech, 1991). Other alternatives for the structure of the inner Oort cloud are summarized and

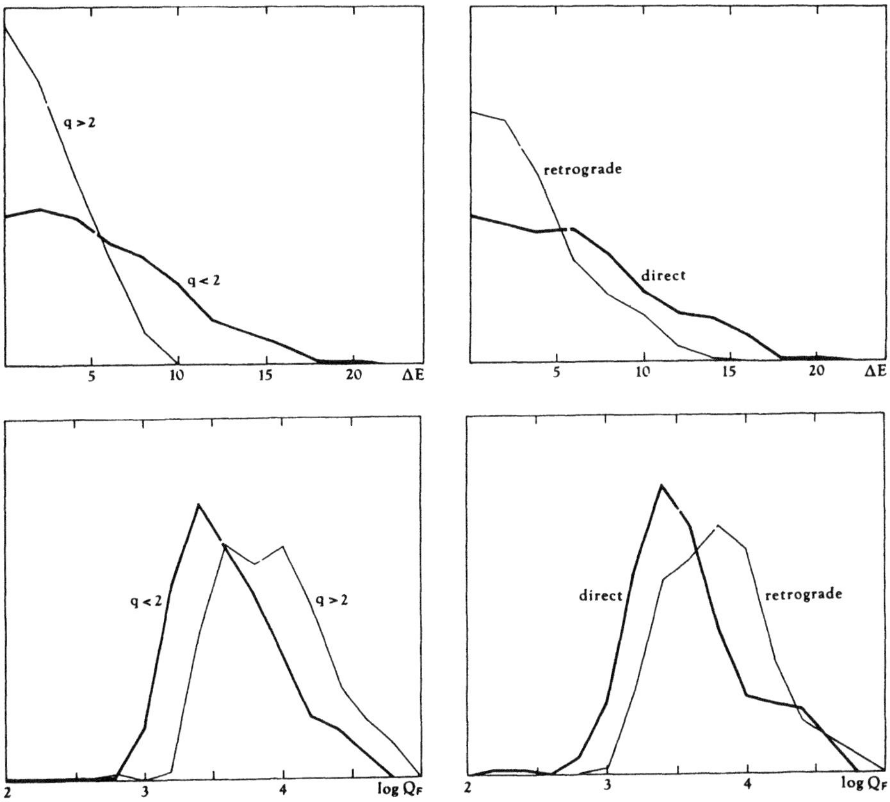

Fig. 3. Distribution of the absolute energy perturbations $\Delta E = \mid \Delta(1/a) \mid$, in 10^{-4} AU^{-1}, and aphelion distances Q (on a logarithmic scale), by their first passage through the inner planetary region. The four plots illustrate the dependence of these quantities on the perihelion distance q and orbital inclination i.

intercompared by Bailey (1990, 1992).

5. Perturbing encounters

The first evolutionary phase of the population of new comets is governed by the efficiency of planetary perturbations around their single perihelion passage. Figure 3 compares this efficiency for two significant parameters : perihelion distance q smaller or greater than 2 AU and inclination i lower or higher than 90° (direct/retrograde). The distributions are derived from all comet orbits listed by Marsden (1992) with the original and future values of $1/a$. The upper half of the figure shows the changes of the binding energy, and the lower half those of the aphelion distance (on a logarithmic scale). One can see that the median changes of the binding energy, marked by small gaps, are almost doubled for $q < 2$, and also for $i < 90°$. The same applies to the reduction of the aphelion distance after the first perihelion passage near the Sun. This implies an acceleration of the next perihelion passage by a factor of 2.5 and 2.7, respectively.

For the further evolution of all populations, the frequency of passages through different heliocentric and jovicentric distances is essential. Figure 4 shows, on the left, the average annual numbers N of comets of different types passing perihelia at $q < 2$, as observed in 1940-1990. Members of the Kreutz group are omitted, and the numbers of new and old comets are multiplied by 1.5 to account for those objects whose original orbits were indeterminate. The ephemeris-aided recoveries and low

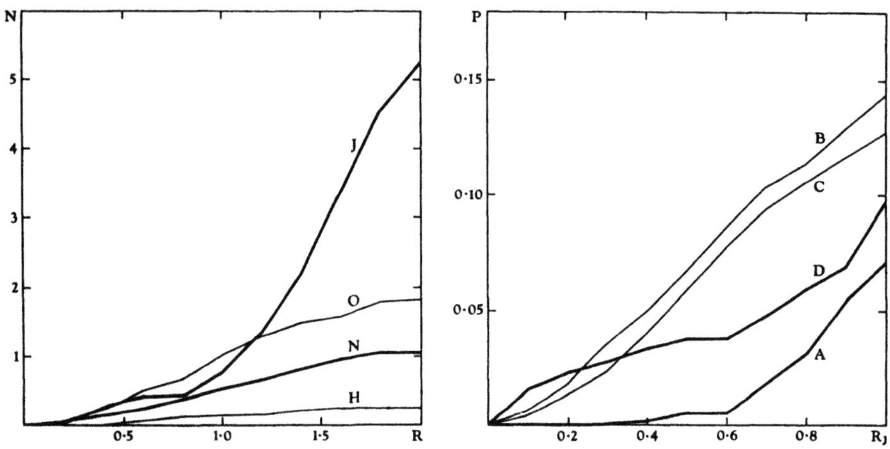

Fig. 4. The average number N of comets belonging to the four basic populations, and passing annualy within different heliocentric distances R (J = Jupiter family, H = Halley type, O = old long-period comets, N = new long-period comets); and probabilities P of passing within different jovicentric distances R_J per revolution for the four subclasses of the Jupiter family (A : $J < 2$, B : $2 < J < 2^{3/2}$, C : $2^{3/2} < J < 3$, D : $J > 3$).

inclinations make the data for the Jupiter family relatively more complete with increasing q. Nevertheless, the differences are definitely too large to be attributed to any observational biases. For the Jupiter family the number of apparitions increases by a factor of 7 when passing from $q < 1$ to $q < 2$, and by a factor of 10 when the peculiar P/Encke is omitted. For all the other populations the corresponding increase is much slower, by a factor of 1.8 to 2.0 ($\simeq 7^{1/3}$). This point is of special interest in connection with the rapid increase of the population of the Jupiter family with q found by Fernández et al. (1992), and the dependence of the aging rate on the perihelion distance.

For these comets the frequency of efficient perturbing encounters with Jupiter is high, and statistically tied with the Tisserand invariant. The plot on the right is based on the long-term integrations by Carusi et al. (1985a), and limited by $R_J = 1$ AU. This corresponds, e.g., to a complete destruction of the compact dust trail behind the comet (Sykes and Walker, 1992; Kresák, 1993). The average frequency of such encounters is once per 14 revolutions for class A ($J < 2$), once per 7 revolutions for class B ($2 < J < 2^{3/2}$), once per 8 revolutions for class C ($2^{3/2} < J < 3$), and once per 10 revolutions for class D ($J > 3$). There is an evident difference between the distribution functions of class A and D, caused by the much lower iovicentric velocities of the latter. These result in a substantial reduction of the perturbed minimum distance against the unperturbed one, and sometimes also in a temporary satellite capture and in a discovery immediately thereafter.

6. Changes in the Jupiter family

The Jupiter family, with 10 members already followed over more than 20 revolutions, is the only population in which the evolutionary changes can be directly observed. The orbital integrations of all such comets of more than one apparition, extending more than four centuries back and forward (Carusi et al., 1985a) were used to determine the frequency of significant processes occuring on time scales comparable with their mean active lifetimes. Based mainly on the cases of disappearance, this was estimated at about 300 $q^{1/2}$ revolutions, or 2500 - 3000 years (Kresák, 1985). However, the detection of their dormant phases (Kresák, 1987), supported recently by the discovery of P/Machholz, P/Takamizawa and P/Hartley 2, suggests that the active lifetime of a comet is not always a single continuous period, but may consist of recurring active phases separated by temporary extinctions. Accordingly, 5000 - 7000 years now seems to be a more reasonable estimate of the total active lifetime. This value is also in better agreement with the results of Weissman (1980), Fernández (1985) and Rickman (1992), obtained by different approaches.

The statistical data are listed in Tables I and II. Table I reflects the asymmetry between the injection of comets into the inner solar system and ejection from it, as already pointed out by Fernández (1985). The disproportion of 2 : 1, referring only to a limited range of revolution periods and perihelion distances, and including the symmetrical libration phenomena, reflects the physical aging of comets and necessity of their supply from outer sources. The total disproportion can be estimated at one order of magnitude. Conversion into asteroid-like objects or total disintegration seems to be the end fate of 90% of these objects, and escape from the Jupiter family only of the rest. This conclusion is supported by the number of small asteroidal objects discovered within the Spacewatch project (Rabinowitz, 1993). There is also a number of larger asteroids the orbits of which are practically indistinguishable from those of the Jupiter family (Kresák, 1985; Kresák and Štohl, 1990), and long-term integrations indicate their similar orbital evolution (Milani et al., 1989).

TABLE I

Mean time spans Δt between significant orbital changes of the comets of the Jupiter family, observed at more than one apparition, in years

Orbital change	Δt (years)
Halley type → Jupiter family	8 000
Jupiter family → Halley type	16 000
$P > P_J \rightarrow P < P_J$	2 800
$P < P_J \rightarrow P > P_J$	3 400
Reduction of q by more than 1 AU	1 800
Increase of q by more than 1 AU	3 000
Entry into libration	2 300
Exit from libration	2 300
Separation of active fragments	600
Splitting into surviving components	6 000
Period of active lifetime	3 000
Total active lifetime	6 000

The frequency of splits of cometary nuclei is given in the lower part of Tab.I. Separation of active fragments surviving not more than 1-2 perihelion passages is a rather frequent phenomen (see P/Biela, P/Brooks 2, P/Giacobini, P/Taylor, P/duToit-Hartley, P/Chernykh). Since there is only one case of formation of a pair of long-lived objects of comparable size, P/Neujmin 3 and P/Van Biesbroeck (Carusi et al., 1985b), the estimated occurrence rate of such events must be taken with reserve. As to the long-period comets, the frequency of splits is higher if referred to their revolutions, but lower if referred to their active lifetimes (Kresák, 1985). The existence of the Kreutz group of sungrazers and of the Taurid complex points to the possibility that disintegrations of extremely large objects may affect substantially the comet populations at a given time, and make them quite different from the long-term average (Steel et al., 1992; Clube, 1992). The existence of objects like 2060 Chiron or P/Schwassmann-Wachmann 1 lends support to the opinion that the disintegration of larger objects may play even a more important role in the replenishment of the inner comet subsystems than the individual dynamical captures.

The chaotic dynamical evolution of short-period comets can be temporarily stabilized by a transient stay in the libration regime. Its efficiency is illustrated by Tab. II, based on the orbit integrations of more-than-one-apparition comets with revolution periods less than that of Jupiter (Carusi et al., 1985a). The table shows that at any time every fourth such comet is librating. The typical duration of one libration cycle, 150 to 200 years, is essentially independent of the resonance ratio. The total duration indicates that a member of the Jupiter family experiences, on the average, two or three libration periods during its active lifetime.

TABLE II
Librations of comets of the Jupiter family about orbital resonances with the planet, at $P < P_J$

Resonance ratio	Objects librating at a time (percentage)	Median duration of one cycle (years)	Mean/median total duration (years)
1/2	11	170	600/400
4/7 or 3/5	4	180	> 700/550
2/3	6	170	> 900/400
3/4 or 4/5	1.5	180	300/250
1/1	1.5	180	250/250
Total	24	170	600/400

Librations about the 5/1, 6/1 and 7/1 resonances with Jupiter play an important role in the dynamical evolution of the comets of Halley type. Their proportion experiencing libration at a time is similar to that in the Jupiter family, but the cycles are twice as long, and the whole libration periods much longer (Carusi et al., 1987).

7. Selection effects

Our present statistical data are obviously very strongly affected by observational biases. One view on this problem is presented by the historical evolution, illustrated by Tabs. III and IV. Table III shows the increase in the number of comets listed in seven main catalogues published since the beginning of orbit computations. Table IV shows the present state for comets which passed their perihelia within the last seven half-centuries. One can see, e.g., that while the number of Halley-type comets increased by a factor of two, that of the Jupiter-family comets increased by a factor of 20 during the last 150 years!

The principal factors on which the discovery or miss of a comet of given size, composition and structure depends are : 1. perihelion distance, determining its maximum heliocentric brightness; 2. observing geometry, determining the maximum geocentric brightness, solar elongation and ecliptical latitude around the perihelion passage; and 3. revolution period, determining the recurrence rate of potential apparitions.

The effect of perihelion distance makes the largest and most stable systems of comets - those with perihelia beyond the orbit of Saturn - entirely unobservable. Further inwards, the detection limit varies with the absolute brightness of the comet and, as far in as the orbit of Mars the discoveries of active comets become fairly complete. The effect of observing geometry is statistically unimportant within the

TABLE III
Cometary orbits listed in different catalogues

	Halley (1705)	Pingré (1783)	Galle (1894)	Baldet (1952)	(1972)	Marsden (1982)	(1992)
S-P comets, all	1	2	29	82	97	121	179
> 1 apparition	1	1	15	35	63	78	102
> 5 apparitions	0	0	5	13	19	28	34
> 10 apparitions	0	0	1	5	8	10	13
> 20 apparitions	0	0	1	2	2	2	2
> 5 revolutions	0	0	8	20	31	42	51
> 10 revolutions	0	0	3	11	19	25	27
> 20 revolutions	0	0	1	3	4	7	10
> 40 revolutions	0	0	0	1	1	1	2
L-P comets, all	21	61	316	453	503	589	672
Sum of orbits	24	67	411	763	924	1109	1353
Sum of objects	22	63	345	535	600	710	851

individual populations. It only prefers the discoveries of short-period comets by the repetition of their returns. This, and the low inclinations in the Jupiter family, enable each of them sooner or later to pass the perihelion near the opposition, where it becomes brighter than at other apparitions, and where also the systematic large-scale surveys concentrate. This is also the reason why more than 90% of them were discovered photographically during the last 50 years, while for the long-period comets the share of photographic discoveries was only 50%. At the time of discovery, the long-period comets were estimated about 100-times brighter, with a median apparent magnitude 10.5 against 15.5. However, since the brightest short-period comets were already known at that time (their ephemeris-aided recoveries are not included in these statistics), and since the subtraction of instrumental effects reduces the difference from 5 to 2–2.5 magnitudes (Kresák and Kresáková, 1990), the apparent brightness alone cannot be taken as a quantitative indicator of the degree of completeness.

During the last 150 years, well covered by observations, most comets of the Jupiter family have made 10 to 30 returns and most of Halley type 1 to 3 returns, but for a majority of long-period comets this time span covers less than 1/100 of their revolution periods. The new comets exceed it even by a factor of 10^4 to 10^5. Assuming a long-term equilibrium, we would have to multiply their observed number by this factor when comparing it with the population of short-period comets. On the other hand, the whole space occupied by new comets is 10^{12}-times larger, which makes their mean spatial density negligible against the Jupiter family.

The increase of the discovery rates of the two main types of comets is illustrated by Fig. 5. The average annual numbers are derived from their totals within 20-

TABLE IV
Cometary apparitions from 50-year intervals, for which orbits of different type and quality are available

	1640 - 1690	1690 - 1740	1740 - 1790	1790 - 1840	1840 - 1890	1890 - 1940	1940 - 1990
All comets :							
apparitions	15	14	39	70	198	249	606
objects	15	14	39	60	161	170	355
Jupiter family :							
apparitions	1	0	5	16	57	121	366
objects	1	0	5	6	20	42	117
Halley type :							
apparitions	1	1	2	5	10	8	13
objects	1	1	2	5	10	8	11
L-P comets :							
all	12	13	32	49	131	120	227
parabolic	11	13	30	42	71	37	66
elliptic	1	0	2	7	48	47	87
hyperbolic	0	0	0	0	12	36	74
old	0	0	0	4	33	39	90
new	0	0	0	0	7	30	64
sungrazers	0	0	0	0	4	0	20

year intervals, with the comets discovered from spacecraft omitted. For the Jupiter family the discovery rate was nearly constant in 1870-1970 at $q < 2$, but since then it increases rapidly. While there is also an increasing proportion of large-q objects, the general trend indicates that the application of modern search techniques has led mainly to discoveries of smaller and absolutely fainter objects, and not to a simple extension of the space covered. For long-period comets the discovery rate - in this case identical with the appearance rate - was relatively stable for a still longer period, since 1840. The subsequent increase was less steep, in particular for comets with $q < 2$.

As a counterpart to Fig. 1, Fig. 6 shows the distributions of perihelion distances for different cometary populations. They are normalized to uniform sums of objects, and reveal substantial differences. Comets of Halley type exhibit a very sharp and symmetrical maximum around $q = 1$, and a void region beyond $q = 2$. For the Jupiter family, a lack of objects with $q < 1$, possibly associated with their shorter survival times, is followed by a less sharp maximum around $q = 1.6$ and a steep drop towards $q = 3$. According to the estimates by Fernández et al. (1992) there is in fact a steep increase for smaller and less active objects. For the long-period comets, the distributions are much smoother on both wings. The only significant difference between new and old comets appears around $q = 4$. A feasible explanation

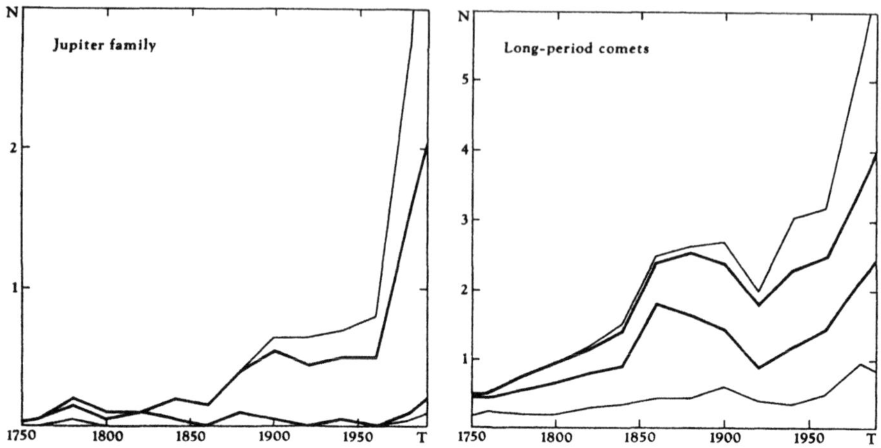

Fig. 5. Changes in the average annual discovery rates N of the members of the Jupiter family (on the left), and long-period comets (on the right) during the last 250 years. The thick lines correspond to $q < 1$ and $q < 2$, and the thin lines to $q < 0.5$ and the total.

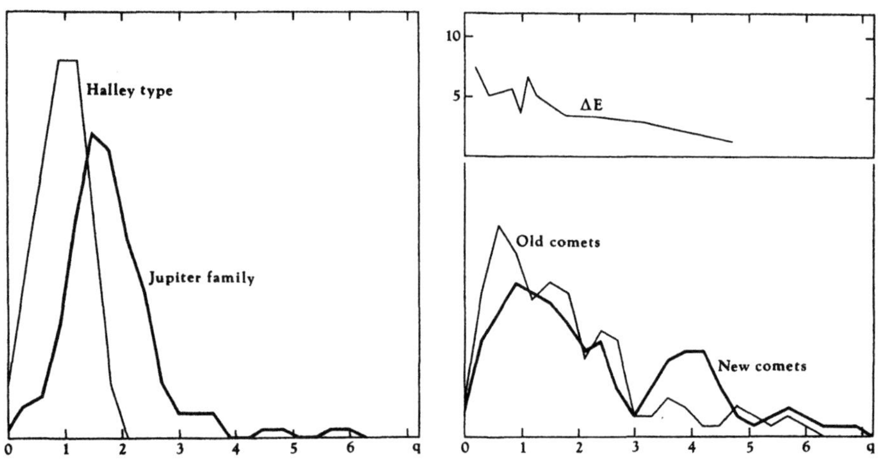

Fig. 6. Normalized distributions of the perihelion distances q of comets belonging to different populations. The plot in the upper right indicates the variations of the median absolute energy perturbations $\Delta E = | \Delta(1/a) |$, in 10^4 AU^{-1}, for the same scale of q.

is an unavoidably wrong classification of those objects for which the planetary perturbations have nearly compensated, resulting in an approximate conservation of the original binding energy $1/a$. This possibility is supported by the dependence of the energy perturbations on perihelion distance. The plot in the upper right gives the medians of the absolute values of ΔE, obtained for 12 equally populated ranges of q from all comets with determined original and future orbits (Marsden, 1992). After a transient decrease of the number of objects at $q = 3$, only the new comets exhibit a secondary maximum, simultaneously with the drop of the typical values of $\mid \Delta(1/a) \mid$ below 0.0002 AU^{-1}. In fact, a transfer of about five comets from new to old (i.e., the same as predicted by the whole distribution pattern and, hence, quite acceptable) would remove the whole discrepancy, and also improve the agreement at small perihelion distances.

Figure 6 demonstrates that significant differences between the individual comet populations do not appear only in the distribution of revolution periods, aphelion distances and inclinations, but also in that of the perihelion distances. This, together with Fig. 5, indicates that there are also significant differences in the distribution of absolute magnitudes and sizes of cometary nuclei. Apparently, this is tied with different dynamical evolution and disintegration processes.

A rough overview of the immense range of sizes and evolutionary time scales, mentioned in the introduction, is presented by Tab. V, in which both of these parameters are scaled by a factor of 1000, with selected illustrative examples. Note that when passing from the linear size to the volume, each step changes into an increase by a factor of 1,000,000,000 !

TABLE V

Size and time scales with a factor of 1000, and a total span of $1 : 10^{18}$

Size/distance (unit $\sim 10^4$ km)	Scale	Time/duration (unit $\sim$ 5 years)
Particle producing a bright meteor	10^{-9}	Meteor outburst
Smallest asteroid observed as yet	10^{-6}	Displacement by the Earth radius at $r = 1$ AU
Nucleus of comet P/Halley	10^{-3}	Comet outburst
The Earth	1	Revolution of a short-period comet
Closest comet approach every few decades	10^3	Total active lifetime of a short-period comet
Kuiper belt	10^6	Revolution of a new comet
Outer Oort cloud	10^9	Age of the solar system

The viewpoints of different investigators on the open problems mentioned in this review are still controversial. Illustrating and stimulating are the extensive round-table discussion records from the Montevideo Workshop, held two years ago (in : Periodic Comets, eds. J.A. Fernández and H. Rickman, 1992, pp. 97-111 and 209-220).

Acknowledgements

This work was supported by the grant No. 492 of the Slovak Academy of Sciences. The author is indebted to Dr. Brian G. Marsden for his careful review and helpful comments.

References

Bailey, M.E. : 1986, "The mean energy transfer to comets in the Oort cloud and implications for cometary origins." *Mon. Not. R. Astron. Soc.*, **218**, 1-30.

Bailey, M.E. : 1990, "Short-period comets : probes of the inner core." In *Asteroids, Comets, Meteors III* (C.-I. Lagerkvist, H. Rickman, B.A. Lindblad and M. Lindgren, Eds.), 221-230, Univ. of Uppsala Press.

Bailey, M.E. : 1992, "Origin of short-period comets." *Celest. Mech.*, **54**, 49-61.

Baldet, F., de Obaldia, G. : 1952, *Catalogue Général des Orbites de Comètes*, CNRS, Paris.

Biermann, L., Huebner, W.F., Lüst, R. : 1983, "Aphelion clustering of new comets : Star tracks through Oort's cloud!" *Proc. Natl. Acad. Sci. USA*, **80**, 5151-5155.

Carusi, A., Valsecchi, G.B. : 1985, "Statistical and numerical studies of the orbital evolution of short-period comets." In *Dynamics of Comets : their Origin and Evolution* (A. Carusi and G.B. Valsecchi, Eds.), 261-278, D. Reidel, Dordrecht.

Carusi, A., Kresák, Ľ., Perozzi, E., Valsecchi, G.B. : 1985a, *Long-term Evolution of Short-period Comets*, A. Hilger, Bristol (together with data for a new edition).

Carusi, A., Kresák, Ľ., Perozzi, E., Valsecchi, G.B. : 1985b, "First results of the integration of motion of short-period comets over 800 years." In *Dynamics of Comets : their Origin and Evolution* (A. Carusi and G.B. Valsecchi, Eds.), 319-340, D. Reidel, Dordrecht.

Carusi, A., Kresák, Ľ., Perozzi, E., Valsecchi, G.B. : 1987, "High-order librations of Halley-type comets." *Astron. Astrophys.*, **187**, 899-905.

Clube, S.V.M. : 1992, "The fundamental role of giant comets in the Earth history." *Celest. Mech.*, **54**, 179-193.

Delsemme, A.H. : 1987, "Galactic tides affect the Oort cloud : an observational confirmation." *Astron. Astrophys.*, **187**, 913-918.

Duncan, M., Quinn, T., Tremaine, S. : 1988, "The origin of short-period comets." *Astrophys. J.*, **328**, L69-L73.

Everhart, E., Raghavan, N. : 1970, "Changes in total energy for 392 long-period comets, 1800-1970." *Astron. J.*, **75**, 258-272.

Fernández, J.A. : 1985, "Dynamical capture and physical decay of short-period comets." *Icarus*, **64**, 308-319.

Fernández, J.A., Rickman, H., Kamél, L. : 1992, "The population size and distribution of perihelion distances of the Jupiter family." In *Periodic Comets* (J.A. Fernández and H. Rickman, Eds.), 143-157, Univ. Montevideo Press.

Galle, J. : 1894, *Verzeichniss der Elemente der bisher berechneten Cometenbahnen*, W. Engelmann, Leipzig.

Halley, E. : 1705, "Tabulae astronomicae." *Phil. Transactions R. Astron. Soc.*, **24**, 1882.

Kresák, Ľ. : 1977, "An alternate interpretation of the Oort cloud of comets?" In *Comets, Asteroids, Meteorites* (A.H. Delsemme, Ed.), 93-97, Univ. Toledo Press, Ohio.

Kresák, Ľ. : 1982, "Dynamical evolution and disintegration of comets." In *Sun and Planetary System* (W. Fricke and G. Teleki, Eds.), 361-370, D. Reidel, Dordrecht.

Kresák, Ľ. : 1985, "The aging and lifetimes of comets." In *Dynamics of Comets : their Origin and Evolution* (A. Carusi and G.B. Valsecchi, Eds.), 279-302, D. Reidel, Dordrecht.

Kresák, Ľ. : 1987, "Dormant phases in the aging of periodic comets." *Astron. Astrophys.*, **187**, 906-908.

Kresák, Ľ. : 1991, "Evidence for physical aging of periodic coemts." In *Comets in the Post-Halley Era* (R.L. Newburn, Jr., M. Neugebauer and J. Rahe, Eds.), 607-628, Kluwer Acad. Publ., Dordrecht.

Kresák, Ľ. : 1992, "Are there any comets coming from interstellar space?" *Astron. Astrophys.*, **259**, 682-691.

Kresák, Ľ. : 1993, "Cometary dust trails and meteor storms." *Astron. Astrophys.*, in press.

Kresák, Ľ., Kresáková, M. : 1990, "Secular variations of absolute magnitudes of periodic comets." *Icarus*, **86**, 82-92.

Kresák, Ľ., Štohl, J. : 1990, " Genetic relationships between comets, asteroids and meteors." In *Asteroids, Comets, Meteors III* (C.-I. Lagerkvist, H. Rickman, B.A. Lindblad and M. Lindgren, Eds.), 379-388, Univ. Uppsala Press.

Kuiper, G.P. : 1951, "Origin of the Solar system." In *Astrophysics* (J.A. Hynek, Ed.), 357-424, McGraw-Hill, New York.

Marsden, B.G. : 1972, *Catalogue of Cometary Orbits* (1st edition), Minor Planet Center, SAO, Cambridge, MA.

Marsden, B.G. : 1982, *Catalogue of Cometary Orbits* (4th edition), Minor Planet Center, SAO, Cambridge, MA.

Marsden, B.G. : 1992, *Catalogue of Cometary Orbits* (7th edition), Minor Planet Center, SAO, Cambridge, MA.

Matese, J.J., Whitman, P.G. : 1992, "A model of the galactic and tidal intercation with the Oort comet cloud." *Celest. Mech.*, **54**, 13-35.

Meech, K.J. : 1991, "Physical aging of comets." In *Comets in the Post-Halley Era* (R.L. Newburn, Jr., M. Neugebauer and J. Rahe, Eds.), 629-669, Kluwer Acad. Publ., Dordrecht.

Meech, K.J., Belton, M.J.S. : 1989, "(2060) Chiron." *IAU Circ.* No. 4770.

Milani, A., Carpino, M., Hahn, G., Nobili, A.M. : 1989, "Project Spaceguard : Dynamics of planet-crossing asteroids." *Icarus*, **78**, 212-269.

Oort, J.H. : 1950, "The structure of the cloud of comets surrounding the Solar System, and a hypothesis concerning its origin." *Bull. Astron. Inst. Neth.*, **11**, 91-110.

Oort, J.H. : 1990, "Orbital distribution of comets." In *Physics and Chemistry of Comets* (W.F. Huebner, Ed.), 235-244, Springer-Verlag, Berlin.

Pingré, P. : 1784, *Cométographie ou Traité historique et théorique des Comètes*, Impr. Royale, Paris.

Quinn, T., Tremaine, S., Duncan, M. : 1990, "Planetary perturbations and the origin of short-period comets." *Astrophys. J.*, **355**, 667-679.

Rabinowitz, D.L. : 1993, "The size distribution of Earth-approaching asteroids." *Astrophys. J.*, in press.

Rickman, H. : 1992, "Structure and evolution of the Jupiter family." *Celest. Mech.*, **54**, 63-69.

Rickman, H., Huebner, W.F. : 1990, "Comet formation and evolution." In *Physics and Chemistry of Comets* (W.F. Huebner, Ed.), 245-303, Springer-Verlag, Berlin.

Rickman, H., Froeschlé, C., Kamél, L., Festou, M.C. : 1991, "Nongravitational effects and the aging of periodic comets." *Astron. J.*, **102**, 1446-1463.

Steel, D., Asher, D. J., Clube, S. V. M. : 1992, "Giant comet origin for the Taurid complex." In *Periodic Comets* (J.A. Fernández and H. Rickman, Eds.), 189-200, Univ. Montevideo Press.

Sykes, M.V., Walker, R.G. : 1992, "Cometary Dust Trails. I. Survey." *Icarus*, **95**, 180-210.

Valsecchi, G.B. : 1992, "Close encounters, planetary masses and the evolution of cometary orbits." In *Periodic Comets* (J.A. Fernández and H. Rickman, Eds.), 81-96, Univ. Montevideo Press.
Weissman, P.R. : 1980, "Physical loss of long-period comets." *Astron. Astrophys.*, **85**, 191-196.
Weissman, P.R. : 1985, "Dynamical evolution of the Oort cloud." In *Dynamics of Comets : their Origin and Evolution* (A. Carusi and G.B. Valsecchi, Eds.), 87-96, D. Reidel, Dordrecht.
Whipple, F.L. : 1978, "Cometary brightness variation and nucleus structure." *Moon and Planets*, **18**, 343-359.
Whipple, F.L. : 1992a, "A new activity index of comets." In *Asteroids, Comets, Meteors 1991* (A.W. Harris and E. Bowell, Eds.), 633-640, Lunar Planet. Inst., Houston.
Whipple, F.L. : 1992b, "The activities of comets related to their aging and origin." *Celest. Mech.*, **54**, 1-11.
Yabushita, S. : 1991, "Maximum nongravitational acceleration due to outgassing cometary nuclei." *Earth, Moon and Planets*, **52**, 87-92.

THE COMET – ASTEROID TRANSITION : RECENT TELESCOPIC OBSERVATIONS

LUCY-ANN MCFADDEN
Department of Astronomy, University of Maryland, College Park, MD 20742
California Space Institute, University of California, San Diego, La Jolla, CA 92093
E–mail mcfadden@astro.umd.edu

Abstract. The phenomenon of a comet-asteroid transition presents the opportunity to study the comet reservoir in comparison to that from which asteroids are derived and to understand the evolution of comets. I discuss six significant areas which have impacted understanding of the comet-asteroid transition from recent years : new discoveries, discovery of dust and volatile comae at 2060 Chiron, observations of comet nuclei, observations of extinct-comet candidates, observations of dust trails associated with short period comets, and discovery of a dust tail on a prediscovery image of 4015 Wilson-Harrington. It is currently possible to recognize an object only as either a comet or an asteroid. There is currently insufficient information to recognize dormant or extinct cometary nuclei, though this is certainly the major goal being sought.

1. Introduction

The objectives of studying the comet-asteroid transition are two-fold, to understand the differences in the composition of the comet and asteroid reservoirs, relating that information back to the formation of the Solar System, and secondly, to understand the evolutionary path of comets in terms of both dynamics and physical properties. I find it useful to review the definitions of comets and asteroids in the manner that was done in Hartmann *et al.* (1987) and Weissman *et al.* (1989). I have reworded the definitions to emphasize what I think is important, the different parts of the Solar System from which the two types of bodies formed.

1. **comet** : a body formed in the outer Solar System. Subsets of comets include the following :
   a. *active comet* : a comet losing volatiles and dust with a detectable coma.
   b. *dormant comet* : a comet with a nucleus which has essentially no volatile loss and hence no detectable coma, though it is active sometimes.
   c. *extinct comet* : a comet that has lost all volatiles.
2. **asteroid** : an interplanetary body that formed in the reservoir between Mars and Jupiter.

Reviews of related topics have been presented in recent years. Wetherill (1991) reviewed the dynamical mechanisms by which comets evolve into dormant or extinct states. The timescales for their evolution into asteroidal orbits is one to two orders of magnitude longer than the time scales of cometary volatile activity, thus making the existence of dormant or extinct comets plausible. Observational scientists are now and have been looking for physical evidence of the evolution of a comet into an object with asteroidal properties with some success but there are also remaining

A. Milani et al. (eds.), Asteroids, Comets, Meteors 1993, 95–110.

questions. I will review the significant observational findings of the past five years concentrating on results since the last ACM meeting in Flagstaff, 1991.

There are six significant aspects of observational studies contributing to our understanding of the phenomenon of the comet-asteroid transition :

1. New discoveries of objects in comet-like orbits which extend our realm of study to the outer Solar System for candidates as extinct or dormant comets.
2. Discovery of dust and volatile comae from 2060 Chiron. Chiron is now known to be a comet.
3. Observational studies of properties of comet nuclei compared to near-Earth asteroids.
4. Studies of individual near-Earth asteroids which are extinct comet candidates.
5. Observations of dust trails associated with short period comets.
6. Discovery of a dust tail on a prediscovery image of 4015 Wilson-Harrington.

2. New Discoveries

Newly discovered objects in the outer Solar System (Figure 1) are logical targets for physical study and comparison with properties of comet nuclei and asteroids. One of these objects, 1991 DA, now named 5335 Damocles, is ~5 km in diameter and is an Earth-approaching object with aphelion beyond Uranus (Steel *et al.* 1992). 5145 Pholus, the outer Solar System object discovered in early 1992 in the course of the Spacewatch Search Program (Scotti *et al.* 1992) crosses the orbits of Saturn, Uranus, and Neptune. These two objects are probably dormant or extinct comets based on their orbital dynamics, though observing efforts to date by Steel *et al.* (1992) for Damocles, and by Hainaut and Smette (1992) for Pholus, revealed no evidence of either dust or gaseous cometary activity. It should be pointed out and we discussed later, that comet-like activity on distant objects is likely to be sporadic. They are far away, relatively small, and thus push the limits of observational techniques, so any one set of negative results is not presently conclusive.

The discovery of more Jupiter-crossing asteroids, for example, 5164 1984 WE_1, provides more candidates with dynamically short life times which might be comets in evolved orbits. Jewitt and Luu (1993a) discovered two distant objects beyond the orbit of Pluto, 1992 QB_1 and 1993 FW which are possibly members of the Kuiper Belt, a postulated source of short-period comets. Four additional objects beyond Neptune were discovered in the fall of 1993 (Luu and Jewitt 1993, Jewitt and Luu 1993b, Williams *et al.* 1993). Future astrometric observations to refine orbits, and photometry to determine their physical properties such as size and color will be directly relevant to understanding the nature of the reservoir of short period comets. The discovered objects both in the outer and inner Solar System increases the available population for the study of the nature of suspected dormant and/or extinct comets.

3. The Discovery of Dust and Volatile Comae from 2060 Chiron

The discovery of cometary activity associated with 2060 Chiron has been pivotal to advancing our understanding of the comet-asteroid transition. Chiron is sometimes active and it is therefore a comet. The story of the discovery of its cometary

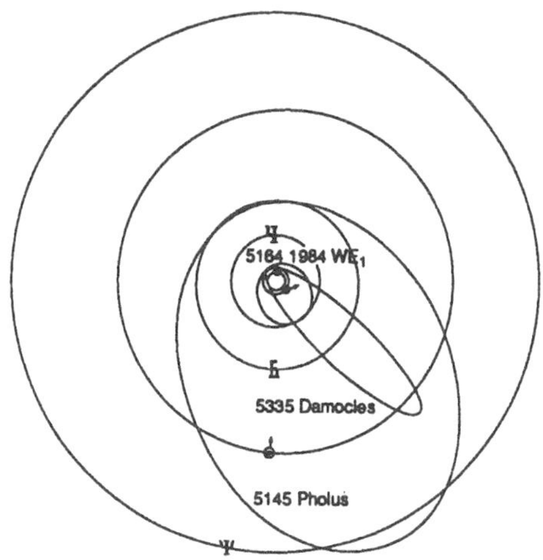

Fig. 1. Plane view of the Solar System with orbits of the planets Earth, Mars, Jupiter, Saturn, Uranus, and Neptune and the recently discovered objects 5145 Pholus, 5335 Damocles, and 5164 1984 WE_1. Their orbits are now or were recently dynamically similar to short-period comets.

activity is illustrative of the interplay of discovery, speculation, and observation, a satisfying example of the process of scientific advancement. Since its discovery in 1977 (Kowal 1979), it was suspected of being a comet. Due to its great distance from Earth, technological limitations of observational instruments and perhaps the limited activity of the comet itself, little was known about it until the wide-spread availability of CCD's in the 1980's. Significant brightness variations as measured with broad-band filters were detected between mid-1987 and 1988 suggesting cometary activity (Hartmann *et al.* 1990). Meech and Belton (1990) first detected a coma in 1989 with CCD imaging. An image of Chiron showing its dust coma is presented in Figure 2. Confirmation of this was reported by Luu and Jewitt (1990a).

A CN emission band was reported by Bus *et al.* (1991) from observations in 1990. Early speculations that Chiron was an inactive comet have now been confirmed by a series of observations. Past brightness variations can be interpreted in terms of cometary activity as Marcialis and Buratti (1993) have recently presented. Buratti and Dunbar (1991) and Luu and Jewitt (1990a) reported short-term impulsive variations. We are now confident because of both independent and repeated observational measurements that Chiron is a comet. Understanding its outbursts and the mechanisms controlling them is the goal of future studies as Chiron comes to perihelion in 1996. The procedure for demonstrating Chiron's cometary nature, detecting both a dust and gas coma can be used to identify other inactive or once dormant comets.

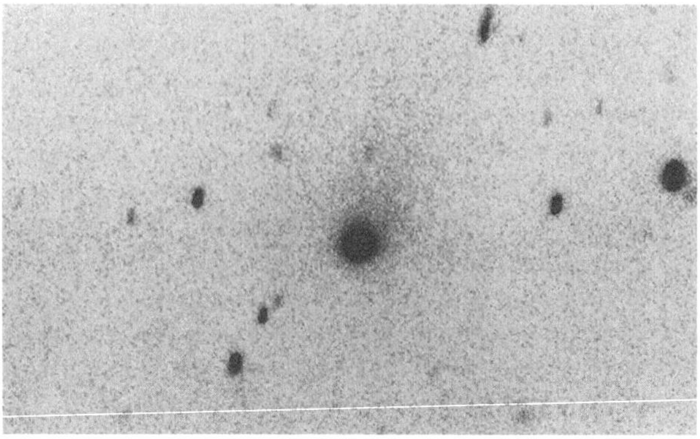

Fig. 2. A CCD image of 2060 Chiron, 1992 January 6, acquired by Karen Meech. The distribution of dust in the coma is asymmetric.

3 1 SEARCH FOR DUST COMAE AROUND OTHER SMALL BODIES

Luu and Jewitt (1992a) searched for faint dust comae among near-Earth asteroids (NEAs) by comparing surface brightness profiles imaged through the R filter with star profiles. In so doing they were able to put upper limits on the mass loss rates ($\dot{M}$) from eleven NEAs of $\leq$ 0.01-0.30 kg sec^{-1} The calculated upper limits are model-dependent; particle density was assumed to be $\rho = 1000$ kg/m^3, mean grain size $\bar{a} = 0.5$ μm, and R, radius, was estimated from $M_v(1,1,0)$ assuming an albedo, p = 0.1. $\dot{M}$ is calculated from equation

$$M_c = \frac{(1.1 \times 10^{-3})\pi\rho\bar{a}\eta R^2_{NEA}}{dr^{0\ 5}\Delta}$$

and can be compared with comets as shown in Figure 3 The parameter, η, is the ratio of the coma cross-section to the nucleus cross-section The angular diameter of the photometry diaphragm, d, is in units of arcsec and r and Δ are in units of AU.

The derived upper limits are 2-4 orders of magnitude less than active comets This is a significant result as long as the parameters of the above equation are reasonable values. An increase in density or mean grain size of the particles would raise the dust production limit though no excess above the star profile was observed. The results of this approach should be made with similar measurements in which the telescope tracks the asteroid instead of the comparison star It is worth continuing to make such observations as opportunities arise because dormant comets are most likely sporadically active.

3 2 SEARCH FOR GAS COMAE

Ground-based and spaceborne instruments have been used to search for OH and CN in some candidate cometary nuclei. Cochran *et al* (1986) searched for CN with

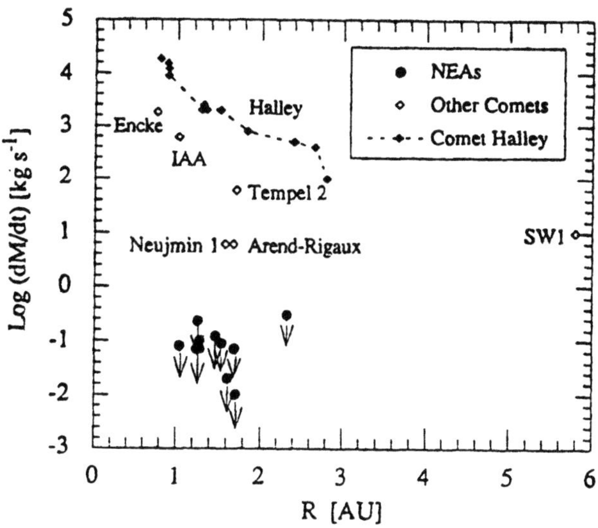

Fig. 3. Derived mass loss rates ($\dot{M}$) for comets and near-Earth asteroids as a function of heliocentric distance (R) from Luu and Jewitt (1992a).

an IDS spectrometer and reported upper limits for a number of cometary candidates. Recent searches for gas comae were presented by Schulz *et al.* (1993) and Chamberlin *et al.* (1993). Three-sigma production rate upper limits were reported for OH at 1566 Icarus, 4015 Wilson-Harrington, and 2201 Oljato (Table I). CN was searched for with a CCD spectrometer from which a 3-σ upper limit was derived for Oljato, and a 2-σ detection was made at 4015 Wilson-Harrington (Table I). That the instrument sensitivities are at the level of the lowest level of measured volatile activity of active comets is particularly frustrating as the results prevent a conclusive difference between volatile activity of comet-candidates among the asteroids and low levels of cometary activity. More sensitive measurements are required.

4. Observations of Comet Nuclei Compared to NEAs

Jewitt (1992) reviewed the physical properties of cometary nuclei in terms of their known rotations, shapes, sizes, and surface reflectance properties. I will describe the general view of comet nuclei, but with the caveat that I am uncomfortable stating their characteristics as representative of all comet nuclei. No one has yet demonstrated that the present distributions are representative of all comet nuclei.

4. 1. SIZES

The six comet nuclei that have been studied directly tend to be small, on order of 2-20 km in effective diameter. But Chiron has a diameter of approximately 300 km (Sykes and Walker 1991, Jewitt and Luu 1992, Campins *et al.* 1992, Campins 1993). Knowing that Chiron is a comet, one cannot accurately say that all comet nuclei are small. I believe we do not yet know the size distribution of comet nuclei.

TABLE I
Upper limits and/or the average of measured gas production rates of selected asteroids and comets

Object	OH (sec^{-1})	CN (sec^{-1})	Reference
2201 Oljato	$\leq$ 2-3 $\times 10^{26}$	$\leq 1 \times 10^{23}$	Schulz *et al.* 1993
1566 Icarus	$\leq 5 \times 10^{25}$		Schulz *et al.* 1993
4015 W.-H.	$\leq$ 2-3 $\times 10^{26}$	1×10^{22}	Chamberlin *et al.* 1993
P/Neujmin 1	7.47×10^{26}	1.66×10^{24}	Campins *et al.* 1987
P/Tempel 2	2.1×10^{27}	3×10^{24}	A'Hearn *et al.* 1989
P/Arend-Rigaux	2.67×10^{26}	1.99×10^{24}	Millis *et al.* 1988

TABLE II
Diameters and albedos of six comet nuclei

Nucleus	D (km)	Albedo	Reference
P/Arend-Rigaux	26 × 16 × 16	0.03	Millis *et al.* 1988
P/Neujmin 1	17.6-21.2	0.03 ± 0.01	Campins *et al.* 1987
P/Encke	4.4-9.8	0.04	Luu and Jewitt 1990b
P/Halley	16.0 × 8.2 × 8.4	0.04	Keller *et al.* 1987
P/Tempel 2	5.9 × 11.8	0.02	A'Hearn *et al.* 1989
2060 Chiron	284 ± 46	0.04 ± 0.03	Campins 1993

4. 2. ROTATION RATE

Belton (1991) thoroughly reviewed the state of the art of determining rotation rates of cometary nuclei. The theory and observations used to determine rotation rates are well-presented in Belton's review, to which I refer anyone who wishes to develop an appreciation for the thought and effort involved in deriving these values and their uncertainties. I suggest that anyone using these values for analysis become familiar with the methods from which the results were derived. These values, as are most physical parameters, are not as certain as reading a number tends to make them appear.

In a study of the rotation rates of NEAs compared to main belt asteroids of similar size range (2-10 km) and to cometary nuclei, Binzel et al. (1992) conclude that the rotation rates of similar-sized main belt asteroids are indistinguishable from the near-Earth asteroids. Thus based on rotation rates alone, the NEAs could have all been derived from the main belt. When one addresses the question of whether or not a contribution from comet nuclei can be distinguished from the NEA population, the measurements of rotation rate of comets are not of uniform reliability and the sample is not of comparable size. Five comet rotation rates were analyzed by Binzel (there are nine now available) compared to 30 measured NEA rotation rates. Using Student's t test, from 0-40% of the NEA population could be made up of a cometary component, assuming that the distribution for the comet nucleus rotation rate and lightcurve amplitude is indeed representative of the comet population. One concludes from this that rotation rate alone is not an adequate discriminator of comets.

4. 3. ALBEDO

The albedos of cometary nuclei studied to date tend to be low, ranging from 0.02-0.04 as currently measured (Jewitt, 1992). We have adopted as reasonable those measurements which have been made on the eight cometary nuclei of low-to-negligible activity and Halley (studied at close range). In the matter of the albedo of Chiron, if one were to believe the albedos derived from the standard thermal model (STM), and/or the isothermal latitude model (ILM), (Campins *et al.* 1992) one would conclude that the albedo of Chiron is higher than any comet nuclei previously measured, with derived albedos of 0.18 and 0.06 respectively. However, when Campins (personal communication, 1993) makes a correction for the contribution of a coma, the resulting albedo is about 0.04. This value seems more likely when one considers the large absorption coefficient of most carbon-bearing materials. If all comets contain some carbonaceous material, it is not surprising that all comet nuclei would be dark. Assuming that our assumptions of the chemical compositions of comets are correct, and that our models from which albedos are derived are reasonable, then the generalization that comet nuclei are dark is reasonable, and maybe even accurate. As a discriminator of a member of the short period comet reservoir, versus the asteroids, a low albedo is not diagnostic. Approximately 60% of all asteroids have albedos of a few percent. So as a signature of an extinct comet, albedo alone is not valid.

4. 4. SPECTRAL REFLECTANCE

Spectral reflectance measurements of NEAs covering 0.4-2.5 μm which contain absorption bands that are diagnostic of mineralogical composition have been obtained by Vilas and McFadden (1992), Bell *et al.* (1988), and Cruikshank *et al.* (1991). If an asteroid or cometary nucleus candidate has a reflectance spectrum with a UV absorption band edge and a 2- and/or 1-μm band with a moderate to high albedo, the surface composition consists predominantly of mafic silicates and the object is most likely derived from the inner edge of the main asteroid belt and is not a remnant of a comet. An asteroid that was once considered a possible comet nucleus by Kresàk (1979), 1866 Sisyphus, can no longer be considered so because its reflectance spectrum as reported in Vilas and McFadden (1992, Figure 4) has the characteristic of an S-type asteroid and is probably derived from the inner asteroid belt. The other two NEAs in Figure 4, 433 Eros and 1036 Ganymed, are the two largest NEAs. Their reflectance spectra also indicate that they most likely were perturbed from the inner main belt based on the spectral criteria mentioned above.

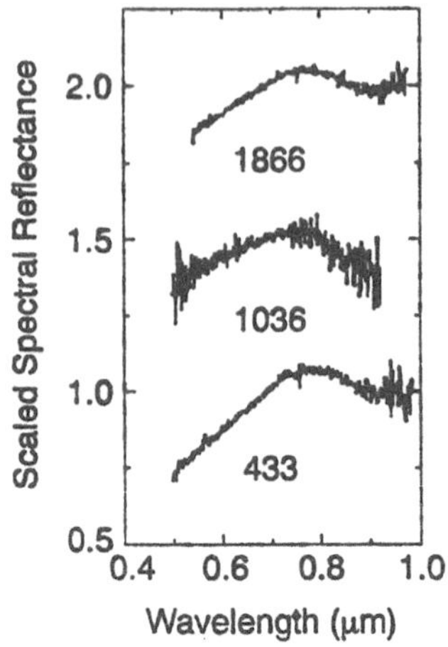

Fig. 4. CCD Spectral reflectance of near-Earth asteroids 1866 Sisyphus, 1036 Ganymed, and 433 Eros from Vilas and McFadden (1992).

Luu (1993) presents spectral reflectance between 0.38-0.7 μm taken with a long-slit CCD spectrometer of both comet nuclei and extinct comet nucleus candidates among the near-Earth asteroids. It is unfortunate that this spectral region contains little information about common rock-forming minerals, no absorption bands down to the instrumental wavelength resolution of 10 Å FWHM, and a wide range of photometric uncertainty (from 5-50%) . In the absence of compositional information, Luu reduces her data to a single parameter called normalized reflectivity gradient S'_λ, expressed in % per 0.1 μm, where positive values imply a reddened spectrum compared to that of the sun,

$$S'_\lambda(\lambda_1, \lambda_2) = (dS/d\lambda) \Big/ S_\lambda$$

$$\lambda = 0.5 \mu m$$

The results are a diversity of values from the highest of 46.1 for 5145 Pholus, to a negative value -3 for 2060 Chiron. The negative slope parameter of 2060 Chiron is a significant exception to the reddening trend that Vilas and Smith (1985) and Hartmann *et al.* (1987) and others believe holds throughout the tremendous distances in the Solar System.

5. Observations of Individual Near-Earth Asteroid, Extinct Comet Candidates

Coordinated observing campaigns and analysis of physical studies of single objects across all available observational wavelengths is another approach to investigating the subject of the comet-asteroid transition. An example of this type of investigation is the study of 2201 Oljato (McFadden *et al.* 1993). Our current conceptualization of Oljato as originating from the asteroid or comet reservoir remains uncertain. If it is an extinct comet, we don't know it because it is not clear what the physical properties of an extinct comet are. Currently, we can't be sure that an object is extinct without prior evidence of cometary activity.

Schulz *et al.* (1993) report upper limits for OH at Oljato from the International Ultraviolet Explorer (IUE) satellite. Oljato is an unusual asteroid. Whether or not it is an extinct or dormant comet, or unusual because of another physical reason, not certain at this time.

6. Dust Trails Associated with Short Period Comets

Dust trails first observed by IRAS and reported by Sykes *et al.* (1986) and Sykes and Walker (1992) consist of large refractory particles (sub-millimeter-sized) ejected at low velocities (on order of m/sec) and have presently been associated only with short period comets. Figure 5 is an illustrative schematic including a comet dust trail.

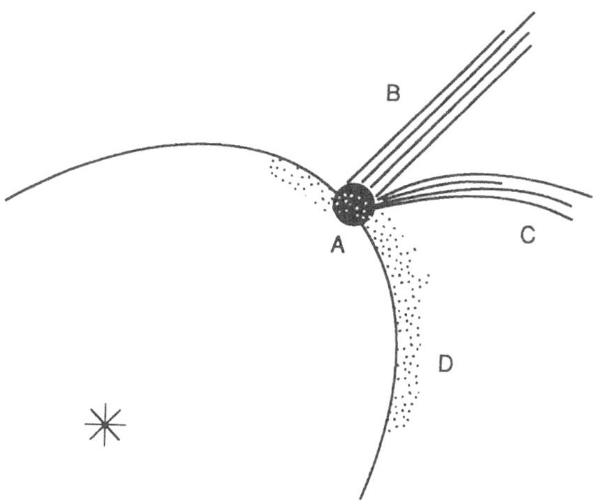

Fig. 5. A schematic drawing of the components of a comet including A) Coma B) Ion tail C) Dust tail and D) Dust trail from Sykes and Walker (1992).

If this association proves unique, that is if no dust trails are associated with asteroids, then the large particle dust trails could be a signature of extinct comet nuclei. In fact, if they are signatures of comets the dust trails will contain a record of the past $10^1 - 10^2$ years of cometary activity of dormant or extinct comets. Though the opinion has been expressed that this is not sufficient time to make dust trails a significant indicator of cometary activity, this may be considered an advantage or a disadvantage. True, there would be no record of cometary activity from thousands of years ago or more, on the other hand, the opportunity to collect information on a scale of centuries is significant compared to the length of an individual's scientific career. It is important to investigate the possibility of the association of dust trails with asteroids to obtain constraints on formation mechanisms of dust trails.

7. Discovery of Dust Tail on Prediscovery Image of 4015 1979 VA

In my opinion the most significant advance in our understanding of the comet-asteroid transition in the past few years is Ted Bowell's discovery of the prediscovery images of 4015 1979VA on the first Palomar Sky Survey prints (Bowell 1992). On plates taken 1949 Nov. 19.1 UT, B.A. Skiff found a slightly fanned tail 2.8' in length on the blue print at position angle 90° ± 5° , the antisunward direction is 76° (Figure 6).

This plate is sensitive to dust tail emissions, whereas the red plate is not sensitive to scattered light from dust. Brian Marsden noted that the 1949 object is comet P/Wilson-Harrington 1949g=1949 III. Reports from the 1949 apparition indicate that the comet was asteroidal in appearance on subsequent Palomar Schmidt exposures through Nov. 25, 1949.

It is important to review the observational data constraining the physical properties of 1979VA, now known as 4015 Wilson-Harrington. My summary is derived from a 1980 presentation (Degewij *et al.* 1980) entitled "1979 VA : Physical Parameters of a Possible Cometary Nucleus." The argument presented at the time, that 1979 VA was a cometary nucleus, was not then persuasive. The evidence available now is.

7. 1. SIZE

During the 1979 apparition, no infrared measurements of this asteroid were obtained from which size information could be derived using the standard radiometric model. Harris and Young estimated a diameter of ~3 km by assuming a low albedo, an assumption made from its UBV colors of B-V=0.66 ± 0.02 and U-B=0.31 ± 0.02. Campins *et al.* (1993) acquired simultaneous visible and near-infrared measurements from the 1992 apparition, deriving a diameter of 2.6-3.8 km.

7. 2. ROTATION RATE

Harris and Young (1983) report a photometric lightcurve, P~4 hours, amplitude=0.06 from two nights' data. In their paper, they note that they have four nights of data but that the photometric quality of the observations was poor and warranted being discarded. Upon the linkage of this asteroid with the periodic comet observed in 1949, Harris checked his data again and reaffirmed that the data quality were poor and that he discarded them for the correct reason, not because he wasn't expecting cometary activity.

7. 3. TAXONOMY

From three nights' observing, Tholen and Tedesco acquired 8-color photometry. Tholen classified it a CF according to his minimal tree cluster analysis (Tholen 1984). It is interesting to note that Tholen speculated that the color represented in *u* and *b* filters of the eight-color system might result from a contribution of CN emission which falls in the bandpass of these two filters.

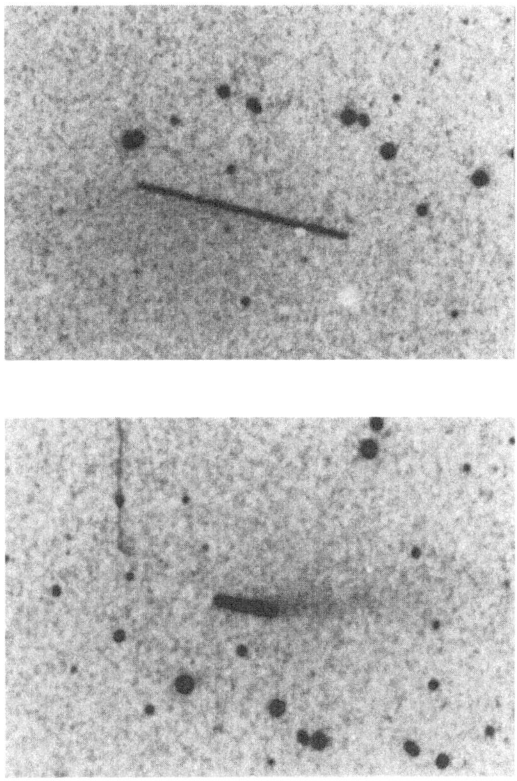

Fig. 6. TOP - Palomar Sky Survey print of 4015 Wilson-Harrington Nov.19.1 1949 UT on red photographic emulsion, 45-min exposure. A very faint tail is reported but not obvious on the print. BOTTOM - Blue emulsion 15-min exposure showing the dust tail with peak apparent brightness of 23 mag/square arcsec.

7 4 ALBEDO

There is no measured albedo from the 1979 apparition. It was always inferred to be low based on the UBV colors and Tholen's taxonomy. Campins *et al.* (1993) derive a J bandpass albedo ranging from 0.02-0.10, depending on the model used. The use of isothermal latitude model, from which the lower albedo is derived, is probably more appropriate to the physical conditions at Wilson-Harrington.

7 5 SPECTRAL REFLECTANCE

Spinrad and Stauffer (Degewij *et al.* 1980) have one night of data covering ~0.4-0.7 μm (Figure 7). Their spectrum combined with Tholen's eight-color asteroid survey data show a UV absorption edge and a flat reflectance (to within 10%) out to 1.6

μm. This type of spectrum is more similar to spectra of C-type asteroids than any of the other comet nuclei observed. Of the observed comet nuclei, they either are quite red (greater than 10% change in reflectance over the observed wavelengths) or don't have much of a UV absorption (e.g. Weissman *et al.* 1989).

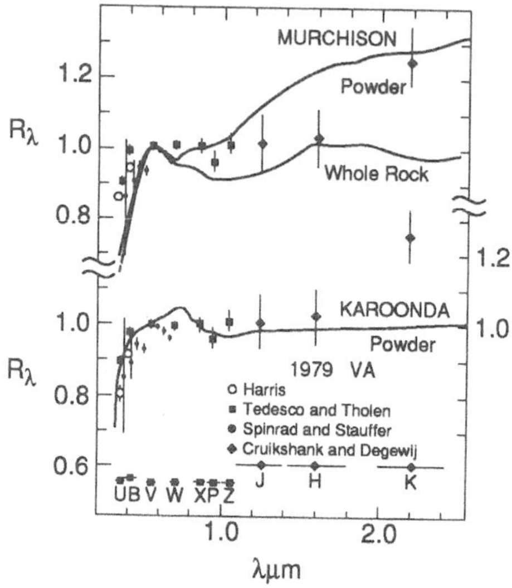

Fig. 7. Spectral reflectance measurements from 1979 apparition of 4015 Wilson-Harrington.

Hartmann *et al.* (1982) obtained JHK and near-simultaneous optical photometry. H-K is high, 0.27 (Figure 8), which fall in a region of color space near J6 (the jovian moon, Himalia). Chiron is not too far away on the V-J, J-K plot. Only comet Slaughter-Burnham, an active comet, has a higher H-K value than 1979 VA, although including the 0.1 mag uncertainty, this high color difference may not be significant. Nevertheless, the region in the near-IR between 1 and 2 μm in comets may be an important region to study and understand better. It is not clear what the physical significance of this result is. Campins *et al.* (1993) did not observe the same phenomenon.

Gaffey provided an interpretation of the available spectra from 1979 (Figure 7) considering possibilities in terms of meteoritic analogues (Degewij *et al.* 1980). His interpretation is of a surface with a moderate iron abundance in a mafic silicate with abundant opaques. The closest similar meteorite type is an evolved (thermally metamorphosed) carbonaceous chondrite, called Karoonda; it is a C4. One can now ask if interpreting the spectrum in terms of meteoritic analogues is appropriate? Doing so is valid, whether or not the interpretation holds must be demonstrated. The red H-K color, which is measured on two different days, is not consistent with the C4 meteoritic analogue.

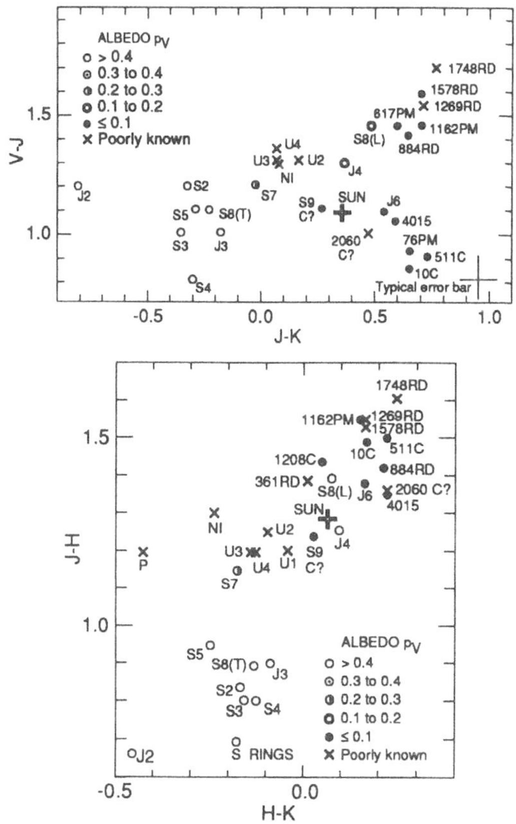

Fig. 8. Near-infrared color plots showing 4015 Wilson-Harrington, some asteroids, comets, and satellites of outer planets.

8. Summary

In summary, active search programs are producing new discoveries of small Solar System bodies. Studying the physical and dynamical properties of these objects will lead us to a better understanding of the composition of the comet and asteroid reservoirs and of the processes by which they evolve into their current orbits.

2060 Chiron, which was referred to as a distant asteroid five years ago, is now known to be a comet. This knowledge provides a new perspective on the dynamical evolution of small bodies and perhaps on our definition of a comet. The discoveries of a dust and gas coma are the first verification of the hypothesized phenomenon of a dormant comet, demonstrating the existence of remnant or sporadic cometary activity. The next step is to characterize the physical properties of those objects known to be dormant or inactive comets with subsequent observations and modelling.

The numerous studies investigating the distribution of physical properties of the nucleus of comets compared to the population of near-Earth asteroids represent a necessary step in the investigation of the comet-asteroid transition. We need to increase the available measurements of the physical properties of comet nuclei, a tremendous amount of work, requiring careful data collection and analysis. At the same time, theoretical studies explaining any significant statistical findings in physical terms ar e required. An example of such work is that of Prialnik and Mekler (1991), in which they model the formation of a dense ice crust below the surface of a comet nucleus as a function of time for comet Halley. Similar studies for objects on different orbits and with different sizes and albedo would be useful.

Efforts directed at studying the physical properties of extinct-comet nucleus candidates have not yet been successful in determining the signature of an inactive object that initially derived from the short period comet reservoir. I have suggested that dust trails from comets might be an indicator of comets, though this matter is currently speculative and may not provide evidence of a cometary origin if trails are also associated with asteroids. It is possible to use existing Infrared Astronomical Satellite (IRAS) data to look for evidence of dust trails associated with asteroids. Success in the effort to find a diagnostic signature of extinct or dormant comets will come from studying those objects known to be dormant and sporadically active. Searches for more dormant comets will contribute to the effort. The association of asteroid 1979 VA with periodic comet P/Wilson-Harrington 1949 III brings the number of known dormant comets to two, Chiron with its sporadic activity, being the other.

Acknowledgements

The author would like to acknowledge conversations with numerous colleagues, Mike A'Hearn, Karen Meech, Humberto Campins, Faith Vilas, Brian Marsden, Ted Bowell, Rick Binzel, Jane Luu, Alan Harris, and Eleonor Helin. Jeff Bytof and Trudy Johnson helped with the manuscript preparation.

References

A'Hearn, M.F., H. Campins, D.G. Schleicher, and R.L. Millis. : 1989, "The nucleus of P/Tempel 2." *Astrophys. J.*, **347**, 1155-1166.

Bell, J.F., P.D. Owensby, B.R. Hawke, and M.J. Gaffey. : 1988, *The 52-color asteroid survey : Final results and interpretation.*, Lunar and Planetary Science (XIX). 57-58. Lunar and Planetary Institute, Houston.

Belton, M.J.S. : 1991, "Characterization of the rotation of cometary nuclei." In *Comets in the Post-Halley Era*, (R.L. Newburn, Jr., M. Neugebauer, J. Rahe, Eds.), 691-721, Kluwer, Boston.

Binzel, R.P., S. Xu, S.J. Bus, and E. Bowell. : 1992, "Origins for the near-Earth asteroids." *Science*, **257**, 779-782.

Bowell, E. : 1992, "(4015) 1979 VA=Comet Wilson-Harrington (1949III)." *IAU Circular*, 5585.

Buratti, B.J. and R.S. Dunbar. : 1991, "Observation of a rapid decrease in the brightness of the coma of 2060 Chiron in 1990 January." *Astrophys. J.*, **336**, 147-149.

Bus, S. J., M.F. A'Hearn, D.G. Schleicher, and E. Bowell. : 1991, "Detection of CN emissions from (2060) Chiron." *Science*, **251**, 774-777.

Campins, H., M.F. A'Hearn, and L.-A. McFadden. : 1987, "The bare nucleus of comet Neujmin 1." *Astrophys. J.*, **316**, 847-857.

Campins, H., D. Jewitt, and C.M. Telesco. : 1992, "Simultaneous visible and thermal infrared observations of (2060) Chiron.[abst.]" *Bull. Amer. Astron. Soc.*, **24**, 993.

Campins, H., B. Osip, B.A.S. Gustafson, G. Rieke, M. Rieke, S. Larson, and D. Schleicher. : 1993, "A multi-wavelength study of the potentially meteorite-producing comet P/Wilson-Harrington (4015 1979VA)." In *Abstracts for Asteroids, Comets, Meteors 1993*, Lunar and Planetary Institute, Houston.

Chamberlin, A.B., L.-A. McFadden, R. Schulz, and D.G. Schleicher. : 1993, "2201 Oljato and 4015 Comet P/Wilson-Harrington : Search for CN(0-0) band emission." In *Abstracts for Asteroids, Comets, Meteors 1993*, Lunar and Planetary Institute, Houston.

Cochran, W.D., A.L. Cochran, and E.S. Barker. : 1986, "Spectroscopy of asteroids in unusual orbits." In *Asteroids, Comets, Meteors II* (C.-I. Lagerkvist, B.A. Lindblad, H. Lundstedt, H. Rickman, Eds.), Uppsala Universitet Reprocentralen.

Cruikshank, D.P., D.J. Tholen, W.K. Hartmann, J.F. Bell, and R. Brown. : 1991, "Three basaltic Earth-approaching asteroids and the source of the basaltic meteorites." *Icarus*, **89**, 1-13.

Degewij, J., J.G. Williams, D.P. Cruikshank, M.J. Gaffey, E.F. Helin, H. Spinrad, J. Stauffer, E.F. Tedesco, and D.J. Tholen. : 1980, "1979 VA : Physical parameters of a possible cometary nucleus [abst.]." *Bull. Amer. Astron. Soc.*, **12**, 665.

Hainaut, O. and A. Smette. : 1992, "1992 AD." *IAU Circular*, 5450.

Harris, A.W. and J.W. Young. : 1983, "Asteroid rotation IV. 1979 Observations." *Icarus*, **54**, 59-109.

Hartmann, W.K., D.P. Cruikshank, and J. Degewij. : 1982, "Remote comets and related bodies : VJHK colorimetry and surface materials." *Icarus*, **52**, 377-408.

Hartmann, W.K., D.J. Tholen, and D.P. Cruikshank. : 1987, "The relationship between active comets, "extinct" comets, and dark asteroids." *Icarus*, **69**, 33-50.

Hartmann, W.K., D.J. Tholen, K. Meech, and D.P. Cruikshank. : 1990, "2060 Chiron : Colorimetry and cometary behavior." *Icarus*, **83**, 1.

Jewitt, D. : 1992, "Observations and physical properties of small Solar System bodies." In *Proceedings of the 30th Liège International Colloquium*, (A. Brahic, J.-C. Gerard, J. Surdej, Eds.), Institute d'Astrophysique, University of Liège, 85-111.

Jewitt, D. and J. Luu. : 1992, "Submillimeter continuum observations of 2060 Chiron." *Astron. J.*, **104**, 398-404.

Jewitt, D. and J. Luu. : 1993a, "Discovery of the candidate Kuiper belt object 1992 QB_1." *Nature*, **362**, 730-732.

Jewitt, D. and J. Luu. : 1993b, "1993 RO." *IAU Circular*, 5865.

Keller, H.U., W.A. Delamere, W.F. Huebner, H.J. Reitsema, H.U. Schmidt, F.L. Whipple, K. Wilhelm, W. Curdt, R. Kramm, N. Thomas, C. Arpigny, C. Barbieri, C.M. Bonnet, S. Cazes, M. Coradini, C.B. Cosmovici, D.W. Hughes, C. Jamar, D. Malaise, K. Schmidt, W. K.H. Schmidt, and P. Siege. : 1987, "Comet P/Halley's nucleus and its activity." *Astron. Astrophys.*, **187**, 807-823.

Kowal, C.T. : 1979, "Chiron." In *Asteroids* (T. Gehrels, Ed.), 436-442, Univ. of Arizona Press, Tucson.

Kresák, L. : 1979, "Dynamical interrelations among comets and asteroids." In *Asteroids* (T. Gehrels, Ed.), 289 - 309, Univ. of Arizona Press, Tucson.

Luu, J.X., and D.C. Jewitt. : 1990a, "Cometary activity in 2060 Chiron." *Astron. J.*, **100**, 913-932.

Luu, J.X., and D.C. Jewitt. : 1990b, "The nucleus of P/Encke." *Icarus*, **86**, 69-81.

Luu, J.X., and D.C. Jewitt. : 1992a, "High resolution surface brightness profiles of near-Earth asteroids." *Icarus*, **97**, 276-287.

Luu, J.X., and D.C. Jewitt. : 1992b, "Near – aphelion CCD photometry of comet P/Schwassmann Wachmann 2." ***Astron. J.***, **104**, 2243-2249.

Luu, J.X. : 1993, "Spectral diversity among the nuclei of comets." ***Icarus***, **104**, 138-148.

Luu, J., and D. Jewitt. : 1993, "1993 RP." *IAU Circular*, 5867.

Marcialis, R.L. and B.J. Buratti. : 1993, "CCD photometry of 2060 Chiron in 1985 and 1991." *Icarus*, **104**, 234-243.

McFadden, L.-A., A. Cochran, E.S. Barker, D.P. Cruikshank, and W.K. Hartmann. : 1993, "The enigmatic object 2201 Oljato : Is it an asteroid or an evolved comet?" *J. Geophys. Res.*, **98(E2)**, 3031-3041.

Meech, K.J. and M.J.S. Belton. : 1990, "The atmosphere of 2060 Chiron." ***Astron. J.***, **100**, 1323-1338.

Millis, R.L., M.F. A'Hearn, and H. Campins. : 1988, "An investigation of the nucleus and coma of comet P/Arend – Rigaux." ***Astrophys. J.***, **324**, 1194-1209.

Prialnik, D. and Y. Mekler. : 1991, "The formation of an ice crust below the dust mantle of a cometary nucleus." ***Astrophys. J.***, **366**, 318-323.

Schulz, R., M.F. A'Hearn, L.-A. McFadden, D.K. Yeomans, M. Haken, and A. Chamberlin. : 1993, "2201 Oljato and 1566 Icarus : Comets and asteroids? A comparison with comet Wilson-Harrington also known as asteroid (4015) 1979 VA." In ***Abstracts for Asteroids, Comets, Meteors 1993***, 264, Lunar and Planetary Institute, Houston.

Scotti, J.V., T. Gehrels, and D.L. Rabinowitz. : 1992, "Automated detection of asteroids in real-time with the spacewatch telescope." In ***Asteroids, Comets, Meteors 1991***, 541-544, Lunar and Planetary Institute, Houston.

Steel, D., R.H. McNaught, and D. Asher : 1992, "1991 DA : An asteroid in a bizarre orbit." In ***Asteroids, Comets, Meteors 1991***, 573-576, Lunar and Planetary Institute, Houston.

Sykes, M.V., L.A. Lebofsky, D.M. Hunten, and F. Low. : 1986, "The discovery of dust trails in the orbits of periodic comets." *Science*, **232**, 1115-1117.

Sykes, M.V. and R.G. Walker. : 1991, "Constraints on the diameter and albedo of 2060 Chiron." *Science*, **251**, 777-780.

Sykes, M.V. and R.G. Walker. : 1992, "Cometary dust trails I. Survey." ***Icarus***, **95**, 180-210.

Tholen, D.J. : 1984, "Asteroid taxonomy from cluster analysis of photometry." Ph. D. Thesis, Univ. of Arizona.

Vilas, F. and B.A. Smith. : 1985, "Reflectance spectrophotometry (~0.5-1.0 μm) of outer-belt asteroids : Implications for primitive, organic Solar System material." ***Icarus***, **64**, 503-516.

Vilas, F. and L.-A. McFadden. : 1992, "CCD reflectance spectra of selected asteroids : Presentation and data analysis considerations." ***Icarus***, **100**, 85-94.

Weissman, P.R., M.F. A'Hearn, L.-A. McFadden, and H. Rickman. : 1989, "Evolution of comets into asteroids." In ***Asteroids II*** (R.P. Binzel, T. Gehrels, M.S. Matthews, Eds.), 880-920, University of Arizona Press, Tucson.

Wetherill, G. W. : 1991, "End products of cometary evolution : Cometary origin of Earth-crossing bodies of asteroidal appearance." In ***Comets in the Post-Halley Era*** (R.L. Newburn, Jr., M. Neugebauer, J. Rahe, Eds.), 537-556, Kluwer, Boston.

Williams, I.P., A. Fitzsimmons, D. O'Ceallaigh. : 1993, "1993 SB and 1993 SC." *IAU Circular*, 5869.

METEOROID STREAMS

DUNCAN STEEL

Anglo-Australian Observatory, Private Bag, Coonabarabran, NSW 2357; and Department of Physics and Mathematical Physics, University of Adelaide, G.P.O. Box 498, Adelaide, SA 5001, Australia

E-mail : DIS@AAOCBN3.AAO.GOV.AU

Abstract. Meteoroid streams, producing meteor showers if some part of the stream has a node near 1 AU, have complex structures which are only just beginning to be understood. The old simplistic idea of a narrow loop being formed about the orbit of a parent comet with one, or possibly two, terrestrial intersection(s) is now being replaced by the recognition that their dynamical evolution may render convoluted and distorted ribbon shapes with eight or more distinct showers being generated. As such the streams are excellent tracers of the sorts of orbital evolution which may be undergone by larger objects (asteroids and comets) in the inner solar system; indeed it is now known that objects presently observed as Apollo-type asteroids may also be the progenitors of streams.

Searches for showers associated with newly-discovered possible parent objects may be carried out either via the calculation of theoretical meteor radiants (which have hitherto been derived using an untenable method), or through searches of catalogues of individual meteor orbits. In order to accomplish the latter, about 68,000 radar, photographic and TV meteor orbits from various programmes in the U.S.A., the former Soviet Union, Canada and Australia are available from the IAU Meteor Data Center, and more than 350,000 orbits of very faint meteors have been determined over the past three years using a new facility in New Zealand.

The discovery amongst IRAS data of dust trails lagging behind comets has opened up a new way in which meteoroid streams may be investigated, although the relationship between these trails and the streams observed as meteor showers at the Earth is by no means clear at this stage. Similarly radar, radio and spacecraft impact observations of meteoroids near cometary nuclei have added to our knowledge.

In spite of the improvement in our understanding of meteoroid streams over the past few years it is clear that there is much still to be done. The words of W.F. Denning in 1923 are still pertinent : "Few astronomers occupy themselves with the observation and investigation of meteors, and yet it is an attractive field of work offering inviting prospects of new discoveries".

1. Historical Introduction

Although meteor showers have been observed since time immemorial, it was only with the great Leonid storms of 1799 and 1833 that meteor studies started to attain any proper scientific understanding. The wonder (and alarm, in some people) provoked by these spectacular events eventually led to the recognition that these great meteor storms occur periodically, and might be linked to a certain comet (P/Tempel-Tuttle in the case of the Leonids), and also that the various seasonal showers with broadly-constant annual activities might similarly have a cometary origin. Thus by 1861 Daniel Kirkwood was able to ask "May not our periodic meteors be the debris of ancient but now disintegrated comets, whose matter has become distributed around their orbits?". The reader is referred to Porter (1952) for a thumbnail sketch of the history of meteor shower studies, and their relation to comets; but it should be noted that the title of his book is semantically incorrect

A. Milani et al. (eds.), Asteroids, Comets, Meteors 1993, 111–126.

(and indeed contrary to later IAU nomenclature recommendations) since strictly-speaking one should refer to 'meteor showers', and 'meteoroid streams'. A more recent discussion of the history of meteor studies has been presented by Beech (1988).

A number of compilations of records of ancient meteor showers have been published; see for example Hasegawa (1992) and references therein for Chinese and Japanese records, and for Arab chronicles see Rada and Stephenson (1992; but see also Kidger, 1993). In the last of those references, historical work on meteor showers observed and recorded by a variety of human cultures is mentioned. Such records are invaluable with regard to providing a check upon theoretical analyses of the dynamical and physical evolution of meteoroid streams : a number of researchers have made use of such tabulations of meteors and fireballs, for example Fox and Williams (1985) and Bailey *et al.* (1990). Records of the activities of well-known streams over the past few centuries are also of great use in aiding an understanding of stream evolution, for example how the rotation of the line of apsides under planetary perturbations brings the node to a heliocentric distance of 1 AU so that a shower is observable for a period of time (*e.g.* in the case of the Geminids, this shower having been observed only since about the middle of the last century : Fox *et al.*, 1982; Hunt *et al.*, 1985). Also noteworthy, regarding the genesis of particular showers/streams, has been work aimed at correlating ancient comets and meteor showers, for example by Kresáková (1987).

2. Modern Studies

Over the past century the dictates of astronomical fashion have led to periods in which interest in meteors and meteoroids has waxed and waned. For example, throughout the 1960's the Harvard Radar Meteor Project in the U.S.A. (Cook *et al.*, 1972) made that nation pre-eminent in the field, but since the demise of that programme there has been comparatively little meteor research done there. This contrasts with the situation in the old eastern European countries, of which the former Soviet Union and Czecho-Slovakia in particular might be mentioned, where very active groups have continued to produce results of import for several decades; indeed, one of the problems faced by this author in preparing this review is that much of the most significant work, judging from the English translations of publication abstracts, has been published only in Russian (and the consequent inaccessibility of the work is the responsibility of this author).

Research groups elsewhere, however, are also making very important contributions to meteor science so that the subject may be characterized as being one full of vigour and rigour (as opposed to *rigor mortis*). Beech (1988) has expressed the view that the study of meteors may be classified as a mature science, but despite that it is by no means true that we have done more than scratch the surface in our quest for an understanding of meteoroid stream behaviour. For example, an unresolved problem remains the mechanism by which streams are dispersed so as to give showers of the observed durations, and this must be tied up with the ejection speeds of the meteoroids from their parent bodies, and the physical mechanisms whereby those speeds are attained. This is the case for the Geminids, which appear to be 'too dispersed' for current models of stream formation and subsequent orbital evolution to accommodate (Fox *et al.*, 1983; Jones, 1985; Jones and Hawkes, 1986; Ryabova, 1989). The nodal motion (or lack of it) in the case of the Geminids has also been an outstanding problem (Fox *et al.*, 1982), with similar questions (with the inclusion of notable mass segregation) being posed by the Quadrantids (Hughes *et al.*, 1981; Froeschlé and Scholl, 1986).

One fundamental benchmark in meteor studies is how the observations of meteors agree with the theory : for example whether meteor storms recur at their predicted times. In many instances the comment of Padevět (1973) is still pertinent : "Even if we use the classical physical theory of meteors, there may still be some possibility for agreement between observations and theory". That comment was made in regard to meteor physics, as opposed to the meteor astronomy which is of interest here, but still it rings true : theory and observations are still a long way apart in many of the outstanding problems of meteoroid streams. The provision of observations of high quality is essential to the testing of any hypothesis, and professionals often do not have the time or the funds to carry out the sorts of observational programmes that are required. Thus the foundation of the International Meteor Organization (Gyssens *et al.*, 1991), with amateur enthusiasts collecting data under controlled conditions, bodes well for the future of this subject.

The scientific study of meteoroid streams, then, remains an area of research with many tantalizing problems yet to be solved. Modern techniques have made many of these amenable to attack, for example the availability of computing power putting large numbers of statistical numerical integrations within the grasp of research groups with quite modest budgets. Over the next few years, with increasing interest in small bodies (asteroids and comets) as potential space mission candidates, there is certain to be added emphasis on the relationship between such objects and the streams that they spawn. The study of meteoroid streams is an exciting field which, after a period of relative quiescence, is now re-awakening to meet the challenges of the spacefaring age.

In this review my intention is only to give an introduction to the field for those who may be newcomers to it, and to point to published work which may be accessed in order to discover the state of the art in various areas of research on meteoroid streams. Due to the lack of space even this intention cannot be fully carried out, and to a large extent only the most recent papers by various researchers on various topics are referenced, or at least those publications which are most easily accessible. This is meant as a review for an ingenue, not an old-hand, although it is hoped that all will find it a suitable compendium for many enquiries.

3. How are meteoroid streams investigated?

In this section I will attempt to give a very simple background sketch about how meteoroid streams are investigated. I will concentrate here on observational/ experimental methods, along with very simple theory, and in section 4 discuss the more intricate theoretical modelling that has been attempted.

3 1 OTHER OBSERVATION TECHNIQUES

Before discussing the standard techniques whereby streams are observed through the showers that they produce, it is worthwhile to mention other techniques that have been applied, and may well be expanded in the future. Impact detectors on spacecraft either in the vicinity of the Earth (Singer and Stanley, 1980) or in deep space (Grün *et al.*, 1993; Grün, this volume) provide stream detection for very small dust particles. Moving to the other end of the mass spectrum, meteoroids with masses to order 1 tonne have been detected hitting the Moon with the Apollo seismometers, and many of these appear to be in streams (Oberst and Nakamura, 1991); and even larger bodies (to 5–10 m) have been telescopically-detected in near-Earth space (Rabinowitz, 1993) and the two nearest-misses to date have been by objects with orbits similar to the Taurid stream, which may extend to kilometre-plus asteroids (Asher *et al.*, 1993). There is also evidence for some meteorites being in streams (Halliday *et al.*, 1990; Terentjeva, 1990; Dodd *et al.*, 1993). Another technique whereby streams might be observed is from the scattering of sunlight by the particles in space, as suggested by Baggaley (1977). At longer wavelengths the IRAS dust trails associated with short-period comets appear to be due to millimetre-sized particles (Sykes and Walker, 1992), although the relationship between these structures and the streams observed impacting the atmosphere is not yet clear; Kresák (1993) discusses the possibility that meteor storms occur when the Earth passes through such trails. At even longer wavelengths, in the millimetre range, meteoroids have been detected near comets through their thermal emission (Crovisier and Schloerb, 1991), and through centimetre-wave radar scattering (Harmon *et al.*, 1989; Campbell *et al.*, 1989).

3 2 METEOR SHOWERS

Our predominant method for sampling any meteoroid stream is by the study of the meteor shower produced when the Earth passes through the stream. Clearly this requires at least part of the stream to have a node near 1 AU. The phenomena produced in the atmosphere may then be studied by ground-based observers, either using optical (visual, photographic, TV) or radar techniques.

The duration of such showers provide information on the transverse stream profile, from the variation in the meteor count rate from day-to-day, and also such factors as the varying mass distribution across the stream. For example, the Geminid shower has a long build-up over about 10 days for the smaller ($\sim$1 mm) particles detected using meteor radars, but a rather shorter climb, over just a few days, for the larger ($\sim$1 cm) optically-detected particles; both then show a rapid fall off in count rates (see *e.g.* the papers cited in section 2 with regard to the Geminids, and Babadzhanov *et al.*, 1992). Other showers have also been studied in this regard; for example the Quadrantids (Porubčan and Cevolani, 1990; Šimek and McIntosh, 1991) and the Perseids (Šimek and Lindblad, 1990).

Whilst day-to-day variations in shower strength render information about the transverse profile through a stream, year-to-year variations can provide information about the lengthways stream profile; these can delineate also the transverse profile since observations on the same day each year will be at slightly different solar longitudes. A good example of how a stream spatial structure may be investigated in this way is given by Bel'kovich (1986).

One confusing factor is the effect of the state of the atmosphere upon meteor count rates, in particular for radar observations. A substantial change in the number of radar-detected meteors will occur if the scale-height at ~100 km alters by only ~10% (*e.g.* see Ellyett and Kennewell, 1980).

Many showers demonstrate consistent count rates from one year to the next, showing that the meteoroids are equably distributed about the stream, and indicating it to be a mature structure : for example see the discussion of the Geminids by Jones (1982). Conversely, some streams are relatively young, or are being dynamically replenished by the parent object, as is the case for the Leonids, spawned by P/Tempel-Tuttle (1866 I). That shower exhibits a 33 year periodicity in its activity, with strong storms occurring when the Earth passes through the stream in the vicinity of the comet (Yeomans, 1981; Williams *et al.*, 1986; Wu and Williams, 1992a). However, the Leonid storms sometimes precede the cometary perihelion passage (*e.g.* in 1799 : Yeomans, 1981) or can occur some years later (an outburst was observed in 1969, three years after the main storm in 1966 : Porubčan and Štohl, 1992)

Another comet-related meteor storm is the October Draconids (or Giacobinids), associated with P/Giacobini-Zinner, with a cyclicity of 13 years (Beech, 1986; Kulikova, 1989; Lebedinets *et al.*, 1990), whilst the great Andromedid storm of 1872 is associated with the break-up of P/Biela some decades before (Babadzhanov *et al.*, 1991b). Since 1988 an increase in the activity of the Perseid shower has been noted, with exceptional displays in 1991 and 1992 leading to the prediction that the parent comet, 1862 III Swift-Tuttle, was in the vicinity, and it was duly re-discovered in 1992 September. In 1993 Perseid activity was spectacular but a storm did not occur; for reports see *IAU Circular 5841* and *5843*. This increase in activity (with perhaps a peak yet to come : Wu and Williams, 1993) is again illustrative of how groupings occur close to the parent (*cf.* Kresák, 1993). For the larger meteoroids (above ~1 cm), producing bright visual meteors, the ejection velocities from the parent are expected to be low, with the initial dispersal in periods, due to the effect of radiation pressure, being only 1% or less. Thus it takes many orbits before the meteoroids released on any particular perihelion passage can become fully dispersed around the orbit; indeed the equable meteor count rates well away from the comet's perihelion passage make the minimum stream age estimable.

Another example of a shower which has undergone outbursts, although in this case the parent comet is not in the vicinity of the Earth, is the April Lyrids (associated with Comet Thatcher, 1861 I). This may be due to some form of mass segregation (Lindblad and Porubčan, 1992; Porubčan and Štohl, 1992) although an explanation based upon orbital resonances with Jupiter has been proposed (Emel'yanenko, 1991).

3.3 EFFECT OF ORBITAL PRECESSION/EVOLUTION UPON SHOWER OCCURRENCE

Whilst certain showers are observed in the present epoch, they were not seen in the past, nor will they be seen at some future time. As the stream orbit precesses under planetary perturbations it may move so as to have both nodes some way away from 1 AU, so that no shower occurs on Earth. For example, as mentioned in section 1, the Geminids have only been observed since the middle of the last century. Historical records, and monitoring of the shower activity now, therefore provide a test of orbital evolution theories. Indeed, in recent years statistical-type integrations of indeterministic cometary or asteroidal orbits have been carried out, with packets of initial orbits with slightly different start elements being used (*e.g.* Hahn and Bailey, 1990). Clearly observations of particles in meteoroid streams represent a real-life portrayal of how slightly different starting elements affect the orbital evolution.

Information on the orbital history of the parent objects can also be gained from a study of their associated streams. For example P/Halley has been observed for almost 2.5 millennia, but this is much shorter than the time for which it has occupied its present general orbit, and any backwards integration becomes indeterminate prior to 1404 B.C. due to a close approach to the Earth in that year (Yeomans and Kiang, 1981). However, the age of the Halleyid stream, producing the Eta Aquarid and Orionid showers, appears to be at least 20,000 years (Jones *et al.*, 1989) so that studies of the orbital evolution of the Halleyids (McIntosh and Jones, 1988; Babadzhanov *et al.*, 1991a; Hajduková and Hajduk, 1991) can provide useful information regarding the comet's history. Stream orbital evolution is more extensively discussed in section 4.

3. 4. SHOWER RADIANTS

Apart from the activity (hourly rates) detected during a shower, information on a stream's structure may be derived from measurements of individual meteor radiants, and orbits. In this section I discuss radiants, and in section 3.5, orbits.

3.4.1. Observed Radiants

The simplest definition of a meteor shower is a concentration of meteors observed to emanate from a particular direction (radiant) at a particular time of year, and with a particular velocity. These specific values then define an orbit for the stream which is producing the shower. Having written that, for many years there has been debate about how a shower may be recognized; whilst there is no argument about the major showers (Perseids, Geminids, Quadrantids, Eta Aquarids, Orionids, Delta Aquarids, Lyrids, Daytime Arietids, Beta Taurids, Zeta Perseids, *etc.*), as soon as the count rates drop to 5 or 10 per hour it is not clear whether a definite shower is being observed, or whether it is a chance alignment, especially for near-ecliptic radiants. The commonly-used list of streams is that presented by Cook (1973), although it is clear that many of those that he tabulates are fictitious, being chance alignments of just five meteors from the Harvard photographic surveys (his major source).

Many catalogues of observed radiants have been published over the past century, with many interpretations being open to doubt. For visual observations the count rates are relatively low, with radars being capable of much higher detection rates. Therefore radiant imaging radars, such as those described by Morton and Jones (1982), Jones and Morton (1982), and by Poole and Roux (1989), are of considerable importance in this regard.

3.4.2. Theoretical Radiants

A theoretical radiant is a value for the radiant calculated from the orbit of a putative parent object, such as an Earth-crossing comet or asteroid. The most-used technique for calculating such radiants was described by Porter (1952), and used by Drummond (1981b, 1982b) and Olsson-Steel (1987) amongst others. The technique involves a determination of the closest approach distance between the orbits of the parent and the Earth (assumed zero eccentricity orbit), and then a rectilinear transposition of the direction cosines of the parent's path to that of the Earth. This is clearly an unphysical treatment since this is *not* how a meteoroid's path alters so as to make Earth-impact possible. Thus the technique will give reasonable results if the actual orbit miss-distance is small, but will give misleading results otherwise.

A far better technique was described by Hasegawa (1990; see also Hasegawa *et al.*, 1992). This involves a rotation of the line of apsides of the comet/asteroid orbit so as to produce a node at 1 AU, the theoretical radiant then being derived in the standard way using the direction cosines for this argument of perihelion-rotated orbit. This is much closer to the actual situation which produces an Earth-intersecting stream, with precession under planetary perturbations being responsible.

However, even this has some drawbacks in that during the stream's evolution the orbital elements (a, e and i) are not constant, so that at the time that a node occurs near 1 AU these elements may have different values to those previously occupied. For example, P/Encke (with $i \simeq 12°$) is known to be the parent of the Taurid meteor showers, but the meteoroids observed hitting the Earth have $i \simeq 2°-6°$; however, when the comet has a node near the Earth it has a reduced inclination too (Whipple and Hamid, 1952; Steel *et al.*, 1991). This is also an excellent example of the fact that meteoroid streams trace out the orbital evolution of their parent bodies (*cf.* section 3.3). Thus any attempt at determining an accurate theoretical radiant must involve a numerical integration of the putative parent's orbit, with times when the nodes reach 1 AU being noted. However, a far simpler method which appears to give reasonable results is the use of the simple integrals of motion discussed by Babadzhanov (1990) and by Babadzhanov and Obrubov (1992b), which allows the orbital elements in epochs of Earth-interception to be determined, and hence the theoretical radiants to be derived.

The next next step in determining a genetic relationship between a hypothetical parent and a meteor shower is to compare the theoretical and observed radiants. Whilst this is not so stringent a test as one in which orbits are directly compared, it is a useful indicator; and theoretical radiants can point observers to useful areas of the sky for examination at certain times of year.

As an example, the genetic relationship between the Eta Aquarid and Orionid showers and P/Halley is clearly established by these showers having radiants close to the theoretical value for the comet. Since there are few sporadic meteoroids with similar velocities (retrograde orbits of large size), and the times of the showers agree with the nodal values of the comet, there is little argument about this link despite the fact that neither shower has an argument of perihelion coincidental with that of the comet (due to the requirement for meteor observation of a node at 1 AU). However, for a low-inclination, moderate-eccentricity orbit the situation would not be so clear cut due to the preponderance of near-ecliptic sporadic meteors (from the Jupiter-family comets).

3.4.3. Twin/Multiple Radiants

It has been discussed above how a stream may precess so as to produce a shower on Earth in this epoch which did not occur some time ago, since at that time it did not have a node at 1 AU. This also suggests that twin showers may occur from a particular stream (*e.g.* the Eta Aquarids and the Orionids from the Halleyid stream); Drummond (1982a) calculated theoretical twin shower radiants for established showers, although for reasons that will become apparent his treatment is insufficiently sophisticated. Even for a stream which precesses whilst preserving its values of (a, e, i), four streams may be produced : at the four combinations of (pre-perihelion leg, post-perihelion leg) and (ascending node, descending node). Thus the Taurid stream produces the (nighttime, pre-perihelion) Northern and Southern Taurids, and the (daytime, post-perihelion) Beta Taurids and Zeta Perseids.

However, an analysis including orbital evolution, again taking the example of the Taurids, shows that 12 distinct showers may be produced by this stream alone (Babadzhanov, 1990; Babadzhanov and Obrubov, 1992b). This is also discussed in section 4.

3. 5. STREAM ORBITS

In order to show with some confidence that a particular macroscopic object (comet or asteroid) has spawned a specific stream, a comparison of orbits is necessary. Indeed, even before this can be done, the stream itself must be demonstrated to exist, and this is best done through the comparison of individual meteor orbits. The tool used to accomplish this has generally been the orbital discriminant, or 'D-criterion', of Southworth and Hawkins (see Drummond, 1981a).

Over the past few years there have been several small-camera measurements of meteor orbits, producing a few hundred such determinations. However, the largest number of meteor orbits generally available are those stored at the IAU Meteor Data Center (IAUMDC) in Sweden, with about 6,000 optical (photographic and TV) and 62,000 radar orbits being archived there (Lindblad, 1991; Steel, 1991). Other radar surveys have been carried out, in particular in the former Soviet Union, from which the data have not been made available. A new survey in New Zealand (Baggaley and Taylor, 1992) has produced more that 350,000 orbits, and these will eventually be archived at the IAUMDC.

These orbits have been used in stream searches too numerous to mention here. Many of the likely stream/comet relations have been tabulated by Cook (1973), and also tested through the simplistic D-criterion test by Drummond (1981a). However, with the orbital evolution discussed in section 3.3 in mind, it is clear that rather more sophisticated testing techniques are required; in particular the parent must be integrated so as to determine its orbit when it is Earth-approaching (*e.g.* see Babadzhanov and Obrubov, 1992b).

Another problem is that of the differentiation of a stream from the background sporadic meteors. To some extent this problem has been addressed by Olsson-Steel (1988) who showed that many Apollo-type asteroids have associated streams, in contradiction to Drummond (1981a, 1982b), but there is still much to be done in this area; see also Porubčan *et al.* (1992).

Whilst it may seem that the number of orbits available should be sufficiently-plentiful for most purposes, in fact for certain showers there is a great sparcity of data. During the 1980's, with the booming interest in P/Halley, there were several studies of the Halleyid stream's orbital evolution (see the papers listed in section 3.3 and references therein). However, Lindblad (1990) showed that the stream orbit listed by Cook (1973), and used by various researchers, was in fact derived from one meteor only. Lindblad found a total of 16 Eta Aquarid orbits amongst the data at the IAUMDC, putting matters on a slightly better footing, but these are now superseded by the 1000+ high-quality orbits from this stream measured in New Zealand (Baggaley *et al.*, 1992, give a preliminary analysis).

4. Stream formation and evolution

Whilst the mechanisms of stream formation remain poorly understood, in particular the source of meteoroid speeds larger than 50 m/sec relative to the parent (as are required by the shower durations discussed in section 2), over the past decade or so our knowledge of the orbital evolution of streams has undergone a great improvement. Before passing on to that subject, some mention of a possible cause of the problematic relative speeds is given.

Gas dynamic drag cannot provide a sufficiently high relative speed, so that some other mechanism is required. However, the foundation of the idea of Whipple (1951) of this gas dynamic drag was that the nucleus of a comet is the source of the outflowing gas in the coma. In fact the data from the spacecraft sent to P/Halley showed that the coma has an extended source : that is, it is evaporation from the lumps of solid material which is the predominant source of gas in the coma. If each of these lumps were to be considered as mini-comets composed to a large degree of volatile material, in accord with our knowledge of cometary nuclei, then clearly the evaporating material will produce a jet force similar in nature to the non-gravitational force that perturbs the orbits of comets themselves. The released mini-comets/meteoroid progenitors will therefore be accelerated relative to the parent comet. Modelling of this effect is commended to any interested reader.

As aforementioned, meteoroids are released from their parent bodies with relatively low speeds, although these may be accelerated by other factors soon after release. Meteor storms occur when the Earth cuts close by the orbit of an active comet, with the comet being in the vicinity (although storms have occurred with no known parent comet being observed; for example, the December Phoenicids in 1956 : see Cook, 1973, but also Ridley, 1962). Such particles will become dispersed around the cometary orbit by the effects of radiation pressure, but complete smearing around that orbit will take some time : see Jones (1982) for a discussion of the Geminids. Other radiative forces, including the Poynting-Robertson effect (Burns *et al.*, 1979) can also cause dispersal and mass segregation.

Another question is that of the dispersal of streams into the sporadic background. This occurs predominantly through the effect of close approaches to Jupiter, at least for the Jupiter-family (period < 20 yr) streams (Olsson-Steel, 1986). Characteristic time-scales are of the order of 10^4 yr for dispersal, and 10^5 yr for physical loss due to impacts with zodiacal dust particles.

There have been many reviews of stream formation and evolution published in the past few years, and here the reader is merely referred to a few suitable papers. These include Lebedinets *et al.* (1990), Williams (1990, 1992), Hajduk (1991), McIntosh (1991), and Babadzhanov and Obrubov (1992b).

Over the past decade work on one stream in particular has provided some insight into the generalities of stream orbital evolution, and this stream is that which produces the strong Quadrantid shower. Soon after the discovery of P/Machholz (1986 VIII) it was realized that this comet was very probably the parent not only of this shower, but also of seven other showers, including the Daytime Arietids, the Delta Aquarids, and the Ursids; however, association of this complex stream with Comet 1491 I cannot be ruled out (Williams and Wu, 1993), which may indicate that the complex was formed from the break-up of a single larger comet. The relationship is particularly surprising in that the orbital elements are very different for these showers and comets (*e.g.* the Quadrantids, Ursids, Delta Aquarids, Daytime Arietids, P/Machholz and 1491 I have $q = 0.977$, 0.968, 0.07, 0.09, 0.127 and 0.761 AU respectively). Recent papers on the orbital evolution of the stream include Zausaev and Pushkarev (1989), McIntosh (1990), Babadzhanov and Obrubov (1992a), Jones and Jones (1992), and Wu and Williams (1992b)

Another stream of particular interest, at least to the author, is the Taurids, which may be related to episodes of terrestrial catastrophism within the past 20,000 years (Steel *et al.*, 1991). As mentioned in section 3.4.3, up to 12 distinct showers may result from this stream (Babadzhanov *et al.*, 1990; Babadzhanov and Obrubov, 1992b), and it is particularly interesting since it it is associated with the shortest-period comet (P/Encke : see *e.g.* Whipple and Hamid, 1952; Jones, 1986) and ten or more known Earth-crossing asteroids (Asher *et al.*, 1993), and also a large fraction of the broad sporadic stream modelled by Štohl and Porubčan (1990, 1992). The breadth of this stream (*i.e.* the duration over which the showers are active) is due in part to its low inclination, leading to a wide range in nodes on the ecliptic.

Most of the above work has been concerned with the relatively short-term evolution of comets of the Jupiter and Halley-type families (*i.e.* those producing most of the recognized meteoroid streams), and have used physical theories based upon simple analytical techniques. Emel'yanenko (1992) has developed methods by which the long-term motion of comets with large eccentricities may be investigated, and thus the diffusion of streams associated with them delineated.

5. The physics of meteoroid streams

In the previous sections of this chapter the main point of discussion has been meteoroid astronomy; to complete this review I will made a brief mention of some aspects of meteoroid physics, such as the masses of streams, their ages, and the physical sources of meteoroids. Although all of these topics have been touched upon above, few suitable references have been given to start the interested reader on the path to an understanding of what is known, what is not known, and what techniques have been applied.

Total stream masses have been computed by Hughes and McBride (1989). That paper reviews work on that topic, and finds that the masses of the streams generally are comparable to the estimated masses of the parent comets. This not only points to the original cometary size, but also provides an estimate of the time since the formation of the stream began. Indeed this technique was used by Jones *et al.* (1989) to derive a range of ages (0.25–6.2×10^4 yr, most probably age 2.3×10^4 yr) for the Orionid (Halleyid) stream. Gustafson (1989) traces back a handful of Geminid meteoroid orbits to their parent asteroid, 3200 Phaethon, and in that ways finds an age of 600–2,000 yr, in accord with several alternate estimates using other techniques (such as the largely-invariant Geminid activity from one year to the next : Jones, 1982).

Whilst the formation and evolution of streams have been addressed in a general sense, the sources of meteoroids have not except to say that they are released from active comets, with some streams from asteroids being known; these may have been active comets when the streams formed (*cf.* the non-detection of P/Encke until two centuries ago, and P/Machholz until 1986, these perhaps being examples of re-activated comets which had undergone a period of quiescence during which they would have appeared as asteroids, if they had been observed). The supply of meteoroidal material was discussed by Whipple (1967), and more recently by, for example, Štohl (1987) and Fulle (1990).

As mentioned in section 4, the residence lifetime of meteoroids in streams is, in general, limited by the effects of close approaches to Jupiter which perturb the particles out of the stream into the sporadic background, but the limiting physical lifetime of meteoroids is due to impacts with zodiacal dust grains (Grün *et al.*, 1985; Olsson-Steel, 1986) which may also remove significant fractions of meteoroids from streams, especially for the longer-period streams which have high-speed encounters with Jupiter and are therefore largely-unaffected by such planetary encounters (*e.g.* the Perseids or the Halleyids). The deduced physical lifetimes of order 10^5 yr are supported by the solar cosmic ray exposure ages of collected dust samples (Bradley *et al.*, 1984; Sandford, 1991). However, the supply of meteoroids and their decay into interplanetary dust remain outstanding problems (Grün *et al.*, 1985).

6. Summary

In the past few years major steps forward have been taken in the search for an understanding of meteoroid streams. In my personal view, the main barrier to a near-complete understanding is our ignorance of how a meteoroid leaves its parent object : if it is accelerated to a large speed, by what mechanism does this acceleration occur?

The dynamics of streams has been a subject of considerable study, with significant breakthroughs occurring. My own personal view puts the following at the top of the list of streams of interest, if a general understanding is to be brought about, for the given reasons :

Geminids : Many individual orbits available, good rate profiles, known parent object, old stream?, very short period, $Q < 4$ AU, so never approaches Jupiter, four associated showers.

Perseids : Many individual orbits available, good rate profiles, old stream, known parent object, long period but never approaches any of the giant planets.

Taurids : Many individual orbits available, known parent (P/Encke but several Apollo asteroids also apparently related), moderate period, twelve associated showers?, orbital evolution dominated by Jovian secular perturbations.

Leonids : Good number of orbits available, known parent, young stream, recurrent storms (one due soon).

Halleyids : Good number of orbits available, known parent, old stream, stream good indicator of the history of the parent comet.

Quadrantids : Good number of orbits available, known parent, fascinating orbital evolution, eight associated showers.

Acknowledgements

This review is dedicated to the memory of Jan Štohl. It is a pleasure to remember discussions with Jan on several of the topics discussed above, and also to give credit to interactions with many of the other meteor astronomers who have contributed to our understanding of this topic. I also acknowledge with thanks reviews by Iwan Williams and Bertil Lindblad. Funding for this work came from the Australian Research Council.

References

Asher, D.J., Clube, S.V.M. and Steel, D.I. : 1993, "Asteroids in the Taurid Complex." ***Mon. Not. Roy. Astron. Soc.***, **264**, 93–105.

Babadzhanov, P.B. : 1990, "Formation of twin meteor showers.". In ***Asteroids, Comets, Meteors III*** (C.-I. Lagerkvist, H. Rickman, B.A. Lindblad and M. Lindgren, Eds.), 497–503, University of Uppsala, Sweden.

Babadzhanov, P.B., Hajduk, A., Obrubov, Yu.V. and Pushkarev, A.N. : 1991a, "Evolution of meteoroid streams of comet Halley." ***Astron. Vestn.***, **25**, 208–216; ***Sol. Sys. Res.***, **25**, 153–160.

Babadzhanov, P.B., Isamutdinov, Sh.O. and Chebotarev, R.P. : 1992, "Some structural features of the Geminid meteor shower." ***Astron. Vestn.***, **26**, 93–97.

Babadzhanov, P.B. and Obrubov, Yu.V. : 1992a, "P/Machholz 1986 VIII and Quadrantid meteoroid stream. Orbital evolution and relationship." In ***Asteroids, Comets, Meteors 1991*** (A.W. Harris and E. Bowell, Eds.), 27–32, Lunar and Planetary Inst., Houston.

Babadzhanov, P.B. and Obrubov, Yu.V. : 1992b, "Evolution of short-period meteor streams." ***Cel. Mech. Dynam. Astron.***, **54**, 111–127.

Babadzhanov, P.B., Obrubov, Yu.V. and Makhmudov, N. : 1990, "Encke meteor showers." ***Astron. Vestn.***, **24**, 18–28; ***Sol. Sys. Res.***, **24**, 12–19.

Babadzhanov, P.B., Wu, Z., Williams, I.P. and Hughes, D.W. : 1991b, "The Leonids, Comet Biela and Biela's associated meteoroid stream." ***Mon. Not. Roy. Astron. Soc.***, **253**, 69–74.

Baggaley, W.J. : 1977, "The possibility of the detection of meteor streams in interplanetary space." ***The Observatory***, **97**, 123–129.

Baggaley, W.J. and Taylor, A.D. : 1992, "Radar meteor orbital structure of southern hemisphere cometary dust streams." In ***Asteroids, Comets, Meteors 1991*** (A.W. Harris and E. Bowell, Eds.), 33–36, Lunar and Planetary Inst., Houston.

Baggaley, W.J., Taylor, A.D. and Steel, D.I. : 1992, "New determinations of radar orbits for meteors from short-period comets." In ***Periodic Comets*** (J.A. Fernández and H. Rickman, Eds.), 115–124, Universidad de la República, Montevideo, Uruguay.

Bailey, M.E., Clube, S.V.M. and Napier, W.M. : 1990, ***The Origin of Comets***, Pergamon, Oxford.

Beech, M. : 1986, "The Draconid meteoroids." ***Astron. J.***, **91**, 159–162.

Beech, M. : 1988, "Meteor astronomy : a mature science?" ***Earth, Moon & Planets***, **43**, 187–194.

Bel'kovich. O.I. : 1986, "On the spatial structure of the Geminid meteor shower." ***Astron. Vestn.***, **20**, 142–151; ***Sol. Sys. Res.***, **20**, 93–101.

Bradley, J.P., Brownlee, D.E. and Fraundorf, P. : 1984, "Discovery of nuclear tracks in interplanetary dust." *Science*, **226**, 1432–1434.

Burns, J.A., Lamy, Ph.L. and Soter, S. : 1979, "Radiation forces on small particles in the solar system." *Icarus*, **40**, 1–48.

Campbell, D.B., Harmon, J.K. and Shapiro, I.I. : 1989, "Radar observations of comet Halley." *Astrophys. J.*, **338**, 1094–1105.

Cook, A.F. : 1973, "A Working List of Meteor Streams.". In *Evolutionary and Physical Properties of Meteoroids, NASA SP-319* (C.L. Hemenway, P.M. Millman and A.F. Cook, Eds.), 183–191, NASA, Washington D.C.

Cook, A.F., Flannery, M.R., Levy, H., McCrosky, R.E., Sekanina, Z., Shao, C.-Y., Southworth, R.B. and Williams, J.T. : 1972, *Meteor Research Program*, NASA CR-2109, Smithsonian Astrophysical Observatory, Cambridge, Mass.

Crovisier, J. and Schloerb, F.P. : 1991, "The study of comets at radio wavelengths." In *Comets in the Post-Halley Era* (R.L. Newburn, Jr., M. Neugebauer and J. Rahe, Eds.), 149–173, Kluwer, Dordrecht.

Dodd, R.T., Wolf, S.F. and Lipschutz, M.E. : 1993, "An H Chondrite Stream : Identification and Confirmation." *J. Geophys. Res.*, **98**, 15105–15118.

Drummond, J.D. : 1981a, "A test of comet and meteor shower associations." *Icarus*, **45**, 545–553.

Drummond, J.D. : 1981b, "Earth-orbit-approaching comets and their theoretical radiants." *Icarus*, **47**, 500–517.

Drummond, J.D. : 1982a, "Theoretical twin meteor showers." *Icarus*, **49**, 135–142.

Drummond, J.D. : 1982b, "Theoretical meteor radiants of Apollo, Amor and Aten asteroids." *Icarus*, **49**, 143–153.

Ellyett, C.D. and Kennewell, J.A. : 1977, "Radar meteor rates and atmospheric density changes." *Nature*, **287**, 521–522.

Emel'yanenko, V.V. : 1991, "Dynamics of the Lyrid meteor swarm." *Astron. Vestn.*, **24**, 308–313; *Sol. Sys. Res.*, **24**, 199–202.

Emel'yanenko, V.V. : 1992, "Dynamics of periodic comets and meteor streams." *Cel. Mech. Dynam. Astron.*, **54**, 91–110.

Fox, K. and Williams, I.P. : 1985, "A possible origin for some ancient December fireballs." *Mon. Not. Roy. Astron. Soc.*, **217**, 407–411.

Fox, K., Williams, I.P. and Hughes, D.W. : 1982, "The evolution of the orbit of the Geminid meteor stream." *Mon. Not. Roy. Astron. Soc.*, **200**, 313–324.

Fox, K., Williams, I.P. and Hughes, D.W. : 1983, "The rate profile of the Geminid meteor shower." *Mon. Not. Roy. Astron. Soc.*, **205**, 1155–1169.

Froeschlé, C. and Scholl, H. : 1986, "Gravitational splitting of Quadrantid-like meteor streams in resonance with Jupiter." *Astron. Astrophys.*, **158**, 259–265.

Fulle, M. : 1990, "Meteoroids from short period comets." *Astron. Astrophys.*, **230**, 220–226.

Grün, E., *et al.* : 1993, "The discovery of jovian dust streams and interstellar grains by the Ulysses spacecraft." *Nature*, **362**, 428–430.

Grün, E., Zook, H.A., Fechtig, H. and Giese, R.H. : 1985, "Collisional balance of the meteoritic complex." *Icarus*, **62**, 244–272.

Gustafson, B.Å.S. : 1989, "Geminid meteoroids traced to cometary activity on Phaethon." *Astron. Astrophys.*, **225**, 533–540.

Gyssens, M., Knöfel, A., Rendtel, J. and Roggemans, P. : 1991, "The International Meteor Organization." In *IAU Colloq. 126 : Origin and Evolution of Interplanetary Dust* (A.C. Levasseur-Regourd and H. Hasegawa, Eds.), 335–338, Kluwer, Dordrecht.

Hahn, G. and Bailey, M.E. : 1990, "Rapid dynamical evolution of giant comet Chiron." *Nature*, **348**, 132–136.

Hajduk, A. : 1991, "Evolution of cometary debris : physical aspects." In *Comets in the Post-Halley Era*, (R.L. Newburn, Jr., M. Neugebauer and J. Rahe, Eds.), 592–606, Kluwer, Dordrecht.

Hajduková, M. and Hajduk, A. : 1991, "Lifetime of meteor streams associated with comet Halley." In *IAU Colloq. 126 : Origin and Evolution of Interplanetary Dust* (A.C. Levasseur-Regourd and H. Hasegawa, Eds.), 323–326, Kluwer, Dordrecht.

Halliday, I., Blackwell, A.T. and Griffin, A.A. : 1990, "Evidence for the existence of groups of meteorite-producing asteroidal fragments." ***Meteoritics*, 25**, 93–99.

Harmon, J.K., Campbell, D.B., Hine, A.A., Shapiro, I.I. and Marsden, B.G. : 1989, "Radar observations of comet IRAS-Iraki-Alcock 1983d." ***Astrophys. J.*, 338**, 1071–1093.

Hasegawa, I. : 1990, "Predictions of the Meteor Radiant Point Associated with a Comet." ***Publ. Astron. Soc. Japan*, 42**, 175–186.

Hasegawa, I. : 1992, "Historical variation in the meteor flux as found in Chinese and Japanese chronicles." ***Cel. Mech. Dynam. Astron.*, 54**, 129–142.

Hasegawa, I., Ueyama, Y. and Ohtsuka, K. : 1992, "Predictions of the Meteor Radiant Point Associated with an Earth-Approaching Minor Planet." ***Publ. Astron. Soc. Japan*, 44**, 45–54.

Hughes, D.W. and McBride, N. : 1989, "The mass of meteoroid streams." ***Mon. Not. Roy. Astron. Soc.*, 240**, 73–79.

Hughes, D.W., Williams, I.P. and Fox, K. : 1981, "The mass segregation and nodal regression of the Quadrantid meteor stream." ***Mon. Not. Roy. Astron. Soc.*, 195**, 625–637.

Hunt, J., Williams, I.P. and Fox, K. : 1985, "Planetary perturbations on the Geminid meteor stream." ***Mon. Not. Roy. Astron. Soc.*, 217**, 533–538.

Jones, J. : 1982, "The annual variation in the activity of the Geminid shower and the theory of the dispersal of meteoroid clusters." ***Mon. Not. Roy. Astron. Soc.*, 198**, 23–32.

Jones, J. : 1985, "The structure of the Geminid meteor stream – I. The effect of planetary perturbations." ***Mon. Not. Roy. Astron. Soc.*, 217**, 523–532.

Jones, J. : 1986, "The effect of gravitational perturbations on the evolution of the Taurid meteor stream complex." ***Mon. Not. Roy. Astron. Soc.*, 221**, 257–267.

Jones, J. and Hawkes, R.L. : 1986, "The structure of the Geminid meteor stream – II. The combined action of the cometary ejection process and gravitational perturbations." ***Mon. Not. Roy. Astron. Soc.*, 223**, 479–486.

Jones, J. and Jones, W. : 1992, "Evolution of the Quadrantid meteor stream." In ***Asteroids, Comets, Meteors 1991*** (A.W. Harris and E. Bowell, Eds.), 269–272, Lunar and Planetary Inst., Houston.

Jones, J., McIntosh, B.A. and Hawkes, R.L. : 1989, "The age of the Orionid meteoroid stream." ***Mon. Not. Roy. Astron. Soc.*, 238**, 179–191.

Jones, J. and Morton, J.D. : 1982, "High resolution radar studies of the Geminid meteor shower." ***Mon. Not. Roy. Astron. Soc.*, 200**, 281–291.

Kidger, M.R. : 1993, "Some Comments on the Identification of Medieval Meteor Showers Recorded by the Arabs." ***Q. Jl. Roy. Astron. Soc.*, 34**, 331–334.

Kresák, Ľ : 1993, "Meteor storms". In ***Meteoroids and their parent bodies*** (J. Štohl and I.P. Williams, Eds.), 147–156, Astron. Inst., Slovak Acad. Sci., Bratislava.

Kresáková, M. : 1987, "Associations between ancient comets and meteor showers." ***Bull. Astron. Inst. Czechoslov.*, 38**, 75–80.

Kulikova, N.V. : 1989, "The Draconid meteor stream (numerical simulation)." ***Astron. Vestn.*, 23**, 289–295; ***Sol. Sys. Res.*, 23**, 175–179.

Lebedinets, V.N., Kulikova, N.V. and Pivnenko, E.A. : 1990, "Effect of reactive deceleration of comets on the structure of meteor swarms." ***Astron. Vestn.*, 24**, 3–17; ***Sol. Sys. Res.*, 24**, 1–11.

Lindblad, B.A. : 1990, "The orbit of the Eta Aquarid meteor stream." In ***Asteroids, Comets, Meteors III*** (C.-I. Lagerkvist, H. Rickman, B.A. Lindblad and M. Lindgren, Eds.), 551–554, University of Uppsala, Sweden.

Lindblad, B.A. : 1991, "The IAU Meteor Data Center in Lund." In *IAU Colloq. 126 : Origin and Evolution of Interplanetary Dust* (A.C. Levasseur-Regourd and H. Hasegawa, Eds.), 311–314, Kluwer, Dordrecht.

Lindblad, B.A. and Porubčan, V. : 1992, "Activity of the Lyrid meteor stream." In ***Asteroids, Comets, Meteors 1991*** (A.W. Harris and E. Bowell, Eds.), 367–370, Lunar and Planetary Inst., Houston.
McIntosh, B.A. : 1990, "Comet P/Machholz and the Quadrantid meteor stream." ***Icarus***, **86**, 299–304.
McIntosh, B.A. : 1991, "Debris from comets : the evolution of meteor streams." In ***Comets in the Post-Halley Era*** (R.L. Newburn, Jr., M. Neugebauer and J. Rahe, Eds.), 557–591, Kluwer, Dordrecht.
McIntosh, B.A. and Jones, J. : 1988, "The Halley comet meteor stream : Numerical modelling of its dynamical evolution." ***Mon. Not. Roy. Astron. Soc.***, **235**, 673–693.
Morton, J.D. and Jones, J. : 1982, "A method of imaging radio meteor radiant distributions." ***Mon. Not. Roy. Astron. Soc.***, **198**, 737–746.
Oberst, J. and Nakamura, Y. : 1991, "A search for clustering among the meteoroid impacts detected by the Apollo lunar seismic network." ***Icarus***, **91**, 315–325.
Olsson-Steel, D. : 1986, "The origin of the sporadic meteoroid component." ***Mon. Not. Roy. Astron. Soc.***, **219**, 47–73.
Olsson-Steel, D. : 1987, "Theoretical meteor radiants of Earth-approaching asteroids and comets." ***Aust. J. Astron.***, **2**, 21–35; see also **2**, 93–101 (1988).
Olsson-Steel, D. : 1988, "Identification of meteoroid streams associated with Apollo asteroids in the Adelaide radar orbit surveys." ***Icarus***, **75**, 64–96.
Padevět, V. : 1973, "Motion of a fragment in disturbed air behind a meteor body." In ***Evolutionary and Physical Properties of Meteoroids, NASA SP-319*** (C.L. Hemenway, P.M. Millman and A.F. Cook, Eds.), 169–174, NASA, Washington D.C.
Poole, L.M.G. and Roux, D.G. : 1989, "Meteor radiant mapping with an all-sky radar." ***Mon. Not. Roy. Astron. Soc.***, **236**, 645–652.
Porter, J.G. : 1952, ***Comets and Meteor Streams***, Chapman and Hall, London.
Porubčan, V. and Cevolani, C. : 1990, "On the activity and structure of the Quadrantid meteor stream." ***Nuovo Cimento***, **13**, 653–662.
Porubčan, V. and Štohl, J. : 1992, "Burst of the 1969 Leonids and 1982 Lyrids." In ***Asteroids, Comets, Meteors 1991*** (A.W. Harris and E. Bowell, Eds.), 469–472, Lunar and Planetary Inst., Houston.
Porubčan, V., Štohl, J. and Vaňa, R. : 1992, "On associations of Apollo asteroids with meteor streams." In ***Asteroids, Comets, Meteors 1991*** (A.W. Harris and E. Bowell, Eds.), 473–476, Lunar and Planetary Inst., Houston.
Rabinowitz, D.L. : 1993, "The size distribution of Earth-approaching asteroids." ***Astrophys. J.***, **407**, 412–427.
Rada, W.S. and Stephenson, F.R. : 1992, "A Catalogue of Meteor Showers in Medieval Arab Chronicles." ***Q. Jl. Roy. Astron. Soc.***, **33**, 5–16.
Ridley, H.B. : 1962, "The Phoenicid meteor shower of 1956 December 5." ***J. Brit. Astron. Assoc.***, **72**, 266–272.
Ryabova, G.O. : 1989, "The effect of secular perturbations and the Poynting-Robertson effect on the structure of the Geminid meteor stream." ***Astron. Vestn.***, **23**, 254–264; ***Sol. Sys. Res.***, **23**, 158–165.
Sandford, S.A. : 1991, "Constraints on the parent bodies of collected interplanetary dust particles." In ***IAU Colloq. 126 : Origin and Evolution of Interplanetary Dust*** (A.C. Levasseur-Regourd and H. Hasegawa, Eds.), 397–404, Kluwer, Dordrecht.
Šimek, M. and Lindblad, B.A. : 1990, "The activity curve of the Perseid meteor stream as determined from short duration meteor radar echoes." In ***Asteroids, Comets, Meteors III***, (C.-I. Lagerkvist, H. Rickman, B.A. Lindblad and M. Lindgren, Eds.), 567–570, University of Uppsala, Sweden.
Šimek, M. and McIntosh, B.A. : 1991, "Cross section of the Quadrantid meteor stream." ***Bull. Astron. Inst. Czechoslov.***, **42**, 124–132.
Singer, S.F. and Stanley, J.E. : 1980, "Submicron particles in meteor streams." In ***IAU Symp. 90, Solid Particles In the Solar System*** (I. Halliday and B.A. McIntosh, Eds.)

329–332, Reidel, Dordrecht.
Steel, D. : 1991, "The orbital distribution and origin of meteoroids." In *IAU Colloq. 126 : Origin and Evolution of Interplanetary Dust* (A.C. Levasseur-Regourd and H. Hasegawa, Eds.), 291–298, Kluwer, Dordrecht.
Steel, D.I., Asher, D.J. and Clube, S.V.M. : 1991, "The structure and evolution of the Taurid Complex." *Mon. Not. Roy. Astron. Soc.*, **251**, 632–648.
Štohl, J. : 1987, "Meteor contribution by short-period comets." *Astron. Astrophys.*, **187**, 933–934.
Štohl, J. and Porubčan, V. : 1990, "Structure of the Taurid meteor complex." In *Asteroids, Comets, Meteors III*, (C.-I. Lagerkvist, H. Rickman, B.A. Lindblad and M. Lindgren, Eds.), 571–574, University of Uppsala, Sweden.
Štohl, J. and Porubčan, V. : 1992, "Dynamical aspects of the Taurid meteor complex." In *IAU Symp. 152 : Chaos, resonance and collective dynamical phenomena in the solar system* (S. Ferraz-Mello, Ed.), 315–324, Kluwer, Dordrecht.
Sykes, M.V. and Walker, R.G. : 1992, "Cometary Dust Trails. I. Survey." *Icarus*, **95**, 180–210.
Terentjeva, A.K. : 1990, "Fireball streams." In *Asteroids, Comets, Meteors III*, (C.-I. Lagerkvist, H. Rickman, B.A. Lindblad and M. Lindgren, Eds.), 579–584, University of Uppsala, Sweden.
Whipple, F.L. : 1951, "A Comet Model. II. Physical Relations for Comets and Meteors." *Astrophys. J.*, **113**, 464–474.
Whipple, F.L. : 1967, "On maintaining the meteoritic complex." In *The Zodiacal Light and the Interplanetary Medium, NASA SP–150* (J.L. Weinberg, Ed.), 409–426, NASA, Washington D.C.
Whipple, F.L. and Hamid, S.E. : 1952, "On the origin of the Taurid meteor streams." *Helwan Obs. Bull.*, **41**, 224–252.
Williams, I.P. : 1990, "The formation of meteor streams." In *Asteroids, Comets, Meteors III* (C.-I. Lagerkvist, H. Rickman, B.A. Lindblad and M. Lindgren, Eds.), 585–594, University of Uppsala, Sweden.
Williams, I.P. : 1992, "The dynamics of meteoroid streams." In *IAU Symp. 152 : Chaos, resonance and collective dynamical phenomena in the solar system* (S. Ferraz-Mello, Ed.), 299–313, Kluwer, Dordrecht.
Williams, I.P., Johnson, C. and Fox, K. : 1986, "Meteor Storms." In *Asteroids, Comets, Meteors II* (C.-I. Lagerkvist, B.A. Lindblad, H. Lundstedt and H. Rickman, Eds.), 559–563, University of Uppsala, Sweden.
Williams, I.P. and Wu, Z. : 1993, "The Quadrantid meteoroid stream and Comet 1491 I." *Mon. Not. Roy. Astron. Soc.*, **264**, 659–664.
Wu, Z. and Williams, I.P. : 1992a, "Formation of the Leonid Meteor Stream and Storm." In *Asteroids, Comets, Meteors 1991*, (A.W. Harris and E. Bowell, Eds.), 661–665, Lunar and Planetary Inst., Houston.
Wu, Z. and Williams, I.P. : 1992b, "On the Quadrantid meteoroid stream complex." *Mon. Not. Roy. Astron. Soc.*, **259**, 617–628.
Wu, Z. and Williams, I.P. : 1993, "The Perseid meteor shower at the present time." *Mon. Not. Roy. Astron. Soc.*, **264**, 980–990.
Yeomans, D.K. : 1981, "Comet Tempel–Tuttle and the Leonid meteors." *Icarus*, **47**, 492–499.
Yeomans, D.K. and Kiang, T. : 1981, "The long-term motion of comet Halley." *Mon. Not. Roy. Astron. Soc.*, **197**, 633–646.
Zausaev, A.F. and Pushkarev, A.N. : 1989, "The orbital evolution of the Quadrantid meteor stream during 4000 years (AD 1950 to 2050 BC)." *Byull. Inst. Astrofiz.*, **80**, 18–24.

ZODIACAL DUST BANDS

S. F. DERMOTT, D. D. DURDA, B. A. S. GUSTAFSON, S. JAYARAMAN
J. C. LIOU AND Y. L. XU
Department of Astronomy,
University of Florida,
Gainesville, FL 32611, U.S.A
E–mail dermott@astro.ufl.edu

Abstract. One of the outstanding problems in solar system science is the source of the particles that constitute the zodiacal cloud. The zodiacal dust bands discovered by IRAS have a pivotal role in this debate because, without doubt, they are the small, tail end products of asteroidal collisions. Geometrical arguments are probably the strongest and the plane of symmetry of the dust bands places their source firmly in the asteroid belt. A cometary source, Comet Encke for example, could exist at the distance of the mainbelt, but the dynamics of cometary orbits makes the formation of cometary dust bands impossible, unless, of course, there is a significant (comparable in volume to the asteroidal families) source of comets interior to the orbit of Jupiter with low (asteroidal) orbital eccentricities. We have suggested that the dust bands are associated with the prominent asteroidal families. The link with the Themis and Koronis families is good but the link with Eos remains to be proved. We show here by detailed modeling that even though the filtered infrared flux in the $25\mu m$ waveband associated with the dust bands is only ~1% of the total signal, this is only the "tip of the iceberg" and that asteroidal dust associated with the bands constitutes ~10% of the zodiacal cloud. This result, plus the observed size-frequency distribution of mainbelt asteroids and the observed ratio of the number of family to non-family asteroids allows us to estimate that asteroidal dust accounts for about one third of the zodiacal cloud. The discovery of the "leading-trailing" asymmetry of the zodiacal cloud in the IRAS data and our interpretation of this asymmetry in terms of a ring of asteroidal particles in resonant lock with the Earth is important for two reasons. (1) The existence of the ring strongly suggests that large (diameter $\geq 12\mu m$) asteroidal particles (or particles with low orbital eccentricities) are transported to the inner solar system by drag forces. (2) The observed ratio of the trailing-leading asymmetry allows an independent estimate of the contribution of asteroidal particles to the zodiacal cloud. These new results have important implications for the source of the interplanetary dust particles (IDPs) collected at the Earth. Because asteroidal particles constitute about one third of the zodiacal cloud and are transported to the inner solar system by drag forces, gravitational focussing by the Earth that results in the preferential capture of particles from orbits with low inclinations and low eccentricities and the possible "funneling" effect of the ring itself, imply that nearly all of the unmelted IDPs collected at the Earth are asteroidal.

1. Introduction

The primary source of particles in the zodiacal cloud is either asteroidal or cometary, but which of these sources is dominant has proved difficult to determine. The work of the Florida group (Dermott et al., 1984, 1985, 1986, 1988, 1989, 1990, 1992, 1993, 1994) has concentrated on modeling the orbital evolution of asteroidal particles. This is partly because of the important role that the IRAS dust bands have in the source debate. It is also true that the asteroidal source is comparatively well-defined and that the small orbital eccentricities of asteroidal particles

A. Milani et al. (eds.), Asteroids, Comets, Meteors 1993, 127–142.

make their dynamics tractable, even in the presence of drag. This is not true for cometary particles. The modeling that we can now do for asteroidal particles cannot, at present, be done for cometary particles. This partly accounts for our lack of work in this area, although some relevant work has recently been completed by Liou (1993). Because of this problem and because asteroids are known to feed the dust bands, the Florida approach is to estimate the contribution of asteroidal particles to the zodiacal cloud. The remainder may primarily be attributed to comets. The cometary trails observed by IRAS are direct evidence of the input of comets to the zodiacal cloud (Sykes and Walker, 1992). Our emphasis on the modeling of an asteroidal source should not be taken to mean that we consider that comets may not have a role in supplying the zodiacal cloud. At present, approximately two thirds of the thermal brightness remains unaccounted for. However, our modeling of the asteroidal dust component is incomplete. The modeling needs to be extended to include (a) a range of particle sizes representing a size distribution and (b) the effects of dust-dust collisions that may increase the effective surface area of the particles as they spiral towards the Sun.

Direct determinations of dust particle orbits using impact detectors or other means are potentially effective ways to discriminate between asteroidal and cometary dust particles. This potential has not yet been realised to the extent of having a major bearing on the question of the origin of the zodiacal cloud. The impact detectors onboard the Galileo and Ulysses spacecraft with effective sensitive areas of $0.1m^2$ can detect particles as small as $4\ 10^{-15}\mu m$ at 20 km/sec (Grün, this volume). Nearly all the detected dust particles have masses below the $10^{-12}g$ mass range , corresponding to diameters in the micron range. This is simply because of the low impact probability of the larger particles. Most of the detected particles correspond to micron or submicron grains that are too small to radiate efficiently in the $25\mu m$ waveband. This size range is also thought to be smaller than those particles that contribute most of the zodiacal light in the visual part of the spectrum by the scattering of sunlight (Röser and Staude, 1978).

Grün et al. (1992) point out that the particle trajectories appear to be consistent with an asteroidal source. The orbits are consistent with a population of particles on low inclination ($I = 0°$ to $30°$) orbits. Inclination is not affected by radiation pressure, so that the larger particles from the same source would also be expected to be in low inclination orbits. However, the low to moderate eccentricities ($e = 0.1$ to 0.5) are affected by radiation forces and may not be indicative of the orbits of the larger particles.

In Section 2, we estimate the contribution that asteroidal dust associated with the dust bands alone makes to the zodiacal cloud. This leads, in Section 3, to an estimate of the contribution from the asteroid belt as a whole. In Section 4, we present evidence that the Earth is embedded in a solar ring of asteroidal particles (or particles with low orbital eccentricities), strongly suggesting that large asteroidal particles are transported from the mainbelt to the Earth by drag forces. In Section 5, we discuss the implications of these new results for the origin of the IDPs collected at the Earth.

The focus of this paper is the zodiacal dust bands, but our conclusions have implications for the origin of the particles in the zodiacal cloud 'as a whole. Other recent reviews on the latter topic that take a much broader viewpoint are by Divine (1993), Greenberg (1991), Grün et al. (1985), Leinert and Grün (1990), and Levasseur-Regourd et al. (1991). Recent reviews and papers on dust bands are by Reach (1991, 1992) and Sykes et al. (1989).

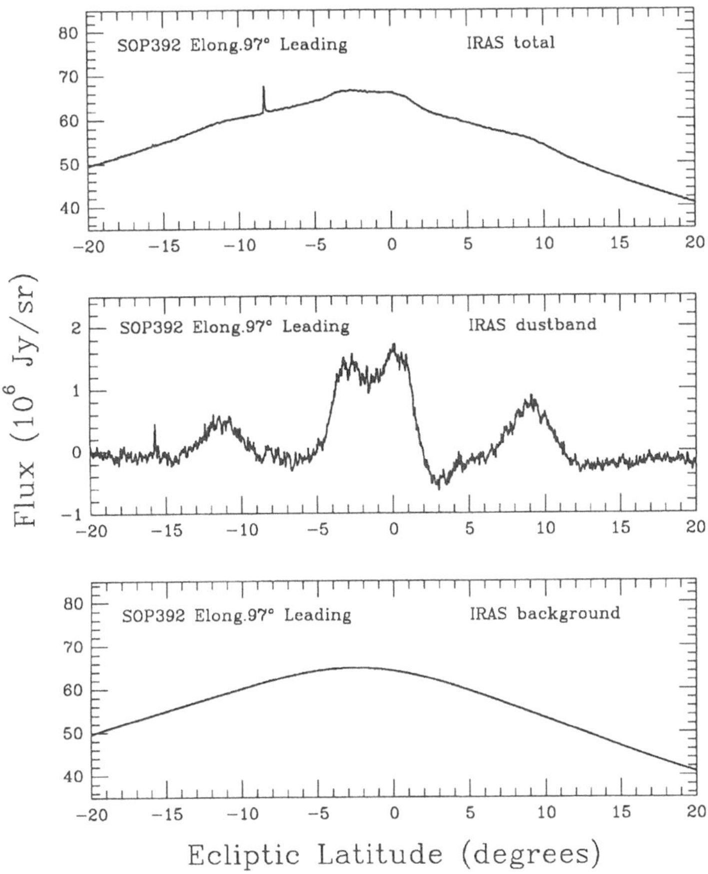

Fig. 1. Top : The total IRAS signal in the $25\mu m$ waveband. Middle : The narrow scale features (dust bands) subtracted from the total signal using a Fourier filter. Bottom : The broad-scale residual (or background) from the filtering process.

2. The IRAS Dust Bands

An example of the data obtained by IRAS in the $25\mu m$ waveband (the galactic signal is minimal in this waveband) is shown in Fig.1. This figure shows the variation of thermal flux with ecliptic latitude for an elongation angle of 97° and for a particular ecliptic longitude of the Earth (SOP 392). Fourier methods are used to separate the dust bands from the broad-scale zodiacal background. It needs to be emphasized that the division between short-scale and long-scale features, that must be specified in the filtering process, is entirely arbitrary and yet this divide determines the amplitudes of the dust bands shown in Fig.1. Thus, in Fig.1 the amplitudes of the "ten-degree" bands are $\sim 1\%$ of the background signal, but these amplitudes could change by a factor as large as ~ 4 simply by changing the filter. We show here how this filtering problem can be bypassed, making possible an estimate of the amount of dust associated with the bands. In 1984, we suggested (Dermott et al., 1984) that the dust bands are produced by the gradual comminution of asteroids in the major families. The confirmation of this hypothesis (an ongoing project) has involved : (1) The development of a new secular perturbation theory that includes the effects of Poynting-Robertson light drag, light pressure and solar wind drag on the evolution of the dust particle orbits (Dermott et al., 1992); (2) The production of a new high resolution Zodiacal History File by IPAC (the Infrared Processing and Analysis Center at Caltech); (3) The development of the SIMUL code : a three-dimensional numerical model that allows the calculation of the thermal flux produced by any particular distribution of dust particle orbits (Dermott et al., 1988). These tools allow the modeling of the asteroidal component of the zodiacal cloud from source to sink.

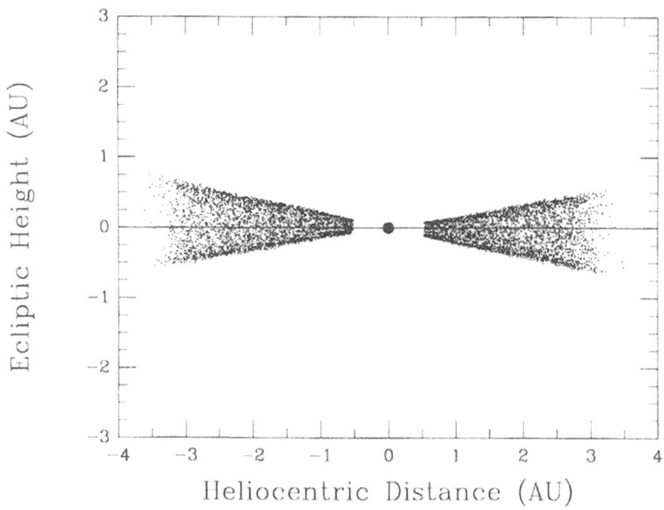

Fig. 2. Vertical slice through a model dust band based on numerical integration of the positions of $9\mu m$ diameter asteroidal particles derived from the Eos family. The integration was stopped at 0.5 A.U.

The cross-sectional structure of a typical dust band is shown in Fig.2. This figure shows the positions of $9\mu m$ diameter particles originating from the Eos family. The particles spiral towards the Sun due to the action of drag forces. During this orbital evolution there are small (calculable) changes in the forced inclinations and nodes.

These forced elements define the effective plane of symmetry of the cloud and vary with particle size and with distance from the Sun. However, the proper inclination remains constant (Dermott et al., 1992) and it is this property that gives rise to an observable pair of dustbands. Because the proper inclination remains constant, there is an enhancement in particle number density at the upper and lower surfaces of the band and this gives rise to an enhancement in flux when the dust band is viewed in a direction close to one of these surfaces (see fig. 2). A pair of dust bands is observed because only the peaks in the flux that rise significantly above that of the broadscale background are observed. Most of the flux arising from the dust associated with the band merges into the background.

The orbital evolution of cometary particles is quite different (Liou, 1993). Because of their higher eccentricities, the variations in the inclinations are not decoupled from those of the eccentricities and the inclination with respect to the plane associated with the forced inclination is no longer constant. Drag forces acting on particles released from Comet Encke with initial inclinations with respect to the forced plane of, for example, 10 degrees experience dispersion and increases up to values as large as 40 degrees (Liou, 1993). This distribution of inclinations may have a role in determining the shape of the background cloud, but it is impossible for such a widely dispersed distribution of inclinations to produce an observable pair of dust bands.

The location of the source of the dust bands can be determined from their observed plane of symmetry. The latitudes of the peaks shown in the middle panel of fig. 1 are closely associated with those particles in the more distant (from the Sun) regions of the dust band. The variation with the longitude of the Earth of the latitudes of these peaks, observed at a constant elongation angle of 90° in the trailing (that is, opposite to the Earth's orbital motion) and in the leading directions, can be used to determine the orientation of the plane of symmetry. Highly precise values of these latitudes can be obtained by exploiting the IRAS observing sequence (Dermott et al. 1993, 1994). To obtain good all-sky coverage, the elongation angle used by IRAS was not held constant at 90°, but was shifted systematically from scan to scan over periods of a few weeks to ensure complete coverage of each "lune" of the sky. For the first two thirds of the mission, the elongation angle, while being kept within the range $\sim 80°$ to $\sim 100°$, was systematically increased from scan to scan for scans in the trailing direction, while for scans in the leading direction the elongation was systematically decreased. These regularities allow us to increase the signal to noise ratio by using all the data in a given lune to obtain (a) the average latitude of a dust band peak for an elongation angle of 90° and (b) the ecliptic longitude of the Earth corresponding to that elongation angle (Dermott et al. 1993, 1994). By plotting these latitudes as a function of the longitude of the Earth, we can find the nodes of the band with respect to the ecliptic. These are the ecliptic longitudes of the Earth at which the latitudes of the mid-points between the North and South peaks in the leading and trailing directions are equal and opposite and equal in magnitude to the forced inclination. Results from Dermott et al. (1992, 1993) are shown in Table 1. The observed plane of symmetry of the "ten-degree" bands is quite distinct from that of the background cloud, but is closely similar to that of the asteroidal families (Eos and Themis, for example) in the outer part of the mainbelt. Other work on the orientation of the zodiacal cloud includes that by Hauser (1988), and Leinert et al. (1980).

Table 1: Orientations of planes of symmetry with respect to Ecliptic.

Feature	Waveband	Ascending		Descending	
		Inclination	Node	Inclination	Node
10 degree bands	$25\mu m$	$1.16 \pm 0.09°$	$99.7 \pm 7.8°$	$1.08 \pm 0.06°$	$265.5 \pm 7.8°$
Eos Family		1.19°	97.1°	1.19°	277.1°
Themis Family		1.22°	97.8°	1.22°	277.8°
Zodiacal Cloud	$25\mu m$	$1.60 \pm 0.02°$	$47.2 \pm 0.8°$	$1.52 \pm 0.02°$	$230.7 \pm 0.8°$

An illustrative example of the method that we use to bypass the filtering problem and to estimate the fraction of the zodiacal cloud that is associated with the dust bands is outlined in fig. 3 shown on the next page. This particular figure shows the results of the modeling that we have done to account for the shape of the dust band shown in fig. 1. Previous work on the modeling of the background zodiacal cloud (Dermott et al. 1992) has shown that the plane of symmetry of the cloud can be accounted for by, for example, asteroidal particles with diameters of $12\mu m$ and densities of $2.7gm/cm^3$ that spiral in from the mainbelt to the Sun without disruption. A more complete model would employ a range of particles sizes and consider the effects of particle breakup due to dust-dust collisions, but our modeling is not yet that sophisticated. In this paper, we use the same single-size asteroidal particles to model the orbital evolution of dust associated with the asteroid families.

The model profile shown in fig. 3a(i) was produced using dust from six prominent families : Themis, Koronis, Eos, Nysa, Dora and Gefion as defined by Zappalà et al. (1990, 1994). The area of dust associated with each family was adjusted to match the observed profile shown in the middle panel of fig. 1 using the following iterative procedure. The raw profile shown in fig. 3a(i) was added to the background profile shown in the bottom panel of fig. 1. The resultant sum was then passed through the same filter that was used to produce the observed dust band. This results in a model dust band that can be compared with the observations - see fig. 3b(i). However, most of the flux seen in fig. 3a(i) goes into the background signal and hence the filtered background exceeds the observed background - see fig. 3b(ii). This excess is subtracted from the background shown in the bottom panel of fig. 1. *before* adding the flux shown in fig. 3a(i). On filtering the resultant sum, we now obtain a model dust band profile as well as a background residual that are the same as those shown in fig. 1 - see fig. 3c. We can show that these results are largely independent of the filter used. Similar modeling has been repeated for many other dust band profiles spanning a wide range of elongation angles (Dermott et al. 1992). Thus, by using the same filter in the modeling process that we use to define the observed dust bands, and iterating, we are able to bypass the arbitrary

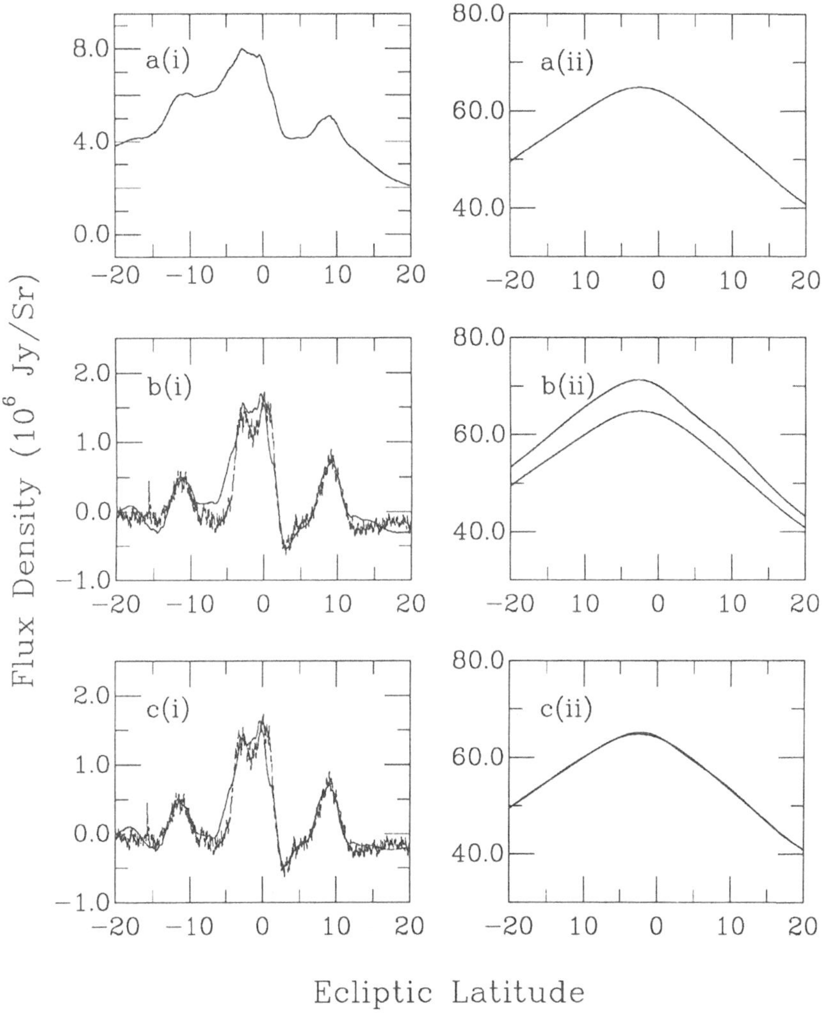

Fig. 3. Outline of the filtering process : (a) Top panel shows the raw model dustband (i) and the observed broadscale back ground (ii); (b) In the first iteration we add a(i) and a(ii) and filter to obtain a model dustband (smooth curve in (i)) similar to observations (noisy curve in (i)) and a background (upper curve in (ii)) which is higher than the observed background (lower curve in (ii)); (c) In the final iteration we subtract the excess flux shown in b(ii) from the background in a(ii) and add a(i) before filtering to obtain the final dustband shown in c(i) and the final background shown in c(ii) that agree with the observations.

divide associated with the filter. This allows us to estimate the true extent of the underlying dust band. Given that the peak flux shown in fig. 3a(i) is about 10% of that shown in fig. 3a(ii), we deduce that about 10% of the dust in the zodiacal cloud is associated with the dust bands and hence with the prominent Zappalà et al. (1990, 1994) asteroid families. We do need to assume that the dust derives from a number of families and not just the two largest (Eos and Themis). In fact, we are not confident that Eos is the dominant contributor to the ten-degree bands, but this will not be certain until we have a more sophisticated model that uses a realistic range of particle sizes. However, these further considerations may change the amount of dust that we associate with particular families but they are unlikely to effect significantly our estimate of the total contribution of the families as a whole to the zodiacal cloud.

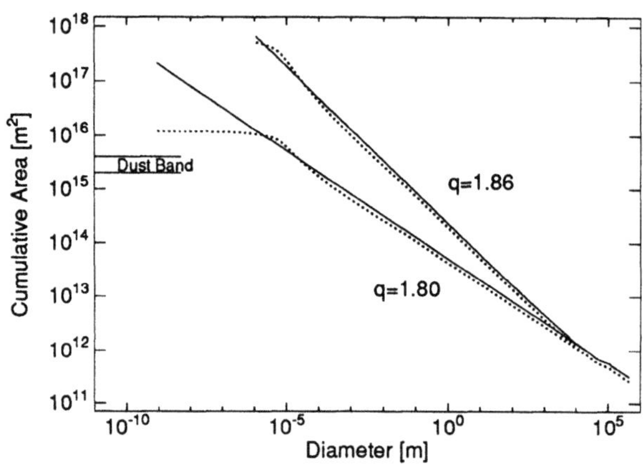

Fig. 4. Cumulative area of dust associated with the Zappalà et al. (1990, 1994) families asteroid families for two values of the size-frequency index q. The solid lines represent the geometrical area, while the dotted lines are the effective area calculated for astronomical silicate using Mie theory. The cumulative area of dust needed to account for the IRAS observations of the dustbands is represented by the banded region on the left.

3. Asteroidal Collisions

The collisional evolution of the asteroid belt has recently been investigated by Durda (1993). The result of Dohnanyi (1969), that the size-frequency index q is 1.833 holds, but only if the strength of the asteroids is independent of their size. Our lack of knowledge of how strength actually depends on size across the micron to 100 km size range (11 decades!) makes the relationship between the total volume of asteroids in collisional "equilibrium" that feed the zodiacal cloud (those asteroids

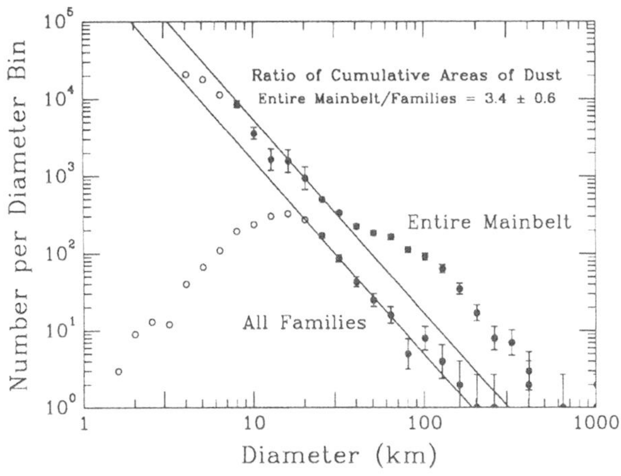

Fig. 5. Size-frequency distribution of (a) all asteroids in the mainbelt (upper line), and (b) the asteroids in the Zappalà et al. (1990, 1994) families (lower line).

with diameters $\leq 30km$) and the effective area of dust that emits radiation in a given waveband somewhat tentative. Thus, even though asteroids are a comparatively well-defined source, we cannot obtain a useful estimate of the effective area of asteroidal dust from first principles. The situation with estimating the contribution of comets to the zodiacal cloud is even worse. Our knowledge of the size-frequency distribution of comets is minimal to non-existent and estimates of the area of dust associated with this source are of little value. In fig. 4, we relate the total volume in all the Zappalà et al. (1990, 1994) asteroid families (Durda, 1993) to the associated effective area of dust in the IRAS $25\mu m$ waveband using two assumed values for q that span the canonical value of 1.833. It would appear that the predicted area of dust may be more than adequate to account for the observations. However, there are many uncertainties. One interesting fact discovered by Durda (1993) is that the value of q may vary from family to family and may be different from that of the general background (the stronger the asteroidal grains, the greater the amount of dust that will be associated with a given source volume). These uncertainties make predictions even more difficult. However, we can use the dust band observations to make a moderately robust estimate of the contribution of the asteroid belt as a whole to the zodiacal cloud. Given that the dust bands are related to the Zappalà et al. (1990, 1994) families, and that the dust associated with these families constitutes 10% of the zodiacal cloud, then the observation that the ratio of family to non-family asteroids in the collisionally evolved asteroidal population is approximately 3.4 : 1 implies that the asteroid belt as a whole is the source of $\sim 1/3$ of the dust in the zodiacal cloud (see fig. 5). Due to the stochastic nature of collisions in the asteroid belt and our lack of knowledge of the number of small, collisionally-evolved asteroids in the outer regions of the asteroid belt, the latter estimate may be uncertain by a factor of two or more (Durda, 1993).

4. The Earth's Asteroidal Ring

The IRAS spacecraft has given us our highest resolution data on the structure of the zodiacal cloud. Dermott et al. (1988) analysed this data and pointed out a marked but peculiar asymmetry, namely that observations of the peak brightness of the zodiacal cloud in the trailing direction (that is, opposite to the Earth's orbital motion) are consistently greater than those in the leading direction. This finding was later confirmed by Reach (1991).

Collisions amongst the asteroids result in the grinding down of the asteroid belt and the production of small dust particles that spiral towards to the Sun due to the action of drag forces. Numerical integrations show that a significant fraction of these particles are temporarily trapped in corotational resonances external to the orbits of each of the terrestrial planets, particularly the Earth (Jackson and Zook, 1989; Dermott et al., 1994) — Figs.6–8. The result is that the Earth is embedded in a ring of asteroidal particles (or, more correctly, a ring of particles with low orbital eccentricities) that is longitudinally near uniform apart from a cavity that contains the planet. The action of drag introduces a phase lag into the equation of motion of these resonant particles. The physical manifestation of this is a phase difference between the longitude of conjunction of the dust particle and the Earth and the apocenter of the dust particle's orbit, with the result that the position of the Earth within the cavity is markedly asymmetric, with the Earth closer to the edge of the cavity in the trailing orbital direction (opposite to the Earth's orbital motion) than in the leading orbital direction (fig. 8). The above discussion certainly applies to mainbelt asteroidal particles. The higher orbital eccentricities of most cometary particles in the vicinity of the Earth (Jackson and Zook, 1992) and of particles deriving from the disruption of near-Earth asteroids make resonant trapping of these particles considerably less likely. We do not deny that some cometary particles in the vicinity of the Earth may have very low orbital eccentricities, but we consider it very unlikely that cometary particles in these orbits constitute a source comparable to that of mainbelt asteroids.

The size-frequency distribution of asteroidal particles is such that the effective area of dust and hence the thermal flux from the ring observed in a given IRAS waveband is dominated by flux from the smaller (but not the smallest) particles in the distribution. These particles have high phase shifts with the result that the dynamics of resonance is in the non-adiabatic regime, for which analytic theory is only a guide and reliable results must be obtained by numerical integration. The drag rates on very small particles can be too high to allow resonance trapping. To obtain a detailed description of the structure of the ring, the dynamics of a wide range of particle sizes need to be considered. Dermott et al. (1994) estimated that asteroidal particles of diameter $12\mu m$ and density $2.7 g/cm^3$ are close to the critical size for trapping into the resonances in significant numbers and estimated the structure of the ring from the dynamics of those particles alone.

On integrating the orbits of 2048 particles originating from the main asteroid belt, we found that up to 20% of the particles are trapped in first-order resonances

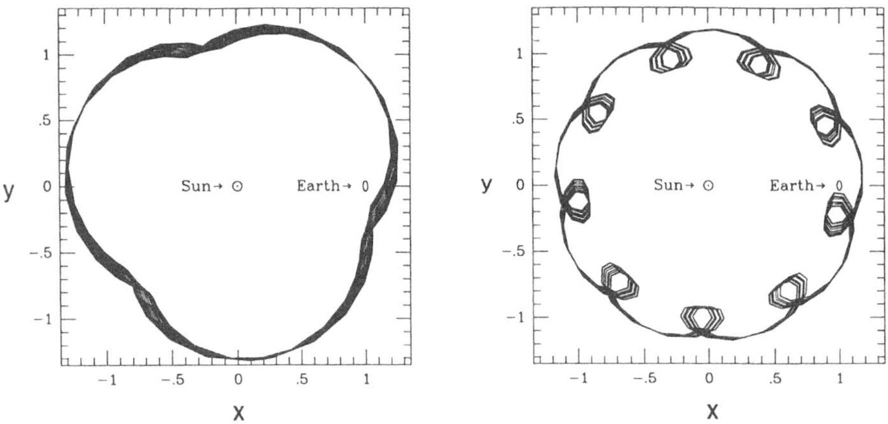

Fig. 6. Orbits of $12\mu m$ asteroidal particles in (a) 3 : 4 resonance and (b) 9 : 10 resonance tracked over one complete libration period and plotted in a frame centred on the sun and rotating with the Earth's mean motion.

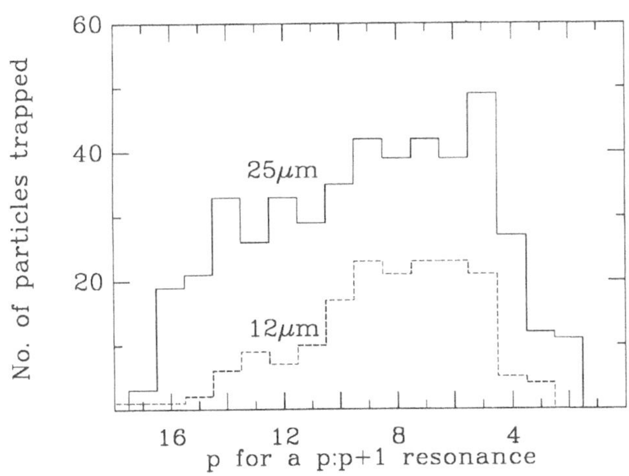

Fig. 7. Relative probabilities of capture into different resonances of asteroidal particles of diameter $12\mu m$ and $25\mu m$.

with $p : p+1$ (the ratio of mean motions) ranging from 3 : 4 to 17 : 18 with occasionally even higher values of p. fig. 7 shows the relative probability of capture into the individual resonances for two sizes of particles.The average trapping time for particles in these resonances was found empirically. Our final model is a simulated 'image' binned in pixels of 0.04 x 0.04 A.U. (fig. 8, left). The resultant image obtained by superposition of the various paths weighted according to the probability of capture and the trapping times shows that, in a rotating reference frame, the trapped particles form a near-uniform ring around the Sun that corotates in inertial space with the Earth.

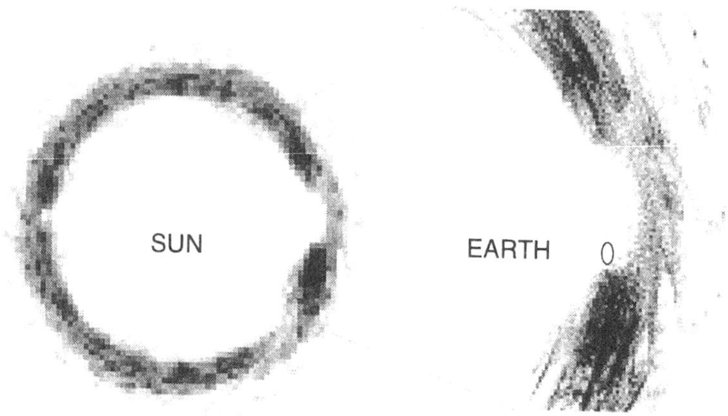

Fig. 8. Numerically simulated images of a heliocentric ring of asteroidal particles corotating with the Earth shown in low resolution on the left and high resolution on the right. The path of the Earth is a small 2 : 1 ellipse that resides in a cavity in the ring.

The peripheral arc-like structures in the ring are artifacts caused by the limited number of single-size particles considered in the model. However, the other noticeable features in fig. 8 are real. Firstly, the existence of a cavity in the ring at the location of the Earth. This is seen more clearly using a higher resolution of 0.01 x 0.01 A.U. The 2 : 1 ellipse in fig. 8 (on the right) is the epicycle of the Earth's orbit in the rotating reference frame. Secondly, fig. 8 also shows that apart from an asymmetry in the mean position of the Earth in the ring's cavity, there is also a marked asymmetry in the longitudinal variation of the particle number density. The increase in resonance strength and the corresponding decrease in phase shift with increasing p value, disperses the longitudes of the pericentres of the particle paths in the rotating frame in the leading direction while concentrating the longitudes of the pericenters in the trailing direction with the result that there is marked enhancement in particle number density behind the Earth in its orbit, as if the Earth had a trailing cloud of dust permanently in its wake. We consider that these asymmetries act together to produce the observed trailing/leading flux ratios of the zodiacal cloud.

We analysed all the IRAS observations of the zodiacal cloud and showed that when viewed at a constant elongation angle of 90°, the ecliptic brightness of the cloud is systematically greater by $\sim 3\%$ in the trailing direction than in the leading

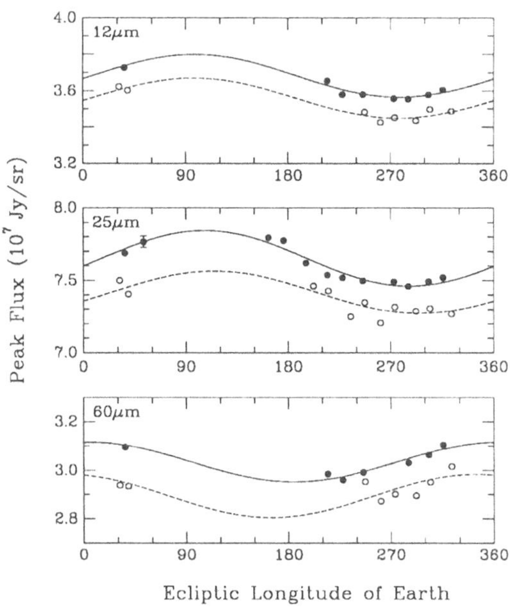

Fig. 9. Variation with the ecliptic longitude of Earth of the peak brightness of the zodiacal cloud as observed by IRAS in the three infrared wavebands at an elongation of 90° in the trailing direction (filled circles and upper curve) and in the leading direction (open circles and lower curve).

direction. This asymmetry is observed at all times of the year and in all three of the IRAS wavebands in which the zodiacal cloud is clearly observed (fig. 9). The existence of asteroidal families and their associated dust bands (also discovered by IRAS) are evidence that the asteroid belt is a significant source of dust in the zodiacal cloud. These new observations will give quantitative information both on the rate of transport of asteroidal dust to the Earth and on the rate of grinding down of the asteroid belt.

We have also calculated that if the zodiacal cloud consisted of $12\mu m$ diameter asteroidal particles alone, with number densities sufficient to account for the observed peak brightness in the IRAS $25\mu m$ waveband, then the trailing/leading asymmetry would be three times higher than that observed. There are two reasons why this must be an overestimate. Small asteroidal particles that exist in the background cloud that cannot be trapped in the resonances because of their high drag rates will act to reduce the trailing/leading asymmetry, as will any cometary particles with comparatively high orbital eccentricities that contribute to the background flux. These calculations and the observed trailing/leading flux ratios imply that, in terms of effective area, the fraction of asteroidal particles that can contribute to the ring is about one third of the total asteroidal particles in the cloud, but more sophisticated calculations are needed here, and more work needs to be done on the distinction of the orbits of cometary and asteroidal particles in the vicinity of the Earth.

5. Source of IDPs collected at the Earth

Without resonant trapping most dust particles spiral past the Earth without striking the planet. However, most particles trapped in resonances are released from these resonances due to close encounters with the Earth. Thus, the ring may act as a funnel through which asteroidal particles are deposited in the Earth's atmosphere. This statement needs to be verified, but could have important implications for the efficiency of the transport of carbonaceous material from the asteroid belt to the Earth.

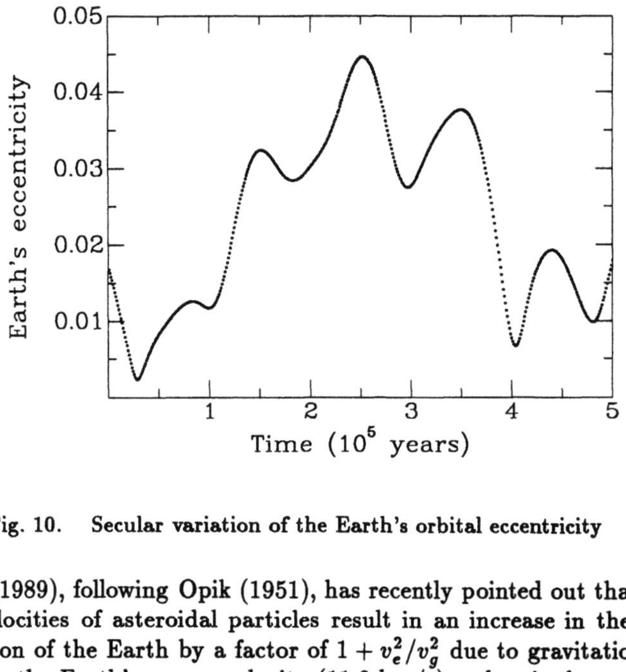

Fig. 10. Secular variation of the Earth's orbital eccentricity

Flynn (1989), following Opik (1951), has recently pointed out that the low geocentric velocities of asteroidal particles result in an increase in the gravitational cross-section of the Earth by a factor of $1 + v_e^2/v_g^2$ due to gravitational focussing, where v_e is the Earth's escape velocity (11.2 km/s) and v_g is the geocentric velocity of the particle. Because asteroidal particles probably constitute one third of the zodiacal cloud and are transported to the inner solar system by drag forces, gravitational focussing by the Earth that results in the preferential capture of particles from orbits with low inclinations and low eccentricities and the possible "funneling" effect of the ring itself, imply that nearly all of the unmelted IDPs collected at the Earth are probably asteroidal. If this is the case, then there will be marked long-term variations in the magnitude of the Earth's resonant ring and in the deposition rate of asteroidal particles on the Earth due to the secular variation of the Earth's orbital eccentricity (fig. 10) and the corresponding variations in the forced eccentricities of the particle orbits. These periodicities should be searched for in

Antarctic ice cores. There will also be annual variations in the deposition rate of asteroidal particles on the Earth due to the Earth's orbital eccentricity that determines the distance of the Earth from the edge of ring cavity (right panel of fig. 8) and also due to the effect of forced inclinations on orbits of low proper inclination (Dermott and Liou, 1994). Strong indications of annual variations exist in the radio observations of meteor echoes (Suzuki, 1991).

Acknowledgements

This work was supported by NASA grants NAGW 1923 and NAG 9–440, by the University of Florida and the IBM Corporation through their Research Computing Initiative at the Northeast Regional Data Center, and by the Geophysics Directorate of the United States Air Force Phillips Laboratory through subcontracts with Mission Research Corporation and General Research Corporation under the Celestial Background Scene Descriptor program.

References

Dermott, S.F., Nicholson, P.D., Burns, J.A., and Houck, J.R. : 1984, "Origin of solar system dustbands discovered by IRAS". *Nature*, **312**, 505–509.

Dermott, S.F., Nicholson, P.D., Burns, J.A., and Houck, J.R. : 1985, "An analysis of the Solar System Dust Bands". In *Properties and Interaction of Interplanetary Dust* (R.H. Giese, P. Lamy, Eds.), 395–409, Reidel, Dordrecht.

Dermott, S.F., Nicholson, P.D., and Wolven, B. : 1986, "Preliminary Analysis of the IRAS Solar System Dust Bands". In *Asteroids Comets and Meteors II* (C-I. Lagerkvist B.A. Lindblad, H. Lundstedt, H. Rickman, Eds.), 583-594, Uppsala Universitet Reprocentralen.

Dermott, S.F., Nicholson, P.D., Kim, Y., Wolven, B., and Tedesco, E.F. : 1988, "The impact of IRAS on Asteroidal Science". In *Comets to Cosmology* (A. Lawrence, Ed.), 3–18, Springer-Verlag, Berlin.

Dermott, S.F., and Nicholson, P.D. : 1989, "IRAS Dustbands and the origin of the Zodiacal Cloud". *Highlights of Astronomy*, **8**, 259–266.

Dermott, S.F., Nicholson, P.D., Gomes, R.S. and Malhotra, R. : 1990 , "Modelling the IRAS solar system dustbands". *Adv. Space Res.*, **10**, 165–172.

Dermott, S. F., et al : 1992 : "Dynamics of the Zodiacal Cloud". In *Chaos, Resonance, and Collective Dynamical Phenomena in the Solar System* (S. Ferraz-Mello,Ed.), 333–347, Kluwer, Dordrecht.

Dermott, S. F., et al. : 1993, "The Origin of IRAS Dustbands". In *Meteroids and Their Parent Bodies* (J. Stohl, I.P. Williams Eds.), 357–366, Astronomical Inst., Slovak. Acad. Sci., Bratislava.

Dermott, S.F., Jayaraman, S., Xu, Y.L. and Liou, J.C. : 1994, "IRAS observations of a ring around the sun : Asteroidal particles in resonant lock with the Earth". *Nature*, (submitted).

Deul, E.R., and Walker, H.J. : 1989, "Responsivity variations in the IRAS survey". *Astron. Astrophys. Suppl. Ser.*, **81**, 207–214.

Divine, N. : 1993, "Five Populations of Interplanetary Meteoroids". *J. Geophys. Res.*, **98**, 17029–17048.

Dohnayi, J.S. : 1969, "Collisional model of asteroids and their debris". *J. Geophys. Res.*, **74**, 2531–2554.

Durda, D.D. : 1993, Ph.D. Thesis, University of Florida.

Flynn, G.J. : 1989, "Atmospheric entry heating : A criterion to distinguish between asteroidal and planetary sources of dust". *Icarus*, **77**, 287–310.
Greenberg, J. M. : 1991, "The Interplanetary medium is thriving". In *Origin and Evolution of Interplanetary Dust* (A. C. Levasseur-Regourd, H. Hasegawa Eds.), 443–451, Kluwer, Dordrecht.
Grün, E., Zook, H. A., Fechtig, H., and Giese, R.H. : 1985, "Collisional balance of the meteoritic complex". *Icarus*, **62**, 244–272.
Grün, E., Fechtig, H. and Kissel, J. : 1985, "Orbits of Interplanteray dust particles insider 1AU as observed by Helios". In *Properties and Interactions of Interplanetary dust* (R. Giese, P. Lamy, Eds.), 101–111, Reidel, Dordrecht.
Grün, E., et al : 1992, "Galileo and Ulysses dust measurements : From Venus to Jupiter". *Geophys. Res. Lett.*, **19**, 1311–1314.
Hauser, M. G. : 1988, "Models for infrared emission from Zodiacal Dust". In *Comets to Cosmology* (A. Lawrence Ed.), 27–39, Springer-Verlag, Berlin.
Jackson, A.A. and Zook, H.A. : 1989, "A Solar System dust ring with the Earth as its shepherd". *Nature*, **337**, 629–631.
Jackson, A.A. and Zook, H.A. : 1992, "Orbital Evolution of Dust particles from Comets and Asteroids". *Icarus*, **97**, 70–84.
Leinert, C., Hanner, M., Ritcher and I., Pitz, E. : 1980, "The plane of symmetry of the interplanetary dust in the inner solar system". *Astron. Astrophys.*, **82**, 328–336.
Leinert, C. and Grün, E. : 1990, "Interplanetary Dust". In *Physics of the Inner Heliosphere I* (R. Schwenn, E. Marsch, Eds.) 207–275, Springer-Verlag, Heidelberg.
Levasseur-Regourd, A.C., Renard, J.B. and Dumont, R. : 1991, "The Zodiacal cloud complex". In *Origin and Evolution of Interplanetary Dust* (A.C. Levasseur-Regourd, H. Hasegawa, Eds.), 131–138, Kluwer, Dordrecht.
Liou, J.C. : 1993, Ph.D. Thesis, University of Florida.
Low. F.J., et al. : 1984, "Infrared cirrcus : New components of the extended infrared emission". *Astrophys. J. Lett.* **278**, L19–L22.
Opik, E.J. : 1951, "Collision probabilities with the planets and the distribution of interplanetary dust". In *Proc. Roy. Irish Acad.*, **54**, Sect A, 165–199.
Reach, W.T. : 1991, "Zodiacal Emission II. Dust near ecliptic". *Astrophys. J.*, **369**, 529–543.
Reach, W.T. : 1992, "Zodiacal Emission III. Dust near Asteroid Belt". *Astrophys. J.*, **392**, 289–299.
Röser, S. and Staude, H.J. : 1978, "The zodiacal light from 1500 Åto 60 micron". *Astron. Astrophys.*, **67**, 381–394.
Suzuki, K. : 1991, "The annual variation of radio meteor echoes observed from 1981 to 1985". In *Origin and Evolution of Interplanetary Dust* (A.C. Levasseur-Regourd, H. Hasegawa, Eds.), 319–322, Kluwer, Dordrecht.
Sykes, M.V., Greenberg, R., Dermott, S.F., Nicholson, P.D., Burns, J.A. and Gautier, T.N. : 1989, "Dust Bands in the Asteroid Belt". In *Asteroids II* (R. Binzel, T. Gehrels, M.S. Matthews, Eds), 336–367, Univ. of Arizona Press., Tucson.
Sykes, M.V. and Walker, R.G : 1992, "Cometary Dust Trails". *Icarus*, **95**, 180–200.
Zappalà, V., Cellino, A. Farinella, P. and Knežević, Z. : 1990, "Asteroid Families I. Identification by hierarchial clustering and reliability assessment". *Astron. J.*, **100**, 2030–2046.
Zappalà, V., Cellino, A. Farinella, P. and Milani, A. : 1994, "Asteroid Families II. Extended emission to unnumbered multiopposition asteroids". *Astron. J.*, **107**, 772–801.

ASTEROID PROPER ELEMENTS : THE BIG PICTURE

ZORAN KNEŽEVIĆ
Astronomska opservatorija, Volgina 7, 11050 Beograd, Yugoslavia

and

ANDREA MILANI
Dipartimento di Matematica, Via Buonarroti 2, 56127 Pisa, Italy
E–mail milani@dm.unipi.it

Abstract. Four perturbation theories presently used to compute asteroid proper elements are reviewed, and their results are briefly discussed (Milani and Knežević, 1990, 1992, 1994, for low to moderate eccentricity/inclination main belt objects; Lemaitre and Morbidelli, 1994, for high e, I objects; Milani, 1993, for Trojans; Schubart, 1982, 1991 for Hildas). The most important recent improvements are described, in particular those pertaining to the upgrades of the previous analytic and semianalytic solutions. The dynamical structure of the asteroid main belt, as defined by the low order mean motion resonances and by linear and nonlinear secular resonances, is considered from the point of view of the effects of these resonances on the accuracy and/or reliability of the computation of proper elements and on the reliability of the identification of asteroid families.

1. What are the proper elements and what they are used for?

Asteroid proper elements are quasi-integrals of motion, stable over very long intervals of time; they represent a sort of "average" characteristics of motion in the sense that they result from a procedure of elimination of short– and long–periodic perturbations. The proper elements are used for two main purposes : (i) they serve as parameters for classification of asteroids into families; (ii) they are employed for studies of the dynamical structure of the asteroid belt.

Asteroid families are groupings of asteroids in the phase space of orbital elements. They were formed by catastrophic breakup of parent bodies, at unknown times in the remote past, and under circumstances (such as geometry of collision, physical properties of the parents, etc.) that are not yet fully understood. The parent body at the breakup might have had osculating elements significantly different from the corresponding proper values; since we do not know the epoch of the collision, the proper elements (due to their constancy in time) become the only tool to reliably recognize families, in particular the small ones consisting of a few members. Until recently, the very identification of the families and their memberships were not very reliable : the proper elements used for this identification were not accurate enough, and/or the classification methods suffered from subjectivity, biases, uncertainties due to insufficient data, etc. The situation is now greatly improved in both respects; the review of the current state of the art regarding proper elements and family classification is the subject of this and some other papers in this book (see e.g. Zappalà and Cellino, Milani, Burbine and Binzel).

The differences of proper orbital elements between family members are related to their relative velocities, averaged over the unknown quantity, the age of the family.

A. Milani et al. (eds.), Asteroids, Comets, Meteors 1993, 143–158.

In order to reliably recognize the families, the changes in the proper elements – over time spans of the order of the family ages– must not exceed an amount given by the typical relative velocities of the family members. The relative velocities of family members are known to be of the order of 100 m/s; for the so called standard metric (Zappalà et al., 1990) this corresponds to $6-7\times10^{-3}$ in terms of the proper elements difference. Hence, proper elements stable –over very long time spans– to about 0.001 in proper eccentricity and sine of inclination and to 0.0005 in proper semimajor axis can be considered as good enough, those stable to 0.003 in $e, \sin I$ should be used with some downweighing, while those with instabilities ≥ 0.01 allow to identify asteroid families, but only in the less densely populated regions of the asteroid belt.

The dynamical structure of the asteroid belt is determined by the mean motion and secular resonances. The resonances split the portion of the phase space corresponding to the asteroid belt in several dynamically distinct zones; they affect the motion of nearby asteroids, give rise to gaps in the distributions of elements, etc. The location in the phase space of the resonances can be determined by computing simple combinations of the so–called fundamental frequencies (the rates of the proper angles), thus the knowledge of the dynamical structure of the belt is essentially equivalent to the ability of listing all the relevant resonances that bound or cut across the asteroid belt region. To our present knowledge, the resonances relevant for the dynamical structure of the main asteroid belt at low to moderate inclinations are the following : (1) the low degree mean motion resonances, like 2/1, 3/1, 4/1, 5/2, 7/3, and others giving rise to Kirkwood gaps; (2) the linear secular resonances $g-g_5$, $g-g_6$, and $s-s_6$ (g, s, g_j, s_j are the frequencies of the proper longitudes of perihelion and node of the asteroid and of the perturbing planets; some authors use another notation, e.g. ν_5, ν_6, ν_{16} for the three linear resonances); (3) nine nonlinear secular resonances that cut across the core asteroid belt ($2.2-3.2\,AU$ at low–to–moderate inclinations) and affect the motion of a significant number of asteroids : $g+s-g_5-s_6$, $z_1=g+s-g_6-s_6$, $g+s-g_5-s_7$, $g-2g_6+g_5$, $g-2g_6+g_7$, $g-3g_6+2g_5$, $s-s_6-g_5+g_6$, $z_2=2g+s-2g_6-s_6$, $z_3=3g+s-3g_6-s_6$. The latter are called nonlinear because the corresponding small divisors represent combinations of more frequencies and appear with terms of higher degree in eccentricity and/or inclination in the equations of motion.

2. Theories and methods for computation of proper elements

Hirayama (1918) discovered asteroid families by looking into the distributions of osculating orbital elements, but he used from the very beginning the concept of proper elements; in his later works (Hirayama, 1923; 1928) he used explicit computations of proper elements by means of a linear theory. Brouwer (1951) used an enhanced Lagrange-type analytical solution of the linear secular perturbation equations and the major planet theory by Brouwer and Van Woerkom (1950) to produce proper elements for all asteroids numbered at that time; his procedure and results were rather advanced with respect to the previous ones and therefore they were used for quite some time (c.f. Anders, 1965; Arnold, 1969; Van Houten et al., 1970; Lindblad and Southworth, 1971). The real breakthrough occurred, however, when Williams (1969) developed a new semianalytic theory, which used no expansion of the perturbing function in eccentricity and inclination of the asteroid, and thus provided results of about the same accuracy regardless of e, I. Proper elements computed by means of this theory have been published by Williams (1979) and by Williams and Hierath (1987), and are sometimes still in use (they can also be found in Williams, 1989). The positions of the surfaces of the linear secular resonances, which are the most important because they bound the main asteroid belt and open large gaps in the distribution of asteroids, were determined using this theory by Williams and Faulkner (1981). The theory of Williams represents a cornerstone in the development of the theories of asteroid motion, and the beginning of the contemporary era in this respect. Afterwards new theories appeared, and there are at present four theories and/or methods in use for computation of asteroid proper elements. These are the purely analytical theory by Milani and Knežević (1990, 1992, 1994), which provides the most accurate results for low to moderate eccentricity/inclination main belt objects, the semianalytic theory by Lemaitre and Morbidelli (1994) which should be used for high e, I objects, a numerical method by Milani (1993) for the Trojans, and a similar one by Schubart (1982, 1991) for the Hildas. All these four catalogues of proper elements are in the public domain, and can be obtained by `anonymous ftp`, as described in Milani et al., this volume. A method to compute proper elements in another case, for asteroids close to some secular resonances, has been developed recently by Morbidelli (1993); the underlying theoretical concept and the results are, however, described in the paper by Froeschlé and Morbidelli in this volume.

2. 1. THEORY BY MILANI AND KNEŽEVIĆ

The theory by Milani and Knežević was developed from the analytical theory of secular perturbations of asteroids by Yuasa (1973), which takes into account terms in the development of the perturbing Hamiltonian up to degree four in eccentricity and inclination, in the first order with respect to perturbing mass, and those of the second degree, in the second order. This theory used the canonical formalism (Hori, 1966) to remove the short– and the long–periodic terms from the Hamiltonian, by means of two successive canonical transformations. Many improvements have been introduced by Milani and Knežević with respect to the original theory. The derivation of the second order terms has been revised, the missing indirect part of the Hamiltonian has been added, the long periodic Hamiltonian has been rearranged to include the second order effects in the linear theory, an iterative procedure has been used to solve for proper elements and frequencies, and last but not least, some errors have been removed from the original equations.

Since it is based on a truncated development of the perturbing Hamiltonian, this theory is particularly suitable for low–to–moderate inclination/eccentricity objects (almost 90% of the main belt asteroids belong to this "core belt"), for which in most cases it provides proper elements with high accuracy. On the other hand, in the high inclination/eccentricity regions and near the main resonances it fails to provide reliable results; the extension of the developments to even higher orders and/or degrees is very difficult and can be achieved by means of the same technique only to some extent (see later, items (b) and (c)).

As claimed by Milani and Knežević (1994), the results achieved by means of the analytical theory are more than good enough for family identification purposes (in the sense described above) for more than 90% of the asteroids belonging to the core belt. This is supported by an extensive set of stability tests, including 35 objects representative of the various regions of the belt and of the families found by Zappalá et al. (1994); the tests consisted of a numerical integration of all the selected objects for 5 Myr (within a dynamical model including the four outer major planets), deriving the time series of proper elements and monitoring their variations over that time span. The results obtained by these tests are summarized in Figure 1, where the square root of the sum of the squares of the RMS of the proper $e, \sin I$ changes over a time span of $5\,Myr$ is plotted versus the proper semimajor axis.

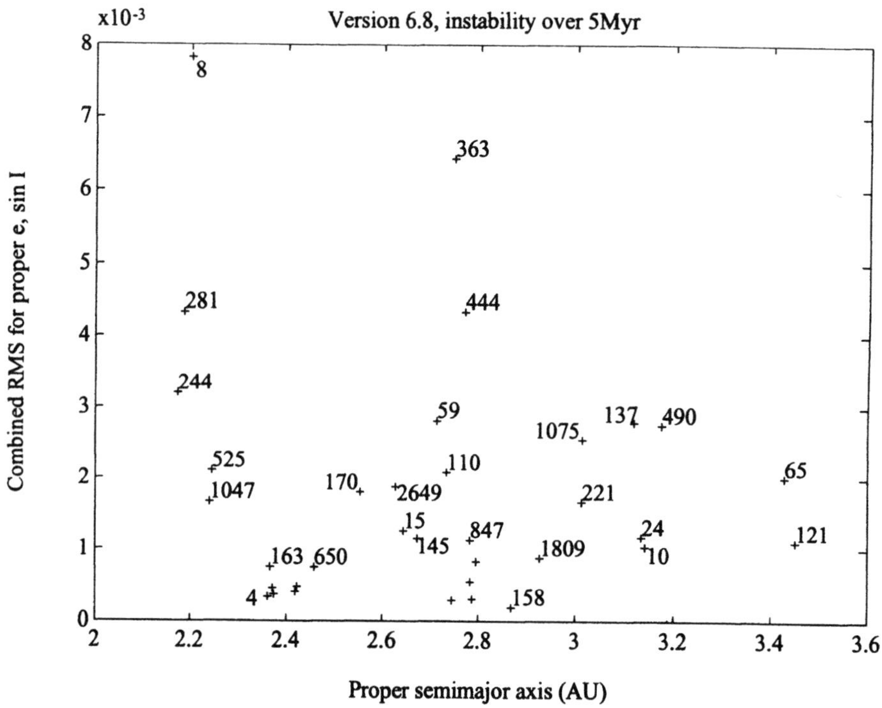

Fig. 1. Summary of the proper elements accuracy tests (from Milani and Knežević, 1994).

There are really but a few cases of seriously degraded accuracy, and they are always due to some resonance (either the 2/1 mean motion resonance or some secular one); it is not even a matter of computing accurate proper elements in these cases, the analytical procedure being divergent, but it is possible to provide a warning and to switch to an adapted theory less sensitive to the resonant effect.

As an illustration of the dynamical structure of the belt, all the most important resonances mentioned in Section 1, affecting the asteroid motion and the determination of proper elements, are plotted (for the average value 0.14 of the proper eccentricity) in a proper inclination vs. proper semimajor axis plane in Figure 2. Contour lines are given for values of the associated divisors of $\pm 0.5\, arcsec/yr$, except for the major $g - g_6$ resonance for which the lines correspond to values of $-2, 0, +2\, arcsec/yr$.

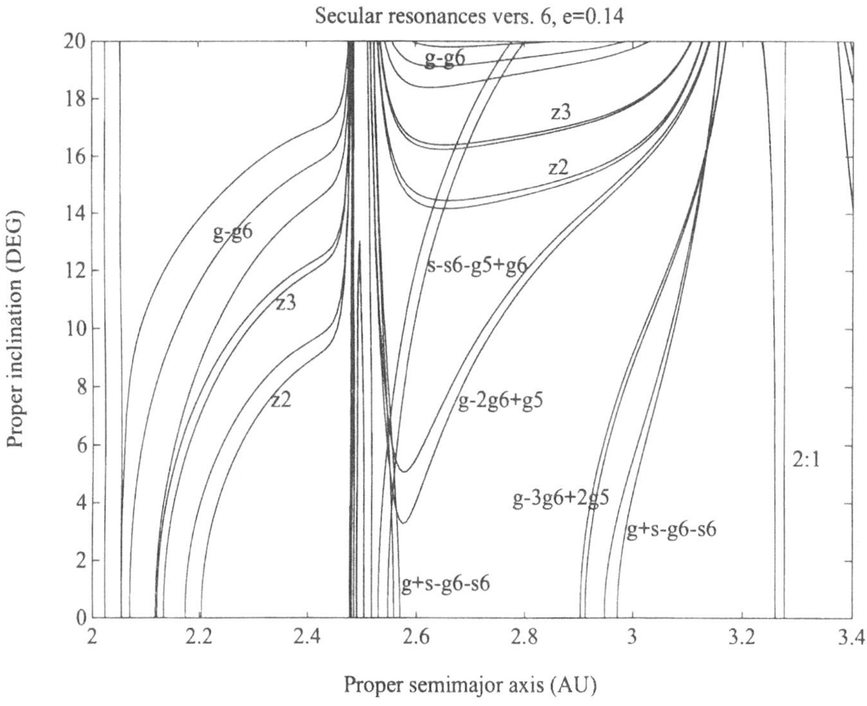

Fig. 2. Positions of secular resonances in the core belt as determined by means of the theory of Milani and Knežević, vers. 6.8.

Among of the nonlinear secular resonances, $g - 2g_6 + g_5$, z_1, and z_2 are the most important both for the effects they cause and for the number of asteroids they affect (since they cut through the most densely populated regions). Many asteroid families are also crossed by the resonances, which in case of a large family (e.g. Eos, see Milani and Knežević, 1992) does not affect the overall recognition of the family, but can affect the list of members; in the case of a small family (e.g. the spurious Lydia family, see Zappalà et al., 1994) their influence can prevent even a reliable identification.

With respect to the version 5.7 of the theory reported at ACM91, the present version 6.8 contains several improvements; since they are described in detail elsewhere (Milani and Knežević, 1994), here we shall only briefly summarize the most important ones.

For the elimination of the short-periodic perturbations and the computation of the asteroid mean elements, the main improvements are as follows. (i) For the removal of the short-periodic effects of Jupiter and Saturn, the previous version of the M&K theory used a development of the perturbing function up to degree four in the first order with respect to the perturbing mass for the direct perturbations, and up to degree three for the indirect perturbations. For reasons of formal completeness of the theory, the terms of degree four were added to the indirect part as well; as shown by Knežević (1992b), this changes the results only by a small amount. (ii) For the perturbation by the inner planets, about 90% of the effects were taken into account by referring the osculating elements to the barycentre of the Sun and the terrestrial planets; this is important because the indirect perturbations of the inner planets can be amplified by small divisors and therefore change the fundamental secular frequencies by a significant amount. (iii) A quality code for mean elements was introduced, to notify the user of the troublesome cases for which the procedure fails to provide accurate mean elements. It allows to automatically detect (and remove from the catalogue) the asteroids locked in the main mean motion resonances. (iv) Technical improvements implemented in the software include the application of a "generic term" technique (Knežević, 1992a), the automatic handling of the reference system changes and of the variable epochs for the input data, the computation of the mean mean anomaly (intended, however, for other purposes, such as asteroid identification), etc.

The improvements implemented in the procedure of removal of the long-periodic perturbations include : the introduction of (a) nonlinear forced terms, (b) second order terms of degree four with squares of the small divisor of the 3/1 resonance, and (c) degree six/eight terms with the z_2 divisor; also (d) an automated switching to an adapted theory has been introduced.

(a) The secular perturbation theory of the major planets contains many terms giving rise to significant forced effects in the motion of asteroids. In the previous versions of the theory of Milani and Knežević (1990, 1992), only the forced terms taken from a linear secular perturbation theory for the planets had been accounted for. Therefore, comparatively large instabilities of the proper elements were found for many asteroids in the vicinity of the resonant surfaces associated with the omitted nonlinear terms; as an example, the results for 1272 *Gefion*, contained an oscillation with frequency $g - 2g_6 + g_5$ and amplitude 4.8×10^{-3} in proper e. In order to remove this source of instability the eight largest forced nonlinear terms from the synthetic theory of major planets by Nobili et al. (1989) have been included in the computation; the result of this simple operation was a significant improvement of the accuracy of proper elements for many test cases previously affected by this problem (the stability for 1272 *Gefion* improved by almost an order of magnitude). However, the theories of the major planets derived from numerical integration include some periodic terms which have not been recognized as integer combinations of fundamental frequencies ("chaotic terms"), and these have not been included, although they have only slightly smaller amplitudes, because their effects cannot be described by an analytical theory.

(b) Although the second order, degree two terms had been included in the previous versions of the theory, the results for some asteroids near the 3/1 mean motion resonance were degraded; this is due to the fact that the most important contribution to the second order effects might come from the degree four part. Among the terms of degree four, in turn, only the terms containing the square of the small divisor are important close to resonance, and these have been added in the present version of the theory. This addition proved to be effective in removing the instabilities from the time series of proper elements for many tested asteroids, the most significant improvement (by a factor of $\simeq 5$) being achieved in the case of 650 *Amalasuntha*. For the 2/1 resonance, on the other hand, the results of adding the same kind of terms were ambiguous, namely, some improvement in some test cases and a slight degradation in others. However, the proper elements for these cases show the features of the chaotic behaviour (intermittency and random walk, positive Lyapunov exponents), and this seems to imply that they are accurate already to the "chaotic limit"; from this we should conclude that no improvement is possible by means of a deterministic theory. Nevertheless the proper elements are, in most of these chaotic cases, stable within a very narrow range; this "stable chaos" phenomenon is far from being fully understood (Milani and Nobili, 1992; Milani, these proceedings; Milani, 1994).

(c) Another only partially successful improvement was the inclusion of the degree six and eight terms with the $z_2 = 2g + s - 2g_6 - s_6$ small divisor. This divisor is the most important one in a cascade of secular resonances that cluster inside the main $g - g_6$ one, and cut across the densely populated Flora region of the core belt. The result of adding this very complex term was rather modest; for the two test cases where the z_2 term was the main source of instability —1047 *Geisha* and 525 *Adelaide*— only about 50% of the corresponding oscillation was removed. Not surprisingly, the Flora region is the only one where family identifications and memberships still have a comparatively low reliability (see Zappalà and Cellino, these proceedings). Further study will be needed to solve this problem.

(d) Since the M&K theory is an iterative one, divergence sometimes occurs, indicating that a long periodic effect contains a divisor close to zero. In a sample of 12,573 asteroids, this occurred in 1,045 cases. The solution applied in such cases is to flag them by means of a special "resonance code", and to try to obtain a convergent result anyway — by switching to an adapted theory from which the nearly resonant term has been removed. Thus, the computation of proper elements becomes more reliable, but somewhat less accurate. In order to identify automatically the resonant term which is the cause of the divergence and therefore has to be eliminated from the theory, a very complicated procedure has been introduced, which now involves an improved iteration scheme (more accurate initial values and tighter convergence control) and checks the size of the terms in three successive iterations, comparing their differences and amplitudes with some predefined threshold values. Because of this complexity the procedure of automatic selection of an appropriate adapted theory cannot be absolutely reliable; sometimes the algorithm fails to identify the responsible resonance, sometimes the procedure diverges even after removal of the supposed "main culprit", etc.; in fact, Milani and Knežević

(1994) estimated that the selection of the suitable adapted theory has a 90% reliability, thus the percentage of cases which have not been properly flagged is less than 1% of the total catalogue of proper elements.

Finally, let us just mention that among the technical improvements to the long periodic software, the most important one is the implementation of the generic term technique (Knežević, 1994), by which a previous 1200 statements computer routine has been replaced by one with 50 statements plus a 1000 lines file storing the information on 300 terms of the literal development of the perturbing Hamiltonian. All the software is placed in the public domain and can be accessed through anonymous `ftp`, at the same server which is used to obtain the proper elements.

It is interesting to note that Milani and Knežević believe that at the present level of development of their analytical theory, they are already beginning to explore the so called "Arnold web" of narrow high degree resonances having small strengths, which sometimes can nevertheless affect the stability of proper elements to some extent. Fundamentally new ideas will be needed if the instability tolerances are to be tightened to values below the present 0.001 level; even more substantial new developments will be needed, if such stability requirements should be extended from the $\simeq 10^7$ *yr* time spans covered by the present stability tests to the lifetime of the solar system.

2. 2. THEORY BY LEMAITRE AND MORBIDELLI

The fundamental difference of the semianalytic theories, such as the one by Lemaitre and Morbidelli (1994), with respect to the analytic theories, such as the one by Milani and Knežević, is due to an observation made already by Kozai (1962). The dynamics at high inclinations and high eccentricities is dominated by the $e^2 i^2 \cos 2\omega$ perturbation, which in such cases must be included in the integrable part of the Hamiltonian in order to obtain meaningful results. Kozai accomplished this by simplifying the dynamical model : by assuming that the perturbing planets are on circular, planar orbits, the secular perturbation Hamiltonian (after averaging out the short periodic perturbations) contains only one angular variable, namely the argument of perihelion ω. Thus this first approximation secular Hamiltonian is integrable, and can be used as a starting point for a perturbative theory. Since this integrable approximation is geometrically very different from the linear theory, there is a big difference in the very definition of the proper elements, at least for the high eccentricities and/or inclinations for which this class of theories should be used.

The first perturbative treatment using the Kozai approximation as a starting point was developed by Williams (1969); in his theory, the portions of the long periodic perturbing function containing e', I' to the first power are accounted for (the primed letters indicate the mean elements of the perturbing planets). The perturbations are computed by numerical averaging of the perturbing function and its derivatives, with respect to the fast variables (the mean anomalies/longitudes of both the asteroid and the perturbing planets). The theory by Williams was very effective, and for many asteroids achieved a level of accuracy comparable to that of the more recent analytical theories; for this reason, Williams' proper elements have been used for a long time. However, there are two main limitations in Williams' theory. First, the theory was developed by means of the non–canonical formalism of classical perturbation theory. To first order in the perturbing masses, the results do not depend upon the formalism; however, it was already difficult to reproduce the results by Williams on the basis of his papers; to go beyond, to a more comprehensive theory, without a more advanced formalism, was impossible. Second, it is by no means true that the effects of higher order in the perturbing mass are always small. For asteroids near (but by no means inside) the main mean motion resonances (such as 2/1, 3/1), a complicated nonlinear effect of the removal of the short periodic terms results in a secular perturbation containing the square of the perturbing masses, but also the inverse square of the small divisor of the near resonance. A theory accounting for this effect can differ from a purely first order theory by as much as 50% in the fundamental frequencies, as in the Themis region. Thus the kind of accuracy needed to identify asteroid families could not be reached in a significant portion of the main belt, unless the theory was extended to the second order in the perturbing masses. Because of the technical limitations of the classical formalism, this was not possible with Williams theory. As a result, the first theories to achieve an accurate treatment of the main second order effects were the analytical theories, first by Yuasa and then by Milani and Knežević, as discussed in Section 2.1. However, this leaved out the region of the main belt with high eccentricity and/or inclination, were the analytical theories cannot work in a satisfactory way.

The first limitation of the semianalytic theories was removed by Henrard (1990), who developed a new semianalytic perturbation scheme within the canonical formalism of a transformation theory (that is, the solution is directly expanded as a series, without going through the expansion of the differential equations in a series). The Henrard method essentially consists of the use of the Arnold–Jost theorem to define action–angle canonical variables for the integrable problem; these variables are theoretically defined by line integrals, and they can be practically computed by numerical quadrature. Then the entire perturbation scheme to be used to compute the solution of a non–integrable problem can be expressed by means of numerical quadratures of the same kind. When this procedure is applied to the Kozai Hamiltonian and to the perturbations of degree one in e', I', Henrard's canonical method reproduces Williams' proper elements and the theory of the location of the secular resonances by Williams and Faulkner (1981); this was shown by Morbidelli and Henrard (1991a, 1991b). Moreover, the formalism is powerful enough to allow the computation of the second order effects, again by numerical quadrature of a more complicated expression. In this way Lemaitre and Morbidelli (1994) have succeeded in computing proper elements with a theory including the second order effects and without truncation in e, I. Clearly, this theory is particularly suitable for objects with high inclination and/or eccentricity, where the analytical theories fail to provide meaningful results. The Lemaitre and Morbidelli method also uses a fixed point iteration to find the proper frequencies and the proper elements from the mean ones (to understand why an iterative algorithm is always needed, see Milani and Knežević, 1990; Milani, 1994). The iteration is divergent near the secular resonance surfaces; however, this method can handle the cases in which the main argument of the Kozai theory, ω, is in libration (divergence can occur near the separatrix), as it is the case for some very high inclination orbits.

Although this theory is still under development, there are already some interesting results that have been reported recently (Lemaitre and Morbidelli, 1994). In the first place, there are some encouraging results on the accuracy achieved with proper elements for a Pallas-like asteroid, which can be considered as a real challenge, being close to the separatrix between circulation and libration of ω. Even by using "version 1" of the theory (with Jupiter as the only perturber, and with only three secular frequencies g_5, g_6, s_6), the changes in the proper eccentricity and sine of inclination both stay within ± 0.015 (within $\pm 1.5\, arcsec/yr$ for the frequencies) over the time span of $2.5\, Myr$ of a numerical integration test. This stability level is good enough for the purpose of family identification in a region with such a sparse background. Indeed Lemaitre and Morbidelli have been able to identify a family around the asteroid *2 Pallas* (see Fig. 4 in Lemaitre and Morbidelli, 1994); such a family had already been proposed by Williams (1989), but with a lower number of members. The accuracy for the less dynamically complex cases can be much better. As an example, for the asteroid *185 Eunike*, located above the $g - g_6$ resonance, the rms instabilities of the proper eccentricity and proper sine of inclination, as inferred from a $4\, Myr$ integration, amount to 0.006 and 0.002, respectively.

TABLE I

Proper elements derived by means of the theory by Lemaitre and Morbidelli, and their differences with respect to those obtained from the theory by Milani and Knežević.

No.	a	e	Δ_e	$\sin I$	$\Delta \sin I$
4	2.36151	0.0882	-0.0105	0.1137	0.0024
15	2.64366	0.1480	-0.0005	0.2261	0.0002
44	2.42276	0.1743	0.0000	0.0555	0.0025
142	2.41867	0.1589	-0.0006	0.0559	0.0002
145	2.67273	0.1685	-0.0006	0.2036	0.0014
158	2.86881	0.0372	-0.0085	0.0372	-0.0003
163	2.36712	0.2078	-0.0024	0.0824	0.0005
170	2.55374	0.0963	-0.0049	0.2600	0.0004
808	2.74515	0.1267	-0.0066	0.0836	-0.0012
847	2.78276	0.0680	0.0012	0.0620	-0.0010
1272	2.78380	0.1298	0.0007	0.1559	-0.0006
1378	2.37482	0.1580	-0.0052	0.0513	-0.0013
1726	2.78741	0.0320	-0.0149	0.0742	-0.0021
1809	2.92742	0.0751	-0.0007	0.0381	0.0015
1932	2.37179	0.1966	-0.0035	0.0363	-0.0032

The theory by Lemaitre and Morbidelli is still evolving quickly. Recently the proper elements "version 2" have become available : they account for the perturbations by Jupiter and Saturn with the frequencies $g_5, g_6, g_7, s_6, 2g_6 - g_5$; some tests have already been performed with s_7, s_8 as well. Two comparisons were made with

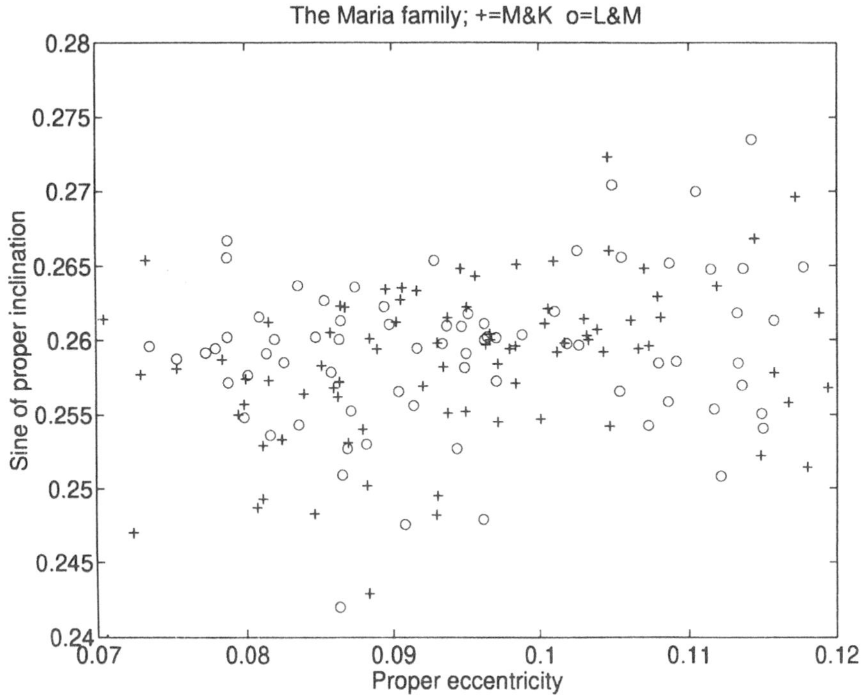

Fig. 3. The Maria family; crosses are data derived by means of the analytical theory by Milani and Knežević, and open circles are obtained from the semianalytical theory by Lemaitre and Morbidelli.

the results by Milani and Knežević, in order to have an idea on the overall reliability of the semianalytic data, and to find out the values of the inclination corresponding to the transition region where the users of the proper elements catalogues are supposed to switch from one theory to another. The data on the latter comparison will be reported elsewhere (Knežević et al., 1994); we shall only state here that the transition region seems to be somewhere between 15° and 17° of inclination, somewhat below the $g - g_6$ resonance. Another comparison is summarized in Table I. Proper elements derived by means of the current version of the theory by Lemaitre and Morbidelli are given together with the differences of these proper values with respect to those by Milani and Knežević, for a number of asteroids in the low-to-moderate inclination zone of the core belt. For most of the objects included in this comparison the agreement is good, but there are still some unexpectedly large discrepancies (4 *Vesta*). Note that in some of these cases the frequencies g, s, as computed by Lemaitre and Morbidelli, are closer to the numerical fits than the corresponding values by Milani and Knežević. Obviously the Lemaitre and Morbidelli proper elements should not be used for low eccentricity and inclination cases, because they are less stable and their computation is less efficient than it is the case for analytic proper elements, but it is important to be able to confirm that, in most cases, they give consistent results. Some of the discrepancies still have to be explained; in particular the stability of the numerical quadrature methods used by the semianalytic methods requires further study to assess the reliability of the computation of proper elements. However, for high inclination and/or eccentricity asteroids the proper elements by Lemaitre and Morbidelli are already at a level of stability and reliability which is useful for the identification of asteroid families, and without doubt the next run of asteroid family classification will use a "composite catalogue" (in the sense of Milani et al., this volume) of proper elements computed by different authors.

An independent test of the significance of the discrepancies in proper elements coming from the two theories is given in Figure 3, which shows the comparatively high inclination region which includes the Maria family, in the proper inclination vs. proper semimajor axis plane, with data from two theories superimposed on the same plot. This is the region of transition between the two theories, in the sense explained above, and proper elements from both theories are supposed to be of about the same accuracy; indeed the agreement of the proper elements and frequencies can be rated as reasonably good.

Two obvious conclusions can be drawn from this figure : the overall shape and identification of the family do not critically depend on the choice of the proper elements theory, while the membership of a particular asteroid (especially at the edge of the family) can be affected. The best idea for this transition region, at least for the moment, might be to use both sets of elements in the classification procedure and to compare the results.

2. 3. METHOD BY MILANI FOR TROJANS

The computation of proper elements for Trojans by means of a synthetic theory, that is, by post-processing the output of a numerical integration, has been first done by Bien and Schubart (1987); they computed the orbits of 41 Trojans for $\simeq 150,000\,yr$, and derived "orbital parameters" for 40 of them. The same technique was improved upon, and applied to a larger set of Trojan orbits, by Milani (1993, and also this volume). The orbits of 184 Trojans have been computed for $1\,Myr$ (for 20 of them the computation has been extended to $5\,Myr$), in a model including the full perturbations by the four giant planets, and with a correction to account for most of the indirect effect of the inner planets. Two files contain the proper elements and frequencies, and their stability as determined by a running box test.

The proper elements as given by Milani are the following : da, a measure of the amplitude of the oscillation of the semimajor axis during one libration period, in AU; as an alternative, also D, a measure of the amplitude of the libration in longitude (in degrees); a proper eccentricity e_p, obtained by filtering out the forced terms; the sine of a proper inclination i_p. The fundamental frequencies f, g, s associated with these three amplitudes are also given. About 87% of the cases have proper elements stable enough for most purposes, with the maximum excursion in the running box test less than $0.001\,AU$ for da, and less than 0.0015 for e_p and $\sin i_p$. Although the dynamical meaning of these "synthetic" proper elements is quite different from the case of an analytic/semianalytic theory, these proper elements are meant to be used exactly in the same way as the others, as an example to identify asteroid families. For any other detail see Milani (1993, and this volume).

2. 4. METHOD BY SCHUBART FOR HILDAS

A synthetic method, based upon the post-processing of numerical integrations, has been developed by Schubart (1982, 1991) to derive "proper parameters" for the Hilda-type asteroids, located in the 3/2 mean motion resonance with Jupiter. The basis of the method is again numerical integration of orbits in the framework of a simplified dynamical model (Sun, Jupiter, Saturn, and solar mass augmented for the masses of the inner planets), covering time spans from $36,500\,yr$ to $109,500\,yr$. Specially defined proper parameters are used in this case, since these asteroids are resonant and no classical secular perturbation is applicable in commensurabilities; the libration gives rise to an oscillation of semimajor axis, for example, and the mean value of this oscillation does not represent a useful quantity, being nearly the same for all Hildas.

The parameters chosen by Schubart to represent the long term characteristics of motion for Hilda asteroids are : $\overline{\sigma}_A$ – a mean amplitude of libration of $\overline{\sigma}$, where the latter is the critical argument in which the argument of perihelion is replaced by its value transformed in such a way to account for the forced effect; $\overline{e}_p$ – a mean value of e_p, where e_p is also a value freed from the forced effect; and i_p – a proper parameter of inclination, a mean value derived from the time series of osculating elements. For the actual derivation of these three parameters, Schubart used a combination of digital filtering and graphical representation of the remaining variations, determining the final results from the respective plots.

Proper parameters were derived in this way for a total of 57 Hildas, two of which, with small eccentricity and a critical argument not in libration (1256 *Normannia* and 4196 1982SA13), are treated in a somewhat different way. Although the proper values obtained by Schubart seem to be fairly reliable (almost the same results have been obtained when parameters for nine objects were recomputed by means of an advanced model and different numerical definitions), certain improvements are possible, as an example extended integrations to filter out the very long periodic effects. The dynamical meaning of Schubart's parameters is again very different from that of either analytical or semianalytical proper elements, nevertheless they can be used to identify asteroid families by translating a difference in proper elements into a relative velocity by means of some metric. At least one family of Hildas has been identified by Schubart (1991).

3. Conclusion

Let us try to summarize this overview of the present state of the art regarding the asteroid proper elements :

- There are four different theories and methods currently in use for the computation of proper elements of different populations or groups of asteroids and contributing to the general catalogue (see Milani et al., this volume) : the analytic theory by Milani and Knežević, the semianalytic theory by Lemaitre

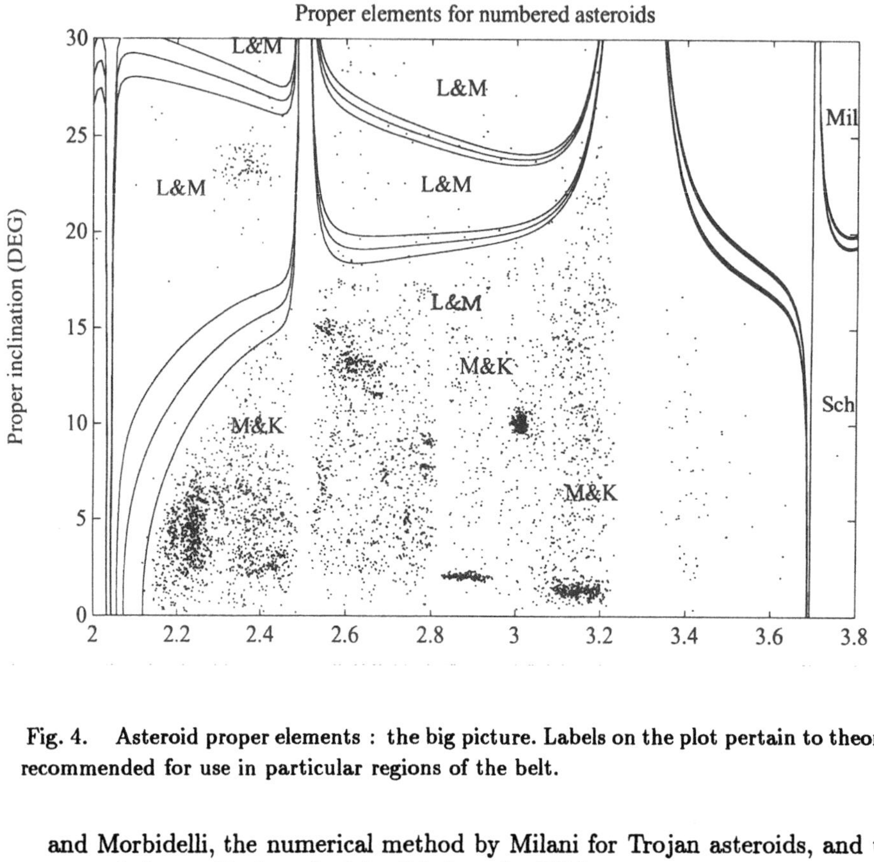

Fig. 4. Asteroid proper elements : the big picture. Labels on the plot pertain to theories recommended for use in particular regions of the belt.

and Morbidelli, the numerical method by Milani for Trojan asteroids, and the numerical-graphical method by Schubart for Hildas.

- Being computed by means of different procedures and even defined in different ways, proper elements are of different accuracy and long term reliability; the user has to bear these differences in mind and has to be particularly cautious when comparing or mixing data from various sources. Most of the data are of accuracy that permits fairly reliable identification of asteroid families against the background, given the number of asteroid orbits presently known.
- As a kind of recipe, one can recommend at the moment to use the theory by Milani and Knežević below 15° of inclination, and the theory by Lemaitre and Morbidelli above 17° (and for special asteroid groupings, like Hungarias and Phocaeas); the region in between can be considered as a transition one in which

the two theories could be used equivalently or alternatively; for resonant Trojans and Hildas there are special methods by Milani and Schubart. In Figure 4. this recipe is summarized on the plot representing almost the entire main asteroid belt — only the proper elements for numbered asteroids are given here, and positions of only three linear secular resonances are indicated; the labels on the plot indicate the theory to be used in each region.

Acknowledgements

We are grateful to P. Farinella and A. Lemaitre for carefully reading the manuscript and suggesting many improvements.

References

Anders, E. : 1965, "Fragmentation history of asteroids." *Icarus* **5**, 399–408.

Arnold, J.R. : 1969, "Asteroid families and jet streams". *Astron. J.* **74**, 1235–1242.

Bien, R. and Schubart, J. : 1987, "Three characteristic parameters for the Trojan group of asteroids", *Astron. Astrophys.* **175**, 292–298.

Brouwer, D. : 1951, "Secular variations of the orbital elements of the principal planets." *Astron. J.* **56**, 9–32.

Brouwer, D. and Van Woerkom, A.J.J. : 1950, "The secular variations of the orbital elements of the principal planets." *Astron. Papers Amer. Ephem. Naut. Almanac* **13**, 81–107.

Henrard, J. : 1990, "A semi-numerical perturbation method for separable hamiltonian systems." *Celest. Mech.* **49**, 43–67.

Hirayama, K : 1918, "Groups of asteroids probably of common origin." *Astron. J.* **31**, 185–188.

Hirayama, K : 1923, "Families of asteroids." *Japan. J. Astron. Geophys.* **1**, 55–93.

Hirayama, K : 1928, "Families of asteroids, second paper." *Japan J. Astron. Geophys.* **5**, 137–162.

Hori, G. : 1966, "Theory of general perturbations with unspecified canonical variables." *Publ. Astron. Soc. Japan* **18**, 287–296.

Knežević, Z. : 1992a, "An efficient algorithm for the analytic computation of planetary short-periodic perturbations." *Bull. Acad. Serbe Sci. Arts (Sci. math.)*, **104**(18), 39–46.

Knežević, Z. : 1992b, "Minor planet short–periodic perturbations : the indirect part of the disturbing function." *Celestial Mechanics*, **55**, 387–404.

Knežević, Z. : 1994, "Generic term representation of the long periodic determining function." *Planet. Space Sci.*, in press.

Knežević, Z., Froeschlé, Ch., Lemaître, A., Milani, A., and Morbidelli, A. : 1994, "Comparison of two theories for calculation of asteroid proper elements.", in preparation.

Kozai, Y. : 1962, "Secular perturbations of asteroids with high inclination and eccentricity." *Astron. J.* **67**, 591–598.

Lemaitre, A., and Morbidelli, A. : 1994, "Calculation of proper elements for high inclined asteroidal orbits." *Celestial Mechanics*, in press.

Lindblad, B.A., and Southworth, R.B. : 1971, "A study of asteroid families and streams by computer technique." In : *Physical Studies of Minor Planets* (T. Gehrels, Ed.), NASA SP-267, U. S. Govt. Print. Off., pp. 337–352.

Milani, A. : 1993, "The Trojan asteroid belt : Proper elements, stability, chaos and families." *Celest. Mech.* **57**, 59–94.

Milani, A. : 1994, "Proper elements and stable chaos." In : *From newton to chaos : modern techniques for understanding and coping with chaos in N-body dynamical systems* (A. E. Roy, B.A. Stevesi, Eds.), Plenum, New York, in press.

Milani, A., and Knežević, Z. : 1990, "Secular perturbation theory and computation of asteroid proper elements." *Celestial Mechanics* **49**, 347–411.

Milani, A., and Knežević, Z. : 1992, "Asteroid proper elements and secular resonances." *Icarus* **98**, 211–232.

Milani, A., and Knežević, Z. : 1994, "Asteroid proper elements and the dynamical structure of the asteroid belt." *Icarus*, in press.

Milani, A., and Nobili, A.M. : 1992, "An example of stable chaos in the Solar System." *Nature* **357**, 569–571.

Morbidelli, A. : 1993, "Asteroid secular resonant proper elements." *Icarus* **105**, 48–66.

Morbidelli, A. and J. Henrard 1991a. Secular resonances in the asteroid belt : Theoretical perturbation approach and the problem of their location. *Celest. Mech.* **51**, 131–168.

Morbidelli, A. and J. Henrard 1991b. The main secular resonances ν_5, ν_6 and ν_{16} in the asteroid belt. *Celest. Mech.* **51**, 169–198.

Nobili, A.M., Milani, A. and Carpino, M. : 1989, "Fundamental frequencies and small divisors in the orbits of the outer planets." *Astron. Astrophys.* **210**, 313–336.

Schubart, J. : 1982, "Three characteristic parameters of orbits of Hilda-type asteroids *Astron. Astrophys.* **114**, 200–204.

Schubart, J. : 1991, "Additional results on orbits of Hilda-type asteroids." *Astron. Astrophys.* **241**, 297–302.

Van Houten, C.J., Van Houten-Groeneveld, I., Herget, P., and Gehrels, T. : 1970, "The Palomar-Leiden Survey of faint minor planets." *Astron. Astrophys. Suppl.* **2**, 339–448.

Williams, J.G. : 1969, "Secular perturbations in the Solar System." Ph.D. Thesis, Univ. California, Los Angeles.

Williams, J.G. : 1979, "Proper orbital elements and family memberships of the asteroids." In : *Asteroids* (T. Gehrels, Ed.), pp.1040–1063, Univ. Arizona Press.

Williams, J.G. : 1989," Asteroid family identifications and proper elements." In : *Asteroids II* (R. P. Binzel, T.Gehrels, M.S. Matthews, Eds.), pp. 1034–1072, Univ. Arizona Press, Tucson.

Williams, J.G.,and Faulkner, J. : 1981, "The position of secular resonance surfaces." *Icarus* **46**, 390–399.

Williams, J.G., and Hierath, J.E. : 1987, "Palomar–Leiden minor planets : Proper elements, frequency distribution, belt boundaries, and family membership." *Icarus* **72**, 276–303.

Yuasa, M. : 1973, "Theory of secular perturbations of asteroids including terms of higher order and higher degree." *Publ. Astron. Soc. Japan* **25**, 399–445.

Zappalà, V., Cellino, A., Farinella, P., and Knežević, Z. : 1990, "Asteroid families I : identification by hierarchical clustering and reliability assessment." *Astron. J.* **100**, 2030–2046.

Zappalà, V., Cellino, A., Farinella, P., and Milani, A. : 1994, "Asteroid families : II. Extension to unnumbered multi-opposition asteroids." *Astron. J.* **107**, 772–801.

THE DYNAMICS OF THE TROJAN ASTEROIDS

ANDREA MILANI
Dipartimento di Matematica, Via Buonarroti 2, I-56127 Pisa, Italia
E–mail milani@dm.unipi.it

Abstract. The state of the art in the dynamics of the Trojan asteroids has progressed rapidly, since it has been possible to perform numerical integrations of many orbits for millions of years. Accurate proper elements are now computed by the synthetic method, that is from the output of a numerical integration; their stability, at least for time spans of a few million years, is good. This has allowed identification of Trojan families with an automated procedure closely mimicking the one used in the main belt. Although the families identified in a reliable way are only four, the occurrence of significant collisional evolution, not unlike that of the main belt, can be confirmed. The dynamical structure of the Trojan cloud, including the location of the main secular resonances, can be deduced from the proper elements and frequencies by a simple fit. However, many problems are not solved : the origin of a significant percentage of chaotic orbits showing no indications of instability; the location of the stability boundary of the Trojan cloud; the origin of the high inclination of most Trojans. We also do not know if there are "Trojans" for some other planets beside Jupiter : only one Mars Trojan has been found.

We are now aware that the Trojan asteroid cloud is not a small feature of the planetary system, but contains a large enough population to be considered an essential component of the Solar System. Thus the widely different level of knowledge and understanding between the main belt and the Trojan cloud is mostly a consequence of the priorities of our research, rather than a reflection of the relevance of the subject. To some extent these priorities are dictated by the observational bias against the discovery and observation of Trojans, which are both farther away and darker. Another reason is that the Trojans are intrinsically a more difficult problem from a dynamical point of view, because their orbits cross the orbit of Jupiter, and close approaches are avoided only because of the 1 : 1 resonance.

Given the striking successes of the last 20 years in the study of the dynamics of the main belt, it is time to ask whether similar results can be obtained for the Trojan cloud, by applying to this problem a suitable amount of research effort. In particular, we would like to know : (1) if it is possible to develop tools equivalent to the proper elements, which have been successfully used for the main belt; (2) if we can detect and study Trojan asteroid families, with the same level of reliability already achieved in the main belt; (3) if we can understand the dynamical structure of the Trojan cloud, including the overall distribution in the phase space with its boundaries and gaps; (4) if we can understand the dynamical stability of the Trojans over time scales comparable to the age of our Solar System, and the interplay between stability, chaos, and the still unknown dissipative processes active during the formation of the Solar System; (5) if there are "Trojans", that is objects in 1 : 1 libration orbits, for some of the other planets besides Jupiter.

This paper reviews the recent progress on all these subjects, and also lists a number of problems which have not yet been solved. For a review of earlier work, see Shoemaker et al. (1989).

A. Milani et al. (eds.), Asteroids, Comets, Meteors 1993, 159–174.

1. PROPER ELEMENTS FOR TROJANS

Proper elements are quantities which are almost constant over a very long time span, e.g. millions of years, for each asteroid moving under the gravitational attraction of the Sun and the major planets. To compute proper elements stable over millions of years requires an accurate theory of the motion of the asteroid applicable over the same time span. For the main belt, such theories do exist, based either on an analytical expansion of the solution into series, or on a semianalytical expansion, using an analytical theory obtained with numerical quadratures (for a more detailed discussion of the definition of proper elements in the different regions of the asteroid belt, see Knežević and Milani, this volume).

For the Trojan-type of orbits, purely analytical theories are very difficult, because the perturbing function is singular at the collisions, and most expansion methods normally used in perturbation theory are strongly divergent for a ratio of semimajor axes close to 1. An analytical theory can be constructed by expanding in powers of small parameters which are related to the distance of the Trojan from the Lagrangian relative equilibria (either L_4, leading Jupiter by 60°, or L_5, trailing at the same distance). The trouble is, the distance of the Trojans from the Lagrangian points is by no means small. The (proper) inclination of a Trojan can be as high as 37°.5, the (proper) eccentricity can be 0.22, the (proper) amplitude of libration in longitude can be 33°.3 : expansions with these quantities as small parameters cannot be a very effective way of computing a theory for the motion of all the Trojans. The most successful analytical theory for the motion of the Trojans was developed by Érdi (1988; and references therein); it is based on the multiple timescale method, and has been pushed to a rather high order in the libration amplitude; however, it only accounts for the perturbations by Jupiter, assumed to be in a fixed elliptic orbit. A semianalytical theory (modelled on the ones by Williams, 1969, and Lemaitre and Morbidelli, 1994) could in principle be computed, but it requires nontrivial modifications to the procedures well established for main belt asteroids. Thus, proper elements for Trojans, of accuracy adequate for the detection of families, were not available until recently.

This unsatisfactory state of affairs was modified when the speed of digital computers become good enough to allow for the numerical integration of a large set of Trojans; Bien and Schubart (1987) computed the orbits of 41 Trojans for a time span of $\simeq 150,000\,yr$ and represented the results in terms of the orbital parameters D, e_p, i_p which played the rôle of the proper elements in an analytical theory. The geometric interpretation of these parameters will be described below; the essential point is that the output of the numerical integration for each asteroid is post–processed by passing through a sequence of coordinate tranformations and digital filters. The output is a best fitting model, which represents in a fairly accurate way the behaviour of the orbital elements of the asteroid over the integration time span. These numerical techniques, pioneered by Schubart (1982), were later improved upon in the so called "synthetic theories" used in *Project LONGSTOP* to describe the behaviour of the outer major planets for $10^8\,yr$ (Carpino et al., 1987).

Recently a modified synthetic theory technique was used by myself (Milani, 1993) to compute proper elements for 174 Trojan asteroids. Apart from technical

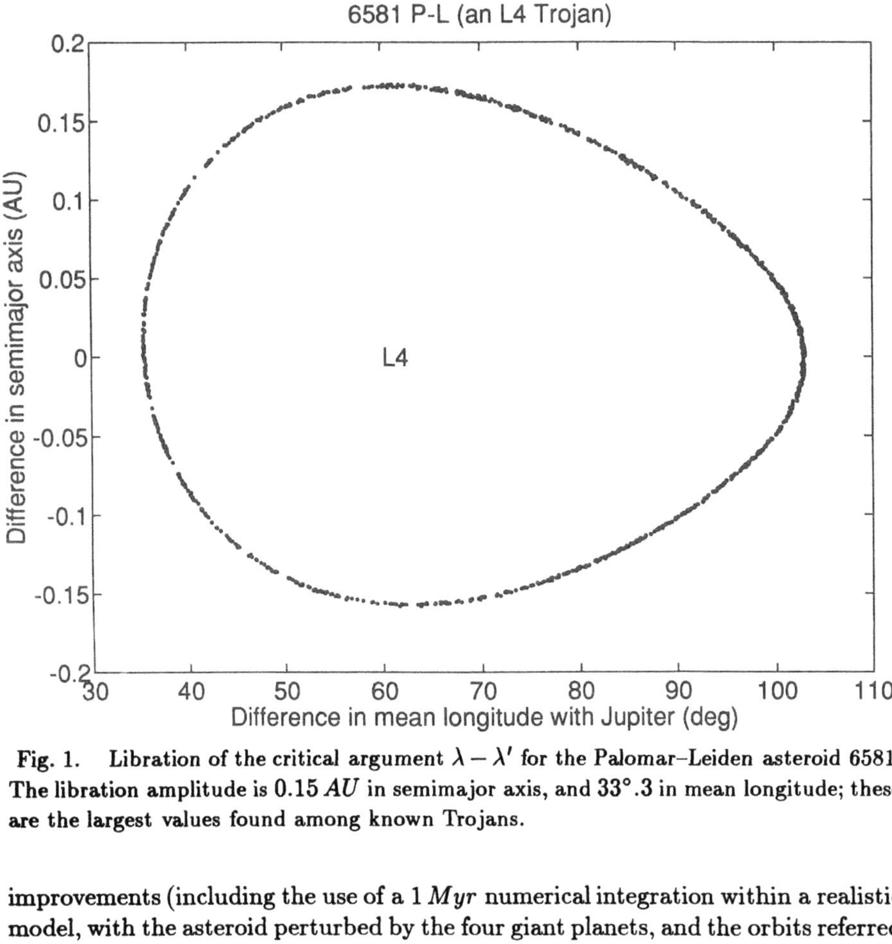

Fig. 1. Libration of the critical argument $\lambda - \lambda'$ for the Palomar–Leiden asteroid 6581. The libration amplitude is $0.15\,AU$ in semimajor axis, and $33^\circ.3$ in mean longitude; these are the largest values found among known Trojans.

improvements (including the use of a $1\,Myr$ numerical integration within a realistic model, with the asteroid perturbed by the four giant planets, and the orbits referred to the barycenter of the inner solar System to account for most of the effect of the inner planets), the method is a modification of the one by Bien and Schubart, and the values of the proper elements are consistent for the cases which have been computed also by them, although the definitions are not the same. The main improvement is that I have derived an estimate of the stability in time of the proper elements, by means of the running box method. Proper elements are derived from the first $500,000\,yr$ of output, then by the data from 50,000 to $550,000\,yr$ into the integration, and so on until the last $500,000\,yr$ are used; this provides 11 different estimates, and the maximum excursion (difference between the maximum and minimum) of these 11 values is used as an estimate of the instability in time of the computed proper elements. The results of this stability test, presented in Table 3 of Milani (1993), were already good, with 87% of the cases with excursions within 0.0025 for $e_p, \sin i_p$ and within $0.001\,AU$ for the proper amplitude of oscillation in semimajor axis da (corresponding to about $0^\circ.2$ for the amplitude of oscillation in longitude). Nevertheless, I have recently recomputed proper elements with an

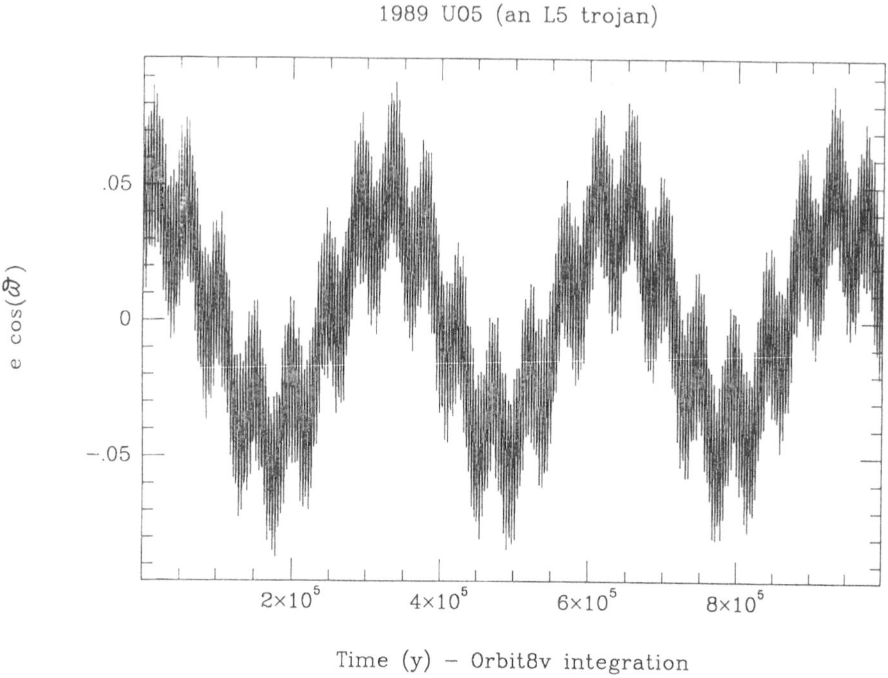

Fig. 2. The variables h, k undergo oscillations with frequencies $g_5, g_6, \ldots$ (the frequencies of revolution of the perihelia of the major planets), and a proper, or free, oscillation with the much faster proper frequency g. The amplitude of the proper mode is, in some cases such as *1989 UO5*, less than the amplitude of the forced g_5 mode : thus the perihelion of the Trojan is locked to the one of Jupiter.

improved algorithm (Version 2.3), which resulted in the same percentage (161 out of 184) of cases with excursions within a reduced range (0.0015 for $e_p, \sin i_p$).

The proper elements version 2.3 are defined as follows. The mean elements $a - a', \lambda - \lambda'$ (difference with Jupiter in semimajor axis and in mean longitude) are obtained by removing the oscillations with periods up to $50\,yr$ (by means of a digital low-pass filter); their behaviour for a large amplitude libration Trojan is shown in Figure 1. A proper element is the first harmonic component (either da, in the $a - a'$ direction, or D in the $\lambda - \lambda'$ direction) with respect to the phase angle θ of the libration. θ is computed as the polar angle around the Lagrangian point, suitably filtered to remove intermediate period oscillations. This definition of D corresponds to half the value given by Bien and Schubart; the latter is not the difference between maximum and minimum of the critical argument $\lambda - \lambda'$, since the libration curve is significantly pear-shaped for large amplitudes (Figure 1).

The proper eccentricity is defined by Fourier analysis of the nonsingular variables $h = e\sin\varpi, k = e\cos\varpi$ (see Figure 2). These variables undergo forced oscillations, with frequencies such as g_5, g_6 (periods $\simeq$ 305,000 and 45,900 yr respectively) and proper oscillations with frequency g and periods ranging from 3,000 to 5,600 yr; given that the frequencies are well separated, it is easy to identify the proper mode by Fourier analysis, and the corresponding amplitude is used as proper eccentricity. To improve the stability of the proper eccentricity, in version 2.3 the Fourier analysis has been performed with respect to the actual value of ϖ, rather than with respect to a linear function of time approximating ϖ; this is closer to the original definition used by Bien and Schubart.

The proper inclination is defined as the Fourier component with frequency s of the proper oscillations in $p = \sin i \sin \Omega, q = \sin i \cos \Omega$; the main problem is that the frequency s changes over a much wider range that it is the case for g (see Section 3), thus for some Trojans it is necessary to use a very long time series to correctly identify the amplitude of the proper oscillation in p, q. To reduce the impact of this problem on the stability of the proper inclination, I have extended the numerical integration to 5 Myr for 20 Trojans.

These results could be improved by carefully analysing the "bad" cases for which the proper elements have a degraded stability, and by introducing a somewhat more complicated synthetic theory to handle most of these. However, the improvements would only concern a few cases; some cases cannot be improved because of the chaotic nature of their orbits (see Section 4). The use of an analytic (or at least semianalytic) theory would also be more informative on the causes of at least some of the less stable cases. However, the brute force work done by means of the numerical integrations such as my own is already good enough to discuss the possible identification of families (Section 2) and the dynamical structure of the cloud, as defined by the lowest degree resonances (Section 3). I hope that somebody will be encouraged by the interesting problems raised by this approach and will try to work in a deeper way on the problem of computing a theory for the long-term motion of the Trojans. The proper elements (as well as the results of the running box stability test) are available by `anonymous ftp`, in the public domain data base described elsewhere in these proceedings.

2. TROJAN ASTEROID FAMILIES

Asteroid families are clusterings in the proper elements space, which are too dense to have occurred by chance, and therefore can be interpreted as the result of the catastrophic disruption of a parent body by collision. Since proper elements are stable for millions of years, the disruption might have occurred very long ago, although we do not know when it occurred. For a more complete discussion, see Zappalà and Cellino (these proceedings). The point which is especially relevant for the discussion of possible families among the Trojans, is that the definition of a family requires not only that the asteroids are collisionally interacting, but also that we must be able to find this out by observing and computing now, millions of years after the event. This is an especially demanding requirement, since we know (from the analysis of the families in the main belt, and also from the identification of the parent body of some class of meteorites, as in the case of the eucrites; see Binzel and Xu, 1993; Farinella et al., these proceedings) that the fragments from the catastrophic disruption of an asteroid can be ejected with relative velocities of several hundreds of meters per second. In the Trojan cloud it is possible that such fragments could be ejected from the stable libration region and would disappear from the Solar System altoghether, after a close approach to Jupiter. It is therefore clear that the total number of collisionally related groupings is much larger than the number of identifiable families. This difficulty is compounded by the lack of physical observations of Trojans, to the point that we can neither confirm nor discard hypothetical families on the basis of similarities/dissimilarities in the spectra and/or the taxonomic types.

The question I would like to address in this Section is therefore the following : can we find in the data on the orbits of the known Trojans reliable evidence for the existence of families? Does this evidence confirm the occurrence, among the Trojans, of a collisional evolution not unlike the one we know in the main belt? Unlike what was done by Shoemaker et al. (1989), I am here interested in finding proof of the existence of at least some families, rather than proposing many possible families.

The methods to be used for this purpose are well established (Zappalà et al., 1990, 1994). We need first to choose a metric in the proper elements space; this distance should be related to the impulsive change of velocity needed to change from one orbit to the other. This correspondence is only approximate (see a discussion in Zappalà et al., 1990; Milani et al., 1992). For the case of the Trojan cloud, I have chosen the simplest possible distance, which has the advantage of giving the same weight to the difference in eccentricity and the difference in inclination, without overweighing the difference in semimajor axis, as it is the case for the main metric used by Zappalà et al.. Therefore this metric preserves the peculiar shape of the Trojan "belt" (it is actually a couple of towers, more than 6 AU in height) :

$$d = \sqrt{\frac{1}{4}(\frac{\delta da}{a})^2 + 2(\delta e)^2 + 2(\delta \sin I)^2}$$

The relative velocity corresponding to this distance is simply $V = d \times V_{Jupiter} \simeq d \times 13\ Km/s$; however, this correspondence has to be understood only as an order of magnitude estimate.

The next step is to compute the distance between any two asteroids, and to sort these distances in ascending order. Then we can apply the Hierarchical Clustering Method : we begin by joining all the couples separated by a distance d less than a given value K; as K is increased step by step (I have used steps of 0.002), a graph is formed, in which the connected components are the candidate groupings. Such a graph is actually shown in Figure 3 (for the "Greeks" around L_4) and Figure 4 (for the "Trojans" around L_5); we have projected on the $e, \sin I$ plane because in the third direction (amplitude of libration in semimajor axis) the Trojan cloud is "thin", and the contribution of the third component to the distances is small.

The procedure needs to be terminated by selecting a maximum value of K, otherwise all the points would be clumped together in a meaningless superfamily. The choice of this maximum K must be such that clumps, of the same size as the ones we are interested in, are unlikely to occur by chance in a population of

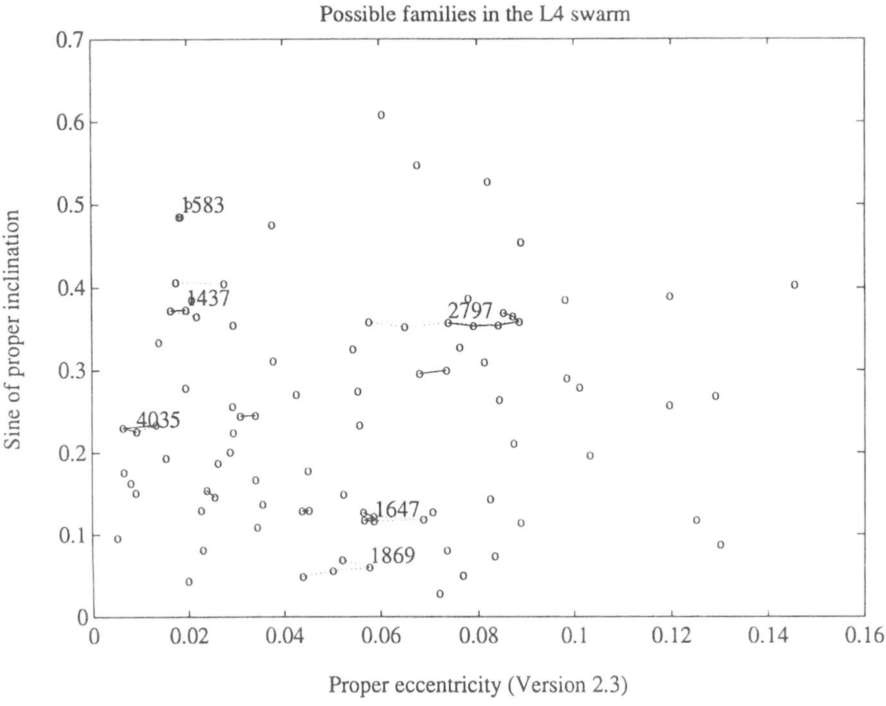

Fig. 3. The L_4 swarm, with a graph indicating the result of the Hierarchical Clustering Method : the solid lines join any two Greeks with a distance $d \leq 0.012$; the groupings of three or more at this distance level should be significant. The dotted lines join couples at a distance $0.012 < d \leq 0.016$; these could be collisionally related, especially if they form large groupings and/or connect to already significant families.

objects not collisionally related. To this purpose, Zappalà et al. (1990) introduced the computation of the Quasi Random Level (QRL), which is the value of K at which groupings of a given size would occur by chance in a population with the same overall distribution, but randomized in such a way that there are no real families. The main problem in applying the same method to the Trojans is that the number of orbits in the catalogues is so much smaller than the one for the main belt, that statistical methods based upon random populations might not work, because of statistical fluctuations over a too small sample.

In an attempt to solve this problem, I have introduced a different method to compute the QRL for very small groupings, with only either 2 or 3 members. I have exploited the property of the Trojan population of being subdivided into two independent swarms, around L_4 and around L_5, which cannot interact collisionally. Then I can use each of the two swarms as a random clone of the other one; if the distances are computed between all the members of both swarms, then the same

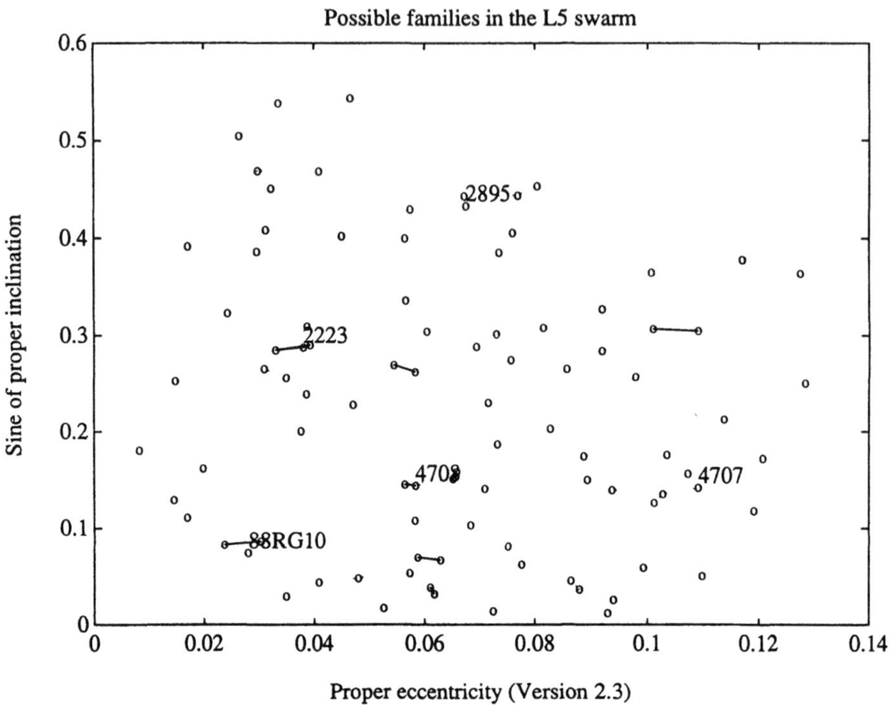

Fig. 4. The L_5 swarm : in the proper eccentricity/sine of proper inclination plane, the neighbouring couples are joined as in Figure 4. The number of statistically significant clusterings is smaller than around L_4.

clustering procedure is used for all together, the spurious groupings (with some members in L_4, some in L_5) play the same rôle of the clusterings found in random samples in the Zappalà et al. method. For the couples, with $K = 0.002$ there is only one couple formed by *1583 Antilochus* and *3801 Thrasimedes*, both in L_4. However, for $K = 0.004$ there are 9 couples, out of which 3 between asteroids in L_4, 6 spurious (one in L_4 and one in L_5), and no one in L_5. The conclusion is that the QRL for couples is less than 0.004, and even the couples as close as 0.006 can easily occur by chance, hence cannot be reliable indicators of a collisional event.

The couple *1583–3801* is very likely to be significant. Actually, the difference in proper elements between *1583 Antilochus* and *3801 Thrasimedes* is small to the point of raising another problem : it corresponds to $d = 0.0006$. When this quantity is translated into relative velocity, it amounts to less than $10\,m/s$. Since the escape velocity from *1583* is $\simeq 65\,m/s$, *3801* needs to be ejected from *1583* at a speed between 65 and $65.7\,m/s$ to escape from the sphere of influence of the larger asteroid, but with such a small relative velocity. Although this is not imposssible, it is unlikely. With P. Farinella I have been studying a non conventional collisional model, involving an intermediate stage as binary Trojan, which could explain this couple with a plausible scenario. Such a model has the interesting feature of predicting an observable property of the couple, namely a long rotation period for both asteroids.

A similar analysis of the occurrence of spurious triples shows that for $K = 0.012$ there are 8 spurious triples, 1 triple in L_5 and 8 triples in L_4. This indicates that the QRL is $\simeq 0.012$, corresponding to $\simeq 150\,m/s$, and also that a significant fraction of the triples in L_4 does not occur by chance. It follows that the groupings of 3 or more Trojans listed below have to be rated as significant, that is are at least very likely to be families, although it is difficult to quantify the statistical confidence level when working with these small numbers.

In the L_4 swarm, the Menelaus family includes, besides the numbered *1647 Menelaus*, *1973SD1, 1973SQ1, 1989AU1*. It is possible that some other asteroids are related to the same family, although the distance is above the QRL : they include the close couple *3548 Eurybates - 6591 PL*, and also *1749 Telamon* and possibly *91HN* (which has, however, very poor proper elements, both because it has not been observed long enough and because of the $s - s_6$ resonance). The Teucer family includes, up to the QRL, *2797 Teucer, 4086, 4946, 2759, 1988BY1* and *1987YU1*; it could have, at a somewhat larger distance, two more members : *3793, 1989AU2*. The triple *4035, 1973SM1, 1986TR6* is marginally significant (a *clump* in Zappalà et al. terminology) because it has exactly 3 members at the QRL for triples. Other groupings, which have been proposed as families by daring authors (including myself), cannot be confirmed by any statistical argument : e.g. Diomedes, Philoctetes.

The L_5 swarm is surprisingly less rich in significant groupings. Whether this is a subtle observational selection effect or a real difference, is an open question. The Sarpedon family includes only 3 members : *2223 Sarpedon, 1988RR10, 1989SC7*; however it is a rather compact triple, with all three mutual distances below 0.012. The Polydoros grouping consists of two couples, *4708 Polydoros - 4792 Lykaon* and *4827 Dares - 4317 PL*, which join one step above the QRL, and therefore does not entirely fulfill the requirement for significance. If they are indeed a family, they might also be related to *4369T3*. Other groupings proposed as families, such as Kryses, Memnon and *1988RG10*, cannot be confirmed.

Does this procedure answer the main problems, or it is only a tutorial on the use of the Hierachical Clustering Method? The reliable results are only two good families, Menelaus and Teucer, plus the odd couple *1583-3801*, in L_4, and only one reasonably good family, Sarpedon, in L_5. All these had already been proposed as families (Shoemaker et al., 1989), together with many others which I am unable to confirm with any statistical argument. However, this is what was expected, given the small numbers available (only 94 Trojans and 90 Greeks). An additional limitation is the narrow range in size of the known Trojans : as we know from the main belt, most families have a larger fragment and the others are much smaller. The surveys performed with the large Palomar Schmidt have produced Trojan orbits too inaccurate to be used for family detection, because they used too short obsrvation time spans. On the contrary the survey performed with a small Schmidt by the Shoemakers has given very accurate and useful orbits, but only for comparatively large asteroids. If we were to look at the main belt at a comparable level of completeness, e.g. if we knew only the orbits of the 603 main belt objects with absolute magnitude $H < 10$, we could identify with certainity only the 3–4 major families. The conclusion is that, given the existence of at least four detectable families among the known Trojans, the collisional evolution of the Trojans is not inconsistent with the one of the main belt. To obtain more information, we need to discover/recover a large number of Trojans; when this will be done, the analytical tools will be ready for investigating the collisional evolution of the Trojan swarm.

3. UNDERSTANDING THE DYNAMICAL STRUCTURE

The proper elements and the corresponding proper frequencies provide also the right coordinate system in which to look to the distribution of the Trojan orbits. The plots in the plane proper eccentricity/sine of proper inclination (as in Figures 3 and 4) do not show any obvious structure, although it is apparent that most Trojans have proper eccentricities $e_p < 0.15$; only one object around L_4, *4835*, and one around L_5, *5144*, have $e_p \simeq 0.22$ (they are the only ones not shown in the figures because of the smaller window). The inclinations are found in a much wider range. The distribution of the amplitude of libration is more interesting (see Figure 5), and the distribution of the frequencies g (of the proper perihelion), s (of the proper node) and f (of the libration phase) clearly indicates the location of some resonances (see Milani, 1993, Figure 11).

The best way to understand the structure of resonances cutting and bounding the Trojan cloud, is to superimpose the asteroid population and the resonance strips computed by some theory providing the frequencies as a function of the proper elements. As an example, this is done in Milani and Knežević (1994), Figures 7–11. However, we do not have a theory to compute the frequencies; we have to resort to a numerical procedure to generate a theory from the experiments (as in Schubart and Bien, 1987). Given the catalogue of proper elements and frequencies for 184 Trojans, I have fit a model in which the frequencies are polynomial expressions in the variables :

$$x = \frac{e_p}{0.15} \quad ; \quad y = \frac{\sin i_p}{0.6} \quad ; \quad z = \frac{da}{0.15\,AU}$$

where the denominators approximate the range of values for each proper element. For the frequency of the proper nodes the best fit gives :

$$\begin{aligned} s(arcsec/year) = & -7.69 - 7.91x^2 + 13.36y^2 - 22.58z^2 + \\ & + 2.37x^2y^2 - 1.77y^4 - 6.33x^2z^2 + 3.96y^2z^2 - 3.51z^4 \end{aligned}$$

where the x^4 (that is, e_p^4) term has been omitted because it is small and poorly determined; only even powers have been used because they are related to the action variables. The fit for s has an RMS of the residuals of $0.065\,arcsec/yr$, and is therefore a very good approximation of the behaviour of the frequency as a function of the proper elements. It shows that the frequency s can change significantly, and indeed for the known Trojans $-33.88 \leq s \leq +3.27$, that is, s can actually be zero. The following secular resonances are therefore possible in the Trojan cloud :

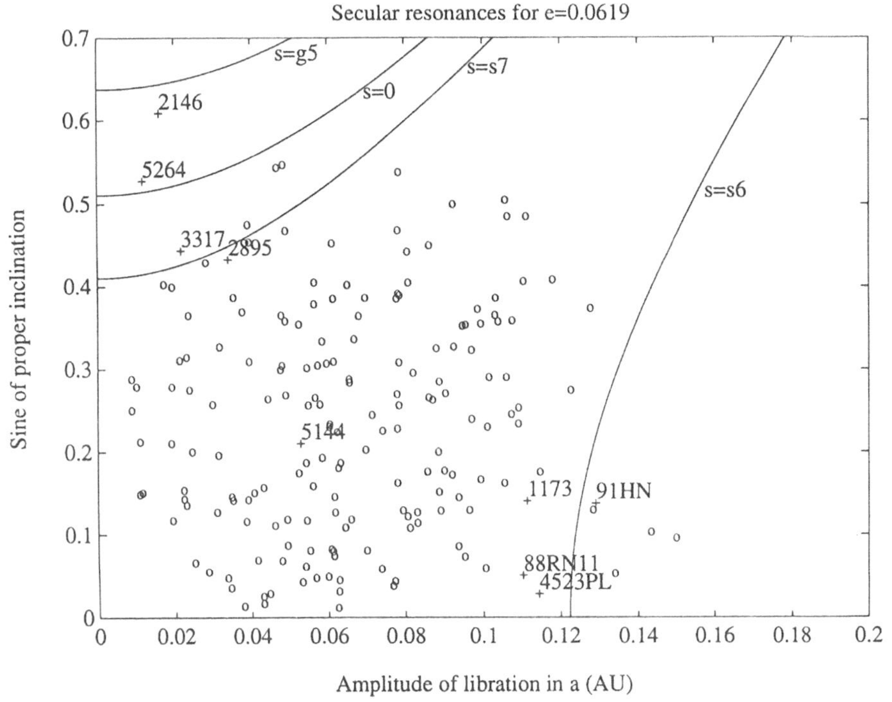

Fig. 5. Secular resonance lines in the plane $da, \sin i_p$, computed for the mean value of e_p; proper elements for the Trojans with a good stability of i_p are plotted as circles. The crosses and the name labels indicate the cases with degraded stability of i_p.

$s - s_6, s - s_7$; the resonance $2s - 2g_5$ could also be relevant (s_i is the proper node frequency, g_i the proper perihelion frequency of planet i). All these resonances have indeed been proposed to explain the behaviour of some specific asteroid (Bien and Schubart, 1986; Nakai and Kinoshita, 1985). Following the main ideas of these authors, it would be possible to produce a very accurate qualitative model of the behaviour of the orbits in these secular resonances; this problem is especially suitable for the application of semianalytic models.

Figure 5 shows, in the plane sine of proper inclination/amplitude of libration in semimajor axis, the actual Trojans superimposed to the lines corresponding to the main secular resonances involving the node, computed for a value of e_p equal to the mean among the real Trojans, that is 0.0619. The asteroids for which the instability of $\sin i_p$ is larger than 0.002 are labelled; they clearly are located near the main resonances (*5144* also has $s \simeq s_6$, but this is not apparent from the Figure, because it has a proper eccentricity very different from the mean value). The Figure also shows that the two main resonances $s - s_6, s - s_7$ bound the region where most Trojans are. Given the small number of cases, it is not clear whether these secular resonances open a gap, or rather define a boundary beyond which there are less asteroids because their dynamical stability is reduced.

Thus the synthetic theory provides us with good clues to understand the dynamical structure, as far as the instability in inclination and the resonances involving the nodes are concerned. The situation is quite different for the eccentricity and the perihelion. A similar fit for the proper perihelion frequency g gives :

$$\begin{aligned} g(arcsec/yr) =& 361.85 - 12.73x^2 - 127.81y^2 + 67.76z^2 + \\ &+ 2.37x^2y^2 + 5.07y^4 - 14.17x^2z^2 - 98.09y^2z^2 - 1.85z^4 \end{aligned}$$

which is also a good fit, with RMS of the residuals of $0.48\,arcsec/yr$. As it is clear from the dominance of the constant term, g cannot change by a large factor : indeed $231.51 \leq g \leq 423.39\,arcsec/yr$ for the known Trojans. This implies that the lowest degree resonances involving the proper perihelion of the Trojan cannot occur : as an example, the Kozai separatrix, defined by $2g = 2s$, beyond which there can be libration of the argument 2ω, cannot occur in the Trojan cloud; also no resonance $g - kg_i$ can occur for any planet i and for any integer $k \leq 8$. For the frequency f of the libration phase :

$$\begin{aligned} f(arcsec/yr) =& 8828.8 + 162.3x^2 - 1069.8y^2 - 448.0z^2 - \\ &- 415.9x^2y^2 + 33.9y^4 + 105.0x^2z^2 + 21.9y^2z^2 - 269.6z^4 \end{aligned}$$

with RMS of the residuals of $20.4\,arcsec/yr$. For known Trojans, $5400 \leq f \leq 8896\,arcsec/yr$, and again no low degree resonance can occur, e.g. $n > kf$ for any $k < 12$, $f > kg$ for any $k < 18$. The three proper frequencies f, g, s being very well separated, no low degree resonance between them is possible. This would seem to indicate that the Trojan cloud is little affected by resonances, and most of the phase space should be filled with regular orbits; this fits well with the good stability of the proper elements as shown by the running box test. However, my experience with the problems of very long-term stability of planetary and asteroidal motions made me quite diffident toward this kind of optimistic statements.

4. OPEN PROBLEMS

As should be clear from this review, what we know about the dynamics of the Trojan cloud is not very much, although we now have examples to work on from the output of systematic numerical investigations. Rather than giving a long list of questions for which we do not know the answer, I would like to stress three main subjects on which we need some progress in the next few years.

The first problem is the origin of chaos in the orbits of Trojans. During my systematic numerical investigation, I have found 31 cases (out of 184) of positive Lyapounov exponent, with values of Lyapounov time (the time in which two neighbouring orbits diverge by a factor exp(1)) ranging between $15,000\,yr$ and 1/6 of the integration time span. Of course it is not possible to deduce from this that the others are regular orbits, but only that the Lyapounov times, if they exist, are longer. The problem is that, out of these 31 chaotic orbits, there is a tentative identifications of the resonance responsible for this behaviour for only a few. 5 cases are near the $s - s_6$, 3 near the $s - s_7$ resonance; in most of these cases, the proper inclination as computed by the synthetic method is not very stable.

On the contrary, for most cases with a detected positive Lyapounov exponent, I could find no resonance of reasonable degree in which they might be locked. In 5 cases, there was some degradation of the stability of the proper amplitude of libration (between 0.001 and 0.002 AU of excursion in the amplitude of oscillation in semimajor axis). In 19 cases, there was no signifcant increase of the instability in any of the three proper elements $da, e_p, \sin i_p$, with respect to what was found in the non–chaotic cases (of course, this statement depends upon the accuracy of the technique used to compute the proper elements : with a more accurate method, smaller instabilities might become detectable). In conclusion, more than 10% of the Trojans exibit some form of *stable chaos*, that is they behave as if the orbits were constrained to remain close to a quasiperiodic orbit, for a time much longer than the Lyapounov time; this behaviour has also been observed in the main belt (Milani and Nobili, 1992; Milani and Knežević, 1994). These orbits are not accurately predictable, beyond a predictability time horizon which is much smaller than the age of the Solar System; but they are approximately predictable, since the proper elements change within a narrow range over a time much longer than the predictability time.

A theoretical explanation for such a strange behaviour is not yet available, although in principle the stochastic region associated to a chaotic orbit can have any size. The only possible explanation is that these orbits are locked into some resonances of "unreasonable" degree, which result into the full range of chaotic phenomena, including the positive Lyapounov exponent, but has too small a strength to produce a large-scale perturbation (Milani, 1994). The problem is not only to find a mechanism responsible for the chaotic behaviour, but also to assess the possibility of chaotic diffusion in the space of proper elements, over an even longer time scale. This cannot be excluded, since very weak resonances can interact between them, and generate by their combined action an Arnold web along which long-range diffusion is possible. The only sound conclusion is that the stability of these "stable chaos" cases can only be guaranteed for the time span of the numerical tests.

Analytical results on the stability of the orbits of the Trojans are not available; however, interesting results have been obtained with some simplifying assumptions. In the circular restricted three dimensional 3–body model, it has been possible to prove rigorously that the orbits librating around the Lagrangian points with amplitudes up to $\simeq 30,000\, Km$ are stable for a time span longer than the age of the solar system (see Celletti and Giorgilli, 1990, and references therein). Although the results of this kind are not directly applicable to real Trojans with a realistic dynamic model, they indicate that chaotic diffusion can be extremely slow to the point of being astrophysically irrelevant. The point is that this is not necessarily the case; more progress in quantitative estimates of diffusion is necessary.

The second problem is to identify the stability boundaries of the Trojan cloud. It is clear that a Trojan with too large an amplitude of libration could be unstable, because of the possibility of a too close approach to Jupiter; the same is true for one with too large an eccentricity. It is also true that chaotic orbits occur more frequently for large libration amplitudes. However, we do not know if there are sharp boundaries, where the Trojan orbits become unstable over timespans much shorter than the age of the solar system, or rather there is a comparatively large transition region from which Trojans can slowly "evaporate" with very long characteristic times, e.g. billions of years. To study the present distribution of the known Trojans is not enough; we need to compute also a number of orbits of hypothetical Trojans, with e_p and/or D larger than the observed values.

The main problem is that the structure of these stability boundaries could be very complicated, becoming closer and closer to a fractal as the instabilities time spans increase; thus it is difficult to design an experiment with random initial conditions in such a way that it provides a reliable answer. Computers are now just becoming powerful enough for this kind of "random sample" experiments over millions of years; therefore I will summarize only some recent work, although earlier work should also be quoted (e.g. Innanen and Mikkola, 1989; Mikkola and Innanen, 1992). Holman and Wisdom (1993) have computed the orbits of about 4,000 test particles, initially near the Lagrange points of all the outer planets, for a time span of 20 Myr. This was possible also because they used a symplectic mapping method (Wisdom and Holman, 1991) with a stepsize of 1 yr; this is of course not an accurate integration, but the advantage, with respect to many previous investigations, is that the approximating dynamical system which is really integrated has the same complexity (e.g., the same number of degrees of freedom, the same possibilities of resonance) of the full problem. They find that the maximum amplitude of libration da for a low inclination Trojan is about $0.2\,AU$, only slightly larger than the maximum value found among the real Trojans. Their experiments also define a large region of stability (for 20 Myr) for initial conditions around the L_4 and L_5 points of Saturn, with the remarkable property that low libration amplitudes ($da < 0.1\,AU$) always result into instability, while intermediate amplitudes ($0.1 < da < 0.2\,AU$) can be stable (this was already known to Innanen and Mikkola, 1989). The origin of this instability is not yet known, although I would suggest that a secular resonance could occur for low libration amplitude. Holman and Wisdom also find considerable stability regions around the Lagrange points of Uranus and Neptune; for the latter, they find a puzzling asymmetry between the leading and the trailing side. They conclude, after a discussion of the observational searches conducted so far, that such a "Trojan" of either Saturn, Uranus or Neptune could have been detected, but the degree of completeness of the surveys is low.

Recently, De la Barre and Kaula (1993) have also investigated the stability of ficticious Trojans, both for Jupiter and Saturn, by using an accurate integration of the full outer solar system. They have followed the orbits remaining stable in the Trojan region for up to 161 Myr; they find a region of stability which is comparable to the range of libration amplitudes of the real Trojans. Their continuing investigation is now focusing on the high inclination orbits. De la Barre and Kaula also computed about 350 orbits of ficticious Saturn "Trojans", 4 of which remained stable for 329 Myr. They confirm the instability for the Saturn Trojans with low libration amplitude, with timescales for large-scale instability of less than 1 Myr; moreover, by computing proper elements (by a synthetic method similar to the one

I have used), they find that the range of proper eccentricities of the stable Saturn Trojans is very small ($e_p < 0.025$), so that all these orbits have a perihelion locked to that of Saturn and revolving around the Sun with frequency g_6, not g (as in the example of Figure 2). A tentative conclusion on the existence of Saturn Trojans is that, when all the orbital elements are considered, the volume of the initial conditions in the phase space for stable Saturn Trojans might be quite small, therefore there could be some, but not many of them.

To conclude the discussion on the existence of "Trojans" for other planets, I need to mention that one such object is known for Mars : asteroid *5261 Eureka*; it has a rather large inclination ($\simeq 20°$) and comparatively low eccentricity, with a perihelion longitude close to that of Mars. Mikkola and Innanen (1993) have investigated the range of initial conditions for stable Mars Trojan orbits (for up to $4.4\,Myr$) and found that only for inclinations $15° < i < 30°$ and $32° < i < 44°$, and for low to moderate eccentricities, stability is possible. Finally, one Earth Trojan has at least tentatively been identified. Milani et al. (1989) report that, in computing the orbit of the asteroid 2063 *Bacchus* in the past, they find that it was in the 1 : 1 resonance with the Earth between $-51,000$ and $-10,000$ years before the present time. Since the orbit of *Bacchus* is highly chaotic in its present state of *Geographos-class* Earth crosser, any orbital computation for such a long time has only a probabilistic value : we can only conclude that such a state is possible for *Bacchus*, and might occur for other Near Earth Objects as well.

The third subject on which I would like to see some progress is the understanding of the origin of the orbital distribution of the Trojans. The main puzzle is the high average inclination of the Trojans, and also the extremely high values found for some of them. It is not enough to say that these orbits appear to be stable : we need to know how they can have arisen in the first place. I know of very little work on this subject in the last few years; the main reference is still Yoder (1979), were we find an explanation based upon the sweeping of the $s - s_6$ secular resonance. However, the discussion by Yoder should be updated taking into account what we now know on proper elements and on the stability boundary.

Acknowledgements

I am grateful to P. Farinella, J.G. Williams, D. Davis and C. Froeschlé for carefully reading the manuscript and suggesting many improvements.

References

Bien, R. and Schubart, J. : 1986, "Two Trojans near a secular resonance : (1173) Anchises and (1871) Astyanax". In *Asteroids Comets Meteors II* (Lagerkvist C.-I., Rickman, H., Eds.), pp. 157–160, Uppsala University.

Bien, R. and Schubart, J. : 1987, "Three characteristic parameters for the Trojan group of asteroids". ***Astron. Astrophys.*** **175**, 292–298.

Binzel, R.P. and Xu, S. : 1993, "Chips off of asteroid 4 Vesta : evidence for the parent body of basaltic achondrite meteorites." *Science*, **260**, 186–190.

Celletti, A. and Giorgilli, A. : 1990, "On the stability of the lagrangian points in the spatial restricted problem of three bodies". *Celest. Mech.* **50**, 31–58.

Carpino, M., Milani, A. and Nobili, A.M. : 1987, "Long–term numerical integrations and synthetic theories for the motion of the outer planets". ***Astron. Astrophys.*** **181**, 182–194.

De la Barre, C. and Kaula, W. : 1993, "An investigation into the existence of Saturnian Trojan asteroids". Division on Dynamical Astronomy, AAS, 1993 Meeting, Santa Barbara, California.

Érdi, B. : 1988, "Long periodic perturbations of Trojan asteroids". ***Celest. Mech.*** **43**, 303–308.

Farinella, P., Gonczi, R., Froeschlé, Ch. and Froeschlé, C. : 1993, "The injection of asteroid fragments into resonances". ***Icarus*** **101**, 174–187.

Holman, M.J. and Wisdom, J. : 1993, "Dynamical stability in the outer solar system and the delivery of short period comets". ***Astron. J.*** **105**, 1987–1999.

Innanen, K.A., and Mikkola, S. : 1989, "Studies on solar system dynamics. I. The stability of Saturnian Trojans." ***Astron. J*** **97** 900–908.

Lemaitre, A., and Morbidelli, A. : 1994, "Calculation of proper elements for high inclined asteroidal orbits". ***Celest. Mech.***, in press.

Mikkola,S. and Innanen, K.A. : 1992, "A numerical exploration of the evolution of Trojan-type asteroidal orbits". ***Astron. J.*** **104**, 1641–1649.

Mikkola, S. and Innanen, K.A. : 1993, "On the stability of martian trojans". Preprint, Turku University.

Milani, A. : 1993, "The Trojan asteroid belt : proper elements, stability, chaos and families". ***Celestial Mechanics*** **57**, 59–94.

Milani, A. : 1994, "Proper elements and stable chaos." In : *From Newton to chaos : modern techniques for understanding and coping with chaos in N–body dynamical systems* (A. E. Roy, B.A. Steves, Eds.), Plenum, New York, in press.

Milani, A., and Knežević, Z. : 1994, "Asteroid proper elements and the dynamical structure of the asteroid main belt." ***Icarus***, in press.

Milani, A., and A. M. Nobili : 1992, "An example of stable chaos in the solar system". ***Nature*** **357**, 569–571.

Milani, A., Carpino, M., Hahn, G. and Nobili, A.M., 1989, Dynamics of planet–crossing asteroids : classes of orbital behaviour, Project SPACEGUARD, ***Icarus*** **78**, 212–269.

Milani, A., Farinella, P. and Knežević, Z. : 1992, "On the search for asteroid families". In *Interelations between physics and dynamics for minor bodies in the solar system*, (D. Benest, C. Froschlé, Eds.), 85–132, Editions Frontières.

Nakai, H. and Kinoshita, H. : 1985, "Secular perturbations of asteroids in secular resonance". ***Celestial Mechanics*** **36**, 391–407.

Shoemaker, E. M., Shoemaker, C. S. and Wolfe, R.F. : 1989, "Trojan asteroids : population, dynamical structure and origin of the L4 and L5 swarms". In ***Asteroids** II* (R.P. Binzel, T. Gehrels, M.S. Mattews, Eds.), pp. 487–523, Arizona University Press.

Schubart, J. 1982, "Three characteristic parameters of orbits of Hilda type asteroids". ***Astron. Astrophys.*** **114**, 200–204.

Schubart, J. and Bien, R. : 1987, "Trojan asteroids : relations between dynamical parameters". ***Astron. Astrophys.*** **175**, 299–302.

Williams, J.G. : 1969, "Secular perturbations in the Solar System". Ph.D. Thesis, Univ. California Los Angeles.

Wisdom, J. and Holman, M.J. : 1991, "Symplectic maps for the N-Body problem." ***Astron. J.*** **102**, 1528–1538.

Yoder, C. F. : 1979 " Notes on the origin of the Trojan asteroids." ***Icarus*** **40**, 341–344.

Zappalà, V., Cellino, A., Farinella, P. and Knežević, Z. : 1990, "Asteroid families I : identification by hierarchical clustering and reliability assessment", ***Astron. J.*** **100**, 2030–2046.

Zappalà, V., Cellino, A., Farinella, P. and Milani, A. : 1994, "Asteroid families : II. Extension to unnumbered multi-opposition asteroids." ***Astron. J.*** **107**, 772–801.

KIRKWOOD GAPS AND RESONANT GROUPS

SYLVIO FERRAZ-MELLO
Instituto Astronômico e Geofísico, Universidade de São Paulo,
Caixa Postal 9638, São Paulo, SP, Brasil.
E-mail SYLVIO@IAG.USP.ANSP.BR

Abstract. This paper is a short review of the dynamics of the asteroidal resonances as currently determined from maps and simulations over $10^6 - 10^7$ years. The main recent results concern the extensive exploration of the phase space to determine domains of initial conditions leading to close approaches to the inner planets, the topological dynamics of the planar Sun-Jupiter-asteroid problem at very high eccentricities and the differences amongst 2/1 and 3/2 resonances able to explain the existence of a gap in the asteroidal belt at the 2/1 resonance and of a group of asteroids in the 3/2 resonance. Current results point to a confirmation of Wisdom's theory for the formation of the gaps by gravitational evolution and scattering by the inner planets.

1. Introduction

Several features in the distribution of the asteroids are associated with resonances. They are groups and gaps appearing in plots where the horizontal axis displays the semi-major axis or the mean motion. The gaps are located at positions corresponding to the 4/1, 3/1, 5/2, 7/3 and 2/1 commensurability between the orbital period of Jupiter and that of the asteroid, while the groups are located in the neighbourhood of the 3/2, 4/3 and 1/1 (fig.1).

The gaps were discovered by Kirkwood in 1867 and are known as Kirkwood gaps. The groups in the neighbourhood of the 3/2 and 4/3 resonances have 57 and 2 members currently known (EMP 1994), respectively. We propose to call them Palisa groups after the name of the astronomer who discovered their two paradigms: (153)Hilda and (279)Thule. The asteroids moving in the neighbourhood of these commensurabilities have similar dynamics and are considered in this review. The dynamics of asteroids in the 1/1 resonance, the Trojans, is very particular and is treated in a separated review (Milani, this volume). The techniques used to study mean-motion resonances have been recently reviewed by Henrard (1988) and Froeschlé and Greenberg (1989). Therefore, the present review is limited to the results directly concerning the occurrence of gaps and groups.

Throughout the 20th century, several hypotheses have been formulated to explain the Kirkwood gaps. One of them, the statistical hypothesis, claims that the gaps are only apparent: asteroids near resonances oscillate about the exact resonance value and expend most of the time far from the commensurabilities. This motion, known as libration or σ-libration, is characterized by the oscillation of the angle

$$\sigma = (r+1)\lambda_{Jup} - r\lambda - \varpi$$

(λ and ϖ denote, respectively, the mean longitude and the longitude of the perihelion and r is a rational number; $(r+1)/r$ is the commensurability ratio) around a fixed value. It was recognized since the early work of Poincaré, and does indeed

A. Milani et al. (eds.), Asteroids, Comets, Meteors 1993, 175–188.

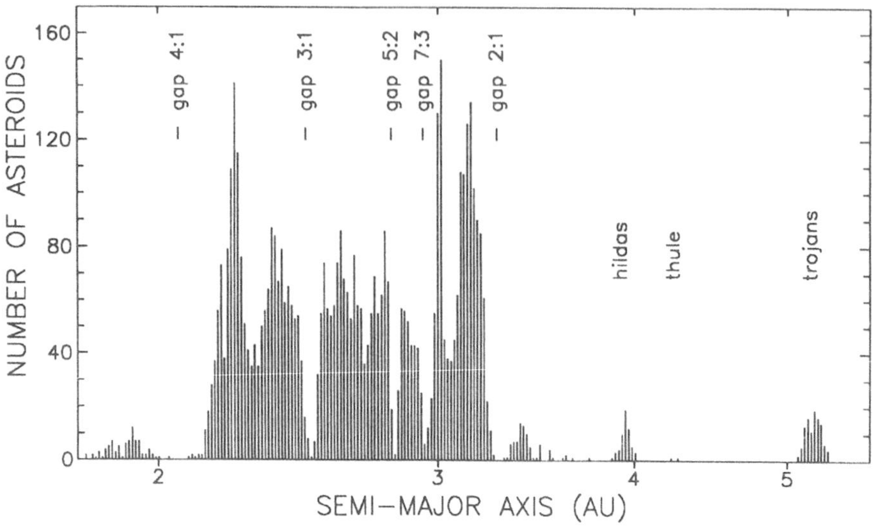

Fig. 1. Histogram of the number of asteroids *vs* semi-major axis showing the location of the main gaps and groups.

exist: it has, for some asteroids, a very large amplitude (up to 0.15 AU, see Ferraz-Mello, 1988). However, it is not large enough to provoke a gap. A complete analysis done by Schweizer (1969) shows that the distribution of the osculating elements in the neighbourhood of the gaps does not significantly differ from the distribution of the mean elements. This is at variance with what should be expected, if the origin of the gaps were of statistical nature. The final argument comes from current numerical simulations over 10^7 years. No regular oscillation has been found with an amplitude large enough to originate a statistical gap as wide as those observed.

The other hypotheses assume that the gaps are real and are either primordial or the result of the orbital evolution of the asteroids. They are usually classed in three groups: cosmogonic, collisional and gravitational. The gravitational theory says that pure gravitational evolution is sufficient to explain the gaps. This theory is difficult to be directly confirmed because of the large time interval elapsed since the origin of the asteroidal belt. However, it stands high in favor since Wisdom's work on the formation of the gap in the 3/1 resonance. Wisdom shows that this gap may be explained by a chaotic diffusion of resonant orbits due to separatrix crossing. This diffusion can drive an asteroid to orbits approaching Mars closely in a very short time-scale ($10^5 - 10^6$ years).

The collisional hypothesis assumes that asteroid collisions are more frequent at resonances. A direct proof or disproof of this hypothesis is difficult. A satisfactory theoretical approach does not exist and numerical simulations are still beyond the existing computational capabilities. Collisions may have played a role in the initial stages of the solar system, but are certainly negligible now, given the small asteroidal masses and space densities. However, if we consider not only collisions amongst asteroids but also with inner planets as well, the collisional hypothesis is part of Wisdom's theory.

At last, there are the cosmogonic hypotheses. They are generally made plausible by the assembling of assumptions in each theory. They generally consist in finding some scenario able to produce gaps in the asteroidal distribution. To the extent that Wisdom's theory of gravitational evolution and scattering of asteroids in resonances is accepted as a general explanation, cosmogonic hypotheses become less important. Indeed, if the efficiency of gravitational evolution mechanisms were as large as pointed out by some recent investigations, the present state of the asteroidal belt at resonances would weakly depend on its primordial state and could provide only a little cosmogonic information. They are not considered in this review.

The crucial point concerning Kirkwood gaps and Palisa groups is that, in spite of their opposite distribution characteristics, they have very similar dynamics. For instance, the theoretical models of the 3/2 and 2/1 resonances are identical and differ only in the numerical value of one parameter (see, for instance, Henrard and Lemaître, 1987; Lemaître and Henrard, 1988). Therefore, in order to be fully accepted, a theory must explain both, the existence of gaps at the 4/1, 3/1, 5/2, 7/3 and 2/1 resonances and the existence of groups at the 3/2 and 4/3 resonances. It must also explain the broadness of the 2/1 and 4/1 gaps, the non-existence of apocentric librators at the 3/2 resonance (while they are abundant at the 2/1 resonance), the existence of some long-lasting librators at the 2/1 gap (while they are absent in other gaps) and the bounded eccentricity range of the 3/2 librators.

2. The 3/1 gap

The first gap to receive an acceptable explanation was the 3/1 gap. Its depletion mechanism was completely unraveled by Wisdom (1982, 1983, 1985). He shows the existence of intermittencies associated with chaos. They appear in this problem as sudden increases – jumps – of the eccentricity (fig.2). Wisdom's simulations show that an asteroid may have a seemingly regular motion for long times and suddenly have a large eccentricity increase due to a separatrix crossing. During these jumps the eccentricity reaches values larger than 0.35 and the asteroid orbit crosses the orbit of Mars. Therefore, the asteroid can have a close approach to the planet. This behaviour is intermittent and the asteroid may escape from close approach many times. However, the close approach eventually occurs and the asteroid exchanges a great amount of energy with Mars, leaving the resonance. The time scale associated with these intermittencies is short ($\sim 10^5$ years) as compared to the age of the solar system. Therefore, the theory is consistent with the absence of permanent asteroids in the 3/1 resonance. The few asteroids presently observed in this resonance are believed to have come from recent captures and are bound to close approaches with the inner planets (Milani *et al.*, 1989).

The work of Wisdom on the 3/1 resonance was extended by Ferraz-Mello and Klafke (1991) and Hadjidemetriou (1992) to very high eccentricities. Using averaged potentials given by semi-numerical expansions, Ferraz-Mello and Klafke mapped the structure of the phase space of the restricted elliptic model in the neighbourhood of this resonance. They show that, besides the region of homoclinic chaos discovered

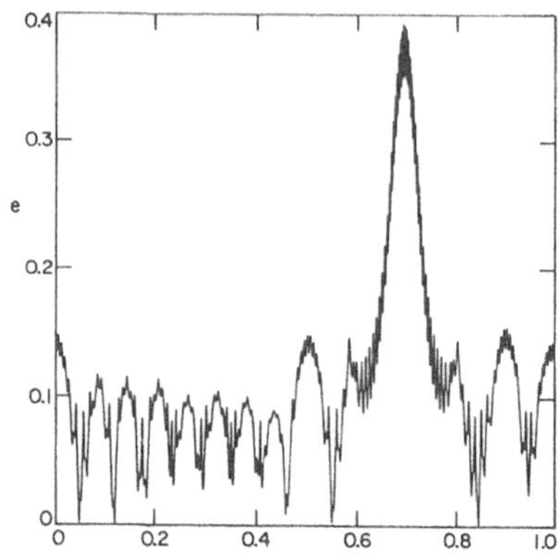

Fig. 2. Intermittencies (jumps) of the eccentricity of a fictitious asteroid in the 3/1 resonance. Total integration time: 200,000 years (Wisdom, 1983).

by Wisdom, there is another at high eccentricities. They also show that, for some values of the energy, a heteroclinic bridge allows the solutions to go from one region to another (fig.3). The intermittencies associated with orbits going through this heteroclinic bridge can drive the asteroidal eccentricity to values as high as 0.9 and then back to a lower eccentricity region. Orbits of fictitious asteroids going through the heteroclinic bridge to high eccentricities were actually computed by Saha (1992). In fact, they are very common: the orbit of (4179) Toutatis lies in this region.

The 3/1 resonance has become a paradigm in the study of asteroidal gaps and has been studied by many authors. In particular we quote Henrard and Caranicolas (1990) who studied the role played by secondary resonances in the complex separatrix-crossing mechanism causing the chaotic behaviour at low eccentricities.

3. The 5/2 and 7/3 gaps

The second-order techniques devised by Wisdom to study the 3/1 resonance use an abridged Hamiltonian that contains only the leading terms of the classical Laplacian expansion of the disturbing potential of Jupiter. The convergence radius of this expansion in the outermost 5/2, 7/3, 2/1 and 3/2 resonances is $e \sim 0.28$, $e \sim 0.26$, $e \sim 0.20$ and $e \sim 0.09$, respectively (it is $e \sim 0.35$ in the 3/1 resonance; see Ferraz-Mello, 1994). When these values are compared to the eccentricities which may lead to collisions with Mars – 0.41, 0.44, 0.49 and 0.58, respectively – we see that theories based on the classical Laplacian expansion can no longer give a verdict on the occurrence, or not, of gravitational mechanisms with amplitude large enough to allow the depletion of a neighbourhood of these resonances.

Wisdom's mapping and perturbative models allowed several authors to show that the resonances 3/1 and 5/2 are quite similar. Šidlichovský and Melendo (1986) show that a low-eccentricity chaotic region exists in the 5/2 resonance and must account for the existence of the gap. Yokoyama and Balthazar (1992) studied the dynamics

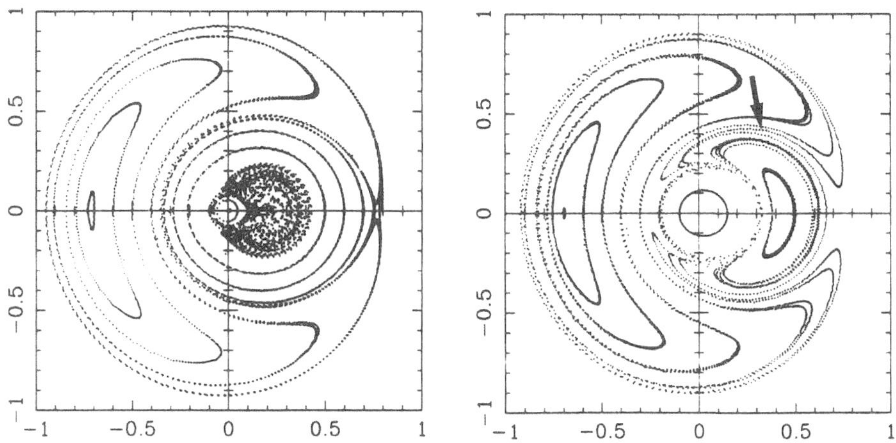

Fig. 3. Poincaré maps ($\sigma = \pi/2$, $\dot{\sigma} < 0$) of the resonance 3/1 in the frame of the planar averaged Sun-Jupiter-asteroid problem at two different energy levels. Coordinates are $x = e.\cos(\varpi - \varpi_{Jup})$, $y = e.\sin(\varpi - \varpi_{Jup})$. At left, the chaotic domain found by Wisdom is seen in the inner part of the figure. At right, the tori confining this region no more exists and a heteroclinic bridge (arrow) is seen (Ferraz-Mello and Klafke, 1991).

of this resonance. They show the separatrix crossing mechanism leading to chaotic behaviour, which is able to transfer one orbit from the low-eccentricity regime (up to 0.1 – 0.2) to an outermost one where it may reach 0.4. The comparison of the results with numerical integrations (Šidlichovský, 1993) shows that Wisdom's techniques, at their limit for this resonance, give a qualitatively correct picture of the solution, in spite of some important quantitative differences (fig 4a). We should also mention some early results of Scholl and Froeschlé (1975) who showed increases of eccentricity up to 0.3 in relatively short time-scales ($\sim 20,000$ years), in agreement with the results reported above.

The chaotic origin of the gap was later confirmed by Ipatov (1990, 1992) by means of an extensive simulation. The simulation shows that the domain of initial conditions corresponding to chaotic orbits coincides with the region depleted in the distribution of the asteroids in the plane eccentricity *vs.* semi-major axis (fig. 4b). Furthermore, using averaged potentials given by semi-numerical expansions, Klafke *et al.* (1992) mapped the structure of the phase space of the restricted elliptic model in the neighbourhood of this resonance. They show that, similar to the 3/1-resonance, there is a chaotic region at high-eccentricities, which is linked to the low-eccentricity region by a heteroclinic bridge for a wide range of energies.

In the case of the 7/3 resonance, Yokoyama and Balthazar (1992) have shown that, in the Sun-Jupiter-asteroid model, there are gravitationally confined low-eccentricity orbits. However, Yoshikawa (1991) studied the distribution of the observed asteroids in the neighbourhood of this commensurability and found that the boundary of the depleted region in the plane eccentricity *vs.* semi-major axis almost

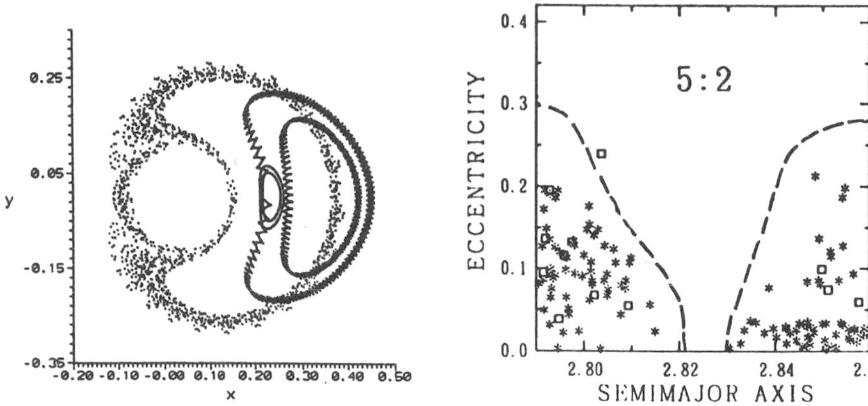

Fig. 4. (a) Comparison of surfaces of section obtained with maps in the 5/2 resonance and real trajectories for high eccentricities. Coordinates as in Fig. 3 (Šidlichovský, 1993). (b) Distribution of the numbered asteroids around the 5/2 resonance when $\sigma = 0$ and $\varpi - \varpi_{Jup} = \pi$ (after Yoshikawa, 1991); the curves show the boundaries of the region corresponding to Mars crossing asteroids with $i = 10°$ (after Ipatov, 1991).

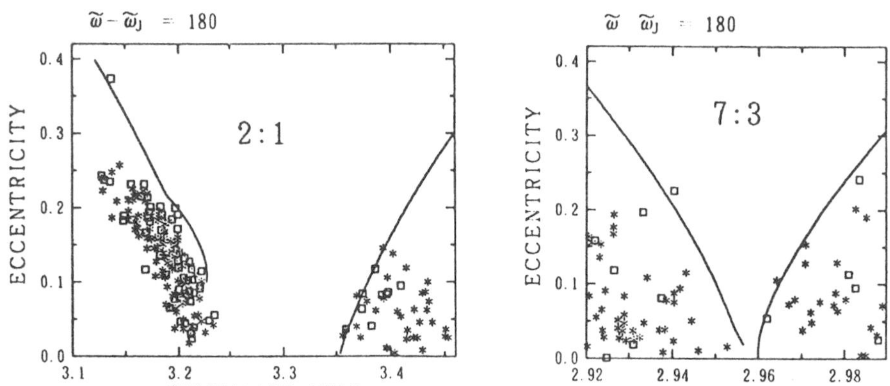

Fig. 5. Distribution of the numbered asteroids around the 7/3 and 2/1 resonances when $\sigma = 0$ and $\varpi - \varpi_{Jup} = \pi$. Two curves drawn in each diagram indicate the boundaries of the resonance region in the planar case. In the diagram of the 2/1 resonance, the asteroids in the lower left region were omitted (asterisk: $i < 15°$, square: $i > 15°$) (Yoshikawa, 1991).

coincides with the boundary between librating and circulating solutions (fig.5). The situation is quite similar to that described below for the 2/1 gap. Even if it does not privilege any particular theory, we may conjecture, by analogy with the 2/1 resonance, that the inclusion of the outer planets in the model may introduce the possibility of intermittencies allowing the low-eccentricity asteroids to migrate to regions of large orbital eccentricity.

4. The 2/1 gap

The existence of low-eccentricity chaotic solutions in the neighbourhood of the 2/1 resonance was first found by Giffen (1973). It was confirmed by Froeschlé and Scholl (1976, 1981), who also showed that those solutions are confined to low-eccentricities by regular motions starting near $e = 0.2$. They were also found by Murray (1986) and Wisdom (1987). Lemaître and Henrard (1990) identified the origin of the chaotic behaviour both in the transition between large-amplitude σ-librations and circulations, at the borders of the resonance, and in the secondary resonances between the libration frequency and the perihelion motion. The topological dynamics of this resonance was recently studied by Moons and Morbidelli (1993) with Poincaré maps up to $e = 0.3$. Their results show the low-eccentricity chaotic region and the quasi-integrable tori confining this region. These results constitute one important part of the current knowledge of the dynamics of the 2/1 resonance, but they are limited to eccentricities smaller than the necessary 0.49 to allow the asteroid orbit to intersect the orbit of Mars.

The difficulties concerning the non-existence of a globally valid expansion of the disturbing function were overcome in different ways.

Morbidelli and Moons (1993) use an approach based on local non-central evaluations of the disturbing function (see Ferraz-Mello and Sato, 1989). They include the two main harmonics of Jupiter's secular resonances, usually denoted as ν_5 and ν_6, and consider inclined orbits. The resulting system has a high number of degrees of freedom and they successively eliminate them by semi-numerical averaging. The final step is a juxtaposition of averages obtained with the separate elimination of the angles associate with ν_5 and ν_6. The solutions are shown on surfaces of section which, as pointed out by the authors, are not Poincaré sections. The results are consistent with the existence of a family of asteroids similar in distribution to that of the Hildas in the resonance 3/2. They do not explain why these asteroids are absent and rather support the need of a cosmogonic hypotheses to explain the 2/1 gap.

We present in figure 6 some Poincaré maps of the planar Sun-Jupiter-asteroid problem obtained with numerical integrations done over 1 Myr and smoothed by filtering out the high frequencies (see Michtchenko and Ferraz-Mello, 1993). These smoothed integrations were interpreted as solutions of an averaged dynamical system with 2 degrees of freedom. The sections show a large region of seemingly ordered motions, even for large libration amplitudes ($\Delta a \sim 0.15$ AU) and eccentricities as large as 0.5.

The chaotic regions identified by Lemaître and Henrard (1990) are seen in (B) and (D). In (B), chaotic motions due to the overlapping of secondary resonances

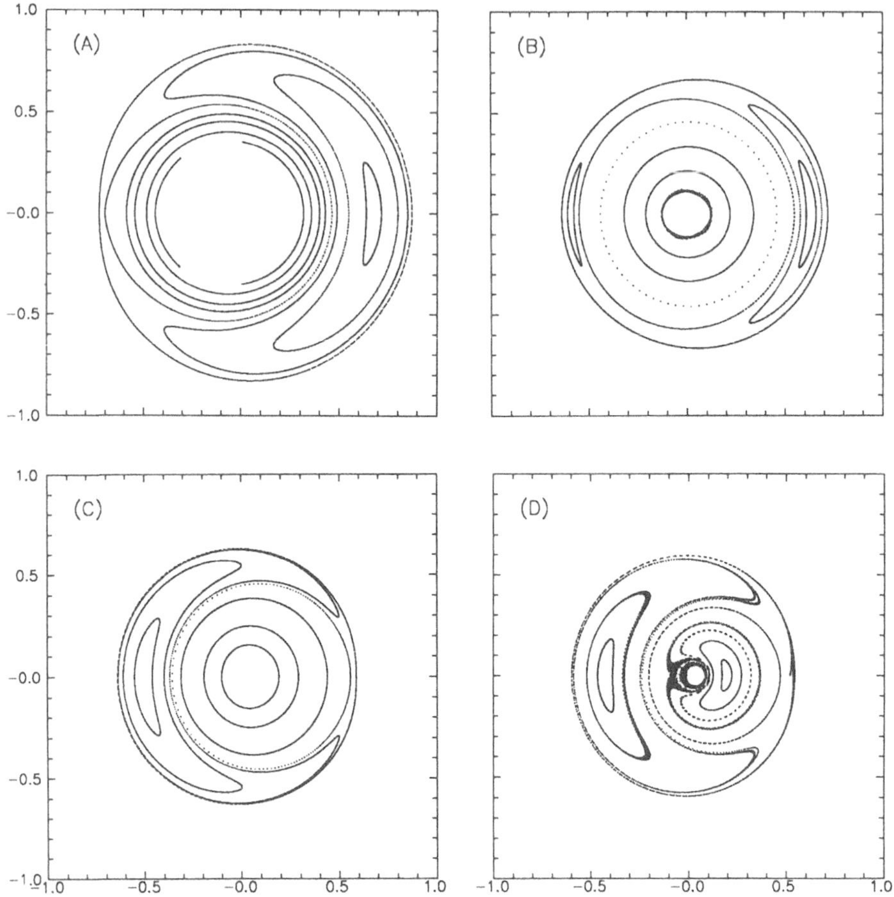

Fig. 6. Poincaré maps ($\sigma = 0,\ \dot{\sigma} > 0$) of the resonance 2/1 in the frame of the planar averaged Sun-Jupiter-asteroid problem. Coordinates as in Fig. 3.

are seen in the middle of the section; in (D), they mark the transition between σ-librations and circulations. These sections extend to high eccentricities the well-known facts that made many authors to find impossible to explain the Kirkwood 2/1 gap by the same mechanisms explaining the 3/1 gap. For instance, in the section (B) one may see that the low-eccentricity chaos is confined by seemingly regular tori, as already mentioned, and is not able to produce intermittencies reaching high-eccentricities. No new chaotic region is visible inside the resonant region.

The results on the Sun-Jupiter-asteroid model do not explain the 2/1 gap. However, when the perturbations due to Saturn are included and the initial conditions assume an inclined orbit ($\imath = 5°$), all solutions become clearly chaotic. The corresponding maximum Lyapunov exponents were estimated from some fifty 5–7 Myr integrations for initial eccentricities in the interval $0.1 < e < 0.4$ and semi-major axes ranging from the middle to the border of the resonance region. The results are generally in the range $10^{-5} - 10^{-3.5}\mathrm{yr}^{-1}$, except for a few orbits starting from the very middle of the resonance. These results show that Saturn triggers the destruction of the regular structures seen in fig. 6. Even if KAM tori may exist, given the high number of degrees of freedom, they no more divide energy manifolds. Some barriers may still exist (see Wiggins, 1990) but, in general, they are not expected to avoid the occurrence of intermittencies allowing orbits to reach high eccentricities and encounter the inner planets.

The actual occurrence of such diffusion in this system, when all the outer planets are considered, is confirmed by numerical simulations. Wisdom (1987) reports the integration of 5 test particles with initial eccentricities 0.05 or 0.1, for 5 Myr. Three of them reached $e \sim 0.4 - 0.5$. Current investigations of Scholl *(personal communication)* show that the range of initial conditions leading to high eccentricities in 12 Myr or less is broad and excludes only orbits close to the libration centers (Fig. 7)

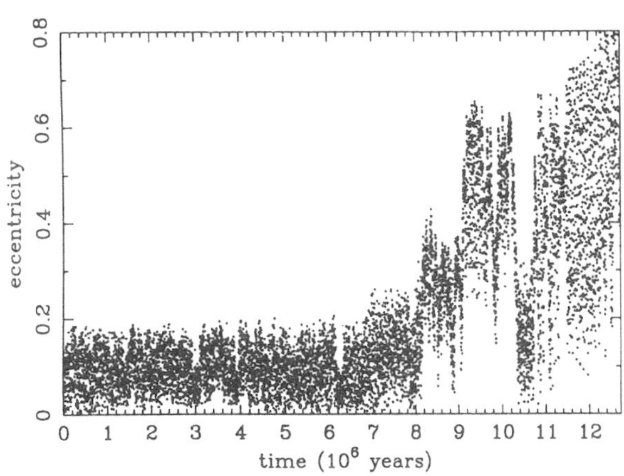

Fig. 7. Evolution of the eccentricity of a fictitious asteroid in the resonance 2/1. (Scholl, *personal communication*)

These results are completed by other evidences of the association of the depleted region with the region where large Lyapunov exponents were found. For instance, Yoshikawa (1991) studied the distribution of the observed asteroids in the neighbourhood of this resonance and found that the boundary of the depleted region in the plane eccentricity *vs.* semi-major axis almost coincides with the boundary of the libration zone (see fig.5).

The above results may be interpreted as responsible for orbital changes able to deplete the resonance at a time rate consistent with the fact that some asteroids are librating inside the 2/1 resonance. Many of them may be considered as temporary but at least one, (3789) Zhonguo, has a smaller eccentricity and seems to be able to remain inside the resonance for a long time. In fact, the long term evolution of this asteroid is not yet known.

However, one should keep in mind that we are dealing with very slow processes, at the limit of the available numerical and analytical techniques. A great deal of investigation is still necessary to confirm the results and to identify all dynamical mechanisms at work in this resonance.

5. The 3/2 and 4/3 groups

The results found in the 2/1 resonance cannot repeat themselves in the 3/2 resonance. Otherwise this resonance should be as depleted as the 2/1 resonance, at variance with the observation of some 60 asteroids moving near the exact commensurability. On the basis of the seemingly regular short-term evolution of some computed orbits in the Sun-Jupiter-asteroid model, Giffen (1973) conjectured that this regularity should be responsible for the existence of this group. A longer integration performed by Wisdom (1987) also shows regular motions in the region where the Hildas are found.

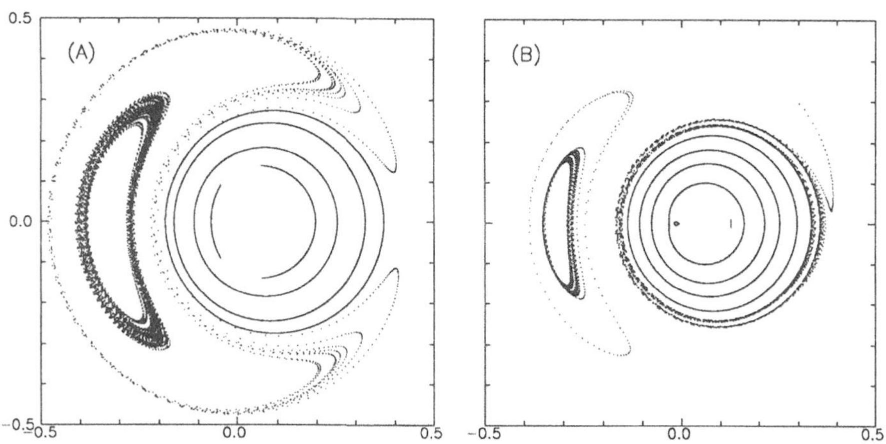

Fig. 8. Poincaré maps ($\sigma = 0$, $\dot{\sigma} > 0$) of the resonance 3/2 in the frame of the planar averaged Sun-Jupiter-asteroid problem. Coordinates as in Fig. 3. Orbits in the perihelion libration lobe are highly chaotic and are bound to close approaches to Jupiter in short times. Orbits in the innermost part remains regular for long terms even when inclined orbits and the perturbations of Saturn are taken into account. The actual Hildas are in the inner region.

Figure 8 shows Poincaré maps of the planar Sun-Jupiter-asteroid problem obtained numerically. They are different from those obtained for the 2/1 resonance in several aspects: (a) the region of perihelion circulation is devoid of chaotic activity (confirmed by Lyapunov exponents tending to zero in numerical integrations over 10 Myr); (b) there is a persistent saddle point at the right side and one lobe of libration of the perihelion at the left side (in the 2/1 resonance there is a reversion of this structure and, then, transition portraits showing two lobes of libration, one at the right and the other at the left side; see section (B) in fig. 6); (c) the bifurcation between the two modes of oscillation of the perihelion is the source of appreciable chaoticity spreading itself over a large part of the perihelion libration lobe and also over the outer circulations.

Figure 8 also shows that other planets are not needed to explain the depletion of the orbits with mean eccentricities larger than ~ 0.3. All sections show a fast diffusion in the libration lobe whose outer orbits are scattered by approaches to Jupiter itself. The inclusion of Saturn only accelerates this phenomenon allowing the orbits to be scattered in less than 1 Myr. The results of Morbidelli and Moons (1993) for this resonance also show an extended chaotic region surrounding seemingly regular motions with $e < 0.25$.

When the perturbations due to Saturn are taken into account and the asteroid is left to move in an inclined orbit, the maximum Lyapunov exponents of the inner regular orbits still tend to very small values. They are in the range $10^{-5.5} - 10^{-7}\mathrm{yr}^{-1}$. Franklin *et al.* (1993) extended to this system some results on the time necessary for the occurrence of sudden orbital transitions, in the N-body asteroidal problem, when the diffusion process in the asteroid eccentricity is assumed as unbounded. They concluded that, in this case, this time is generally much larger than the age of the Solar System and that there is no contradiction in the fact that most of the real Hildas are formally chaotic.

The reported results show many similarities in the dynamics of the 2/1 and 3/2 resonances. In many respects, the 2/1-gap orbits appear as more regular than those of the 3/2 resonance. The only instance in which this fact is reversed is when Lyapunov exponents are computed in the Sun-Jupiter-Saturn-asteroid problem. The fact that the maximum Lyapunov exponents of the orbits in the 2/1 gap are at least 2 orders of magnitude larger than those in the 3/2 group is, likely, one reason for the observed distribution differences.

One open question concerns the existence of chaotic low-eccentricity orbits close to the exact commensurability. For low eccentricities, it is possible to calculate the locus of secondary resonances similar to those responsible for the low-eccentricity chaos of the 2/1 resonance (Lemaître and Henrard, 1988; Michtchenko, 1993). Schubart (1990) computed 3 orbits in this region and found no special features other than secondary resonances. Michtchenko (1993) computed a large number of orbits in this region and analyzed them by means of Fourier Transforms. Her results point to the existence of thin layers of chaotic orbits with very small eccentricities. Some of these layers are clearly associated with secondary resonances.

The Palisa group at the 4/3 resonance is recognized as a group, in spite of its only 2 members. They are the only known asteroids with a mean period between 8 years (period of the Hildas) and 11.8 years (period of the Trojans and Jupiter). The structure of the 4/3 resonance was recently studied by Ries (1993) in the frame of the planar restricted three-body problem. Her results point toward the survival of low-eccentricity objects ($e < 0.075$).

6. The 4/1 gap

In this review, we have not considered the particular problems related to the 4/1 gap. The dynamics in the neighbourhood of this commensurability is complex even when Jupiter is kept moving in a fixed ellipse (see Klafke *et al.*, 1992). Moreover, this resonance is situated at 2.06 AU, just where the secular resonances ν_6 and ν_{16}, as derived from general secular theories, occurs for low-eccentricity asteroids (fig. 9). Scholl and Froeschlé (1991) have shown that the overlapping of these resonances is responsible for a complex dynamic behaviour in the whole 2.0 – 2.1 AU region. In this region, the necessary eccentricity for close approach to Mars is only 0.2. Thus, even without a complete dynamical study, one can say that this is certainly a reason for the huge observed depletion.

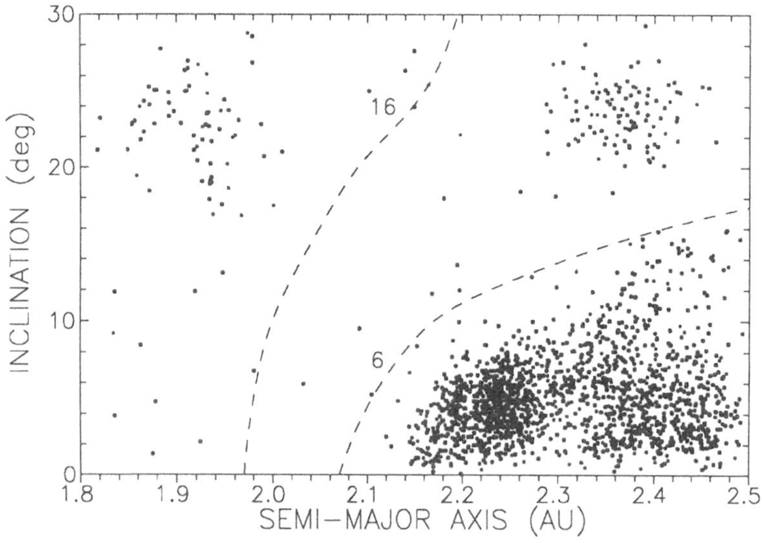

Fig. 9. Distribution of the asteroids in the neighbourhood of the 4/1 resonance showing the wide gap at 2.0 – 2.1 AU. The locus of the secular resonances ν_6 and ν_{16}, as derived from general secular theories, is also shown.

7. Conclusion

We reviewed the main period resonances. Whenever possible, we emphasized the results concerning the topological dynamics of these resonances rather than those that are only generic simulations. The large quantity of simulations conducted in the past 20 years has already shown that asteroids can be found in these places only if some protective mechanisms related to σ-libration or secular resonance are acting.

The collected results refer to current maps and simulations extending over 10^6 – 10^7 years. They show that several conclusions, obtained in the past on the basis of simulations over $10^4 - 10^5$ years, are not correct. This fact may serve to prevent us from assuming that current conclusions are final. They are certainly an improvement on the previous scenario but we do not know what will be unraveled when our theories become able to show evolutions over a time span as long as $10^8 - 10^9$ years.

References

Efemeridi Malikh Planet na 1994 god (EMP 1994), Inst. Teoret. Astron., St. Peterburg, Russia (1993).

Ferraz-Mello, S. : 1988, "The high-eccentricity libration of the Hildas". *Astron. J.* **96**, 400-408.

Ferraz-Mello, S. : 1994, "The convergence domain of the Laplacian expansion of the disturbing function". *Celest. Mech. Dyn. Astron.* **58**.

Ferraz-Mello, S. and Klafke, J.C. : 1991, "A model for the study of very-high-eccentricity asteroidal motion. The 3: 1 resonance". In *Predictability, Stability and Chaos in N-Body Dynamical Systems* (A.E.Roy, ed.), Plenum Press, New York, 177-184.

Franklin, F., Lecar, M. and Murison, M. : 1993, "Chaotic orbits and long-term stability : an example from asteroids of the Hilda group". *Astron. J.* **105**, 2336-2343.

Froeschlé, C. and Greenberg, R. : 1989, "Mean-motion resonances". In *Asteroids II* (R.P.Binzel *et al.*, eds.), Univ. Arizona Press, Tucson, 827-844.

Froeschlé, C. and Scholl, H. : 1976, "On the dynamical topology of the Kirkwood gaps". *Astron. Astrophys.* **48**, 389-393.

Froeschlé, C. and Scholl, H. : 1981, "The stochasticity of peculiar orbits in the 2/1 Kirkwood gap". *Astron. Astrophys.* **93**, 62-66.

Froeschlé, Ch. and Scholl, H. : 1986, "The secular resonance ν_6 in the asteroidal belt". *Astron. Astrophys.* **166**, 326-332.

Giffen, R. : 1973, "A study of commensurable motion in the asteroidal belt". *Astron. Astrophys.* **23**, 387-403.

Hadjidemetriou, J. : 1992, "The elliptic restricted problem at the 3: 1 resonance". *Celest. Mech. Dyn. Astron.* **53**, 151-183.

Henrard, J. : 1988, "Resonances in the planar restricted elliptic problem". In *Long-term Dynamical Behaviour of Natural and Artificial N-Body Systems* (A.E.Roy, ed.), Plenum Press, New York, 405-425.

Henrard, J. and Caranicolas, N. : 1990, "Motion near the 3/1 resonance of the planar elliptic restricted three body problem". *Celest. Mech. Dynam. Astron.* **47**, 99-121.

Henrard, J. and Lemaître, A. : 1987, "A perturbative treatment of the 2/1 jovian resonance". *Icarus* **69**, 266-279.

Ipatov, S.I. : 1990, "Variations in orbital eccentricities of asteroids near the 5: 2 resonance". *Sov. Astron. Letters* **15**, 324-328.

Ipatov, S.I. : 1992, "Evolution of asteroidal orbits at the 5: 2 resonances". *Icarus* **95**, 100-114; **97**, 309.

Klafke, J.C., Ferraz-Mello, S. and Michtchenko, T. : 1992 "Very-high-eccentricity librations at some higher-order resonances". In *IAU Symposium 152* (S.Ferraz-Mello, ed.), Kluwer, Dordrecht, 153-158.

Lemaître, A. and Henrard, J. : 1988, "The 3/2 resonance". *Celest. Mech.* **43**, 91-98.

Lemaître, A. and Henrard, J. : 1990, "Origin of the chaotic behaviour in the 2/1 Kirkwood gap". *Icarus* **83**, 391-409.

Michtchenko, T.A. : 1993, *Dr. Thesis*, University of São Paulo.

Michtchenko, T.A. and Ferraz-Mello, S. : 1993, "The high-eccentricity libration of the Hildas. II. Synthetic-theory approach". *Celest. Mech. Dynam. Astron.* **56**, 121-129.

Milani, A., Carpino, M., Hahn, G. and Nobili, A.M. : 1989, "Project Spaceguard : Dynamics of planet-crossing asteroids. Classes of Orbital Behaviour". *Icarus* **78**, 212-269.

Moons, M. and Morbidelli, A. : 1993, "The main mean-motion commensurabilities in the planar circular and elliptical problem". ***Celest. Mech. Dynam. Astron.*** **57**, 99-108.

Morbidelli, A. and Moons, M. : 1993, "Secular resonances in mean-motion commensurabilities. The 2/1 and 3/2 cases". ***Icarus*** **102**, 316-332.

Murray, C.D. : 1986, "Structure of the 2: 1 and 3: 2 Jovian Resonances". ***Icarus*** **65**, 70-82.

Ries, J.G. : 1993, "Numerical exploration of the 4: 3 resonance in the Elliptic Restricted Problem". ***Bull. Amer. Astron. Soc.*** (*in press*).

Saha, P. : 1992, "Simulating the 3: 1 Kirkwood gap". ***Icarus***, **100**, 434-439.

Scholl, H. and Froeschlé, C. : 1974, "Asteroidal motion at the 3/1 commensurability". ***Astron. Astrophys.*** **33**, 455-458.

Scholl, H. and Froeschlé, C. : 1975, "Asteroidal motion at the 5/2, 7/3 and 2/1 resonances". ***Astron. Astrophys.*** **42**, 457-463.

Scholl, H. and Froeschlé, Ch. : 1991, "The ν_6 secular resonance region near 2 AU : A possible source of meteorites". ***Astron. Astrophys.*** **245**, 316- 336.

Schubart, J. : 1990, "The low-eccentricity gap at the Hilda group of asteroids". In ***Asteroids, Comets, Meteors III*** (C.-I. Lagerqvist *et al.*, eds.), Reprocentralen HSC, Uppsala, 171-174.

Schweizer, F. : 1969, "Resonant asteroids in the Kirkwood gaps and statistical explanation of the gap". ***Astron. J.*** **74**, 779-788.

Šidlichovský, M. : 1993, "Chaotic behaviour of trajectories for the fourth and third order asteroidal resonances". ***Celest. Mech. Dyn. Astron.*** **56**, 143-152.

Šidlichovský, M. and Melendo, B. : 1986, "Mapping for the 5/2 asteroidal commensurability". ***Bull. Astron. Inst. Czechoslov.*** **37**, 65-80.

Wiggins, S. : 1990, ***Chaotic Transport in Dynamical Systems***, Springer-Verlag, New York.

Wisdom, J. : 1982, "The origin of Kirkwood gaps: A mapping for asteroidal motion near the 3/1 commensurability". ***Astron. J.*** **85**, 1122-1133.

Wisdom, J. : 1983, "Chaotic behaviour and the origin of the 3/1 Kirkwood gap". ***Icarus*** **56**, 51-74.

Wisdom, J. : 1985, "A perturbative treatment of motion near the 3/1 commensurability". ***Icarus*** **63**, 279-282.

Wisdom, J. : 1987, "Chaotic dynamics in the Solar System". ***Icarus*** **72**, 241-275.

Yokoyama, T. and Balthazar, J.M. : 1992, "Application of Wisdom's perturbative method for 5: 2 and 7: 3 resonances". ***Icarus***, **99**, 175-190.

Yoshikawa, M. : 1990, "Motions of asteroids at the Kirkwood gaps. I. On the 3: 1 resonance with Jupiter". ***Icarus*** **87**, 78-102.

Yoshikawa, M. : 1991, "Motions of asteroids at the Kirkwood gaps. II. On the 5: 2, 7: 3 and 2: 1 resonances with Jupiter". ***Icarus*** **92**, 94-117.

THE SECULAR RESONANCES IN THE SOLAR SYSTEM

CHRISTIANE FROESCHLE
O.C.A.Laboratoire G.D. Cassini, CNRS URA 1362 B.P.229 F-06304 Nice Cedex 4, France
E–mail Froesch@obs-nice.fr

ALESSANDRO MORBIDELLI
O.C.A.Laboratoire Cerga, CNRS URA 1360 B.P.229 F-06304 Nice Cedex 4, France
E–mail Morby@obs-nice.fr

Abstract. In the last three years new studies on secular resonances have been done. The second–order and fourth–degree secular perturbation theory of Milani and Knežević allowed to point out the effect of mean motion resonances on the location of the linear and non linear secular resonances. Moreover this theory improved the knowledge of the exact location of the $g = g_6$ (i.e. ν_6) resonance at low inclination. Morbidelli and Henrard revisited the semi–numerical method of Williams, taking into account the quadratic terms in the perturbing masses. They computed not only the location of secular resonances, but also provided a global description of the resonant dynamics in the main secular resonances namely $g = g_5$ (i.e. ν_5), $g = g_6$ (i.e. ν_6) and $s = s_6$ (i.e. ν_{16}). The resonant proper element algorithm developed by Morbidelli allows to identify the dynamical nature of resonant objects, and is a powerful tool to study the mechanisms of meteorite transport to the inner Solar System. Purely numerical experiments have been done, which show : (i) the complexity of the dynamics when two resonances overlap; (ii) the efficiency of successive crossings of non linear resonances in pumping up the inclination of small bodies; (iii) the efficiency of the secular resonance ν_6 as a source of meteorites up to 2.4 AU.

1. Introduction.

According to secular perturbation theories, the orbital elements of the planets change with periods ranging from thousands to millions of years. If one restricts to the Sun–Jupiter–Saturn system, these changes are quasi–periodic with three basic frequencies : g_5 (the average precession rate of Jupiter's longitude of perihelion), g_6 (the average precession rate of Saturn's longitude of perihelion) and s_6 (the precession rate of both nodes); additional frequencies appear when the full solar system is taken into account. The planets exert also secular perturbations on any small body orbiting around the Sun and force the precession of their orbits; we denote by g the precession frequency of the asteroid's longitude of perihelion, and s the precession frequency of its node. These secular perturbations give particularly large effects when a secular resonance occurs, namely when the frequency of precession of the small body g or s (or a combination of these frequencies) becomes nearly equal to an eigenfrequency (or a combination of eigenfrequencies) of the planetary system. In the following we will call *linear* secular resonances those involving only *one* asteroid frequency and *one* planetary frequency.

In the present review (since previous ones have been published, Scholl et al. 1989, Froeschlé and Scholl 1989) we will just mention the most important steps which

A. Milani et al. (eds.), Asteroids, Comets, Meteors 1993, 189–204.

have been done in the field of secular resonances up to 1989 and we will devote this paper to the new results obtained after 1989.

The phenomena of secular resonances have been known since the end of the last century (LeVerrier 1855, Tisserand 1882, Poincaré 1892). The first modern approach to study secular resonances was done by Williams (1969). He developed a semi–numerical theory based on Gauss averaging method, which can be applied to derive proper elements of asteroidal orbits even for large values of eccentricity and/or inclinations, and mapped the contours of secular resonances in the main asteroid belt as a function of the proper elements a, e and i. This method was later refined (Williams and Faulkner 1981) to derive detailed graphical maps of the surfaces which denote the location of the three strongest secular resonances, in the proper element space with semi major axis ranging between 1.25 and 3.5 AU.

Since the development of a new generation of high speed computers, orbital evolution of resonant and near resonant asteroids have been studied. These purely numerical experiments were based mainly on the four body model, i.e. Sun–Jupiter–Saturn–Asteroid (see Froeschlé and Scholl 1989). The numerical experiments yielded new quantitative results and confirmed the qualitative behaviour of orbital evolution at secular resonances obtained previously by semi–analytical theories. Moreover they revealed the existence of chaotic motions.

These numerical works revived the interest for the study of secular resonances and, as a consequence, new improved theories appeared such as the ones by Yoshikawa (1987) and Šidlichovský (1989), for the $g = g_6$ resonance, and by Nakai and Kinoshita (1985), for the $s = s_6$ resonance (see Froeschlé and Scholl 1989 for a review of these works). The more recent theories were developed by Milani and Knežević (1990), and also by Morbidelli and Henrard (1991a).

Milani and Knežević (1990) developed an analytical theory to compute proper elements, where only Jupiter and Saturn are considered. Their work is based on an explicit expansion of the hamiltonian of the model in powers of the eccentricity and inclination of both the asteroid and the planets; such expansion is truncated at degree 4 for what concerns the part which is linear in the masses of the planets, and degree 2 for what concerns the quadratic part. All the terms of order larger than 2 in the masses are also neglected. As a result they located in the proper element space not only the linear secular resonances but also the ones of degree 4 in eccentricity and inclination. Retaining the quadratic terms in the perturbing masses they pointed out the strong effects imposed by mean motion resonances on the location of secular ones. Knežević et al.(1991), generalizing the previous theory by considering the perturbation of the four major planets, determined the location of the linear secular resonances from 2 to 50 AU. Milani and Knežević (1992) published an improved version of their theory, taking into account the secular perturbation of the four outer planets and part of the effect of the inner planets. They studied the effects of non linear secular resonances upon asteroid families. Finally, Milani and Knežević (1994) analyzed in detail the dynamical behaviour of several asteroids close to some non linear secular resonances.

Šidlichovský (1990) extending his non linear theory developed in 1989 studied the problem of overlapping of secular resonances. The simple Chirikov overlapping criterion was applied to the hamiltonian where both terms corresponding to secular resonances $g = g_5$ and $g = g_6$ are taken into account. The overlap criterion yields a critical value of the quantity $P = \sqrt{(1-e^2)}.(1-\cos i)$ depending on the semimajor axis of the asteroid. When P is greater than the critical value, resonance overlap occurs and chaotic motion appears. A mapping of the hamiltonian is presented which allows to calculate rapidly trajectories.

All these theories, based on the classical expansion of the perturbation in power series of the asteroid's eccentricity and inclination, have necessary a poor accuracy in determining the location and the dynamics of secular resonances at large eccentricity and inclination. Therefore, Morbidelli and Henrard (1991a) revisited the semi–numerical method first developed by Williams (1969), avoiding the expansion in power series of asteroid's e and i. Introducing suitable action–angle variables, they take completely into account the strongly non–linear dynamics related to the motion of the perihelion argument of the small body, which is dominant at high inclination, as shown by Kozai (1962). Furthermore they improved Williams' results on the location of resonances retaining the quadratic terms in the mass ratio (neglected by Williams). Moreover they studied in detail the dynamics of the three principal secular resonances $g = g_6$ (i.e. ν_6 in Williams' notations), $g = g_5$ (i.e. ν_5) and $s = s_6$ (i.e. ν_{16}) (Morbidelli and Henrard, 1991b).

Recently Morbidelli (1993) developed an efficient algorithm for the computation of the dynamical evolution of asteroids which are inside or close to a secular resonance. This algorithm is able to identify the dynamical nature of resonant objects. Since it requires a short CPU time for its execution, it is a powerful tool to study the dynamics of many fictitious objects in order to investigate the mechanisms of meteorite transport to the Earth. In addition, this work has pointed out the dynamical peculiarities of the $g = g_6$ resonance.

In the last years new numerical experiments have been performed (Scholl and Froeschlé 1991, Froeschlé and Scholl 1992, Farinella et al. 1993a) including not only Jupiter and Saturn but also Mars and the Earth. These new results show the role played by the secular resonances –linear and non linear ones– in the distribution of the asteroids in the inner belt and also the importance of the secular resonance $g = g_6$ for the delivery of meteorites to the Earth.

In the following we review in Section 2 the new theories developed since 1990. Then we discuss the new results obtained by numerical experiments in section 3.

2. Theories.

The recent works on secular resonances can be classified in two groups : the analytical theory which consists in expanding the perturbation with respect to the eccentricity and the inclination of the small body, and the semi–numerical one which avoids any expansion in series of the small body's e and i, and therefore require the numerical evaluation of integrals.

2. 1. MILANI– KNEŽEVIĆ APPROACH (MK).

In order to compute proper elements (i.e. quasi–integrals of motion) the MK theory was first developed in 1990, and improved in 1992 and 1993 by adding the new perturbation terms; it is based on the Lie series formalism and uses an iterative algorithm. In the disturbing function which is averaged to eliminate short periodic terms, they include the first order term in the masses expanded up to the degree 4 in eccentricity and inclination (both for the asteroid and the planets), and the quadratic term in the masses expanded up to the second degree. Their results on the location of secular resonances are a byproduct of their proper elements work.

Indeed secular resonances enter the proper elements theory as small divisors which contain combinations of the fundamental frequencies g and s of the asteroid and the fixed frequencies g_j and s_j of the planets. Therefore, in order to compute the location of secular resonances one has to compute the values of the frequencies g and s as functions of the proper orbital elements and search for those that satisfy a resonance condition; in particular linear secular resonances (the strongest ones) are given by the relations $g = g_j$ and $s = s_j$. As it is well known, classical linear theories provide two basic frequencies $g = g_\circ$ and $s = s_\circ$, with $s_\circ = -g_\circ$, functions of the semi–major axis only; $g_\circ$ and $s_\circ$ are also called "free oscillation frequencies". In Milani and Knežević work, g and s are given by $g_\circ$ and $s_\circ$ corrected by considering the contribution of the quadratic term in the masses (which makes $s_\circ \neq -g_\circ$) plus a correction coming from the terms of degree 4 in e and i; this makes the final frequencies to become functions also of the proper eccentricity and inclination.

Milani and Knežević work points out the effect of mean motion resonances on the location of secular ones. This is due to the fact that the frequencies are corrected by the quadratic term in the masses which contains small denominators corresponding to the main mean motion resonances. On the other hand, since the quadratic term in the masses is computed up to degree 2 only in e and i, only the effect of mean motion resonances of the form $p+1 : p$ and $p+2 : p$ is detected.

An example of their results is reported in Figure 1 where the location of secular resonances is plotted in the proper elements space (a, i) for $e = 0.1$.

This work has been generalized by Knežević et al. (1991) who determined the location of the linear secular resonances up to the outer solar system. The basic result is that none of the considered secular resonances exist beyond 50 AU, so that one can conclude that these secular resonances are not effective for transporting inwards comets belonging to a possible Kuiper belt.

For what concerns the study of the resonant dynamics, Milani and Knežević (1992) have done a detailed study only on the effects of the resonances $g+s = g_6+s_6$ and $g+s = g_5+s_7$, which both cut the Eos family. Even if these resonances have a minor effect on the capability to identify the family members, they strongly influence the long term dynamics of Eos family. Also the small Lydia family is strongly perturbed by these resonances. As it is underlined by the authors, the effects of non linear secular resonances on a small family can result in the loss of some family members and even in the impossibility of identifying the family in a reliable way.

More recently Milani and Knežević (1994) analyzed the dynamical behaviour of several asteroids close to non–linear secular resonances, by numerical integrating their orbits for several million years and by computing adaptive proper elements (i.e. proper elements computed by dropping the resonant term) as function of time. The asteroids they have investigated are the following ones : 8 Flora ($3g+s = 3g_6+$

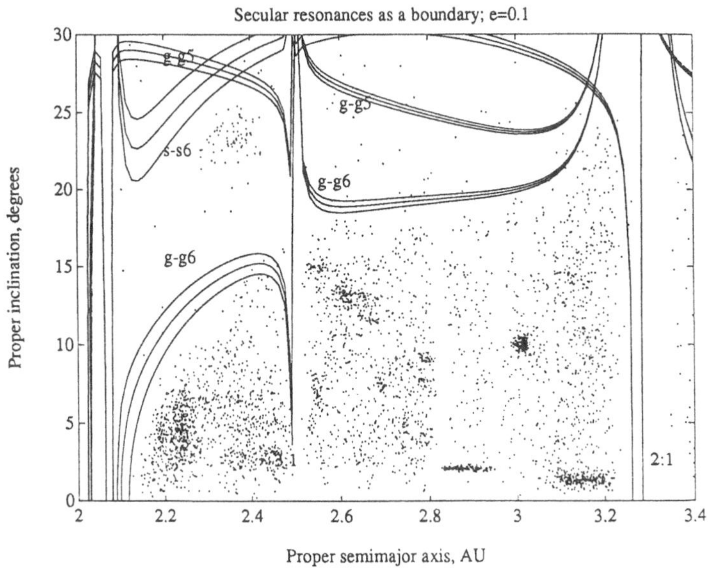

Fig. 1. The location of the main linear secular resonances for proper $e = 0.1$ according to Milani and Knežević (1990). The location of the resonances is strongly affected by the interaction with the mean motion resonances 3/1 and 2/1. The contour lines around the secular resonances correspond to a width of +/− 1"/yr.

s_6); 1047 Geisha and 244 Sita ($2g+s = 2g_6+s_6$); 2649 Oongaq ($s = g_5-g_6+s_6$); 59 Elpis, 244 Oceana and 214 Aschea ($g = 2g_6-g_5$); 1809 Prometheus ($g = 3g_6-2g_5$).

The main limit of Milani and Knežević work is due to their perturbation expansion. First of all, since the hamiltonian is expanded in power series of the asteroid's eccentricity and inclination, their results loose accuracy at large e or $\imath$. Moreover, since the term $e^2\imath^2\cos(2\omega)$ is considered only as a "perturbation", the region at large inclination cannot be studied properly, since this term (as pointed out by Kozai) becomes dominant, giving a strongly non–linear dynamics. Nevertheless their results have a good accuracy for the low to moderate e and $\imath$.

2 2 MORBIDELLI AND HENRARD APPROACH (MH)

The MH theory -close to that of Williams (1969)- avoids any truncation of powers of the asteroid's eccentricity and inclination. The averaged hamiltonian –as in Williams' method– is expanded in power series of the planetary eccentricities and inclinations,(i.e. e' and $\imath'$ are assumed implicitly to be smaller than whatever the asteroid's e and $\imath$). The quadratic terms in the perturbing masses (which were completely neglected in Williams' work) are taken into account.

Introducing suitable angle–action variables their method can be applied all over the phase space, i.e. in regions of circulation and also libration of the perihelion argument. In order to study the dynamics of the secular resonances, Morbidelli and Henrard develop a local perturbation study in the neighbourhood of a resonant orbit. Moreover, they deduce from the local study suitable integrable global models for the description of the dynamics of the three main secular resonances $g = g_5$, $g = g_6$ and $s = s_6$.

The main steps for what concerns the localization of secular resonances are the following (Morbidelli and Henrard, 1991a) :

- The hamiltonian is split into an integrable part (hamiltonian of the 2-body problem) and a perturbation one. The perturbation part is globally of order Jupiter mass $\sim 10^{-3}$, and is time-dependent through the planetary elements which are considered as given functions of time with fixed proper frequencies g_j and s_j.
- the hamiltonian is averaged with respect to short periodic terms by applying the Lie algorithm and truncated at order two in the masses. The linear term in the masses is computed without any expansion in powers of e and i; conversely the quadratic term in the masses is computed up to degree 4 only. More recently, Lemaitre and Morbidelli (1994) computed the quadratic term in the masses in a semi-numerical way, i.e. avoiding expansions.
- the averaged hamiltonian is expanded in power series of e' and i', which are assumed to be the new perturbation parameters, denoting K_m the term of degree m in e', i'. Therefore, the main part of the averaged hamiltonian turns out to be the one given by the planets assumed on coplanar circular orbits, called K_0; this part is integrable, although highly non-linear, and describes completely the dynamics related to the motion of the argument of perihelion, which is dominant at large inclination, as pointed out by Kozai (1962). In particular at large inclination, the argument of perihelion may librate, locked in the "Kozai resonance".
- Suitable action-angle variables are introduced in order to eliminate the dependence of K_0 on the argument of perihelion. In the new variables, K_0 is characterized by two fixed frequencies, which are the *proper frequencies* g and s of the system. The comparison of these frequencies with those of the planetary system gives the location of secular resonances. If the quadratic term in the masses is neglected, Morbidelli and Henrard recover completely the result of Williams and Faulkner (1981). The inclusion of the quadratic terms, confirms the results of Milani and Knežević on the influence of mean motion resonances on the location of secular ones. However, the authors pointed out that the results depend critically on the degree of truncation of the computation of the quadratic term, especially close to the 2/1 mean motion commensurability (3.0 – 3.2 A.U.) (see Figure 2.). Recently Morbidelli et al. (1993b), applying the method of successive elimination of harmonics, have computed the proper frequencies g and s in this region without passing through the computation of higher order terms in the masses, therefore obtaining much more reliable results (see Fig. 3).

For what concerns the description of the resonant dynamics, Morbidelli and Henrard's work (1991b) provides a global description of the motion in the main linear resonances $g = g_6$, $s = s_6$, and partly also in $g = g_5$.

Basically, the linear term in e', i' (called K_1) is written in the action-angle variables introduced in order to normalize K_0; furthermore, assuming to be in presence of only one isolated resonance, only the corresponding resonant term is

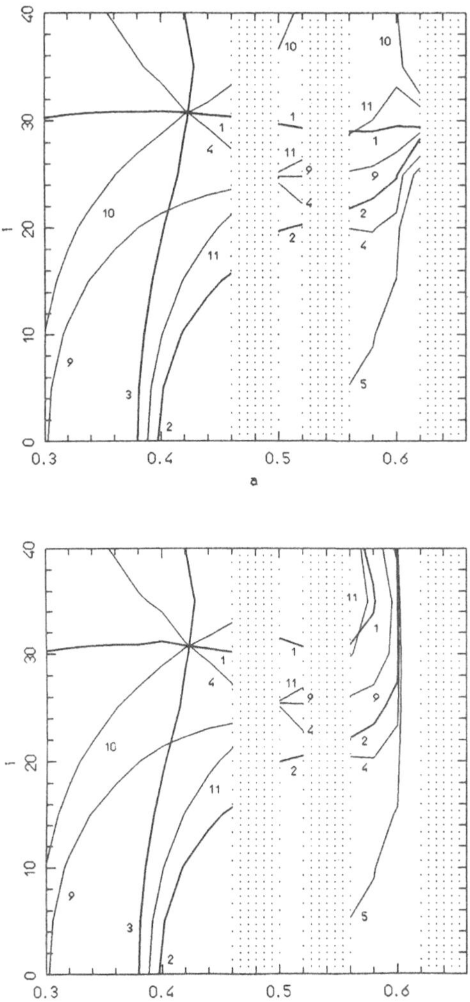

Fig. 2. The location of secular resonances for proper e=0.1 according to Morbidelli and Henrard (1991a). The semi major axis is expressed in units of Jupiter ($a_J = 5.2026$ A.U.) The enumeration of the resonances is the following : **1** $g = g_5$, **2** $g = g_6$; **3** $s = s_6$; **4** $g = s = g_5 + s_6$; **5** $g + s = g_6 + s_6$; **9** $2g = g_5 + g_6$; **10** $g - s = g_5 - s_6$; **11** $g - s = g_6 - s_6$. In the picture above the quadratic term in the masses is truncated at order 2, in the picture below at order 4. The comparison shows how such a computation is sensitive to the truncation order, at least in the region at large inclination, in between 2.9 and 3.2 AU. The dotted bands cover the mean motion commensurabilities 3/1, 5/2 and 2/1 which are forbidden regions for the theory since the computation of the quadratic term of the masses is singular.

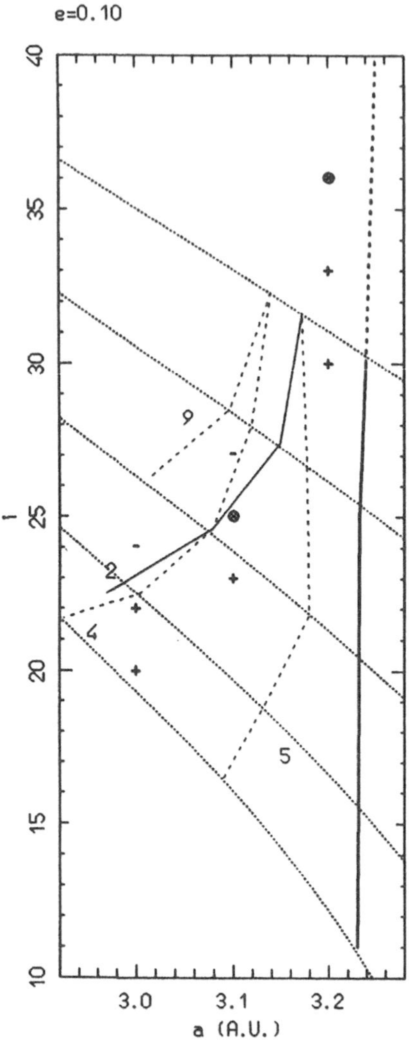

Fig. 3. The location of secular resonances at proper $e = 0.1$ and a between 2.9 and 3.3 AU according to the more sophisticated theory by Morbidelli et al. (1993b). The enumeration of the resonances is the same as in figure 2. The vertical line at 3.24 AU denotes the separatrix of the 2/1 mean motion commensurability. Secular resonances are bent to larger inclination approaching the 2/1 commensurability. This is confirmed by numerical integrations : the + denotes the initial conditions of test orbits which turn out to be below the $g = g_6$ secular resonance (2); the − denotes those which are above the resonance; the $\otimes$ denotes those which are inside the secular resonance.

retained in K_1, while the other terms are dropped. This gives an integrable model (similar to the one usually called "second fundamental model", see Henrard and Lemaitre, 1983) of the considered resonant dynamics. The main results can be shortly summarized as follows :

- $g = g_6$: the global pictures of the resonant phase space are only in partial agreement with those of the previous work by Yoshikawa (1987), but are confirmed by numerical simulations;
- $s = s_6$: the results confirm the previous ones by Nakai and Kinoshita (1985), in the region at moderate inclination where ω (perihelion argument) circulates. Moreover the ν_{16} resonance is found also in the region of ω–libration. In Froeschlé et al. (1991), the theory is applied in order to give a complete explanation of the orbital behaviour of the asteroid 2335 James;
- $g = g_5$: this resonance is located near a critical threshold where the coefficient of the corresponding resonant term in K_1 changes the sign; therefore the uncertainty on the global resonant picture is great, due mainly to the poor accuracy of the quadratic term in the masses computed, at this step, from truncated power series of e and i.

2. 3. QUANTITATIVE RESONANT DYNAMICS.

The MH theory on the dynamics in the main secular resonances has been recently improved from the quantitative point of view and implemented in the resonant proper elements program by Morbidelli (1993). The basic approach is the following :

1) the perturbation term K_1 is expanded in Fourier series of the secular angles;
2) all the *non-resonant* terms in K_1 are eliminated via a canonical transformation similar to that used for the computation of usual proper elements; this gives the "semi-proper elements" of the asteroid;
3) in the semi-proper elements space, the hamiltonian is thus reduced to an integrable resonant one of the kind $K_0 + k_1 \cos\sigma$, where σ is the critical angle of the secular resonance (for example $\varpi - g_6 t$ for the $g = g_6$ resonance), and k_1 is the coefficient of the resonant term in the Fourier expansion of K_1;
4) the semi-proper elements of the asteroid are the initial conditions for the integration of such resonant hamiltonian : the solution gives the long term dynamical evolution of the considered body. In particular the resonant proper elements are defined as the values of e and i while $\sigma = 0°, 180°$.

Figure 4 shows an example of application of the resonant proper elements algorithm on a fictitious asteroid in the $g = g_6$ resonance. On the left hand side the global picture of the resonant phase space is plotted, computed as level curves of $K_0 + k_1 \cos\sigma$; the dot marks the actual semi-proper elements of the asteroid, i.e. the present position of the asteroid on such a phase space. On the right hand side, the full numerical integration of the secular behaviour of the object is plotted, for a check. The coordinates are $x = e\cos\sigma$ and $y = e\sin\sigma$. As one sees, the resonant proper elements algorithm describes accurately the long term dynamical evolution of the object, averaging out all the "short" periodic oscillations due to the non-resonant secular terms. Additional examples, referring also to the case where the eccentricity blows up, can be found in Morbidelli (1993).

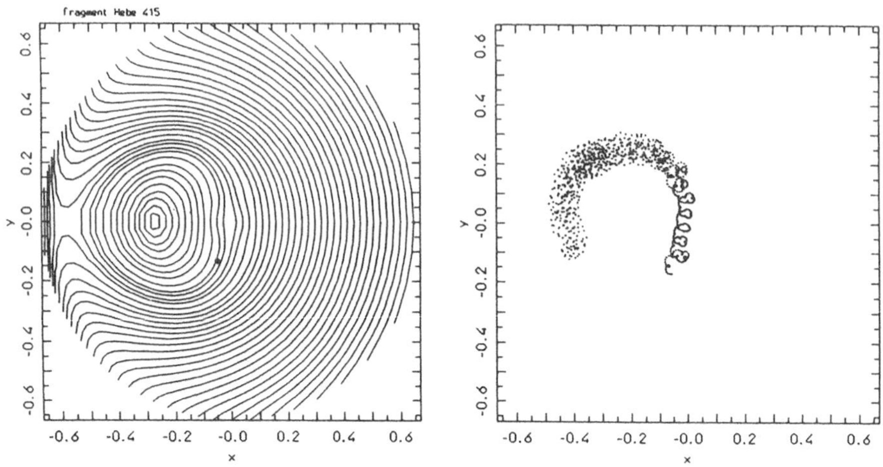

Fig. 4. On the left, the phase space computed by the resonant proper element algorithm for a fictitious asteroid. The dot marks the present position of the body in the semi-proper elements space. On the right, the evolution of the orbit of the body as integrated numerically. The coordinates are $x = e\cos\sigma$ and $y = e\sin\sigma$, where σ is the critical angle of the $g = g_6$ secular resonance.

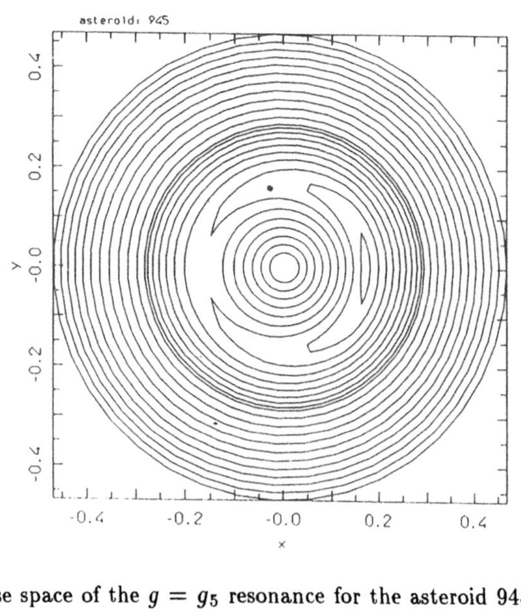

Fig. 5. The phase space of the $g = g_5$ resonance for the asteroid 945 Barcelona. The dot marks the present position of the asteroid in the semi-proper elements space.

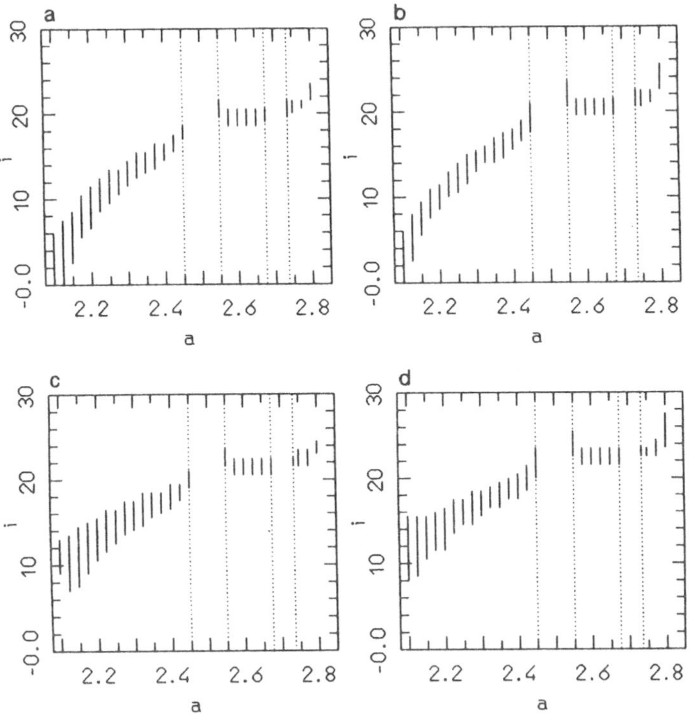

Fig. 6. The sets of the initial conditions in the $g = g_6$ resonance with $e = 0.15$, which lead to $e > 0.4$. Figure a (top left) is made for $\omega = 0°$ and $\Omega = 124.19°$ ($\sigma = 0°$); figure b (top right) is for $\omega = 180°$ and $\Omega = 124.19°$ ($\sigma = 180°$); figure c (bottom left) is for $\omega = 180°$ and $\Omega = 304.19°$ ($\sigma = 0°$); figure d (bottom right) is for $\omega = 0°$

The quantitative improvement with respect to the original MH theory on secular resonant dynamics is due to two facts :

a – instead of neglecting all non–resonant terms in K_1, one eliminates them via the suitable variable transformation which gives the semi–proper elements;

b – the quadratic term in the masses is taken into account and computed in a semi–numerical way which allows to avoid expansion in powers of the asteroid's eccentricity and inclination.

As a consequence, the resonant proper elements algorithm can be applied with reliability also to the $g = g_5$ resonance (see fig. 5 for the asteroid 945 Barcelona).

The relevant advantage of the resonant proper elements algorithm with respect to pure numerical integration is the CPU cost : to construct a global picture of the phase–space as in figures 4 and 5 (which corresponds to computing the dynamical evolution for millions of years) takes less than 1 minute of CPU time on a HP710. This allows, on the one hand, to study thousands of real/fictitious bodies, as in Morbidelli et al. (1993a) for the investigation of dynamical mechanism of meteorite transport; on the other hand, a systematic exploration of the resonant dynamics, by analyzing hundreds of initial conditions chosen on a suitable grid. This allowed, for example, to point out the dynamical peculiarities of the $g = g_6$ resonance, which is the *only* secular one able to pump up the asteroid's eccentricity up to extremely large values ($e > 0.8$). Figure 6 shows the sets of initial conditions in the (a, i) mean elements space with $e = 0.15$ and different values of the angles, which lead to $e > 0.4$; in other words, the dangerous regions of the $g = g_6$ resonance.

The main limit of the resonant proper elements algorithm is that it does not take into account the inner planets; as a consequence close approaches, which are the relevant dynamical features of any orbit with large eccentricity, are not taken into account in the model. Therefore, one can apply the algorithm in order to study the dynamical evolution of an object only up to the time when the first close approach occurs.

3. Numerical Experiments.

As pointed out in the introduction, a lot of numerical experiments have been done since the 1980's. Most of them investigated the orbital evolution of fictitious and real asteroids located in or near the principal resonances. These quantitative experiments gave a detailed picture of the dynamical properties of the secular resonances. Transitions between libration, inner and outer circulations have been found, indicating the occurrence of chaotic motion.

Froeschlé et al. (1991) have pointed out the complex dynamical behaviour of the asteroid 2335 James (first found located by Williams in the $s = s_6$ resonance). A backwards integration showed that James was temporarily located in the $g = g_5$ and $s = s_6$ resonances. In a Sun–Mars–Jupiter–Saturn model, the numerical integration showed close encounters of the asteroid with Mars, without ejecting James from the $s = s_6$ resonance.

Numerical experiments on fictitious small bodies with initial eccentricities e=0.1 have been performed in the overlapping region of the 3/1 mean motion resonance and of the $g = g_6$ secular resonance (Froeschlé and Scholl 1993). The dynamics in this region is very complex as suggested by the recent theories. The analysis of the dynamical behaviour of orbits in the $g = g_6$ secular resonance shows that the dynamical picture of the phase space changes approaching the 3/1 mean motion commensurability : for example, the libration motion is reversed. Inside the 3/1 resonance region the secular resonance $g = g_5$ is the dominant one, and some secondary secular resonances as $g - s = g_6 - s_6$ and $2g = g_5 + g_6$ are present.

Froeschlé and Scholl (1992) studied the effects of linear and non-linear secular resonances in the inner belt $a \leq 2.4$ AU, which appears to be depopulated at inclinations larger than 12°. This region is surrounded by the three main resonances $g = g_5$, $g = g_6$, $s = s_6$ and is crossed by non–linear ones. Numerical experiments of fictitious bodies show that the inclinations are pumped up by successive crossings through the non linear resonances. Bodies located at the border of the $g = g_6$ resonance with semi major axis $a \leq 2.4$ AU become Earth–crossers on a time scales of 1 Myr.

Many studies have shown the potential importance of secular resonances for the delivery of meteorites and Aten/Apollo/Amor objects to planet–crossing orbits,

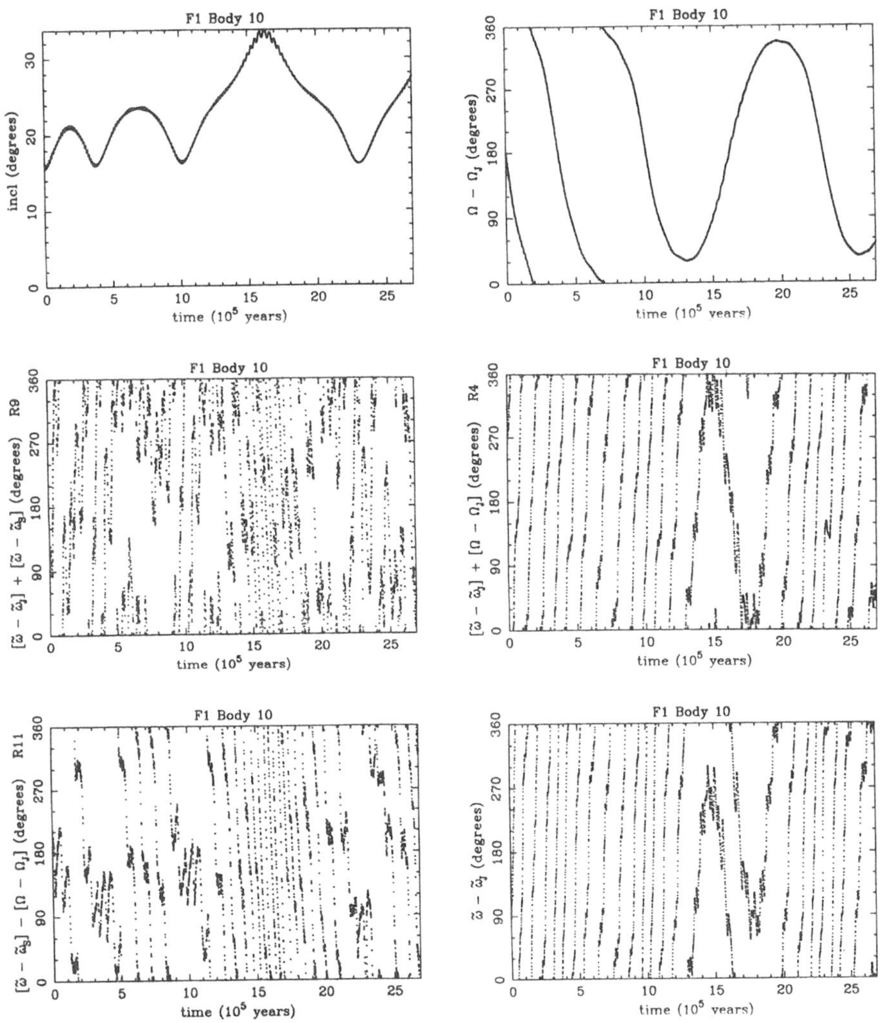

Fig. 7. An unpublished numerical simulation by Froeschlé and Scholl (see also Froeschlé and Scholl, 1992) which shows a fictitious body the inclination of which is pumped up from 16 to 34 degrees by chaotic diffusion through secular resonances. The orbit first crosses the secular resonances $2g = g_5 + g_6$ (critical argument $\tilde{\omega} - \tilde{\omega}_J + \tilde{\omega} - \tilde{\omega}_S$) and $g - s = g_6 - s_6$ (critical argument $\tilde{\omega} - \tilde{\omega}_S + \Omega - \Omega_J$) and is finally captured in the resonance $s = s_6$ (critical argument $\Omega - \Omega_J$) and, temporary, in the $g = g_5$ (critical argument $\tilde{\omega} - \tilde{\omega}_J$).

often using Monte–Carlo techniques to model the chaotic orbital evolution of asteroid fragments (Wetherill 1977, 1979, 1988, Wetherill and Williams 1979). Milani et al. (1989) suggested, on the basis of a numerical investigation, that the region of secular resonances around 2 AU could be at least as important as the Kirkwood gaps as a route for Earth crossing.

Scholl and Froeschlé (1991) showed that the proximity between the $g = g_6$ secular resonance and the 4/1 mean motion resonance broadens significantly the Earth–crossing region in the inner belt. In the Sun–Mars–Jupiter–Saturn model, integrations of bodies with initial eccentricity equal to 0.05 and 0.1 have been performed for several semimajor axes $a \leq 2.13$ AU. Due to the proximity of the resonance $g = g_6$ the eccentricity is pumped up inducing close encounters with Mars. At that time random walk of the semimajor axis starts and the body can be trapped either in the resonance 4/1, which makes the eccentricity to increase up to 0.5 after 10^4 years, or in the $g = g_6$ resonance; in the latter case the time scale to become Earth–crosser is at least 10^5 years.

Farinella et al. 1993a have quantitatively modeled the chance insertion of collisional fragments into the $g = g_6$ and 3/1 resonances, through which they can achieve Earth–crossing orbits. They show that the two resonances are potentially effective channels for fragment collection and delivery. The efficiency of the $g = g_6$ is comparable to the efficiency of the 3/1 resonance. Moreover the $g = g_6$ resonance is found to be an efficient fragment collector not only near the inner edge of the asteroid belt, but also for several asteroids with semimajor axes about 2.4 and 2.7 AU, see Farinella et al., this volume.

Numerical integrations of 18 fictitious fragments ejected from the asteroid 6 Hebe located very close to the $g = g_6$ resonance at semimajor axe of 2.42 AU have been performed by Farinella et al. 1993b. For five of these fragments the authors found that the resonance $g = g_6$ pumps up the eccentricity to value ≥ 0.6 which result into Earth–crossing within a time scale of 1 Myr, several close encounters with our planet cause chaotic orbital evolution. Some more fragments become Mars–crossers albeit not Earth–crossers. Two bodies are injected into the 3/1 mean motion resonance with Jupiter, and display also chaotic behaviour leading to Earth–crossing.

4. The new frontiers

The many numerical and theoretical works which have been done since the beginning of this decade have thrown a new light onto the secular evolution of the orbits in the main asteroid belt pointing out the relevant dynamical role of secular resonances.

However, some problems of non–negligible importance for a deeper comprehension of the features of the solar system, are still to be investigated.

First of all, we stress that most of the works described above are devoted to isolated secular resonances. Just few papers (Šidlichovský, 1990; Froeschlé and Scholl, 1992, Milani and Knežević, 1994) point out the importance of interactions among secular resonances. In particular, overlapping among non–linear secular resonances could give rise to important phenomena of diffusion through the asteroid belt; this is the way by which faint resonances could induce macroscopic phenomena. Conversely, isolated secondary resonances could give rise to chaotic phenomena (with positive Liapunov exponent) but with no importance on macroscopic stability, in the sense that the action variables do not show significant changes.

Moreover, the secular theories and the numerical experiments should be extended to the outer solar system, following the road opened by Knežević et al. (1991) who, on the other hand, did not provide any information on the strength of secular resonances and, consequently, on their dynamical role.

A new subject of work is the role of secular resonances inside mean motion commensurabilities. A recent paper by Morbidelli and Moons (1993) shows that secular resonances exist and are responsible of the existence of large chaotic zones. This seems to be a promising way to explain the existence of Kirkwood gaps.

References

Farinella, P., R. Gonczi, Ch Froeschlé and Cl. Froeschlé 1993a. The Injection of Asteroids Fragments into Resonances. *Icarus* **101**, 174–197.

Farinella, P., Ch. Froeschlé and R. Gonczi 1993b. Meteorites from the asteroid 6 Hebe. *Celest. Mech.* **56**,287–305.

Froeschlé, Ch., A. Morbidelli and H. Scholl 1991. Complex Dynamical Behaviour of The Asteroid 2335 James associated with the Secular Resonances ν_5 and ν_{16} : Numerical Studies and Theoretical Interpretation. *Astron. Astrophys.* **24**, 553–562.

Froeschlé, Ch. and H. Scholl 1989. The three principal secular resonances ν_5 ν_6 and ν_{16} in the asteroidal belt.*Celest. Mech.* **46**, 231–251.

Froeschlé, Ch. and H. Scholl 1992. The Effect of Secular Resonances in the Asteroid Region between 2.1 and 2.4 AU. *Asteroids Comets and Meteors 1991*, 205–209.

Froeschlé, Ch. and H. Scholl 1993. Numerical experiments in the 3/1 and ν_6 overlapping region. *Celest. Mech.* **56**, 163–176.

Henrard, J., and A. Lemaitre 1983. A second fundamental model for resonance. *Celest. Mech.* **30**, 197–218.

Knežević, Z., A. Milani, P. Farinella, Ch. Froeschlé and Cl. Froeschlé 1991. Secular Resonances from 2 to 50 AU. *Icarus* **93**, 316–330.

Kozai, Y. 1962. Secular perturbations of asteroids with high inclination and eccentricities. *Astron. J.* **67**, 591–598.

Lemaitre, A. and A. Morbidelli 1994. Calculation of proper elements for high inclined asteroidal orbits. *Celest. Mech.*, in press.

LeVerrier, U. J. 1855. Développement de la fonction qui sert de base au calcul des mouvements des planètes. *Ann. Obs. Paris* **1**, 258–342.

Milani, A., M. Carpino, G. Hahn and A. M. Nobili 1989. Dynamics of planet-crossing asteroids : classes of orbital behavior. Project SPACEGUARD. *Icarus* **78**, 212–269.

Milani, A. and Z. Knežević 1990. Secular perturbation theory and computation of asteroid proper elements. *Celestial Mechanics* **49**, 247–411.

Milani, A. and Z. Kneževič 1992. Asteroid proper elements and secular resonances. *Icarus* **98**, 211–232.

Milani, A. and Z. Kneževič 1994. Asteroid proper elements and the dynamical structure of the asteroid main belt. *Icarus*, in press.

Morbidelli, A. and J. Henrard 1991a. Secular resonances in the asteroid belt : Theoretical perturbation approach and the problem of their location. *Celest. Mech.* **51**, 131–168.

Morbidelli, A. and J. Henrard 1991b. The main secular resonances ν_5, ν_6 and ν_{16} in the asteroid belt. *Celest. Mech.* **51**, 169–198.

Morbidelli, A. 1993. Asteroid secular resonant proper elements. *Icarus* **105**, 48–66.

Morbidelli, A. and M. Moons 1993. Secular resonances in mean motion commensurabilities : the 2/1 and 3/2 cases. *Icarus* **102**, 316–332.

Morbidelli, A., R. Gonczi, Ch. Froeschlé, and P. Farinella 1993a. Delivery of meteorites through the ν_6 secular resonance. ***Astron. Astrophys.***, in press.

Morbidelli, A., H. Scholl and Ch. Froeschlé 1993b. The location of secular resonances close to the 2/1 commensurability. ***Astron. Astrophys.* 278**, 644,653.

Nakai,H., and H. Kinoshita 1985. Secular perturbations of asteroids in secular resonances. ***Celest. Mech.* 36**, 391–407.

Poincaré, H. 1892. Les Méthodes nouvelles de la Mécanique Céleste. Gauthier Villars, Paris.

Šidlichovský, M. 1989. Secular resonances and the second fundamental model. ***Bull. Astron. Inst. Czech.* 40**, 92–104.

Šidlichovský, M. 1990. The existence of a chaotic region due to the overlap of secular resonances ν_5 and ν_6. ***Celest. Mech.* 32**, 177–196.

Scholl, H., Ch. Froeschlé, M. Yoshikawa, H. Kinoshita and G. H. Williams 1989. Secular Resonances. In ***Asteroids II*** (R.P. Binzel, T.Gehrels, M.S. Matthews, Eds.), 845–861, Univ. of Arizona Press.

Scholl, H. and Ch. Froeschlé 1991. The ν_6 Secular Resonance Region near 2.AU : a Possible Source of Meteorites. ***Astron. Astrophys.* 245**, 316–321.

Tisserand, M. F. 1882. Mémoire sur les mouvements séculaires des plans des orbites des trois planètes, ***Ann. Obs. Paris* 16**, E1–E57.

Wetherill, G. W. 1977. In ***Comets, Minor planets and Meteorites*** (A.H. Delsemme ,Ed.), 283–292, Univ. of Toledo Press.

Wetherill, G. W. 1979. Steady state populations of Apollo–Amor objects. ***Icarus* 37**, 96–112.

Wetherill, G. W. 1988. Where do the apollo objects come from? ***Icarus* 76**, 1–18.

Wetherill, G. W. and J. G. Williams 1979. In ***Origin and Distribution of the Elements*** (L. H. Ahrens, Ed.), 19–31, Pergamon Oxford/ New York.

Williams, J. G. 1969. Secular perturbations in the Solar System. Ph.D. Dissertation, Univ. California Los Angeles.

Williams, J.G. and J. Faulkner 1981. The positions of secular resonances surfaces. ***Icarus* 46**, 390–399.

Yoshikawa, M. 1987. A simple analytical model for the secular resonance ν_6 in the asteroidal belt. ***Celest. Mech.* 40**, 233–272.

METEORITE DELIVERY AND TRANSPORT

PAOLO FARINELLA
Dipartimento di Matematica, Università di Pisa,
Via Buonarroti 2, I–56127 Pisa, Italy, and
Observatoire de la Côte d'Azur, B.P. 229, F–06304 Nice Cedex 4, France
E–mail : TWIN2@VM.CNUCE.CNR.IT, ZAPPALA@OBS–NICE.FR

CLAUDE FROESCHLÉ AND ROBERT GONCZI
Observatoire de la Côte d'Azur, B.P. 229, F–06304 Nice Cedex 4, France

Abstract. Understanding how meteorites and near–Earth asteroids reach their Earth–crossing orbits starting as fragments from main–belt asteroids is a basic prerequisite to identifying the original parent bodies of these objects and building a self–consistent cosmogonical interpretation of the observed properties of meteorites. We review the recent progress made in this area and the most important remaining open problems. These concern the physics of asteroidal collisions, the size distribution of small main–belt asteroids, the efficiency of different dynamical routes, and the relationships between asteroid taxonomic types based on spectrophotometry data and meteorite classes having different thermal histories and compositions.

1. Introduction

Every day, tons of interplanetary material fall into the Earth's atmosphere. Occasionally fragments heavier than a few kilograms partially survive atmospheric crossing, to reach the Earth's surface and become meteorites. The fall of larger bodies also occurs, with a frequency roughly proportional to the -2 power of the chosen threshold diameter — corresponding to the typical size distribution of the interplanetary complex. The typical interval between two impacts is a few centuries for 50–meter sized bodies, causing a 10 Mton of TNT explosion like that which in 1908 devastated the Siberian *taiga* at Tunguska; a few times 10^5 yr for km–sized objects; and $\approx 10^8$ years for an asteroid or comet 10 km across, causing a global climatic and ecological catastrophe, such as that recorded at the K–T boundary. Averaged over geologic time, a total influx of about 1.7×10^8 kg/yr hits the Earth over the mass range from 10^{-21} to 10^{15} kg (Ceplecha 1992).

Where does this material come from? Until 200 years ago, the fall of stones from outer space was widely believed by scientists to be a matter of superstition; on the other hand, at the beginning of this century, meteorites were thought to be interstellar objects travelling on hyperbolic orbits. Today, we know that cometary nuclei and debris contribute to the interplanetary complex (though apparently not to the small solid bodies reaching the Earth's surface), and there is evidence that a tiny fraction of the meteorites come from the Moon and Mars. However, the birthplace of most meteorites has been recognized to lie in the main asteroid belt, where the growth of a full–sized planet was aborted during the early evolution of the solar system (see Lipschutz et al. 1989, for a recent review). Therefore, for scientists interested in the earliest history of the Earth and the planetary system,

A. Milani et al. (eds.), Asteroids, Comets, Meteors 1993, 205–222.

meteorites are veritable Rosetta stones that include a wide variety of materials which have survived almost unaltered since 4.5×10^9 years ago.

However, until recently serious problems remained in understanding the asteroid–meteorite connection. As a consequence of interasteroidal impact events, fragments are usually ejected at velocities of the order of 100 m/s, far too small for directly causing the drastic orbital changes needed to achieve planet–crossing orbits. On the other hand, impulsive velocity increments of several km/s would induce in the meteoritic material shock and thermal modifications which in general are not observed. Thus the interplay between collisions and subtle dynamical effects is required to transport material from the asteroid belt to the Earth. A long–standing issue in planetary science is that of identifying and estimating the efficiency of these dynamical routes. Also, the meteorite delivery models should be consistent with the observed properties of meteorites, such as orbits, fall times and cosmic–ray exposure ages, and with the inferred (dis)similarities in composition and thermal history between the meteorite and asteroid classes. As we shall see, in recent years significant progress has been achieved in addressing these problems — albeit of course a number of puzzles are still awaiting a convincing solution.

The remainder of this paper is organized as follows. In Secs. 2 and 3 we will discuss the ejection of fragments from main–belt asteroids, in particular the evidence concerning ejection velocities and the way the fragment yield depends on size of the source object. Sec. 4 will be devoted to dynamics, both of fictitious asteroid fragments and of real near–Earth objects. The overall efficiency of the meteorite transport process will be estimated in Sec. 5, and the evidence provided by asteroid spectrophotometry will be shortly reviewed in Sec. 6. Sec. 7 will summarize the open problems and the most important areas of future work.

2. Fragment ejection velocities

Interasteroidal collisions provide an effective mechanism not only to generate the observed distribution of asteroid sizes and spins (Davis et al. 1989), but also to eject fragments out of the surfaces and even the interiors of existing asteroids. When fragments exceeding several km in size are ejected in dynamically "quiet" regions of the asteroid belt, the resulting orbital clusterings may remain detectable for billions of years as *dynamical families* (Milani et al. 1992; Zappalà and Cellino, this volume). On the other hand, when fragments are injected into chaotic zones of the phase space, they may end up as planet–crossing asteroids/meteoroids.

However, a crucial quantitative problem concerns the typical ejection speeds of asteroid fragments. Since the orbital velocities of main belt–asteroids are of the order of 20 km/s, velocity increments of hundreds of m/s are required just to achieve a 1% change in the orbital elements. Moreover, it is easy to estimate that the escape velocity of an asteroid of (mean) radius R is $V_{esc} = (60\,\mathrm{m/s}) \times (R/50\,\mathrm{km})$ (we have assumed here a density $\rho \approx 2.5$ g/cm^3 — note that $V_{esc} \propto \sqrt{\rho}$, so that uncertainties in the density do not affect much V_{esc}). Therefore, ejection velocities ≈ 100 m/s are needed just to escape "to infinity" from any sizable asteroid.

In fact, most current models of the outcomes of asteroidal collisions, involving either catastrophic break–up or cratering of the target (Greenberg and Chapman

1983; Wetherill 1987; Davis et al. 1989; Farinella et al. 1993a,b; Petit and Farinella 1993) assume that the magnitude of the fragment ejection velocity V_{ej} follows a power–law distribution with a lower cutoff. This is a reasonable approximation of the shape of the distribution observed after laboratory experiments on hypervelocity impacts (see Gault et al. 1963, and Stöffler et al. 1975, for craters; Davis and Ryan 1990, and Nakamura and Fujiwara 1991, for break–up events). Then the number of fragments ejected with speeds in the interval $(V_{ej}, V_{ej} + dV_{ej})$ is

$$dN(V_{ej}) = C\ V_{ej}^{-\alpha}\ dV_{ej}\ , \tag{1}$$

for $V_{ej} > V_{min}$, while $dN(V_{ej}) = 0$ for $V_{ej} < V_{min}$ (implying that V_{min} is the lower–cutoff velocity). For the exponent α the value 3.25 is roughly consistent with experimental results quoted above (the total kinetic energy of the fragments is finite only provided $\alpha > 3$). Provided $\alpha > 2$, the mean value $< V >$ of the distribution is $\frac{(\alpha-1)}{(\alpha-2)} V_{min}$ ($= 1.8\, V_{min}$ for $\alpha = 3.25$). Note that Eq. (1) implies that the distribution of the relative velocity "at infinity" V is given by :

$$dN(V) = CV(V^2 + V_{esc}^2)^{-(\alpha+1)/2}\ dV\ , \tag{2}$$

for $V_{min} < V_{esc}$; otherwise, $dN = 0$ whenever $V < (V_{min}^2 - V_{esc}^2)^{1/2}$. The peak of (2) occurs at $V = V_{esc}/\sqrt{\alpha}$, namely at a relative velocity smaller than, but comparable to the target's escape velocity; however, this is true only provided $V_{min} < V_{esc}$, otherwise the peak is at $V = (V_{min}^2 - V_{esc}^2)^{1/2}$.

Clear evidence on the typical values of $< V >$ or V_{min} is provided by asteroid families. From the escape velocity of their largest members and the observed dispersion of proper elements, typical relative velocities of a few hundreds of m/s are easily inferred (Zappalà et al. 1984, 1990; Bendjoya et al. 1991). However, such values are definitely *not* consistent with the results of laboratory experiments. In all the hypervelocity impact fragmentation experiments performed so far, typical fragment ejection velocities have been found to be only ≈ 10 m/s. As for the cratering events, the experiments with basalt targets carried out by Gault et al. (1963) indicate $V_{min} \approx 50$ m/s, while lower values were observed for "softer" targets. Again, asteroid families (in particular, the Vesta family as recently studied by Binzel and Xu 1993, see Sec. 6) provide convincing evidence that in giant impact cratering events large fragments (up to ≈ 10 km) can be ejected at velocities of several hundreds of m/s. Such values are confirmed by Vickery's (1986) analysis of the distribution of secondary craters surrounding some large impact craters on the Moon and Mars (though she pointed out the alternative possibility that large secondary craters are formed by clusters of small ejecta rather than single fragments).

Thus, it is worth stressing that we have a serious problem here, as the development of working meteorite delivery (and family formation) models requires the assumption that large–scale asteroidal collisions are much more efficient than the small–scale laboratory impacts in accelerating the fragments. As the physical reason for this discrepancy is not understood, further experimental and theoretical research in this area is urgently needed.

A related problem is that no correlation between fragment size (or mass) and ejection velocity, such as indicated by recent experiments (Nakamura and Fujiwara

1991; Nakamura et al. 1992) has been included in the fragment ejection models so far. This is an important limitation, as results by Petit and Farinella (1993) imply that such a correlation can affect in a significant way the overall amount and the size/velocity distribution of the escaping fragments.

3. Fragment yield vs. size of source asteroid

Meteorites represent just the end–product of a "collisional cascade", namely they are probably multi–generation fragments separated by a number of impact/ejection events — occurred both in the main asteroid belt and in planet–crossing orbits — from their primordial precursor asteroids (which are usually referred to as "parent bodies"). This is a complex process, for which realistic models have not yet been developed. However, important questions are whether collisional fragments are currently generated more effectively by large or small main–belt asteroids, and by cratering or fragmentation events. To address these questions, a simple model has been developed by Farinella et al. (1993a); while it provides useful first–approximation results, it also shows where the critical areas of uncertainty are.

The model assumes that : (i) the collision rate against a target asteroid of radius R is $P_i R^2 N_{pr}$, where P_i is a constant (the average intrinsic collision probability, see Farinella and Davis 1992) and N_{pr} is the assumed number of projectiles; (ii) the projectile population has a power–law cumulative size distribution with a negative exponent $-b$, i.e., the number of projectiles larger than r is $N(>r) = Kr^{-b}$, where K is a constant (b is assumed to be < 3, so that the total mass in the distribution is finite); (iii) catastrophic fragmentation of the target occurs whenever the ratio between projectile and target size exceeds a critical threshold ζ, independent of target size; (iv) otherwise, a crater is created with an excavated mass proportional to the projectile's kinetic energy through a constant coefficient L; (v) the largest possible crater corresponds to an excavated mass given by a factor γ times the target mass [if the average impact velocity is V_i, it follows from assumptions (iii) and (iv) that $\gamma = LV_i^2\zeta^3/2$]; (vi) only fragments with $V_{ej} > V_{esc}$ escape reaccumulation, so that if we define $R_{min} = 83\,km\ (V_{min}/100\,m/s)$ as the target radius for which $V_{esc} = V_{min}$, namely the minimum radius for which part of the fragments are reaccumulated, then for $R > R_{min}$ the escape velocity is $V_{min}R/R_{min}$ and the fraction of escaping fragments is $(R/R_{min})^{1-\alpha}$. Note that here we have assumed that V_{min} does not depend on fragment size; different assumptions would lead to a much more complicated treatment of the fragment reaccumulation process (see Petit and Farinella 1993). Let us now consider separately (1) impact fragmentation and (2) cratering events.

(1) The frequency of catastrophic collisions for a target of radius R is $P_i R^2 N(>\zeta R) = P_i K\zeta^{-b}R^{2-b}$. Since in any break–up event the whole target mass is converted into fragments, the mass ejection rate is $(4\pi\rho K P_i/3)\zeta^{-b}R^{5-b}$ (ρ being the target density). For $R > R_{min}$, this has to be multiplied times $(R/R_{min})^{1-\alpha}$ to account for fragment reaccumulation.

(2) If we assume that craters of all sizes up to the largest possible one have been formed on the target, the total excavated mass is just L times the total delivered kinetic energy, which can be computed by integrating over projectile sizes from

0 to ζR. This yields a mass ejection rate $\frac{2\pi b}{3(3-b)}KP_i\rho V^2 L\zeta^{3-b}R^{5-b}$. Again, for $R > R_{min}$ the additional reaccumulation factor $(R/R_{min})^{1-\alpha}$ is needed.

Thus, within the simplifying assumptions described above, the same scaling rule applies to both cratering and fragmentation : the mass ejection rate is $\propto R^{(5-b)}$, and the gravitational reaccumulation of fragments can be taken into account by changing the scaling factor into $R^{(6-b-\alpha)}$ for $R > R_{min}$. For each target asteroid, a relative *fragment delivery efficiency* can thus be derived by multiplying the fraction of ejected fragments falling into resonances by a scaling factor proportional to $(R/25\,\mathrm{km})^{(5-b)}$ for $R < R_{min}$, and to $(R/25\,\mathrm{km})^{(5-b)}(R/R_{min})^{(1-\alpha)}$ for $R > R_{min}$. In this way the fragment delivery efficiency is normalized so as to be just equal to the percentage of delivered fragments for $R = 25$ km, and scaled for different sizes according to the rule derived above.

It is worth noting that this simple collisional model predicts that the mass ratio between crater ejecta and break–up fragments is independent of target size, and is simply given by $\frac{b}{2(3-b)}LV_i^2\zeta^3 = \frac{b\gamma}{3-b}$. For $\gamma = 0.1$ (see Davis et al. 1989; Petit and Farinella 1993) and b ranging between 2 and 2.5 (see later), this ratio ranges between 0.2 and 0.5. These values are somewhat increased if we are interested in the crater/break–up ejecta mass ratio for fragments up to some given size $\bar{r}$, smaller than the size of the largest body included in the crater distribution. As a consequence, for asteroids such as 4 Vesta, which probably have never been shattered but have undergone giant cratering events close to the break–up threshold (Davis et al. 1985), the fragment delivery efficiency given by the scaling rule described above is overestimated only by a factor 2 or 3. Anyway, this argument shows that the contributions to the production of asteroid fragments (and to the meteorite yield) of bodies originated in cratering and break–up events are of the same order, with a possible dominance of break–up fragments by a factor of a few.

We stress that the scaling rules would be different if one assumed size–dependent impact response parameters, such as implied by the scaling theory for collisional fragmentation developed by Housen and Holsapple (1990). Their nominal model (see Fig. 4 in their paper) predicts a variation of the impact strength (hence of the cube of our coefficient ζ) of the order of a factor 100 over the asteroid size range. Thus the number of projectiles capable of shattering any given target may differ by a factor up to $\approx 100^{b/3}$ with respect to the predictions of the model adopted above, and so does the target's lifetime vs. fragmentation and the mass delivered by break–up events per unit time. However, the current strength scaling models are not reliable enough for basing on them a more refined analysis, at least until new experimental or numerical modelling work will have provided independent evidence. But clearly this is another critical area of model uncertainty.

A further serious problem is related to the projectile size distribution index b. For Earth–crossing objects, the lunar cratering record is consistent with $b \approx 2$ (Baldwin 1971; Neukum et al. 1975). Cellino et al. (1991) have analysed different zones of the belt and different size ranges (but always exceeding a suitable "completeness threshold" at a few tens of km, to avoid discovery biases) using the IRAS data base of asteroid albedos and diameters, and have found that b varies in the range from ≈ 1 to ≈ 3. For main–belt asteroids smaller than about 20 km, relevant data come from the Palomar–Leiden Survey (PLS; Van Houten et al. 1970) and from the crater counts on the surface of 951 Gaspra (see chapter by Chapman in this volume) : for PLS asteroids, typically several km in size, b appears to range between about 2 and 2.5, whereas the projectiles impacting Gaspra (between 20 and 60 m in diameter) display a steeper index, $b \approx 3.3$. Thus there is evidence that b is actually variable in a complex way, rather than constant at the 2.5 equilibrium value predicted by Dohnanyi (1969) for a collisionally evolving population with size–independent collisional response parameters.

The unknown size distribution of small asteroids affects in a critical way the relative fragment production efficiencies of asteroids of different sizes. This can be seen by recalling that for constant b, at small sizes the efficiency scaling factor is $R^{(5-b)}$, and since the number of bodies in an interval $(R, R + dR)$ is proportional to $R^{-b-1}\,dR$, for $b = 2$ all equal size intervals would provide the same contribution, and small asteroids would not be very important. On the other hand, for $b = 2.5$ one would get a logarithmic growth of the overall efficiency when the lower limit of the considered size range is decreased, hence the contribution of small asteroids would become dominant (e.g., fragments generated by asteroids with diameters between 100 m and 10 km would be as abundant as those coming from all larger bodies). Thus, in order to assess the contribution of small main–belt asteroids to the flux of Earth–crossing fragments, one should determine their size distribution in a more reliable way than it is possible with currently available data.

4. Resonant routes

After the pioneering studies of the 70's and the early 80's (e.g., Williams 1973; Scholl and Froeschlé 1977; Wetherill and Williams 1979), much work has been devoted by celestial mechanicians in recent years to understand and numerically explore the dynamics of the resonances present in the main asteroid belt and near its boundaries. Two different types of resonances are relevant here : *mean motion resonances*, occurring for nearly–commensurable values of the orbital periods of the minor body and Jupiter, and *secular resonances*, involving commensurability (actually, equality for the three main resonances) between the nodal or apsidal precession rates of the orbit of a minor body (perturbed by the planets) and the eigenfrequencies of the orbits of the planets themselves (perturbing each other). As far as meteorite transport is concerned, the 3/1 mean motion resonance (Wisdom 1983, 1985; Yoshikawa 1990) and the ν_6 secular resonance (Froeschlé and Scholl 1987, 1992; Scholl and Froeschlé 1991; Morbidelli et al. 1993) provide the most effective routes. The other main secular resonances do not cross or approach the most densely populated regions of the main belt (Knežević et al. 1991); and among the other mean motion resonances, the 4/1 resonance (Dahlgren et al. 1992) appears just to contribute to the large chaotic zone lying in the semimajor axis range between 2 and 2.1 AU due to secular resonances, whereas the 5/2 resonance (Hahn et al. 1991; Ipatov 1992) inserts fragments into Jupiter–approaching, comet–like orbits whose lifetime vs. ejection from the solar system or impact onto Jupiter is much less than the typical time needed to hit the inner planets.

The dynamics of fictitious asteroid fragments injected from real asteroids into the resonances has been recently studied both by direct numerical integration of

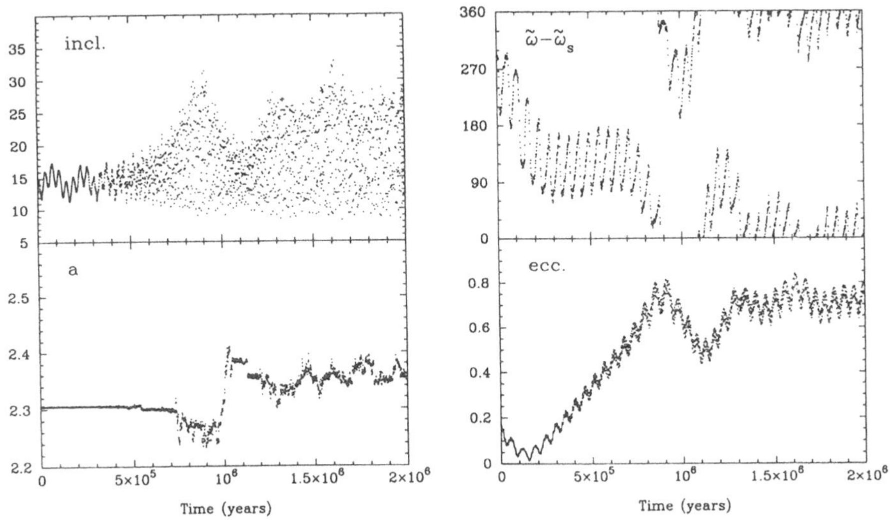

Fig. 1. Orbital evolution of a fragment from 6 Hebe, injected into the ν_6 secular resonance. The figure shows the inclination i (degrees), the semimajor axis a (AU), the ν_6 critical argument $\varpi - \varpi_{Saturn}$ (degrees), and the eccentricity e vs. time over the integration time span of 2 Myr.

their orbits (Farinella et al. 1993b), and through an *ad hoc* analytical theory for the long–term behavior of ν_6-resonant objects (Morbidelli et al. 1993). Of course the former method is more accurate in individual cases, but the latter is much faster (by a factor of several hundreds) and allows to carry out statistical studies on samples of tens of thousands of bodies with reasonable computing time requirements.

The behaviour of two fairly typical resonant fragments from the large asteroid 6 Hebe, as determined by Farinella's et al. (1993b) numerical integrations, is shown in Figs. 1 and 2. It is interesting to note that several real Earth and Mars–crossing asteroids have orbits similar to those of the integrated fictitious fragments. Also, it can be noted that one of the three photographically recorded meteorite falls, named Příbram, yielded the following orbital elements : $a = 2.40$ AU, $e = 0.67$ and $i = 10°.5$ (Ceplecha 1977). Both the semimajor axis and the inclination are very close to the corresponding elements of 6 Hebe. Of course it is not possible to trace back the origin of a meteorite or a near-Earth asteroid by dynamical arguments only, since for chaotic orbits numerical integrations over very long spans of time provide only qualitative information. However, the results shown in Fig. 1, where a Hebe fragment ends up with a Příbram like orbit within less than 1 Myr, support the idea that this meteorite (an H5 ordinary chondrite) may be a fragment ejected from Hebe, whose eccentricity was pumped up by the ν_6 resonance.

The statistical results obtained with the analytical algorithm of Morbidelli et al. (1993) are summarized in Table I. For each of 20 asteroids close to the ν_6 resonance, the Table columns give the number, the diameter, the taxonomic type and the percentages of fictitious fragments injected into the 3/1, 5/2 and ν_6 resonances

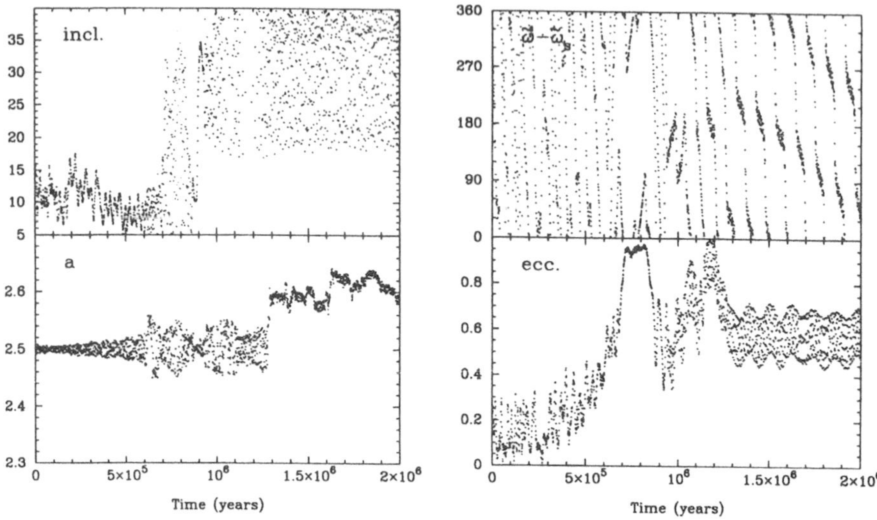

Fig. 2. The same as Fig. 1, but for a Hebe fragment injected into the 3/1 mean motion resonance.

(in the latter case, the maximum eccentricity achieved due to secular resonant perturbations was used to distinguish between Mars- and Earth-crossers). The fragments were assumed to be ejected isotropically at random times from the source body, with a velocity distribution computed from Eq. (1) and $V_{min} = 100$ m/s.

The Table shows that several of these asteroids are likely to be efficient sources of meteorites. This applies in particular to the largest ones, such as 6 Hebe, 304 Olga, 739 Mandeville and 759 Vinifera, because the percentages given in the Table should be multiplied by the scaling factor discussed in Sec. 3. Assuming $b = 2.5$, this implies for instance that Hebe's 3.3% of Earth-crossing fragments (adding up the contributions of the ν_6 and the 3/1 resonances) corresponds to a meteorite yield 11 times that of 1468 Zomba and 980 times that of 2368 Beltrovata. Both b and the diameters of small asteroids are uncertain, but these ratios show clearly that sheer size is very important in determining meteorite yields. Other objects deliver a substantial fraction of their fragments into the 3/1 (623 Chimaera, 1892 Lucienne) and 5/2 (631 Philippina, 907 Rhoda, 1222 Tina) Kirkwood gaps. As we will discuss later, these asteroids should become the targets of detailed spectrophotometric investigations, to assess their similarity to some of the known meteorite classes.

An important remark is that, over a time scale of the order of 10^8 yr, many Mars-crossing (or even Mars-grazing) fragments are likely to end up into Earth-crossing orbits, as a consequence of repeated Mars encounters changing their location in a random-walk fashion with respect to the ν_6 resonance (Wetherill and Williams 1979). While this indirect mechanism requires a longer time for transporting fragments into Earth-crossing orbits, its efficiency with respect to direct resonant transport is increased by the fact that most source asteroids deliver a much larger proportion of Mars-crossers than of Earth crossers (see Table I). A

TABLE I
Percentages of resonant fragments from 20 real asteroids.

No.	D (km)	Type	3/1	5/2	ν_6 Mars-cr.	ν_6 Earth-cr.
6	192	S	2.4	0.0	7.6	0.9
304	68	C	0.9	0.0	33.2	2.6
623	46	EMPC	18.0	0.0	5.2	0.4
631	60	S	0.3	18.8	1.9	0.6
739	110	EMP	0.0	2.3	35.8	3.5
759	53		1.0	0.3	30.0	8.4
907	66	C	0.4	28.7	1.4	0.6
930	39		3.1	0.0	10.0	1.5
963	11	S	0.1	0.0	14.6	0.7
1222	14		0.0	29.0	19.4	2.5
1468	18		0.1	0.0	99.2	80.0
1892	17		15.1	0.0	2.0	0.4
1916	4	S	0.0	0.0	100.0	13.4
2033	9		0.0	0.0	85.4	3.6
2368	3	SQ	0.1	0.0	99.1	79.3
2548	12		0.7	0.2	31.1	0.8
2604	18		0.4	0.0	72.4	2.4
3402	3		0.0	0.0	97.1	25.8
3833	3		0.0	0.0	96.8	70.9
4095	5		0.0	0.0	47.7	3.3

quantitative assessment of this mechanism requires an algorithm accounting for the effects of close encounters between fragments and planets and of fragmentation *en route* to the Earth, in addition to resonances. This is also needed to model the further evolution of fragments once they have reached Earth–crossing orbits. Since direct numerical integrations of thousands of bodies over 10^8 yr in the inner solar system are still out of reach, refined analytical models for the resonant effects will have to be coupled with Monte–Carlo statistical treatments of close encounters and collisions, building upon the pioneering work of Wetherill (1985, 1987, 1988).

What about the dynamical evolution of real near–Earth asteroids? For a general study and a classification of their amazingly complex behaviour, we refer the reader to Milani et al. (1989). However, due to computing time constraints and the parallel structure of their computer (requiring a fixed stepsize integration method), these authors could only treat the close encounters in an approximate fashion, and the integration time span (2×10^5 yr) was not long enough to detect secular resonances. We have thus repeated some of these integrations, using a variable stepsize method and extending the time span to a few Myr. The results will be discussed in detail in a forthcoming paper (Froeschlé et al. 1994). Figs. 3 and 4 show two interesting examples, involving two near-Earth asteroids whose spectral properties have been recently interpreted as suggesting a relationship to some meteorite classes.

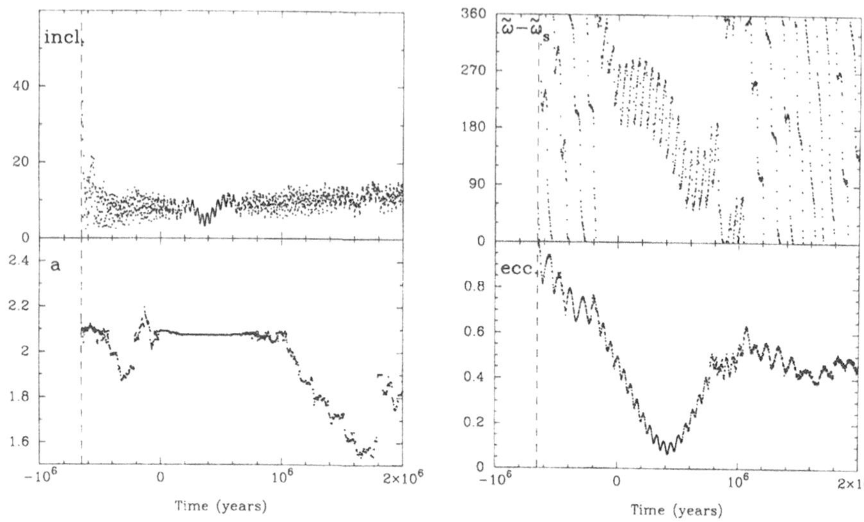

Fig. 3. Orbital evolution of asteroid 3551 (1983 RD) over 3 Myr. The integration backward in time is stopped (dashed line) when the asteroid's distance from the solar centre becomes $< R_{\odot}$.

Asteroids 3551, 3908 and 4055 have all V–type near–infrared spectra very similar to those of Vesta and of the basaltic achondrite meteorites (Cruikshank et al. 1991). Their osculating orbits are fairly similar, leading to the suggestion that at least 3551 and 3908 are members of an "association" or "stream" which includes also some bright meteors (Drummond and Wisniewski 1990; Drummond 1991), and was generated by the break–up of a common precursor. However, the results of the numerical integrations show that over a few Myr the dynamical evolution of these three objects is very different, suggesting that unless the three objects have been separated very recently ($< 10^5$ yr ago, a very unlikely possibility), the current similarity of the three orbits is coincidental. As shown in Fig. 3, the orbit of 3551 is currently locked into the ν_6 resonance, and as a result the eccentricity reaches Earth–crossing values both in the past and in the future. Close encounters within 10^5 km occur both with the Earth and with Venus, and at $t = -656,000$ yr the eccentricity gets so close to 1 that the asteroid hits the Sun! Of course, this does not imply that the real asteroid is younger than that, nor it proves that it really passed very close to the Sun, as the orbit is chaotic and integrations over so long time spans are not "deterministic" — but it is a good example of the drastic orbital changes caused by the coupling between encounter and resonance effects.

Fig. 4 shows the behavior of asteroid 3103, which has been suggested by Gaffey et al. (1992) to be related to the enstatite achondrite meteorites and to the E–type

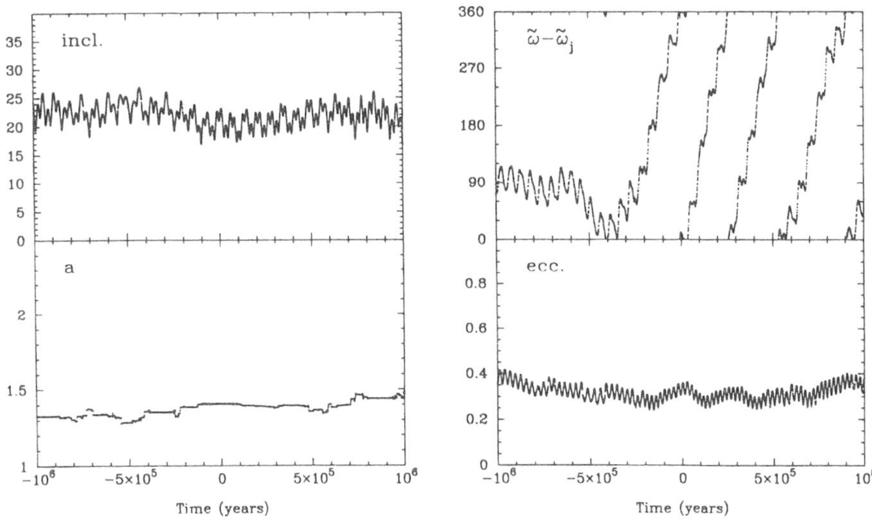

Fig. 4. Orbital evolution of asteroid 3103 (1982 BB) over 2 Myr. The evolution of the ν_5 critical argument $\varpi - \varpi_{Jupiter}$ of the ν_5 is shown here in the right upper plot.

asteroids present in the Hungaria region (at the innermost edge of the belt, with semimajor axes $\approx$ 1.9 AU). In the classification of Milani et al. (1989), this is a fairly typical Geographos-type orbit, whose encounter-driven evolution is slowed down by the large inclination. The 3/5 mean motion resonance with the Earth detected by Milani et al. appears to be short-lived; however, some 0.5 Myr in the past the orbit gets locked into the ν_5 secular resonance. There is no sign that this object may have evolved from the Hungaria region, but our integration time span is still much shorter than the impact lifetime of the asteroid.

These numerical experiments suggest that the orbital evolution of Earth-approaching asteroids and meteorites is diverse, complex and caused by several dynamical mechanisms acting over different time scales. In particular, secular resonances often play an important role. Due to the properties of chaos, there is no way of "going back in time" to the original fragment ejection event in the main belt just by integrating backwards the equations of motion. But the relative importance of the different dynamical pathways can be probably assessed with more confidence than it has been done so far.

5. Efficiency of the delivery and transport process

Is the total flux of Earth-impacting objects predicted by fragment delivery models similar to that observed? As argued by Wetherill (discussion following Greenberg and Nolan 1989), achieving an order-of-magnitude agreement with the observed flux is an important constraint for meteorite delivery models, in spite of the considerable uncertainty on the value of the flux itself. Although Farinella et al. (1993a) modeled only the fragment injection into resonances, and did not deal with their further dynamical and collisional evolution, some order–of–magnitude estimates can be based on their results. According to them, the total fragment mass injected into the resonant strips is of the order to 100 times the yield of a single asteroid 50 km across. Let us assume that the average lifetime of such a body is of $\approx 2 \times 10^9$ yr, derived from an average intrinsic collision probability of 2.85×10^{-18} km^{-2} yr^{-1} for main–belt asteroids (Farinella and Davis 1992) and a population of 250,000 potential disrupting projectiles. Although probably uncertain by plus or minus a factor two, due to our poor knowledge of asteroid impact strengths and of the relevant projectile flux, this lifetime is consistent with Wetherill's (1985) Eq. (6) and with the current estimates for small asteroids (Farinella et al. 1992). Of the 8×10^9 kg/yr of material thus injected in the resonances, only a few percent will reach the Earth, due to collisional comminution *en route* (the observed ratio of about 10 between the lifetimes of planet–crossing objects against impacts on the planets and the typical exposure ages of ordinary chondrites suggests that most asteroid fragments are broken up again while in planet–crossing orbits) and other possible outcomes of the dynamical evolution (impact against other planets, ejection from the solar system). This small "survival factor" is in rough agreement with Wetherill's (1985, Tables 4 and 5) Earth impact efficiencies of $\approx 10^{-4}$ for random ejecta from asteroids whose fragments can reach the 3/1 resonance, because in Farinella's et al. (1993a) model the average fraction of fragments delivered to the resonances by all the asteroids, weighted to account for the source size as described in Sec. 3, is of the order of a few percent. Thus, within the assumptions stated above, the model predicts an Earth flux of asteroid fragments of the order of 10^8 kg/yr, in good agreement with the estimate of $\approx 1.7 \times 10^8$ kg/yr derived by Ceplecha (1992) for all the interplanetary bodies hitting the Earth (including those of cometary origin). Note that the flux of bodies in the meteorite size range (up to about 1 meter) is about two orders of magnitude smaller (Halliday et al. 1984; Ceplecha 1988); this corresponds to the fact that only $\approx 1\%$ of the mass of asteroid fragments lie in the meteorite size range, according to the typical size distribution of fragments from hypervelocity impacts.

6. Evidence from spectrophotometry

In the last few years, a growing amount of work has been dedicated to comparing asteroid and meteorite properties, in order to determine the likely composition and thermal history of asteroids and to find out candidate sources for the different meteorite classes. In particular, the comparison of asteroid and meteorite spectra has been based on new, high–quality CCD observations (Vilas and McFadden 1992; see also the chapter by Burbine and Binzel in this volume). Principal analysis classification of the two data samples has been performed (Britt et al. 1992), and the implications of such comparisons have been carefully assessed (Fanale et al. 1992; McSween 1992; Keil et al. 1992; Clark et al. 1992). Laboratory work on meteorite spectra has explored such specific problems as the effects of mixing different mineralogic types (Hiroi et al. 1993) and using powdered meteorite samples (Salisbury et al. 1991). Specific comparisons between individual asteroid and meteorite classes, based on spectral data, has shown that a number of intriguing similarities indeed exist, involving both main-belt (Hiroi and Takeda 1990; Burbine et al. 1992) and near-Earth asteroids (Cruikshank et al. 1991; Gaffey et al. 1992). For more details on this work, we refer the reader to the reviews by Wetherill and Chapman (1988), Lipschutz et al. (1989) and Gaffey et al. (1989, 1993). In this context, two issues have important implications on the meteorite delivery and transport studies : the new findings on the Vesta family and the origin of basaltic achondrites, and the ongoing debate on the identity of the ordinary chondrite parent bodies.

Owing to its peculiar spectral features, it has long been suggested that 4 Vesta is the parent body of basaltic achondrites (e.g., Drake 1979; Greenberg and Chapman 1983). As we already mentioned, Cruikshank et al. (1991) have recently found that three km-sized near-Earth asteroids display reflectance spectra nearly identical to that of Vesta (hence they have been classified as V-types, a class whose only sizable main-belt member is Vesta itself), and this supports the idea that fragments from Vesta can be inserted into planet-crossing orbits. The main difficulty with this possibility has always been the high ejection velocity — at least some 600 m/s — needed to reach a resonant zone starting from Vesta; this led Wetherill (1987) to the conclusion that Vesta's Earth-impact efficiency must be very low. Farinella et al. (1993a) have confirmed that the average ΔV of fictitious Vesta fragments reaching the 3/1 resonance is about 700 m/s (and exceeds 1 km/s for ν_6); but they have also found that unless the size distribution index b is large and V_{min} is small, its large size makes Vesta a relatively efficient fragment deliverer. With $V_{min} = 100$ m/s, some 2.4% of the fragments escaping from Vesta fall in the 3/1 resonance and may become Earth-crossers; as a consequence, Vesta's fragment delivery efficiency accounts for 1% to 6% of the total from the main asteroid belt. Although the delivery efficiency might be overestimated by a factor ≈ 3 if Vesta has never been shattered (see Sec. 3), in order of magnitude these results are consistent with the observed abundance of basaltic achondrite falls.

From the point of view of cratering mechanics, it is not clear how unshocked fragments can be accelerated to speeds of many hundreds of m/s (see Greenberg and Chapman 1983, p. 475). However, both Zappalà et al. (1990) and Bendjoya et al. (1991) found that Vesta is the major body of a dynamical family, which includes many smaller members between 5 and 10 km in size. Recent observations by Binzel and Xu (1993) have revealed that the spectra of these minor family members (and also of some neighbouring objects, located between the Vesta family and the 3/1 Kirkwood gap) have the distinctive absorption bands of basaltic achondrites. This implies that large fragments have indeed been ejected from Vesta during a giant cratering event, at speeds well exceeding the escape velocity of ≈ 300 m/s. Note also that extrapolating down to 1 km diameter the family size distribution with $b = 2.5$ (typical for the families, according to Cellino et al. 1991) implies that thousands of km-sized fragments were produced in the family-forming event, hence several tens of them could reach the 3/1 resonance and become Earth-crossers. A similar mechanism might have worked for other families found to lie in the vicinity of resonances (Cellino and Zappalà 1993).

On the other hand, a convincing and unequivocal solution is still missing for the so-called "spectrophotometric paradox" (Wetherill and Chapman 1988), namely the fact that ordinary chondrites, which account for $\approx 75\%$ of all the observed meteorite falls, cannot be matched with the typical reflectance spectra of any common asteroid taxonomic type (and in particular of S–types, most abundant in the inner belt and earlier considered as the most plausible parent bodies). Two alternative possibilities are currently favoured. Farinella et al. (1993a,b) have suggested that most ordinary chondrites may come from just a few sizable S–type asteroids (such as 6 Hebe) having high fragment delivery efficiencies and chondritic compositions, unlike those of most other S–types. This explanation would be consistent with the evidence from meteoritic groupings, which suggests that almost all ordinary chondrites come from a small number — possibly as few as three — of original asteroids (Lipschutz et al. 1989), and also with their diameters inferred from meteorite cooling histories (between 100 and 300 km; see Pellas and Storzer 1981; Pellas and Fiéni 1988). The origin of many ordinary chondrites in a few large–scale collisional events is also suggested by the observed clustering of the cosmic–ray exposure ages (Crabb and Schultz 1981; Marti and Graf 1992).

An obvious observational test for this explanation is that of obtaining for the candidate asteroids quoted above new, detailed spectrophotometry data, capable of clearly constraining their composition, as in the work reviewed by Gaffey et al. (1989) on 8 Flora, 15 Eunomia and some other bodies. Interestingly, Gaffey et al. (1994) have recently reported that a small subtype of the S–types asteroids (named S(IV), and accounting for just $\approx 10\%$ of all S–types) display silicate assemblages consistent with, although not specifically diagnostic of, undifferentiated ordinary chondritic assemblages. The S(IV) subtype includes 6 Hebe, and is the only S subtype showing a significant concentration in the semimajor axis range between 2.3 and 2.6 AU, where both the 3/1 and the ν_6 resonances provide efficient transport routes to Earth–crossing orbits (see Sec. 4). However, the reason for this preferential location is quite puzzling, if the S(IV) subgroup includes the only S–type asteroids having escaped differentiation.

The alternative explanation is that ordinary chondrites come from the numerous main–belt population with diameters smaller than a few tens of km, for which few spectral data exist (Bell et al. 1989; Fanale et al. 1992), assuming that they have escaped the primordial heating events that led to (partial or total) melting of the larger asteroids. Binzel et al. (1993) have recently found that the 7–km asteroid 3628 Božěmcová, located just outside the 3/1 Kirkwood gap, has the distinctive spectral signature of ordinary chondrites (even more than 1862 Apollo and a few other Earth–approachers, classified in the Q taxonomic type). However, this finding does not really support the idea that ordinary chondrites come from "primitive" small asteroids. In the first place, no other small asteroid in a sample of 80 surveyed by Binzel et al. shares the ordinary chondrite spectral features, and this contrasts with the large abundance of this class among the meteorite falls. Secondly, asteroids of size ≈ 10 km have collisional lifetimes much shorter than the age of the solar system (Farinella et al. 1992; see also the chapter by Chapman in this volume). If so, where are the sources for the small ordinary chondrite asteroids?

A compromise solution between the two explanations may be that the surfaces of large ordinary chondrite parent bodies are somewhat camouflaged, and only when one observes fresh fragments excavated from them the material with the diagnostic spectral features can come into view. However, more observational evidence is clearly needed to settle this important issue.

7. Open problems and future work

Future work on the subject of this paper should include a number of different issues and techniques. First, the real efficiency of the different dynamical routes between the asteroid belt and the Earth should be tested by extensive numerical experiments, both to explore in a more systematic way the relevant parts of the phase space and to determine the fate of fictitious fragments ejected — according to some plausible model of collisional outcomes — from specific asteroids, which look as promising candidate sources for meteorites and near-Earth asteroids. Second, the same asteroids should be the target of further, detailed astronomical observations, aimed at determining their physical properties, surface morphology and, in particular, their likely composition and its consistency with the assumed meteorite counterparts. Another problem whose crucial importance is apparent for the understanding of many issues related to the evolution of asteroids, and therefore should be addressed by new, *ad hoc* observational surveys, is the size distribution of small (say, km–sized) main–belt asteroids. Finally, experimental and modelling work is still needed in the field of collisional physics, in order to better understand the mechanisms of fragment acceleration and the way some basic impact-response parameters (strength, ejecta velocity distribution, crater excavation efficiency) may vary with size and material properties.

Acknowledgements

We are grateful to R.P. Binzel, Z. Ceplecha, D.R. Davis, Ch. Froeschlé, G. Hahn, Z. Knežević, A. Milani, A. Morbidelli, G.W. Wetherill and V. Zappalà for many useful discussions and comments. P.F. has performed much of the work described here while staying at the Observatoire de la Côte d'Azur (Nice, France), thanks to the "G. Colombo" fellowship of the European Space Agency.

References

Baldwin, R.B. : 1971, "On the history of the lunar impact cratering : The absolute time scale and the origin of planetesimals", *Icarus*, **14**, 36–52.

Bell, J.F., Davis, D.R., Hartmann, W.K. and Gaffey, M.J. : 1989, "Asteroids : The big picture". In *Asteroids II* (T. Gehrels, Ed.), 921–945, Univ. of Arizona Press, Tucson.

Bendjoya, Ph., Slezak, E. and Froeschlé, C. : 1991, "The wavelet transform : a new tool for asteroid family determination", *Astron. Astrophys.*, **251**, 312–330.

Binzel, R.P. and Xu, S. : 1993, "Chips off Vesta and a near–resonance source for basaltic achondrite meteorites", *Science*, **260**, 186–190.

Binzel, R.P., Xu, S., Bus, S.J., Skrutskie, M.F., Meyer, M.R., Knezek, P. and Barler, E.S. : 1993, "Discovery of a main–belt asteroid resembling ordinary chondrite meteorites", *Science*, **262**, 1541–1543.

Britt, D.T., Tholen, D.J., Bell, J.F. and Pieters, C.M. : 1992, "Comparison of asteroid and meteorite spectra : Classification by principal components analysis", *Icarus*, **99**, 153–166.

Burbine, T.H., Gaffey, M.J. and Bell, J.F. : 1992, "S-asteroids 387 Aquitania and 980 Anacostia : Possible fragments of the breakup of a spinel-bearing parent body with CO3/CV3 affinities", ***Meteoritics***, **27**, 424-434.

Cellino, A. and Zappalà, V. : 1993, "Asteroid 'clans' : Super-families or multiple events?", ***Celest. Mech.***, **57**, 37-47.

Cellino, A., Zappalà, V. and Farinella, P. : 1991, "The asteroid size distribution from IRAS data", ***Mon. Not. R. astr. Soc.***, **253**, 561-574.

Ceplecha, Z. : 1977, "Fireballs photographed in central Europe", ***Bull. Astron. Inst. Czech.***, **28**, 328-340.

Ceplecha, Z. : 1988, "Earth's influx of different populations of sporadic meteoroids from photographic and television data", ***Bull. Astron. Inst. Czech.***, **39**, 221-236.

Ceplecha, Z. : 1992, "Influx of interplanetary bodies onto Earth", ***Astron. Astrophys.***, **263**, 361-366.

Clark, B.E., Fanale, F.P. and Salisbury, J.W. : 1992, "Meteorite-asteroid spectral comparison : The effects of comminution, melting, and recrystallization", ***Icarus***, **97**, 288-297.

Crabb, J. and Schultz, L. : 1981, "Cosmic-ray exposure ages of ordinary chondrites and their significance for parent-body stratigraphy", ***Geochem. Cosmochem. Acta***, **45**, 2151-2160.

Cruikshank, D.P., Tholen, D.J., Hartmann, W.K., Bell, J.F. and Brown, R.H. : 1991, "Three basaltic Earth-approaching asteroids and the source of basaltic meteorites", ***Icarus***, **89**, 1-13.

Dahlgren, M., Hahn, G., Lagerkvist, C.-I. and Lundstrom, M. : 1992, "The orbital evolution of real asteroids near the 4/1 mean motion resonance with Jupiter". In ***Asteroids, Comets, Meteors 1991*** (A.W. Harris and E. Bowell, Eds.), 141-144, Lunar and Planetary Institute, Houston.

Davis, D.R. and Ryan, E. : 1990, "On collisional disruption : Experimental results and scaling laws", ***Icarus***, **83**, 156-182.

Davis, D.R., Chapman, C.R., Weidenschilling, S.J. and Greenberg, R. : 1985, "Collisional history of asteroids : Evidence from Vesta and the Hirayama families", ***Icarus***, **62**, 30-53.

Davis, D.R., Farinella, P., Paolicchi, P., Weidenschilling, S.J. and Binzel, R.P. : 1989, "Asteroid collisional history : Effects on sizes and spins". In ***Asteroids II*** (T. Gehrels, Ed.), 805-826, Univ. of Arizona Press, Tucson.

Dohnanyi, J.W. : 1969, "Collisional model of asteroids and their debris", ***J. Geophys. Res.***, **74**, 2531-2554.

Drake, M.J. : 1979, "Geochemical evolution of the eucrite parent body : Possible nature and evolution of asteroid 4 Vesta". In ***Asteroids*** (T. Gehrels, Ed.), 765-782, Univ. of Arizona Press, Tucson.

Drummond, J.D. : 1991, "Earth-approaching asteroid streams", ***Icarus***, **89**, 14-25.

Drummond, J.D. and Wisniewski, W.Z. : 1990, "The rotational poles and shapes of 1580 Betulia and 3908 (1980 PA) from one apparition", ***Icarus***, **83**, 349-359.

Fanale, F.P., Clark, B.E. and Bell, J.F. : 1992, "A spectral analysis of ordinary chondrites, S-type asteroids, and their component materials : Genetic implications", ***J. Geophys. Res.***, **97**, 20,863-20,874.

Farinella, P. and Davis, D.R. : 1992, "Collision rates and impact velocities in the main asteroid belt", ***Icarus***, **97**, 111-123.

Farinella, P., Davis, D.R., Cellino, A. and Zappalà, V. : 1992, "The collisional lifetime of asteroid 951 Gaspra", ***Astron. Astrophys.***, **257**, 329-330.

Farinella, P., Gonczi, R., Froeschlé, Ch. and Froeschlé, C. : 1993a, "The injection of asteroid fragments into resonances", ***Icarus***, **101**, 174-187.

Farinella, P., Gonczi, R. and Froeschlé, Ch. : 1993b, "Meteorites from the asteroid 6 Hebe", ***Celest. Mech.***, **56**, 287-305.

Froeschlé, Ch. and Scholl, H. : 1987, "Orbital evolution of asteroids near the secular resonance ν_6", ***Astron. Astrophys.***, **179**, 294-303.

Froeschlé, Ch. and Scholl, H. : 1992, "The effect of secular resonances in the asteroid region between 2.1 and 2.4 *AU*". In ***Asteroids, Comets, Meteors 1991*** (A.W. Harris and E. Bowell, Eds.), 205–209, Lunar and Planetary Institute, Houston.

Froeschlé, Ch., Gonczi, R., Farinella, P. and Morbidelli, A. : 1994, "Orbital evolution of near–Earth asteroids affected by secular resonances", in preparation.

Gaffey, M.J., Bell, J.F. and Cruikshank, D.P. : 1989, "Reflectance spectroscopy and asteroid surface mineralogy". In ***Asteroids II*** (R.P. Binzel, T. Gehrels, and M.S. Matthews, Eds.), 98–127, Univ. of Arizona Press, Tucson.

Gaffey, M.J., Reed, K.L. and Kelley, M.S. : 1992, "Relationship of E–type Apollo asteroid 3103 (1982 BB) to the enstatite achondrite meteorites and the Hungaria asteroids", ***Icarus*, 100**, 95–109.

Gaffey, M.J., Burbine, T.H. and Binzel, R.P. : 1993, "Asteroid spectroscopy : Progress and perspectives", ***Meteoritics*, 28**, 161–187.

Gaffey, M.J., Bell, J.F., Brown, R.H., Burbine, T.H., Piatek, J.L., Reed, K.L. and Chaky, D.A. : 1994, "Mineralogical variations within the S-type asteroid class", ***Icarus*, 106**, 573–602.

Gault, D.E., Shoemaker, E.M. and Moore, H.J. : 1963, "Spray ejected from the lunar surface by meteoroid impact", *NASA Tech. Note D–1767.*

Greenberg, R. and Chapman, C.R. : 1983, "Asteroids and meteorites : Parent bodies and delivered samples", ***Icarus*, 55**, 455–481.

Greenberg, R. and Nolan, M.C. : 1989, "Delivery of asteroids and meteorites to the inner solar system". In ***Asteroids II*** (R.P. Binzel, T. Gehrels, and M.S. Matthews, Eds.), 778–804, Univ. of Arizona Press, Tucson.

Hahn, G., Lagerkvist, C.-I., Lindgren, M. and Dahlgren, M. : 1991, "Orbital evolution studies of asteroids near the 5/2 mean motion resonance with Jupiter", ***Astron. Astrophys.*, 246**, 603–618.

Halliday, I., Blackwell, A.T. and Griffin, A.A. : 1984, "The frequency of meteorite falls on the Earth", ***Science*, 223**, 1405–1407.

Hiroi, T. and Takeda, H. : 1990, "A method to determine silicate abundances from reflectance spectra with applications to asteroid 29 Amphitrite associating it with primitive achondrite meteorites", ***Icarus*, 88**, 205–227.

Hiroi, T., Bell, J.F., Takeda, H. and Pieters, C.M. : 1993, "Modeling of S–type asteroid spectra using primitive achondrite and iron meteorites", ***Icarus*, 102**, 107–116.

Housen, K.R., and Holsapple, K.A. : 1990, "On the fragmentation of asteroids and planetary satellites", ***Icarus*, 84**, 226–253.

Ipatov, S.I. : 1992, "Evolution of asteroidal orbits at the 5/2 resonance", ***Icarus*, 95**, 100–114.

Keil, K., Bell, J.F. and Britt, D.T. : 1992, "Reflection spectra of shocked ordinary chondrites and their relationship to asteroids", ***Icarus*, 98**, 43–53.

Knežević, Z., Milani, A., Farinella, P., Froeschlé, Ch. and Froeschlé, C. : 1991 "Secular resonances from 2 to 50 AU", ***Icarus* 93**, 316–330.

Lipschutz, M.E., Gaffey, M.J. and Pellas, P. : 1989, "Meteoritic parent bodies : Nature, number, size and relation to present–day asteroids". In ***Asteroids II*** (R.P. Binzel, T. Gehrels, and M.S. Matthews, Eds.), 740–777, Univ. of Arizona Press, Tucson.

Marti, K. and Graf, T. : 1992, "Cosmic–ray exposure history of ordinary chondrites", ***Ann. Rev. Earth. Planet. Sci.*, 20**, 221–243.

McSween, H.Y.Jr. : 1992, "Redox effects in ordinary chondrites and implications for asteroid spectrophotometry", ***Icarus*, 95**, 239–243.

Milani, A., Carpino, M., Hahn, G. and Nobili, A.M. : 1989, "Dynamics of planet–crossing asteroids : Classes of orbital behavior — Project SPACEGUARD", ***Icarus*, 78**, 212–269.

Milani, A., Farinella, P. and Knežević, Z. : 1992, "On the search for asteroid families". In ***Interrelations between Physics and Dynamics for Minor Bodies in the Solar System*** (D. Benest and C. Froeschlé, Eds.), 85–132, Editions Frontières, Gif–sur–Yvette, France.

Morbidelli, A., Gonczi, R., Froeschlé, Ch. and Farinella, P. : 1993, "Meteorite delivery through the ν_6 resonance", ***Astron. Astrophys.***, in press.

Nakamura, A. and Fujiwara, A. : 1991, "Velocity distribution of fragments formed in a simulated collisional disruption", ***Icarus***, **92**, 132–146.

Nakamura, A., Suguiyama, K. and Fujiwara, A. : 1992, "Velocity and spin of fragments from impact disruptions, I. An experimental approach to a general law between mass and velocity", ***Icarus***, **100**, 127–135.

Neukum, G., König, B., Fechtig, H. and Storzer, D. : 1975, "Cratering in the Earth–Moon system : Consequences for age determination by crater counting", ***Lunar Planet. Sci. Conf.***, **VI**, 2597–2620.

Pellas, P. and Fiéni, C. : 1988, "Thermal histories of ordinary chondrite parent asteroids", ***Lunar Planet. Sci. Conf.***, **XIX**, 915–916.

Pellas, P. and Storzer, D. : 1981, "^{244}Pu fission track thermometry and its application to stony meteorites", ***Proc. R. Soc. Lond. A***, **374**, 253–270.

Petit, J.–M. and Farinella, P. : 1993, "Modelling the outcomes of high–velocity impacts between small solar system bodies", ***Celest. Mech.***, **57**, 1–28.

Salisbury, J.W., D'Aria, D.M. and Jarosewich, E. : 1991, "Midinfrared (2.5 – 13.5 μm) reflectance spectra of powered stony meteorites", ***Icarus***, **92**, 280–297.

Scholl, H. and Froeschlé, C. : 1977, "The Kirkwood gaps as an asteroidal source of meteorites". In ***Comets, Asteroids, Meteorites*** (A.H. Delsemme, Ed.), 293–295, Univ. of Toledo Press, Toledo.

Scholl, H. and Froeschlé, Ch. : 1991, "The ν_6 secular resonance region near 2 AU : A possible source of meteorites", ***Astron. Astrophys.***, **245**, 316–321.

Stöffler, D., Gault, D.E., Wedekind, J. and Polkowski, G. : 1975, "Experimental hyper-velocity impact into quartz sand : Distribution and shock metamorphism of ejecta", ***J. Geophys. Res.***, **80**, 4062–4077.

Van Houten, C.J., Van Houten–Groeneveld, I., Herget, P. and Gehrels, T. : 1970, "The Palomar–Leiden Survey of faint minor planets", ***Astron. Astrophys. Suppl.***, **2**, 339–448.

Vickery, A.M. : 1986, "Size–velocity distribution of large ejecta fragments", ***Icarus***, **67**, 224–236.

Vilas, F. and McFadden, L.A. : 1992, "CCD reflectance spectra of selected asteroids, I. Presentation and data analysis considerations", ***Icarus***, **100**, 85–94.

Wetherill, G.W. : 1985, "Asteroidal source of ordinary chondrites", ***Meteoritics***, **20**, 1–21.

Wetherill, G.W. : 1987, "Dynamical relations between asteroids, meteorites and Apollo–Amor objects", ***Phil. Trans. R. Soc. Lond. A***, **323**, 323–337.

Wetherill, G.W. : 1988, "Where do the Apollo objects come from?", ***Icarus***, **76**, 1–18.

Wetherill, G.W. and Chapman, C.R. : 1988, "Asteroids and meteorites". In ***Meteorites and the Early Solar System*** (J.F. Kerridge and M.S. Matthews, Eds.), 35–67, Univ. of Arizona Press, Tucson.

Wetherill, G.W. and Williams, J.G. : 1979, "Origin of differentiated meteorites". In ***Origin and Distribution of the Elements*** (L.H. Ahrens, Ed.), 19–31, Pergamon Press, Oxford.

Williams, J.G. : 1973, "Meteorites from the asteroid belt?", ***Eos***, **54**, 233.

Wisdom, J. : 1983, "Chaotic behavior and the origin of the 3/1 Kirkwood gap", ***Icarus***, **56**, 51–74.

Wisdom, J. : 1985 "Meteorites may follow a chaotic route to earth", ***Nature***, **315**, 731–733.

Yoshikawa, M. : 1990, "Motions of asteroids at the Kirkwood gaps, I. On the 3/1 resonance with Jupiter", ***Icarus***, **87**, 78–102.

Zappalà, V., Farinella, P., Knežević, Z. and Paolicchi, P. : 1984, "Collisional origin of the asteroid families : Mass and velocity distributions", ***Icarus***, **59**, 261–285.

Zappalà, V., Cellino, A., Farinella, P. and Knežević, Z. : 1990, "Asteroid families, I. Identification by hierarchical clustering and reliability assessment", ***Astron. J.***, **100**, 2030–2046.

DYNAMICS OF COMETS : RECENT DEVELOPMENTS AND NEW CHALLENGES

JULIO A. FERNÁNDEZ
Departamento de Astronomía, Facultad de Ciencias,
Tristán Narvaja 1674, 11200 Montevideo, URUGUAY
E–mail julio@fisica.edu.uy

Abstract. There is a broad consensus that long-period comets come from a huge reservoir surrounding the solar system, as proposed originally by Oort. Yet, the classical picture of the Oort cloud has substantially changed during the last decade. In addition to passing stars, the tidal force of the galactic disk and giant molecular clouds have also been identified as major perturbers of the Oort cloud. In particular, the latter may be responsible for limiting the size of the stable Oort cloud to no more than $\approx 10^4 AU$, i.e. about one tenth of the classical Oort's radius.

Most comets are injected into the planetary region by the quasi-steady action of the tidal force of the galactic disk. The concentration of aphelion points of dynamically young comets toward mid-galactic latitudes is a consequence of its dominant influence. The frequency of comet passages into the inner planetary region could experience significant fluctuations with time as the Oort cloud meets random strong perturbers. The observed ordered pattern of most comet aphelia, associated with the galactic structure, argues against a recent strong perturbation of the Oort cloud.

The origin of the Jupiter family has become another point of intense debate. Jupiter family comets may come from a transneptunian comet belt -the Kuiper belt- from where they can reach the planetary region through chaotic motion. The Kuiper belt has become accessible to large telescopes, as shown by the recent discoveries of 1992QB1 and 1993FW, possibly belt members. The major challenge will be to explore the region usually inaccessible to external perturbers that goes from $\sim 30AU$ to a few thousand AU. A significant mass may have been locked there from the beginnings of the solar system, giving rise to an inner core that feeds the outer or classical Oort cloud. Our aim will be to briefly discuss some of the topics summarized here.

1. Steady and random perturbers of the Oort cloud. Comet showers

Comets in the Oort cloud evolve dynamically under the action of several external perturbers. Distant - background - stars passing at distances greater than, say $2-3\times 10^4 AU$, and tides of the galactic disk exert a quasi-steady perturbing action. From time to time, the Oort cloud is strongly perturbed by stars that come very close to the Sun or by encounters with giant molecular clouds (GMCs). These sporadic encounters seem to have major effects in the dynamical stability of the Oort cloud, the injection rate of comets into the planetary region and, perhaps, in the evolution of the planetary atmospheres, life and the impact cratering rate.

The action of stellar perturbations on Oort comets has already been well studied by several authors (e.g. Rickman 1976, Weissman 1980), so we will not repeat the analysis of their dynamical effects here. For a general review see, for instance, Fernández and Ip (1991).

The tidal force of the galactic disk is very effective in changing the angular momentum or perihelion distance q of Oort comets (e.g. Byl 1983, Heisler and

A. Milani et al. (eds.), Asteroids, Comets, Meteors 1993, 223–240.

Tremaine 1986). However, its overall effect on the comet's energy is negligible. This is because the energy's perturbation along the outgoing leg of the comet's orbit roughly compensates for that along the incoming leg, so that its net effect is very small, typically of order (q/a), where a is the comet's semimajor axis (Byl 1983). Therefore the perihelia of Oort comets can be rather frequently injected into the planetary region, but the comets themselves can hardly become hyperbolic by the action of this force. Let us consider an Oort cloud comet of semimajor axis a and aphelion direction at a galactic latitude ϕ. The change in its transverse velocity due to the action of the galactic tidal force acting during an orbital revolution P is

$$(\Delta v_T)_{tide} = 3\pi G \rho a P \cos\alpha \sin 2\phi, \quad (1)$$

where α is the angle between the orbital plane and the plane perpendicular to the galactic disk containing the radius Sun-comet (assumed to keep the same direction along the àpsidal line through the orbital revolution, which is more or less fulfilled for near-parabolic comets).ρ is the density of the galactic disk in the Sun's neighborhood, which is about $0.15 M_\odot pc^{-3}$ (Tremaine 1993).

A population of comets thermalized by external perturbers will have randomly oriented velocity vectors so all the directions will be possible, except those falling very close to the solar direction since these comets will be quickly removed by planetary perturbations. Consequently, there will be an empty region in the velocity phase space around the solar direction known as the *loss cone* (Hills 1981).

Should the loss cone be completely empty, the Oort comets would then stay away from the planetary region. Now, the combined effect of stellar perturbations and tides of the galactic disk will cause a steady refilling of the loss cone. At the end, losses by planetary perturbations will reach an equilibrium with refilling by external perturbers, so a fraction $f \leq 1$ of the loss cone will be kept filled at any time. The fraction f will depend on the semimajor axis a of Oort comets. For larger a the action of external perturbers will be stronger, so an ever increasing fraction of the loss cone will be filled; the increase with a goes as $f \propto a^7$ (Fernández 1992). There will be a limiting semimajor axis $a = a_{fill}$ such that $f = 1$, so comets with $a > a_{fill}$ will have their loss cones permanently filled. Consequently, the influx rate of Oort comets with $a > a_{fill}$ will be constant throughout the time, no matter the random action of strong perturbers. Showers will be produced by comets with $a < a_{fill}$.

Strictly speaking, the tidal force of the galactic disk depends on the galactic latitude ϕ (cf. eq.(1)); it is maximum for $\phi = 45°$ and goes down to zero for $\phi = 0°$ or $\pm 90°$. Therefore, this force is very efficient in injecting Oort comets at mid-galactic latitudes but it cannot do that near the galactic equator or poles. In this regard the distribution of aphelion points of LP comets shows a clear concentration at mid-galactic latitudes (Delsemme 1987). Galactic tidal forces fill the loss cones of Oort cloud comets at mid-galactic latitude ($\phi \sim 30° - 60°$) for $a \gtrsim 3.3 \times 10^4 AU$, while stellar perturbations do that for comets with $a \gtrsim 4 \times 10^4 AU$ (if the latter were the only acting perturber). These results show the predominance of galactic tides in the injection rate of Oort cloud comets in the planetary region. Only at low or high galactic latitudes ($|\phi| \lesssim 15°$ or $|\phi| \gtrsim 75°$), stellar perturbations overcome galactic tidal forces (Fig.1).

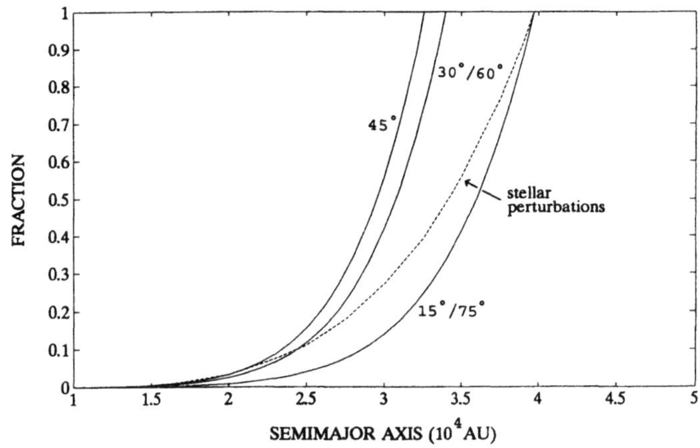

Fig. 1. Fraction of loss cone filled with comets as a function of the semimajor axis, as due to the action of stellar perturbations (dashed curve) and tides of the galactic disk (solid curves) at the galactic latitudes indicated beside each curve (adapted from Fernández 1992).

A very close stellar passage or a penetrating encounter with a GMC can perturb the inner portions of the Oort cloud where loss cones are empty, or at least partially empty, so they will be suddenly refilled. A sharp increase in the injection rate of Oort comets in the planetary region will ensue that will last until their loss cones are emptied, which is of the order of their typical orbital period P (about $10^6 yr$). Afterwards the Oort cloud will relax to the state previous to the perturbation. Hills (1981) called such a sudden enhancement in the cometary flux a *comet shower*.

The intensity of a comet shower will depend on the external perturber and on the degree of central condensation of the Oort cloud. For a heavily concentrated Oort cloud, a close stellar passage at $D_{\odot} \sim 10^4 AU$ can trigger a comet shower $\sim 10^2$ times as intense as the background comet flux at average intervals of a few 10^7 years (see numerical results by Heisler 1990 and Fernández 1992). Similar effects can be reached by penetrating encounters with GMCs.

Whether clues to past comet showers can be found in the impact cratering record is a very controversial issue. The problem is that the terrestrial cratering rate is dominated by Earth-crossing asteroids, while the steady-state flux of comets contribute to no more than $\sim 10\%$ (e.g. Bailey 1991, Weissman 1991).A simple estimate shows that the intensity of a comet shower should be at least several 10^2 times greater than the steady-state comet flux to show up in crater statistics.

The evidence that most new comets seem to be deflected to the inner planetary region by tidal torques of the galactic disk -a steady perturber- suggests that the frequency of comet passages is currently near its quiescent level. Comets injected during a shower might greatly exceed the steady supply of Oort comets to the point of erasing or severely weakening the galactic signature in the distribution of aphelion points. Indeed, there are some small aphelion clusterings (e.g. Biermann et al. 1983), suggesting that one or more stars penetrated the Oort cloud in the recent past leading to weak showers, though these seem to contribute to only a minor fraction of the overall aphelion sample.

The invoked association of aphelion clusterings with the solar antapex (e.g. Oja 1975) is another point that awaits elucidation. The major difficulty seems to be the lack of a sound physical reason to explain it. Among some possible explanations we can mention : the capture of a cloud of interstellar comets with low relative velocities in the recent past (Valtonen and Innanen 1982), and Chandrashekar's "dynamical friction" in which the massive Sun -but not the light comets- loses kinetic energy to deflected interstellar particles, thus giving comets a net acceleration in the direction of the solar motion (Brunini 1993). There is the possibility that just by chance a clustering caused by a close stellar passage (Biermann et al. 1983) happens to lie close to the antapex.

2. The Jupiter-Saturn barrier

As mentioned, under the tidal force of the galactic disk Oort comets will mainly change their angular momentum H or their perihelion distance. A change in the transverse velocity Δv_T will be associated to a change in the angular momentum $\Delta H = r \times \Delta v_T$, where Δv_T is given by eq.(1). For a comet in a near-parabolic orbit $H \simeq (2GM_\odot q)^{1/2}$, whereby $\Delta H = H/2 \times \Delta q/q$. By taking a time-average distance $r = 1.5a$, we obtain the relative change in the comet's perihelion distance after one orbital revolution as

$$\Delta q/q \simeq 12.7\pi^2 M_\odot^{-1} \rho q^{-1/2} a^{7/2} \cos\alpha \sin 2\phi. \tag{2}$$

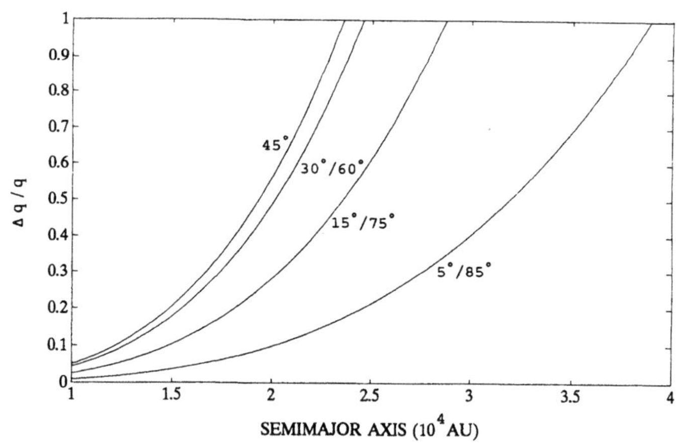

Fig. 2. Relative change in the perihelion distance of Oort comets with original $q = 7.5AU$, as caused by the tidal force of the galactic disk, as a function of their semimajor axes and for the galactic latitudes indicated beside each curve.

Computed values of the relative change $\Delta q/q$ as a function of the comet's semimajor axis a are shown in Fig.2. Note that the ratio $\Delta q/q$ depends on q. We will adopt $q = q_b = 7.5AU$ (Fernández 1992), which represents approximately the boundary between Oort comets strongly perturbed by the planets ($q < q_b$) and those weakly perturbed ($q > q_b$); in other words, the boundary between average energy changes per perihelion passage greater or smaller than $10^{-4}AU^{-1}$, respectively. We see from eq.(2) that Oort comets with rather small values of a, say $\lesssim 1.8 \times 10^4 AU$, have $\Delta q/q \lesssim 0.5$ for all galactic latitudes. Consequently, they will drift in q very smoothly under the action of galactic tidal torques. These results are in reasonable agreement with those obtained analytically by Matese and Whitman (1992) and the orbital calculations of the change in q for the observed sample of near-parabolic orbits carried out by Yabushita (1989).

Since planetary perturbations in the region of Uranus and Neptune are very weak, Oort comets with $a \lesssim 2 \times 10^4 AU$ can actually move back and forth between the Oort cloud and the outer planetary region several times without changing significantly either a or q (Fernández 1981, Weissman 1985). Once an Oort comet reaches the Jupiter-Saturn region during its slow drift in the q phase space, planetary perturbations are already strong enough to remove it from the Oort cloud. It seems then that there is no way for Oort comets with $a \lesssim 2 \times 10^4 AU$ to leap over the Jupiter-Saturn barrier and reach the region of the terrestrial planets where they become observable (Weissman 1985). The situation changes for Oort comets with larger semimajor axes, since they can experience changes in q large enough to leap over the Jupiter-Saturn barrier from the Uranus-Neptune region or beyond to the observable region (say a sphere centered on the Sun of about $3AU$ radius within which most of the discovered comets have their perihelia).

The observed distribution of the original reciprocal semimajor axis $(1/a)_{orig}$ shows a concentration in the range $2 \times 10^{-5} < (1/a)_{orig} < 5 \times 10^{-5} AU^{-1}$ (Fig.3), i.e. semimajor axes in the range $2 - 5 \times 10^4 AU$. The histogram of Fig.3 only takes into account comets with $q > 2AU$ to avoid as much as possible uncertainties in $(1/a)_{orig}$ due to nongravitational forces. The lower limit agrees well with the minimum a derived above for Oort comets to be able to leap over the Jupiter-Saturn barrier. The upper limit may indicate the distance at which the Oort cloud population has decreased substantially, as comets there have ever shorter dynamical lifetimes.

3. The stability limit of the Oort cloud. The inner core

Comets may be removed from the Oort cloud either because they enter the planetary region where they are perturbed by the planets, or because they get enough energy from external perturbers to overcome the gravitational field of the Sun. The latter will be the only loss mechanism for Oort comets whose perihelia lie outside the planetary region. It has become clear that passing stars are not the major external perturbers responsible for pumping up energies of Oort comets to escape velocities. Biermann (1978) pointed out that encounters with GMCs may have a dramatic dynamical influence, causing the disruption of the outer layers of the Oort cloud (e.g. Napier and Staniucha 1982). Typical masses of GMCs are of the order of $1 - 2 \times 10^5 M_{\odot}$ with mean diameters of $\sim 45pc$ (Blitz 1993). The number of penetrating encounters of the Sun with GMCs during the solar system age is ~ 5, with an uncertainty of a factor ~ 2 (Bailey 1983, Weinberg et al. 1987).

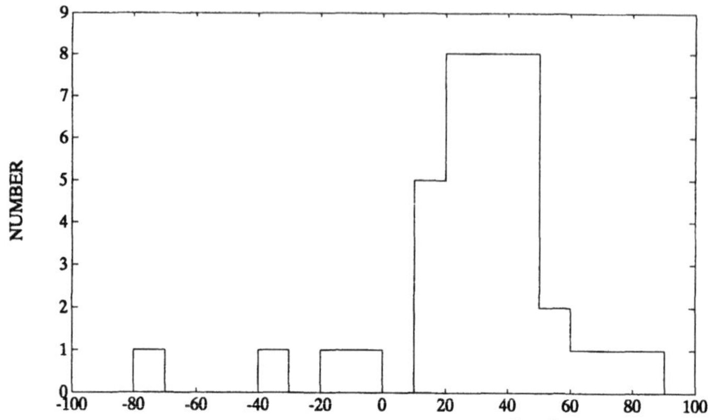

Fig. 3. Distribution of the original reciprocal semimajor axes of the observed new comets with perihelion distances $q > 2AU$ taken from Marsden and Williams (1992) catalogue. Only those comets with good orbit determinations are plotted, which the authors define as being of quality classes 1A and 1B.

GMCs are not uniform entities, but they are composed of numerous dense clumps with a wide range of masses, the largest ones being of the order of a few $10^3 M_{\odot}$ (Blitz 1993). Actually, during a penetrating encounter with a GMC, the largest dynamical effect on the Oort cloud may be expected to occur during close approaches to some of the most massive clumps (Stern 1990). Soft (nonpenetrating) encounters with GMCs and the more numerous penetrating encounters with less massive molecular clouds will increase somewhat the cumulative velocity perturbation and thus decrease the stability boundary.

Weinberg et al. (1987) showed that binary stars with separations $a \sim 8000AU$ have dynamical lifetimes of the order of the solar system age ($4.6 \times 10^9 yr$)(Fig.4). Hut and Tremaine (1985) found a half-life of $3 \times 10^9 yr$ for comets with $a = 25000AU$ perturbed either by stars or by GMCs; their combined effect will decrease the semimajor axis to values of $\sim 10^4 AU$ over the solar system age. Indeed, some observations tend to confirm the conclusion that the stability boundary is at about $10^4 AU$. For instance, the maximum separations of wide binary stars are of the order of $0.1pc$ (e.g. Latham et al. 1991), and they are even smaller for the presumably older late-type stars. Poveda (1988) concludes that the classical Oort cloud should not be stable over time scales longer than the solar system age for distances greater than a few $10^3 AU$. We then come to the conclusion that the outer or classical Oort cloud -from where Oort comets are driven into the observable region (cf. Fig.3)- must not be primordial. It must have been steadily replenished from a dense inner core throughout the solar system age to make up for the dynamical losses.

The realization that the stability limit of the Oort cloud is smaller than the range of semimajor axes ($2 - 5 \times 10^4 AU$) for which tidal torques of the galactic disk are effective in bringing comets into the planetary region leads to a new concept : the existence of a dense inner core of the Oort cloud. During penetrating encounters

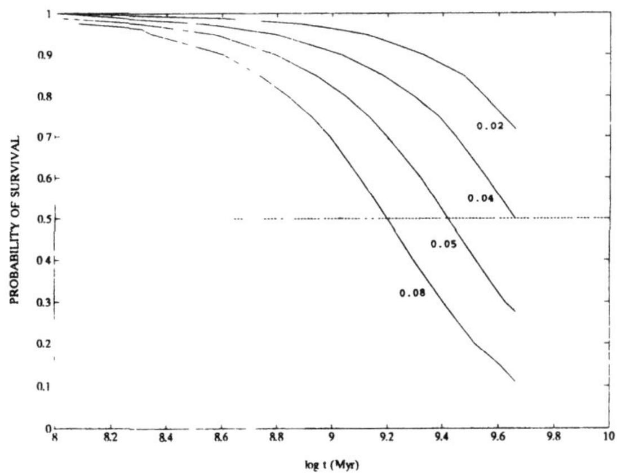

Fig. 4. Probability of survival of binaries with initial semimajor axes indicated beside each graph (in parsecs), subject to perturbations by stars and GMCs. The lifetime $t_{1/2}$ is given by the intersection of these curves with the $P = 0.5$ locus (adapted from Weinberg et al. 1987).

with GMCs or very close stellar passages some comets of the inner core gain enough energy to be transferred to the classical Oort cloud, while others will be directly injected into the planetary region causing a comet shower. The capture of transient Oort clouds from interstellar clouds has been proposed by Clube and Napier (1984) as an alternative to the inner core hypothesis. Yet, serious difficulties with capture theories have been hovering around for quite a long time, such as the very low capture efficiency of interstellar comets that would lead to extremely high mass densities of interstellar comets, or anomalously low encounter velocities, in order to produce enough captures (e.g. Valtonen and Innanen 1982, Zheng et al. 1990), and the lack of observed comets with clearly hyperbolic original orbits (Kresák 1992).

4. Multiple-step capture process of Oort cloud comets by the Jovian planets

Let us now analyze the later evolution of Oort comets injected into the observable region as *new* comets. This evolution can be approximately described as an one-dimensional random-walk in the energy space. The binding energy x of a comet is proportional to the reciprocal semimajor axis $(1/a)$, so we can take $x = (1/a)$. After a passage by the planetary region the comet will be either lost to the interstellar space (i.e. it gets an energy $x < 0$), or it remains bound to the solar system with a new energy $x' = x + \Delta x > 0$. For long-period (LP) comets the orbital energy x is the parameter that experiences the greatest variation during a passage by the planetary region. The changes in the other orbital parameters are much smaller, though their long-term effects may be non-negligible as we will see below for the inclination.

A fraction of the comets coming from the Oort cloud (i.e. energies $x \simeq 0$) can reach a certain energy level $x_f > 0$ under the action of planetary perturbations after a certain number of passages; the rest will be lost along hyperbolic orbits. The probability of capture, p_c, of a LP comet starting in a parabolic orbit to an orbit with energy x_f (assumed to be elliptic, i.e. $x_f > 0$) is given by (Fernández and Gallardo 1993)

$$p_c \simeq 0.5\sigma/x_f, \tag{3}$$

where σ is the typical energy change per passage, which can be expressed by the standard deviation of the distribution function of energy changes $\Psi(\Delta x)$, assumed to be symmetrical with respect to $\Delta x = 0$. We note that Everhart (1969) showed that the distribution of energy changes has asymmetric tails of large values of Δx; nevertheless, the symmetry hypothesis used before is still a good approximation for small Δx. For low-inclination LP comets reaching the observable region, say $q \lesssim 2-3AU$, typical energy changes are : $\sigma \sim 15\times10^{-4}AU^{-1}$, while for retrograde orbits $\sigma \sim 6\times10^{-4}AU^{-1}$. Introducing these results in eq.(3) we find that the capture probability is $\sim 2-3$ times greater for comets in direct orbits than for retrograde ones. The difference in σ between comets in direct and retrograde orbits is much larger in the outer planetary region (cf. Fernández 1981), which implies a much larger difference in their capture probabilities. We will come back to this point in the next Section. As eq.(3) shows, the capture probability is also inversely proportional to the energy level reached by the comets.

Second-order changes in the inclination may become significant after hundreds of revolutions. The main effect will be a tendency of near-perpendicular comets ($i \sim 90°$) to shift toward retrograde orbits (Fernández and Gallardo 1993). This mechanism can explain the Kreutz family of sungrazers (Bailey et al. 1992). The combined effects of the dependence of the capture probability on the inclination and the shift in inclination, make that the greater losses of retrograde orbits in small-q comets by the first effect are roughly offset by the gains by the second one, so the balance between direct and retrograde orbits is more or less kept throughout the dynamical evolution (Fig.5). When we allow for physical losses by taking a limiting number of revolutions, the resulting i-distribution shows a clear depletion of retrograde orbits in better agreement with the observed i-distribution of *old* LP comets and intermediate-period (IP) comets (say, comets with periods $20 < P < 1000yr$) and perihelion distances $q < 2AU$. These results suggest that the observed i-distribution of small-q LP comets arises from a combination of dynamical causes and physical losses. The observed depletion of old, retrograde LP comets suggests average lifetimes of a few hundred revolutions before disintegration or deactivation.

Computer results by Fernández and Gallardo (1993) suggest that the steady-state population of IP comets with $q < 2AU$ may be of about 300 comets. The current number of comets detected in this class is 21, so the degree of completeness of the sample may be less than 10%. The difficulties at detecting IP comets are great owing to their extreme faintness; some IP comets may have become inactive, looking asteroidal as may be the case of 1991DA. The capture probability of IP comets into short-period (SP) orbits ($P < 20yr$) turns out to be very low, of the

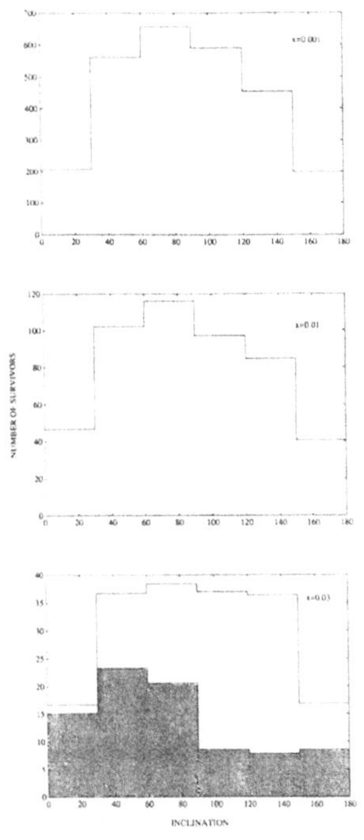

Fig. 5. Computed inclination-distributions of LP comets evolving through successive passages by the planetary region to larger energies x. The shaded histogram is for a simulation that includes physical losses by setting a limiting number of 400 revolutions (Fernández and Gallardo 1993).

order of 10^{-2}, leading to steady-state populations of captured SP comets one-two orders of magnitude too small as compared to the observed population.

5. The Jupiter family. Possible source regions

SP comets ($P < 20yr$) or Jupiter family comets form a very peculiar group clearly distinguished from a dynamical viewpoint from the other comets of longer periods. First, they have low-inclination orbits and their values of the Tisserand constant T are mainly concentrated in the range $2.5 \lesssim T \lesssim 3$, whereas the observed comets with longer periods have lower T values; for $q < 2AU$ these fall in the range $-2 < T < 2$ (Fig.6). The aphelia of SP comets tend to be concentrated around Jupiter's orbit and, in particular, around its perihelion which shows the dominant role of Jupiter in their dynamical evolution. There is a question of nomenclature that is convenient to analyze very briefly. Sometimes the term "short-period" is applied to all comets with $P < 200$ yr. Now, there is a clear distinction around $P \sim 20yr$ in the sense that comets with periods $20 < P < 200yr$, classically called "intermediate-period" comets, or more recently Halley-type comets, have Tisserand constants $T < 2$ like the ones of LP comets. On the other hand, SP comets or Jupiter family comets have $T > 2$, with only three exceptions so far : P/Tuttle, P/IRAS and P/Machholz.

Therefore, the classification of comets according to their orbital periods continues to be very useful and with some dynamical basis, adding perhaps the Tisserand constant to discriminate some uncertain cases, as Valsecchi (1992) suggested.

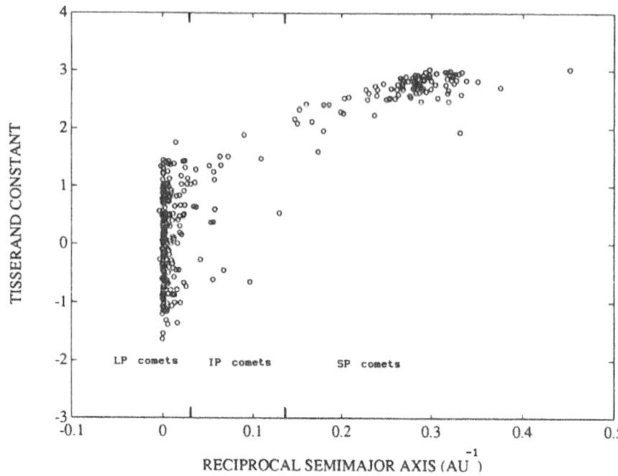

Fig. 6. Plot of the Tisserand constant T versus the reciprocal semimajor axis of the observed LP comets with well determined osculating values of $1/a$, IP comets and SP comets. We only considered comets with $q < 2AU$ for all the dynamical classes.

Much attention has been paid to the dynamical evolution of SP comets and their physical decay, which is of paramount importance to assess their population size, distribution of perihelion distances and their interrelationship with the Apollo-Amor asteroids and the meteoroidal complex. Tancredi and Rickman (1992) have carried out long-term orbital integrations of the observed Jupiter-family comets over $2000yr$ centered on the present time. They find that the mean values of q, a and i are minimum near the present. The minimum of the mean q (and mean a) may possibly be due to observational selection effects, since we tend to discover SP comets when they get closer to the Sun. The slope of the mean-q values of SP comets is steeper in the past than in the future, which was interpreted by Fernández (1985) in terms of an average physical lifetime of $\sim 10^3$ revolutions ($\sim 10^4 yr$). Tancredi and Rickman found a less pronounced past-future asymmetry which would imply a longer physical lifetime. Orbital integrations of all the known SP comets over $\pm 10^7 yr$ by Levison and Duncan (1993a) fully corroborated the minimum in the mean $\imath$ at present. Why the system is so flat at present is interpreted by Levison and Duncan as due to a recent capture of comets from a flat source. As the system evolves -either into the past or into the future- it becomes much less flat as SP comets tend to occupy all the possible dynamical niches but, as the authors argue, such an evolved system of SP comets can never be observed due to their short physical lifetimes as compared to their dynamical ones.

The origin of the Jupiter family has been visualized as the dynamical end product of Oort comets captured by Jupiter through many revolutions (e.g. Everhart 1972).

There are however several independent results in conflict with this hypothesis as, for instance, the failure to produce the right number of SP comets (e.g. Joss 1973, Fernández and Ip 1983) and some computed i-distributions that yield a significant fraction of SP comets in retrograde orbits (Duncan et al. 1988, Wetherill 1991). Computed distributions of T and q are also in conflict with the observed ones (Fernández and Gallardo 1993). For instance, computed q-distributions tend to be uniform within Jupiter's orbit, while the observed one suggests a very steep increase in the number of SP comets with q, at least up to $q \sim 2.5AU$ (Fernández et al. 1992).

Given the above difficulties one has to ask whether there may be more suitable comet sources for the Jupiter family. Comets like P/Schwassmann-Wachmann 1 in the Jupiter-Saturn region may be suitable progenitors. It moves on a low-inclination ($i = 9^\circ.4$), low-eccentricity ($e = 0.045$) orbit and has a Tisserand constant $T = 2.98$, i.e. within the range of most SP comets. Indeed P/S-W 1 may belong to a large transient population of about 30,000-100,000 comets (Delsemme 1973), large enough to keep the population of SP comets in steady-state. Such comets with $T \approx 3$ have very low relative velocities with Jupiter, so captures as temporary satellites can occur rather frequently (Carusi and Valsecchi 1981). The recently discovered comet P/Shoemaker-Levy 9(1993e) on an orbit similar to Jupiter's (IAU Circular No.5800) can be a nice example of such a kind of objects.

Since comets in the Jupiter-Saturn region have rather short dynamical lifetimes (Gladman and Duncan 1990), they must be continuously replenished from a source in the outer planetary region. Chiron may be representative of this more distant population (the sample of discovered outer solar system objects is growing very fast). The question is whether suitable storage places of outer solar system bodies exist within the planetary region. From several surveys, Shoemaker et al. (1993) have found that the number of Trojans with diameters $> 15km$ associated to the Lagrange $L4$ and $L5$ points should be of several thousand. The Trojan population down to a radius $\sim 1km$ would be of some 10^5 bodies. It is then possible that some comet-like bodies in the Jupiter's zone with a Tisserand constant $T \sim 3$ are indeed escaped Trojans as already argued by Rabe (1972). Trojan-like swarms might be associated to the Lagrange $L4$ and $L5$ points of the other Jovian planets, for which preliminary integrations tend to confirm long dynamical survivals (Holman and Wisdom 1993). Whether these might be suitable sources to provide low-inclination comets, via slow dissipation from the swarms, has to be assessed through future surveys and studies of their dynamical stability.

A possibility is that the transient populations between the giant planets mainly come from the capture of Oort comets by Saturn, Uranus and Neptune (Bailey 1986). The capture efficiency of these planets should be studied in more detail as well as their dependence on the orbital inclination; in other words, its capability to sort out high-inclination comets in order to produce a flat population of SP comets. Some preliminary results (Duncan et al. 1988, Wetherill 1991) seem to indicate that a fraction of comets with initial random inclinations in Neptune-crossing orbits will be captured in retrograde SP orbits with $q < 1.5AU$, which would argue against this source.

The results presented by Duncan et al. (1988) have been criticized by Stagg and

Bailey (1989) because of the use of artificially high masses for the giant planets in order to speed up the computations. We also note that Duncan et al. start to follow the capture process to SP orbits from comets with perihelia in the range $20 < q < 30AU$ and semimajor axis $a = 50AU$(i.e. an energy $x = 0.02AU^{-1}$), for which they assume an isotropic, prograde distribution of inclinations (i.e. uniform in $\cos i$ between 0 and 1). Now, it is very likely that the inclination distribution of comets that reach the energy level $x = 0.02AU^{-1}$ has already departed very substantially from an isotropic distribution (cf. Section 4). As eq.(3) shows, the capture probability to a certain energy state is proportional to the typical energy change σ. Now, in the outer planetary region σ decreases very sharply with the inclination, in such a way that : $\sigma(i \sim 0) \sim 10\sigma(i \sim 180°)$ (Fernández 1981), so the capture of Oort comets to the energy level $x = 0.02AU^{-1}(a = 50AU)$ will be strongly biased toward low-inclination orbits. Wetherill (1991) has used Öpik's two-body formalism to study the capture of parabolic comets with perihelia in the range $20 - 30AU$ into Jupiter-family comets with $q < 1.5AU$. He also finds that a fraction of the computed Jupiter-family comets will have retrograde orbits. Even though Wetherill is able to simulate the capture process without need of artificially increasing the masses of the Jovian planets, the problem is that Öpik's method can only handle strong perturbations in close encounters, which is not well suited to follow the dynamical evolution of comets starting in near parabolic orbits.

The above difficulties strongly favors to review the whole subject of comet capture by the Jovian planets in a more realistic way, namely adopting for the Jovian planets their actual masses and following the orbital evolution from the beginning as near parabolic comets to their final capture as SP comets.

Another interesting possibility -perhaps the most promising one- is that most Jupiter-family comets come from a comet reservoir located beyond Neptune. On cosmogonic grounds, Kuiper (1951) explained such a reservoir on the basis of residual planetesimals left after the formation of Neptune in the region between $30 - 50AU$. The *Kuiper belt* -as it was named- would have a strongly flattened, ring-shaped structure, which would explain the i-distribution of the observed SP comets. The recent discoveries of objects 1992QB1 and 1993FW located at about $40AU$ (Jewitt and Luu 1993 and IAU Circ.5370), may represent the first direct detections of bodies belonging to the Kuiper belt (for more information see Luu, this volume).

6. Kuiper belt dynamics

The question now is what might be the mechanism(s) responsible for the scattering of Kuiper belt comets to Neptune-crossing orbits. Once belt comets become Neptune-crossers, they can be handed down to the control of the other Jovian planets, so a fraction of them may end up incorporated to the Jupiter family (Fernández 1980). The existence of a number of lunar-sized objects within the belt might provide the required stirring effect (Fernández 1980, Ip and Fernández 1991). Based on the tilt of Uranus's orbit, the binary system Pluto-Charon and the retrograde orbit of Triton, Stern (1991) also argues that a primordial population of $10^3 - km$ size bodies existed in the $20 - 50AU$ region. The problem is that no such large objects have so far been discovered through several sky surveys (e.g. Luu and

Jewitt 1988, Kowal 1989, Levison and Duncan 1990), which put severe constraints on the size of the largest objects that might remain undetected in the belt. The conclusion seems to be that if such large bodies existed in the early solar system, all of them -except Pluto- are by now gone.

A major issue to be solved concerns the orbital stability of bodies stored in the Kuiper belt. The dynamical evolution of small, gravitationally noninteracting objects in the outer solar system was examined numerically by Torbett (1989) and Gladman and Duncan (1990) among others. Torbett found that the orbits of bodies initially located at radial distances less than $50AU$ will become chaotic and be quickly perturbed to Neptune-crossing orbits within $10^8 yr$, though this result will depend on their initial eccentricities and inclinations. The distinction between chaotic and regular orbits was set by determining whether the Lyapunov exponent tended to zero or to a certain positive constant within the computed period. Gladman and Duncan also concluded that orbits of small bodies that start on circular orbits within $34AU$ will become Neptune-crossers in about $20Myr$. These results suggest that the inner edge of the Kuiper belt should be detached from Neptune's orbit by several AU.

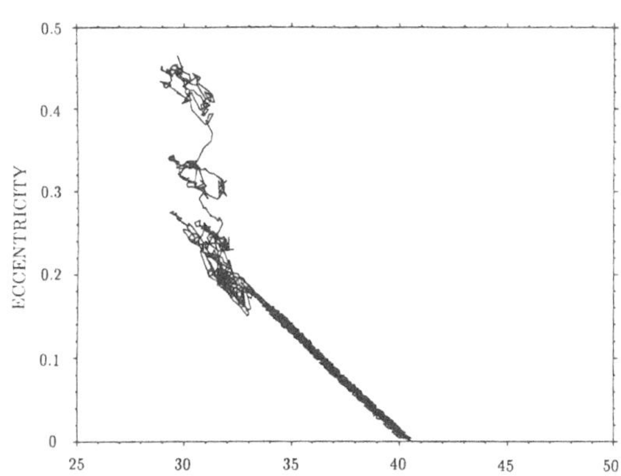

Fig. 7. Dynamical evolution in the parametric plane (q, e) of a hypothetical Kuiper-belt body initially on a circular orbit located in the invariable plane. Lines of constant semi-major axis are diagonal (Holman and Wisdom 1993).

Much more extensive numerical integrations over periods up to $10^9 yr$ have very recently been undertaken by Holman and Wisdom (1993) and Levison and Duncan (1993b) by means of a very efficient symplectic integration technique. These studies help to shed some light on the dynamical structure of the Kuiper belt. They show that the inner edge of the belt is depleted as comets become Neptune-crossers. The erosion rate of the belt follows a rather complex pattern with the heliocentric distance, where bumps are noted at the 3 : 2 and 2 : 1 mean motion resonances with Neptune (at about $40AU$ and $48AU$, respectively). It is very likely that the depletion within $\sim 35AU$ over the solar system age is essentially complete, and that the belt has been very heavily eroded up to $\sim 45AU$. Holman and Wisdom depict very nicely how belt comets on initially near-circular orbits beyond the planetary region evolve until they become Neptune-crossers (see Fig.7). Their results show that belt comets first tend to evolve in eccentricity and perihelion distance, keeping their semimajor axes more or less constant (represented by a diagonal in the plane (q, e)). When they become Neptune-crossers the evolution mainly proceeds in a and e, keeping in this case q more or less constant (a vertical line in the plane (q, e)). Thus, belt comets seem to diffuse through chaotic zones associated to mean motion and secular resonances in a dynamical process resembling the delivery of meteorites to the Earth's zone.

7. Summary and conclusions

In short, we can mention as some of the major advances in our understanding of comet dynamics the following :

1) Tides of the galactic disk exert a dominant influence on the evolution of the angular momentum of Oort comets and thus on the rate at which their perihelia are injected into the planetary region. This reflects in the concentration of the aphelion points of new and dynamically young comets at mid-galactic latitudes. Background stars play a minor role : they can overcome galactic tides only close to the galactic equator or the galactic poles ($|\phi| \lesssim 15°$ or $|\phi| \gtrsim 75°$).

2) The rather smooth injection of Oort comets by the quasi-steady action of galactic tidal torques and background stars is punctuated from time to time by sudden enhancements in the injection rate triggered by penetrating encounters with GMCs or very close stellar passages. There is no evidence of a shower occurring at present, though weak showers in the recent past cannot be ruled out as some small aphelion clusterings suggest.

3) The Oort cloud is dynamically unstable over the solar system age for $a \gtrsim 10^4 AU$. Penetrating encounters with GMCs and close stellar passages are responsible for disrupting the outer layers of the Oort cloud. Since most observed new comets have original semimajor axes in the range $2 - 5 \times 10^4 AU$, they may have been placed there from an inner core over time scales shorter than the solar system age. Hence the inner core appears as a necessary device to maintain the unstable, classical Oort cloud.

4) The Jupiter-Saturn barrier can be overshot by Oort comets with $a \gtrsim 2 \times 10^4 AU$ at mid-galactic latitudes (say, $\phi \sim 30° - 60°$), for which galactic tidal torques are able to sharply decrease the perihelion distances to the observable range after one revolution.

5) The observed LP comets and IP comets can be explained by a multiple-step capture process by Jupiter. The strong depletion of retrograde orbits among the old LP comets ($P \lesssim 10^3 yr$) and IP comets of perihelion distances $q < 2AU$ may be due to physical losses, implying average physical lifetimes of a few 10^2 revolutions.

6) The origin of SP or Jupiter-family comets is still controversial. The capture of comets by Jupiter from a spherical source (either the Oort cloud or the inner core) leads to a fraction of SP comets in retrograde orbits, which is not observed.

The computed values of the Tisserand constant, the distribution of the perihelion distances and the capture efficiency are also in disagreement. Additional captures of Oort comets by Saturn, Uranus and Neptune might not solve the previous discrepancies, though this is a point that deserves further study. A few storage places might still exist within the planetary region around the Lagrange $L4$ and $L5$ points of the Jovian planets (i.e. Jupiter's Trojan swarms and similar swarms for the other Jovian planets). An alternative and promising view is that SP comets come from a transneptunian comet belt, known as the Kuiper belt.

7) The problem of the dynamical stability of comets in the Kuiper belt and the rate at which they can be transferred to the planetary region is still open. Some preliminary numerical results seem to indicate that the belt has been completely depleted by planetary perturbations within $\sim 35AU$. The dynamical time scales for belt comets originally on near-circular orbits to become Neptune-crossers are of the order of the solar system age in the range $\sim 35 - 50AU$. Therefore, this may be the region of the belt undergoing the stronger depletion at present and, at the same time, the most suitable place to provide SP comets.

Acknowledgements

I thank G. B. Valsecchi and another anonymous referee for their useful comments on the manuscript.

References

Bailey, M.E. : 1983 "The structure and evolution of the Solar System comet cloud." ***Mon. Not. Roy. Astron. Soc.*** **204**, 603–633.

Bailey, M.E. : 1986 "The near-parabolic flux and the origin of short-period comets." ***Nature*** **324**, 350–352.

Bailey, M.E. : 1991 "Comet craters versus asteroid craters." ***Adv. Space Res.*** **11**, 43–60.

Bailey, M.E., Chambers, J.E., and Hahn, G. : 1992 "Origin of sungrazers : a frequent cometary end-state." ***Astron. Astrophys.*** **257**, 315–322.

Biermann, L. : 1978 "Dense interstellar clouds and comets." In ***Astronomical Papers dedicated to Bengt Strömgren*** (A. Reiz and T. Anderson, Eds.), 327–336, Copenhagen Observatory.

Biermann, L., Huebner, W.F., and Lüst, R. : 1983 "Aphelion clustering of "new" comets : Star tracks through Oort's cloud." ***Proc. Natl. Acad. Sci. (USA)*** **80**, 5151–5155.

Blitz, L. : 1993 "Giant molecular clouds." In ***Protostars and Planets III*** (E.H. Levy and J.I. Lunine, Eds.), 125–161, Univ. of Arizona Press, Tucson.

Brunini, A. : 1993 "Dynamical friction induces perturbations on Oort cloud comets." ***Astron. Astrophys.*** **273**, 684–694.

Byl, J. : 1983 "Galactic perturbations on near-parabolic cometary orbits." ***Moon and Planets*** **29**, 121–137.

Carusi, A., and Valsecchi, G.B. : 1981 "Temporary satellite captures of comets by Jupiter." ***Astron. Astrophys.*** **94**, 226–228.

Clube, S.V.M., and Napier, W.M. : 1984 "Comet capture from molecular clouds : a dynamical constraint on star and planet formation." ***Mon. Not. Roy. Astron. Soc.*** **208**, 575–588.

Delsemme, A.H. : 1973 "Origin of the short-period comets." ***Astron. Astrophys.*** **29**, 377–381.

Delsemme, A.H. : 1987 "Galactic tides affect the Oort cloud : an observational confirmation." *Astron. Astrophys.* **187**, 913–918.

Duncan, M., Quinn, T., and Tremaine, S. : 1988 "The origin of short-period comets." *Astrophys. J. Letts.* **328**, L69–L73.

Everhart, E. : 1969 "Close encounters of comets and planets." *Astron. J.* **74**, 735–750.

Everhart, E. : 1972 "The origin of short-period comets." *Astrophys. Lett.* **10**, 131–135.

Fernández, J.A. : 1980 "On the existence of a comet belt beyond Neptune." *Mon. Not. Roy. Astron. Soc.* **192**, 481–491.

Fernández, J.A. : 1981 "New and evolved comets in the solar system." *Astron. Astrophys.* **96**, 26–35.

Fernández, J.A. : 1985 "Dynamical capture and physical decay of short-period comets." *Icarus* **64**, 308–319.

Fernández, J.A. : 1992 "Comet showers." In *Chaos, Resonance and Collective Dynamical Phenomena in the Solar System* (S. Ferraz-Mello, Ed.), , IAU Symp. No.152, 239–254, Kluwer, Netherlands.

Fernández, J.A., and Gallardo, T. : 1993 "The transfer of comets from parabolic orbits to short-period orbits : numerical studies." *Astron. Astrophys*, in press.

Fernández, J.A., and Ip W.-H. : 1983 "Dynamical origin of the short-period comets." In *Asteroids, Comets, Meteors* (C.-I. Lagerkvist and H. Rickman, Eds.), 387–390 , Uppsala University.

Fernández, J.A., and Ip W.-H. : 1991 "Statistical and evolutionary aspects of cometary orbits." In *Comets in the Post-Halley Era* (R.L. Newburn, M. Neugebauer, and J. Rahe, Eds.), IAU Coll. No. 121, 487–535, Kluwer, Netherlands.

Fernández, J.A., Rickman, H., and Kamél, L. : 1992 "The population size and distribution of perihelion distances of the Jupiter family." In *Workshop on Periodic Comets* (J.A. Fernández and H. Rickman, Eds.), 143–157 , Univ. República, Montevideo.

Gladman, B. and Duncan, M. : 1990 "On the fates of minor bodies in the outer solar system." *Astron. J.* **100**, 1680–1696.

Heisler, J. : 1990 "Monte Carlo simulations of the Oort comet cloud." *Icarus* **88**, 104–121.

Heisler, J. and Tremaine, S. : 1986 "The influence of the galactic tidal field on the Oort comet cloud." *Icarus* **65**, 13–26.

Hills, J.G. : 1981 "Comet showers and the steady-state infall of comets from the Oort cloud." *Astron. J.* **86**, 1730–1740.

Holman, M.J., and Wisdom, J. : 1993 "Dynamical stability in the outer solar system and the delivery of short period comets." *Astron. J.* **105**, 1987–1999.

Ip, W.-H., and Fernández, J.A. : 1991 "Steady-state injection of short-period comets from the trans-Neptunian cometary belt." *Icarus* **92**, 185–193.

Jewitt, D., and Luu, J. : 1993 "Discovery of the candidate Kuiper belt object 1992QB1", *Nature* **362**, 730–732.

Joss, P.C. : 1973 "On the origin of short-period comets." *Astron. Astrophys.* **25**, 271–273.

Kowal, C.T. : 1989 "A solar system survey." *Icarus* **77**, 118–123.

Kresák, L. : 1992 "Are there any comets coming from interstellar space?." *Astron. Astrophys.* **259**, 682–691.

Kuiper, G.P. : 1951 "On the origin of the solar system." In *Astrophysics* (J.A. Hynek, Ed.), 357–427 , McGraw-Hill, New York.

Latham, D.W., Mazeh, T., Davis, R.J., Stefanik, R.P., and Abt, H.A. : 1991 "A search for wide binaries in the north galactic pole using precise radial velocities." *Astrophys. J.* **101**, 625–636.

Levison, H.F., and Duncan, M.J. : 1990 "A search of proto-comets in the outer regions of the solar system." *Astron. J.* **100**, 1669–1675.

Levison, H.F., and Duncan, M.J. : 1993a Submitted to *Icarus*.

Levison, H.F., and Duncan, M.J. : 1993b "The gravitational sculpting of the Kuiper belt." *Astrophys. J.* **406**, L35–L38.

Luu, J.X., and Jewitt, D. : 1988 "A two-part search for slow-moving objects." ***Astron. J.*** **95**, 1256–1262.

Marsden, B.G., and Williams, G.V. : 1992 ***Catalogue of Cometary Orbits, Seventh Edition***, Smithsonian Astrophys. Center.

Matese, J.J., and Whitman, P.G. : 1992 "A model of the galactic tidal interaction with the Oort comet cloud." ***Cel. Mech. Dyn. Astron.*** **54**, 13–35.

Napier, W.M., and Staniucha, M. : 1982 "Interstellar planetesimals - I. Dissipation of a primordial cloud of comets by tidal encounters with massive nebulae." ***Mon. Not. Roy. Astron. Soc.*** **198**, 723–735.

Oja, H. : 1975 "Perihelion distribution of near-parabolic comets." ***Astron. Astrophys.*** **43**, 317–319.

Poveda, A. : 1988 "Distant components in double and multiple stars." ***Astrophys. Space Sci.*** **142**, 67–78.

Rabe, E. : 1972 "Orbital characteristics of comets passing through the 1 : 1 commensurability with Jupiter." In ***The Motion, Evolution of Orbits, and Origin of Comets*** (G.A. Chebotarev, E.I. Kazimirchak-Polonskaya and B.G. Marsden, Eds.), IAU Symp. No. 45, 55–60, Kluwer, The Netherlands.

Rickman, H. : 1976 "Stellar perturbations of orbits of long-period comets and their significance for cometary capture." ***Bull. Astron. Inst. Czech.*** **27**, 92–105.

Shoemaker, E.M., Shoemaker, C.S., and Levison, H.F. : 1993 "Survey of the Jupiter Trojans." Presented at ***Asteroids, Comets, Meteors 1993***, IAU Symp.160, Belgirate, Italy.

Stagg, C.R., and Bailey, M.E. : 1989 "Stochastic capture of short-period comets." ***Mon. Not. Roy. Astron. Soc.*** **241**, 507–541.

Stern, S.A. : 1990 "External perturbations on distant planetary orbits and objects in the Kuiper belt." ***Cel. Mech. Dyn. Astron.*** **47**, 267–273.

Stern, S.A. : 1991 "On the number of planets in the outer solar system : evidence of a substantial population of 1000-km bodies." ***Icarus*** **90**, 271–281.

Tancredi, G., and Rickman, H. : 1992 "The evolution of Jupiter family comets over 2000 years." In ***Chaos, Resonance and Collective Dynamical Phenomena in the Solar System*** (S. Ferraz-Mello, Ed.), , IAU Symp. No.152, 269–274, Kluwer, Netherlands.

Torbett, M.V. : 1989 "Chaotic motion in a comet disk beyond Neptune : the delivery of short-period comets." ***Astron. J.*** **98**, 1477–1481.

Tremaine, S. : 1993 "The distribution of comets around stars." In ***Planets around Pulsars*** (J.A. Phillips, S.E. Thorsett and S.R. Kulkarni, Eds.), Astron. Soc. Pacific Conf. Series, in press.

Valsecchi, G.B. : 1992 Round Table on Dynamics of Periodic Comets, In ***Workshop on Periodic Comets*** (J.A. Fernández and H. Rickman, Eds.), 98–100 , Univ. República, Montevideo.

Valtonen, M.J. and Innanen, K.A. : 1982 "The capture of interstellar comets." ***Astrophys. J.*** **255**, 307–315.

Weinberg, M.D., Shapiro, S.L., and Wasserman, I. : 1987 "The dynamical fate of wide binaries in the solar neighborhood." ***Astrophys. J.*** **312**, 367–389.

Weissman, P.R. : 1980 "Stellar perturbations of the cometary cloud." ***Nature*** **288**, 242–243.

Weissman, P.R. : 1985 "Dynamical evolution of the Oort cloud." In ***Dynamics of Comets : Their Origin and Evolution*** (A. Carusi and G.B. Valsecchi, Eds.), IAU Coll. No. 83, 87–96, Reidel, The Netherlands.

Weissman, P.R. : 1991 "The cometary impactor flux at the Earth." In ***Global Catastrophes in Earth History*** (V. Sharpton and P. Ward , Eds.), GSA Special Papers 247, 171–180.

Wetherill, G.W. : 1991 "End-products of cometary evolution : cometary origin of Earth-crossing bodies of asteroidal appearance." In ***Comets in the Post-Halley Era*** (R.L. Newburn Jr., M. Neugebauer and J. Rahe, Eds.), Proc. IAU Coll. No. 121, 537–556, Kluwer, The Netherlands.

Yabushita, S. : 1989 "On the orbital elements of nearly parabolic comets at previous returns as perturbed by the galactic tidal field." ***Astron. J.*** **97**, 262–264.

Zheng, J.-Q., Valtonen, M.J., and Valtaoja, L. : 1990 "Capture of comets during the evolution of a star cluster and the origin of the Oort cloud." *Cel. Mech. Dyn. Astron.* **49**, 265–272.

A REVIEW OF COMETS AND NONGRAVITATIONAL FORCES

D. K. YEOMANS
Jet Propulsion Laboratory,
California Institute of Technology,
4800 Oak Grove Dr, Pasadena,
CA 91109, USA
E-mail DKY@NAIF.SPAN

Abstract. A review is given on the recent advances in understanding the nature of the nongravitational forces that affect the motions of nearly all active comets.

1. Introduction

One of the primary reasons for Whipple's introduction of his icy conglomerate model for the cometary nucleus was to explain the so-called nongravitational accelerations that were evident in the motions of many active periodic comets (Whipple 1950, 1951). That is, even after all the gravitational perturbations of the planets were taken into account, the observations of many active comets could not be well represented without the introduction of additional so-called nongravitational effects into the dynamical model. These effects are brought about by cometary activity when momentum is transferred to the nucleus by the sublimating ices. Whipple noted that for an active, rotating, icy cometary nucleus, a thermal lag between cometary noon and the time of maximum outgassing would introduce transverse accelerations in a comet's motion. In an attempt to model these effects, Marsden (1968, 1969) first introduced a semi-empirical nongravitational acceleration model using what are now termed Style I nongravitational parameters.

Style II parameters were added when Marsden et al. (1973) introduced what has become the standard, or symmetric, nongravitational acceleration model for cometary motions; a rotating cometary nucleus is assumed to undergo vaporization from water ice that acts symmetrically with respect to perihelion. That is, at the same heliocentric distance before and after perihelion, the cometary nucleus experiences the same nongravitational acceleration. The cometary equations of motion are written :

$$\frac{d^2\mathbf{r}}{dt^2} = -\mu\frac{\mathbf{r}}{r^3} + \frac{\partial R}{\partial \mathbf{r}} + A_1 g(r)\hat{\mathbf{r}} + A_2 g(r)\hat{\mathbf{T}} \qquad (1)$$

where $g(r) = \alpha(r/r_o)^{-m}(1 + (r/r_o)^n)^{-k}$.
The acceleration is given in astronomical units/(ephemeris day)^2, μ is the product of the gravitational constant and the solar mass, and R is the planetary disturbing function. The scale distance r_o is the heliocentric distance inside which the bulk of solar insolation goes to sublimating the comet's ices. For water ice, $r_o = 2.808$ AU and the normalizing constant $\alpha = 0.111262$. The exponents m, n, and k equal 2.15, 5.093 and 4.6142, respectively. The nongravitational acceleration is represented by a radial term, $A_1\ g(r)$, and a transverse term, $A_2\ g(r)$, in the equations of motion.

A. Milani et al. (eds.), Asteroids, Comets, Meteors 1993, 241–254.

The radial unit vector ($\hat{\mathbf{r}}$) is defined outward along the Sun-comet line, while the transverse unit vector ($\hat{\mathbf{T}}$) is directed normal to $\hat{\mathbf{r}}$, in the orbit plane, and in the direction of the comet's motion. An acceleration component normal to the orbit plane is likely also present for most active comets, but its periodic nature makes a meaningful solution for it difficult in these computations because we are solving for an average nongravitational acceleration effect over three or more apparitions. If the comet's nucleus were not rotating, the outgassing in this model would always be toward the Sun and the resulting nongravitational acceleration would act only in the anti-solar direction. However, the rotation of the nucleus, coupled with a thermal lag angle between the nucleus subsolar point and the point on the nucleus where there is maximum outgassing, introduces a transverse acceleration component in either the direction of the comet's motion or contrary to it – depending upon the nucleus rotation direction.

Equation 2 represents the time derivative of the comet's orbital semi-major axis (a) as a result of radial and transverse perturbing accelerations (R_p, T_p).

$$\frac{da}{dt} = \frac{2}{n(1-e^2)^{1/2}}\left[(e\sin\nu)R_p + \frac{p}{r}T_p\right] \tag{2}$$

In this equation, n, e, ν, and r denote, respectively, the orbital mean motion, eccentricity, true anomaly, and the comet's heliocentric distance, while p is the orbital semi-latus rectum, $a(1-e^2)$. Because of the thermal lag angle, a comet in direct rotation will have a positive transverse nongravitational acceleration component, and from equation 2, it is apparent that the comet's orbital semi-major axis will increase with time (its orbital energy will increase). Likewise, a comet in retrograde rotation will be acted upon by a negative Tp and its semi-major axis will decrease with time. Because the nongravitational acceleration is assumed to act symmetrically with respect to perihelion, the time-averaged effect of the periodic radial acceleration cancels out.

When introducing the standard model, Marsden *et al.* (1973) included possible time dependences in the transverse parameter (A_2). Currently, however, the standard nongravitational acceleration model is most often used solving only for the radial and transverse parameters (A_1 and A_2) over data intervals short enough so that neglected time dependences do not cause systematic trends in the residuals. Solutions for the nongravitational parameters usually require data from at least three apparitions and by comparing the nongravitational parameters determined from several of these short arc solutions, one can determine their change with time.

Largely because of its success in allowing accurate ephemeris predictions, the "standard" nongravitational force model has been in use for two decades. More recently, it has become understood that, while this model is successful in representing the astrometric observational data and allowing the computation of accurate ephemeris predictions, the standard model does not represent a completely accurate representation of the actual processes taking place in the cometary nucleus.

More than a century and a half ago, Bessel (1836) noted that a comet expelling material in a radial sunward direction would suffer a recoil force and if the expulsion of material did not take place symmetrically with respect to perihelion, there would be a shortening or lengthening of the comet's period depending upon whether the comet expelled more material before or after perihelion (see equation 2). Although Bessel did not identify the physical mechanism with water vaporization from the nucleus, his concept of cometary nongravitational forces was essentially correct. This review will focus upon this redirection of thought from a view whereby transverse nongravitational effects arise from a rotating comet that outgasses symmetrically with respect to perihelion to a view whereby the radial nongravitational effects are more important and outgassing from surface vents acts asymmetrically with respect to perihelion. For a more comprehensive outline of the earlier work on cometary nongravitational forces, the reader is directed to previously published reviews (Marsden 1968, 1969; Marsden *et al.* 1973; Marsden 1985; Yeomans 1991). Section 2 will focus upon the recent attempts to discern physical characteristics of comets using the results of nongravitational force modeling and Section 3 will outline the efforts to improve upon the standard model for modeling the motions of active comets. Section 4 will examine the long-term behavior of nongravitational effects and the constraints they provide upon the physical characteristics of the cometary nucleus. Section 5 is a summary of these discussions.

2. Inferring Nucleus Characteristics Using Nongravitational Parameters

Rickman (1986) pointed out that radial outgassing forces that act asymmetrically with respect to perihelion were the likely cause of the nongravitational effects upon comets Halley and Kopff and he went on to make estimates of their masses and bulk densities. He noted that the water production curves for Halley and other comets show an asymmetry with respect to perihelion so that the effect of the radial component, integrated over one orbital period, does not cancel out. The nucleus masses were estimated by comparing nongravitational parameters with the rocket-like forces expected from the gas production curves. In turn, the gas production curves were determined from the light curves using an empirical relationship developed by Festou. The bulk density for comet Halley was estimated to be 0.1–0.2 $g\ cm^{-3}$ and that for Kopff was lower still. Rickman *et al.* (1987b) continued this type of analysis and estimated masses for 29 short period comets. The change in the total orbital period per revolution is the sum of the contributions from the radial and transverse rocket effects. The mass of each comet was determined as a function of its estimated thermal and rotational properties. As a group, the bulk densities of these objects were estimated to be less than 0.5 $g\ cm^{-3}$ suggesting that the cometary nucleus is a very porous structure. This type of analysis depends upon the assumption that there is a correlation between the light curve and the assumed gas production curve, that thermal lag angles are present, and that the surface of each object has an un-mantled free sublimating area. Using a similar approach for comet Halley, Sagdeev *et al.* (1988) estimated a bulk density of 0.6 $g\ cm^{-3}$ with error bars of +0.9 and –0.4 $g\ cm^{-3}$. After a rather complete discussion of the method and the uncertainties involved in this type of analysis, Peale (1989) concluded that it is difficult to meaningfully constrain the bulk density of comet Halley.

Festou *et al.* (1990) established a statistical correlation between the nongravitational parameters and the asymmetry of the gas production rates with respect to perihelion. A linear relationship was devised between the change in the comet's orbital period (ΔP) due to nongravitational effects and the difference (E) between the integrated gas production rate before and after perihelion. A linear relationship should exist between ΔP and E if the radial nongravitational effect is dominant and if the maximum outgassing rate divided by the cometary mass is reasonably constant from comet to comet. Using an early version of the curve of Δ P vs E, Festou *et al.* (1989) predicted a correction to the predicted time of perihelion passage time (ΔP) for P/Brorsen-Metcalf of between −30 and −15 days. When the comet was recovered by Eleanor Helin on July 4, 1989 it soon became evident that the predicted time of perihelion passage required a correction of −15.6 days. In retrospect, Festou *et al.* (1990) noted that their prediction was fortuitous because the prediction should have been −6 to −15 days and the 1989 light curve did not display the same asymmetry they assumed for the 1919 apparition. There remains a question as to whether the magnitude of a perihelion time correction can be accurately predicted from a comet's light curve characteristics. In this regard, one should make note of comet Crommelin. The light curve of this comet reaches a maximum ten days before perihelion yet the motion of this comet suggests that its nongravitational effects are very small. Whether or not the magnitude of a comet's nongravitational effects can be discerned from its light curve, it is evident that the light curve can, in most cases, be used to indicate whether orbital energy is being added or subtracted as a result of the cometary outgassing.

Sekanina (1993a) examined the effects of discrete outgassing sources upon the motions of periodic comets. In a series of arguments, Sekanina pointed out the complex effects that discrete active areas on the nucleus can have upon the cometary nongravitational accelerations. In Sekanina's model, the nongravitational effects depend upon the spin-vector orientation and upon the location of the active areas on the nucleus. These active areas can introduce both radial and transverse nongravitational effects even when the thermal lag angle is assumed to equal zero. Active areas can easily introduce seasonal effects whereby a source is more active before (or after) perihelion. Unlike the standard model, the sign of A_2 has no correlation with the direction of nucleus rotation. The initiation of new active areas, or the dying out of existing areas, is used to explain the time dependence or erratic behavior of the transverse nongravitational parameter (A_2) with time. For example, the transverse nongravitational parameter for periodic comet Giacobini-Zinner, after remaining nearly constant from 1900 through 1959, became erratic thereafter and for comet Comas-Solá, A_2 went from positive to negative over the 1935–1987 interval (see Figure 1).

One of the results of Sekanina's work (1993a) was to define more clearly the deficiencies of the current standard nongravitational force model and to point out how the large set of standard (Style II) nongravitational parameters might be used to provide useful information for modeling comets under the new paradigm. At any given time, the radial, transverse, and normal components of the standard model do not necessarily represent the actual nongravitational accelerations resulting from the outgassing of one or more discrete active sources. However, a successful orbital solution for the standard nongravitational parameters should approximately account for the nongravitational perturbations when these perturbations are integrated completely around the cometary orbit for three or more returns. Attempts to solve for nongravitational parameters over time scales of less than one orbital period can be expected to improve the observation residuals but these parameters are probably meaningless for interpreting the physical nature of the cometary nucleus.

Sekanina noted that the correlation between the sign of A_2 and the perihelion asymmetry of the water production rate is valid for all possible combinations of the rotation parameters as long as only a single region is active on the nucleus. This correlation indicates an insignificant heat transfer lag in the sublimation process. This being the case, there is no correlation between the sign of A_2 and the sense of nucleus rotation. Assuming no heat transfer lags, the sign of A_2 is then to be interpreted in terms of the distribution of active areas with respect to the spin vector orientation. Sekanina noted that for active comets that have been seen at 10 or more apparitions, there is an equal number of comets with A_2 less than zero and those with A_2 greater than zero. The largest absolute values of the A_2 parameter seem to be associated with periodic comets having large perihelion distances (i.e. large q's). From Figure 2, we note that periodic comets Gunn (q = 2.47 AU), Schwassmann-Wachmann 2 (2.07), and Brooks 2 (1.84) have the largest magnitudes for their A_2 parameters. Sekanina notes that this may be due to use of an incorrect value for the scale distance r_o in the standard model.

When interpreted in light of Sekanina's model, both negative and positive values of A_1 are realistic; the former corresponds to an advancement in the orbital line of apsides while the latter suggests a regression of this line. For many well observed comets and asteroids that have small semi-major axes and large eccentricities, care must be taken to include the effects of general relativity in the equations of motion because these effects also introduce a non negligible radial acceleration toward the sun (Sitarski, 1983, 1992a).

A nongravitational acceleration acting normal to the comet's orbit plane will affect the longitude of the ascending node and the orbital inclination but neither of these perturbations are secular. Since these perturbations are modulated by either sine or cosine functions of the true anomaly, much of the nongravitational perturbations upon the two orbital elements would average to zero even if the normal perturbative forces remain positive or negative throughout the orbit. Sekanina (1993b) noted that a meaningful solution for the normal nongravitational parameter (A_3) would be possible only for the special case where the perturbations upon the ascending node and the inclination yield a similar value of A_3. Solutions for the A_3 parameter are not usually successful but there are some exceptions. Meaningful values of A_3 were obtained for the 1808–1988 apparitions of P/Grigg- Skjellerup, the 1906–1991 apparitions of P/Metcalf-Brewington and the 1958–1983 apparitions of P/Kopff (Sitarski 1991, 1992b; Rickman *et al.*, 1987a). Over the four returns of periodic comet Clark, Nakano (1992) found a value for A_3 with a formal uncertainty of only 3% and Sekanina (1993b) suggested that for this comet, the effective nongravitational perturbations on the ascending node and the orbital inclination are about equal. The rotation axis at perihelion is located in the plane defined by the sun-comet line at perihelion and perpendicular to the comet's orbit plane with the axis inclined about 45 degrees with respect to the orbit plane. As seen from the comet, the rotation axis at perihelion would be pointing in the general direction of the sun but about 45 degrees above it. In this configuration for the spin axis of comet Clark, the cometary outgassing produces a non-canceling nongravitational thrust in a normal direction so that the solution for A_3 would be valid.

3. Modeling the Nongravitational Motion of a Comet

Several efforts have been made to improve upon the standard nongravitational force model either by attempting to change the dynamical model to more closely reflect the physical model of the nucleus or by altering the mechanism by which nongravitational accelerations are introduced.

Froeschlé and Rickman (1986) and Rickman and Froeschlé (1986) used theoretical calculations to examine the secular evolution of the nongravitational parameters as a function of the heliocentric distance for various kinds of short period comets and different assumed thermal inertias. In general, their values of these parameters did not correspond to those computed from the standard model. In fact, there was such a wide variation in the respective behavior of the A_1, A_2, and A_3 parameters that no generally applicable model for the nongravitational effects was suggested. They noted that improved models would likely have to include the effects of rotation pole orientation and seasonal heat flows.

Using the asymmetric light curve of comet Halley, Yeomans (1984) attempted to employ the nucleus rotation parameters introduced by Sekanina (1981) to improve the nongravitational force model for comet Halley. For this latter model, the outgassing is assumed to result from a sub-solar active area which is defined by its cometocentric solar longitude at perihelion (P), the obliquity of the nucleus equator with respect to the orbit plane (I), and by a thermal lag angle measured from the cometary subsolar point to the point of maximum outgassing. Although the optimum lag angle and obliquity turned out to be small, in apparent agreement with subsequent results, the orbital solution did not improve upon the standard nongravitational force model.

For his investigation of the two apparitions of P/Metcalf-Brewington, Sitarski (1992b) found that his nongravitational force model employing a single nongravitational acceleration parameter (A) together with the lag angle, obliquity (I) and cometocentric longitude of the sun at perihelion (P) gave nearly identical results when compared to an orbital solution using A_1, A_2, and A_3 of the standard model. For the investigation of the motion of P/Grigg-Skjellerup and P/Swift-Gehrels, Sitarski (1991) and Bielicki and Sitarski (1991) introduced their model using A, the lag angle, I, and P but found it necessary to also introduce dI/dt and dP/dt for the model of P/Swift-Gehrels' motion over the 1889–1991 interval and dI/dt, dP/dt, and a center-of-light/center-of-mass offset for the motion of P/Grigg-Skjellerup over the 1808–1988 interval.

Yeomans and Chodas (1989) modified the standard nongravitational acceleration model to allow the water vaporization curve to peak a certain number of days either before or after perihelion. They found that for several periodic comets that exhibit asymmetric light curves, an asymmetric outgassing model can improve upon the orbital representation of the astrometric data. The asymmetric model, which more accurately mimics the comet's outgassing history, often yields values for the radial and transverse nongravitational parameters that are completely different from the corresponding values derived from using the standard symmetric model. For P/d'Arrest the water vaporization curve (as modeled by the asymmetric nongravitational acceleration function) and the visual light curve reach maxima very close to one another (40 days post perihelion). For P/Giacobini-Zinner, the optimum values in the modeled perihelion offset approximately mimic the observed light curve history with both reaching a maximum at perihelion prior to 1959, then a maximum at 25 days after perihelion in 1959, back to a maximum at perihelion in 1972 and finally to a maximum 15 days prior to perihelion in 1985. For P/Kopff, the computed value of A_2 does not pass through zero with the asymmetric model whereas it does for the symmetric model. Recent results comparing comet Halley orbital solutions using the asymmetric and standard nongravitational acceleration models are given in Table 1.

4. The Time Dependence of Nongravitational Effects

From Figures 1 and 2, it is clear that a number of comets have rather constant values for their A_2 parameters and an equal number of comets have rather time dependent A_2 parameters. In the latter case, the smoothly varying time dependence, and especially the passage of the A_2 parameter through zero, is likely due to major precessional motion of the rotation pole with time. A number of comets like Finlay, Giacobini-Zinner, Brooks 2, and Whipple have rather erratic values of A_2 as a function of time that are perhaps due to the cessation of old active areas or the initiation of new ones (Sekanina 1993a). As an illustration of this process, imagine that the rotation pole of a nucleus is in its orbital plane and aligned perpendicular to the sun-comet line at perihelion. Further imagine that there is an active area on one polar cap so that on the way into perihelion this polar region is active and produces a radial acceleration that would reduce the orbital semi-major axis (see equation 2). On the way out from perihelion, this active polar region is in shadow and thus inactive. Likewise if only the opposite polar region was active, the determined value of A_2 would have an opposite sign because the orbital semi-major axis (and orbital period) would then increase rather than decrease.

If the time dependence of A_2 is smoothly varying, some success has been achieved by introducing time dependencies directly into the nongravitational force model (for example, see Marsden *et al.*, 1973; Sitarski, 1991). However, for so-called erratic comets, solutions have normally been made over short enough time intervals that the nongravitational parameters can be assumed constant. By making several solutions over different time intervals, the changing values of the nongravitational parameters can be noted. The information plotted in Figures 1 and 2 was determined in this way.

For a few of the less active periodic comets (Arend-Rigaux, Neujmin 1, Schwassmann-Wachmann 1), no nongravitational effects could be detected in the orbital solutions. Among the relatively few objects exhibiting time independent nongravitational effects are comets Tempel 2, Tuttle, d'Arrest, and Halley. Carusi *et al.* (1991) pointed out an early 1678 apparition of periodic comet d'Arrest and noted that the nongravitational effects for this comet were time independent for nearly 50 revolutions. Yeomans and Chodas (1989) found that the recent astrometric data for this comet could be represented most accurately with an asymmetrical nongravitational acceleration model that reached a maximum some 40 days after perihelion. The light curve of this comet consistently reaches a maximum about this same time. For Halley, Yeomans and Kiang (1981) found the comet's motion to be consistent with time independent nongravitational parameters over two millennia. In addition the absolute magnitude of comet Halley has shown no obvious changes during its two thousand year observational interval (Broughton 1979; Bortle and Morris 1984)

Table 1 outlines a recent orbit determination for comet Halley including the orbital elements for its last four apparitions and its next return in 2061. The asymmetric nongravitational force model of Yeomans and Chodas (1989) was employed and various dates were input for when the modelled outgassing reached its maximum value; the optimum value (in the sense that the weighted rms residual reached a minimum) occurred when the modelled outgassing maximum was taken as approximately 35 days after perihelion. In addition to a solution for the six orbital elements, a solution was made for an offset between the measured photometric center of the comet's image and its center of mass. This offset is assumed to vary along the comet-sun line with an inverse square dependence on the heliocentric distance. The offset at one AU from the sun was determined to be 849 km sunward. This result is in accordance with Medvedev's (1993) estimate of about 1000 km for Halley's post-perihelion offset in the photocenter.

For the standard model, the transverse nongravitational parameter (A_2) is very well determined and it is this transverse acceleration that causes the nongravitational effects. However, for the asymmetric model, the transverse nongravitational parameter (A_2) is often indeterminate and it is the well-determined radial nongravitational parameter (A_1) and the fact that the maximum outgassing is assumed to reach a maximum 35 days after perihelion that introduces the nongravitational effect. Comet Halley's consistent behavior from apparition to apparition is responsible for an increase in the comet's orbital period (4 days) over what might be expected if only gravitational perturbations were taken into account. Various modeling efforts have concluded that the spin state of comet Halley is not in simple rotation. Although not well understood, the rotation characteristics are likely to include both a rotation about its long axis and precession about an axis nearly perpendicular to the rotation axis. However, the long term constancy of Halley's nongravitational parameters do not allow for major changes of these axes with time. Spin axis precession and nutation are likely to be evident but chaotic tumbling over time scales of about two thousand years certainly can be excluded. Over the same time scale, there cannot have been major changes in the location or activity of the outgassing vents.

Yau *et al.* (1993) found the observations of comet Swift-Tuttle in 69 B.C., A.D. 188, 1737, 1862, and 1992-93 to be consistent with no nongravitational effects in this comet's motion and, as was the case for comet Halley, there have been no

TABLE I

Based on the asymmetric nongravitational acceleration model by Yeomans and Chodas (1989), an orbital solution for comet Halley was computed from 7525 observations over the interval 1759 January 23 through March 18, 1991. The root mean square (RMS) weighted residual reached a minimum (best solution) when the modelled water vaporization curve was assumed to peak 35 days after perihelion (DT = +35 days). A sunward offset of the photometric center-of-light from the comet's center-of-mass was assumed to be operative at each return; at a heliocentric distance of one AU, this offset (S_o) was determined to be some 849 km. Planetary perturbations were computed at each time step using planetary ephemeris DE200 (with outer planet masses improved using the results from the Voyager spacecraft flybys). Full relativistic equations of motion were employed. The final weighted, unweighted, and normalized RMS residuals are given. For comparison, the nongravitational parameters, center-of-light/center-of-mass offset and RMS residuals are also given for a solution using the standard nongravitational force model.

	Asymmetric model	Standard Model
A_1 (10^{-10}AU day^{-2})	3.6038 (± 0.1043)	1.9704 (± 0.2789)
A_2 (10^{-10}AU day^{-2})	0.0532 (± 0.0525)	1.55443 (± 0.00003)
S_o in km	849.2 (± 17)	830.8 (± 17)
DT in days	+35	0
RMS (weighted)	1.141	1.154
RMS (unweighted, arcsec)	5.351	5.562
RMS (normalized)	1.027	1.039

Orbital elements (J2000) from above asymmetric model orbit

Epoch (TDB)	e	q(AU)	ω	Ω	i	T (TDB)
1759 Mar 21.0	0.96768749	0.584473925	110.708762	57.245864	162.372379	1759 Mar 13.059370
1835 Nov 18.0	0.96739544	0.586568623	110.704026	57.518405	162.258718	1835 Nov 16.439845
1910 May 9.0	0.96730219	0.587212031	111.737103	58.562661	162.218514	1910 Apr 20.178242
1986 Feb 19.0	0.96727580	0.587103940	111.865650	58.860054	162.242195	1986 Feb 9.458966
2061 Aug 4.0	0.96657663	0.592780500	112.052286	59.392434	161.965091	2061 Jul 28.719901

obvious changes in this comet's absolute magnitudes over two millennia. Yau *et al.* (1993) list the entire set of orbital elements for this comet from 703 B.C. through A.D. 2392 and Marsden *et al.* (1993) present orbital elements over the 69 B.C.–A.D. 3302 interval. For the similar epochs of osculation, the orbital elements of Yau *et al.* (1993) and Marsden *et al.* (1993) are similar and both groups point out that, over the long term, there are no nongravitational effects evident for this comet.

Because comet Swift-Tuttle's absolute magnitude has not changed significantly over two millennia and there is a lack of significant nongravitational effects over the same period, additional constraints can be placed upon the model for this comet's nucleus. At one AU from the sun, the outgassing activity of Swift-Tuttle is comparable with that of comet Halley at the same heliocentric distance (A'Hearn 1993). Yet comet Halley experiences an increase in its orbital period of 4 days per revolution due to nongravitational effects while Swift-Tuttle has no perceptible change in its period. If the mass of Swift-Tuttle were significantly larger than Halley's, one would not expect to be able to detect a nongravitational acceleration in its orbital motion. Alternatively, if the outgassing of Swift-Tuttle were directed radially toward the sun and if this outgassing were symmetric with respect to perihelion, there would be no nongravitational effects evident in its orbital motion. However, for this latter case, the rotation axis as well as the size and location of the active regions would have been constant over its entire observational interval.

If comet Swift-Tuttle could be shown to outgas preferentially before or after perihelion, it would strongly imply that the mass of this comet is far larger than that of comet Halley. In this case, although the massive comet would experience a substantial nongravitational force, its nongravitational acceleration would remain undetectable. In the analysis by Green (1993), there was no obvious asymmetry in the comet's visual light curve. Even so, either the string of conditions mentioned in the last two sentences of the previous paragraph are true or the comet is substantially more massive than comet Halley. The latter case seems far more likely. Based upon an analysis of their respective meteor stream characteristics, Hughes and McBride (1989) concluded that the mass of comet Swift-Tuttle is about ten times larger than comet Halley.

5. Summary

Although the notion of an icy conglomerate model for a cometary nucleus (Whipple 1950, 1951) is still in basic agreement with the observations, there have been a number of recent modifications and refinements to this model. Largely as a result of the impressive image of comet Halley's nucleus taken by the Giotto spacecraft, the picture of an outgassing sunlit hemisphere has been replaced by a "vent" model whereby the outgassing activity takes place from discrete active areas.

The observation that the sign of the transverse nongravitational parameter (A_2) is strongly correlated with a comet's outgassing asymmetry with respect to perihelion suggests that the nongravitational accelerations are due to radial outgassing toward the sun. This concept would then replace the earlier notion whereby the nongravitational effects were viewed as resulting from a transverse acceleration introduced by a significant heat lag between the subsolar point of the cometary nucleus and the point of maximum outgassing. Although the widely used nongravitational acceleration model introduced by Marsden *et al.* (1973) is based upon the older model and assumes symmetric outgassing with respect to perihelion, it has been successful in providing improved ephemerides for many comets.

Since a large majority of the orbits for active periodic comets have been computed with the nongravitational acceleration model of Marsden *et al.* (1993), it is important to be able to interpret the nongravitational parameters (A_1, A_2, A_3) from this model in terms of the more recent "vent" model. For this latter model, the radial nongravitational parameter (A_1) is the more important nongravitational parameter and it can be either positive or negative. For the older standard model, it was the transverse nongravitational parameter (A_2) that was dominant and a positive value of A_1 was expected. With a few notable exceptions, a solution for the normal nongravitational parameter (A_3) is not appropriate in either the standard or the newer "vent" model.

For the many comets whose nongravitational parameters are time dependent, one can introduce time dependencies directly into the equations of motion if the nongravitational effects vary smoothly with time. Otherwise, sets of nongravitational parameter solutions can be made over short time intervals and the resulting parameters compared to note the time dependencies. Under the older model, the time varying nongravitational parameters were most often interpreted in terms of a precessing rotation axis. For example, a uniform precession of the rotation axis though the comet's orbital plane would result in a smooth sign change in the transverse nongravitational parameter. For the vent model, these time dependencies are still interpreted in terms of rotation axis motion whereas erratic time dependencies are interpreted as the cessation of old vents or the initiation of new ones.

For a comet, like Halley, that is both active and shows no significant time dependencies in its nongravitational effects, one must assume that the same active areas have maintained about the same activity over the comet's observed interval and that its rotation axis has not suffered a major change in orientation. Since an active vent on the surface of comet Halley's nucleus is likely to sublimate tens of meters of ice during each apparition, the vents would appear to get quite deep without changing their outgassing characteristics. This enigmatic behavior is worthy of additional study as is the contrast between the observed, complex rotation state of comet Halley and the apparent long-term stability of its rotation axes.

For a comet like Swift-Tuttle that is active yet whose motion is not subjected to significant nongravitational effects, there are additional constraints upon the characteristics of its nucleus. Either a comet of this type is so massive that nongravitational forces do not affect the comet's orbital motion to any significant degree or the comet's outgassing is directed radially toward the sun and symmetric with respect to perihelion. In the latter case, the activity and location of the active vents, as well as the rotation pole orientation, must have remained rather constant over its entire observed interval.

Acknowledgements

The author would like to acknowledge the helpful suggestions provided by Z. Sekanina and P.R. Weissman as well as the assistance of K. Yau. The research described in this paper was carried out at the Jet Propulsion Laboratory, California Institute of Technology, under contract with the National Aeronautics and Space Administration.

References

A'Hearn, M. : 1993, Personal communication.

Bessel, F.W. : 1836, "Bemerkungen über mögliche Unzulänglichkeit der die Anziehungen allein berücksichtigenden Theorie der Kometen. *Astron. Nach.*, **13**, 345–350.

Bielicki, M. and Sitarski, G. : 1991, "Nongravitational Motion of Comet P/Swift-Gehrels." *Acta Astron.*, **41**, 309–323.

Bortle, J.E. and Morris, C.S. : 1984, "Brighter Prospects for Halley's Comet." *Sky and Telescope*, Jan. 1984, pp. 9–12.

Broughton, R.P. : 1979, "The Visibility of Halley's comet." *J.R.A.S. Canada*, **73**, 24–36.

Carusi, A., Kresák, L., Kresáková, M., and Valsecchi, G.B. : 1991, "Observations of Periodic Comet d'Arrest in 1678 and Implications for its Evolutionary History." *Astron. Astrophys.*, **252**, 377–384.

Festou, M., Rickman, H., and Kamél, l. : 1989, *IAU Circ.* No. 4804.

Festou, M., Rickman, H., and Kamél, l. : 1990, "Using Light-curve Asymmetries to Predict comet Returns." *Nature*, **345**, 235–238.

Froeschlé, C. and Rickman, H. : 1986, "Model Calculations of Nongravitational Forces on Short Period Comets. I. Low-Obliquity Case." *Astron. Astrophys.*, **170**, 145–160.

Green, D.W.E. : 1993, "The Brightness of P/Swift-Tuttle in 1992-1993." *International Comet Quarterly*, **15**, 182.

Hughes, D.W. and McBride, N. : 1989, "The Mass of Meteoroid Streams." *Mon. Not. R. Astr. Soc.*, **240**, 73–79.

Marsden, B.G. : 1968, "Comets and Nongravitational Forces." *Astron. J.*, **73**, 367–379.

Marsden, B.G. : 1969, "Comets and Nongravitational Forces II." *Astron. J.*, **74**, 720–734.

Marsden, B.G. : 1985, "Nongravitational Forces on Comets : The First Fifteen Years." In *Dynamics of Comets : Their Origin and Evolution* (A. Carusi and G.B. Valsecchi, Eds.), pp. 343–352, D. Reidel Publishing Company, Astrophysic and Space Science Library, vol. 115.

Marsden, B.G., Sekanina, Z., and Yeomans, D.K. : 1973, "Comets and Nongravitational Forces. V." *Astron. J.*, **78**, 211–225.

Marsden, B.G., Williams, G.V., Kronk, G.W., and Waddington, W.G. : 1993, "Update on Comet Swift-Tuttle." *Icarus*, **105**, 420–426.

Medvedev, Yu.D. : 1993, "On the Dust Cloud in a Comet's Head and its Role in Forming the Photocentre Shift Phenomena." *Asteroids, Comets, Meteors 1993 poster paper*, Belgirate, Italy, June, 1993.

Nakano, S. : 1992, "Periodic Comet Clark." *Minor Planet Circular* 20122.

Peale, S.J. : 1989, "On the Density of Halley's Comet." *Icarus*, **82**, 36-49.

Rickman, H. : 1986, "Masses and Densities of Comets Halley and Kopff." In *The Comet Nucleus sample Return Mission Proc. Workshop*, pp. 195–205, Canterbury, UK, 15–17 July 1986, ESA SP-249.

Rickman, H. and Froeschlé, C. : 1986, "Model Calculations of Nongravitational Forces on Short Period Comets. II. High Obliquity Case." *Astron. Astrophys.*, **170**, 161–166.

Rickman, H., Sitarski, G., and Todorovic-Juchniewicz, B. : 1987a, "Nongravitational motion of comet P/Kopff during 1958–1983." *Astron. Astrophys.* **188**, 206–211.

Rickman, H., Kamél, L., Festou, M.C., and Froeschlé, C. : 1987b, "Estimates of Masses, Volumes and Densities of Short-Period Comet Nuclei." *Symposium on the diversity and similarity of comets* (E.J. Rolfe and B. Battrick, Eds.), pp. 471–481, ESA SP-278.

Sagdeev, R.Z., Elyasberg, P.E., and Moroz, V.I. : 1988, "Is the Nucleus of Comet Halley a Low Density Body?" *Nature*, **331**, 240–242.

Sekanina, Z. : 1981, "Rotation and Precession of Cometary Nuclei." *Annual Review of Earth and Planetary Science*, **9**, 113–145.

Sekanina, Z. : 1993a, "Effects of Discrete-Source Outgassing on Motions of Periodic Comets and Discontinuous Orbital Anomalies." *Astron. J.*, **105**, 702–735.

Sekanina, Z. : 1993b, "Nongravitational Motions of Comets : Component of the Recoil Force Normal to Orbital Plane." *Astron. Astrophys.*, **277**, 265–282.

Sitarski, G. : 1983, "Effects of General Relativity in the Motions of Minor Planets and Comets." *Acta Astron.*, **33**, 295–304.

Sitarski, G. : 1991, "Linkage of all the Apparitions of the Periodic Comet Grigg-Skjellerup during 1808–1988." *Acta Astron.*, **41**, 237–253.

Sitarski, G. : 1992a, "On the Relativistic Motion of (1566) Icarus." *Astron. J.*, **104**, 1226–1229.

Sitarski, G. : 1992b, "Motion of Comet P/Metcalf-Brewington (1906 VI = 1991a)." *Acta Astron.*, **42**, 49–57.

Whipple, F.L. : 1950, "A Comet Model. I. The Acceleration of comet Encke." *Astron. J.*, **111**, 375–394.

Whipple, F.L. : 1951, "A Comet Model II. Physical Relations for Comets and Meteors." *Astrophys. J.*, **113**, 464–474.

Yau, K., Yeomans, D.K., and Weissman, P.R. : 1993, "The Past and Future Motion of P/Swift-Tuttle." *Mon. Not. R. Astr. Soc.* (in press).

Yeomans, D.K. : 1984, "The Orbits of Comets Halley and Giacobini-Zinner." *Cometary Astrometry* (D.K. Yeomans, R.M. West, R.S. Harrington, and B.G. Marsden, Eds.), pp. 167–175, JPL Publication 84–82.

Yeomans, D.K. : 1991, "Cometary Orbital Dynamics and Astrometry." In *Comets in the Post-Halley Era*, vol. 1 (R.L. Newburn, Jr., M. Neugebauer, and J. Rahe, Eds.), pp. 3–17, Kluwer Academic Publishers.

Yeomans, D.K. and Kiang, T. : 1981, "The Long-Term Motion of Comet Halley." *Mon. Not. R. Astr. Soc.*, **197**, 633–646.

Yeomans, D.K. and Chodas, P.W. : 1989, "An Asymmetric Outgassing Model for Cometary Nongravitational Accelerations." *Astron. J.*, **98**, 1083–1093.

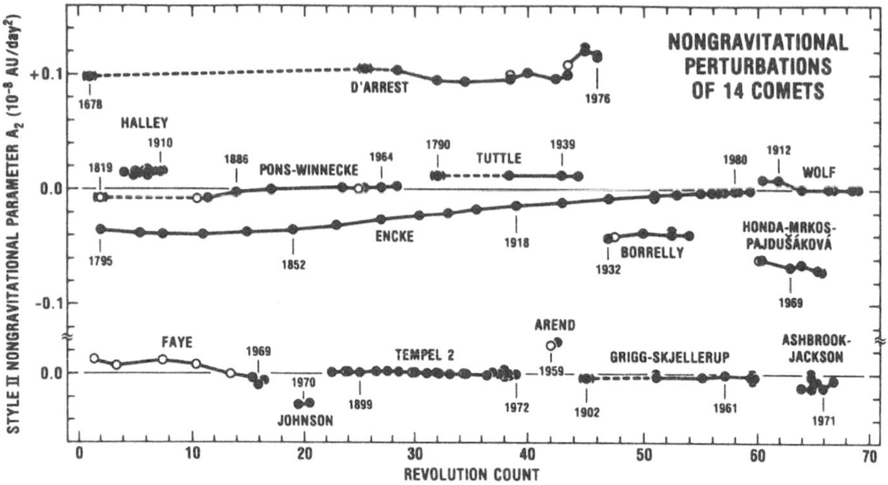

Figure 1. Fourteen Comets whose Nongravitational Effects are either Constant or Slowly Varying with Time. Figures 1 and 2 Courtesy of Z. Sekanina (1993a).

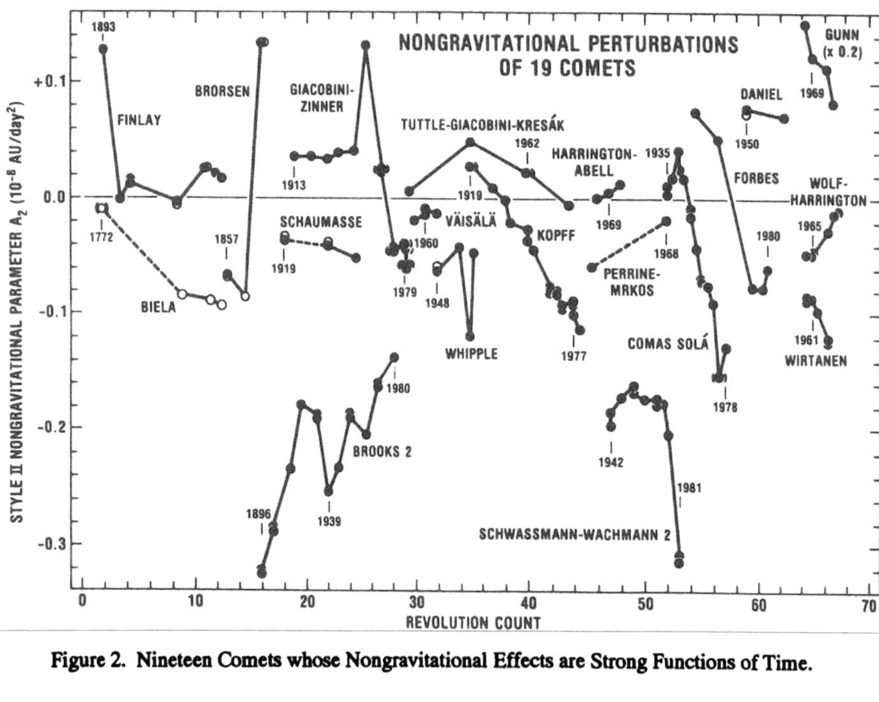

Figure 2. Nineteen Comets whose Nongravitational Effects are Strong Functions of Time.

ASTEROID SPECTROSCOPY AND MINERALOGY

THOMAS H. BURBINE AND RICHARD P. BINZEL
Department of Earth, Atmospheric and Planetary Sciences, MIT, Cambridge, MA 02139, U.S.A.
E-mail THB@ASTRON.MIT.EDU
E-mail RPB@ASTRON.MIT.EDU

Abstract. Spectroscopic observations of asteroids of smaller sizes and in the near-infrared have led to many exciting discoveries that have increased substantially our knowledge of the mineralogy of asteroids. These discoveries include the identification of hydrated M and E-asteroids, the mineralogical diversity of the S-class, anomalous S-asteroids, a possible source body of the aubrites, Vesta-like objects in the main belt, possible ordinary chondrite-like objects and one of the reddest objects in the solar system.

1. Introduction

Since the publication of *Asteroids II* in 1989, there has been a considerable amount of work in asteroid spectroscopy that has increased substantially our knowledge of the mineralogical diversity of asteroids within and beyond the asteroid belt. Asteroid types that were just postulated to exist have been discovered in the past few years. These discoveries coupled with the identification of some objects not even considered to exist have led to a very exciting time in asteroid spectroscopy. This review paper will discuss seven of these discoveries and they include

1) the hydrated M and E-asteroids,
2) the mineralogical diversity of the S-class,
3) the two anomalous S-asteroids (387 Aquitania and 980 Anacostia) with features uncharacteristic of any other type of S-asteroid,
4) the linkage of the earth-crossing E-asteroid 3103 Eger with the aubrites,
5) the Vesta-like objects in the Vesta family and also extending out to the Jovian 3 : 1 resonance region,
6) a possible ordinary chondrite-like object in the main belt and
7) the outer solar system asteroid 5145 Pholus that is one of the reddest objects in the solar system.

In addition to this paper, readers should also refer to more extensive recent reviews by Gaffey *et al.* (1993b) and Pieters and McFadden (1994).

2. Asteroid Taxonomy and Mineralogy

The most widely used taxonomy today is the one developed by Tholen (1984) using seven colors (normalized to the visual filter) from the Eight-Color Asteroid Survey (ECAS) (Zellner *et al.* 1985), which observed 589 asteroids using eight broadband filters from 0.34 to 1.04 μm. Tholen created fourteen asteroid classes with eleven of the classes distinguished by their ECAS spectra alone. Three classes (E, M and P) had degenerate ECAS spectra and could only be distinguished by visual albedo. Later taxonomies by Barucci *et al.* (1987) and Tedesco *et al.* (1989) used different

A. Milani et al. (eds.), Asteroids, Comets, Meteors 1993, 255–270.

classification methods on the ECAS data set in conjunction with IRAS albedos. The resulting taxonomies that they produced yielded classifications similar to Tholen's.

Mineralogical analysis of an asteroid's spectrum is done by comparing it to the spectral characteristics of laboratory measured materials. [Detailed reviews of asteroid and mineral spectroscopy can be found in Gaffey *et al.* (1989) and Gaffey *et al.* (1993b).] The mineralogical information that can be obtained from the analysis of a visual wavelength spectrum is the presence or absence of an ultraviolet absorption feature due to Fe^{+2}, the presence or absence of an absorption feature due to olivine and/or pyroxene that appears longward of 0.7 μm and is centered approximately between 0.9 and 1.1 μm (usually called the 1 μm feature) and the slope of the spectrum longwards of 0.55 μm that is due to presence or absence of some reddening agent (e.g., organics, NiFe). Also the visual albedo can be used to discriminate between different mineralogical types that have similar visual spectral characteristics. Mineralogical analysis of ECAS spectra is hampered by the low resolution of the spectra with reflectances at only eight wavelengths from 0.3 to 1.1 μm, the coverage of only part of the 1 μm absorption feature (when present) that can extend out into the near-infrared and the lack of any information on the presence or absence of a 2 μm absorption feature usually due to pyroxene.

The analysis of the ECAS data, in conjunction with a number of other data sets (e.g., radar, near-infrared photometry) that were done for much smaller numbers of asteroids, led to a picture of the asteroid belt up to 1989 that is best described by the Bell superclass paradigm (Bell 1986, Bell *et al.* 1989) (Table I). This paradigm groups the asteroid classes according to the degree of metamorphic heating that they are believed to have received. A very general mineralogic interpretation is given for each class plus the meteorites that are possible analogs to asteroids in that class. However the seven discoveries that are discussed in this paper will show that many of the asteroid classes are now known to contain very mineralogically diverse assemblages and that Table I gives only a very general framework of the asteroid belt. Recent revisions of the Bell superclass paradigm have been done by Burbine and Bell (1993a, 1993b) to try to better represent the compositional diversity within the classes.

3. Asteroid Spectroscopy

In addition to the progress in mineralogic analyses, substantial advances have been made in the observational techniques for asteroid spectroscopy at both visual and near-infrared wavelengths. The primary factor in these advances has been the incorporation of charge-coupled devices (CCDs) into asteroid spectroscopic observing programs. Because CCDs are area detectors with high quantum efficiencies, complete visual spectra within all or part of the range from 0.4 to 1.0 μm can be recorded in a single exposure with a much higher spectral resolution (~10 Å) than filter photometry. Simultaneous recording of all parts of the spectrum (as well as the sky) makes the measurements insensitive to flux variations caused by the irregular shape of a rotating asteroid. At visual wavelengths, several researchers have led the movement into using CCDs for asteroid spectroscopy (e.g., Luu and Jewitt 1990, Vilas and McFadden 1993). Improved quantum efficiencies at near-infrared

TABLE I
Bell Superclasses

Superclass	Class	Inferred Minerals	Analogous Meteorites
Primitive	D	clays, organics	(none)
	P	clays, organics	(none)
	C	clays, C, organics	CI, CM chondrites
	K	ol, pyx, carbon	CV and CO chondrites
Metamorphic	T	?	?
	B+G+F	clays, opaques	altered carb. chondrites
	Q	pyx, ol, gray NiFe	H, L, LL chondrites ?
Igneous	V	plag, pyx, ol	basaltic achondrites
	R	ol, pyx	ol-rich achondrites?
	S	pyx, ol, red NiFe	pallasites, lodranites, irons ?
	A	ol	brachinites
	M	NiFe	irons
	E	Fe-free pyx	aubrites

wavelengths have allowed some important asteroid discoveries (discussed below) and have facilitated observations being pushed to fainter and smaller asteroids (e.g., Clark 1993).

The Small Main-Belt Asteroid Spectroscopic Survey (SMASS), initiated at the Massachusetts Institute of Technology in 1990 (Binzel and Xu 1990, Xu *et al.* 1993, Xu 1994) provides an example of the potential for asteroid spectroscopy using CCDs. Figure 1 shows the approximate size limits for asteroids that were well-sampled by previous surveys, such as the Eight-Color Asteroid Survey by Zellner *et al.* (1985). Use of CCDs with modern spectrographs on two-meter telescopes allows the limit for visual spectra to be pushed to well below 10 km diameter asteroids in the inner main-belt. The effective limit for SMASS, which utilizes the 2.4 m Hiltner telescope of the Michigan-Dartmouth-MIT Observatory, is shown in Figure 1. This new capability to spectroscopically sample nearly all of the numbered asteroids has lead to several important discoveries (discussed below) and promises a tremendous potential for further advances during the 1990s.

Another important observational technique is rotationally resolved spectroscopy where spectra are taken at different times throughout the asteroid's rotational period. These rotationally resolved spectra allow for compositional differences to be measured across an asteroid. This technique has been used to make a compositional map of Vesta (Gaffey 1983), to determine the igneous nature of 8 Flora (Gaffey 1984) and to characterize compositional differences across 243 Ida (Barucci and Lazzarin 1993).

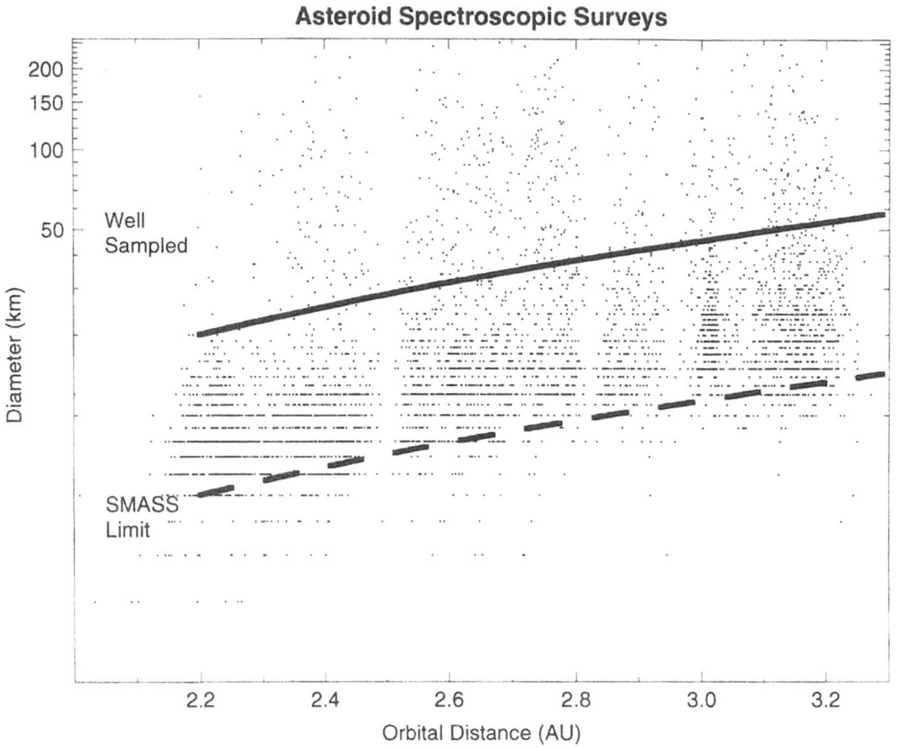

Fig. 1. Orbital distribution of asteroids plotted versus their diameters. The top line depicts the approximate limit where previous surveys have obtained reasonable sampling statistics among the larger asteroids. With the advent of CCDs, programs such as the Small Main-Belt Asteroid Spectroscopic Survey (SMASS, lower dashed line) can routinely sample the spectral properties of main-belt asteroids having diameters of about 10 km or less.

4. Hydrated M and E-asteroids

Spectra from approximately 2.5 to 3.5 μm allow the determination of the presence or absence of a broad absorption feature centered at approximately 3 μm that is due to hydrated silicates on the surface of an asteroid. The presence of hydrated silicates indicates that the original parent bodies, believed to have been composed of anhydrous silicates and ice, underwent aqueous alteration due to a mild heating episode that raised temperatures in the interior above the freezing point of water. The most detailed observational work in this area was done by Jones *et al.* (1990), who analyzed 3 μm observations of 46 asteroids. They found that approximately two-thirds of the C-type asteroids had hydrated silicate surface compositions.

Observations were also done on "igneous" asteroids (asteroids whose surface mineralogies are believed to have formed from a melt) to try to confirm that 3 μm absorption features are not characteristic of these supposedly "anhydrous" asteroids. Observations of three S-asteroids and 4 Vesta confirmed the absence of hydrated silicates on the surfaces of these asteroids. However two of the three M-class asteroids observed (55 Pandora and 92 Undina) were found to have 3 μm features that seem to indicate the presence of hydrated silicates and not the metallic iron compositions that are the mnemonic corresponding to the classification letter "M." [Undina was originally given the designation of X (EMP) by Tholen (1984) due to the asteroid's lack of a visual albedo at that time, but was given the designation of M based on its IRAS albedo of 0.20 in other taxonomies (Barucci *et al.* 1987, Tedesco *et al.* 1989).] The 3 μm observations of these two asteroids have been repeated to try to confirm or reject the presence of hydrated silicates (Howell personal communication, Rivkin *et al.* 1994). Preliminary observations of 55 Pandora (Howell personal communication) are not consistent with the observations of Jones *et al.* (1990) and appear to indicate an anhydrous assemblage. However observations of 92 Undina (Rivkin *et al.* 1994) confirm the presence of hydrated silicates on this asteroid. The observed M-asteroid without a 3 μm feature is 16 Psyche, which is generally thought to have a considerable metallic iron component due to its high radar albedo (Ostro *et al.* 1985).

Observations were also made of the E-type asteroid 44 Nysa. Nysa and other E-type asteroids have been thought to have surfaces dominated by a nonabsorbing, spectrally neutral phase such as enstatite, forsterite or some other iron-poor silicate due to their high albedos and relatively featureless spectra (Gaffey *et al.* 1989). These asteroids are believed to have surfaces that are the crusts or exposed mantles of differentiated parent bodies with E-chondrite bulk compositions (Zellner *et al.* 1977). Nysa also has a weak 0.85-0.90 μm pyroxene feature, not evident in other E-type asteroid spectra, that is consistent with low-iron enstatite seen in some meteorites (Gaffey *et al.* 1989). However 3 μm observations of Nysa (Howell personal communication, Rivkin *et al.* 1994) appear to indicate the presence of hydrated silicates on this asteroid. Either the E or M-types with 3 μm features are not igneous or some currently unknown post-igneous aqueous alteration processes have operated on these asteroids. Thus, these observations appear to indicate that some percentage of the classified M and E-asteroids may have compositions similar to the C-type asteroids. Just as importantly, the apparent discovery of hydrated silicates among asteroids classified as "M" and "E" reminds us that taxonomy is based on the characteristics of spectral features and does not automatically carry with it an interpretation of the mineralogy.

5. Diversity of the S-class

High-resolution spectra from 0.8 to 2.5 μm allow for a better determination of the 1 μm feature, the degree to which the spectrum is reddened, the identification of the presence or absence of absorption features such as the 1.1 μm feature due to feldspar, a 2 μm feature (due to minerals such as pyroxene and spinel) and a 2.2 μm feature attributed to organics (Cruikshank *et al.* 1991a). The most extensive observing program in this wavelength range was done by Bell *et al.* (1988) who observed 119 asteroids (predominately of the S-class) as part of the 52-channel (also known as the 52-color) survey. Analyses of the 52-channel data have led to the creation of the K-class (Bell 1988), which are objects with S-like ECAS spectra but flat reflectances in the near-infrared, and the creation of a preliminary asteroid taxonomy using a neural network (an artificial intelligence program that will recognize patterns in a data set) (Howell *et al.* 1994). This taxonomy supports the separation of the K-class from the S-class and also defines two sub-classes within the S-class (So and Sp). So-asteroids are olivine-rich with a red continuum slope while Sp-asteroids are pyroxene-rich with a less red continuum slope. The remaining S-asteroids show intermediate olivine-pyroxene compositions with a moderately red continuum slope.

The 52-channel data, in conjunction with ECAS and/or 24-filter data, yield coverage of both the 1 and 2 μm absorption features thereby allowing much better mineralogical analyses of asteroids (A, S and V-classes) that have features predominately due to olivine and/or pyroxene. From the initial classification by Tholen (1984), the S-asteroids were known to be a very compositionally diverse spectral class, but no subdivision of the S-class was attempted. The spectral range and resolution of the 52-channel data allow for a much better determination of the mineralogic characteristics of the olivine and/or pyroxene absorption features that appear to be crucial for sub-dividing the S-class. For olivine-orthopyroxene mixtures, the relative areas of the 1 and 2 μm features (called the band area ratio) and the center of the 1 μm feature (called the band I center) have been found to be related to the olivine-orthopyroxene abundance (Cloutis *et al.* 1986). Gaffey *et al.* (1993a) used these parameters to group 39 S-asteroids with 52-channel spectra into seven compositionally distinct subtypes [S(I)-S(VII)]. The reflectance spectra of S(I)-asteroids are indicative of monomineralic olivine assemblages, the spectra of the S(II) to S(VI)-asteroids are indicative of varying olivine/pyroxene abundances and the spectra of S(VII)-asteroids are indicative of pyroxene or pyroxene- feldspar assemblages.

These spectral parameters (band area ratio and band I center) were also compared to the spectral parameters derived from the analysis of the spectra of ordinary chondrites. Except for the S(IV)-subclass, the S-asteroids have very different spectral parameters from the ordinary chondrites, which implies different silicate (olivine and pyroxene) mineralogies and no relationship between these two sets of objects. The S(IV)-asteroids (e.g., 3 Juno, 6 Hebe, 7 Iris) have silicate mineralogies consistent with ordinary chondrites and appear to be the only viable ordinary chondrite parent body candidates among the large main-belt S-asteroid population. However the silicate mineralogies of the S(IV)-asteroids are also consistent with the mineralogy of objects that have undergone small degrees of partial differentiation (primitive achondrites), so these objects need to be more intensely studied before any definite conclusions can be made. It is interesting to note that the S(IV)-asteroids appear to cluster at semimajor axes between 2.3 and 2.6 AU (Gaffey *et al.* 1993a), just where the Jovian 3 : 1 and ν_6 secular resonances both provide viable dynamical routes to Earth-crossing orbits (e.g., Farinella *et al.* 1993).

6. Anomalous S-asteroids

The 52-channel data also have allowed for the identification of some asteroids that appear distinctly different when compared to other S-types. Burbine *et al.* (1992) identified S-asteroid 387 Aquitania and SU-asteroid 980 Anacostia as having absorption features not characteristic of assemblages of olivine and/or pyroxene. The reflectance spectra of assemblages of olivine and/or pyroxene always have a 1 μm feature stronger than the corresponding 2 μm feature. In contrast, these two asteroids each have a 2 μm feature stronger than the corresponding weak to non-existent 1 μm feature. This feature has been interpreted as being either due to spinel (Burbine *et al.* 1992) or chromite (a member of the spinel group) (Cloutis and Gaffey 1993). The presence of spinel would appear to indicate a surface composed of CO/CV carbonaceous material enriched in calcium-aluminum (CAI) inclusions while the presence of chromite would indicate an achondritic surface enriched in chromite that maybe similar to the composition of two unique chromite-rich achondrites that have just been recently described (Meteorite Working Group 1993, Takeda *et al.* 1993).

7. E-asteroid 3103 Eger and the aubrites

As stated before, E-asteroids, because of their high albedos and relatively featureless spectra, have been interpreted as having surfaces dominated by a nonabsorbing, spectrally neutral phase such as enstatite or forsterite. These E-asteroids have been thought to be the probable parent bodies of the aubrites or enstatite achondrites (e.g., Zellner *et al.* 1977). The Hungaria asteroid region, which is located between 1.79 AU and 1.98 AU, is dominated by E-types and contains approximately half of all E-types known in the asteroid belt.

The only Earth-crossing E-type asteroid known is 3103 Eger (provisionally designated as 1982 BB), which was classified as an E-type (Veeder et al. 1989) due to its relatively neutral visual colors (Wisniewski 1987) and high albedo (0.53-0.63) (Veeder *et al.* 1989). Observations by Gaffey *et al.* (1992) in the visible and near-infrared confirmed that Eger has a featureless spectrum consistent with its classification as an E-type. Gaffey *et al.* (1992) have proposed that this object is the actual near-Earth parent body of the aubrites. The lines of evidence include the rareness of E-type asteroids in Earth-approaching orbits, the time of day of aubrite falls that indicates a limited range of source orbits with many of those orbits similar to Eger, the clustering of cosmic ray exposure ages (which records the time that a meteorite spent within a meter of the surface of its original parent body or as a meter-size fragment in interstellar space) for the aubrites and similarities in mineral compositions that suggest that most aubrites are derived from a single parent body (e.g., Keil 1989). The aphelion distance of Eger (1.905 AU) falls within the Hungaria zone and this fact, in conjunction with this asteroid's designation as an E-type, appears to indicate that Eger was most probably derived from that region. These results represented the first plausible direct link between a meteorite group (aubrites) and a particular source region of the asteroid belt.

8. Vesta-like Asteroids

Asteroid 4 Vesta is unique in being the only large main-belt asteroid in the V-class. Vesta's reflectance spectrum also is known to be very similar to the spectra of basaltic achondritic meteorites (McCord *et al.* 1970, Feierberg *et al.* 1980, Feierberg and Drake 1980). This spectral similarity argued for Vesta being the source of the basaltic achondrites, however cratering mechanics argued against Vesta being the source due to the belief (Wetherill 1987) that ejection velocities would not be high enough to allow fragments to reach some of the dynamically favored meteorite source regions such as the Jovian 3 : 1 and ν_6 secular resonances. But the identification of a statistically robust family around Vesta (Zappalá *et al.* 1990) showed that fragments could be ejected from Vesta's surface at velocities well exceeding the escape velocity (300 m/s) and could obtain velocities ($>$ 600 m/s) that were high enough to reach the Jovian 3 : 1 resonance (Farinella *et al.* 1993). Three basaltic Earth approaching asteroids were also found to exist (Cruikshank *et al.* 1991b), however it was still very difficult to persuasively argue that Vesta was the source of the basaltic achondrites.

Recent CCD spectroscopy of small asteroids (diameters $\leq$ 10 km) near Vesta appears to have established Vesta as indeed being the source of the basaltic achondrite meteorites. These asteroids were part of a spectroscopic survey of Xu (1994) whose goal was to determine the spectral characteristics of these small main-belt asteroids so as to reveal possible difference in composition between the large and small asteroid populations and hopefully establish new links in the asteroid-meteorite connection. Binzel and Xu (1993) identified twenty-one main-belt asteroids that have CCD spectra from 0.5 to 1.0 μm that are distinctly similar to Vesta. Thirteen of these asteroids were found within the previously predicted Vesta family, while the other eight bridge the orbital space between Vesta and the Jovian 3 : 1 resonance (Figure 2). It appears very likely that all of these fragments were initially derived from Vesta through one or more impact excavations.

Binzel and Xu (1993) noticed that fifteen of these asteroids had spectra very similar to Vesta (and to eucrites) with absorption band centers near 0.94 μm and classified them as V-types (Figure 3). However the remaining six asteroids had spectra that peaked more sharply at a shorter wavelength and showed a much deeper and less broad 0.90 μm absorption band (Figure 4). These characteristics are suggestive of diogenites and these asteroids were classified with the letter J (mnemonic for the diogenite Johnstown). A principal component analysis (Burbine and Xu personal communication) shows that the J class does indeed separate from the V class. A true mineralogical distinction between the V and J classes requires near-infrared spectral data. Specifically, diogenites are distinguished from eucrites by the lack of a weak feldspar absorption band near 1.2 μm.

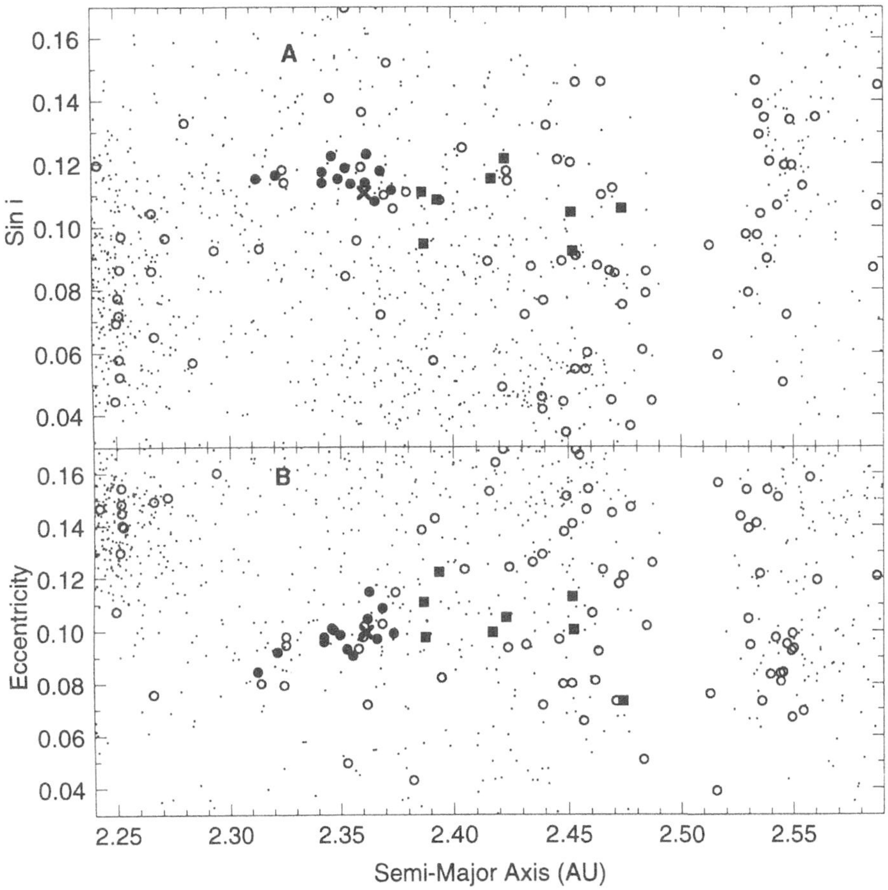

Fig. 2. Distribution of asteroids in proper orbital element space in the inner main belt for (**A**) semimajor axis versus sin *i* (inclination) and (**B**) semimajor axis versus eccentricity. Vesta's position is denoted by **X**. The symbol • indicates the positions for the thirteen Vesta family asteroids that we have observed to display basaltic achondrite spectra. Positions for asteroids that we have also observed, but which do not have basaltic achondrite spectra are denoted by o. Eight additional discovered basaltic achondrite asteroids (filled squares), bridge the region between Vesta and the Jovian 3 : 1 resonance (2.50 AU).

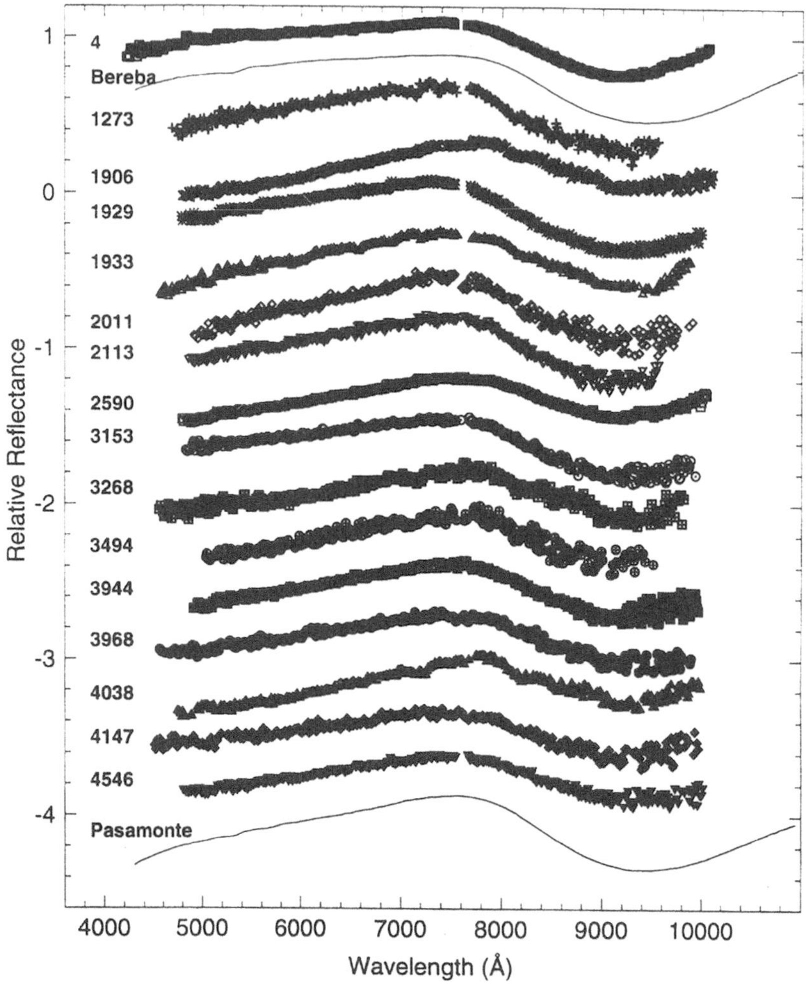

Fig. 3. Optical reflectance spectra for fifteen asteroids that display a deep absorption band near 0.9 μm, characteristic of basaltic achondrites. A spectrum of Vesta, obtained with the same instrumentation, and spectra of eucrites Bereba and Pasamonte (solid lines) (Gaffey 1976) are included for comparison. All spectra have been normalized to unity at 0.55 μm and have been offset vertically for clarity.

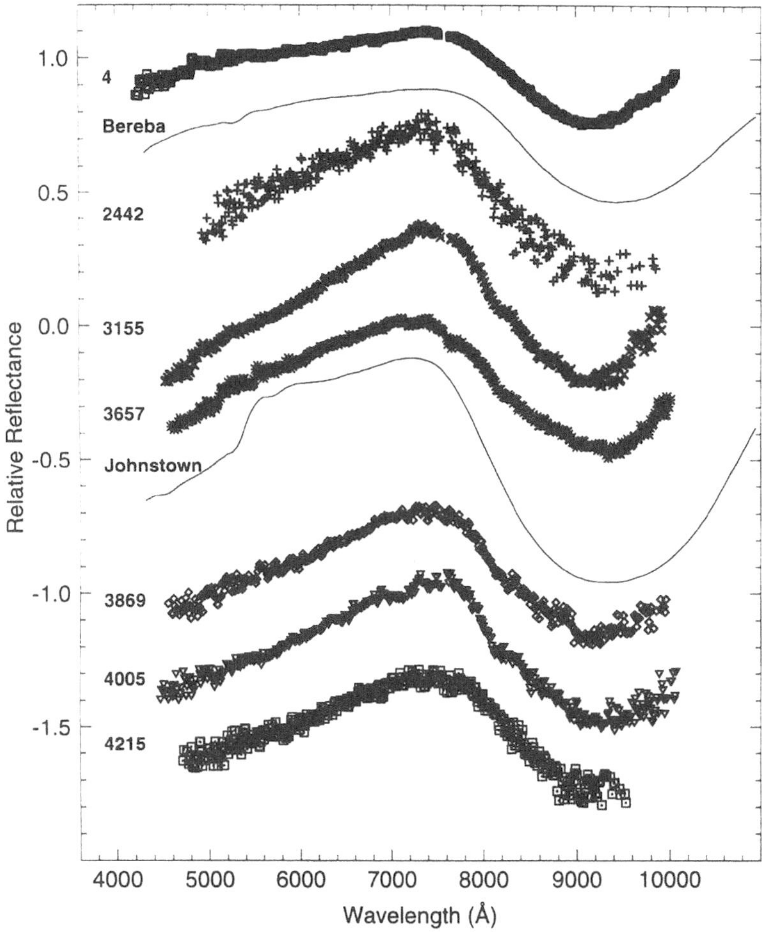

Fig. 4. Optical reflectance spectra for six additional basaltic achondrite main-belt asteroids. Vesta and the eucrite Bereba (Gaffey 1976) are included for comparison. Compared with eucrites, these asteroids show steeper slopes below 0.75 μm, sharper peaks and deeper 0.9 μm absorption bands centered at a shorter wavelength. Their spectral properties are more analogous to the diogenites. A spectrum of the diogenite Johnstown (Gaffey 1976) is included for comparison. All spectra have been normalized to unity at 0.55 μm and have been offset vertically for clarity.

9. Possible Main Belt Ordinary Chondrite Asteroid

Also part of the CCD spectroscopic survey of Xu (1994) was the discovery of the spectrally anomalous characteristics of asteroid 3628 Božněmcová (diameter $\approx$ 7 km) (Binzel *et al.* 1993). As shown in Figure 5**A**, Božněmcová has a CCD spectrum significantly different than the spectra from 0.5 to 1.0 μm of all other known types of asteroids. The figure clearly shows that its spectrum is distinct from the Q-type asteroid 1862 Apollo. However the CCD spectrum of this asteroid appears very similar to the corresponding spectra of ordinary chondrites (Figure 5**B**). The only near-infrared data available for this object are from JH (1.2 and 1.6 μm) photometry that appear roughly consistent with the ordinary chondrite spectra within the 1 σ error bars. Much higher resolution near-infrared data are needed to confirm the existence of an ordinary chondrite-like surface for this asteroid.

In addition to Božněmcová's spectral similarity to ordinary chondrites, it is also located near the Jovian 3 : 1 resonance, a location which many researchers have predicted as an ordinary chondrite source region. It has been argued that many more ordinary chondrite-like objects should exist in this region and have not been seen only because of their small sizes (Bell *et al.* 1989). These postulated "small" ordinary chondrite asteroids could either be primordial objects or the relatively young fragments from larger precursor objects. However of the approximately 70 small asteroids (diameters $\leq$ 20 km) observed in this region, only one of them had any spectral similarity to ordinary chondrites. Thus at present, the discovery of one possible ordinary chondrite asteroid does not resolve the problem for identifying the source(s) for this most common meteorite type.

10. 5145 Pholus

Another spectrally anomalous asteroid is 5145 Pholus, which was discovered on 1992 January 9 just inside the orbit of Saturn (Scotti 1992). Pholus has a perihelion distance of 8.7 AU and an aphelion distance of 32.3 AU. CCD spectra from 0.5 to 1.0 μm (Fink *et al.* 1992; Binzel 1992), BVRI colors (0.43 to 0.8 μm) (Mueller *et al.* 1992) and JHK colors (1.2, 1.6 and 2.2 μm) (Davies and Sykes 1992) show Pholus to be one of the reddest objects in the solar system. Its spectrum is unlike the spectrum of any known asteroid.

The spectrum of Pholus is best matched by the spectra of tholins (Cruikshank *et al.* 1991a), which are the nonvolatile residues produced by the irradiation of C-bearing ices and gas mixtures. Due to its anomalous spectrum and distance from the sun, it has been postulated that Pholus is a "comet" that has yet to have "turned on" and started cometary activity (Fink *et al.* 1992, Mueller *et al.* 1992, Binzel 1992). Pholus has been tentatively classified as a Z-type asteroid (Mueller *et al.* 1992)

11. Future Work

The exciting new discoveries in asteroid spectroscopy and taxonomy within just the last few years appear to foretell of a promising and interesting future for the 1990s.

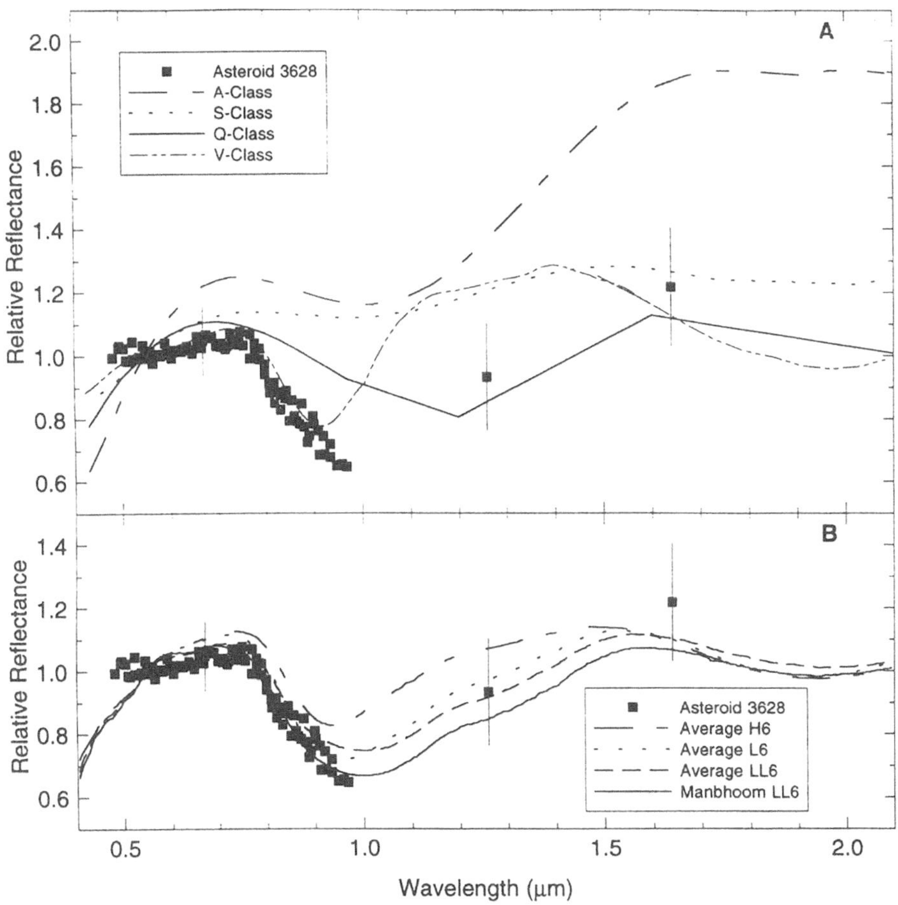

Fig. 5. Visible and near-infrared reflectance spectrum of the main-belt asteroid 3628 Božněmcová compared with (**A**) spectra of other asteroids and (**B**) spectra of ordinary chondrites. Error bars are affixed to the points denoting simultaneous measurements at 0.67, 1.26 and 1.64 μm. The spectral features of Božněmcová are presently unique among all measurements of main-belt asteroids. In comparison with meteorite spectra (**B**), the measurements of Božněmcová are most consistent with L6 and LL6 ordinary chondrites. All spectra have been normalized to unity at 0.55 μm.

Continued directions toward surveying smaller asteroids and for obtaining spectra out into the near-infrared promise to illuminate many of the current problems in the field. Importantly the trend appears away from just assigning a taxonomic letter to individual asteroids. Instead, the trend is toward more thoroughly understanding the mineralogy of individual asteroids and making comparisons based on mineralogy. We are taking our first steps toward what might be called "asteroid geology."

Acknowledgements

This paper is dedicated to Roger Burns who passed away just as this paper was being completed. Roger's work on the spectroscopy of minerals is the basis for much of the mineralogical analyses of asteroids today. We would like to thank Shui Xu for the preparation of some of the figures. We would also like to thank Paolo Farinella and Vincenzo Zappalá for their reviews.

References

Barucci, M. A. and Lazzarin, M. : 1993, "Koronis family : Reflectance spectra of 243 Ida, 1442 Corvina and 2226 Cunitza." *Planet. Space Sci*, **41**, 641–646.

Barucci, M. A., Capria, M. T., Coradini, A. and Fulchignoni, M. : 1987, "Classification of asteroids using G-mode analysis." *Icarus*, **72**, 304–324.

Bell, J. F. : 1986, "Mineralogical evolution of meteorite parent bodies (abstract)." *Lunar Planet. Sci.*, **XVII**, 985–986.

Bell, J. F. : 1988, "A probable asteroidal parent body for the CV or CO chondrites (abstract). *Meteoritics*, **23**, 256–257.

Bell, J. F., Davis, D. R., Hartmann, W. K. and Gaffey, M. J. : 1989, "Asteroids : The big picture." In *Asteroids II* (R. P. Binzel, T. Gehrels and M. S. Matthews, Eds.), 921–945. University of Arizona Press.

Bell, J. F., Owensby, P. D., Hawke, B. R. and Gaffey, M. J. : 1988, "The 52–color asteroid survey : Final results and interpretation (abstract)." *Lunar Planet. Sci.*, **XIX**, 57–58.

Binzel, R. P. : 1992, "The optical spectrum of 5145 Pholus." *Icarus*, **99**, 238–240.

Binzel, R. P. and Xu, S. : 1990, "Asteroid 2965 Surikov : A piece of the missing mantle? (abstract)." *Bull. Am. Astron. Soc.*, **22**, 1115–1116.

Binzel, R. P. and Xu, S. : 1993, "Chips off of asteroid 4 Vesta : Evidence for the parent body of basaltic achondrite meteorites." *Science*, **260**, 186–191.

Binzel, R. P., Xu, S., Bus, S. J., Skrutskie, M. F., Meyer, M. R., Knezek, P. and Barker, E. S. : 1993. "Discovery of a main-belt asteroid resembling ordinary chondrite meteorites." *Science*, **262**, 1541–1543.

Burbine, T. H. and Bell, J. F. : 1993a, "Asteroid taxonomy : Problems and proposed solutions (abstract)." *IAU Symposium 160 : Asteroids, Comets, Meteors 1993*, 49.

Burbine, T. H. and Bell, J. F. : 1993b, "How diverse is the asteroid belt? (abstract)." *Lunar Planet. Sci.*, **XXIV**, 223-224.

Burbine, T. H., Gaffey, M. J. and Bell, J. F. : 1992, "S-asteroids 387 Aquitania and 980 Anacostia : Possible fragments of the breakup of a spinel-bearing parent body with CO3/CV3 affinities." *Meteoritics*, **27**, 424–434.

Clark, B. E. : 1993, "Spectral Reflectance Studies and Optical Surface Alteration in the Search for Links Between Meteorites and Asteroids." Ph. D. Thesis, University of Hawaii.

Cloutis, E. A. and Gaffey, M. J., : 1993, "The constituent minerals in calcium-aluminum inclusions : Spectral reflectance properties and implications for CO carbonaceous chondrites and asteroids." *Icarus*, **105**, 568–579.

Cloutis, E. A., Gaffey, M. J., Jackowski, T. L. and Reed, K. L. : 1986, "Calibration of phase abundance, composition, and particle size distribution for olivine-orthopyroxene mixtures from reflectance spectra." *J. Geophys. Res.*, **91**, 11,641–11,653.

Cruikshank, D. P., Allamandola, L. J., Hartmann, W. K., Tholen, D. J., Brown, R. H., Matthews, C. N. and Bell, J. F. : 1991a, "Solid C≡N bearing material on outer solar system bodies." *Icarus*, **94**, 345–353.

Cruikshank, D. P., Tholen, D. J., Hartmann, W. K., Bell, J. F. and Brown, R. H. : 1991b, "Three basaltic Earth-approaching asteroids and the source of the basaltic meteorites." *Icarus*, **89**, 1–13.

Davies, J. K. and Sykes, M. V. : 1992, "(5145) 1992 AD." *IAU Circular*, 5480.

Farinella, P., Gonczi, R., Froeschlé, Ch. and Froeschlé, C. : 1993, "The injection of asteroid fragments into resonances." *Icarus*, **101**, 174–187.

Feierberg, M. A. and Drake, M. A. : 1980, "The meteorite-asteroid connection : The infrared spectra of eucrites, shergottites, and Vesta." *Science*, **209**, 805-807.

Feierberg, M. A., Larson, H. P., Fink, U. and Smith, H. A. : 1980, "Spectroscopic evidence for two achondrite parent bodies : Asteroids 349 Dembowska and 4 Vesta." journalGeochim. Cosmochim. Acta, **45**, 971–981.

Fink, U., Hoffman, M., Grundy, W., Hicks, M. and Sears W. : 1992, "The steep red spectrum of 1992 AD : An asteroid covered with organic material?"*Icarus*, **97**, 145–149.

Gaffey, M. J. : 1976, "Spectral reflectance characteristics of the meteorite classes." *J. Geophys. Res.*, **81**, 905–920.

Gaffey, M. J. : 1983, "The asteroid (4) Vesta : Rotational spectral variations, surface material heterogeneity, and implications for the origin of the basaltic achondrites (abstract)." *Lunar Planet. Sci.*, **XIV**, 231–232.

Gaffey, M. J. : 1984. "Rotational spectral variations of asteroid (8) Flora : Implications for the nature of the S-type asteroids and for the parent bodies of the ordinary chondrites." *Icarus*, **60**, 83–114.

Gaffey, M. J., Bell, J. F., Brown, R. H., Burbine, T. H., Piatek, J. L., Reed, K . L. and Chaky, D. L. : 1993a, "Mineralogic variations within the S-type asteroid class." *Icarus*, **106**, 573–602.

Gaffey, M. J., Bell, J. F. and Cruikshank, D. P. : 1989, "Reflectance spectroscopy and asteroid surface mineralogy." In *Asteroids II* (R. P. Binzel, T. Gehrels and M. S. Matthews, Eds.), 98–127. University of Arizona Press.

Gaffey, M. J., Burbine, T. H. and Binzel, R. P. : 1993b, "Asteroid spectroscopy : Progress and perspectives." *Meteoritics*, **28**, 161-187.

Gaffey, M. J., Reed, K. L. and Kelley, M. S. : 1992, "Relationship of E-type Apollo asteroid 3103 (1982 BB) to the enstatite achondrite meteorites and the Hungaria asteroids." *Icarus*, **100**, 95–109.

Howell, E. S., Merényi, E. and Lebofsky, L. A. : 1994, "Classification of asteroid spectra using a neural network." *Jour. Geophys. Res.*, in press.

Jones, T. D., Lebofsky, L. A., Lewis, J. S. and Marley, M. S. : 1990, "The composition and origin of the C, P, and D asteroids : Water as a tracer of thermal evolution in the outer belt." *Icarus*, **88**, 172–192.

Keil, K. : 1989, "Enstatite meteorites and their parent bodies." *Meteoritics*, **24**, 195–208.

Luu, J. X. and Jewitt, D. C. : 1990, "Charge-coupled device spectra of asteroids. I. Near-Earth and 3 : 1 resonance asteroids." *Astron. J.*, **99**, 1985–2011.

McCord, T. B., Adams, J. B. and Johnson, T. V. : 1970, "Asteroid Vesta : Spectral reflectivity and compositional implications." *Science*, **168**, 1445–1447.

Meteorite Working Group, : 1993. *Antarctic Meteorite Newsletter* (R. Score and M. Lindstrom, Eds.), **16**, 15.

Mueller, B. E. A., Tholen, D. J., Hartmann, W. K. and Cruikshank, D. : 1992, "Extraordinary colors of asteroidal object (5145) 1992 AD." *Icarus*, **97**, 150–154.
Ostro, S. J., Campbell, D. J. and Shapiro, I. I. : 1985, "Mainbelt asteroids : Dual-polarization radar observations." *Science*, **229**, 442–446.
Pieters, C. M. and McFadden, L. A. : 1994, "Meteorite and asteroid reflectance spectroscopy : Clues to early solar system processes." *Ann. Rev. of Earth and Plan. Science*, in press.
Rivkin, A. S., Britt, D. T., Howell, E. S. and Lebofsky, L. A. : 1994, "Hydrated E-class and M-class asteroids (abstract)." *Lunar Planet. Sci.*, **XXV**, in press.
Scotti J. V. : 1992, "1992 AD." *IAU Circular*, 5434.
Takeda, H., Saika, K., Otsuki, M. and Hiroi, T. : 1993, "A new Antarctic meteorite with chromite, orthopyroxene and metal with reference to a formation model of S asteroids (abstract)." *Lunar Planet. Sci.*, **XXIV**, 1395–1396.
Tedesco, E. F., Matson, D. L., Veeder, G. J., Gradie, J. C. and Lebofsky, L. A. : 1989, "A three-parameter asteroid taxonomy." *Astron. J.*, **97**, 580–606.
Tholen, D. J. : 1984, "Asteroid Taxonomy from Cluster Analysis of Photometry." Ph. D. Thesis, University of Arizona.
Veeder, G. J., Hanner, M. S., Matson, D. L., Tedesco, E. F., Lebofsky, L. A. and Tokunaga, A. T. : 1989, "Radiometry of near-Earth asteroids." *Astron. J.*, **97**, 1211–1219.
Vilas, F. and McFadden, L. A. : 1993, "CCD reflectance spectra of selected asteroids. I. Presentation and data analysis considerations." *Icarus*, **100**, 85–94.
Wetherill, G. W. : 1987, "Dynamical relations between asteroids, meteorites and Apollo-Amor objects." *Phil. Trans. R. Soc. Lond. A*, **323**, 323–337.
Wisniewski, W. Z. : 1987, "Photometry of six radar target asteroids.", *Icarus*, **90**, 117–122.
Xu, S. : 1994, "CCD Photometry and Spectroscopy of Small Main-Belt Asteroids." Ph. D. Thesis, Massachusetts Institute of Technology.
Xu, S., Binzel, R. P., Burbine, T. H. and Bus, S. J. : 1993, "Small main-belt asteroid spectroscopic survey (abstract)." *Bull. Am. Astron. Soc.*, **25**, 1135.
Zappalá, V., Cellino, P., Farinella, P. and Kneževič, Z. : 1990, "Asteroid families. 1. Identification by hierarchical clustering and reliability assessment." *Astron. J.*, **100**, 2030–2046.
Zellner, B., Leake, M., Morrison, D. and Williams, J. G. : 1977, "The E asteroids and the origin of the enstatite chondrites." *Geochim. Cosmochim. Acta*, **41**, 1759–1767.
Zellner, B., Tholen, D. J. and Tedesco, E. F. : 1985, "The eight-color asteroid survey : Results for 589 minor planets." *Icarus*, **61**, 355–416.

COHERENT BACKSCATTERING BY SOLAR SYSTEM DUST PARTICLES

KARRI MUINONEN

Observatory, P.O. Box 14
FIN-00014 University of Helsinki, Finland
E-mail karri@ceres.helsinki.fi

Abstract. I review progress in interpreting the opposition effect and negative linear polarization observed for solar system dust particles. The so-called coherent backscattering mechanism has recently been introduced to explain the observations. However, fundamental difficulties in theoretical modeling still prevent quantitative interpretation. I also review some of the key observations that questioned the hitherto widely accepted mutual-shadowing explanation for the opposition effect. I summarize previous theoretical and experimental work on the opposition effect and negative polarization.

1. Introduction

Close to astronomical opposition, there are two intriguing light scattering phenomena observed for solar system dust particles : the *opposition effect* and *negative linear polarization.* As for the increase of brightness, the behaviour in both narrow and wide ranges of solar phase angle is discussed. The phase angle is the angular distance between the Sun (light source) and the observer (detector) as seen from the object (sample). The wide-angle brightening from large phase angles down to about 10° is common for dark rough particulate surfaces and can be termed the *strong backscattering effect.* The narrow-angle brightening, a rapid increase in brightness towards zero phase angle, observed for phase angles lower than 10°, is generally called the opposition effect. In the present review, I do not distinguish between the opposition effect and opposition spike, the term sometimes used for "anomalously" narrow opposition effects.

The sunlight incident on solar system bodies is known to be unpolarized to great precision. The degree of linear polarization of scattered light, or briefly linear polarization, is defined as the ratio

$$P = \frac{I_\perp - I_\parallel}{I_\perp + I_\parallel} , \tag{1}$$

where $I_\parallel$ and $I_\perp$ refer to the intensity components parallel and perpendicular to the scattering plane, here defined by the Sun, the object and the observer. Negative polarization is observed when the parallel component is greater than the perpendicular. This is opposite to first-order Rayleigh scattering or Fresnel reflection, which result in perpendicular, positive polarization. Typically, the observed angular ranges for negative polarization vary between 15° and 25°.

None of the existing treatments of electromagnetic scattering by inhomogeneous media is generally applicable to solar system dust. The present review is partially

A. Milani et al. (eds.), Asteroids, Comets, Meteors 1993, 271–296.

based on the following material : Muinonen (1990) on coherent backscattering, and the opposition effect and negative linear polarization, Pellicori (1971) on the history and milestones of polarimetry, Kravtsov and Saichev (1982) on double–passage effects in inhomogeneous media, Bowell et al. (1989) on the Hapke and Lumme–Bowell scattering laws, McGurn (1990) on the opposition effect and Anderson localization, Barabanenkov et al. (1991) on enhanced (or coherent) backscattering, van Tiggelen (1992) on multiple scattering and localization of light, Mishchenko and Dlugach (1993) on coherent backscattering, and Shkuratov et al. (1994) on the theoretical models for negative polarization.

Since 1887 it has been known that the brightness of the Saturnian planet–ring system peaks sharply at opposition. Using Müller's observations (1893), Seeliger (1887) assigned the phenomenon to the rings and gave the classical mutual shadowing explanation : ring particles hide their own shadows at exact backscattering geometry, which leads to opposition brightening. Lyot (1929) reported the reversal of linear polarization for the rings, but offered only rather general ideas for the origin of the phenomenon.

The strong backscattering effect was reliably observed for the Moon by Herschel (1847), who could not explain the brightening and concluded that there must have been a systematic error in the measurements. However, Herschel's observations were later confirmed by Russell (1916a), who was aware of Seeliger's (1887) theoretical considerations for Saturn's rings. Russell (1916b) further proposed that the lunar surface was covered with fragments of rock. At the moment, the main reason for the strong backscattering effect is thought to be macroscopic shadowing, or rough–surface shadowing, due to structures formed by the regolith dust particles.

The coherent backscattering mechanism has been pointed out by Watson (1969) in studies on multiple scattering of electromagnetic waves in an underdense plasma, and by de Wolf (1971) in studies on electromagnetic reflection from a dielectric turbulent medium. The relevance of the coherent backscattering mechanism to the opposition effect of the Moon was mentioned but ignored in studies of electromagnetic scattering (e.g., Kuga and Ishimaru 1984, O'Donnell and Mendez 1987). In the solar system context, it was introduced as a possible explanation for the opposition effect and negative polarization by Shkuratov (1988b, 1989) and Muinonen (1989ab, 1990). The coherent backscattering mechanism does not require any specific assumptions of particle or regolith geometry.

The coherent backscattering mechanism is described in Figure 1a, where an electromagnetic plane wave (solid line) with wavenumber $k = 2\pi/\lambda$ (λ being the wavelength) is scattered in the backward direction via several scattering elements located of the order of the wavelength to tens of wavelengths apart. The scattering elements can be individual particles, subparticles in an aggregate particle, facets on a rough solid surface, or other irregularities. But there is the companion electromagnetic wave (dashed line) that propagates through the same scattering elements in the opposite (or time–reversed) direction. These two waves always interfere constructively in the exact backward direction, but not necessarily in other directions. Averaging over a distribution of scattering element locations will result in a backscattering peak, called enhanced or coherent backscattering. The backscattering peak tends to be narrower for higher–order interactions, the average distance be-

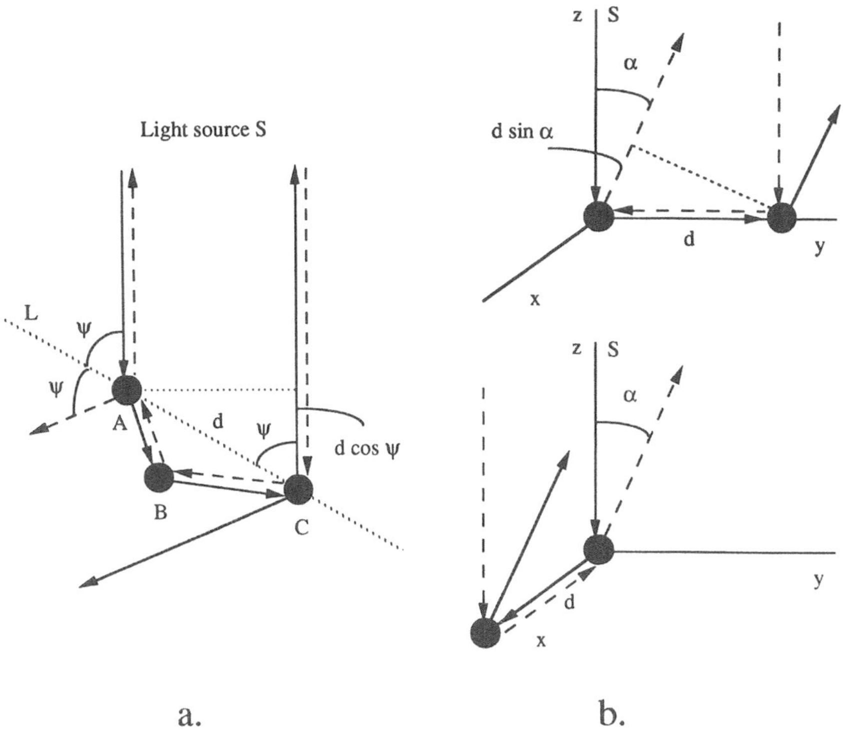

Fig. 1. (a) Opposition effect due to the coherent backscattering mechanism. The multiply scattering electromagnetic waves propagating in opposite directions (solid and dashed lines) always interfere constructively at the backward direction (phase angle $\alpha = 0°$) and in directions forming a cone with axis L. In other directions, the interference varies between constructive and destructive depending on the wavelength, and the distance and orientation of the first and last scattering elements. (b) Negative linear polarization due to the coherent backscattering mechanism. For non-zero phase angles and second-order scattering, the interference favors negative polarization : in the yz-plane in the scattering geometry leading to positive polarization the interference depends on the phase difference $\delta = kd \sin\alpha$ (upper panel), whereas the interference is always constructive in the geometry causing negative polarization (lower panel). Averaging over the element locations will result in a backscattering peak and net negative polarization.

tween the end elements increasing with increasing order of interaction.

Note that the interference is constructive in the directions of a cone defined by rotating the light source direction around the L-axis (Figure 1a) that joins the two end scattering elements—this is the key to the negative polarization mechanism. In Figure 1b, an electromagnetic plane wave is scattered via two scattering elements at a distance d from each other. The negative polarization can be understood by calculating the phase differences in the yz-plane in the two scattering geometries (upper and lower panels). Since first-order scattering is predominantly positively polarized (e.g., Rayleigh scattering and Fresnel reflection) the scattering elements sufficiently far away from each other ($kd \gg 1$) interact mainly with the electric field vector perpendicular to the plane defined by the source and the scattering elements. The observer in the yz–plane will measure positive polarization from the geometry in the upper panel of Figure 1b, and negative polarization from the geometry in the lower panel. However, the positive polarization suffers from the phase difference $\delta = kd \sin \alpha$, whereas the phase difference for the negative polarization is zero for all phase angles. Isotropic averaging over positions of the scattering centers will result both in an increase of the brightness and in a reversal of polarization near the backward direction (at exactly zero phase angle the polarization goes to zero). Scattering orders higher than the second can also experience preferential interaction geometries, and lead to net negative polarization.

As an example, Figure 2 shows the backscattering peak and negative linear polarization in total diffuse scattering for a simple scattering system of a small dipole particle close to a semi–infinite electromagnetically homogeneous medium.

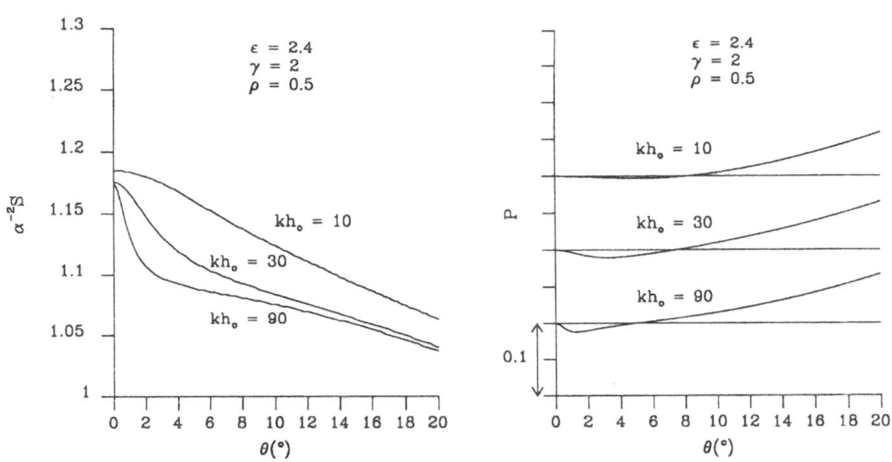

Fig. 2. Total diffuse scattering by a small dipole particle close to a material interface (Muinonen et al. 1991). (a) The coherent backscattering peak, and (b) the accompanying negative linear polarization. ϵ is the electric permittivity of the particle and the semi–infinite medium, γ and h_o are the power and the mean in the Gamma distribution for the distance between the particle and the interface, and ρ is the standard deviation for the Gaussian random slope of the scattering system.

Another phenomenon related to brightness variations of asteroids and other atmosphereless solar system bodies is the color opposition effect (Mikhail 1970). For example, the Moon is bluest at the time of its full phase. In fact, if the optical parameters of the surface coincide for two wavelengths, the coherent backscattering mechanism predicts a narrower opposition effect for the shorter wavelength. This is in agreement with the lunar color opposition effect. However, the color opposition effect of the solar system dust particles needs further observational verification.

The coherent backscattering peak and negative linear polarization generally depend on refractive index. For small real refractive indices, the multiple scattering contribution is driven toward high scattering orders, and one can expect sharper backscattering phenomena than for larger real refractive indices. Increasing absorption decreases the lengths of the interaction paths, thus broadening the backscattering peak. On the other hand, for metallic materials with complex refractive indices, the dependence becomes more difficult, since the backscattering phenomena can sometimes be affected by electromagnetic surface wave (or surface polariton) interactions, besides the multiple reflections between different surface elements.

The coherent backscattering mechanism is a possible explanation for some observations by Sir Isaac Newton in 1730. In attempting to understand light scattering by rough surfaces, Newton faced the very same difficulties as did researchers to come (Newton 1952) :

> *There is no Glass or Speculum how well soever polished, but, besides the Light which it refracts or reflects regularly, scatters every way irregularly a faint light, by means of which the polish'd Surface, when illuminated in a dark room by a beam of the Sun's Light, may be easily seen in all positions of the eye. There are certain Phaenomena of this scatter'd Light, which when I first observed them, seem'd very strange and surprizing to me.*

In one of Newton's numerous experiments, the Sun shone into his darkened chamber through a hole in a white opaque paper chart, one third of an inch wide. The sunlight was then normally incident on a spherically concave mirror, with the radius of curvature of five feet eleven inches, and with a quick–silver coating on the convex side. The distance between the paper chart and the mirror surface was about six feet. Newton writes about the scattered light he noticed in his detector, i.e., on the opaque chart :

> *...there was, in their common center [of rainbows] a white round Spot of faint Light, something broader than the [specularly] reflected beam of Light, which beam sometimes fell upon the middle of the Spot, and sometimes by a little inclination of the Speculum receded from the middle, and left the Spot white to the center.*

The Spot around the backscattering direction can be coherent backscattering from the rough interface between the glass and quick silver. It can result from Anderson localization of polaritons by the roughness of the surface. There is debate whether the enhanced backscattering by certain metallic surfaces is predominated by multiply reflected electromagnetic waves, cyclically propagating surface polaritons (leading to coherent backscattering), or double–passage of individual surface polaritons.

In Section 2, I summarize the observations of the opposition effect and negative linear polarization for solar system objects. Section 3 reviews theories and experiments in the era before the introduction of the coherent backscattering mechanism. Research on the coherent backscattering mechanism is then reviewed in Section 4, and discussion follows in Section 5.

2. Observations

The opposition effect and negative linear polarization have been confirmed for the planets Mercury and Mars, the Moon, and the Martian, Jovian, and Saturnian satellites. The phenomena have been observed for asteroids, Saturn's rings, and for the interplanetary dust, in which case the opposition brightening has been called the Gegenschein. Additionally, the opposition effect has been observed for the Uranian satellites, and the negative polarization for cometary dust. The review of observations is here constrained to the Moon, Saturn's rings, asteroids (44) Nysa, (64) Angelina, (165) Loreley, (419) Aurelia, and the Galilean satellite Europa. Both the opposition effect and negative linear polarization have been experimentally verified in laboratory measurements for rough and particulate surfaces.

The opposition effects and negative polarizations are shown collectively in Figures 3 and 4 in the order of downward steepening slope of the linear part of the magnitude phase dependence. The slope can be related to the amount of multiple scattering so, upward in Figures 3 and 4, the objects are principally brighter. However, Saturn's rings probably consist of particles whose geometric albedos exceed those for the asteroids (44) Nysa and (64) Angelina, and also, there are hints that (64) Angelina can be slightly brighter than (44) Nysa. Unfortunately, there are no published polarization observations for (165) Loreley and (419) Aurelia. The reader is encouraged to consult the original references for information on the geometric albedos, and spectral bands for the magnitudes and polarizations.

As for Saturn's rings, the phase angle range covered with Earth-based observations is rather narrow, and the reversal of linear polarization into positive values is not seen. There is discussion whether the polarization tends to zero at zero phase angle for Saturn's rings. The sharp polarization surge for Saturn's rings can be called the polarization opposition effect (Mishchenko 1993).

Gehrels (1956) confirmed the nonlinear opposition brightening (magnitude scale) for the asteroid (20) Massalia, and introduced the term "opposition effect". The lunar opposition effect was then pointed out by Gehrels et al. (1964) and by van Diggelen (1964), although it was already discovered by Barabashev (1922) and Markov (1924), and could be distinguished in Rougier's extensive observations (1933). In short, the Moon is roughly twice as bright at opposition than on the day just before or just after opposition. Hapke (1966) interpreted the effect using the mutual shadowing mechanism, after constructing a new shadowing function for the lunar strong backscattering effect (1963). The negative linear polarization was discovered for the Moon by Lyot (1929).

Brown and Cruikshank (1983) published intriguingly sharp opposition surges for the Uranian satellites, the surge amplitudes closing a factor of two (the maximum enhancement factor from mutual shadowing or coherent backscattering as separate

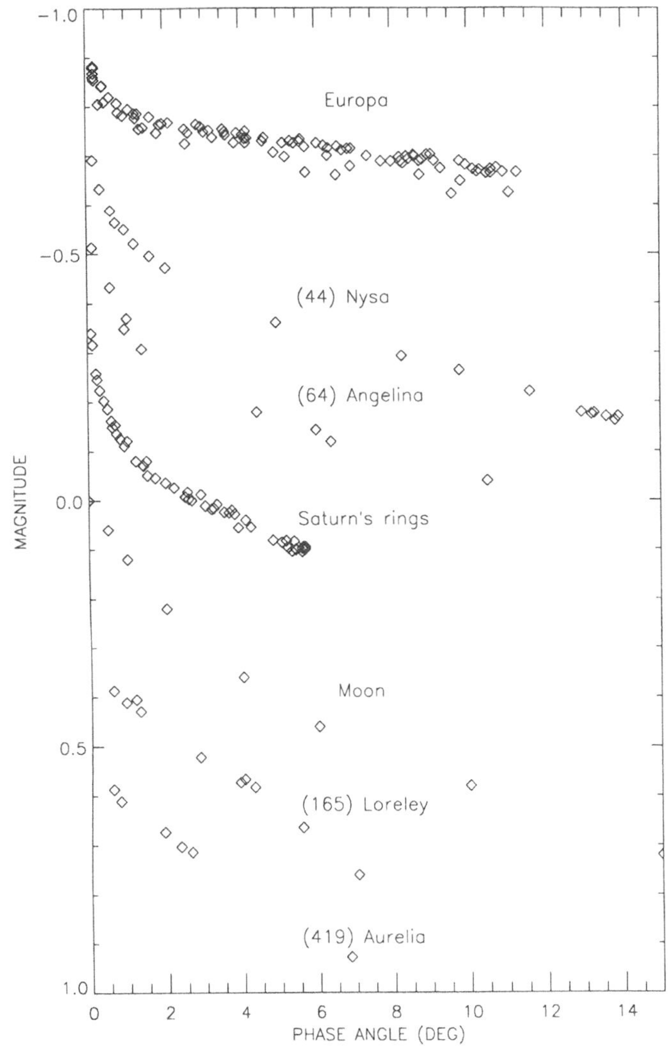

Fig. 3. The opposition effects for the Galilean satellite Europa (Thompson and Lockwood 1992), the E-type asteroids (44) Nysa and (64) Angelina (Harris et al. 1989), Saturn's rings (Franklin and Cook 1965), the Moon (Rougier 1933, Bowell et al. 1989), (165) Loreley (Harris and Young 1988), and (419) Aurelia (Harris and Young 1989). For clarity, the phase curves have been shifted using the following constants in magnitude scale (from Europa downwards) : −5.9, −7.6, −8.1, 0.09, −0.85, −7.55, and −7.95.

mechanisms). Prior to these observations, only Saturn's rings had been known to exhibit a comparable brightness surge.

Thompson and Lockwood (1992; see also Domingue et al. 1991) discovered the

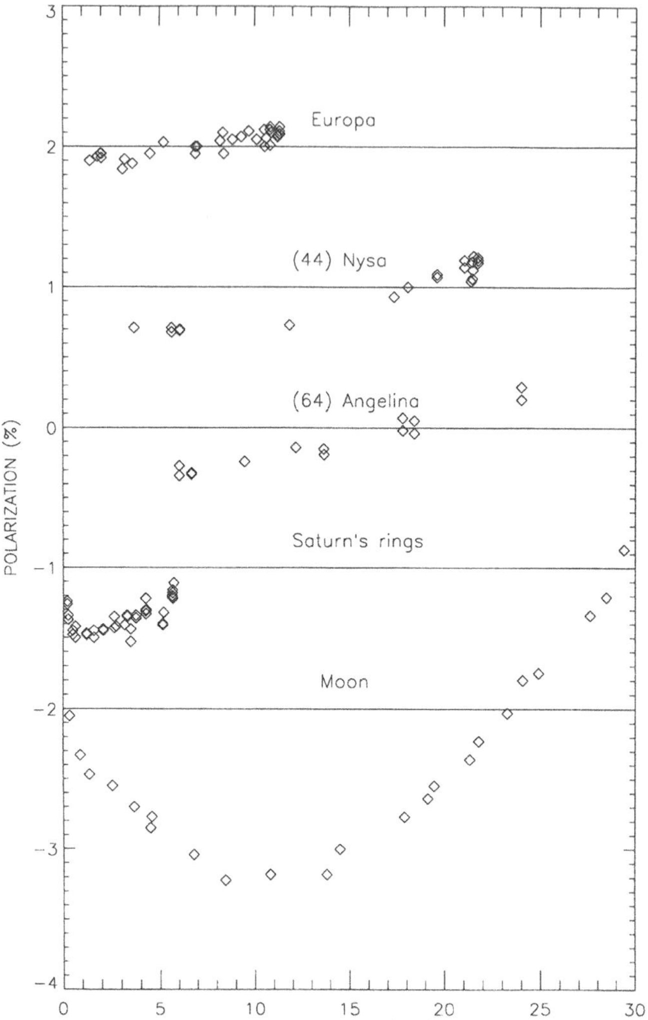

Fig. 4. The linear polarizations for the Galilean satellite Europa (Dollfus 1975), the E–type asteroids (44) Nysa and (64) Angelina (Zellner and Gradie 1976), Saturn's rings and the Moon (Lyot 1929). For clarity, the polarization phase curves have been shifted using the following constants (percentage points, from Europa downwards) : 2.0, 1.0, 0.0, −1.0, and −2.0.

record–setting sharp opposition surge for the Galilean satellite Europa in late 1986. An important observation was made by Harris et al. (1989), who measured an unusually sharp opposition effect for the bright E–type asteroid (44) Nysa. The other E–type asteroid (64) Angelina was found to exhibit a sharp opposition effect, too.

These observations raised another problem for a pure mutual shadowing explanation : how could a particle volume fraction of the order of 1% be understood for the regolith of an asteroid? In fact, these opposition effects resemble that for Saturn's rings (Harris et al. 1989).

Harris and Young (1988) have reported on dark asteroids with very small opposition effects. The observations of the dark asteroids (165) Loreley and (419) Aurelia are shown in Figure 3 and exhibit practically no opposition effect. Note the steeper slope of the strong backscattering effect for the darker objects, mainly due to the decreased multiple scattering. It is probable that the main contribution to the phase curves of (165) Loreley and (419) Aurelia comes from shadowing due to surface irregularities, which hints that, perhaps, no narrow–angle opposition effect can be expected from shadowing in the case of regoliths of atmosphereless bodies.

Objects exhibiting the opposition effect also exhibit negative polarization. Lyot (1929) discovered the negative polarization for Saturn's rings and reported it simultaneously with the linear polarization for the Moon. Since then, the polarization of Saturn's rings has been observed, for example, by Johnson et al. (1980). Further studies of the lunar polarization have been carried out by Dollfus and Bowell (1970), Shkuratov and Opanasenko (1992), and by Shkuratov et al. (1992). Asteroid polarization curves have been extensively observed by Zellner et al. (1974) and by Zellner and Gradie (1976).

Unfortunately, the polarizations of (165) Loreley and (419) Aurelia have not been recorded. It would be important to observe very dark asteroids and search for saturation of polarization, the stage where the negative polarization begins to disappear with decreasing single–particle albedo. It has already been found in laboratory experiments by Zellner et al. (1977) for mixtures of hydrated magnesian silicate and small particles of carbon black.

Qualitative interpretation of the opposition effects in Figure 3 can be carried out on the basis of the coherent backscattering mechanism. For the dark objects (165) Loreley and (419) Aurelia, the absence of a nonlinear brightening can be due to the absence of multiply scattered light. For the brighter Moon, a non–linear surge begins to show up and could be ascribed to increased multiple scattering. For the even brighter (44) Nysa and (64) Angelina, as for Saturn's rings, the amount of multiple scattering is considerable, causing a pronounced opposition effect. Furthermore, Europa is the brightest object in this sample and, in agreement with the coherent backscattering mechanism, exhibits the sharpest opposition effect.

The polarization observations also agree with the predictions of the coherent backscattering mechanism. The bright satellite Europa, asteroids (44) Nysa and (64) Angelina, and Saturn's rings have a weaker polarization because of increased proportion of unpolarized multiple scattering from orders higher than the second. Note that the theoretical calculations also support the idea that the negative polarization extends to larger phase angles for larger refractive indices.

By far the most significant dilemma in adopting the coherent backscattering explanation for the opposition phenomena is the often largely differing angular widths of the observed opposition effect and negative polarization. In that light, one can understand the backscattering phenomena of Europa and Saturn's rings as caused by coherent backscattering, but has difficulties in explaining the entire negative polarization ranges for the Moon, and the asteroids (44) Nysa and (64) Angelina (Figures 3 and 4). Further theoretical and experimental studies will have to be carried out to resolve whether coherent backscattering in multi-scale inhomogeneous media (including scattering from rough interfaces) could lead to such broad polarization effects.

3. Theories and Experiments

Here I review the theoretical and experimental work before the coherent backscattering era in studies of light scattering by solar system dust particles. Mutual shadowing was the generally accepted explanation for the opposition effect. It was the most popular explanation for the negative linear polarization, too, even without solid theoretical or experimental support. Partly because of that, numerous negative polarization models or mechanisms have appeared in the literature.

3. 1. INTENSITY

As mentioned before, the first steps in the interpretation of the opposition effect of Saturn's rings were taken by Seeliger (1887) on the basis of Müller's observations (1893). However, if one assumes the classical many–particle–thick ring model and that the opposition effect is related to mutual shadowing, the volume density (or volume fraction) for the ring particles would have to be much lower than predicted from ring dynamics studies (L. W. Esposito 1988, private communication). This is a serious contradiction between two different approaches to determining the volume density, as was emphasized by Irvine et al. (1988), who showed that a very small amount of trough retroreflecting particles in Saturn's rings would explain the opposition effect and relax the requirement of a many–particle–thick ring system. (Trough and corner retroreflection are second–order and third–order backward reflection from two and three perpendicular facets, respectively.) The same result would follow if the opposition effect could be addressed to coherent backscattering in individual ring particles.

Bobrov (1940) investigated the phase curve of Saturn's rings and later generalized Seeliger's shadowing calculations for particles of unequal size (1961). The theoretical considerations of Seeliger and Bobrov were reviewed by Irvine (1966), who also briefly analyzed the possible applications of the mutual shadowing calculations. Hämeen–Anttila and Vaaraniemi (1975) derived a photometric function for Saturn's rings assuming a monolayer ring model. Esposito (1979) extended the applicability of the classical shadowing calculations to higher particle volume densities by means of a van der Waals correction in the particle distribution function. He showed that the contribution from second–order shadowing is very small compared to the dominant first–order one. The most recent mutual shadowing calculations have been carried out by Shkuratov (1988a), Peltoniemi and Lumme (1992), and Peltoniemi (1993).

Hapke (1963) constructed a shadowing model for the lunar strong backscattering effect, preceded by photometric studies of complicated surfaces (Hapke and van Horn 1963). The lunar opposition effect was confirmed by Gehrels et al. (1964), and by van Diggelen (1964). Soon thereafter, Hapke (1966) extended his shadowing calculations to explain the lunar opposition effect, too. Since then, Hapke (1981) has been refining his theoretical calculations, treating light scattering by regoliths using radiative transfer equation. Theoretical work has been followed by experiments and interpretation of observations (Hapke and Wells 1981), and a correction for macroscopic roughness (Hapke 1984). In his relatively recent work (Hapke 1986), he still emphasized the opposition effect as being due to mutual shadowing.

Lumme (1971) derived a mutual shadowing model to minimize the restrictive assumptions of the earlier calculations, and found good agreement between observations and his theoretical considerations. Later, Lumme and Bowell (1981a) developed a radiative transfer model including new shadowing calculations and considerations of multiple scattering. The theoretical model was followed by an interpretation of phase curves (1981b), including the interpretation of the observed asteroid phase curves (e.g., Harris and Young 1989). The model was also applied to the lunar phase curve (Lumme and Irvine 1982). In connection to the Phobos space mission, Lumme et al. (1990) further developed their treatment of macroscopic surface roughness.

Heated discussion has sometimes taken place when the Hapke and Lumme–Bowell models have been compared to each other (Hapke 1982, Lumme and Bowell 1982). Both models have been criticized for unphysical parametrization. A recent review of the models is presented in Bowell et al. (1989).

Whitaker (1979) compared the phase relations of several asteroids, Mercury, and the Moon, and noted the similarity of the opposition effects. He expressed doubts concerning the mutual shadowing explanation. The first–order similarity for the opposition effect of asteroids was also noticed by Scaltriti and Zappalà (1980). They extracted the linear part of the phase curves (magnitude scale) and showed the similarity of the remaining nonlinear effects. However, they stated that the resulting rather similar nonlinear surges might have been due to the restricted range of geometric albedos in their sample of asteroids.

Opposition effect mechanisms other than mutual shadowing or coherent backscattering have been considered every now and then. The corner, trough and lense retroreflection mechanisms, which have their basis on the shape and structure of the dust particles, were summarized by Trowbridge (1978, 1984), and also analyzed by Akimov (1980).

Muinonen (1989a) and Muinonen et al. (1989) carried out investigations of internal corner and trough retroreflection from randomly oriented crystal–shaped particles. Strong backscattering peaks were confirmed when the particles contained facets precisely perpendicular to each other, but the peaks were rather sensitive to small changes from the rectangular forms. Adding the electromagnetic phase in the computations would produce coherent backscattering, and thus sharpen the backscattering peaks due to trough and corner retroreflection. However, the assumption of a sufficient number of specific geometries among the surface particles may be correct in some cases, but seems unlikely to be generally true. The same difficulty rises when trying to explain the observed opposition effects with the glory phenomenon of light scattered by spheres (e.g., Franklin and Cook 1965). (Note, however, that the glory phenomenon can be partly understood as coherent backscattering.)

3. 2. LINEAR POLARIZATION

For negative linear polarization, Öhman (1955) suggested that the negative polarization could be due to trough retroreflection. He was aware of the fact that these kinds of surface structures would also lead to an opposition effect. Although the lunar samples do not support the widespread occurrence of rectangular troughs in the dust particles, this mechanism has a sound physical basis, is worthy of an examination, and may occasionally contribute to the observed linear polarization. The trough retroreflection mechanism for polarization was studied further by Muinonen (1989a) and Muinonen et al. (1989) in the context of crystal scattering with a result parallel to that for the intensity : negative polarization appeared for rectangular crystal particles, but disappeared for non–rectangular shapes.

Hopfield (1966) pointed out that negative linear polarization could be explained by a Sommerfeld diffraction mechanism (Born and Wolf 1970) connected to mutual shadowing. Behind the edge of a thin half–plane of material with high electric conductivity, to be taken as an opaque dust particle, the scattered electromagnetic field is negatively polarized. According to Hopfield, this component would then be reflected from underlying structures and be registered by the observer. The requirement of a specific diffraction and reflection geometry and material of high electric conductivity restricts the applicability of this interpretation. Hopfield emphasized that it is natural to search for a common explanation of the opposition effect and negative polarization. Lumme et al. (1980) made laboratory experiments to investigate the mechanism but were not convinced of its relevance in the context of solar system observations.

McCoyd (1967) related the negative polarization to transmission and total internal reflection, in an assumed coating on the scattering dust particles. Being only one–dimensional, the model is limited, though it leads to correct qualitative results. Furthermore, the assumption of a coating seems to make this interpretation unrealistic for solar system objects. Lumme (1979) introduced an explanation based on transmission through transparent dust particles. Negative polarization is also observed for opaque materials, and thus cannot be fully explained with transmitted wave components.

Steigman (1978) suggested a model, where second–order reflection in flat–bottomed cylindrical pits results in negative polarization. Steigman's model is one of those models, in which the specific geometric requirements decrease the applicability. However, it indicates the ambiguity of the inversion problem : knowing only the linear polarization it is impossible to derive unambiguous information about the surface conditions.

Wolff (1975) presented a second–order reflection explanation based on a hypothesis of selective shadowing of negatively and positively polarized rays. No mathematically consistent theoretical verification of this mechanism has been presented, though it has undoubtedly been the most popular one. The mechanism has been experimentally verified only for non–isotropic surfaces covered with parallel grooves (Geake et al. 1984). The extension to realistic isotropic surfaces is far from obvious. However, Geake et al. concluded that second–order interactions are needed to produce negative polarization and that diffraction effects may play a significant role when small–scale surface textures are involved.

Gehrels (1977) supported Wolff's mechanism, and presented a geometric flat-bottomed pit model that, according to him, resulted in a net negative polarization near the backward direction. Muinonen et al. (1992) carried out some simple geometric calculations for shadowing by cylindrical pits and bumps of varying height-to-width ratios, and concluded that Wolff's mechanism was at least of an order of magnitude too weak to be significant.

Kolokolova (1990) offered complicated multiparameter geometric considerations for random rough surfaces that seemed to support Wolff's mechanism. In contrast, Kolokolova et al. (1993) analyzed the experiments by Geake and Geake (1990) and concluded that Wolff's mechanism could not explain the linear polarization of subwavelength particles but, rather, the coherent backscattering mechanism had to be invoked to explain the measurements.

Using Wolff's mechanism, Shkuratov (1982) and Shkuratov et al. (1988b) modeled negative polarization for particulate media and for opaque rough surfaces. Theoretical calculations by Peltoniemi et al. (1989) for stochastically rough particles and by Peltoniemi and Lumme (1992) for closely packed scattering media do not confirm a purely geometric explanation for the negative polarization. Moreover, theoretical studies of rough surface shadowing (Muinonen et al. 1990) show that, e.g., at normal incidence, a surface has to be extremely rough to eliminate the emergence of rays into small phase angles. The supporters of Wolff's shadowing mechanism are invited to present solid theoretical and experimental evidence for the mechanism—currently, no such evidence is available in the literature.

Lumme and Bowell (1985) studied light scattering by interplanetary dust, and derived the single particle phase functions for the parallel and perpendicular polarizations. Comparing these phase functions with the observations of planetary regoliths, they then concluded that the linear polarization of the interplanetary dust particles strongly resembled that of the typical regolith particles. Light scattering by cometary dust (e.g., Levasseur-Regourd et al. 1993) further adds to the complementary set of scattering problems in the solar system.

3. 3. LABORATORY MEASUREMENTS

Oetking's (1966) laboratory experiments indicated that most terrestrial substances, including standard diffusing surfaces, showed a prominent rise in reflectivity near the backscattering direction. He verified, for example, that fine micron-sized particles of aluminum oxide developed much higher backscattering peaks than coarser particles of the same material. Extensive experimental polarization studies of simulated lunar surfaces have been carried out by Egan (1967), Dollfus et al. (1971), Dollfus and Titulaer (1971), and by Bowell et al. (1972), who include particularly good documentation of the measured samples.

4. Coherent Backscattering Mechanism

It is useful to summarize the present state of research in the field of electromagnetic scattering by random media. There are several reasons why the existing theoretical treatments are not directly applicable to light scattering by solar system dust particles. The dust particles, though plausibly exhibiting wavelength-sized irregularities, are themselves thought to be very large compared to the wavelength. Also, there are dust particles of various materials in regoliths, and the particles themselves can be inhomogeneous aggregates. The regolith dust is probably closely packed with a volume density about 40%. Thus solar system regoliths are multi–scale inhomogeneous random media with strongly fluctuating electromagnetic conditions (i.e., having strong contrasts in the electric permittivity).

Coherent backscattering has been studied mainly for two kinds of scattering media : rough surfaces that are interfaces of two electromagnetically homogeneous media; and volumes of discrete randomly distributed scatterers. At the moment, both lines of research are relevant for light scattering by solar system dust particles. In particular, there has been recent activity in studies of rough surface scattering, which can be partly understood by the very fundamental nature of rough interfaces in all scattering problems.

In addition to coherent backscattering by rough surfaces and volumes of particles, there is coherent backscattering by single particles, which, excluding Mie scattering for a moment, has not been much studied. Such backscattering is of utmost relevance in the solar system context, where the irregular particles are mostly much larger than the wavelength of light. The single particles can sometimes be taken as clusters of subparticles; the wavelength–sized or smaller subparticles are probably not uniformly distributed, for example, in planetary regoliths. Ultimately, one needs to take ensemble averages (e.g., orientation averages) of individual particles and thus approaches the volume scattering problem. The basic mechanism itself can nevertheless be due to the single particles themselves, and not due to the interactions between these particles.

There are nomenclature problems when one discusses radiative transfer and electromagnetic theories in the same context. Examples of such difficulties are "homogeneous media" and "rough surfaces". The reader is urged to consult the literature to overcome possible nomenclature problems.

4. 1. ROUGH SURFACES

The profound experiments by O'Donnell and Mendez (1987) confirmed the existence of the backward enhancement for random rough surfaces. They also measured negative linear polarization, although they used different terminology and were not familiar with the observations of solar system dust particles. They noticed that, for normal incidence on an isotropic surface, the secondary maxima near the backward direction behaved differently in the planes parallel and perpendicular to the electric vector of the incident field. Experiments have also been carried out by Gu et al. (1989). They compared the results with the theory by McGurn et al. (1985) and by Celli et al. (1985), which stated that, in certain conditions, the backward enhancement can be related to the localization of surface polaritons by the random surface roughness. The measured backscattering enhancement exceeded the theoretical predictions.

Theoretical studies of scattering by random rough surfaces were further carried out by Bahar and Fitzwatēr (1989) and by Soto–Crespo and Nieto–Vesperinas (1989). Bahar and Fitzwater related the backward enhancement to first–order scattering and, on the other hand, Michel et al. (1989) showed that the phenomenon rose from multiple scattering. The approaches by Soto–Crespo and Nieto–Vesperinas and by Michel et al. are restricted to one–dimensional random rough surfaces.

Ishimaru (1990) pointed out that two different enhancement phenomena are mainly observed for rough surfaces. The enhancement seems to occur both for large heights and steep slopes, and also for small height variations. The latter can be understood with the help of surface wave modes, but the former needs more examination. Enhanced backscattering through a deep random phase screen has been investigated by Jakeman (1988) and by Tapster et al. (1989). The book edited by Nieto–Vesperinas and Dainty (1990) summarizes open questions of electromagnetic scattering in volumes and surfaces.

Investigations of rough surface backscattering have been continued by, e.g., Kim et al. (1990), Soto–Grespo et al. (1990), and Bruce and Dainty (1991ab). McGurn (1990) provides a review of rough surface scattering with an emphasis on Anderson localization of surface polaritons.

4. 2. SINGLE PARTICLES

A fact that has been almost ignored in recent studies of coherent backscattering is the coherent nature of backscattering by spherical particles (van de Hulst 1957). van de Hulst treats scattering by spheres in the geometric optics approximation, including the electromagnetic phase, thus accounting for coherent backscattering. In general, it is impossible to distinguish strictly between single particles and aggregates of single particles, and thus some many–particle calculations are reviewed in this section.

How do irregular single particles exhibit coherent backscattering? The mechanism of cyclic passage of waves inside an individual scatterer and the subsequent coherent backscattering is indicated for an arbitrary particle by Muinonen (1989a), when introducing the Kirchhoff approximation (and its modified version) for light scattering by single particles. Preliminary results for tetrahedral crystals in the full Kirchhoff approximation (Muinonen 1991) showed weak signs of coherent backscattering in both intensity and polarization.

The simplest possible electromagnetic scattering problem of two particles is that of two dipole (or Rayleigh) particles, and was solved analytically by Muinonen (1989b). The negative linear polarization was verified to accompany the backscattering enhancement in second–order scattering, and the qualitative explanation for the coherent backscattering mechanism for the negative polarization was given. However, due to the small scattering cross section, the coherent second–order backscattering peak and negative polarization do not show up in total angular scattering in the validity region of the electric dipole approximation.

Lindell et al. (1991) and Muinonen et al. (1991) considered scattering by small objects close to an interface using the Exact Image Theory formulation. A generalized Green function was derived to account for the interface close to the scattering particle, assumed to be small compared to the wavelength. Intuitively, it is clear that the multiple scattering contributions increase when the second single dipole scatterer in Muinonen (1989b) is replaced with a dielectric half–space. Indeed, the co–existence of the negative polarization and the backscattering enhancement in the total diffuse scattered field component are confirmed through simulations of a rough surface covered with dipolar inhomogeneities (Figure 2). This averaging is only a rough approach to simulate natural conditions, and is not to be understood as a model for solar system observations. In further verifications, the first–order Fresnel reflection component should also be taken into account.

Muinonen and Lumme (1991) showed results for coherent backscattering from two spherically curved surface elements, and demonstrated how the backscattering peak and negative polarization vanished when the electromagnetic phase was excluded from the calculations. In general, the calculations by Muinonen et al. can be criticized for not being readily applicable to the interpretation of solar system observations.

Lumme and Rahola (1994; see also Hage 1991) computed scattering by porous single particles using an improved numerical method for the discrete–dipole approximation by Purcell and Pennypacker (1973). In searching for coherent backscattering phenomena, they could not conclude whether they saw backscattering enhancement or negative linear polarization caused by coherent backscattering. Since the discrete–dipole approximation is well suited for studies of interference phenomena, I encourage the continuation of search efforts.

4. 3. VOLUMES OF PARTICLES

Pre–1982 backscattering enhancement studies have been well reviewed by Kravtsov and Saichev (1982). They discussed the fluctuation effects in double passage of waves propagating through the same inhomogeneities of a random medium, touching also the topic of coherent backscattering. The mechanism of coherent backscattering was presented by Watson (1969), who studied multiple scattering of electromagnetic waves in an extended underdense plasma. Watson emphasized that for calculating radar backscattering, the coherence of multiply scattering waves propagating in opposite directions had to be taken into account. de Wolf (1971) brought up the phenomenon in his treatment of electromagnetic scattering by an extended dielectric turbulent medium.

The coherent backscattering phenomenon has been experimentally verified in controlled light scattering measurements for densely distributed latex and polystyrene microspheres by Kuga and Ishimaru (1984), van Albada and Lagendijk (1985), and by Wolf and Maret (1985). In these experiments, the widths of the backscattering peaks were less than 1°. After these verifications, the interest in the phenomenon has increased rapidly in the field of electromagnetic scattering. van Albada and Lagendijk, and Wolf and Maret investigated the coherent backscattering phenomenon in connection to weak localization of photons in disordered dielectric media. This field of research is reviewed by MacKintosh and John (1988), and by McGurn (1990).

Tsang and Ishimaru (1984, 1985) interpreted the measurements on the basis of the so–called diffusion approximation. They used a scalar diagrammatic theory for randomly distributed, isotropically scattering point–like particles, and found that the angular width of the peak was related to the transport length in the medium. Also, they concluded that multiple scattering contributions from orders higher than the second play an important role in the peak formation. Although this approach clarifies the essential physics in coherent backscattering, the analysis can be criticized for its scalar nature, and for the non-realistic assumption of isotropic single scattering. The latter assumption appears analogous to Chandrasekhar's work on radiative transfer (1960). Later, the approach was extended to non-isotropic scatterers (Ishimaru and Tsang 1988).

Ma et al. (1988) treated electromagnetic scattering by discrete random medium in a distorted Born approximation using transition-matrix (T-matrix) formalism, and emphasized that the scatterers in the aforementioned experiments were large or comparable to the wavelength, a fact which should be taken into account. The formalism enabled the consideration of the size and orientation distributions and physical properties of the scatterers (Varadan et al. 1983, Varadan et al. 1987). They interpreted the measurements by van Albada and Lagendijk (1985) and indicated the CPU time problems encountered in the numerical analysis. Also, they brought up the fundamental question of whether the applied pair statistics satisfactorily approximate the real higher order statistics in a discrete random medium.

Bahar and Fitzwater (1988) studied the backscattering enhancement from random distributions of finitely conducting spherical and random rough particles using radiative transfer equation and the so-called full-wave approach. In their treatise on random medium, they related the backward enhancement to single scattering from either smooth or rough spheres. Evidently, the full-wave approach leads to interesting results.

The coherent backscattering mechanism has been recently investigated by Shkuratov (1988b, 1989), Shkuratov et al. (1989), and Shkuratov and Muinonen (1991) using both scalar and vector formulations of radiative transfer and including the electromagnetic phase. One of their goals has been to derive simple formulae describing the main features of the observations and experiments. However, they can be criticized for just this philosophy : over-simplification may yield misleading results.

Coherent backscattering was invoked to explain the unusual radar characteristics of outer planet satellites by Hapke (1990). Hapke et al. (1993) carried out a circular polarization experiment to verify whether the lunar opposition effect was caused by shadow-hiding or coherent backscattering. They studied the linear polarization and circular polarization ratios, defined as the ratios of the intensity scattered with the sense of polarization not expected upon reflection from a smooth surface to the intensity scattered with the sense that would be expected. They noticed that the linear polarization ratio decreased as the circular polarization ratio increased for the lunar samples toward the backward direction or opposition. They then stated that the experiment provided "unequivocal proof" that coherent backscattering was dominating the opposition effects for the measured samples. However, the lack of quantitative theoretical interpretation of their laboratory results renders their proof speculative. Their experimental results do hint strongly at a multiple scattering explanation for the opposition phenomena, but the results await a quantitative theoretical explanation just as the lunar opposition effect and negative polarization.

Mishchenko and Dlugach (1991) analyzed the amplitude of the coherent backscattering peak using a vector multiple-scattering theory. They assumed that the enhancement was an exact multiple of two compared to the radiative transfer contribution from scattering orders higher than the first. The amplitude of the opposition effect was then defined as the ratio of the total coherent and incoherent scattering to the total incoherent scattering. The amplitude for the opposition effect varied from about 1.3 to about 1.6, resembling the opposition effects of solar system objects.

The copolarized and depolarized backscattering enhancement factors were studied by Mishchenko (1991) for Rayleigh scatterers and spherical latex particles in water. Mishchenko does not calculate the angular dependence of the coherent backscattering peak but assumes the shape of the peak from other contexts (e.g., Ozrin 1992ab). This "shortcut" is a deficiency in his otherwise thorough work that derives from radiative transfer theory through the aforementioned assumption of the amplitude of the coherent backscattering peak.

Mishchenko and Dlugach (1992) studied the opposition effect of Saturn's rings in the context of coherent backscattering. Mishchenko (1993) then studied the negative linear polarization of Saturn's rings, and concluded that coherent backscattering cannot explain the entire range of negative polarization, but rather, mechanisms such as Wolff's shadowing mechanism (Wolff 1975) had to be incorporated.

Mishchenko and Dlugach (1993) studied the coherent backscattering mechanism for the E–type asteroids (44) Nysa and (64) Angelina (see also Shkuratov and Muinonen 1991), and gave a brief review of the present status of coherent backscattering studies. Kolokolova et al. (1993) supported Wolff's mechanism as a general explanation for the negative polarization, but stated that the coherent backscattering mechanism needed to be introduced to explain the negative polarization of samples of subwavelength–sized particles.

Peters (1992) has proposed a vector formulation accounting for polarization and absorption effects and small or large scatterers. His work is a generalization of the studies by MacKintosh and John (1988). However, Peters was bound to introduce approximations that were later criticized by Mishchenko and Dlugach (1993), and the validity of the approximations still need to be verified via more thorough theoretical and experimental work.

Ozrin (1992ab) developed a vector formulation for coherent backscattering by samples of Rayleigh scatterers. He considered a semi–infinite random medium composed of nonabsorbing point–like scatterers. For normally incident linearly polarized light, he found the angular distribution of the backward scattered intensity. However, the deficiency in Ozrin's work was his assumption that the diffuse background radiation was independent of the phase angle near the backward direction.

Recently, Shkuratov et al. reported strong negative polarization for samples that they considered relevant for the interpretation of the solar system observations (Shkuratov and Akimov 1987; Shkuratov et al. 1987, 1988a). Furthermore, Shkuratov and Opanasenko (1992) and Shkuratov et al. (1992) reported studies of the lunar negative and positive polarization, accompanied by supporting laboratory simulations. I am currently involved in designing and constructing a new scatterometer for experiments, in particular, close to the backscattering direction (Piironen and Muinonen 1993).

5. Discussion

Unambiguous quantitative inversions of solar system scattering problems are usually not possible, which is evident from the fact that many of the mechanisms for the opposition effect or negative polarization described above could work in special cases. However, the inversion problems can be approached through an analysis that attempts to make as few assumptions as possible. Adopting this approach for the backscattering studies, we should concentrate our research on the coherent backscattering mechanism and, further, on its relation to mutual shadowing.

Many theoretical aspects of the coherent backscattering phenomenon are still not well understood. For example, the coherent internal backscattering by single particles cannot be omitted in the case of bright materials. This could be investigated by treating the scattering by stochastically rough particles in the Kirchhoff approximation. Examinations of second–order external reflection from random rough surfaces should be continued, which would give further insight, especially, into the origin of the negative polarization.

The Exact Image Theory (e.g., Lindell et al. 1991) can also be applied to scatterers that are not necessarily small compared to the wavelength. The main advantage of the method is that it uses the already existing solutions for isolated objects. The possible inhomogeneity of the individual dust particles could be taken into account, perhaps, by applying the Exact Image Theory to scatterers that are inhomogeneities of an otherwise homogeneous half–space medium. These kinds of studies are needed in the investigation of light scattering by solar system dust particles, since it is not well understood where the main contributions to coherent backscattering come from. Investigations could also be continued by analyzing a three–particle scattering problem in the electric dipole approximation. This would give insight into the intensity and polarization effects in third–order scattering. These solutions could also be important for further improvements of the discrete–dipole approximation (e.g., Lumme and Rahola 1994).

Future laboratory experiments should first be concentrated on controlled measurements in which all the relevant parameters of the rough or particulate surface are known. For example, in the case of rough surfaces, detailed knowledge of the surface height statistics is necessary for theoretical treatments. Unfortunately, a part of the laboratory work in the past must be classified as pure laboratory observations rather than as controlled experiments. It is the intelligent synthesis of theory and experiments that finally leads us to understand the opposition effect and negative linear polarization.

Acknowledgements

I am grateful to Kari Lumme, Edward Bowell, Bruce Hapke and Anny–Chantal Levasseur–Regourd for their impact on the original manuscript. I would like to thank Yurij Shkuratov for guiding me to the treasures of Sir Isaac Newton's Opticks.

References

Akimov, L. A. : 1980, "O prirode oppozicionnogo effekta". *Vestnik Kharkovskogo Univ.* **204**, 3–12.

Bahar, E., and Fitzwater, M. A. : 1988, "Enhancement of backscattered diffuse specific intensities from random distributions of finitely conducting particles with rough surfaces". *J. Opt. Soc. Am. A* **5**, 89–98.

Bahar, E., and Fitzwater, M. A. : 1989, "Depolarization and backscatter enhancement in light scattering from random rough surfaces : comparison of full–wave theory with experiment". *J. Opt. Soc. Am. A* **6**, 33–43.

Barabanenkov, Yu. N., Kravtsov, Yu. A., Ozrin, V. D., and Saichev, A. I. : 1991, "Enhanced backscattering in optics". In *Progress in Optics XXIX* (E. Wolf, Ed.), 67–197, Elsevier.

Barabashev, N. P. : 1922, "Bestimmung der Erdalbedo und des Reflexionsgesetzes für die Oberfläche der Mondmeere. Theorie der Rillen". *Astron. Nachr.* **217**, 445–452.

Bobrov, M. S. : 1940, "On the physical interpretation of the phase curve of Saturn's rings". *Astron. Zh.* **17**, 1–8.

Bobrov, M. S. : 1961, "A generalisation of the theory of the shadow effect on Saturn's rings for the case of particles of unequal size". *Astron. Zh.* **38**, 669–680.

Born, M., and Wolf, E. : 1970, *Principles of Optics* (4th edition), Pergamon.

Bowell, E., Dollfus, A., and Geake, J. E. : 1972, "Polarimetric properties of the lunar surface and its interpretation. Part 5 : Apollo 14 and Luna 16 samples". In *Proc. 3rd Lunar Sci. Conf.* (Supplement 3, Geochimica et Cosmochimica Acta) **3**, 3103–3126, The M.I.T. Press.

Bowell, E., Hapke, B., Lumme, K., Harris, A. W., Domingue, D., and Peltoniemi, J. I. : 1989, "Application of photometric models to asteroids". In *Asteroids II* (R. P. Binzel, T. Gehrels, and M. Matthews, Eds.), 524–556, Univ. of Arizona Press.

Brown, R. H., and Cruikshank, D. P. : 1983, "The Uranian satellites : surface compositions and opposition brightness surges". *Icarus* **55**, 83–92.

Bruce, N. C., and Dainty, J. C. : 1991a, "Multiple scattering from random rough surfaces using the Kirchhoff approximation". *Journal of Modern Optics* **38**, 579–590.

Bruce, N. C., and Dainty, J. C. : 1991b, "Multiple scattering from rough dielectric and metal surfaces using the Kirchhoff approximation". *Journal of Modern Optics* **38**, 1471–1481.

Celli, V., Maradudin, A. A., Marvin, A. M., and McGurn, A. R. : 1985, "Some aspects of light scattering from a randomly rough metal surface". *J. Opt. Soc. Am. A* **2**, 2225–2239.

Chandrasekhar, S. : 1960, *Radiative Transfer*, Dover.

de Wolf, D. A. : 1971, "Electromagnetic reflections from an extended turbulent medium : cumulative forward–scatter single backscatter approximation". *IEEE Trans. Antennas Propag.* **19**, 254–262.

Dollfus, A. : 1975, "Optical polarimetry of the Galilean satellites of Jupiter". *Icarus* **25**, 416–431.

Dollfus, A., and Bowell, E. : 1970, "Polarimetric properties of the lunar surface and its interpretation. Part I. Telescopic observations," *Astron. Astrophys.* **10**, 29–53.

Dollfus, A., and Titulaer, C. : 1971, "Polarimetric properties of the lunar surface and its interpretation. Part III. Volcanic samples in several wavelengths". *Astron. Astrophys.* **12**, 199–209.

Dollfus, A., Bowell, E., and Titulaer, C. : 1971, "Polarimetric properties of the lunar surface and its interpretation. Part II. Terrestrial samples in orange light." *Astron. Astrophys.* **10**, 450–466.

Domingue, D. L., Hapke, B. W., Lockwood, G. W., and Thompson, D. T. : 1991, "Europa's phase curve : implications for surface structure". *Icarus* **90**, 30–42.

Egan, W. G. : 1967, "Polarimetric measurements of simulated lunar surfaces". *J. Geophys. Res.* **72**, 3233–3245.

Esposito, L. W. : 1979, "Extensions to the classical calculation of the effect of mutual shadowing in diffuse reflection". *Icarus* **39**, 69–80.

Franklin, F. A., and Cook, A. F. : 1965, "Optical properties of Saturn's rings. II. Two-color phase curves of the two bright rings". *Astron. J.* **70**, 704–720.

Geake, J. E., and Geake, M. : 1990, "A remote-sensing method for sub-wavelength grains on planetary surfaces by optical polarimetry". *Mon. Not. Royal Astron. Soc.* **245**, 46–55.

Geake, J. E., Geake, M., and Zellner, B. H. : 1984, "Experiments to test theoretical models of the polarization of light by rough surfaces". *Mon. Not. R. Astr. Soc.* **210**, 89–112.

Gehrels, T. : 1956, "Photometric studies of asteroids. V. The light-curve and phase function of 20 Massalia". *Ap. J.* **123**, 331–338.

Gehrels, T. : 1977, "The physical basis of the polarimetric method for deriving asteroid albedos". In *The Interrelated Origin of Comets, Asteroids, and Meteorites* (A. H. Delsemme, Ed.), 253–256, Univ. of Toledo Publs.

Gehrels, T., Coffeen, T., and Owings, D. : 1964, "Wavelength dependence of polarization. III. The lunar surface". *Astron. J.* **69**, 826–852.

Gu, Z.-H., Dummer, R. S., Maradudin, A. A., and McGurn, A. R. : 1989, "Experimental study of the opposition effect in the scattering of light from a randomly rough metal surface". *Appl. Opt.* **28**, 537–543.

Hage, J. : 1991, "The optics of porous particles and the nature of comets". PhD-thesis, University of Leiden.

Hapke, B. : 1963, "A theoretical photometric function for the lunar surface". *J. Geophys. Res.* **68**, 4571–4586.

Hapke, B. : 1966, "An improved theoretical lunar photometric function". *Astron. J.* **71**, 333–339.

Hapke, B. : 1981, "Bidirectional reflectance spectroscopy : 1. Theory". *J. Geophys. Res.* **86**, 3039–3054.

Hapke, B. : 1982, "The Lumme-Bowell photometric parameters : reality or fantasy?" *Bull. Amer. Astron. Soc.* **14**, 726.

Hapke, B. : 1984, "Bidirectional reflectance spectroscopy : 3. Correction for macroscopic roughness". *Icarus* **59**, 41–59.

Hapke, B. : 1986, "Bidirectional reflectance spectroscopy : 4. The extinction coefficient and the opposition effect". *Icarus* **67**, 264–280.

Hapke, B. : 1990, "Coherent backscatter and the radar characteristics of outer planet satellites". *Icarus* **88**, 407–417.

Hapke, B., and van Horn, H. : 1963, "Photometric studies of complex surfaces, with applications to the moon". *J. Geophys. Res.* **68**, 4545–4570.

Hapke, B., and Wells, E. : 1981, "Bidirectional reflectance spectroscopy : 2. Experiments and observations". *J. Geophys. Res.* **86**, 3055–3060.

Hapke, B., Nelson, R. M., and Smythe, W. D. : 1993, "The opposition effect of the Moon : the contribution of coherent backscatter". *Science* **260**, 509–511.

Harris, A. W., and Young, J. W. : 1988, "Two dark asteroids with very small opposition effects". *Lunar Planet. Sci. XIX* **18**, 447–448.

Harris, A. W., and Young, J. W. : 1989, "Asteroid lightcurve observations from 1979–1981". *Icarus* **81**, 314–364.

Harris, A. W., Young, J. W., Contreiras, L., Dockweiler, T., Belkora, L., Salo, H., Harris, W. D., Bowell, E., Poutanen, M., Binzel, R. P., Tholen, D. J., and Wang, S. : 1989, "Phase relations of high albedo asteroids : the unusual opposition brightening of 44 Nysa and 64 Angelina". *Icarus* **81**, 365–374.

Herschel, J. : 1847, "Results of astronomical observations made at the Cape of Good Hope". 353–374 (London, 1847).

Hopfield, J. J. : 1966, "Mechanism of lunar polarization". *Science* **151**, 1380–1381.

Hämeen–Anttila, K. A., and Vaaraniemi, P. : 1975, "A theoretical photometric function of Saturn's rings". *Icarus* **25**, 470–478.

Irvine, W. M. : 1966, "The shadowing effect in diffuse reflection". *J. Geophys. Res.* **71**, 2931–2937.

Irvine, W. M., Muinonen, K., and Lumme, K. : 1988, "Is the mutual shadowing explanation for the opposition effect of Saturn's rings still valid?" *Bull. Amer. Astron. Soc.* **20**, 853.

Ishimaru, A. : 1990, "Experimental and theoretical studies on enhanced backscattering from scatterers and rough surfaces". In *Scattering in Volumes and Surfaces* (M. Nieto–Vesperinas and J. C. Dainty, Eds.), 1–15, North Holland.

Ishimaru, A., and Tsang, L. : 1988, "Backscattering enhancement of random discrete scatterers of moderate sizes". *J. Opt. Soc. Am. A* **5**, 228–236.

Jakeman, E. : 1988, "Enhanced backscattering through a deep random phase screen". *J. Opt. Soc. Am. A* **5**, 1638–1648.

Johnson, P. E., Kemp, J. C., King, R., Parker, T. E., and Barbour, M. S. : 1980, "New results from optical polarimetry of Saturn's rings". *Nature* **283**, 146–149.

Kim, M.–J., Dainty, J. C., Friberg, A. T., and Sant, A. J. : 1990, "Experimental study of enhanced backscattering from one– and two–dimensional random rough surfaces". *J. Opt. Soc. Am. A* **7**, 569–577.

Kolokolova, L. O. : 1990, "Dependence of polarization on optical and structural properties of the surfaces of atmosphereless bodies". *Icarus* **84**, 305–314.

Kolokolova, L. O., Mishchenko, M. I., and Wolff, M. : 1993, "On the negative polarization of light scattered by subwavelength regolithic grains". *Mon. Not. R. Astron. Soc.* **260**, 550–552.

Kravtsov, Y. A., and Saichev, A. I. : 1982, "Effects of double passage of waves in randomly inhomogeneous media". *Sov. Phys. Usp.* **25**, 494–508.

Kuga, Y., and Ishimaru, A. : 1984, "Retroreflectance from a dense distribution of spherical particles". *J. Opt. Soc. Am. A* **1**, 831–835.

Levasseur–Regourd, A. C., Dumont, R., and Renard, J. B. : 1990, "A comparison between polarimetric properties of cometary dust and interplanetary dust particles". *Icarus* **86**, 264–272.

Lindell, I. V., Sihvola, A. H., Muinonen, K. O., and Barber, P. W. : 1991, "Scattering by a small object close to an interface. I. Exact Image Theory formulation". *J. Opt. Soc. Am. A* **8**, 472–476.

Lumme, K. : 1971, "Interpretation of the light curves of some non–atmospheric bodies in the solar system". *Astrophys. Space Sci.* **13**, 219–230.

Lumme, K. : 1979, "A model for the polarization of atmosphereless bodies". *Bull. Amer. Astron. Soc.* **11**, 562.

Lumme, K., and Bowell, E. : 1981a, "Radiative transfer in the surfaces of atmosphereless bodies. I. Theory". *Astron. J.* **86**, 1694–1704.

Lumme, K., and Bowell, E. : 1981b, "Radiative transfer in the surfaces of atmosphereless bodies. II. Interpretation of phase curves". *Astron. J.* **86**, 1705–1721.

Lumme, K., and Bowell, E. : 1982, "A reply to Hapke's criticism of the Lumme–Bowell photometric theory". *Bull. Amer. Astron. Soc.* **14**, 726.

Lumme, K., and Bowell, E. : 1985, "Photometric properties of zodiacal light particles". *Icarus* **62**, 54–71.

Lumme, K., and Irvine, W. M. : 1982, "Radiative transfer in the surfaces of atmosphereless bodies. III. Interpretation of lunar photometry". *Astron. J.* **87**, 1076–1082.

Lumme, K., and Rahola, J. : 1994, "Light scattering by porous dust particles in the discrete–dipole approximation". *Ap. J.*, in press.

Lumme, K., Bowell, E., and Zellner, B. : 1980, "The negative polarization of light scattered from rough surfaces seems to be largely due to diffraction". *Bull. Amer. Astron. Soc.* **12**, 663.

Lumme, K., Peltoniemi, J. I., and Irvine, W. M. : 1990, "Diffuse reflection in stochastically bounded semi-infinite media". ***Trans. Theory and Stat. Phys.*** **19**(3–5), 317–332.

Lyot, B. : 1929, "Recherches sur la polarisation de la lumière des planètes et de quelques substances terrestres". ***Ann. Obs. Paris*** **8**(1), 1–161.

Ma, Y., Varadan, V. V., and Varadan, V. K. : 1988, "Scattered intensity of a wave propagating in a discrete random medium". ***Appl. Opt.*** **27**, 2469–2477.

MacKintosh, F. C., and John, S. : 1988, "Coherent backscattering of light in the presence of time–reversal–noninvariant and parity–nonconserving media". ***Phys. Rev. B*** **37**, 1884–1897.

Markov, A. : 1924, "Les particularités dans la réflexion de la lumière par la surface de la Lune". ***Astron. Nachr.*** **221**, 65–78.

McCoyd, G. C. : 1967, "Polarization properties of a simple dielectric rough–surface model". ***J. Opt. Soc. Am.*** **57**, 1345–1350.

McGurn, A. R. : 1990, "Enhanced retroreflectance effects in the reflection of light from randomly rough surfaces". ***Surf. Sci. Rep.*** **10**, 357–410.

McGurn, A. R., Maradudin, A. A., and Celli, V. : 1985, "Localization effects in the scattering of light from a randomly rough crating". ***Phys. Rev. B*** **31**, 4866–4871.

Michel, T., Maradudin, A. A., and Mendez, E. R. : 1989, "Enhanced backscattering of light from a non–Gaussian random metal surface". ***J. Opt. Soc. Am. B*** **6**, 2438–2446.

Mikhail, J. S. : 1970, "Colour variations with phase of selected regions of the lunar surface". ***The Moon*** **2**, 167–201.

Mishchenko, M. I. : 1992, "Polarization characteristics of the coherent backscatter opposition effect". ***Earth, Moon, Planets*** **58**, 127–144.

Mishchenko, M. I. : 1993, "On the nature of the polarization opposition effect exhibited by Saturn's rings". ***Ap. J.*** **411**, 351–361.

Mishchenko, M. I., and Dlugach, J. M. : 1991, "The amplitude of the opposition effect due to weak localization of photons in discrete disordered media". ***Astrophys. Space Sci.*** **189**, 151–154.

Mishchenko, M. I., and Dlugach, J. M. : 1992, "Can weak localization of photons explain the opposition effect of Saturn's rings?" ***Mon. Not. Royal Astron. Soc.*** **254**, 15p–18p.

Mishchenko, M. I., and Dlugach, J. M. : 1993, "Coherent backscatter and the opposition effect for E–type asteroids". ***Planet. Space Sci.*** **41**, 173–181.

Muinonen, K. : 1989a, "Scattering of light by crystals : a modified Kirchhoff approximation". ***Appl. Opt.*** **28**, 3044–3050.

Muinonen, K. : 1989b, "Electromagnetic scattering by two interacting dipoles". In ***Proc. 1989 URSI Electromagnetic Theory Symp.*** (Stockholm, Sweden), 428–430.

Muinonen, K. : 1990, "Light scattering by inhomogeneous media : backward enhancement and reversal of polarization". PhD–thesis, University of Helsinki.

Muinonen, K. : 1991, "Scattering of light by crystals : a full Kirchhoff approximation". ***Optical Society of America, 1991 Annual Meeting, Technical Digest*** (San Jose, California, U.S.A.), 119.

Muinonen, K., and Lumme, K. : 1991, "Light scattering by solar system dust : the opposition effect and the reversal of linear polarization". In ***IAU Colloquium 126, Origin and Evolution of Interplanetary Dust*** (A.–C. Levasseur–Regourd and H. Hasekawa, Eds.), 159–162, Kluwer Academic Press.

Muinonen, K., Lumme, K., Peltoniemi, J. I., and Irvine, W. M. : 1989, "Light scattering by randomly oriented crystals". ***Appl. Opt.*** **28**, 3051–3060.

Muinonen, K., Lumme, K., Zhukov, B., Peltoniemi, J. I., Kaasalainen, M. and Irvine, W. M. : 1990, "Statistical photoclinometric methods for the analysis of surface topography". Space Research Institute, Academy of Sciences, USSR, Preprint 1631, 1–16.

Muinonen, K. O., Sihvola, A. H., Lindell, I. V., and Lumme, K. A. : 1991, "Scattering by a small object close to an interface. II. Study of backscattering". ***J. Opt. Soc. Am. A*** **8**, 477–482.

Muinonen, K., Peltoniemi, J. I., and Lumme, K. : 1992, "Is the negative linear polarization of solar system dust caused by shadowing or coherent backscattering?" In ***Workshop on Polarization III, Light Scattered by Irregular Dust Grains, with Emphasis on Cometary and Interplanetary Dust*** (Paris, France), 4pp.

Müller, G. : 1893, "Helligkeitsbestimmungen der grossen Planeten und einiger Asteroiden". ***Publ. Astrophys. Obs. Potsdam Nr. 30*, 8**, 193–389.

Newton, I. : 1952, ***Opticks***, Dover. (Based on the Fourth Edition, London, 1730.)

Nieto–Vesperinas, M., and Dainty, J. C., Eds. : 1990, ***Scattering in Volumes and Surfaces***, North–Holland.

O'Donnell, K. A., and Mendez, E. R. : 1987, "Experimental study of scattering from characterized random surfaces". ***J. Opt. Soc. Am. A* 4**, 1194–1205.

Oetking, P. : 1966, "Photometric studies of diffusely reflecting surfaces with applications to the brightness of the moon". ***J. Geophys. Res.* 71**, 2505–2513.

Ozrin, V. D. : 1992a, "Exact solution for coherent backscattering from a semi–infinite random medium of anisotropic scatterers". ***Phys. Lett. A* 162**, 341–345.

Ozrin, V. D. : 1992b, "Exact solution for coherent backscattering of polarized light from a random medium of Rayleigh scatterers". ***Waves Random Med.* 2**, 141–164.

Öhman, Y. : 1955, "A tentative explanation of the negative polarization in diffuse reflection". ***Stockholm Obs. Ann.* 18**, 8, 1–10.

Pellicori, S. F. : 1971, "Polarizing properties of pulverized materials with special reference to the lunar surface". ***Appl. Opt.* 10**, 270–285.

Peltoniemi, J. I. : 1993, "Light scattering in planetary regoliths and cloudy atmospheres". PhD–thesis, University of Helsinki.

Peltoniemi, J. I., and Lumme, K. : 1992, "Light scattering by closely packed particulate media". ***J. Opt. Soc. Am. A* 9**, 1320–1326.

Peltoniemi, J. I., Lumme, K., Muinonen, K., and Irvine, W. M. : 1989, "Scattering of light by stochastically rough particles". ***Appl. Opt.* 28**, 4088–4095.

Peters, K. J. : 1992, "Coherent–backscatter effect : a vector formulation accounting for polarization and absorption effects and small or large scatterers". ***Phys. Rev. B* 46**, 801–812.

Piironen, J., and Muinonen, K. : 1993, "Backscattering of light by snow : laboratory experiments". ***IAU Symposium 160 : Asteroids, Comets, Meteors 1993*** (Belgirate, Italy), 248.

Purcell, E. M., and Pennypacker, C. R. : 1973, "Scattering and absorption of light by nonspherical dielectric grains". ***Ap. J.* 186**, 705–714.

Rougier, A. : 1933, "Photométrie photoélectrique globale de la Lune". ***Ann. Obs. Strasbourg* 2**, 205–339.

Russell, H. N. : 1916a, "The stellar magnitudes of the sun, moon, and planets". ***Ap. J.* 43**, 103–129.

Russell, H. N. : 1916b, "On the albedo of the planets and their satellites". ***Ap. J.* 43**, 173–196.

Scaltriti, F., and Zappalà, V. : 1980, "The similarity of the opposition effect among asteroids". ***Astron. Astrophys.* 83**, 249–251.

Seeliger, H. von : 1887, "Zur Theorie der Beleuchtung der grossen Planeten, insbesondere des Saturn". ***Abh. Bayer. Akad. Wiss. Math. Naturwiss.***, Kl. **16**, 405–516.

Shkuratov, Yu. G. : 1982, "A model for the negative polarization of light scattered by cosmic bodies deprived of atmospheres". ***Astron. Zh.* 59**, 817–822.

Shkuratov, Yu. G. : 1988a, "Shadow component of phase brightness function of atmosphereless cosmic bodies". ***Kin. Phys. Neb. Tel* 4**, 60–66.

Shkuratov, Yu. G. : 1988b, "Diffractional model of the brightness surge of complex structure surfaces". ***Kin. Phys. Neb. Tel* 4**, 33–39.

Shkuratov, Yu. G. : 1989, "New mechanism of the negative polarization of light scattered by atmosphereless cosmic bodies". ***Astron. Vestnik* 23**, 176–180.

Shkuratov, Yu. G., and Akimov, L. A. : 1987, "Laboratory studies of negative polarization of light scattered by complex structure surfaces. Some sequences for atmosphereless cosmic bodies. I". ***Kin. Phys. Neb. Tel* 3**, 22–27.

Shkuratov, Yu. G., and Muinonen, K. : 1991, "Interpreting asteroid photometry and polarimetry using a model of shadowing and coherent backscattering". In ***Asteroids, Comets, Meteors 1991*** (A. W. Harris and E. Bowell, Eds.), 549–552, Lunar and Planetary Institute.

Shkuratov, Yu. G., and Opanasenko, N. V. : 1992, "Polarimetric and photometric properties of the Moon : Telescope Observation and Laboratory Simulation. 2. The positive polarization". ***Icarus* 99**, 468–484.

Shkuratov, Yu. G., Akimov, L. A., Stankevich, N. P., Melkumova, L. Ya., Latynina, I. I., and Bogdanova, T. B. : 1987, "Laboratory studies of negative polarization of light scattered by complex structure surfaces. Some sequences for atmosphereless cosmic bodies. II". ***Kin. Phys. Neb. Tel* 3**, 32–37.

Shkuratov, Yu. G., Melkumova, L. Ya., and Badyukov, D. D. : 1988a, "Laboratory studies of negative polarization of light scattered by complex structure surfaces. Some sequences for atmosphereless cosmic bodies. III". ***Kin. Phys. Neb. Tel* 4**, 11–18.

Shkuratov, Yu. G., Opanasenko, N. V., and Melkumova, L. Ya. : 1988b, "The model of negative polarization of light scattered by opaque rough surfaces". Preprint No 366, IRE AN USSR, Kharkov, 1–26.

Shkuratov, Yu. G., Opanasenko, N. V., and Melkumova, L. Ya. : 1989, "Interference surge of backscattering and negative polarization of light reflected by complex structure". Preprint No 361, IRE AN USSR, Kharkov, 1–26.

Shkuratov, Yu. G., Opanasenko, N. V., and Kreslavsky, M. A. : 1992, "Polarimetric and photometric properties of the Moon : Telescope Observations and Laboratory Simulations. 1. The negative polarization". ***Icarus* 95**, 283–299.

Shkuratov, Yu. G., Muinonen, K., Bowell, E., Lumme, K., Peltoniemi, J. I., Kreslavsky, M. A., Stankevich D. G., Tishkovetz, V. P., Opanasenko, N. V., and Melkumova, L. Y. : 1994, "A critical review of theoretical models for the negative polarization of light scattered by atmosphereless solar system bodies". Submitted to ***Earth, Moon, and Planets.***

Soto–Crespo, J. M., and Nieto–Vesperinas, M. : 1989, "Electromagnetic scattering from very rough random surfaces and deep reflection gratings". ***J. Opt. Soc. Am. A* 6**, 367–384.

Soto–Grespo, J. M., Nieto–Vesperinas, M., and Friberg, A. T. : 1990, "Scattering from slightly rough random surfaces : a detailed study on the validity of the small perturbation method". ***J. Opt. Soc. Am. A* 7**, 1185–1201.

Steigman, G. A. : 1978, "A polarimetric model for a dust–covered planetary surface". ***Mon. Not. R. Astr. Soc.* 185**, 877–888.

Tapster, P. R., Weeks, A. R., and Jakeman, E. : 1989, "Observation of backscattering enhancement through atmospheric phase screens". ***J. Opt. Soc. Am. A* 6**, 517–522.

Thompson, D. T., and Lockwood, G. W. : 1992, "Photoelectric photometry of Europa and Callisto". ***J. Geophys. Res.* 97**, 14761–14772.

Trowbridge, T. S. : 1978, "Retroreflection from rough surfaces". ***J. Opt. Soc. Am.* 68**, 1225–1242.

Trowbridge, T. S. : 1984, "Rough–surface retroreflection by focusing and shadowing below a randomly undulating interface". ***J. Opt. Soc. Am. A* 1**, 1019–1027.

Tsang, L. and Ishimaru, A. : 1984, "Backscattering enhancement of random discrete scatterers". ***J. Opt. Soc. Am. A* 1**, 836–839.

Tsang, L., and Ishimaru, A. : 1985, "Theory of backscattering enhancement of random discrete isotropic scatterers based on the summation of all ladder and cyclical terms". ***J. Opt. Soc. Am. A* 2**, 1331–1338.

van Albada, M. P., and Lagendijk, A. : 1985, "Observation of weak localization of light in a random medium". ***Phys. Rev. Lett.* 55**, 2692–2695.

van de Hulst, H. C. : 1957, ***Light Scattering by Small Particles***, Wiley.
van Diggelen, J. : 1964, "The radiance of lunar objects near opposition". ***Planet. Space Sci.*** **13**, 271–279.
van Tiggelen, B. A. : 1992, "Multiple scattering and localization of light". PhD–thesis, University of Amsterdam.
Varadan, V. K., Vringi, V. N., Varadan, V. V., and Ishimaru, A. : 1983, "Multiple scattering theory for waves in discrete random media and comparison with experiments". ***Radio Sci.*** **18**, 321–327.
Varadan, V. V., Varadan, V. K., Ma, Y., and Steele, W. A. : 1987, "Effects of non-spherical statistics on EM wave propagation in discrete random media". ***Radio Sci.*** **22**, 491.
Watson, K. M. : 1969, "Multiple scattering of electromagnetic waves in an underdense plasma". ***J. Math. Phys.*** **10**, 688–702.
Whitaker, E. A. : 1979, "Implications for asteroidal regolith properties from comparisons with the lunar phase relation and theoretical considerations". ***Icarus*** **40**, 406–417.
Wolf, P.-E., and Maret, G. : 1985, "Weak localization and coherent backscattering of photons in disordered media". ***Phys. Rev. Lett.*** **55**, 2696–2699.
Wolff, M. : 1975, "Polarization of light reflected from rough planetary surface". ***Appl. Opt.*** **14**, 1395–1405.
Zellner, B., and Gradie, J. : 1976, "Minor planets and related objects. XX. Polarimetric evidence for the albedos and compositions of 94 asteroids". ***Astron. J.*** **81**, 262–280.
Zellner, B., Gehrels, T., and Gradie, J. : 1974, "Minor planets and related objects. XVI. Polarimetric diameters". ***Astron. J.*** **79**, 1100–1110.
Zellner, B., Lebertre, T., and Day, K. : 1977, "Laboratory polarimetry of dark carbon-bearing silicates and the asteroid albedo scale". ***Proc. 8th Lunar Sci. Conf.***, Preprint Nr. 770006, 1–8.

COMETARY NUCLEI

HANS RICKMAN
Astronomiska observatoriet, Box 515, S-75120 Uppsala, Sweden
E-mail HANS@ASTRO.UU.SE

Abstract. After a brief summary of current concepts on cometary nuclei, resulting from recent advances in observational techniques and laboratory experiments on cometary analogue materials, some of the outstanding problems facing theoretical modelling and interpretations of the new data is described. Emphasis is placed on the outgassing of species more volatile than water over the history of cometary nuclei, the structure of the nuclei in terms of porosity and tensile strength on a variety of different scales, and the volumetric importance and distribution of the refractory "dust" component. Near-surface interactions of dust grains and outflowing gas may be associated with the formation of large dust aggregates and mantling of the nuclei.

1. Introduction

Our picture of cometary nuclei has changed a great deal in recent years, in response to the stimulus of new, partially space-borne observations and laboratory experiments on simulated cometary materials. Nevertheless, the picture appears far from complete. While some aspects have become relatively well-established, others remain uncertain or controversial. This paper will discuss some of the issues which, in spite of recent progress, remain to be clarified. As a background to this discussion, let us summarize some current concepts about cometary nuclei, on which most authors agree.

- Cometary nuclei are icy bodies. The composition of the ice is dominated by H_2O, but other volatiles are retained as well even in comets that have come of age, like P/Halley. Examples are CO_2 and CO.
- The H_2O ice is initially amorphous and thus able to trap the other volatiles at temperatures far above their standard sublimation levels. This, however, implies that the nuclei were formed and have been preserved at very low temperatures.
- As solar heating drives a crystallization front inward with release of trapped gases, considerable vapour pressures may build up, leading to fracture or splitting of the nuclei and/or to irregular bursts of outgassing.
- Compared with the ice and vapours, the "dust" component of the nuclear material is of equal or even greater importance by mass. It is composed of silicates and organic compounds in an apparently intricate mixture that may reflect presolar core-mantle grains.
- The structure of the nuclei is fluffy, implying bulk densities well below 1 g/cm^3 and a volumetrically important network of pores, perhaps from micron-scale up to macroscopic voids.
- The gas-dust interactions next to the surface involve a complicated scenario that leads to the entailment of grains with a wide size spectrum in the outflowing gas but also to the formation of "dust mantles" that reduce or quench the gas flow.

A. Milani et al. (eds.), Asteroids, Comets, Meteors 1993, 297–312.

- Due to the common occurrence of dust mantles, the outgassing from cometary nuclei is concentrated to "active regions" often referred to as local spots or vents. The nuclei evolve by opening up or closure of active spots, and a common end state may be complete mantling.
- The nuclear surface layers of new Oort cloud comets, or comets that have never before been close to the Sun, should bear witness of Gyrs of cosmic-ray exposure and so might have special optical and chemical properties.
- The size spectrum of cometary nuclei is very wide, extending upward to Chiron-size objects or larger and downward to sub-km radii. Their spins are relatively slow and the shapes are irregular, often elongated.

These concepts, all consistent with Whipple's (1950) basic model, will be subject to critical scrutiny in the following sections, attempting to identify the weak points of our understanding and the pieces of information – both observational, experimental and theoretical – that are yet to be sought.

2. Volatile outgassing

Let us start by a statement that is hard to challenge : H_2O is the dominant driver of cometary activity close to the Sun (heliocentric distance $r \lesssim 2-3$ AU). This is most clearly borne out by the fact that nuclear production rates of other gases are always found to be at levels of several percent or less with respect to H_2O (Mumma *et al.* 1993; Crovisier 1993). We can safely conclude that the volatile composition of cometary nuclei is H_2O-dominated. However, depending on the thermal properties and the detailed energetics of the surface layers, the H_2O sublimation rate is bound to fall off rapidly with r, starting from within 5 AU, if solar heating is the only source of energy (Rickman 1992a). And yet, one often observes grain comae or tails at $r > 5$ AU despite the fact that H_2O sublimation cannot be expected to lift the grains off the nuclei. Reasons to conclude on other outgassing mechanisms in distant comets have been reviewed by Meech (1993), the basic argument being that large quantities of grains are driven out from the nuclei at distances ($r \gtrsim 10$ AU) where water ice has to remain inert. Fig. 1, from Meech (1993), shows several examples of long-period (LP) comets exhibiting a continuous decline of the activity level with increasing r, and comet P/Halley is seen to have returned, temporarily, to the typical LP comet activity level during its recent outburst.

The driver of the distant activity, unless an extremely strong internal heat source is invoked, has to be the outgassing of species more volatile than H_2O. Moreover, since these are just minor constituents, we are led to the concepts of subsurface production and diffusion of vapours to the surface through the porous icy matrix of the nuclei. *A priori*, one may then consider two different alternatives :

- individual ices sublimating independently at different depths; or
- release of gases, trapped in amorphous H_2O ice (Bar-Nun *et al.* 1987), as the ice crystallizes.

The first of these scenarios requires us to explain how volatile ices, e.g. CO, can be retained in the P/Halley nucleus, and the second raises the question of survival of amorphous ice. The problem of retention of CO ice involves two main

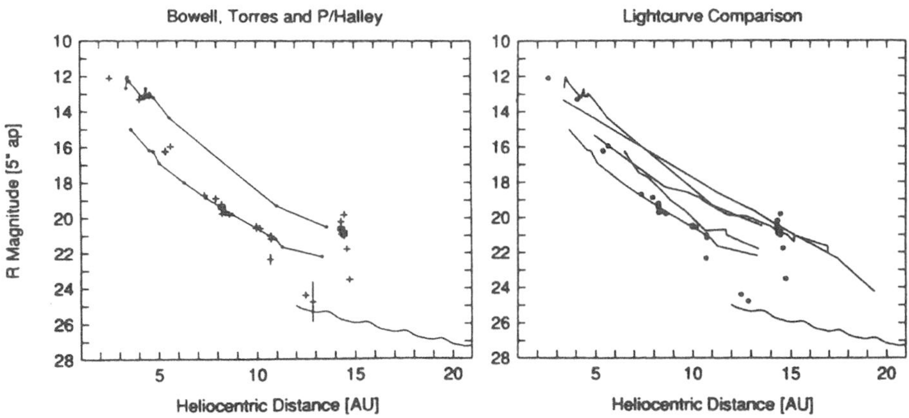

Fig. 1. *Left :* Light curves of comets 1982 I Bowell (upper curve), 1987 V Torres (lower curve) and 1986 III P/Halley (crosses); the curve to the lower right is P/Halley's theoretical bare nucleus light curve. *Right :* Light curve comparison for P/Halley (filled circles and lower-right curve) and five LP comets (solid curves). ¿From Meech (1993).

issues. First, the radiogenic heating due to the decay of ^{40}K, ^{232}Th, ^{235}U and ^{238}U included into the silicate fraction of the dust component causes a rise of the central temperature (T_c) which, depending on the size and thermal diffusivity of the nucleus, may reach the level ($T_c \sim 30$ K) where CO ice sublimates rapidly. This would lead to an outward flow of CO vapour and recondensation of the ice closer to the surface (Whipple and Stefanik 1966), and consequently, upon capture into a short-period (SP) orbit, the CO ice might be concentrated in the outer layers of the nucleus which are the first to be eroded away. Thus the second issue is whether any CO ice remains after a decrease to, say, half the initial radius following surface insolation and H_2O ice sublimation. Model calculations are needed in order to solve this problem.

It seems fair to say that the choice between the two above scenarios, and the preference for the second one among most authors, is largely governed by considerations of how and where cometary nuclei were formed. The ice in molecular clouds is known to be amorphous (see Greenberg 1989a) and extrapolation of laboratory data (Schmitt *et al.* 1989) indicates that the crystallization time scale exceeds the age of the Solar System at $T \lesssim 75$ K. Thus the assumption that cometary nuclei formed out of primordial grains of the presolar cloud leads directly to the concept of amorphous ice. Even if the grains were thermally processed, e.g. upon entry into the solar nebula (Lunine *et al.* 1991), recondensation of ice at $T \lesssim 50$ K would still form the amorphous phase (see, however, Shul'man 1992 for a different point of view). The problem then reduces to that of the thermal history of the nucleus and the survival of amorphous ice in the presence of radiogenic and solar-induced heating. One important difference between this problem and that of CO ice retention is that crystallization, in contrast to CO ice sublimation, is an exothermic process.

Recently models were computed by Haruyama *et al.* (1993) for the thermal evolution of cometary nuclei over the age of the Solar System in response to radiogenic heating by the above-mentioned isotopes. The nuclei were assumed to consist of a mixture of amorphous ice and dust in the form of core-mantle grains (Greenberg 1989b). For the thermal conductivity of amorphous ice, the experimentally determined value by Kouchi *et al.* (1992) was used, which is roughly a factor 10^4 lower than earlier, theoretical estimates (e.g. Klinger 1980). This has an important consequence in facilitating runaway crystallization, caused by the latent heat release of 90 J/g, since the heat becomes essentially locked up in sizeable nuclei for the age of the Solar System. Indeed, as an example, it was found that for an optimal (though realistic) dust/ice ratio of unity and an *ad hoc* (though realistic) assumption that the porous grain matrix has a bulk conductivity 100 times less than the individual grains, all nuclei larger than 2 km radius would undergo runaway crystallization.

This is an interesting indication, highlighting a potential problem in explaining the preservation of amorphous ice over the age of the Solar System. More realistic models should now be devised, incorporating gas release and recondensation and exploring a range of values for the amorphous ice conductivity as well as for the silicates/organics ratio of the dust component, assuming chondritic abundances of radioisotopes only for the silicate fraction.

For dynamically old comets, especially short-period ones, an additional problem arises in clarifying what happens in response to solar heating, assuming the amorphous ice indeed survives until the entry into SP orbits. Early models adapted to the case of P/Halley (Prialnik and Bar-Nun 1987) indicated a series of major crystallization bursts that in due time would reach the central region of the nucleus. The amplitude of the events proved to be a function of the model parameters chosen (Prialnik and Bar-Nun 1988), but a progressive crystallization of the whole nucleus remains a likely outcome. Furthermore, it has been suggested (e.g. Klinger 1983) that, for typical Jupiter family comets (perihelion distance $q \simeq 1.5$ AU; orbital period $P \simeq 7$ yr), orbital mean temperatures in excess of the critical range where fast crystallization ensues should imply a rapid phase transition of the whole nucleus into the crystalline state. The models by Espinasse *et al.* (1989; 1991), tailored to the case of P/Churyumov-Gerasimenko and utilizing a complete set of equations for the coupled gas and heat diffusions in a porous nucleus with a propagating crystallization front associated with gas release, indeed indicated runaway behaviour by extrapolation beyond the 35 years spanned by the calculations.

However, similar models were recently developed by Tancredi *et al.* (1993), using a different numerical approach and exploring more realistic values of physical and chemical parameters over a longer time span. As a result, treating a nucleus with equal amounts of dust and ice and with 10% trapped CO in the amorphous H_2O ice, the crystallization was found to start out very rapidly in agreement with Espinasse *et al.*, but it gradually slows down so that the speed of the front approaches that of surface erosion. Consequently, in this case extrapolation beyond 500 years indicates a time scale of $\sim 10^4$ years for complete crystallization (Fig. 2). This is of the same order as the dynamical lifetime of observable Jupiter family comets. Instrumental for the reduction of the crystallization rate is the absorption of part of the latent heat of crystallization by the inert dust component as well as the additional expense of heat for sublimating the CO ice that condenses below the front.

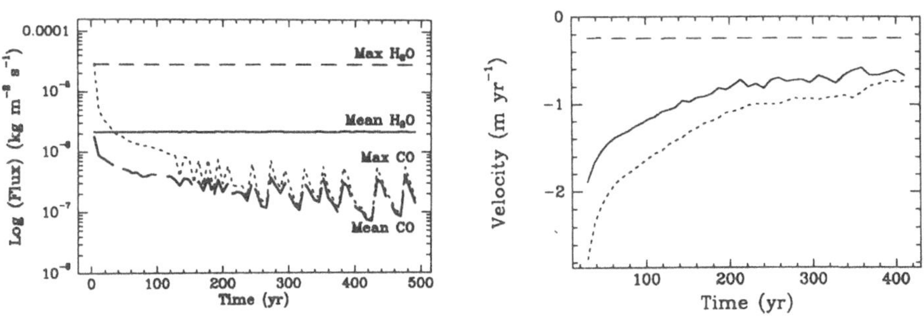

Fig. 2. *Left :* H_2O and CO production rates for a model nucleus with 10% trapped CO moving in a Jupiter family orbit; thin curves indicate the maximum rates (at perihelion for H_2O) and thick curves indicate the mean values over one orbital revolution. *Right :* Surface erosional velocity (dashed curve), velocity of the crystallization front (solid curve), and velocity of the recondensed CO ice zone (dotted curve) for the same model. From Tancredi *et al.* (1993).

A common feature of most models so far computed is that crystallization proceeds in bursts. Such bursts may apparently occur over a wide range of heliocentric distance, and they have thus been invoked as a tentative explanation for the P/Halley outburst at $r \simeq 14$ AU (Prialnik and Bar-Nun 1993; Schmitt *et al.* 1993). Moreover, for the case of trapped CO, they cause large variations of the CO production rate between successive apparitions. For $q \simeq 1.5$ AU this is always several orders of magnitude below the peak H_2O production rate of an unmantled, icy nucleus. Interestingly, though, the situation appears quite different for comets such as P/Gehrels 3 or P/Smirnova-Chernykh with $q \simeq 3.5$ AU. Even for this kind of orbit crystallization is quite fast, at least initially (Tancredi *et al.* 1993), and one would expect that the activity of large-q Jupiter family comets might be largely CO-driven – an indication that remains to verify observationally.

3. Structural properties

An important observational constraint on the porosity of cometary nuclei comes from mass determination via the nongravitational effects in the orbital motions. However, some of the relevant parameters cannot be observed and must hence be estimated on the basis of physical modelling. The determination and analytical representation of nongravitational effects in orbital solutions is reviewed by Yeomans (this volume). For the purpose of mass determination, use is made only of the principal effect, i.e., the perturbation of the mean motion ($n = 2\pi/P$) or orbital period (P). Integrated over one orbital revolution this amounts to :

$$\Delta P = \frac{a^{5/2}}{(GM_\odot)^{3/2}} \cdot 6\pi m \langle u_g \rangle \cdot \frac{Q_m}{M} \Big\{ \langle C_r \rangle E + \langle C_t \rangle T \Big\} \tag{1}$$

where a is the orbital semimajor axis, G is the gravitational constant, $M_\odot$ and M are the masses of the Sun and the comet, respectively, m is the mean mass of the gas molecules flowing out from the nucleus, $\langle u_g \rangle$ is an orbital average of the mean outflow velocity, and Q_m is the maximum gas production rate from the cometary nucleus. The curly bracket features the orbital averages of the direction cosines (C_r and C_t) of the force vector along the radial and transverse directions in the orbital plane, and E and T are parameters characterizing the gas production curve $Q(t)$ (Rickman 1992b). E measures the perihelion asymmetry of the gas production curve and T measures its peakedness; it should be noted that $\langle C_r \rangle$ may not be meaningful unless C_r is nearly constant.

Solving for M from Eq. (1) requires knowledge of the gas production curve and the velocity vector of the average gaseous outflow. The incompleteness of such knowledge translates into error bars on the derived value of M. Considering the velocity modulus factor, $\langle u_g \rangle$, its value depends on the geometry of the outflow from the nucleus, and modelling of this involves 2-D hydrodynamical simulations for an expanding gas-dust mixture (Kitamura 1986). Special issues that have been discussed include the averaging of the vertical velocity component in the Knudsen layer next to the nucleus (Bisikalo *et al.* 1989), the return flux due to molecular collisions and partial thermalization of the velocity distribution (Crifo 1987), and the additional recoil momentum of gas molecules returning to the surface after accelerating the dust grains (Peale 1990). The result is a velocity vector perpendicular to the local nuclear surface, whose magnitude can be written : $u_\ell = \zeta_\ell v_{th}(T_s)$ where $v_{th} = \sqrt{8kT_s/\pi m}$ is the thermal speed of the gas molecules and ζ_ℓ is a local momentum transfer coefficient. Until now the values used for ζ_ℓ by different authors (Sagdeev *et al.* 1988; Rickman 1989; Peale 1990) have been crude estimates based on simplified arguments, and realistic model calculations remain to be done.

The mean outflow velocity $\mathbf{u}_g$ is a global average of the local velocity vector $\mathbf{u}_\ell$ over the whole gas-producing area during the complete spin of the nucleus, weighted by the local gas flux. Its modulus is u_g and its direction cosines are $-C_r$ and $-C_t$. The vectorial averaging implies that u_g should be smaller than the global average of u_ℓ, so we can formally write : $u_g = \zeta_g u_\ell$, introducing $\zeta_g < 1$ as a global momentum transfer coefficient. We are currently faced with different models for the nuclear spin of comets as well as the activity distribution around their nuclear surfaces, influencing the estimates of ζ_g, C_r and C_t (see Fig. 3). On the one hand, the spin is sometimes modelled as pure rotation around a fixed axis or one precessing due to the torque exerted by asymmetric outgassing from a non-spherical nucleus (e.g. Whipple and Sekanina 1979; Sekanina 1984; Sekanina 1992) and sometimes as a complex motion involving rotation, precession and nutation excited by the jet torque (Sekanina 1987; Rickman 1989; Banaszkiewicz and Szutowicz 1990). On the other hand, there is also an unsettled issue about the relative contributions to cometary outgassing from isolated jets emerging from local active spots and from broad background fans with roots seated all around the nuclei. In P/Halley the jets appear to dominate (Keller 1990), but it appears premature to extrapolate this result to all comets. Concerning the spin model, we also have good evidence for a complex state of rotation for P/Halley (Belton 1991), and it would seem natural to apply such models in general.

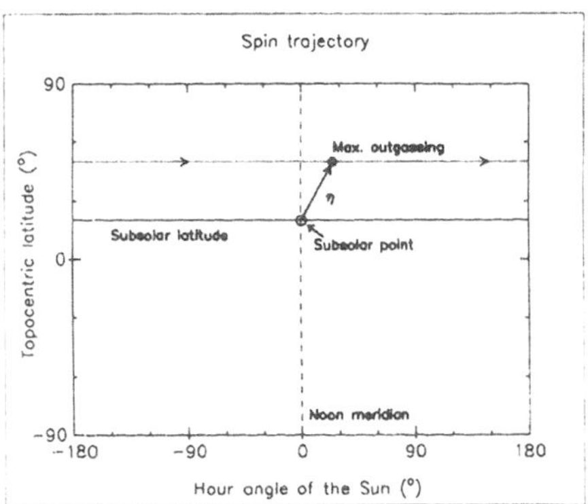

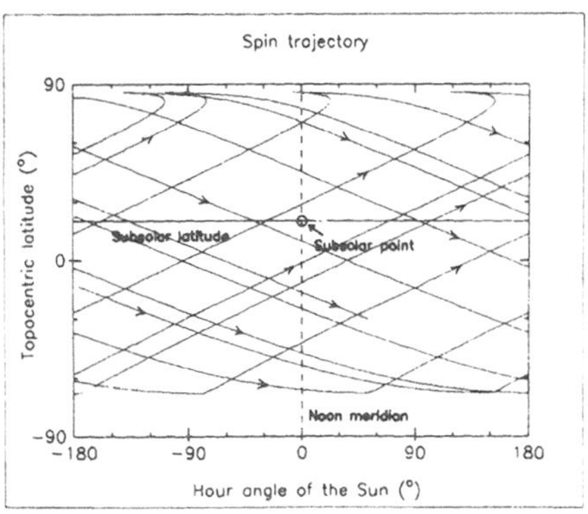

Fig. 3. *Left :* Map of a cometary nucleus with a fixed spin axis; an active spot describes the horizontal trajectory with arrows indicating the direction of rotation, and a thermal lag causes the point of maximum outgassing to be displaced from the subsolar meridian. *Right :* Corresponding trajectory during one month for an active spot on a spheroidal nucleus rotating around the long body axis, which in turn precesses around a fixed angular momentum axis. The maximum outgassing in this case tends to occur closer to the subsolar latitude. From Rickman (1989).

However, Eq. (1) shows that the interpretation of the nongravitational effect goes hand in hand with that of the perihelion asymmetry of the gas production curve, i.e., the light curve asymmetry. Recent work (Yeomans and Chodas 1989; Festou *et al.* 1990; Sekanina 1992) has convincingly demonstrated the validity of Bessel's (1836) conjecture about the importance of the 'radial' term in this equation. As yet, the jet outflow model with pure rotation offers the only straightforward explanation to the frequent occurrence of strong light curve asymmetries with both preperihelion and postperihelion peaks among short-period comets by means of the seasonal variation of insolation onto the dominant active spot. For comet P/Halley the coexistence of an important postperihelion outgassing excess and a rotational motion involving significant wobbling remains unexplained, and the constancy of the nongravitational effect over 30 orbital revolutions appears enigmatic.

The lack of knowledge about the distribution of outgassing activity around the nuclear surface and the rotational state for any given comet naturally transforms into an error bar on ζ_g, but unless extreme cases are considered, this one is unimportant compared to that on ζ_ℓ. Of much greater consequence is the fact that one-spot models with pure rotation (Sekanina 1992) imply a general correlation between E and $\langle C_t \rangle$ so that the radial and transverse terms in Eq. (1) tend to act in concert and to contribute comparably to the total effect. More complex models would generally attach less significance to the transverse term (Rickman 1989); and even a small, though unknown, thermal lag may upset the above result for comets with nearly symmetric gas production curves. To summarize the current situation, nuclear masses may be estimated for comets with large values of $|E|$ from Eq. (1) using a model calculation of $\langle u_g \rangle$ and observational data on Q_m and E, taking as an *ad hoc* estimate : $\langle C_r \rangle E + \langle C_t \rangle T \approx E$. As a consequence, we are limited to comets with well-observed, asymmetric gas production curves for which Q_m and E are known with decent accuracy. Even so, an unavoidable error of a factor $2-3$ comes from the uncertainties of the above-described estimates of $\langle u_g \rangle$ and $\langle C_r \rangle E + \langle C_t \rangle T$. Only in exceptional cases, such as that of P/Halley, can this modelling uncertainty be reduced. However, better knowledge of the gas production curves would lead to some improvement, e.g. when considering the typical mean density of a statistical sample of cometary nuclei. From the cases analyzed as yet (Rickman 1989; Rickman *et al.* 1992), it appears that cometary nuclei are very porous (mean density $\rho \lesssim 0.5$ g/cm^3).

Porosity means a fluffy structure, but do we know over which scales the fluffiness extends? Donn (1990) advocates a fractal build-up of cometary nuclei resulting from formation by low-velocity random accretion. Numerical simulations (Daniels and Hughes 1981) yield open structures with a fractal dimension $D \approx 2$, as illustrated by Fig. 4a from Hughes (1988), i.e., a density decreasing with linear size (R) as $\rho \propto R^{-1}$. This obviously cannot hold over many orders of magnitude in real nuclei, whose bulk packing factors appear to be ~ 0.1 judging from the above density estimate. Hence a great deal of compaction must have occurred during the accretion process (Donn 1990). Information on the formation conditions is hence stored in the inhomogeneity of the nuclei, in particular the average variation of the packing factor with the volume of the sample considered. Related to this is the tensile strength spectrum, i.e. the variation of the typical tensile strength of the material with the sample area considered, since this may be affected by the collisional merging of particles and dependent on the velocities involved.

In addition to the apparently fluffy structure, some evidence for the random low-velocity accretion scenario is present in the nuclear photometric observations indicating generally elongated shapes (Jewitt and Meech 1988), as expected from numerical simulations (Fig. 4b), and perhaps also in the waist-like feature on the P/Halley nucleus (Keller 1990) that appears to mark the intersection of two large cometesimals. But affirmative evidence for a fractal-like build-up, and characterization of the associated porosity and tensile strength spectra, are very difficult to realize and offer a great challenge for future cometary exploration. Near-surface *in situ* studies by landers and penetrators may be insufficient as long as they sample only the transformed, crystalline crust (Sect. 2) where sintering may have modi-

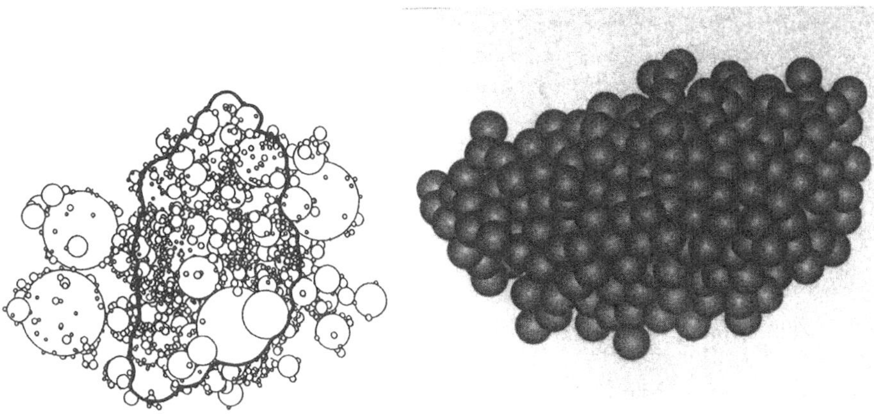

Fig. 4. *Left* : Fractal model of a newly accreted cometary nucleus, self-similar from micrometric to kilometric scales; the thick contour indicates the shape after considerable surface erosion has occurred. After Hughes (1988). *Right* . Simulated cometary nucleus formed by 1000 randomly accreted spheres of equal size. From Jewitt and Meech (1988).

fied the microstructure (Kochan *et al.* 1989). Recently modifications of the pore structure by the pressure of released gases at the crystallization front were given a preliminary treatment (Prialnik *et al* 1993), indicating the possibility of catastrophic tensile failure that would probably cause fragments to be split off the nucleus.

Analysis of cometary splittings (Sekanina 1982) offers an interesting possibility to probe the tensile strength spectrum over a wide range, if supplemented by further data on cometary dust trails or meteor streams and radar or microwave observations of the near-nucleus regions of comets of different degrees of activity. Comet 1993e P/Shoemaker-Levy 9 is a particularly fascinating case, having split up as it passed within Jupiter's Roche limit in July 1992 to an as yet unknown miss distance (Carusi *et al.* 1993). On this occasion tensile failure occurred at least on a large scale, yielding the $\sim$ 20 large fragments now seen as separate comets. If some pieces should just barely escape collision with Jupiter in July 1994, their splitting behaviour should be monitored as closely as possible.

4. The dust component

How much refractories do cometary nuclei contain in relation to the volatiles? We are interested in finding the value, or the range of values, of the dust/ice ratio. This is not amenable to direct observations, but one often considers dust/gas ratios in cometary comae based on ratios of continuum brightness to band emissions as indicators. However, it is important to note that there is a long way to go from this indicator to the desired quantity. The scattered light from the dust coma is dominated by a relatively narrow range of grain sizes, which apparently does not contribute a great deal to the total mass. It is well known that comets vary a lot in appearance and that some appear dusty, while others appear gassy, but one cannot infer with certainty a difference in nuclear composition from this.

In situ measurements of the grain size distribution are available for the P/Halley coma from the S/C encounters in March 1986. Fig. 5, from Curdt and Keller (1989), summarizes these data. The grains with radii $a \sim 1$ μm that dominate the visual scattered light show a relatively rapid fall-off (cumulative power-law mass index $\alpha \simeq 0.9$) toward larger sizes, as measured by the PIA and DID dust detectors, and by simple extrapolation one would find a rather small dust/gas production ratio ~ 0.2, close to the pre-encounter expectation from analyses of dust comae and tails in other comets (Divine *et al.* 1986). But accounting for the large particles that shook the Giotto spacecraft (Curdt and Keller 1989) and the total dust fluence that decelerated it (Edenhofer *et al.* 1987), one finds a much flatter distribution ($\alpha \simeq 0.2$) at $a \gtrsim 0.1$ mm, and as a result the dust/gas ratio is considerably increased. In fact the result is strongly dependent on the maximum grain size involved. This general property seems confirmed by the GEM-DIDSY detections at P/Grigg-Skjellerup (McDonnell *et al.* 1993) and the radio science indication of an even larger impacting particle (Pätzold *et al.* 1993).

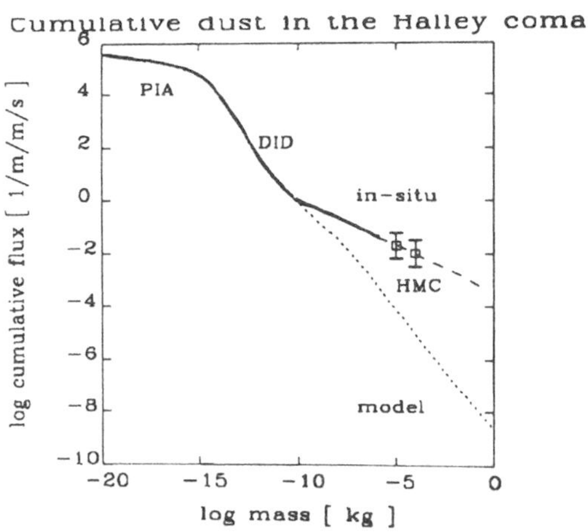

Fig. 5. Cumulative flux of dust grains of mass $> m$ registered in the P/Halley coma by PIA and DID during the last two minutes before closest approach, depicted by the solid curve. The two data points to the right are derived from S/C attitude changes measured by the HMC camera. The dotted line shows the pre-encounter model distribution. From Curdt and Keller (1989).

A total dust/gas production ratio of order unity is indicated, but the exact value cannot yet be inferred. Nonetheless, this is an important matter, since on cosmochemical grounds the presolar grains should have a dust/ice ratio close to 1 (Greenberg 1989b), and the finding of a value far from this would indicate a severe problem for the current understanding of comet formation. Values of $2-3$ for the dust/gas production ratio correspond to a cutoff at the "maximum liftable size" (mass $\sim$ 1 kg; Keller 1989), but even though the existence of such large particles in the P/Halley coma is indicated by radar observations (Campbell *et al.* 1989), the inference is still risky since the meaning of the concept is not clear (see below). Furthermore, there may be large variations of the amount of dust produced as large grains during the course of the apparition, as temporary dust mantles are formed and blown away, so it appears premature to conclude that cometary nuclei contain substantially more dust than ice. Nevertheless, the issue is of great importance for several reasons. Not only would a dust/ice ratio of 2 – 3 cast doubt on the reference picture of comet formation, but it would call upon a new picture of cometary material as a refractory matrix with icy inclusions – an "icy dirt ball" (Keller 1989). This might have important consequences for the understanding of dust mantle development (Kührt and Keller 1993).

Of considerable interest is also the ratio of organics to silicates in the dust component. The total production rate of complex, carbon-bearing molecules cannot yet be determined by remote-sensing techniques, even though it is probably related to the strength of the emission near 3.4 μm (e.g. Baas *et al.* 1986), and though such molecules may provide a source for the most conspicuous radicals observed in visual spectra as indicated by the CN jets (A'Hearn *et al.* 1986) or C_2 jets (Cosmovici *et al.* 1988) in P/Halley. We are thus limited to the snapshot picture of relative CHON grain abundance offered by the PIA instrument during the Giotto flyby (Langevin *et al.* 1987), indicating a production rate of organics comparable to that of silicates.

A very important but still elusive property of cometary material is the size distribution of dust grains residing inside the nuclei. There is an apparent conflict between the presence of coma grains with a very wide size spectrum and the reference picture of cometary formation, where the starting point is a unimodal distribution of presolar grains of submicron size with refractory cores and icy mantles. The most natural solution may be the agglomeration of these submicron refractory cores into larger structures in the near-surface layers as a consequence of the outgassing process. Laboratory sublimation experiments, like KOSI (Grün *et al.* 1991), have provided a wealth of useful information relevant to the understanding of such processes, but theoretical modelling is lagging behind and thus it remains difficult to predict the outcome in real comets exposed to varying insolation conditions as they move around their orbits.

In previous modelling work it has generally been assumed that the whole observed size spectrum of dust grains resides "ready-made" in the nuclear ice, whereupon one applies concepts like the critical radius a_c of the maximum liftable grains for the outgassing flux Z, initially introduced by Whipple (1950) :

$$a_c = \frac{C_D u_\ell m Z}{\frac{8}{3}\rho_g R\left(\frac{4}{3}\pi G\rho - \Omega^2 \cos^2\beta\right)} \tag{2}$$

where C_D is the drag coefficient, ρ_g is the density of the grain, R is the radius of the nucleus, Ω is its spin angular velocity and β is the latitude (Rickman *et al.*

1990). Using this, one models the accumulation of grains with $a > a_c$ on a freely outgassing surface until a critical level of coverage is reached (e.g. Shul'man 1972) such that a dust mantle is formed. A physical theory for this dust mantle (e.g. Shul'man 1981; Horányi *et al.* 1984; Rickman and Fernández 1986) is then used to follow its evolution and to check whether it remains stable against the gas pressure building up by ice sublimation underneath, or whether mantle blowoff occurs.

There are several problems plaguing this kind of models. First, their predictive power is weak since many parameters are badly constrained and a wide range of outcomes remains possible. For instance, the build-up of a dust mantle appears very slow according to the results of Fanale and Salvail (1984), whereas Rickman *et al.* (1990) found mantles to be formed and destroyed on very short time scales. Secondly, as noted above, there is an inconsistency in the scenario used, since the problem of forming the large dust aggregates with $a > a_c$ is shortcut. A step toward a treatment of this problem was taken in the framework of the "friable sponge" model (see Horányi and Kecskeméty 1983) using percolation theory to predict the grain size distribution. A further-reaching idea that remains to be explored is to model the gas-grain interactions in the boundary layer where the H_2O vapour feeding the outflow is produced (Rickman 1992a). In this layer there should be a gradual stripping of the icy mantles off the grains and a vigorous gas flow causing the grains to move relative to each other and accrete into larger structures but also tending to tear these structures apart.

It seems unlikely to have mm- or cm-sized aggregates formed in this way though their existence is indisputable based on observations of cometary fireballs or dust trails. An attractive picture is that these form as a result of mantle blowoff, thus requiring to have unstable mantles of rather frequent occurrence. Thus the large grains involved in the above estimate of the total dust/gas production ratio may be expelled from the nucleus in a peculiar manner, which casts doubt on the application of the critical radius as given in Eq. (2). To improve our understanding of the evolution of cometary nuclei as well as their composition and the origin of the meteoroidal and zodiacal dust complex, it is an urgent task to develop models for the physics of the near-surface layers of the nuclei, and to present a self-consistent picture of the formation of large dust aggregates and mantles. To this end the experience recently gained through laboratory experiments should be exploited, but we may need even more input from such work, e.g. concerning the cohesive strength of dust aggregates.

5. Conclusion

The reference picture of cometary nuclei sketched in the introduction represents a real achievement, when compared with what was usually thought before the age of space exploration of comets, but it is not at all finished - in fact, the picture is just beginning to emerge. There are still many tasks that have to be better fulfilled and matters that one should learn much more about. Needless to say, nothing appears basically wrong about the reference picture, but most aspects remain quite uncertain. Thus our knowledge of the true nature of cometary nuclei is still very limited, and the picture can only be accepted or discarded in favour of something better after further advances have been made. This is a matter of both observations, experiments and theoretical models, and it may be a necessary step to have future cometary space missions such as those now under study, e.g. *Rosetta*.

References

A'Hearn, M.F., Hoban, S., Birch, P.V., Bowers, C., Martin, R. and Klinglesmith III, D.A. : 1986, "Cyanogen jets in comet Halley". *Nature*, **324**, 649–651.

Baas, F., Geballe, T.R. and Walther, D.M. : 1986, "Spectroscopy of the 3.4 micron emission feature in comet Halley". *Astrophys. J. Lett.*, **311**, L97–L101.

Banaszkiewicz, M. and Szutowicz, S. : 1990, "Thermal Modelling of Precessing Cometary Nuclei". In *Asteroids, Comets, Meteors III* (C.-I. Lagerkvist, H. Rickman, B.A. Lindblad, M. Lindgren, Eds.), 239–242, Uppsala Univ. Reprocentralen.

Bar-Nun, A., Dror, J., Kochavi, E. and Laufer, D. : 1987, "Amorphous water ice and its ability to trap gases". *Phys. Rev.*, **B 35**, 2427–2435.

Belton, M.J.S. : 1991, "Characterization of the Rotation of Cometary Nuclei". In *Comets in the Post-Halley Era*, vol. 2 (R.L. Newburn, Jr., M. Neugebauer, J. Rahe, Eds.), 691–721, Kluwer, Dordrecht.

Bessel, F.W. : 1836, "Bemerkungen über mögliche Unzulänglichkeit der die Anziehungen allein berücksichtigenden Theorie der Kometen". *Astron. Nachr.*, **13**, 345–350.

Bisikalo, D.V., Marov, M.Ya., Shematovich, V.I. and Strel'nitskij : 1989, "The flow of the subliming gas in the near-nuclear (Knudsen) layer of the cometary coma". *Adv. Space Res.*, **9**, (3)53–(3)58.

Campbell, D.B., Harmon, J.K. and Shapiro, I.I. : 1989, "Radar observations of comet Halley". *Astrophys. J.*, **338**, 1094–1105.

Carusi, A., Marsden, B.G. and Valsecchi, G.B. : 1993, "The probable collision of P/Shoemaker-Levy 9 (1993e) with Jupiter in 1994". *Planet. Space Sci.*, in press.

Cosmovici, C.B., Schwarz, G., Ip, W.-H. and Mack, P. : 1988, "Gas and dust jets in the inner coma of comet Halley". *Nature*, **332**, 705–709.

Crifo, J.F. : 1987, "Improved gas-kinetic treatment of cometary water sublimation and recondensation : application to comet P/Halley". *Astron. Astrophys.*, **187**, 438–450.

Crovisier, J. : 1993, "Molecular abundances in comets". *Planet. Space Sci.*, in press.

Curdt, W. and Keller, H.U. : 1989, "Large Dust Particles along the Giotto Trajectory". *MPAE Report*, MPAE-W-55-89-19.

Daniels, P.A. and Hughes, D.W. : 1981, "The Accretion of Cosmic Dust – A Computer Simulation". *Mon. Not. R. Astr. Soc.*, **195**, 1001–1009.

Divine, N., Fechtig, H., Gombosi, T.I., Hanner, M.S., Keller, H.U., Larson, S.M., Mendis, D.A., Newburn Jr., R.L., Reinhard, R., Sekanina, Z. and Yeomans, D.K. : 1986, "The Comet Halley dust and gas environment". *Space Sci. Rev.*, **43**, 1–104.

Donn, B.D. : 1990, "The formation and structure of fluffy cometary nuclei from random accumulation of grains". *Astron. Astrophys.*, **235**, 441–446.

Edenhofer, P., Bird, M.K., Brenkle, J.P., Buschert, H., Kursinski, E.R., Mottinger, N.A., Porsche, H., Stelzried, C.T. and Volland, H. : 1987, "Dust distribution of comet P/Halley's inner coma determined from the Giotto Radio-Science Experiment". *Astron. Astrophys.*, **187**, 712–718.

Espinasse, S., Klinger, J., Ritz, C. and Schmitt, B. : 1989, "A method of estimating temperature profiles and chemical differentiation in the near surface layers of porous comet nuclei – first results for comet P/Churyumov-Gerasimenko". In *Physics and Mechanics of Cometary Materials*, ESA SP-302 (J. Hunt, T.D. Guyenne, Eds.), 185–190, ESA Publ. Div.

Espinasse, S., Klinger, J., Ritz, C. and Schmitt, B. : 1991, "Modeling of the Thermal Behavior and of the Chemical Differentiation of Cometary Nuclei". *Icarus*, **92**, 350–365.

Fanale, F.P. and Salvail, J.R. : 1984, "An idealized short-period comet model : Surface insolation, H_2O flux, dust flux, and mantle evolution". *Icarus*, **60**, 476–511.

Festou, M.C., Rickman, H. and Kamél, L. : 1990, "Using comet light-curve asymmetries to predict comet returns". *Nature*, **345**, 235–238.

Greenberg, J.M. : 1989a, "Interstellar Dust : an Overview of Physical and Chemical Evolution". In *Evolution of Interstellar Dust and Related Topics* (A. Bonetti, J.M. Greenberg, S. Aiello, Eds.), 7–52, North Holland.

Greenberg, J.M. : 1989b, "From interstellar dust to comet dust and interplanetary matter". In *Highl. of Astron.* (D. McNally, Ed.), **8**, 241–250, IAU.

Grün, E., Bar-Nun, A., Benkhoff, J., Bischoff, A., Düren, H., Hellmann, H., Hesselbarth, P., Hsiung, P., Keller, H.U., Klinger, J., Knölker, J., Kochan, H., Kohl, H., Kölzer, G., Krankowsky, D., Lämmerzahl, P., Mauersberger, K., Neukum, G., Oehler, A., Ratke, L., Roessler, K., Schwehm, G., Spohn, T., Stöffler, D. and Thiel, K. : 1991, "Laboratory simulation of cometary processes : Results from first KOSI experiments". In *Comets in the Post-Halley Era*, vol. 1 (R.L. Newburn, Jr., M. Neugebauer, J. Rahe, Eds.), 277–297, Kluwer, Dordrecht.

Haruyama, J., Yamamoto, T., Mizutani, H. and Greenberg, J.M. : 1993, "Thermal History of Comets during Residence in the Oort Cloud : Effect of Radiogenic Heating in Combination with the Very Low Thermal Conductivity of Amorphous Ice". *J. Geophys. Res. Planets*, in press.

Horányi, M. and Kecskeméty, K. : 1983, "Cometary dust". In *Asteroids, Comets, Meteors* (C.-I. Lagerkvist, H. Rickman, Eds.), 269–272, Uppsala Univ. Reprocentralen.

Horányi, M., Gombosi, T.I., Cravens, T.E., Körösmezey, A., Kecskeméty, K., Nagy, A.F. and Szegö, K. : 1984, "The friable sponge model of a cometary nucleus". *Astrophys. J.*, **278**, 449–455.

Hughes, D.W. : 1988, "Cometary Magnitude Distribution and the Ratio between the Numbers of Long- and Short-Period Comets". *Icarus*, **73**, 149–162.

Jewitt, D.C. and Meech, K. : 1988, "Optical properties of cometary nuclei and a preliminary comparison with asteroids". *Astrophys. J.*, **328**, 974–986.

Keller, H.U. : 1989, "Comets – dirty snowballs or icy dirtballs?". In *Physics and Mechanics of Cometary Materials*, ESA SP-302 (J. Hunt, T.D. Guyenne, Eds.), 39–45, ESA Publ. Div.

Keller, H.U. : 1990, "The Nucleus". In *Physics and Chemistry of Comets* (W.F. Huebner, Ed.), 13–68, Springer-Verlag.

Kitamura, Y. : 1986, "Axisymmetric Dusty Gas Jet in the Inner Coma of a Comet". *Icarus*, **66**, 241–257.

Klinger, J. : 1980, "Influence of a Phase Transition of Ice on the Heat and Mass Balance of Comets". *Science*, **209**, 271–272.

Klinger, J. : 1983, "Classification of Cometary Orbits Based on the Concept of Orbital Mean Temperature". *Icarus*, **55**, 169–176.

Kochan, H., Roessler, K., Ratke, L., Heyl, M., Hellmann, H. and Schwehm, G. : 1989, "Crustal strength of different model comet materials". In *Physics and Mechanics of Cometary Materials*, ESA SP-302 (J. Hunt, T.D. Guyenne, Eds.), 115–119, ESA Publ. Div.

Kouchi, A., Greenberg, J.M., Yamamoto, T. and Mukai, T. : 1992, "Extremely low thermal conductivity of amorphous ice : relevance to comet evolution". *Astrophys. J. Lett.*, **388**, L73–L76.

Kührt, E. and Keller, H.U. : 1993, "The formation of cometary surface crusts". *Icarus*, in press.

Langevin, Y., Kissel, J., Bertaux, J.-L. and Chassefière, E. : 1987, "First statistical analysis of 5000 mass spectra of cometary grains obtained by PUMA 1 (Vega 1) and PIA (Giotto) impact ionization mass spectrometers in the compressed modes". *Astron. Astrophys.*, **187**, 761–766.

Lunine, J., Engel, S., Rizk, B. and Horányi, M. : 1991, "Sublimation and Reformation of

Icy Grains in the Primitive Solar Nebula". *Icarus*, **94**, 333–344.

McDonnell, J.A.M., McBride, N., Beard, R., Bussoletti, E., Colangeli, L., Eberhardt, P., Firth, J.G., Grard, R., Green, S.F., Greenberg, J.M., Grün, E., Hughes, D.W., Keller, H.U., Kissel, J., Lindblad, B.A., Mandeville, J.-C., Perry, C.H., Rembor, K., Rickman, H., Schwehm, G., Turner, R.F., Wallis, M.K. and Zarnecki, J.C. : 1993, "Dust Impact Detection System : first results from the Giotto Extended Mission encounter with Comet Grigg-Skjellerup". *Nature*, **362**, 732–734.

Meech, K.J. : 1993, "Observations of the cessation of activity in Comets P/Halley, Torres (1987 V), and Bowell (1982 I) and implications for the mechanism of the P/Halley outburst". In *Workshop on the Activity of Distant Comets* (W.F. Huebner, H.U. Keller, D. Jewitt, J. Klinger, R. West, Eds.), 12–20, Southwest Research Institute.

Mumma, M.J., Weissman, P.R. and Stern, S.A. : 1993, "Comets and the Origin of the Solar System : Reading the Rosetta Stone". In *Protostars and Planets III* (E.H. Levy, J.I. Lunine, M.S. Matthews, Eds.), 1177-1252, Univ. of Arizona Press, Tucson.

Pätzold, M., Edenhofer, P., Bird, M.K. and Volland, H. : 1993, "The Giotto encounter with comet P/Grigg-Skjellerup : first results from the Giotto Radio-Science Experiment". *Astron. Astrophys. Lett.*, **268**, L13–L16.

Peale, S.J. : 1990, "On the Density of Halley's Comet". *Icarus*, **82**, 36–49.

Prialnik, D. and Bar-Nun, A. : 1987, "On the evolution and activity of cometary nuclei". *Astrophys. J.*, **313**, 893–905.

Prialnik, D. and Bar-Nun, A. : 1988, "The Formation of a Permanent Dust Mantle and Its Effect on Cometary Activity". *Icarus*, **74**, 272–283.

Prialnik, D. and Bar-Nun, A. : 1993, "Distant Outbursts of Comet P/Halley Caused by Crystallization of Amorphous Ice". In *Workshop on the Activity of Distant Comets* (W.F. Huebner, H.U. Keller, D. Jewitt, J. Klinger, R. West, Eds.), 100–109, Southwest Research Institute.

Prialnik, D., Egozi, U., Bar-Nun, A., Podolak, M. and Greenzweig, Y. : 1993, "On Pore Size and Fracture in Gas-Laden Comet Nuclei". *Icarus*, in press.

Rickman, H. : 1989, "The nucleus of Comet Halley : surface structure, mean density, gas and dust production". *Adv. Space Res.*, **9**, (3)59–(3)71.

Rickman, H. : 1992a, "Cometary nuclei – recent gas flux modelling and applications". *Ann. Geophysicae*, **10**, 157–168.

Rickman, H. : 1992b, "Physico-Dynamical Evolution of Aging Comets". In *Interrelations between Physics and Dynamics for Minor Bodies in the Solar System*, Goutelas 1991 (D. Benest, C. Froeschlé, Eds.), 197–263, Ed. Frontières.

Rickman, H. and Fernández, J.A. : 1986, "Formation and Blowoff of a Cometary Dust Mantle". In *Comet Nucleus Sample Return Mission*, ESA SP-249 (O. Melita, Ed.), 185–194, ESA Publ. Div.

Rickman, H., Fernández, J.A. and Gustafson, B.Å.S. : 1990, "Formation of stable dust mantles on short-period comet nuclei". *Astron. Astrophys.*, **237**, 524–535.

Rickman, H., Festou, M.C., Tancredi, G. and Kamél, L. : 1992, "Twentieth century light curves and the nucleus of comet P/Tempel 2". In *Asteroids, Comets, Meteors 1991* (A.W. Harris, E. Bowell, Eds.), 509–512, Lunar and Planetary Institute, Houston.

Sagdeev, R.Z., Elyasberg, P.E. and Moroz, V.I. : 1988, "Is the nucleus of Comet Halley a low-density body?". *Nature*, **331**, 240–242.

Schmitt, B., Espinasse, S., Grim, R.J.A., Greenberg, J.M. and Klinger, J. : 1989, "Laboratory studies of cometary ice analogues". In *Physics and Mechanics of Cometary Materials*, ESA SP-302 (J. Hunt, T.D. Guyenne, Eds.), 65–69, ESA Publ. Div.

Schmitt, B., Espinasse, S. and Klinger, J. : 1993, "A possible explanation of the outburst of Comet P/Halley at 14 astronomical units from the Sun". In *Workshop on the Activity of Distant Comets* (W.F. Huebner, H.U. Keller, D. Jewitt, J. Klinger, R. West, Eds.), 124–132, Southwest Research Institute.

Sekanina, Z. : 1982, "The problem of split comets in review". In *Comets* (L.L. Wilkening, Ed.), 251–287, Univ. Arizona Press.

Sekanina, Z. : 1984, "Precession model for the nucleus of periodic comet Kopff". ***Astron. J.***, **89**, 1573–1586.
Sekanina, Z. : 1987, "Nucleus of Comet Halley as a Torque-Free Rigid Rotator". ***Nature***, **325**, 326–328.
Sekanina, Z. : 1992, "Effect of discrete-source outgassing on motions of periodic comets and discontinuous orbital anomalies". ***Astron. J.***, **105**, 702–735.
Shul'man, L.M. : 1972, "The evolution of cometary nuclei". In ***The Motion, Evolution of Orbits, and Origin of Comets*** (G.A. Chebotarev, E.I. Kazimirchak-Polonskaya, B.G. Marsden, Eds.), 271–276, D. Reidel Publ. Co.
Shul'man, L.M. : 1981, "A two-layer model of the cometary nucleus". ***Astrometria i Astrofisica***, **45**, 21–34.
Shul'man, L.M. : 1992, "On the Intrinsic Source of Energy in Cometary Nuclei : Amorphous Ice or Hydrated Ions?". ***Bull. Amer. Astron. Soc.***, **24**, 1012 (Abstract).
Tancredi, G., Rickman, H. and Greenberg, J.M. : 1993, "Thermochemistry of cometary nuclei. I. The Jupiter family case". ***Astron. Astrophys.***, in press.
Whipple, F.L. : 1950, "A comet model. I. The acceleration of Comet Encke". ***Astrophys. J.***, **111**, 375–394.
Whipple, F.L. and Sekanina, Z. : 1979, "Comet Encke : Precession of the spin axis, nongravitational motion and sublimation". ***Astron. J.***, **84**, 1894–1909.
Whipple, F.L. and Stefanik, R.P. : 1966, "On the physics and splitting of cometary nuclei". ***Mém. Soc. Roy. Sci. Liège***, Sér. 5, **12**, 33–52.
Yeomans, D.K. and Chodas, P.C. : 1989, "An asymmetric outgassing model for cometary nongravitational accelerations". ***Astron. J.***, **98**, 1083–1093.

MOLECULAR ABUNDANCES IN COMETS

JACQUES CROVISIER
Observatoire de Paris, Section de Meudon, F-92195 Meudon, France
E-mail CROVISIE@FRMEU51.BITNET or MESIOA : : CROVISIE

Abstract. The molecular composition of cometary volatiles is a basic information on the nature of comets and a clue to their formation mechanisms. It is only recently that direct identifications of cometary volatiles were obtained through in situ exploration as well as from remote sensing at UV, IR and radio wavelengths. An inventory of known cometary volatiles is presented, with a critical review of the evaluations of their abundances.

1. Introduction

The molecular composition of cometary volatiles is a basic information on the nature of comets and an important clue to their formation mechanisms. For a long time, the only information on this composition was coming from the signatures of dissociation products in the visible spectrum. It is only recently that we obtained direct identifications of cometary volatiles by in situ exploration as well as by remote sensing at UV, IR and radio wavelengths. An inventory of known cometary volatiles will be presented, with a critical review of the evaluation of their abundances. Previous reviews on this topic were made recently by Krankowsky (1991) and Mumma *et al.* (1993b) and as part of the comprehensive review of Festou *et al.* (1993). The implications for the nucleus composition, structure, and evolution, as well as the possible link to the interstellar medium, are only briefly mentioned here, as they are treated elsewhere in this volume (Rickman 1994; Greenberg and Shalabiea, 1994). Cometary isotopic ratios, which also contain a strong cosmogonic message, will not be discussed due to lack of space. A recent review is given by Krankowsky (1991) : in summary, the [D]/[H] ratio is enriched; the other known ratios (such as $[^{18}O]/[^{16}O]$ and $[^{13}C]/[^{12}C]$) are close to terrestrial.

2. The methods and their reliability

There are now several ways to measure cometary volatile abundances, each of them having its own advantages and drawbacks. A synopsis is given in Table I.

A decade ago, if one excepts the CO molecule detected in the UV, no cometary parent molecule was directly observed. One had to *guess* the nature of the parent volatiles from the presence of the radicals, atoms and ions observed in the visible spectrum of comets. With the opening of new spectral domains, several parent molecules are now reliably identified. The existence of others, however, has still to be inferred from educated guesses.

Most of the electronic and fundamental vibrational bands of interest for cometary parent molecules are emitted through resonant fluorescence excited by the Sun (Yamamoto 1982; Crovisier and Encrenaz, 1983; Weaver and Mumma, 1984). Their excitation rates (the so-called *g-factors*) depend only upon the solar radiation and the molecular constants. It is thus straightforward to derive column densities from

A. Milani et al. (eds.), Asteroids, Comets, Meteors 1993, 313–326.

TABLE I
Methods for the evaluation of molecular abundances in comets.

Method	Example	Problems and reliability
Rotational lines (individual)	HCN (radio)	modelling of rotational distribution required
Rovibrational lines (individual)	CH_4 (IR)	idem
Vibrational bands (total)	CH_3OH (IR)	blending and identification
Electronic bands	CO (UV)	generally not available for parent molecules
Mass spectroscopy	N_2	mass ambiguity, chemical modelling required
Observation of dissociation products	NH_3 (from NH)	chemical modelling and space distribution required

band intensities, at least for optically thin lines. Problems, however, may arise for the identification of these bands from low-resolution spectra.

Individual rotational lines and rovibrational lines observed at high resolution lead to unambiguous identifications. It is more tricky to retrieve production rates, however, because one then needs to know the rotational population distribution. In some cases when it is possible to observe simultaneously several lines and to measure this distribution (e.g. CH_3OH at radio wavelengths), reliable results may be obtained. In other cases, one has to rely on models. Molecules are expected to evolve to *fluorescence equilibrium*, governed by the balance between spontaneous decay and fluorescence excitation of the vibrational or electronic bands (Crovisier 1984, 1987; Weaver and Mumma, 1984). However, radio and IR telescopes generally sample the inner coma where this equilibrium is not yet reached. More sophisticated models must be used, in which the coma kinetic temperature and collision rates are important (but unfortunately ill-known) parameters.

Unique information was obtained, during the *Giotto* fly-by of P/Halley, by the Neutral Mass Spectrometer (*NMS*; Krankowsky *et al.* 1986), the Ion Mass Spectrometer (*IMS*; Balsiger *et al.* 1986) and the *PICCA* instrument (Mitchell *et al.* 1992). These data are not easy to interpret. The main reason is that these instruments measured mass per charge (m/q) ratios with a low resolution and could not distinguish between species with the same atomic mass (i.e. CO, N_2 and C_2H_4) without further information. In several cases, a useful guideline for the interpretation of mass spectra was provided by complementary information from remote-sensing spectroscopy. Retrieving the abundance of parent molecules from ion densities requires a careful chemical modelling, which is hampered by our incomplete knowledge of ion-molecule reaction rates and of the actual coma chemical composition. Some of the preliminary interpretations were erroneous and the detailed analyses are still going on.

These observations lead to determinations of column densities (of densities for mass spectroscopy), which in turn have to be converted into production rates to get finally abundances. This requires the knowledge of the spatial distribution, modelling of which involves the molecular lifetimes and expansion velocities. In the best cases, information on the scale lengths can be obtained from brightness profiles, on the expansion velocities, from the radio line shapes. In the worst cases, one does not even know whether the species under study is coming from the nucleus, or from a distributed source, which results in a large uncertainty upon the production rate (see the discussions on CO and H_2CO in the next Section). The derivation of relative abundances is difficult because observations are rarely simultaneous and cometary activity may be variable. In addition to these problems, most observations of cometary parent molecules are made with state-of-the-art techniques which may be subject to calibration difficulties.

3. Molecular identifications and abundances

Our present knowledge of the cometary volatile abundances is summarized in Table II. In the following, all abundances will be given relative to water.

3. 1. WATER

Water is well studied through its dissociation products : through the OH radical which can be observed in the near UV from the *IUE* satellite or even from the ground, and in the radio at 18 cm wavelength; through space observations of the Lyman α line of hydrogen; through observation of the prompt emission of the forbidden lines of oxygen. The problems in deriving water production rates from these indirect observations were discussed in a special session of a preceding *Asteroids Comets Meteors* conference (Lagerkvist *et al.* 1990). Up to now, water was directly observed in only two comets, through its IR vibrational bands : in P/Halley from the *KAO* (Mumma *et al.* 1986) and from the *VEGA IKS* and *TKS* IR spectrometers (Combes *et al.* 1988; Krasnopolsky *et al.* 1986), in comet Wilson 1987 VII from the KAO (Larson *et al.* 1988). It is expected that future space observations in the submillimetre and far-infrared ranges (e.g. with *ISO*, *SWAS* and *FIRST*) will easily detect the fundamental rotational lines of water. Water was, of course, observed by the *Giotto NMS* (Krankowsky *et al.* 1986). It was found to constitute 80 % (by number) of the whole cometary gas, thus confirming its status of dominant cometary volatile.

3. 2. CARBON MONOXIDE AND CARBON DIOXIDE

Carbon monoxide was first detected in comet West 1976 VI by rocket UV observations (Feldman and Brune, 1976), then observed in several comets with the *IUE*, sounding rockets or the Space Shuttle (Feldman 1991; Feldman *et al.* 1991). Its fundamental vibrational band at 4.7 μm was marginally detected by the *VEGA-1* infrared spectrometer in P/Halley (Combes *et al.* 1988) and one of its rovibrational lines was tentatively detected in Austin 1990 V (DiSanti *et al.* 1992b). CO was also studied by the *NMS* in P/Halley (Eberhardt *et al.* 1987). Highly different CO abundances were observed : from 2 % in Bradfield 1979 X to 15–20 % in comets West 1976 VI and P/Halley. Analyses of UV observations made with different fields of view as well as the CO density evolution observed in situ by *NMS* show that the

TABLE II
A review of the volatile composition of comets.

species	abundance	comets	methods	references
H_2O	100.	many	photodiss. prod.	
		P/H, Wilson	IR	1, 2, 3, 4, 5
		P/H	NMS	6
CO	2.–20.	several	UV	7, 8, 9
		P/H, Austin	IR	2, 10
		P/H	NMS	11
CO_2	2.7	P/H	IR	2
	3.5	P/H	NMS	6
	6.	P/Hartley 2	CO prompt em.	12
H_2CO	0.03–1.5	several	radio	13, 14, 15, 16
	$< 1.$	several	IR	17
	4.	P/H	IR	2
	> 3.8	P/H	NMS, IMS	18, 19, 20
CH_3OH	1.–7.	Austin, Levy, P/S-T	radio	16, 21, 22, 23, 24
		several	IR	25, 26
		P/H	NMS, IMS	19, 27
C_2H_5OH	< 0.5	Levy	radio	28
HCOOH	< 0.2	Austin, Levy	radio	28
CH_4	< 0.5–2.	P/H, Wilson, Levy	IR	5, 29, 30, 31
	2.	P/H	IMS	32
NH_3	0.1–0.3	several	NH, NH_2 (vis.)	33
	0.4–1.0	P/H	NH (UV)	34
	0.5–2.	P/H	IMS, NMS	32, 35
HCN	0.05–0.2	several	radio	15, 23, 24, 36, 37
		P/H	NMS, IMS	35
HC_3N	< 0.02	P/H, Levy	radio	28, 36
N_2	0.2–0.02	P/H, Bradfield	N_2^+ (vis.)	38, 39
	< 1.5	P/H	NMS, IMS	11, 19
H_2S	0.1–0.3	Austin, Levy, P/S-T	radio	21, 22, 24, 40
		P/H	IMS, NMS	19, 41
CS_2	0.1	several	CS (UV)	9
S_2	0.05	IRAS-A-A	UV	42, 43
OCS	< 0.3	Levy	radio	40
	< 0.5	P/H, Austin	IR	2, 44
H_2CS	< 0.1	Levy	radio	28
SO_2	< 0.001	P/H, Bradfied	UV	45
He	$< 3.$	Austin	UV	46
Ar	$< 17.$	Austin, Levy	UV	8, 46

Notes to Table II :

Comets : IRAS-Araki-Alcock 1983 VII, P/Halley 1986 III, Wilson 1987 VII, Bradfield 1987 XXIX, Austin 1990 V, Levy 1990 XX, P/Hartley 2 1991 XV, P/Swift-Tuttle 1992t.
References : **1** Mumma et al. 1986; **2** Combes et al. 1988; **3** Krasnopolsky et al. 1986; **4** Weaver et al. 1986; **5** Larson et al. 1988; **6** Krankowsky et al. 1986; **7** Feldman and Brune 1976; **8** Feldman et al. 1991; **9** Feldman 1991 and references therein; **10** DiSanti et al. 1992b; **11** Eberhardt et al. 1987; **12** Weaver et al. 1994; **13** Snyder et al. 1989; **14** Colom et al. 1992; **15** Schloerb and Ge 1992; **16** Bockelée-Morvan et al. 1994c; **17** Reuter et al. 1992; **18** Krankowsky 1991; **19** Geiss et al. 1991; **20** Meier et al. 1993; **21** Bockelée-Morvan et al. 1990; **22** Bockelée-Morvan et al. 1991; **23** Bockelée-Morvan et al. 1994b; **24** Despois et al. 1994; **25** Hoban et al. 1991; **26** Hoban et al. 1994; **27** Eberhardt et al. 1991; **28** Crovisier et al. 1993; **29** Drapatz et al. 1986; **30** Kawara et al. 1988; **31** Brooke et al. 1991; **32** Allen et al. 1987; **33** Wyckoff et al. 1991b; **34** Feldman et al. 1993; **35** Ip et al. 1990; **36** Crovisier and Schloerb 1991 and references therein; **37** Wootten et al. 1994; **38** Wickoff et al. 1991a; **39** Lutz et al. 1993; **40** Crovisier et al. 1991; **41** Marconi et al. 1990; **42** A'Hearn et al. 1983; **43** Budzien and Feldman 1992; **44** DiSanti et al. 1992a; **45** Kim and A'Hearn 1991; **46** Stern et al. 1992.

origin of CO is twofold : one part is coming from the nucleus region, another one from a distributed source with a scale length of the order of 10^4 km.

Carbon dioxide, which cannot be observed from the ground due to the absence of allowed rotational lines and to strong telluric absorption of its IR bands, was strongly suspected from the existence of CO_2^+. CO_2 was observed only in P/Halley through its strong ν_3 band at 4.3 μm by *VEGA-1* (Combes *et al.* 1988) and through mass spectroscopy by *NMS* (Krankowsky *et al.* 1986). In the UV spectrum of comet P/Hartley 2 1991 XV observed with the *HST*, Weaver *et al.* (1994) identified bands of the CO Cameron system around 200 nm that they attributed to CO prompt emission following photodissociation of CO_2. They estimated a carbon dioxide abundance of 6 % in this comet, and a CO abundance < 2 %.

3. 3. FORMALDEHYDE

The detection of formaldehyde was first claimed in P/Halley through its IR bands at 3.56 μm (Combes *et al.* 1988; subsequently reanalyzed by Mumma and Reuter, 1989) and at 6 cm radio wavelength (Snyder *et al.* 1989). The derived abundance was 4 %. Formaldehyde was then detected at 226 and 352 GHz in P/Brorsen-Metcalf, Austin 1990 V, Levy 1990 XX and P/Swift-Tuttle (Colom *et al.* 1992; Schloerb and Ge 1992; Bockelée-Morvan *et al.* 1994c), with much smaller abundances (0.03 to 1 %) assuming direct release from the nucleus. Reuter *et al.* (1992) failed to detect H_2CO from the ground in the IR of several comets, with limits typically smaller than one percent (for direct release from the nucleus). On the basis of these results, Bockelée-Morvan and Crovisier (1992) questioned the detections and abundance determinations of formaldehyde in P/Halley.

H_2CO was also observed in situ in P/Halley by the *NMS* and *IMS* instruments (Krankowsky 1991; Geiss *et al.* 1991; Meier *et al.* 1993). Both observations show that formaldehyde is released from a distributed source. The abundance retrieved by *NMS* is 3.8 % at least (with 3 % produced inside the r < 4700 km region) and the distribution is consistent with *no* formaldehyde directly released from the nucleus. The determinations given above from radio and IR observations, based upon a nucleus release, are thus underestimates of the true H_2CO abundances. A reliable determination will await the knowledge of the space distribution and of the origin of H_2CO. The difference, however, between the high abundance of H_2CO in P/Halley and those in the other comets cannot be reconciled by the space distribution. H_2CO may not be a primary cometary volatile and therefore may be irrelevant for cosmogonic considerations. Meier *et al.* (1993) suggested that formaldehyde polymers on grains would sublimate as H_2CO monomers. In turn, photodissociation of H_2CO could contribute to part of the distributed source required for CO.

3. 4. METHANOL

Methanol was unambiguously identified in comets Austin 1990 V, Levy 1990 X and P/Swift-Tuttle through its rotational lines at millimetre and submillimetre wavelengths (Bockelée-Morvan *et al.* 1990, 1991, 1994b). Several rotational lines can be observed simultaneously with the same spectrometer : this allows one to derive the rotational distribution and to estimate the methanol abundance almost independently of any assumption on the excitation. The methanol ν_3 band was tentatively identified to a 3.52 μm emission feature in comet Austin by Hoban *et al.* (1991). This band was then retrieved in archive IR spectra and in spectra of subsequent comets (Bockelée-Morvan *et al.* 1994a; Hoban *et al.* 1994; Davies *et al.* 1993). Methanol is also detected from its protonated form $CH_3OH_2^+$ in the *Giotto* mass spectra (Geiss *et al.* 1991; Eberhardt *et al.* 1991). Methanol abundances from both radio and IR observations are reliable and in good agreement. They range from 1 to 7 %, depending upon the comet, and the variation from comet to comet seems to be real. This makes methanol one of the most abundant parent molecules. It must be noted that the photodissociation of CH_3OH cannot explain the distributed source of formaldehyde, because the dissociation product is presumably CH_3O rather than H_2CO, and because the lifetime of methanol is too large (10^5 s) to be reconciled with the H_2CO observations.

3. 5. METHANE

Methane was searched for through individual lines of its ν_3 band at 3.3 μm with the *KAO* or from the ground (Drapatz *et al.* 1986; Kawara *et al.* 1988; Larson *et al.* 1988; Brooke *et al.* 1991). The retrieved upper limits depend upon the assumptions made on the rotational temperature. They range from 0.2 to 2 % in comets P/Halley and Levy 1990 XX. Larson *et al.* (1988) reported a possible detection that exceeds the 3–σ level (corresponding to an abundance 1.4–4.5 %) in comet Wilson 1987 VII, but too marginal to constitute a definite detection of cometary methane. Allen *et al.* (1987) attributed a peak in the *IMS* mass spectra to CH_3^+ coming from methane and deduced an abundance of 2 % of CH_4 for P/Halley. This value is subject to caution since they neglected the contribution of methanol and other methyl compounds in their chemical model. Haider *et al.* (1993), from another model, revised the abundance to 0.5 %.

3. 6. NITROGEN SPECIES

Ammonia was never directly observed if one excepts an unconfirmed detection at centimetre wavelengths (Altenhoff *et al.* 1984). From the bands of NH and NH_2 in the visible, Wyckoff *et al.* (1991b) estimated abundances of 0.1–0.3 % in several comets, including P/Halley. Using a different approach, Feldman *et al.* (1993) estimated a higher abundance (0.4–1.0 %) in P/Halley and seven other comets from the NH bands observed by the *IUE* and *ASTRON* satellites. Still higher abundances (1–2 %) were retrieved by Allen *et al.* (1987) through chemical modelling of the *Giotto IMS* data (since NH_4^+ is blended with water-group ions, the determination is quite indirect). However, Ip *et al.* (1990) retrieved only 0.5 % from the same data, and Geiss *et al.* (1991) argued that these abundances might be overestimated, due to flaws in the chemical modelling (see also Haider *et al.* 1993). It may be anticipated that direct detection of the fundamental submillimetre rotational lines of NH_3 will be possible by observing from space and will permit a more reliable assessment of its abundance.

First claimed to be detected in comet Kohoutek 1973 XII (Huebner *et al.* 1974), hydrogen cyanide was definitely identified at millimetre wavelengths in P/Halley (Crovisier and Schloerb 1991 and references therein) and several subsequent comets (Bockelée-Morvan *et al.* 1990, 1994c; Schloerb and Ge, 1992). Several rotational transitions (J 1–0, 3–2 and 4–3) have now been observed. Although the observations are not always simultaneous, or are made with different beams or telescopes, it may be expected that a careful modelling of these data will constrain the excitation conditions of the molecule and lead to a reliable determination of its abundance. Abundances presently observed range from 0.05 to 0.2 %. Ip *et al.* (1990) derived an abundance upper limit of 0.02 % for HCN in P/Halley from the *IMS* data; this might be due to flaws in their chemical model. Whether or not the observed HCN abundance can explain all the CN radicals observed in the visible is still a debated question. From first evaluations this is not the case (Crovisier and Schloerb, 1991). Other potential CN-parents (HC_3N, CH_3CN...) were unsuccessfully searched for at radio wavelengths (Bockelée-Morvan *et al.* 1987; Crovisier and Schloerb, 1991; Crovisier *et al.* 1993), and an early radio detection of CH_3CN (Ulich and Conklin, 1974) could never be confirmed. The possibility that CN could be released from organic grains has been invoked (see the discussion in Greenberg and Shalabiea, 1994).

Molecular nitrogen N_2 has no favourable spectral transition to be searched for. Wyckoff *et al.* (1991a) estimated its abundance to 0.02 % in P/Halley from the spectral signal of the $N_2{}^+$ ion in the visible. Observations of comet Bradfield 1987 XXIX by Lutz *et al.* (1993) suggest an abundance an order of magnitude higher. No stringent upper limit could be obtained from the *Giotto* mass spectrometers due to mass ambiguity with CO (Balsiger *et al.* 1986; Eberhardt *et al.* 1987). The $[N_2]/[NH_3]$ ratio appears to be much smaller in comets than in star-forming regions (where N_2 is traced by $N_2{}^+$; Womack *et al.* 1992). This may be interpreted either by a complete conversion of N_2 into NH_3 in the presolar nebula, or by the incapacity of comet nuclei to retain the highly volatile N_2.

Like molecular nitrogen, other very volatile species such as molecular hydrogen [which was searched for in the IR by Smyth *et al.* (1989) and in the UV as reported by Feldman (1991)] and noble gases [searched for in the UV by Feldman *et al.* (1991) and Stern *et al.* (1992)] are depleted in comets.

3. 7. SULPHUR SPECIES

H_2S was detected at millimetre wavelengths in comets Austin 1990 V, Levy 1990 XX (Crovisier *et al.* 1991) and P/Swift-Tuttle (Despois *et al.* 1994). Assuming a rotational excitation temperature similar to that of CH_3OH leads to abundances 0.1 to 0.3 %, depending on the comet. H_2S is also suspected in P/Halley from a peak at 35 amu/e in the data of the *PICCA* and *IMS* instruments on *Giotto* (Marconi *et al.* 1990; Geiss *et al.* 1991). S_2 was identified in *IUE* UV spectra uniquely in comet IRAS-Araki-Alcock 1983 VII during an outburst (A'Hearn *et al.* 1983; Budzien and Feldman, 1992). The abundance is 5×10^{-4}, but this species may come from processing of the nucleus surface rather than being primordial.

No other sulphur parent molecule has yet been identified, but it seems very likely that the CS radical, currently observed in *IUE* UV spectra, comes from the CS_2 molecule, with abundances of the order of 0.1 % (Feldman 1991). Significant upper limits were obtained for OCS, H_2CS, SO and SO_2 (Table II). Whether or not sulphur is depleted with respect to oxygen in cometary volatiles is still a debated question (Azoulay and Festou 1986; Crovisier *et al.* 1991; Mumma *et al.* 1993b). The abundance of the sulphur molecules may not be enough to explain the production rate of the S atoms observed in the UV and other sulphur compounds may exist in comets.

3. 8. AROMATICS

The presence of aromatics is suggested by their characteristic emission feature at 3.28 μm. This feature was especially conspicuous in the spectrum of comet Levy 1990 XX (Davies *et al.* 1991), but it is apparently variable from comet to comet (Bockelée-Morvan *et al.* 1994a). Its intensity may be explained by a relatively small abundance of aromatics, if it is emitted after electronic excitation in the UV followed by internal conversion to vibrational excitation, as is invoked for the emission of similar bands of PAHs in the interstellar medium, rather than by resonant fluorescence : Bockelée-Morvan *et al.* (1994a) retrieved abundances of the order of 10^{-5}. Moreels *et al.* (1994) identified phenanthrene, a three-cycle PAH, in the visible spectrum of P/Halley recorded by the *VEGA-TKS* instrument, for which they estimated a production rate of 0.15% relative to water. Infrared and visible aromatics production rates are therefore controversial.

4. Pending problems

4. 1. WHAT SPECIES REMAIN TO BE IDENTIFIED ?

Water represents only 80 % (by number) of the cometary gases coming from the nucleus. It is suggested from Table II that the observed species cannot fully account for the remaining 20 %, and that a significant fraction (possibly 5 to 10 %) consists of species still to be identified. There are evidences of mass spectroscopy peaks at medium and high m/q which are still to be identified.

The problem of the origin of the cometary radicals observed in the visible is not fully solved : C_2 and C_3 are orphan radicals for which parents are still to be found. The CN radical production rate is not completely explained by HCN. It must be also noted that numerous features in visible cometary spectra are still unidentified (Arpigny *et al.* 1993); most of them are presumably due to radicals and ions.

The origin of the 3.4 μm cometary emission has been a puzzle since its discovery (Encrenaz *et al.* . 1988; Encrenaz and Knacke, 1991). Thermal emission of small grains with an organic mantle and/or fluorescence of large molecules excited by solar UV were invoked because they could be caused by a small amount of matter and we were not expecting a large fraction of cometary volatiles besides water. This situation has to be reconsidered since we know now that methanol is present in comets with a large abundance (more than 5 % in some objects). However, methanol alone can only explain about half of the 3.4 μm emission (Reuter 1992; Bockelée-Morvan *et al.* 1994a). The remaining emission, which peaks at 3.43 μm, could be explained by a few percent of *CHO* species, but no definite identification could yet be proposed. Some species like HCOOH or C_2H_5OH are ruled out by radio observations (Table II).

A related problem may be that of the 2.8 μm excess emission (Tokunaga *et al.* 1987). In addition to hot vibrational bands of water (as was formerly proposed by Bockelée-Morvan and Crovisier 1989), the OH stretching-mode of other *CHO* species may be contributing to this emission.

4. 2. OUTGASSING FROM GRAINS

One of the new ideas on cometary processes that emerged from the exploration of Halley's comet was that a significant fraction of cometary volatiles could directly come from dust grains. Evidences for such a process are coming from the discovery of an important fraction of dust particles composed of *CHON* elements, thus presumably of organic molecules, and the fact already mentioned above that species such as CO and H_2CO are coming, at least partly, from a distributed source. From the space distribution of CN and possibly of other radicals, it was also suspected that part of these radicals may not come from gas-phase parent molecules, although the relation of such a source to dust is not yet clear.

The release of molecules and radicals from grains is an ill-known mechanism. That water and other volatiles could come from icy grains is not so new an idea. Such grains, however, would have a very short lifetime (at least at $r = 1$ AU). Their expected scale length is such that molecules outgassing from such grains should be indistinguishable from molecules directly coming from the nucleus by ground-based observations. It thus seems unreasonable to explain, for instance, the CO distribution (which needs a source with a scale length of the order of 10^4 km) by CO-coated grains. Photosputtering is presumably inefficient to explain large production rates. A more realistic process could be the progressive sublimation of *semi refractories* such as some polymers (like polyoxymethylene) or small PAHs, or the pyrolysis of organic refractories (Greenberg and Shalabiea, 1994).

4. 3. VARIATIONS FROM COMET TO COMET

It is very important to study whether all comets show similar molecular abundances. Variations could betray different origins or different histories. The data for parent molecules are still very sparse, but a high comet-to-comet variability is suggested for methanol or formaldehyde. Schleicher (1994) reports a systematic study of the relative production rates of radicals : a rather homogeneous [CN]/[OH] ratio is found, but the existence of two populations is suspected from the $[C_2]/[OH]$ ratio. Extreme cases are comet Yanaka 1988 XXIV, dramatically depleted in C_2 and CN, but with conspicuous NH_2 bands (Fink 1992; see Greenberg *et al.* 1993 for a discussion) and P/Wolf-Harrington, depleted in C_2 and C_3 compared to CN and OH (Schleicher *et al.* 1993).

4. 4. VARIATION WITH HELIOCENTRIC DISTANCE; ACTIVITY OF DISTANT COMETS

The activity of comets at large heliocentric distances has been known for a long time (see Table 1 in Meech 1991). This topic dramatically came back with the observation of an outburst of P/Halley at 14 AU from the Sun (West *et al.* 1991) and was recently the subject of a workshop (Huebner *et al.* 1993). Since water can not likely sublimate at more than 4–5 AU from the Sun, more volatile species must be responsible for outgassing at these large distances. Very little could be learned from the cometary spectra, because of the very low fluorescence rates at large geocentric distances. The P/Halley spectrum during its outburst was featureless. The CN band was detected in the spectrum of Chiron at $r = 10$ AU (Bus *et al.* 1991). CN and CO^+ were detected in P/Schwassmann-Wachmann 1 at $r = 6$ AU (Cochran and Cochran, 1991). These detections are puzzling, since HCN, the plausible parent for CN, is not expected to sublimate at Chiron's distance, and the ionization rate of CO at the heliocentric distance of P/Schwassmann-Wachmann 1 should be excessively small. Carbon monoxide is the most plausible volatile to explain cometary activity at large distance, but its direct identification will not be easy.

All the present determinations of cometary volatile abundances rely on the gas phase coma observed at heliocentric distances of the order of 1 AU. One cannot infer the nucleus volatile composition without taking into account the important sublimation fractionation effects (Espinasse *et al.* 1991; Prialnik 1994; Rickman 1994). The coma composition of distant comets may be surprisingly different.

5. Conclusion

We cannot yet propose a "standard" cometary composition, because we are still lacking data on several key molecules which can difficultly be observed from the ground, and also because this composition may highly vary from comet to comet. Table II summarizes our present knowledge of the subject. The composition is strikingly similar to that observed in interstellar matter (Yamamoto 1991; Greenberg and Shalabeia, 1994), but it is still premature to draw a firm conclusion on the origin of comets. Awaiting accurate analyses from future space missions such as *Rosetta*, we can expect important progresses on our knowledge of cometary composition from the development of the microwave and infrared spectroscopic techniques in the near future.

References

A'Hearn, M.F., Feldman, P.D. and Schleicher, D.G. : 1983, "The discovery of S_2 in comet Iras-Araki-Alcock 1983d." ***Astrophys. J. Lett.***, **274**, L99–L103.

Allen, M., Delitsky, M., Huntress, W. et al. : 1987, "Evidence for methane and ammonia in the coma of P/Halley." ***Astron. Astrophys.***, **187**, 502–512.

Altenhoff, W.J., Batrla, W., Huchtmeier, W.K. et al. : 1984, "Radio observations of comet 1983d." ***Astron. Astrophys. Lett.***, **125**, L19–L22.

Arpigny, C., Rahe, J., Donn, B., Dossin, B. and Wyckoff, S. : 1994, ***Atlas of Cometary Spectra***, Reidel, Dordrecht, in press.

Azoulay,G. and Festou, M.C. : 1986, "The abundance of sulphur in comets." In ***Asteroids, Comets, Meteors II*** (C.-I. Lagerkvist, B.A Lindblad, H. Lunds, H. Rickman, Eds.), 273–277, Uppsala Univ. Reprocentralen.

Balsiger, H., Altwegg, K., Buhler, F. et al. : 1986, "Ion composition and dynamics at comet Halley." ***Nature***, **321**, 330–334.

Bockelée-Morvan, D., Crovisier, J., Despois, D. et al. : 1987, "Molecular observations of comets P/Giacobini-Zinner 1984e and P/Halley 1982i at millimetre wavelengths." ***Astron. Astrophys.***, **180**, 253–262.

Bockelée-Morvan, D. and Crovisier, J. : 1989, "The nature of the 2.8 μm feature in cometary spectra." ***Astron. Astrophys.***, **216**, 278–283.

Bockelée-Morvan, D., Crovisier, J., Colom, P., Despois, D. and Paubert, G. : 1990, "Observations of parent molecules in comets at millimetre wavelengths : HCN, H_2S, H_2CO and CH_3OH." *ESA–SP315*, 143–148.

Bockelée-Morvan, D., Colom, P., Crovisier, J., Despois, D. and Paubert, G. : 1991, "Microwave detection of hydrogen sulphide and methanol in comet Austin (1989c1)." ***Nature***, **350**, 318–320.

Bockelée-Morvan, D. and Crovisier, J. : 1992, "Formaldehyde in comets : II. Excitation of the rotational lines." ***Astron. Astrophys.***, **264**, 282–291.

Bockelée-Morvan, D., Brooke, T.Y. and Crovisier, J. : 1994a, "On the origins of the 3.2-3.6 μm emission feature in comets." ***Icarus***, submitted.

Bockelée-Morvan, D., Crovisier, J., Colom, P. and Despois, D. : 1994b, " The rotational lines of methanol in comets Austin 1990 V and Levy 1990 XX." ***Astron. Astrophys.***, in press.

Bockelée-Morvan, D., Padman, R., Davies, D.K. and Crovisier, J. : 1994c, "Observations of submillimetre lines of CH_3OH, HCN and H_2CO in comet P/Swift-Tuttle with the James Clerk Maxwell Telescope." ***Planet. Space Sci.***, in press.

Brooke, T.Y., Tokunaga, A.T., Weaver, H.A., Chin, G. and Geballe, T.R. : 1991, "A sensitive upper limit on the methane abundance in comet Levy (1990c)." ***Astrophys. J. Lett.***, **372**, L113–L116.

Budzien, S.A. and Feldman, P.D. : 1992, "Upper limits to the S_2 abundance in several comets observed with the International Ultraviolet Explorer." ***Icarus***, **99**, 143–152.

Bus, S.J., A'Hearn, M.F., Schleicher, D.G. and Bowell, E. : 1991, "Detection of CN emission from (2060) Chiron." ***Science***, **251**, 774–777.

Cochran, A.L. and Cochran, W.D. : 1991, "The first detection of CN and the distribution of CO^+ gas in the coma of comet P/Schwassmann-Wachmann 1." ***Icarus***, **90**, 172–175.

Colom, P., Crovisier, J., Bockelée-Morvan, D., Despois, D. and Paubert, G. : 1992, "Formaldehyde in comets : I." ***Astron. Astrophys.***, **264**, 270–281.

Combes, M., Moroz, V.I., Crovisier, J. et al. : 1988, "The 2.5 to 12 μm spectrum of comet Halley from the IKS-VEGA experiment." ***Icarus***, **76**, 404–436.

Crovisier, J. : 1984, "The water molecule in comets : fluorescence mechanisms and thermodynamics of the inner coma." ***Astron. Astrophys.***, **130**, 361–372.
Crovisier, J. : 1987, "Rotational and vibrational synthetic spectra of linear parent molecules in comets." ***Astron. Astrophys. Suppl.***, **68**, 223–258.
Crovisier, J. and Encrenaz, T. : 1983 : "Infrared fluorescence of molecules in comets : the general synthetic spectrum." ***Astron. Astrophys.***, **126**, 170–182.
Crovisier, J. and Schloerb, F.P. : 1991, "The study of comets at radio wavelengths." In ***Comets in the Post-Halley Era*** (R.L. Newburn Jr., M. Neugebauer, J. Rahe, Eds.), 149–173, Kluwer, Dordrecht.
Crovisier, J., Despois, D, Bockelée-Morvan, D., Colom, P. and Paubert, G. : 1991, "Microwave observations of hydrogen sulfide and searches for other sulfur compounds in comets Austin (1989c1) and Levy (1990c)." ***Icarus***, **93**, 246–258.
Crovisier, J., Bockelée-Morvan, D., Colom, P., Despois, D. and Paubert, G. : 1993, "A search for parent molecules at millimetre wavelengths in comets Austin 1990 V and Levy 1990 XX : upper limits for undetected species." ***Astron. Astrophys.***, **269**, 527–540.
Davies, J.K., Green, S.F. and Geballe, T.R. : 1991, "The detection of a strong 3.28 μm emission feature in comet Levy." ***Mon. Not. Roy. Astron. Soc.***, **251**, 148–151.
Davies, J.K., Mumma, M.J., Reuter, D.C., et al. : 1993, "The 3.2-3.6 micron spectrum of comet P/Swift-Tuttle." ***Mon. Not. Roy. Astron. Soc.***, **251**, 1022–1026.
Despois, D., Paubert, G., Colom, P., Bockelée-Morvan, D. and Crovisier, J. : 1994, "Observations of comet P/Swift-Tuttle 1992t at IRAM." ***Planet. Space Sci.***, in press.
DiSanti, M.A., Mumma, M.J. and Lacy, J.H. : 1992a : "A sensitive upper limit to OCS in comet Austin (1989c1) from a search for ν_2 emission at 4.85 μm." ***Icarus***, **97**, 155–158.
DiSanti, M.A., Mumma, M.J., Lacy, J.H. and Parmar, P. : 1992b, "A possible detection of infrared emission from CO in comet Austin (1989c1)." ***Icarus***, **96**, 151–160.
Drapatz, S., Larson, H.P. and Davis, D.S. : 1987, "Search for methane in comet P/Halley." ***Astron. Astrophys***, **187**, 497–501.
Eberhardt, P., Krankowsky, D., Schulte, U. et al. : 1987, "The CO and N_2 abundance in comet P/Halley." ***Astron. Astrophys.***, **187**, 481–484.
Eberhardt, P., Meier, R., Krankowsky, D. and Hodges, R.R. : 1991, "Methanol abundance in comet P/Halley from in situ measurements." ***B.A.A.S.***, **23**, 1161.
Encrenaz, T., d'Hendecourt, L. and Puget, J.L. : 1988, "The interpretation of the 3.2–3.5 μm emission feature in the spectrum of comet P/Halley : abundances in the comet and in interstellar matter." ***Astron. Astrophys.***, **207**, 162–173.
Encrenaz, T. and Knacke, R.F. : 1991, "Carbonaceous compounds in comets : infrared observations." In ***Comets in the Post-Halley Era*** (R.L. Newburn Jr., M. Neugebauer, J. Rahe, Eds.), 107–137, Kluwer, Dordrecht.
Espinasse, S., Klinger, J., Ritz, C. and Schmitt, B. : 1991, "Modelling of the thermal behaviour and of the chemical differentiation of cometary nuclei." ***Icarus***, **92**, 350–365.
Feldman, P.D. : 1991, "Ultraviolet spectroscopy of cometary comae." In ***Comets in the Post-Halley Era*** (R.L. Newburn Jr., M. Neugebauer, J. Rahe, Eds.), 139–148, Kluwer, Dordrecht.
Feldman, P.D. and Brune, W.H. : 1976, "Carbon production in comet West (1975n)." ***Astrophys. J. Lett.***, **209**, L45–L48.
Feldman, P.D., Davidsen, A.F., Blair, W.P. et al. : 1991, "Observations of comet Levy (1990c) with the Hopkins ultraviolet telescope." ***Astrophys. J. Lett.***, **379**, L37–L40.
Feldman, P.D., Fournier, K.B., Grinin, V.P. and Zvereva, A.M. : 1993, "The abundance of ammonia in comet P/Halley derived from ultraviolet spectrophotometry of NH by ASTRON and IUE." ***Astrophys. J.***, **404**, 348–355.
Festou, M.C., Rickman, H. and West, R.M. : 1993, "Comets." ***Astron. Astrophys. Rev.***, **4**, 363-447 and **5**, 37–163.
Fink, U. : 1992, "Comet Yanaka (1988r) : a new class of carbon-poor comet." ***Science***, **257**, 1926–1929.

Geiss, J., Altwegg, K., Anders, E. et al. : 1991, "Interpretation of the ion mass spectra in the mass per charge range 25-35 amu/e obtained in the inner coma of Halley's comet by the HIS-sensor of the Giotto IMS experiment." ***Astron. Astrophys.***, **247**, 226–234.

Greenberg, J.M. and Shalabiea, O. : 1994, "Comets as a reflection of interstellar medium chemistry." This volume.

Greenberg, J.M., Singh, P.D. and de Almeda, A.A. : 1993, "What is new about the new comet Yanaka (1988r) ?" ***Astrophys. J. Lett.***, **414**, L45–L48.

Haider, S.A., Bhardwaj, A. and Singhal, R.P. : 1993, "Role of auroral and photoelectrons on the abundance of methane and ammonia in the coma of comet Halley." ***Icarus***, **101**, 234–243.

Hoban, S., Mumma, M., Reuter, D. et al. : 1991, "A tentative identification of methanol as the progenitor of the 3.52 μm feature in several comets." ***Icarus***, **93**, 122–134.

Hoban, S., Reuter, D.C., DiSanti, M.A., Mumma, M.J. and Elston, R. : 1994, "Infrared observations of methanol in comet P/Swift-Tuttle." ***Icarus***, in press.

Huebner, W.F., Keller, H.U., Jewitt, D., Klinger, J. and West, R., Eds : 1993, ***Workshop on the Activity of Distant Comets***, Southwest Research Institute, San Antonio, Texas.

Huebner, W.F., Snyder, L.E. and Buhl, D. : 1974, "HCN radio emission from comet Kohoutek (1973f)." ***Icarus***, **23**, 580–585.

Ip, W.-H., Balsiger, H., Geiss, J. et al. : 1990, "Giotto IMS measurements of the production rate of hydrogen cyanide in the coma of comet Halley." ***Ann. Geophys.***, **8**, 319–326.

Kawara, K., Gregory, B., Yamamoto, T. and Shibai, H. : 1988, "Infrared spectroscopic observation of methane in comet P/Halley." ***Astron. Astrophys.***, **207**, 174–181.

Kim, S.J. and A'Hearn, A.F. : 1991, "Upper limits for SO and SO_2 in comets." ***Icarus***, **90**, 79–95.

Kim, S.J., A'Hearn, M.F. and Larson, S.M. : 1990, "Multi-cycle fluorescence : application to S_2 in comet Iras-Araki-Alcock 1983d." ***Icarus***, **87**, 440–451.

Krankowsky, D. : 1991, "The composition of comets." In ***Comets in the Post-Halley Era*** (R.L. Newburn Jr., M. Neugebauer, J. Rahe, Eds.), 855–877, Kluwer, Dordrecht.

Krankowsky, D., Lammerzahl, P., Herrwerth, I. et al. : 1986, "In situ gas and ion measurements at comet Halley." ***Nature***, **321**, 326–329.

Krasnopolsky, V.A., Gogoshev, M., Moreels, G. et al. : 1986, "Spectroscopic study of comet Halley by the Vega 2 three-channel spectrometer." ***Nature***, **321**, 269–271.

Lagerkvist, C.-I., Rickman, H., Lindblad, B.A. and Lindgren, M., Eds. : 1990, ***Asteroids Comets Meteors III***, Uppsala Univ. Reprocentralen.

Larson, L., Weaver, H.A., Mumma, M.J. and Drapatz, S. : 1988, "Airborne infrared spectroscopy of comet Wilson (1986l) and comparisons with comet Halley." ***Astrophys. J.***, **338**, 1106–1114.

Lutz, B.L., Womack, M. and Wagner, R.M. : 1993, "Ion abundances and implications for photochemistry in comets Halley (1986 III) and Bradfield (1987 XXIX)." ***Astrophys. J.***, **407**, 402–411.

Marconi, M.L., Mendis, D.A., Korth, A. et al. : 1990, "The identification of H_3S^+ with the ion of mass per charge (m/q) 35 observed in the coma of comet Halley." ***Astrophys. J. Lett.***, **352**, L17–L20.

Meech, K.J. : 1991, "Physical aging in comets." In ***Comets in the Post-Halley Era*** (R.L. Newburn Jr., M. Neugebauer, J. Rahe, Eds.), 629–669, Kluwer, Dordrecht.

Meier, R., Eberhardt, P., Krankowsky, D. and Hodges, R.R. : 1993, "The extended formaldehyde source in comet P/Halley." ***Astron. Astrophys.***, **277**, 677–690.

Mitchell, D.L., Lin, R.P., Carlson, C.W. et al. : 1992, "The origin of complex organic ions in the coma of comet Halley." ***Icarus***, **98**, 125–133.

Moreels, G., Clairemidi, J., Hermine, P., Brechignac, P. and Rousselot, P. : 1994, "Detection of a polycyclic aromatic molecule in P/Halley." ***Astron. Astrophys.***, in press.

Mumma, M.J. and Reuter, D.C. : 1989, "On the identification of formaldehyde in Halley's comet." ***Astrophys. J.***, **344**, 940–948.

Mumma, M.J., Weaver, H.A., Larson, H.P., Davis, D.S. and Williams, M. : 1986, "Detection of water vapor in Halley's comet." *Science*, **232**, 1523–1528.
Mumma, M.J., Weissman, P.R., and Stern, S.A. : 1993b, "Comets and the origin of the solar system : reading the Rosetta stone." In ***Protostars and Planets III*** (E.H. Levy, J.I Lunine, Eds.), 1177–1252, University of Arizona Press, Tucson.
Prialnik, D. : 1994, "Modelling the structure and activity of comet nuclei." This volume.
Reuter, D.C. : 1992, "The contribution of methanol to the 3.4 μm emission feature in comets." ***Astrophys. J.***, **386**, 330–335.
Reuter, D.C., Hoban, S. and Mumma, M.J. : 1992, "An infrared search for formaldehyde in several comets." ***Icarus***, **95**, 329–332.
Rickman, H. : 1994, "Comet nuclei." This volume.
Schleicher, D.G. : 1994, "Comet taxonomy and evolution." This volume.
Schleicher, D.G., Bus, S.J., Osip, D.J. : 1993, "The anomalous molecular abundances of comet P/Wolf-Harrington." ***Icarus***, **104**, 157–166.
Schloerb, F.P. and Ge, W. : 1992, "Submillimeter line observations of comet Levy (1990c)." In ***Asteroids, Comets, Meteors 1991,*** (A.W. Harris, E. Bowell, Eds.), 533–536, Lunar and Planetary Institute, Houston.
Smyth, W.A., Wolstencroft, R.D. and Lutz, B.L. : 1989, "The 2.4 micron spectrum of comet Halley : a search for H_2 emission." ***The Observatory***, **109**, 19–23.
Snyder, L.E., Palmer, P. and de Pater, I. : 1989, "Radio detection of formaldehyde emission from comet Halley." ***Astron. J.***, **97**, 246–253.
Stern, S.A., Green, J.C., Cash, W. and Cook, T.A. : 1992, "Helium and argon abundance constraints and the thermal evolution of comet Austin (1989c1)." ***Icarus***, **95**, 157–161.
Tokunaga, N.T., Nagata, T. and Smith, R.G. : 1987, "Detection of a new emission band at 2.8 μm in comet P/Halley." ***Astron. Astrophys.***, **187**, 519–522.
Ulich, B.L. and Conklin, E.K. : 1974, "Detection of methyl cyanide in comet Kohoutek." ***Nature***, **248**, 121–122.
Weaver, H.A., Feldman, P.D., McPhate, J.B. et al. : 1994, "Detection of CO Cameron band emission in comet P/Hartley 2 (1991 XV) with the Hubble Space Telescope." ***Astrophys. J.***, in press.
Weaver, H.A. and Mumma, M.J. : 1984, "Infrared molecular emissions from comets." ***Astrophys. J.***, **276**, 782–797.
Weaver, H.A., Mumma M.J. and Larson, H.P. : 1991, "Infrared spectroscopy of cometary parent molecules." In ***Comets in the Post-Halley Era*** (R.L. Newburn Jr., M. Neugebauer, J. Rahe, Eds.), 93–106, Kluwer, Dordrecht.
Weaver, H.A., Mumma, M.J., Larson, H.P. and Davis, D.S. : 1986, "Post-perihelion observations of water in comet Halley." ***Nature***, **324**, 441–444.
West, R.M., Hainaut, O. and Smette, A. : 1991, "Post-perihelion observations of P/Halley. III. An outburst at r = 14.3 AU." ***Astron. Astrophys. Lett.***, **246**, L77–L80.
Womack, M., Wickoff, S. and Ziurys, L.M. : 1992, "Observational constraints on solar nebula nitrogen chemistry : N_2/NH_3." ***Astrophys. J.***, **401**, 728–735.
Wootten, A., Latter, W. and Despois, D. : 1994, "HCN emission from comet P/Swift-Tuttle." ***Planet. Space Sci.***, in press.
Wyckoff, S.W., Tegler, S.C. and Engel, L. : 1991a, "Nitrogen abundance in comet Halley." ***Astrophys. J.***, **367**, 641–648.
Wyckoff, S.W., Tegler, S.C. and Engel, L. : 1991b, "Ammonia abundances in four comets." ***Astrophys. J.***, **368**, 279–286.
Yamamoto, T. : 1982, "Evaluation of infrared line emission from constituent molecules of cometary nuclei." ***Astron. Astrophys.***, **109**, 326–330.
Yamamoto, T. : 1991, "Chemical theories on the origin of comets." In ***Comets in the Post-Halley Era*** (R.L. Newburn Jr., M. Neugebauer, J. Rahe, Eds.), 361—376, Kluwer, Dordrecht.

COMETS AS A REFLECTION OF INTERSTELLAR MEDIUM CHEMISTRY

J. MAYO GREENBERG AND OSAMA M. SHALABIEA
Huygens Laboratory, University of Leiden, Leiden, The Netherlands
E–mail MAYO@RULHL1.LEIDENUNIV.NL

Abstract. A brief summary of the observed infrared and other properties of interstellar dust is given. Chemical, physical and morphological criteria are discussed concerning the degree to which there are constraints relating comets to interstellar dust chemistry representative of the presolar nebula. Results of theoretical modelling of dust and gas evolution in dense clouds are used to compare with observed dust composition . Sources of the distribution of simple as well as complex molecules in the coma are related to what is presently known about the volatile ices in interstellar dust and to processes leading to evaporation of organic "refractory" grain mantle material represented by laboratory residues produced by photoprocessing of ices. The criterion of preservation of interstellar volatiles in comets leads to the further criterion that the ice in comets is amorphous. Criteria for relating interstellar dust volatiles to asteroids are discussed.

1. Introduction

The chemistry of the interstellar medium is seen in the distribution of the gaseous molecules and in the molecular composition of the dust. The former has been primarily obtained by the use of infrared in the 2.5 μm to 15 μm range. Both of these kinds of observations are direct in the sense that the molecules detected are the molecules present. On the other hand, the chemistry of comets; i.e., the chemical composition of the nuclei of comets, is based on deduction and induction from observations of the volatile and refractory components in the comet coma, both of which have generally undergone considerable changes from their original nucleus abundances resulting from physical and chemical processes after leaving the nucleus. The original or parent comet molecules are seldom observed directly in the coma. Even the most abundant water molecule is mostly studied via its dissociation products OH and H and has been directly observed in only two comets : Halley and Wilson 1987 VII (see Crovisier 1994, this volume and Mumma et al., 1993). Thus the relative abundances of most of the commonly assumed parent molecules are not simply obtained. A major exception is H_2O which is presently believed to be known with sufficient reliability that it is generally used as the reference molecule for all the components of the comet.

There is no general agreement on the degree to which the interstellar dust has been evaporated or modified prior to comet formation. Not all alternatives are reviewed here. Emphasis will be given to those properties of dust and comets, such as the degree of amorphicity of solid H_2O which may impose a strict constraint on whether or to what degree the interstellar dust has evaporated and recondensed.

Even were we to assume that the comet nucleus composition is *exactly* that of the presolar interstellar medium, the question remains what *is* the presolar interstellar medium? There exist neither observations of an interstellar cloud at the stage of

A. Milani et al. (eds.), Asteroids, Comets, Meteors 1993, 327–342.

collapse just *before* a star is born nor, as yet, any good theories for the chemistry of the gas and dust in such a very dense cloud. And, as already noted, even if the precometary medium composition were known, the comet nucleus composition is not, at least until there has been a successful comet nucleus sample return mission.

The problem, then, of establishing a precise correspondence between presolar chemistry and comet composition involves, in the former, a forward extrapolation from observed conditions before star formation or a backward extrapolation of the medium after star formation, and for the latter, a time reversed extrapolation of the composition in the comet coma.

The correspondence between molecular abundance ratios in comet nuclei and in the interstellar medium out of which they are formed is thus fraught with potential pitfalls. *Both* the chemistry *and* the morphology of the interstellar dust are needed to provide the connections to comet nuclei.

In this paper we consider both refractory and volatile components of the comet nucleus. It is important to do this because they are not entirely separate. First, there is a wide range of volatility in the so-called refractory dust components and secondly, some of the simple molecules of the gaseous coma are derivatives of complex dust organic "refractory" molecules and not daughters of frozen cometary volatiles. With this caveat in mind we shall begin by summarizing some of the known properties of interstellar gas and dust with emphasis on those aspects which seem most likely to bear on comet chemistry.

One of the clues indicating the proper approach to be taken is the H_2O to CO ratio in molecular clouds and in comets. While it is well known that CO is by far the most abundant gas phase molecule in the interstellar medium (obviously excepting H_2) it is substantially less abundant in the comet coma than H_2O. On the other hand *solid* H_2O is the most abundant mantle molecule on the interstellar grains, generally exceeding CO by a factor of 5 or more. It is therefore logical to place particular emphasis on the interstellar *dust* molecules both as observed and as deduced from dust/gas cloud chemistry as an initial basis for considering comet composition.

2. Interstellar Cloud Chemistry

We shall first summarize some of the observational evidence for the molecular abundances in grains. Next we shall give an example of a dust/gas chemistry scheme which provides a basis for explaining simultaneously the dust mantle and gas phase abundances.

2. 1. INTERSTELLAR DUST OBSERVATIONS

2.1.1 Refractory components :

Silicates :

In diffuse interstellar clouds the dust consists of at least three and possibly four separate components differing in both size and chemical composition. The major fraction by mass is in the particles whose mean size is in the order of one tenth micron. These particles consist of a core of silicates which is both amorphous and elongated- a seeming contradiction in properties. Nevertheless, observations of the excess polarization produced by interstellar dust in the 9.7 μm Si-O stretch band as seen, for example, in the Orion BN source (see Whittet, 1992 for a review) and in other sources as well show that the silicate is elongated (certainly nonspherical). Furthermore this band has the width characteristic of a generalized amorphous silicate (Day, 1979). It has been conjectured that, although the material is indeed amorphous, it is elongated by virtue of consisting of two or several particles which somehow stick together, when formed in the atmosphere of cool evolved oxygen rich stars (Greenberg, 1989). The fact that the H_2O ice band (see later) also exhibits excess linear polarization indicates that it forms as a mantle on the silicate (Dyčk and Lonsdale, 1981). Observations of the abundances of Si, Mg, and Fe in the clouds indicates that of the order of 95% of these elements are found in the amorphous silicates. Although the 9.7 μm absorption is always observed where interstellar extinction exists, the cosmic (solar system) abundances of the rocky elements (Si, Mg, Fe) is insufficient to provide, as particulates, the observed total extinction (Spitzer, 1978; Greenberg, 1991a). An additional component is required (see next).

The organic refractory component :

In diffuse clouds an insufficiency of rocky elements to produce the observed amount of interstellar extinction by silicates alone, suggests that mantles of the organics - OCN elements - must be involved. If one examines the spectrum toward the galactic center, the silicate feature far outweighs any evidence for mantles - certainly the ice band is absent. But the apparently insignificant (with respect to the "normally" strong ice bands) feature at 3.4 μm characteristic of CH stretches in organic mixtures is *the* evidence for the organic mantle needed. Spectroscopically it compares well with that of the organic residue which results from long term laboratory photoprocessing of ices (Schutte and Greenberg, 1986; Sandford et al, 1991). These molecules are presumed to be created in molecular clouds by photoprocessing of the ices. It was predicted that such a mantle material should be observed wherever the ice band is not present (Greenberg, 1973) and has now been observed not only towards the galactic center but also towards the object for which it was originally predicted, namely VI Cygni# 12 (Sandford et al. 1991). In order to account for the observed continuum properties of interstellar dust using silicates and organic refractories alone in the diffuse interstellar medium a mass ratio of organic to silicates of the order of one is required (Greenberg, 1982b). One would expect the organic refractory dust mantle to be little changed when the diffuse cloud dust accretes a volatile icy mantle in the molecular cloud phase. The absence or reduction of the 3.4 μm feature in the extended line of sight toward protostellar objects (Allamandola et al. 1992) applies to molecular cloud material which has undergone various degrees of processing (heating, etc.) following star formation and it would certainly be a puzzle if it could also be demonstrated for *pre*stellar molecular cloud dust as seen toward, say, Elias 16.

Small carbonaceous : The smaller ($\leq$ 0.01 μm) particulates which are variously suggested to provide primarily the ultraviolet extinction are all carbon rich (Jenniskens et al, 1993). Their abundances are not well established but it has been suggested that between $\simeq$ 5 and 10% of the total available carbon is contained in large polycyclic aromatic hydrocarbons (PAH's) (Allamandola et al., 1989) and as much as $\simeq$ 20% of the carbon is required to account for the 220 nm extinction (absorption) bump (Greenberg, 1982a).

2.1.2 Volatile ice components :

The volatile components of interstellar dust are no more uniformly observed than the gaseous components in molecular clouds. However, just as CO is a ubiquitous gas phase molecule whose abundance in a variety of clouds is high and rather uniform so H_2O ice is the most generally observed volatile molecule in the dust. In the following we will briefly summarize the observational evidence for the ice mantle molecules.

H_2O :

Water ice has now been observed in a wide variety of astrophysical regions. It was first observed in the Orion BN-KL source and since then in many other regions of new star formation as well as in more normal clouds. Its most generally observed feature is at about 3.07 μm. The 6 μm feature is not as frequently observed and the 12 μm feature is blended with the 9.7 μm silicate feature making it difficult, but not impossible, to derive molecular abundances. Not all the interpretations of the observations agree with respect to the actual abundance of the H_2O molecule. This is partially because the effects of mixtures of other molecules on the 3.07 μm H_2O absorption strength and also the effect of the *particle shape* on the band *absorption shape* are not usually considered, as, for example in van de Bult et al. (1984). The excess polarization at 3.07 μm shows in Elias 16 as in BN that the ice has a non spherical form. The optical depth relative to the extinction of $\tau_{3.07}/A_v = 0.059 \pm 0.003$ obtained by Smith et al. (1993) in the Taurus dark cloud yields an $[H_2O/H]$ ratio of $\simeq 5 \times 10^{-5}$, using $A_v/N_H = 0.5 \times 10^{-21}$, $\sigma_{H_2O} \simeq 6 \times 10^{-19}$ cm^2. The solid H_2O optical depth at 3.07 μm is reduced in an ice mixture because of the diluting presence of nonpolymeric H_2O. This reduction has been estimated to be as much as by a factor of the order of 0.5 for a fractional ice composition of 60% (Greenberg, 1982b) so that the solid H_2O estimate in the above cloud could be as high as 10^{-4}. The H_2O seen through the cool molecular cloud towards Elias 16 is well characterized by an ice band which matches that for non-spherical grain models with amorphous ice mantles (van de Bult, Greenberg and Whittet, 1984). Its abundance is $[H_2O/H] = 7 \times 10^{-5}$ is compared with 2×10^{-5} for gaseous CO along the same line of sight and with 1×10^{-5} solid CO (Tielens et al, 1991). We do not here generally consider protostars or YSO (young stellar objects) spectra as representative of interstellar dust for comets because there is considerable modification of the relative proportions of the various components caused by heating following star formation. This is particularly true for CO which evaporates much more readily than the H_2O. Although, as pointed out by Tielens and Allamandola (1987), the optical depth at 3.07 μm may be less reliable than that at 6 μm for obtaining H_2O abundances it is much more frequently observed and provides a far greater data base. In any case, the fact that H_2O is the most abundant solid phase molecule turns out to be unambiguously clear. The generally accepted amorphicity of the ice in comets provides a constraint on its origin as interstellar ice mantles. It has been demonstrated that if the dust ice is evaporated and recondensed in the solar nebula it would be crystalline even at relatively low temperatures (Kouchi et al, 1993).

CO :

Again considering the Taurus molecular cloud, the solid CO abundance is less than the solid H_2O abundance, as already noted by Whittet et al. (1988) and as since confirmed by many later observations (see Whittet and Duley, 1992 for a review). It is also very clear that in such objects as in Orion near young hot stars (Geballe, 1986) and in W33A the solid CO is substantially reduced (Lacy et al., 1984) undoubtedly as a result of heating of the dust (Schmitt et al., 1989). This will be shown in the later discussion of the OCN^- ion. In Elias 16 the solid CO abundance is $[CO/H] \approx 1 \times 10^{-5}$ so that the water to CO ratio is $[H_2O/CO] \approx 7$; i.e. about 15% CO. In no case has a solid CO abundance exceeded the solid H_2O whereas in the gas phase $[CO/H_2O] \geq 10$. We should point out that the gas phase H_2O is exceedingly difficult to measure. The low value is, however, confirmed by theoretical dust/gas modelling as well as by pure gas phase interstellar cloud modelling (van Dishoeck et al., 1993) as will be shown explicitly shown later.

CO_2 :

The detection of the solid CO_2 by its CO stretch is made difficult because of the atmospheric absorption by gaseous CO_2. However its bending mode absorption at 15.2 μm is not so obscured. Nevertheless, being on the edge of the broad H_2O 12 μm absorption has prevented its clear detection. By a very difficult subtraction procedure using the IRAS Low Resolution Spectra d'Hendecourt and de Muizon (1989) made the first discovery of the solid CO_2 by comparison with laboratory spectra. While the detection has not been made for normal molecular cloud dust, there is abundant laboratory data on the ultraviolet irradiated interstellar ice samples containing H_2O and CO, which show that CO_2 *must* be present in molecular cloud dust (d'Hendecourt et al. 1985; Greenberg and d'Hendecourt, 1985). Its abundance by laboratory data as well as by the limited observation is possibly as high as CO.

CH_3OH :

Observations of the 3.53 μm and 9.8 μm CH_3OH bands toward various embedded protostellar objects give widely varying abundances of CH_3OH relative to H_2O in interstellar ices; i.e., between 4% and 40% (Grim et al, 1991; Schutte et al, 1991; Allamandola et al, 1992; Skinner et al, 1992, Schutte et al, 1994). Thus although the presence of methanol in dust is well demonstrated its abundance in precometary dust is not yet demonstrated. Note that methanol has not yet been discovered in *normal* molecular clouds but this would be an observational selection effect because it is made in or on the dust (see section on dust/gas modelling). In fact theoretical dust/gas modelling (see Table II) indicates that methanol is probably not very abundant in *fairly* dense clouds which are still at a stage of *pre*-protostellar cloud contraction.

H_2CO :

It is only very recently that formaldehyde has been detected in interstellar dust even though it, like methanol and carbon dioxide, must be created by photoprocessing of dust ice mantle mixtures or possibly by surface reactions (Tielens and Hagen, 1982). Its abundance in the source of AFGL 2136 has been estimated as about comparable with CH_3OH; i.e. about 3% (Schutte et al, 1994). The region of the observation is not a simple molecular cloud because the source is a deeply embedded object, a YSO.

NH_3, CH_4 :

In spite of the fact that NH_3 and CH_4 are expected to be substantially represented in the interstellar ice mantles in molecular clouds, particularly in young clouds, neither one can claim to be *quantitatively* detected. Early claims for the NH_3 abundance were not justified by later interpretation of the spectral features at about 2.95 μm although the expected induced long wavelength wing of the 3.07 μm ice band is consistent with a fractional abundance of about 10% (van de Bult et al., 1984). The methane molecule has only been observed in deeply embedded sources and, in the solid phase, may there be even more abundant than CO. But note that CO has a higher vapour pressure and for heated dust a starting ratio of $[CO/CH_4] > 1$ could be inverted. Since CO is relatively depleted in W33A it is not surprising that for this source CH_4 is significantly more abundant (Lacy et al., 1991). In any case, the fact that CH_4 can even be comparable with CO in the solid shows that it is produced most likely on the dust rather than in the gas phase where $[CH_4/CO]$ is typically $\approx 10^{-3}$. Thus CH_4 is like H_2O, produced predominantly as a solid whereas CO is produced primarily in the gas and *then* accreted.

OCS :

This sulphur molecule has been observed in W33A - again a deeply embedded object rather than a dense molecular cloud (Geballe et al., 1985; Grim and Greenberg, 1987a). Nevertheless, the deduction from the laboratory comparisons would indicate that this molecule should also be present in significant amounts in molecular clouds. In laboratory experiments H_2S was used to produce OCS (Grim and Greenberg, 1987a) but the original evidence for H_2S has been shown to be incorrect (Allamandola et al, 1992).

OCN^-, NH_4^+ :

These ions are put together because they probably form as counter ions as a result of photoprocessing of dust mantles. The OCN^- : CO ratio is much higher in W33A (a heated region) than in NGC7538 IRS9 (Grim and Greenberg, 1987b; Grim et al., 1989). The latter is probably closer to prestellar dust.

2. 2. INTERSTELLAR DUST/GAS : THEORETICAL AND LABORATORY MODELLING

A series of paper originating in Leiden have examined the combined time dependent evolution of gas and dust (Tielens and Hagen, 1982; d'Hendecourt et al., 1985; Breukers, 1991) taking into account surface reactions, ultraviolet photoprocessing of the dust and explosive desorption of volatile dust mantles. The most recent effort in this direction has explicitly examined photoprocessing of the interstellar mantle molecules with the aim of following the development of such molecules as H_2CO and CH_3OH as well as the more common molecules (Shalabiea and Greenberg, 1993). Not only are the predicted gas phase molecules in better agreement with observations but the dust mantle molecules are also well represented. We repeat the caveat already expressed earlier that although the calculations have been performed for fairly dense interstellar clouds they have not yet been carried through the phase in which the cloud collapses toward the protosolar nebula phase. Nevertheless the results are encouraging both in comparison with interstellar dust observations *and* comet coma observations. We shall summarize below the results of laboratory and theoretical studies of interstellar dust and interstellar gas.

2.2.1 Molecular Gas and Dust : Theoretical modelling

The basic physical processes considered, in addition to the usual gas phase interactions, are surface reactions, accretion, photoprocessing and dust explosive desorption. A total of 217 molecular species and 2074 individual reactions are included. The cloud density is $n_o = n_H + 2n_{H_2} = 2 \times 10^4$ cm^{-3} where most of the hydrogen winds up as H_2. Solid state photoproduction of three crucial molecules CO_2, H_2CO and CH_3OH are specifically included basing their photoproduction on laboratory results. In table I are given some of the calculated gas phase species compared with their observed values and with values calculated without including dust. This table is presented to show how the theory, including the dust as we have done in the total chemical scheme, provides a more reliable set of values than that using gas phase reactions alone.

We note in particular, that the gas phase abundances of H_2CO, CH_3OH and CH_4 are an order of magnitude or more greater when the dust is included in the reaction scheme and that they are then closer to the observed values. A detailed discussion of all parameters involved in the chemical scheme is beyond the scope of this paper. For details see Shalabiea and Greenberg (1993). Of more direct relevance is the dust mantle composition. The theoretical derived dust mantle components are presented in table II for the two dust/gas models (low photoproduction and high photoproduction) represented by their gas phase abundances in the two right columns of Table I.

Table I : Some steady state molecular gas abundances using pure gas phase reactions and combined dust/gas reactions. Cloud density $n_o = 2 \times 10^4$cm^{-3}, $T_d = T_G = 10$K. Low and high p.p. refer to high and low solid state photoproduction rates estimated from laboratory results. Two cosmic ray ionization rates, ζ, are used.

	Observed		Gas alone		dust/gas (ζ= 5(-17))	
	L134N[a]	BN[b]	ζ=1(-17)	=5(-17)	low p.p.	high p.p.
CO	8(-5)	1(-4)	1.5(-4)	1.6(-4)	7.0(-5)	6.8(-5)
H_2O	–	1(-5)	3.2(-6)	9.7(-6)	5.8(-6)	6.1(-6)
H_2CO	2(-8)	>6(-8)	1.1(-9)	2.3(-9)	2.9(-8)	2.7(-7)
CH_3OH	2(-9)	0.1-1(-6)	7.4(-11)	3.9(-10)	2.6(-9)	1.3(-7)
NH_3	3(-8)	1(-7)	5.1(-9)	1.5(-8)	1.5(-7)	1.5(-7)
CH_4	–	1(-7)	9.7(-9)	1.3(-8)	1.3(-7)	1.0(-7)
HCN	1(-8)	–	6.5(-9)	3.0(-9)	2.0(-8)	3.0(-8)

[a] see van Dishoeck et al., 1993 for the cloud L134N
[b] see van Dishoeck et al., 1993 for Orion BN/ext. ridge

We see that the predicted ratio of the CO to H_2O is $\leq 1/4$ in the solid while, in the gas, the predicted ratio is > 10. Furthermore the formaldehyde and methanol fractions- relative to H_2O - are generally an order of magnitude greater in the dust than in the gas *and* since the H_2O in the solid is much higher than in the gas this represents at least *several* orders of magnitude improvement as compared with the values observed in the comet coma. Note that the methanol to water ratio is much lower and its CO to water ratio is much higher than in W33A but also note that if *this* dust (in table II) were to be heated not only would the CO be reduced but a substantial portion of the formaldehyde would likely be converted to methanol (Zhao, 1990). Also of interest is the CH_4 : CO ratio in the dust in relation to that observed in the comet and that predicted by solar nebula chemistry, assuming all the dust volatiles have evaporated before comet nuclei formation. From table II, $CH_4/CO = 0.01$ to 0.0007 which is outside the observed ratio range for comet Halley (0.03 - 0.17) but is orders of magnitude closer than the solar nebula disequilibrium calculations prediction of $CH_4/CO < 10^{-7}$ (Prinn and Fegley, 1989; Fegley and Prinn, 1989).

Table II. Solid state dust mantle abundance fractions of several key molecular species resulting from theoretical time dependent chemistry at steady state in a cloud with $n_o = 2 \times 10^4$ cm^{-3} and ionization rate $\zeta = 5 \times 10^{-17}$ s^{-1}

	low p.p	high p.p
CO	0.18	0.15
H_2O	0.70	0.58
H_2CO	0.01	0.15
CH_3OH	0.001	0.02
NH_3	0.008	0.008
CH_4	0.002	0.0001

2.2.2 Laboratory Produced Organic Residues :

A laboratory analog of the photoprocessing of interstellar ices lead, after warmup, to organic residues which are presumed to resemble the organic refractories which are observed by their 3.4 μm absorption. The chemical characteristics of the laboratory organics is taken as a first approximation to interstellar organic mantles. GCMS analysis shows that it consists in part of many prebiotically significant molecules (Briggs et al., 1992), the most abundant having been determined to be approximately 7% ethylene glycol ($HOCH_2CH_2OH$), 30% glycolic acid ($HOCH_2CO_2H$), 21% 2-hydroxyacetimine ($HCCH_2CONH_2$), 3% urea (NH_2CONH_2), 6% glycerol ($HOCH_2CH(OH)CH_2OH$), 7% glyceric acid ($HOCH_2CH(OH)CO_2H$) and 19% glyceramide ($HOCH_2(OH)CONH_2$). Mass spectroscopic analysis of a residue has revealed masses up to $\sim$ 1000 and identifications of both aliphatic and aromatic hydrocarbons have been obtained. These new results indicate the presence of highly unsaturated, aromatic hydrocarbons (see table III). This agrees qualitatively with what has been found for meteoritic material (Grady et al., 1983; Gilmour and Pillinger, 1985) and cometary material (Kissel and Krueger, 1987; Mukhin et al., 1989). The most refractory part of the organic residue material consists of relatively high molecular weight ($\geq$ 400 AMU) polymers, which have not yet been completely characterized.

Finally, we also studied the volatile part by means of a gas collector (Mendoza-Gómez, 1992). Once the initial gas mixture (H_2O : CO : NH_3= 5 : 5 : 1) was irradiated at 12K, the sample was slowly warmed up, and the material coming off was trapped with a gas collector between given temperatures. The results found for materials coming off from 100 K to room temperature are shown in table IIIb. It is interesting to note how little nitrogen is contained in the intermediate masses shown here compared with the compounds discussed earlier (Briggs et al., 1992). Coincidentally, nitrogen appears to be underabundant in comet volatiles (Wyckoff et al., 1991) and may possibly be embedded in the highly polymerized fraction which in our residue has not yet been analyzed. These new results further confirm comet origins as being the interstellar dust.

Table III. (a) High-resolution fast atom bombardment mass-spectral data of the organic refractory residues (U.V. irradiated H_2O : CO : NH_3=5 : 5 : 1), (b) High resolution electron mass-spectral data of the gas material collected from 100K to room temperature

Mass no. (a)	elemental composition	Mass no. (b)	elemental composition
219	$C_{14}H_{19}O_2$	44	CO_2
203	$C_{14}H_{19}O$	44	C_2H_4O
189	$C_{13}H_{17}O$	70	C_4H_6O
173	$C_{13}H_{17}$	84	C_5H_8O
159	$C_{12}H_{15}$	98	$C_6H_{10}O$
157	$C_{12}H_{13}$	103	C_7H_5N
		115	C_7H_17N
		116	C_9H_8

It is possible that in some astrophysical situations, when the dust is heated in the presence of protostellar sources, these intermediate size molecules are evaporated and become mixed with the gas.

Some of the molecules identified using high resolution (elemental compositions), MS/MS mass spectral data (mass spectra of individual peaks) and direct comparison with mass spectra of model compounds are shown below in Figure 1.

Note how cyclic compounds seem to be well represented. This is encouraging because polycyclic aromatics appear to require a source in the interstellar medium (Greenberg et al., 1992) rather than circumstellar as often suggested. Indeed there is also evidence for a polycyclic aromatic phenanthrene (Moreels et al., 1993), in comet Halley spectra which would be similar to the m/z = 214 labeled molecule in Figure 1.

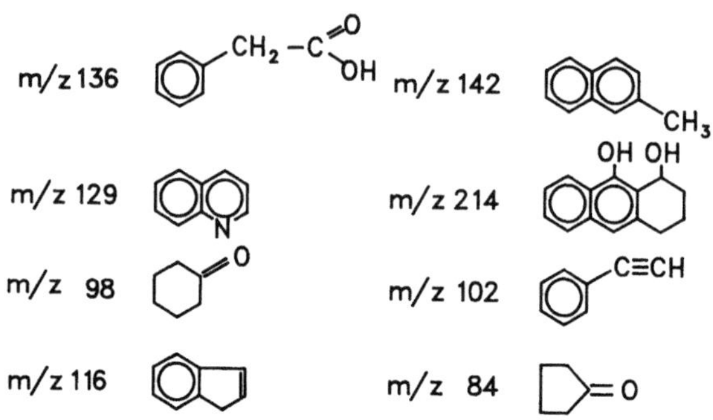

Fig. 1. Identified cyclic molecules in organic residue.

3. Physical Description of Comet Dust

3. 1. POROSITY OF INTERSTELLAR DUST AGGREGATES DERIVED FROM COMET HALLEY DATA

A detailed calculation was performed for the dust of comet Halley (Greenberg and Hage, 1990)in which it was shown that the interstellar dust model for the coma consisting of silicate core-organic refractory mantles (volatiles like H_2O etc. have been depleted in the coma dust) can both simultaneously and quantitatively fulfill the observational constraints of : (1) the observed strength of the 3.4 and 9.7 μm emission bands; (2) the *shape* of the 9.7 μm band; (3) the relative amounts of silicates to organic materials measured in situ; (4) the mass distribution of the dust *if* the coma dust is a fluffy aggregate at least as porous as $P = 0.93$ and with a likely porosity as high as 0.975; i.e. 97.5% empty space. The organic refractories not only provide a source of the 3.4 μm emission, they are ***absolutely required*** to heat the dust, ***otherwise*** the 9.7 μm emission would not exist because silicates do not absorb enough solar radiation to achieve the ***above*** black body temperature needed for the excess ***spectral*** emission. The key words here are ***above black body temperature***! Very fluffy comet dust can approach the absorption/emission properties of a ***cloud*** of tenth micron core-mantle dust particles whose temperatures can greatly exceed the temperatures of the aggregated particles considered as compact. One should not automatically assume that such dust has no rigidity because some rigidity is already shown by aerogels with porosity as high as or higher than 0.99. The coma dust temperature for a basic core-mantle grain with partial organic refractory evaporation (Si/O.R ≈ 2) at 0.98 au from the sun is 560K as compared with about 290K black body temperature. A fluffy aggregate of such particles with $P = 0.95$ and overall size 2.5 μm would be at $T \approx 500$K whereas as a *solid* particle with the same mass its temperature would be only ≈ 300K. However, for particles as large as 50 μm (mass $\approx 5 \times 10^{-8}$ g) the temperature is down again to $T \approx 310$K.

3. 2. SOURCES OF SIMPLE COMA SPECIES IN THE DUST ORGANICS

While the emphasis in the past was to try to identify the volatile molecules as parents of coma molecules it is now well recognized that a significant number of daughter molecules have as their source the relatively nonvolatile organic species which are part of the dust. What is the source of C_2 and C_3 which Crovisier (this volume) denotes as orphan molecules for which no specific parent has been identified? We believe that the answer must lie in the existence of the organic refractory interstellar dust mantles. We shall discuss this both from the point of view of observed infrared properties of comet dust as well as the laboratory analog results on simulated interstellar organics. There is also a recent and important observation of a new comet Yanaka 1988r for which the conspicuous absence of C_2 and CN may be used to infer *why* they must arise from dust organic refractories. We shall first present a brief summary of an interstellar dust model of comet dust from which it may be shown *physically* how the organics provide a source of volatiles in the comet coma which is continuously distributed as the dust clouds expands away from the nucleus. We shall show also how some of the identified laboratory produced organics may be evaporated, dissociated or pyrolized to produce the observed (range of) abundances of such species as C_2, CN as well as CO.

The smaller components in the organic refractory component generally have higher vapor pressures than the hydrocarbon species detected by mass spectrometry, some of which are large polymers. In general the temperatures at which pyrolysis occurs in these components is typically $T \geq 410$K (Briggs et al, 1992). Since sufficiently large particles - even fluffy ones - approach the black body temperature , we expect that only a limited size (mass) range of particles in the comet dust distribution can contribute substantially to distributed volatiles. The fact that there is a substantial fraction of complex organics in comets (and comet dust) is thus a *necessary* but not *sufficient* condition for providing a distributed source of CO, C_2, CN etc. This question of the required porosity was first semi- quantitatively considered by Greenberg and Hage (1990) in the production of CO. By imposing the lower limit $T \geq 410$K we can refine this by noting that : 1) the total possible CO mass fraction source in the volatile organics is about 0.3 (see the list of molecules identified in the laboratory residues as the organic refractories which contain CO groups), and 2) the volatile organics constitute (originally) about 1/3 of the total dust mass. Using these refractories we find - based on the dust temperature calculations of Greenberg and Hage (1990) - that the maximum *distributed* CO production rate is $\dot{M}(CO) \approx 0.3 \times 10^5$ g s^{-1} for M_d ($<10^{-8}$); i.e. a factor of at least 30 times more dust than that limited to Md $< 10^{-8}$g would be needed to account for the required CO distributed production rate of 8×10^5 g s^{-1} as in Eberhardt et al (1987). This requires particles with $P \geq 0.975$ to provide the temperature $T \geq 400$ K for the particle masses up to $m \approx 3 \times 10^{-7}$ g which would *just* provide the distributed CO. Thus in order to make a good case for organic (semi)refractories being the source of such distributed molecules as CO, CN, etc., high porosity is critical to achieve the high temperatures needed to evaporate the parent organic molecules as a sufficient source of the distributed molecules. Furthermore the farther out in the coma one looks, the higher the mean dust temperature because of more and more small particles so that a substantial fraction of the total dust could be a temperature approaching that of individual core-mantle particles. This makes for distributed molecules from more refractory organics being made possible; e.g., CN is generally in more refractory components than is CO.

3. 3. DUST PROPERTIES NEAR A NEW COMET NUCLEUS DERIVED FROM COMET YANAKA 1988R :

We have comet Yanaka (1988r) as an example of what happens to the source of distributed molecules if the dust is limited to large cool particles. A deficiency in the observed abundances of C_2 and CN in the comet may be explained in terms of the properties of its organic refractory components plus the fact that it is a new comet *and* the small aperture (very close to the nucleus) with which it was observed. First of all, the parent source of C_2 and CN is relatively tightly bound within the comet dust whose outer layer has been processed by cosmic rays while in the Oort cloud (Strazzulla and Johnson, 1991). Secondly, the cosmic-ray-processed dust is less fragile (is better glued together because the volatile ices are converted to organics) than that of a periodic comet so that its size distribution is weighted toward larger cooler particles than those of an old comet. The coma dust from the outer crust is probably also less porous because it has far less volatiles to be evaporated. Finally, the small central observation diaphragm further exaggerates the size distribution effect by not allowing time for fragmentation to occur. The consequent low surface-to-mass ratio and the relative coolness of the dust *within* the regions of observation of comet Yanaka (1988r) combined in a nonlinear way to inhibit severely the "usual" sources of C_2 and CN. It is worth noting here that the weakness of the 3.4 and 9.7 μm excess emissions in another dynamically new comet also observed with small aperture (Hanner et al, 1993) are consistent with the suggestions for the dust properties of comet Yanaka(1988r).

4. Asteroids and meteorites

The connection between interstellar chemistry and asteroids is at least one step removed from that with comets; and it is a further step to meteorites which are asteroidal debris. Among other things the region where asteroids are formed and their subsequent temperature history precludes that water could have survived in solid form. The evidence that liquid water was incorporated in asteroids is suggested as a reason for the layer-lattice silicates in those IDP's (interplanetary dust particles) which appear to be debris of asteroids, such silicates resulting from hydration in the presence of liquid water (Sandford, 1991). Was this water melted interstellar ice or was it water which had formed in the solar nebula after the dust evaporated? The answer to this may depend on the apparent correlation between the abundance of carbonates and layer-lattice silicates in meteorites (Greenberg, 1991b) . The formation of carbonates implies the presence of a significant amount of CO_2 in solution. The question is where does the required amount of CO_2 originate? The abundance of CO_2 in interstellar dust is substantial - probably of the order of that of CO (about 5% ,or more, relative to that of H_2O)- but in models of solar nebula chemistry it is at least ten times less(Fegley,1993; Prinn,1993). On this basis one would need to attribute the carbonic acid as forming from dissolved interstellar

CO_2 brought in with the interstellar ice which has melted in the interior of the asteroid. A further connection has been suggested by Cronin and Chang (1993) via the organic rich CI1 and CM2 carbonaceous chondrites which contain minerals also (as in the IDP's) believed to indicate hydrous activity in the meteorite parent body. The organic chemistry of these meteorites and the substantial deuteration of their organic compounds as well as enrichment in ^{15}N and in ^{13}C relative to terrestrial matter are "consistent with a formation scheme in which (1) a parent body was formed from volatile-rich icy planetesimals containing interstellar organic matter, (2) warming of the parent body led to an extensive aqueous phase in which the interstellar organics underwent various reactions, and (3) residual volatiles were largely lost leaving behind the suite of nonvolatile compounds that now characterize these meteorites.." Thus a simultaneous correlation of substantial organics, layer-lattice silicates, carbonates and isotopic enrichments would go far toward establishing an asteroid-interstellar dust connection.

5. Conclusions

It has been shown that not only the volatiles (ices) of interstellar dust are related to comet coma molecules but also the organic refractory components. Preservation of volatiles in comets in approximate proportions to those observed or inferred for prestellar interstellar dust appears to preclude their evaporation before comet nucleus formation. Further evidence is given showing that the criteria for distributed molecules from the refractory component are that the comet dust and, by inference, the comet nucleus be extremely porous. A requirement of $P > 0.975$ on the comet dust leads to a nucleus density of < 0.3 cm^{-3} consistent with the values derived by Rickman (this volume). The amorphous property of H_2O ice in comet nuclei can be attributed *only* to the amorphous ice in interstellar dust because ice condensed in the solar nebula would have to be crystalline.

Acknowledgements

This work has been supported by NASA grant NGR 33-018-148 and by a grant from the Netherlands Space Research Organization (SRON). One of us (OMS) would like to thank the World Laboratory for a fellowship. We thank Dr. W. Schutte for some suggestions. We also thank an unknown referee for some constructive criticisms.

References

Allamandola, L.J., Tielens, A.G.G.M. and Barker, J.R. : 1989, " Interstellar polycyclic aromatic hydrocarbons : the infrared emission bands, the excitation/emission mechanism, and the astrophysical implications." ***Astrophys. J. Suppl. Ser.***, **71**, 733–775.

Allamandola, L.J., Sandford, S.A., Tielens, A.G.G.M. and Herbst, T.M. : 1992, " Infrared spectroscopy of dense clouds in the C-H stretch region : methanol and diamonds." ***Astrophys. J.***, **399**, 134–146.

Breukers, R.J.L.H. : 1991, ***Thermal and chemical processes in the evolution of interstellar dust and gas*** , Ph. D. Thesis, Univ. of Leiden.

Briggs, R., Ertem, G., Ferris, J.P., Greenberg, J.M., McCain, P.J., Mendoza-Gómez, C.X. and Schutte, W. : 1992, "Comet Halley as an aggregate of interstellar dust and further evidence for the photochemical formation of organics in the interstellar medium." *Origins of life and evolution of the biosphere*, **22**, 287–307.

Cronin, J.R. and Chang, S. : 1993, "Organic matter in meteorites : molecular and isotopic analysis of the Murchison meteorite." In *The Chemistry of Life's Origins* (J.M. Greenberg, C.X. Mendoza-Gómez, V. Pirronello, Eds.), 259–299, Kluwer, Dordrecht.

Crovisier, J. : 1994, "Molecular abundances in comets." This volume.

Day K.L. : 1979, "Mid-infrared optical properties of vapor-condensed magnesium silicates." *Astrophys. J.*, **234**, 158–161.

d'Hendecourt, L.B., Allamandola, L.J. and Greenberg, J.M. : 1985, "Time dependent chemistry in dense interstellar clouds. I. Grain surface reactions, gas/grain interactions and infrared spectroscopy." *Astron. Astrophys.*, **152**, 130–150.

d'Hendecourt, L.B. and Jourdain de Muizon, M. : 1989, "The discovery of interstellar carbon dioxide." *Astron. Astrophys.*, **223**, L5–L8.

Dyck, H.M., Lonsdale, C.J. : 1981, "Polarimetry of infrared sources." In *Infrared Astronomy* (C.G. Wynn-Williams, D.P., Cruikshank, Eds.), 223–236, Reidel, Dordrecht.

Eberhardt, P.D., Krankowsky, D., Schulte, U. et al. : 1987, " The CO and N_2 abundance in comet P/Halley." *Astron. Astrophys.*, **187**, 481–484.

Fegley,B. : 1993, "Chemistry of the solar nebula." In *The Chemistry of Life's Origins*, (J.M. Greenberg, C.X. Mendoza-Gómez, V. Pirronello, Eds.), 75–147, Kluwer, Dordrecht.

Fegley, B. and Prinn, R.G. : 1989, "Solar nebula chemistry : implications for volatiles in the solar system." In *The formation and evolution of planetary systems* (H.A. Weaver, L. Danly, Eds.), 171–205, Cambridge Univ. Press.

Geballe, T.R. : 1986, "Absorption by solid and gaseous CO towards obscured infrared objects." *Astron. Astrophys.*, **162**, 248–252.

Geballe, T.R., Baas, F., Greenberg, J.M. and Schutte, W. : 1985, " New infrared absorption features due to solid phase molecules containing sulphur in W33A." *Astron. Astrophys.*, **146**, L6–L8.

Gilmour, I. and Pillinger, C.T. : 1985, *Org. Geochem.*, **8**, 421.

Grady, M.M., Wright, I.P., Fallick, A.E. and Pillinger, C.T. : 1983, In *Proc. 8th symp. Antarctic Meteorites*, p. 289.

Greenberg, J.M. : 1973, "Chemical and physical properties of interstellar dust." In *Molecules in the galactic environment* (M.A. Gordon, L.E. Snyder, Eds.), 94–124, Wiley.

Greenberg, J.M. : 1982a, "Dust in dense clouds. One stage in a cycle." In *Submillimetre Wave Astronomy* (J.E. Beckman, J.P. Phillips, Eds.), 261–306, Cambridge Univ. Press.

Greenberg, J.M. : 1982b, "What are comets made of ? - a model based on interstellar dust." In *Comets* (L.L. Wilkenning, Ed.), 131–163, Univ. of Arizona Press, Tucson.

Greenberg, J.M. : 1989, "The core-mantle model of interstellar grains and the cosmic dust connection." In *Interstellar dust* (L.J. Allamandola, A.G.G.M. Tielens, Eds.), 345–355, Reidel, Dordrecht.

Greenberg, J.M. : 1991a, "Physical, chemical and optical interactions with interstellar dust." In *Chemistry in Space* (J.M. Greenberg, V. Pirronello, Eds.), 227–261, Kluwer, Dordrecht.

Greenberg, J.M. : 1991b, "The interplanetary medium is thriving." In *Origin and evolution of interplanetary dust* (A.C. Levasseur-Regourd, H. Hasegawa, Eds.), 443–451, Kluwer, Dordrecht.

Greenberg, J.M. and d'Hendecourt, L.B. : 1985, "Evolution of ices from interstellar space to the solar system." In *Ices in the solar system* (J. Klinger, D. Benest, A. Dollfus, R. Smoluchowski, Eds.), 185-104, Reidel, Dordrecht.

Greenberg, J.M. and Hage, J.I. : 1990, "From interstellar dust to comets : a unification of observational constraints." *Astrophys. J.*, **361**, 260–274.

Greenberg, J.M., Mendoza-Gómez, C.X., de Groot, M.S. and Breukers, R. : 1992, "Laboratory dust studies and gas-grain chemistry." In ***Dust and Chemistry in Astronomy*** (T.K. Millar, D.A.Williams, Eds.), 265–288, IOP Publ.

Grim, R.J.A., Baas, F., Geballe, T.R., Greenberg, J.M. and Schutte, W. : 1991, " Detection of solid methanol toward W33A." ***Astron. Astrophys.***, **243**, 473–477.

Grim, R.J.A. and Greenberg, J.M. : 1987a, "Photoprocessing of H_2S in interstellar grain mantles as an explanation for S_22 in comets." ***Astron. Astrophys.***, **181**, 155–168.

Grim, R.J.A. and Greenberg, J.M. : 1987b, "Ions in grain mantles : the 4.62 micron absorption by OCN^- in W33A." ***Astrophys. J. Lett.***, **321**, L91–L96.

Grim, R.J.A., Greenberg, J.M., Schutte, W. and Schmitt, B. : 1989, "Ions in grain mantles : a new explanation for the 6.86 micron absorption in W33A." ***Astrophys. J. Lett.***, **341**, L87–L90.

Hanner, M.S., Russell, R.W., Lynch, D.K. and Brooke, T.Y. : 1993, " Infrared spectroscopy and photometry of comet Austin 1990V." ***Icarus***, **101**, 64–70.

Jenniskens, P., Baratta, G.A., Kouchi, A., de Groot, M.S., Greenberg, J.M. and Strazzulla, G. : 1993," Carbon dust formation on interstellar grains." ***Astron. Astrophys.***, **273**, 583–600.

Kissel, J. and Krueger, F.R. : 1987, "The organic component in dust from comet Halley as measured by the PUMA mass spectrometer on board Vega 1." ***Nature***, **326**, 755–760.

Kouchi, A., Yamamoto, T., Kozasa, T., Kuroda, T. and Greenberg, J.M. : 1993 "Conditions for condensation and preservation of amorphous ice and crystallinity of astrophysical ices." ***Astron. Astrophys.***, submitted.

Lacy, J.H., Baas, F., Allamandola, L.J., Persson, S.E., McGregor, P.J., Lonsdale , C.J., Geballe, T.R. and van de Bult, C.E.P.M. : 1984, "4.6 micron absorption features due to solid phase CO and cyano group molecules toward compact infrared sources." ***Astrophys. J.***, **276**, 533–543.

Lacy, J.H., Carr, J.S., Evans, N.J., Baas, F., Achtermann, J.M. and Arents, J.F. : 1991, "Discovery of interstellar methane : observations of gaseous and solid CH_4 absorption toward young stars in molecular clouds." ***Astrophys. J.***, **376**, 556–560.

Mendoza-Gómez, C.X. 1992 : "Complex irradiation products in the interstellar medium." Ph.D thesis, Leiden.

Moreels, G., Clairemidi, J., Hermine, P., Brechignac, P. and Rousselot, P. : 1993 "Detection of a polycyclic aromatic molecule in comet P/HALLEY." ***Astron. Astrophys.***, in press.

Mukhin, L.M., Dikov, Y.P., Evlanov, E.N., Fomenkova, M.N., Nazarov, M.A., Priludsky, O.F., Sagdeev, R.Z. and Zubkov, B.U. : 1989. In ***Lunar Planetary Science conf. XX***, p. 733, Houston.

Mumma, M.J., Stern, S.A. and Weissman, P.R. : 1993, "Comets and the origin of the solar system : Reading the Rosetta stone." In ***Planets and Protostars III*** (E.H. Levy, J.I. Lunine, M.S. Matthews, Eds.), 1177–1252, Univ. of Arizona Press, Tucson.

Prinn, R.G. : 1993, "Chemistry and evolution of gaseous circumstellar disks." In ***Protostars and Planets III*** (E.H. Levy and J.I. Lunine, Eds.), 1005–1028, Univ. of Arizona Press, Tucson.

Prinn, R.G. and Fegley, B. : 1989, "Solar nebula chemistry : origin of planetary, satellite, and cometary volatiles." In ***Origin and Evolution of Planetary and Satellite Atmospheres*** (S.K. Atreya, J.B. Pollack, M.S. Matthews, Eds.), 78–136, Univ. of Arizona Press, Tucson.

Rickman, H. : 1994, "Cometary nuclei." This volume.

Sandford, S.A. : 1991, "Constraints on the parents bodies of collected interplanetary dust particles." In ***Origin and evolution of interplanetary dust*** (A.C. Levasseur-Regourd, H. Hasegawa, Eds.).

Sandford, S.A., Allamandola, L.J., Tielens, A.G.G.M., Sellgren, K., Tapia, M. and Pendleton, Y. : 1991, "The interstellar C-H stretching band near 3.4 microns : constraints on the composition of organic material in the diffuse interstellar medium." ***Astrophys.***

J., **371**, 607–620.
Schmitt, B., Greenberg, J.M. and Grim, R.J.A. : 1989, "The temperature dependence of the CO infrared band strengths in CO : H_2O ices." *Astrophys. J. Lett.*, **340**, L33–L36.
Schutte, W.A., Gerakines, P.A., van Dishoeck, E.F., Greenberg, J.M. and Geballe, T.R. : 1994, "Possible detection of solid formaldehyde towards the embedded source GL 2136." In *Physical chemistry of molecules and grains in space*, 50th Internat. Meeting, Div. Chimie Physique, Soc. Française de Chimie, Mont Sainte-Odile (France), 6-10 Sept. 1993.
Schutte, W. and Greenberg, J.M. : 1986, "Formation of organic molecules on interstellar dust particles." In *Light on dark matter* (Proc. IRAS Symp., Noordwijk 10-14 June 1985) (F. Israel, Ed.), 229-232, Reidel, Dordrecht.
Schutte, W.A., Tielens, A.G.G.M. and Sandford, S.A. : 1991, "10 micron spectra of protostars and the solid methanol abundance." *Astrophys. J.*, 382-529.
Shalabiea, O.M. and Greenberg, J.M. : 1993, "Photoprocessing of grain mantles and explosive desorptions. Two key processes in dust/gas chemical modelling." *Astron. Astrophys.*, submitted.
Skinner, C.J., Tielens, A.G.G.M., Barlow, M.J. and Justtanont, K. : 1992 "Methanol ice in the protostar GL 2136." *Astrophys. J. Lett.*, **399**, L79–L82.
Smith, R.G., Sellgren and Brooke, T.Y. : 1993, "Grain mantles in the Taurus dark cloud." *Mon. Not. Roy. Astron. Soc.*, **263**, 749–766.
Spitzer, L. : 1978, *Physical processes in the interstellar medium*, Wiley.
Strazzulla, G. and Johnson, R.E. : 1991, "Irradiation effects on comets and cometary debris." In *Comets in the Post Halley Era* (R.L. Newburn, M. Neugebauer, J. Rahe, Eds.), 243–276, Kluwer, Dordrecht.
Tielens, A.G.G.M. and Allamandola, L.J. : 1987, In *Physical Processes in Dense Clouds*, (G.E. Morfill, M. Scholer, Eds.), 333, Reidel, Dordrecht.
Tielens, A.G.G.M. and Hagen, W. : 1982, "Model calculations of the molecular composition of interstellar grain mantles." *Astron. Astrophys.*, **114**, 245–260.
Tielens, A.G.G.M., Tokunaga, A.T., Geballe, T.R. and Baas, F. : 1991, " Interstellar solid CO : polar and nonpolar interstellar ices." *Astrophys. J.* **381**, 191–199.
van de Bult, C.E.P.M., Greenberg, J.M. and Whittet, D.C.B. : 1984, "Ice in the Taurus molecular cloud : modelling of the 3μm profile." *Mon. Not. Roy. Astron. Soc*, **214**, 289–305
van Dishoeck, E.F., G.A. Blake, B.T. Draine and Lunine, J.I. : 1993, "The chemical evolution of protostellar and protoplanetary matter." In *Protostars and Planets III*, (E.H. Levy, J.I. Lunine, M.S. Matthews, Eds.), 163–241, Univ. of Arizona Press, Tucson.
Whittet, D.C.B. : 1992, "Dust in The Galactic Environment." IOP Publishing.
Whittet, D.C.B., Bode, M.F., Longmore, A.J., Adamson, A.J., McFadzean, A.D., Aitken,D.K. and Roche, P.F. : 1988, "Infrared spectroscopy of dust in the Taurus dark clouds : ice and silicates." *Mon. Not. Roy. Astr. Soc.*, **233**, 321–336.
Whittet, D.C.B. and Duley, W.W. : 1992, " Carbon monoxide frosts in the interstellar medium." *Astron. Astrophys. Rev.*, **2**, 167–189.
Wyckoff,S., Tegler, S.C. and Engel, L. : 1991, "Ammonia abundances in four comets." *Astrophys. J.*, **368**, 279–286.
Zhao, Nansheng : 1990, "Photochemistry of interstellar and cometary ices." Ph. D. Thesis, Univ. of Leiden.

METEOROID PROPERTIES FROM PHOTOGRAPHIC RECORDS OF METEORS AND FIREBALLS

ZDENĚK CEPLECHA

Academy of Sciences of the Czech Republic, 251 65 Ondřejov Observatory

Abstract. Statistical criteria based on the single-body theory enabled the distinction of different composition groups of meteoroids in the past. The new single-body model proposed by Pecina and Ceplecha (1983, 1984) is capable of determining the individual values of ablation coefficients, which has proved to be a better tool for separating meteoroids of different ablation properties. However, a significant fraction of fireballs exhibit a time-dependence of residuals, when the single-body model is applied to their photographic observations. This was recently explained by assuming sudden fragmentation at a point (gross-fragmentation). The proposed gross-fragmentation model was checked in exceptional cases, when splitting of a fireball was directly visible on the photographs. The new fragmentation model was then applied to the best photographic records of Prairie Network fireballs (PN). Least-squares fit of computed to observed distances along a meteoroid trajectory determines uniquely the ablation coefficient, the shape-density coefficient, the position of the gross-fragmentation point and the amount of fragmented material relative to the main body mass. This enabled not only a better classification according to ablation coefficient (composition groups), but also a recognition of different strength categories according to dynamic pressure at the fragmentation point. Except for composition groups (types) I, II, IIIA, IIIB, each meteoroid with precise photographic data on its fireball can be classified as NF (no-fragmenting), 1F (with one point of fragmentation) and MF (with many points of fragmentation). The fragmenting meteoroids (1F and MF) can moreover be sorted into several categories (a, b, c, d, e) according to dynamic pressure at the fragmentation point. Thus the classification became two dimensional, separating meteoroid composition from structure. Values of ablation coefficients and bulk-densities were revised using this model. The amount of fragmented mass relatively to the main body was also determined. Typical sudden fragmentation for almost half of all fragmenting meteoroids is equivalent to stripping away slightly more than half of the mass.

1. Classification from preheating and ablation.

When a meteoroid collides with the Earth's atmosphere, most of the mutual kinetic energy is freed in interaction processes. The meteor phenomenon is thus very much dependent on the composition and structure of the meteoroid. Even the preheating is a sensitive indicator of the physical properties of the material from which the meteor body is composed.

The beginning heights of the luminous trajectories can thus be used for classification of meteoroids (Ceplecha 1967, 1968). This was done in the past for bodies in the mass range from 2×10^{-8} kg to 0.5 kg by recognizing different discrete levels of beginning heights. Four groups denoted A, B, C, and D can be specified. The A group meteors start their luminous trajectory much lower down than the D group meteors. The ratio of the air density at the beginning height between the two most populated groups A and C is a factor of 5, which corresponds to a difference in the normalized beginning heights of 11 km between A and C groups. Almost the same

A. Milani et al. (eds.), Asteroids, Comets, Meteors 1993, 343–356.

difference is obtained between C and D group, thus the total difference between A and D group is 21 km reflecting the difference of the material constants (heat conductivity, heat capacity, bulk-density and radiation emissivity). The levels are schematically presented in Fig. 1. Details on all previous work are summarized in Ceplecha (1988).

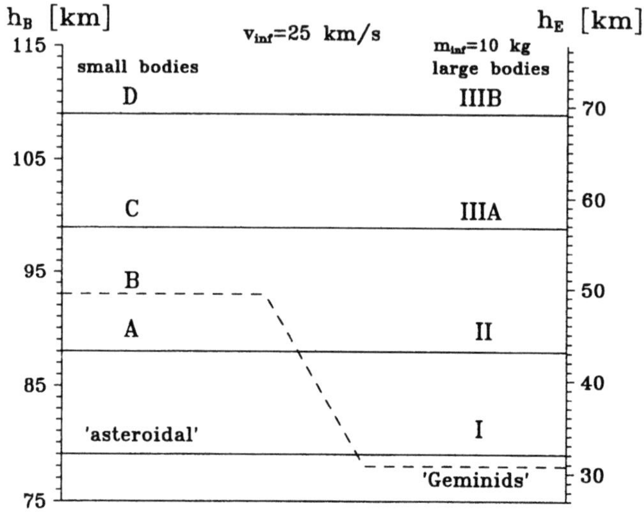

Fig. 1. Schematic representation of discrete levels of meteor beginning heights, h_B, and of meteor terminal heights, h_E, and their mutual relationship.

Bodies larger than 0.1 kg can be better classified according to their ability to penetrate the atmosphere. These bodies produce fireballs and our knowledge of them comes from photographic observations of fireball networks (EN, PN, MORP; Ceplecha, 1988; Ceplecha and McCrosky, 1969; McCrosky *et al.*, 1971; Halliday *et al.*, 1984, 1989). Except for very early portions, the ablation part of their trajectories is practically identical with their luminous trajectories. Thus terminal heights reflect the ablation ability and are an excellent indicator of the composition and structure of these larger bodies (Ceplecha and McCrosky, 1976). Terminal heights have to be normalized to the same velocity, to the same mass and to the same inclination of the trajectory to the horizon, when we want to use them for classification of fireball bodies. These normalized terminal heights for large meteor bodies cover a huge range of 1 : 1000, if expressed as ratios of the air densities at the terminal point.

The fireball networks collected data on meteoroids in the mass range from 0.1 kg to 5×10^4 kg; the upper limit corresponds to a size of 7 m. In this mass range we can again recognize several discrete levels of these normalized terminal heights : members of these levels were denoted as group I, II, IIIA and IIIB. The terminal height levels for fireballs are schematically related to the beginning height levels for

TABLE I
Survey of populations of large meteoroids (ρ = bulk density, σ = ablation coefficient).

population (type)	ρ g/cm^3	σ s^2/km^2	characteristic orbit for 0.1 - 1 m size a	e	i	assumed composition
I	3.7	.017	2.4	.68	6°	stony
II	2.0	.041	2.3	.61	5°	carbonaceous
IIIA	0.75	.10	2.4	.82	4°	cometary
IIIAi	0.75	.10	$\approx\infty$	.99	random	cometary
IIIA[C3]	0.75	.10	2.7	.67	random	cometary
IIIB	0.27	.21	3.0	.70	13°	soft cometary

faint photographic meteors in Fig. 1. A summary of all the fireball groups with data on average ablation coefficients and bulk densities are given in Table I. Enormous variety of compositions and structures exceeding our "surface experience" from studying meteorites is evident : meteorites are just samples of the strongest part of interplanetary matter coming to the Earth, which survived the atmospheric entry (groups I and II). The group I level belongs to stony meteorite falls as proved by Příbram, Lost City and Innisfree meteorite falls. The asteroidal origin of these bodies is evident. Group II is suspected to be composed of several systems of carbonaceous bodies of both asteroidal and cometary origin. The presence of carbon in spectra of meteoroids of this group is one of several verifications of the proposed composition. Group IIIA and IIIB are groups with meteoroids of cometary origin as it is evident from individual coincidences with meteor streams. The low density cometary material similar to the meteoroids from P/Giacobini-Zinner belongs to group IIIB. Recent statistical studies of bodies in the mass range from 10^4 to 10^7 kg (sizes of the order of 10 m) based partly on extrapolations of fireball data (Ceplecha, 1988) and partly on Spacewatch telescope discoveries of very small "asteroidal" bodies (Rabinowitz, 1993), yielded evidence that most of these bodies belong to IIIB group (Ceplecha, 1992). Thus the 10 m size bodies in the solar system may be mostly inactive material of cometary composition and structure. Alternatively, they may also belong to some not yet known system of bodies, which is relatively very strong at 10 m sizes.

Statistical criteria and studies based on the single body theory enabled distinction of all these different composition groups in the past (Ceplecha and McCrosky, 1976; Ceplecha and Borovička, 1992; Sekanina, 1983; Wetherill and ReVelle, 1981a,b). The biggest disadvantage of statistical procedures was the impossibility of separating the value of ablation coefficient from the bulk density of meteoroids : this could only be done in a statistical way (Table I contains these "statistical" values).

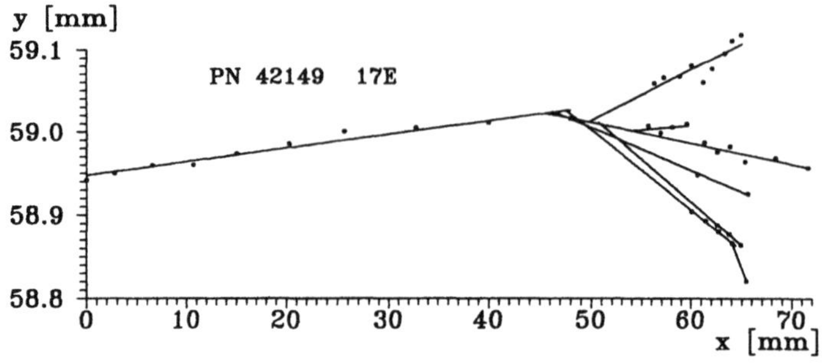

Fig. 2. Splitting trails of fireball PN 42149. The x and y are rectangular coordinates defined by the measuring device (Ascorecord). The individual measured points of different splitting trails are given as dots. Least-squares fits of straight lines to these points define trails of individual fragments. The y-axis is greatly enlarged to better visualize the separation of individual fragments. The main trail is terminated at the largest x-value.

2. Gross fragmentation

A new single body model was proposed by Pecina and Ceplecha (1983, 1984). This model was applied to individual well-observed meteoroids (multi-station photographs) and yielded the possibility of separating the ablation coefficient from the bulk density. This method was recently used to all available data on PN and EN fireballs with the result of better individual values of ablation coefficients (Ceplecha and Borovička, 1992). At the same time a new phenomenon was revealed : gross-fragmentation, *i.e.* a sudden instantaneous fragmentation (Ceplecha and McCrosky, 1992). When the time dependencies of residuals received from single body solutions were inspected by simple graphical means, the gross-fragmentation seemed to be important for about 25% of all observed fireballs.

Recently a new model of gross-fragmentation was proposed and checked on by means of several cases, where the fragmentation is directly visible as splitting trails on the photographs (Fig. 2). The gross-fragmentation model gives the distance flown by the meteoroid along its trajectory, l , as a function of time : $l = l(t)$. This distance is also directly derived as l_{obs} from measured values on the multi-station meteor photographs, *i.e.* from apparent distances along the trail measured for each time mark (velocity is an indirectly derived value and moreover it is the first derivative of the measured distances). The mathematics belonging to this model, the procedures used, and many details and results of application to PN fireballs, can be found in Ceplecha *et al.* (1993). Together with the observed heights, h_{obs}, which are available for each time mark, t, the observed distances, l_{obs}, determine uniquely the ablation coefficient, the shape-density coefficient, the position of the gross-fragmentation point, and the amount of fragmented material relatively to the main body mass just by means of the least squares fit of l to l_{obs}. (Fig. 3).

A computer program for automatic searching for gross-fragmentation points was prepared and applied to a sample of 80 records of PN fireballs with precise values of the measured distances and heights. From those 21 proved to be without gross-

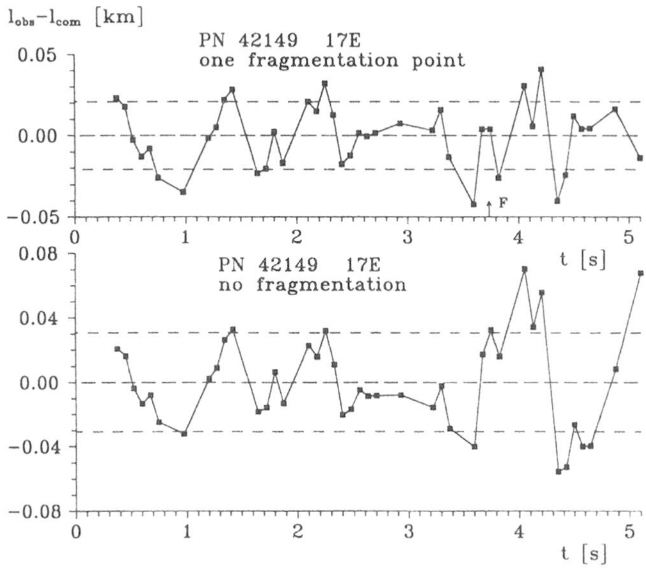

Fig. 3. Residuals, $l_{obs} - l_{com}$, of solutions without fragmentation and with one fragmentation point. The position of the fragmentation point (denoted by F) is derived from the least squares solution and can be compared with the geometrically determined fragmentation point (see Fig. 2). Residuals for the one-fragmentation-point solution are smaller and almost time-independent.

fragmentation (Table IV : class NF), 19 having one point with an overwhelming amount of fragmentation (class 1F), and 11 with many points (at least two) of comparable amount of fragmentation (class MF). 29 records had too low accuracy (class LA) to decide among these classes. Thus we have a new tool for classification of such meteoroids for which we have very precise data available, *i.e.* classification according to their ability to withstand fragmentation and their ability to then be broken suddenly in the atmosphere at a distinct height. The classification thus becomes two dimensional in separating at least partly the influence of composition from the influence of the structure of the body.

2. 1. STRENGTH CATEGORIES

The dynamic pressure $p = \rho v^2$ was computed at the fragmentation points of the 1F and MF cases : see histogram in Fig. 4. The precision of the individual values is better than the size of the histogram bins. Preference for some pressure values seems suggested. Thus we can define several strength categories of fireball meteoroids and determine average values of pressures at which meteoroid gross-fragmentation takes place, as proposed in Table II.

We also studied the maximum dynamic pressure acting on the NF fireballs. These pressures were mostly below 12 Mdyn/cm^2 with 4 exceptions of very strong bodies, *i.e.* PN 38737, PN 39521, PN 39775B, which survived as a single body up

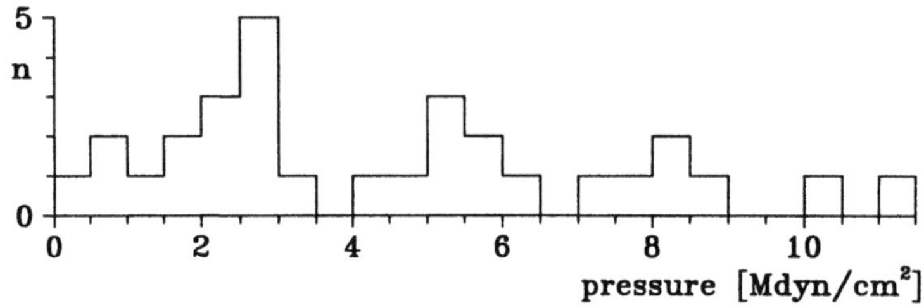

Fig. 4. Histogram of the dynamic pressures at the fragmentation points of 1F and MF fireballs.

TABLE II
Definition of strength categories.

strength category	interval of p [Mdyn/cm^2]	average p [Mdyn/cm^2]
a	$p < 1.4$	0.8
b	$1.4 \leq p < 3.9$	2.5
c	$3.9 \leq p < 6.7$	5.3
d	$6.7 \leq p < 9.7$	8.0
e	$9.7 \leq p < 12.$	11.

to 15 Mdyn/cm^2, and PN 39935A, which survived up to 50 Mdyn/cm^2.

2. 2. ABLATION

One of the parameters determined by the model is the ablation coefficient, σ. One value of σ, which was determined for all 80 studied fireballs, is closer to the value after the fragmentation point, denoted σ_2. Histograms of the ablation coefficients determined for different fragmentation styles are given in Fig. 5. The ablation coefficients for the 1F and MF fireballs sorted into individual groups (I, II, IIIA and IIIB) proved to be smaller than the previously published statistical values (Ceplecha, 1988), but quite close to the values derived for NF fireballs. But when the 1F+MF fireballs were computed as NF cases (*i.e.* as single body neglecting their fragmentation), the values of the ablation coefficients were larger and closer to their previous statistical values. This is one more justification for the validity of this fragmentation model. The resulting average values of the ablation coefficients σ are given in Table III for different groups and different fragmentation styles. The statistics are quite small, since we chose only the most precise data available on fireballs. On the other hand, individual values within these statistics are rather precise values disclosing *properties of individual meteoroids*.

Table III also reveals that the gross-fragmentation is more frequent for stronger

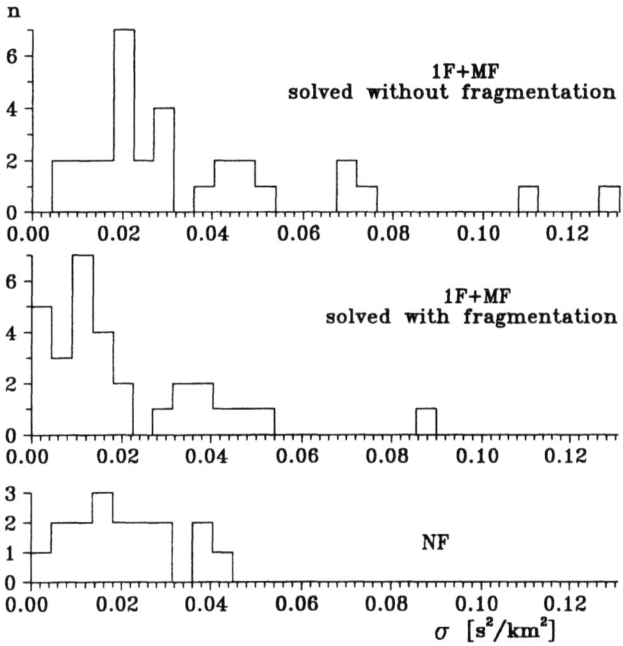

Fig. 5. Histograms of ablation coefficients show the difference between neglecting the gross fragmentation (top : higher values) and taking it into account in the fragmentation model (middle). Genuine NF cases as ascertained by the fragmentation model (bottom) show distribution similar to the 1F and MF cases (middle).

TABLE III

Average ablation coefficient σ in s^2/km^2 for different fireball groups (types). n = number of cases, PSV = previous statistical values (Ceplecha, 1988).

type	n	1F+MF assumed as NF	1F+MF	n	NF	PSV
I	21	0.023	**0.011**	12	0.015	0.017
II	8	0.060	**0.038**	6	0.035	0.041
IIIA	1	0.128	0.088	2	0.15	0.10
IIIB				1	0.59	0.21

types of fireballs. For group I, there are almost twice as many 1F+MF as NF whereas for group II, these two categories occur almost equally. And the only IIIB case we were able to compute with the fragmentation model is NF. But one should have in mind that the ablation in both the single-body and the fragmentation model contains continuous fragmentation as well. Thus the IIIB type (PN 39973) ablated only in form of continuous fragmentation without any sudden gross-fragmentation. The maximum pressure it suffered was only 0.088 $Mdyn/cm^2$ at the terminal height of 75 km with a terminal velocity of 15.5 $\pm .3$ km/s and with an ablation coefficient of 0.59 $\pm .02$ s^2/km^2; its initial velocity was computed as 16.555 $\pm .008$ km/s and the bulk density as 0.4 g/cm^3. Other extreme of NF is PN 39935A, a type I fireball, which survived pressure of 50 $Mdyn/cm^2$ without fragmentation. Its very low ablation coefficient 0.0010 ± 0.0005 guarantees that continuous fragmentation was almost completely absent.

The two dimensional classification of fireballs is also evident from Fig. 6, where the ablation coefficient σ is plotted against the dynamic pressure at the fragmentation point. Horizontal lines divide the groups, vertical lines the strength categories. All computed 1F and MF fireballs are presented : the MF cases are all in the group I region, but covering the whole b to e category. Thus the multiple fragmentation is typical for the type I fireballs with $p > 2$ $Mdyn/cm^2$. About one half of the type I gross-fragmentation is realized at several points for one fireball. Type II fireballs with gross-fragmentation have usually just one fragmentation point. The numbers of cases in different fragmentation styles and for different types of fireballs are given in Table IV. Using the fragmentation model solutions, individual σ values separate type I fireballs better from type II fireballs than using statistical criteria or individual no-fragmentation solutions.

TABLE IV

Number of cases according to fragmentation styles and fireball types.

type	NF	1F	MF
I	12	11	10
II	6	7	1
IIIA	2	1	0
IIIB	1	0	0

2. 3. BULK DENSITY

Bulk densities can be also computed for each fireball individually providing we have a photometrically well determined mass. Fewer fireballs have had such data determined and only quite rough statements can be made. All values of bulk densities computed without fragmentation for 1F+MF cases are significantly lower than when computed with the fragmentation model (for the Geminids : see also Ceplecha and McCrosky, 1992). On the other hand, the bulk density values computed for all 17 (1F+MF) fireballs with relevant data by using the fragmentation model came close to the statistical values of bulk densities (Ceplecha, 1988). This revision of

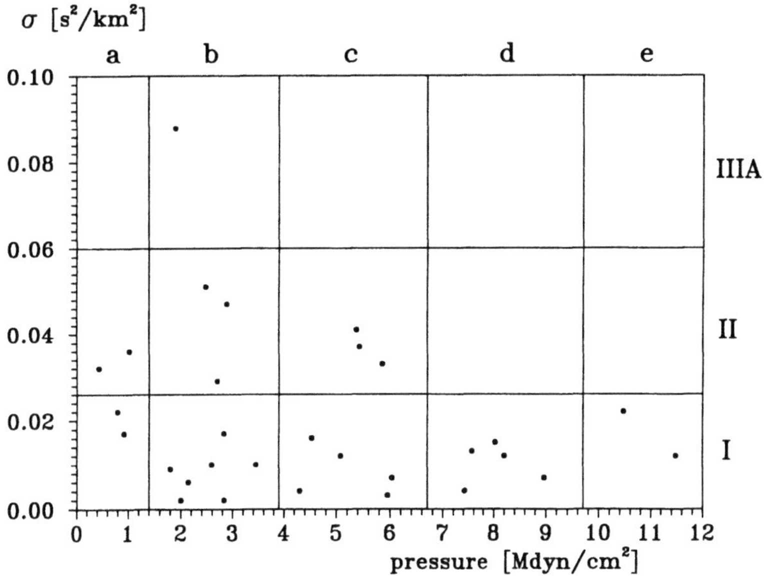

Fig. 6. Ablation coefficient σ plotted against pressure at the fragmentation point. Each point belongs to one fireball. The grid lines are the dividing lines among different types and categories of fragmenting fireballs (ablation classes or types I, II, IIIA, and strength categories a, b, c, d, e).

the average ablation coefficients and bulk densities for individual fireball groups does not change the values of the average statistical densities for groups I and II, but it makes the ablation coefficients smaller. The reason for this may be the statistical method used for the separation of σ from K, because only the product σK was determined from observations by Ceplecha and McCrosky (1976). Thus the statistical values of bulk densities for group I is $\rho = 3.7$ g/cm^3 and for group II is $\rho = 2.0$ g/cm^3, and these seem to be close to reality. The one IIIA class fireball for which we were able to compute the bulk density gave 0.75 g/cm^3, identical to the statistical value. The one of IIIB classification gave 0.4 g/cm^3, higher than the statistical value, but probably a preferable value, because its photometry was precisely calibrated.

2. 4. FRAGMENTED MASS

Another parameter also determined in the fragmentation model is the amount of fragmented mass, m_f, relatively to the mass immediately before gross-fragmentation. A histogram is given in Fig. 7. The amount of fragmentation for 1F class is typically 60%, which corresponds to breaking the body into approximately two halves with some accompanying small fragments. The second most common value is between 95% and 99%, which corresponds to almost complete disruption of the body. The distribution of m_f for the MF class is more random than for the 1F class. A typical sudden fragmentation for almost half of all fragmented meteoroids is equivalent to stripping away slightly more than half of the mass.

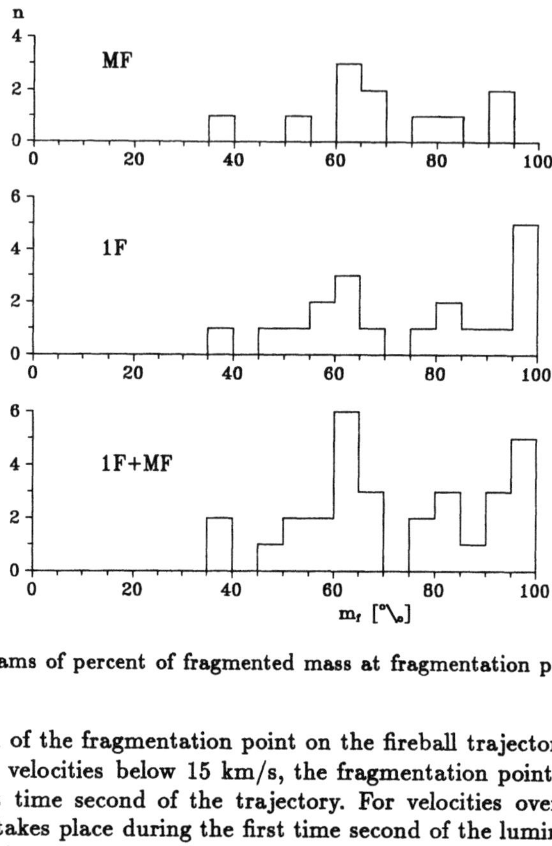

Fig. 7. Histograms of percent of fragmented mass at fragmentation points for 1F and MF fireballs.

The position of the fragmentation point on the fireball trajectory is dependent on velocity. At velocities below 15 km/s, the fragmentation point usually occurs during the last time second of the trajectory. For velocities over 20 km/s, the fragmentation takes place during the first time second of the luminous trajectory. For higher velocities, gross-fragmentation might even cause the onset of luminosity, but there is not much change of velocity at the beginning point for such large bodies to decide, if this is the case (see for comparison Geminids with $v \approx 36$ km/s in Ceplecha and McCrosky, 1992). Application of the fragmentation model to fainter meteors (small masses) with enough change of velocity shortly after the beginning of the luminous trajectory may be another possibility for understanding how the beginning of a luminous trajectory is related to sudden fragmentation.

3. Precision from independent records

The precision of results in applying the fragmentation model may be demonstrated by independent results on the same fireball from two independent records taken at two different stations. 4 fireballs with records of good precision of dynamic data from both stations were computed. The fragmentation model was thus applied independently to two records of the same fireball and comparison of the results gave excellent agreement in all 4 cases (Table V).

TABLE V

Fragmentation model applied to 4 cases of the same fireball with independent records from two different stations. v_∞ = initial no-atmosphere velocity; σ = ablation coefficient; m_∞ = initial mass; m_E = terminal mass; ε_l = standard deviation for one case of observed distance along the trajectory.

fireball no. + st.	class	v_∞ [km/s]	σ [s^2/km^2]	m_∞ [kg]	m_E [kg]	ε_l [m]
38856 15E	NF	19.757 ±.006	0.0254 ±.0005	0.036	0.0009	±16
38856 16W	NF	19.751 ±.011	0.0248 ±.0006	0.033	0.0007	±21
39055 8S	1F	16.013 ±.003	0.0469 ±.0005	0.78	0.004	±14
39055 14E	1F	16.017 ±.002	0.0457 ±.0002	0.75	0.005	± 9
39065 16E	NF	17.321 ±.005	0.0303 ±.0005	1.3	0.037	±16
39065 7S	NF	17.343 ±.010	0.030 ±.002	1.2	0.12	±20
39820 14S	MF	24.624 ±.002	0.0147 ±.0006	2.7	0.03	±26
39820 15W	1F	24.644 ±.003	0.0129 ±.0005	2.0	0.02	±23

Fireball PN 39820 is just a borderline case in the fragmentation classification, but it is more likely 1F class, because the more precise record from station 15W classifies it as 1F. This fireball actually demonstrates differences of results to be expected for borderline cases with uncertain classification between 1F and MF. At this occasion it is useful to stress that also the NF class cases, when classified as NF cases by using our gross-fragmentation model, demonstrate the precision of the model. If 1F or MF cases are dealt with by the single body model for the same fireball and the results for two independent records compared, the differences are significantly greater than those in Table V.

4. Proposal

Generally speaking, with the precision of the photographic fireball data (PN, EN, MORP), only the best records can be used for a detailed study of the gross-fragmentation dynamics. For most of these cases, only the model with one-fragmentation point gave reasonable results. This invokes a proposal for putting into operation some system of cameras with much longer focal distances and sufficiently large fields of view. Precision in observed distances along the trajectory should be of the order of several meters. The focal length of the cameras should be around 1 m with large fields of view. But the biggest problem could be the realization of time marks with sufficiently high precision for fireballs photographed with such cameras.

Very recently, Hills and Goda (1993) presented a new elaborate model of meteoroid (asteroid) fragmentation during the atmospheric penetration. Even if bodies larger than the meteoroids accessible to precise photographic records are the primary concern of this paper, the smallest sizes covered by the theoretical model of Hills and Goda correspond to the largest sizes of the recorded meteoroids. Thus, some predictions of this new fragmentation theory of very large bodies could be partly checked on by photographic observations of fireballs (Prairie Network data).

Acknowledgements

My sincere thanks are due to Dr. R.E. McCrosky for his excellent Prairie Network photographs of fireball splittings and for his kindness in lending me such films from the Smithsonian Astrophysical Observatory archive located at the Oak Ridge Observatory, Harvard, MA, USA. The tedious measurements of these films with fireball splittings were realized at the Ondřejov Observatory of the Academy of Sciences of the Czech Republic by J. Keclíková using a Zeiss Ascorecord, for which work I want to express my special thanks. My thanks are due to C.-Y. Shao for his help with searching in the PN archive and for his enormous work connected with the original measurements of the PN films. I am very much indebted to Dr. J. Borovička and Dr. P. Spurný for their fragmentation computations and for very valuable scientific discussions.

This work has been supported by grant 30319 from the Czechoslovak Academy of Sciences.

References

Ceplecha, Z. : 1967, "Classification of meteor orbits." *Smithson. Contr. Astrophys.*, **11**, 35.

Ceplecha, Z. : 1968, "Discrete levels of meteor beginning height." *Smithson. Astrophys. Obs.Spec. Rep.*, **279**, 1.

Ceplecha, Z. : 1988, "Earth's influx of different populations of sporadic meteoroids from photographic and television data." *Bull. Astron. Inst. Czechosl.*, **39**, 221.

Ceplecha, Z. : 1992, "Influx of interplanetary bodies onto Earth." *Astron. Astrophys.*, **263**, 361.
Ceplecha, Z. and Borovička, J. : 1992, "Meteors." In *Interrelations between Physics and Dynamics for Minor Bodies in the Solar System* (D. Benest, Cl. Froeschlé, Eds.), 309–368, Editions Frontières, Gif-sur-Yvette, France.
Ceplecha, Z. and McCrosky, R.E. : 1976, "Fireball end heights : a diagnostic for the structure of meteoric material." *J. Geophys. Res.*, **81**, 6257.
Ceplecha, Z. and McCrosky, R.E. : 1992, "Gross-fragmentation of meteoroids and bulk density of Geminids from photographic fireball records." In *Asteroids, Comets, Meteors 1991* (A.W. Harris and E. Bowell, Eds.), 109–112, Lunar and Planetary Institute, Houston, USA.
Ceplecha, Z., Spurný, P., Borovička, J. and Keclíková, J. : 1993, "Atmospheric fragmentation of meteoroids." *Astron. Astrophys.*, **279**, 615.
Halliday, I., Blackwell, A.T. and Griffin, A.A. : 1984, "The frequency of meteorite falls on Earth." *Science*, **223**, 1405.
Halliday, I., Blackwell, A.T. and Griffin, A.A. : 1989, "Detailed records of many unrecovered meteorites in western Canada for which further searches are recommended." *J. Roy. Astron. Soc. Canada*, **83**, 49.
Hills, J.G. and Goda, M.P. : 1993, "The fragmentation of small asteroids in the atmosphere." *Astron. J.*, **105**, 1114.
McCrosky, R.E. and Ceplecha, Z. : 1969, "Photographic networks for fireballs." In *Meteorite Research* (P.E. Millman, Ed.), 600, Reidel, Dordrecht.
McCrosky, R. E., Posen, A., Schwartz, G. and Shao, C.-Y. : 1971, "Lost City meteorite : its recovery and a comparison with other fireballs." *J. Geophys. Res.*, **76**, 4090.
McCrosky, R. E., Shao, C.-Y. and Posen, A. : 1977, "Prairie Network Fireball Data." *Center for Astrophysics, Preprint Series*, Nos. 665, 721.
Pecina, P. and Ceplecha, Z. : 1983, "New aspects in single-body meteor physics." *Bull. Astron. Inst. Czechosl.*, **34**, 102.
Pecina, P. and Ceplecha, Z. : 1984, "Importance of atmospheric models for interpretation of photographic fireball data." *Bull. Astron. Inst. Czechosl.*, **35**, 120.
Rabinowitz, D. : 1993, "The size distribution of the Earth-approaching asteroids". *Astroph. J.*, **407**, 412.
Sekanina, Z. : 1983, "The Tunguska event : no cometary signature in evidence". *Astron. J.*, **88**, 1382.
Wetherill, G.W. and ReVelle, D.O. : 1981a, "Which fireballs are meteorites ? A study of the Prairie Network Photographic Meteor Data". *Icarus*, **48**, 308.
Wetherill, G.W. and ReVelle, D.O. : 1981b, "Relationships between comets, large meteors, and meteorites". In *Comets* (L. Wilkening, Ed.), 297, Univ. of Arizona Press, Tucson, USA.

Late note on Strength categories (2.1.)

The"preference for some pressure values" was questioned by one of the referees asking for more statistics in this respect. The pattern in Fig. 4 is not random : at least it shows a systematic decrease with the increasing pressure. The fact that the highest pressure is 11.5 Mdyn/cm^2 has also a statistical significance. If the histogram is constructed up to, say, 100 Mdyn/cm^2, the distribution is clearly not random. From the fact that the histogram is plotted from 0 to 11.5 Mdyn/cm^2 one cannot say it should be random inside this interval.
The observational precision of individual values of pressure is better than the chosen pressure interval of the histogram : statistical spread in such a case corresponds to the natural distribution of the events.
We should not handle each pressure interval of the histogram as an isolated case : it is the change of the numbers of cases from one interval to the neighboring interval, which makes the systematic part of the distribution. Not only the maxima, but also the minima of the distribution

are important. The assumption that the pattern is completely random inside the plotted interval yields the average number of cases in each box 1.3 ±1.3 (the standard deviation), while assumption of a linear change of number of cases with the increasing pressure yields ±0.7 with the *b*, *c*, *d* maxima being significant (over one standard deviation). The distribution is not known a priori : a comparison of maxima and minima significance in Fig 4 is given in Table VI for several assumed distributions. There is no doubt about significance of the *b*, *c* maxima and *b*/*c* and *c*/*d* minima. The *d* maximum and *a*/*b* and *d*/*e* minima are significant only if the decrease of the number of cases with the increasing pressure is accounted for. The *a* region is defined in Table II only as all cases lower than 1.4 Mdyn/cm^2 : there are 4 cases in this region; this is significantly more than what could be expected from a possible admixture of the *b* distribution. The *e* region is formed by the two largest pressure values observed. The significance of the *e* region is clear just from the fact that there are no larger values : these two values could hardly be a statistical wing of the *d* distribution.

If interpolation polynomial solutions are used for the whole distribution pattern in Fig. 4, then the systematic part of errors using a first order polynomial is 82%; using a second order polynomial the systematic part of errors is 80%; a polynomial of the order 9 yields a solution with 65% of systematic errors still left. The big change occurs first, when the order of the polynomial is chosen as 10 : the systematic error decreases to 25% only, and this polynomial yields maxima at 0.5, 2.4, 5.5, 8.4, 10.5 Mdyn/cm^2, and minima at 1.2, 4.0, 7.0, 9.9, quite close to the values declared in Table II.

Summarizing : the *b* maximum in Fig. 4 is the only certain. The *c* and *d* maxima are also over one standard deviation, but to a lesser extent than *b* : their reality should be checked on in the future, when more data of very high precision – necessary for this analysis – will be available. The *a* and *e* regions are included, because of low probability that they belong to the neighboring *b* or *d* distributions, respectively. The extreme possibility of a single Gaussian distribution over the whole interval is in this paper superseded by the other extreme possibility that there are 5 independent Gaussian distributions inside this interval.

The following Table VI contains deviations of the maxima and the minima of Table II from the corresponding average values. These deviations are expressed in units of one standard deviation. The maxima and minima with values ≥1.0 are statistically significant. The values of standard deviations are : ±1.3 cases in a box for the random distribution; ±1.1 cases in a box for the single Gaussian distribution; ±0.7 cases in a box for the linear distribution; ±0.8 cases in a box for the quadratic and the power 4 distribution; ±0.5 cases in a box for the quintuplet Gaussian distribution used in this paper.

TABLE VI

notation of maximum or minimum	distribution				
	random	single Gaussian	polynomial		
			linear	quadratic	power 4
a	0.5	1.5	-	-	-
a/b	0.2	0.5	1.4	1.2	1.3
b	2.8	3.5	4.8	4.3	3.7
b/c	1.0	1.5	2.4	2.3	2.6
c	1.3	0.9	2.4	1.9	2.1
c/d	1.0	1.6	1.8	1.7	1.4
d	0.5	0.8	1.6	1.3	1.6
d/e	1.0	0.5	1.1	1.0	0.9
e	0.2	0.8	0.6	0.8	0.5

THE *GALILEO* ENCOUNTERS WITH GASPRA AND IDA

CLARK R. CHAPMAN
Planetary Science Institute,
620 N. 6th Avenue, Tucson, Arizona 85705
E–mail cchapman@nasamail.nasa.gov

Abstract. The Galileo spacecraft encounters with 951 Gaspra and 243 Ida have provided the first close-up pictures and measurements of asteroids. These two small, S-type asteroids are both irregular in shape, confirming generalized pre-encounter interpretations from groundbased data. Gaspra is lightly cratered by small, fresh craters whereas Ida's surface is heavily covered by craters of all sizes and in all stages of degradation. Unless there are major differences in strength between these two bodies, Ida may be about 10 times older than Gaspra – approaching the age of the solar system. Both asteroids have grooves, although not as prominent as on Phobos. Ida has a population of boulders, particularly near its ends. While Ida seems to have a deep regolith, Gaspra is more nearly in a state of erosion, although there is evidence for an older megaregolith. The data are thus far not conclusive about the geophysical properties of these objects (e.g. whether they are rubble piles) and there are as yet no firm conclusions about how asteroid families are produced by catastrophic collisions. Interesting spectral data relevant to the S-type asteroid controversy (e.g. spatial variations on Gaspra) may lead to some useful generalizations after the remaining Ida data are returned and analyzed in spring 1994. Unexpected magnetic anomalies observed in the vicinities of both asteroids are being studied.

1. Introduction

Until recently, our insights about asteroids came from a very unbalanced set of data. First, there is a great body of astronomical data, obtained from telescopic observations of unresolved, point sources (radar images, which will become more important in the future, are a rare exception to the rule that groundbased data have no spatial resolution). Second, we have a wealth of hand-samples excavated from asteroid surfaces, the meteorites. But until the Galileo spacecraft, enroute to Jupiter, flew past the inner main-belt asteroid 951 Gaspra on 29 October 1991, we had no data at intermediate scales of resolution. For the first time, Galileo provided us with a picture of an asteroid, directly showing its shape, its geological features, and variations in color and other properties across its surface. This encounter thus provided a potential linkage with the other disparate data sources. In addition, the inferences, extrapolations, and theoretical models derived from astronomical data can now be compared with "ground truth," thus calibrating the confidence with which we can derive valid inferences from astronomical data of the countless asteroids that will never be visited by spacecraft. Also, the speculative "meteorite parent bodies," derived from cosmochemical inferences from meteoritical data, can be compared for the first time with real asteroids.

Then, on 28 August 1993, Galileo made its second and final encounter with an asteroid, 243 Ida, one of the larger members of the Koronis family. Only a small portion of the Ida data have been returned to Earth at the time of this writing (nearly all of the remainder are planned for playback in spring 1994). Therefore,

A. Milani et al. (eds.), Asteroids, Comets, Meteors 1993, 357–365.

this review only briefly comments on some of the preliminary results for Ida. Ida, of course, provides a basis for understanding the degree to which the intermediate-resolution studies of Gaspra represent typical or anomalous aspects of asteroids. Since both Gaspra and Ida are main-belt members of the S spectral type and they are both roughly the same size (Ida is less than three times as large as Gaspra), these fly-by encounters represent only a first step in the close-up study of asteroids. Continuing improvements in groundbased radar and application of cheap, generic spacecraft encounters – like the forthcoming Clementine studies of 1620 Geographos – will provide some additional insights about different kinds of asteroids at these intermediate scales. However, the excellent quality of the Galileo data for Gaspra and Ida provides a strong rationale for the use of state-of-the-art instrumentation specifically designed for small body studies. And the kinds of questions that remain unanswered after the Galileo encounters argue that a dedicated program of asteroid exploration will require rendezvous, in situ measurements, and sample return.

Since part of Galileo's instrument package was designed for the Galilean satellites, some of the instruments are well suited to these first asteroid flyby observations. So far, the important scientific results have come primarily from the Solid State Imaging (SSI) camera, the Near-Infrared Mapping Spectrometer (NIMS), and – surprisingly – the magnetometer. Spacecraft encounters at both asteroids were restricted to distances beyond any expectation for existence of dangerous debris trapped in orbit around the targets. Those distances were nearly optimal for imaging because of uncertainties in target ephemeris and the need to sample large areas of sky to be assured of capturing the asteroids in the mosaics; if Galileo had flown much closer, there would have been inadequate time to take mosaics large enough to guarantee showing the targets at larger phase angles, best for studying geologic features. For both fly-bys, the data were recorded on the spacecraft tape recorder. Due to failure of Galileo's high-gain antenna, data return is complicated and time-consuming. In the case of Gaspra, much of the data was played back shortly before the second Earth-encounter, when the low-gain antenna was sufficient. For Ida, the best SSI image plus associated data from other instruments, were returned at 40 bits per second (bps) in September 1993; it is expected that most of the rest of the recorded data can be played back in spring 1994 (also at 40 bps) prior to planned observations of the impact of fragments of Comet Shoemaker-Levy 9 into Jupiter in July 1994.

Preliminary reports interpreting the images of Gaspra (Belton *et al.*, 1992) and the apparent magnetic anomaly near Gaspra (Kivelson *et al.*, 1993) have appeared in Science. A collection of more detailed reports from the imaging team is scheduled to be published in Icarus in early 1994. A preliminary report on the data returned from Ida during September 1993 will be available by the time this review appears.

Briefly, the close-up pictures of Gaspra and Ida did not produce stunningly unexpected revelations. In some ways, the interpretations of groundbased and meteoritical data, and predictions by theorists, had not overlooked anything terribly fundamental. But for the asteroid research community, there were more than enough surprises from both encounters. We now know what asteroids (at least two of them) look like, we have new ideas about the size distribution of smaller objects in the asteroid belt (from the cratering record on Gaspra and Ida), we have information that begins to constrain heretofore highly speculative models of regolith evolution on small bodies, we have magnetometer evidence of unexpected interactions between asteroids and the interplanetary space environment, and we have interesting hints relevant to the longstanding scientific controversy concerning the nature of S-type asteroids and their relationships to differentiated and undifferentiated meteorites.

2. 951 Gaspra

Gaspra is one of the most irregular bodies ever seen from close range in the solar system, unless Ida proves to be even more irregular (Fig. 1). Gaspra's principal diameters are 18.2 x 10.5 x 8.9 km (Veverka *et al.*, 1993a), with an average diameter of 12.2 ± 0.8 km. In the closest image, it presents a planar surface or facet to the spacecraft, and other facets are visible on the edges. One facet, named Dunne Regio, is 5 x 7 km and is flat to within 200 m. Subsequent playback of earlier pictures of other sides of Gaspra revealed a lumpy visage, with a smaller lump attached to a larger lump. Gaspra's shape is consistent with ideas that asteroids might be rubble piles, composed of multiple bodies following a size distribution in which the largest component contains most of the mass. It is also consistent with Gaspra being the core fragment from a catastrophically disrupted larger precursor body, in which case each planar surface may mark the locus of spallation of other outer fragments.

Davies *et al.* (1993) have determined that Gaspra's pole is located at RA 9°.5 ±0.9, Dec +26°.7 ±1.2 (J2000), similar to the less precise orientation determined from pre-encounter analysis of telescopic lightcurves (Magnusson *et al.*, 1992), and they confirm the groundbased spin period of about 7.04 hours (prograde). No satellite of Gaspra has been found in the 150-image data set; 57 images contained at least a part of Gaspra itself. Helfenstein *et al.* (1993) find Gaspra's geometric albedo to be ~0.22 (with regional variations 10% or less), very high for an S-type (but Gaspra was known to be an outlier from typical S types). The bolometric Bond albedo is 0.12 ± 0.03.

The most prominent characteristic of Gaspra is its smoothed surface, peppered by a sprinkling of small, fresh craters. (I will address the craters in more detail below.) Other features have been recognized including grooves and compressional features (Veverka *et al.*, 1993b), in addition to the more general ridges, facets, and lumps that make up Gaspra's general shape. Thomas *et al.* (1993) argue that there is a continuity of structural elements, or "fabric", suggesting that Gaspra is a monolithic body, rather than a gravitationally-bound assemblage of smaller pieces. I would note, however, that the structural elements have been resolved only for one side of Gaspra, from high-resolution pictures taken under similar lighting geometries; furthermore, such a conclusion is not inconsistent with the expectation (from typical fragment size distributions) that a rubble pile would be dominated by one large piece and only a few significant subordinate pieces.

Images taken through different filters through the visible and near-IR, as well as NIMS spectra of different parts of Gaspra's surface, indicate slight spatial variations in composition. In some places (especially ridges), Gaspra's albedo is slightly brighter, the prominent olivine absorption band is relatively stronger, and the overall reflection spectrum is slightly less red, while Gaspra's facets tend to be redder but have lower spectral contrast (perhaps they are slightly more metal-rich). Granahan *et al.* (1993) have intercompared the two data sets and find a spread in the olivine/pyroxene ratio comparable to that separating the H, L, and LL groups of ordinary chondrites – but, in Gaspra's case, even the olivine-poor end-member is more olivine rich than the most olivine-rich ordinary chondrites (the LL's). This evidence supports the idea that Gaspra is a differentiated body. However, large units of substantially differentiated composition are not evident anywhere on Gaspra; the color variations are only about 5%. The smaller differences that are seen may best be attributed to episodic segregation of materials, perhaps by downslope movement following seismic shaking by rare, larger impacts.

Analysis of more than 600 craters (Chapman *et al.*, 1993) from the highest resolution image of Gaspra reveals that most craters are very fresh and follow an unexpectedly (to some) steep size-frequency distribution (see Fig. 2). The crater production function has a differential power-law exponent (population index) of about -4.3 over the range 0.2 to 0.6 km diameter, considerably steeper than the -3.5 characteristic of collisional fragmentation processes (Dohnanyi, 1971) and steeper than the range of possibilities evaluated before encounter by Namiki and Binzel (1991). All craters together, including the modest numbers of shallow, "soft", degraded craters, actually follow a slightly shallower size distribution (slope = $\sim$3.8), but the degraded craters appear to represent an earlier generation of larger craters (typically greater than 500 m diameter) that peek through the last episode of general crater erasure.

Although the -4.3 slope for the production function was surprising to some, it is identical to the production function for craters of similar sizes on the lunar surface; the equivalent slope has also been reported (although it remains controversial) for small bodies in Earth-crossing orbits (Rabinowitz, 1993). Recent work on asteroid collisional evolution, reported at the Gubbio Workshop (cf. Campo Bagatin *et al.*, 1993), suggests that the -3.5 slope may not be sacred, and there may be ways to explain the steeper slope on Gaspra without having to invoke such special circumstances as a recent spray of small particles from a catastrophic fragmentation in the Flora family region.

The steeply sloping production function has implications for the small scale structure of Gaspra's surface, such as its putative regolith. Although perhaps half of the ejecta may be retained from larger cratering events on Gaspra (Housen, 1992), the visible craters could have distributed only a few meters of material around Gaspra (Carr *et al.*, 1993; Chapman *et al.*, 1993). The smaller cratering events would excavate and scour the surface very efficiently, since the majority of ejecta would escape Gaspra from each of these events. In fact, Gaspra's surface must be in net erosion and there can be no continuous production of (or reworking of) regolith on its surface (Chapman *et al.*, 1993). Thus there is little opportunity for optical alteration of surface materials on Gaspra analogous to processes on the lunar surface. The indirect arguments for slight alteration and for minor downslope movement of particulates (invoked, for example, to explain the correlation of color differences with topography) probably pertain to occasional episodes, perhaps induced by seismic shaking due to large impacts, during which the ancient megaregolith is briefly

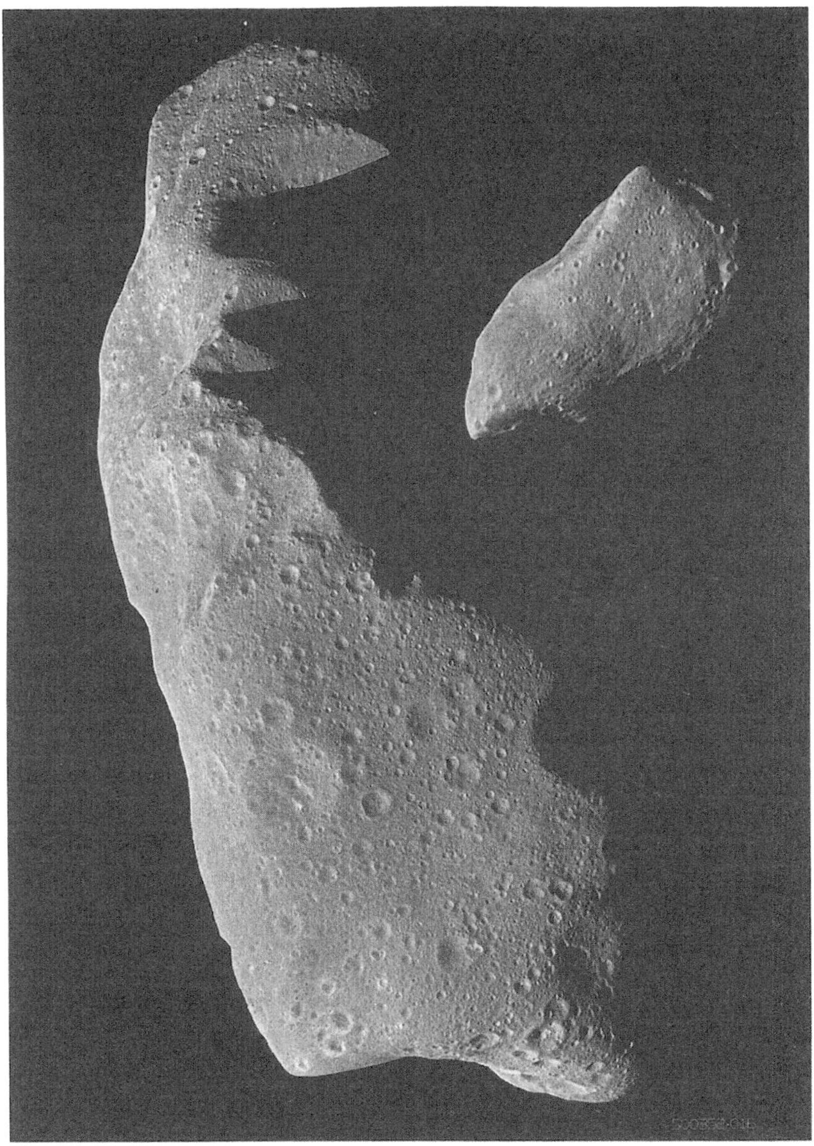

Fig. 1. Galileo images of Gaspra and Ida, showing their approximate relative sizes.

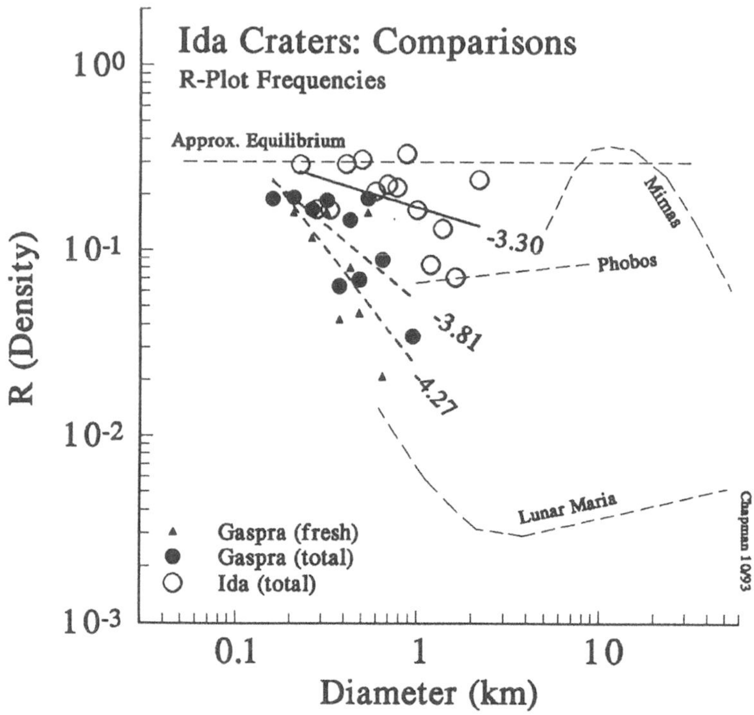

Fig. 2. Crater size-frequency data for Gaspra and Ida, with comparisons. These "R-plot" consists of differential frequency data, normalized by dividing by D^{-3}. Modified from a figure in Chapman *et al.* (1993).

mobilized after which it then remains quiescent for another long duration.

It is plausible that the largest impacts on Gaspra, and on its precursor body, would have developed a poorly consolidated megaregolith. Such megaregolithic structure at depth is probably required to explain the subdued larger craters, the generally softened appearance of Gaspra's profile and surface, and the grooves. The megaregolith may have been formed, along with the planar facets, when Gaspra was formed by the catastrophic disruption (e.g. by spallation) of a larger precursor body in the Flora family zone of the asteroid belt. Perhaps there have been a few subsequent episodes of shaking and blanketing due to sub-catastrophic impacts, although no craters larger than 3 km diameter have been found on Gaspra. Possibly a large crater remains hidden in the small, unimaged part of Gaspra near its southern pole.

The cratering age of Gaspra (i.e. the age of exposure of the visible, smoothed surface, to the production function, as represented by the fresh crater population) has been estimated (Belton *et al.*, 1992) as 200 My, assuming that Gaspra responds to cratering impacts as a body with rocky or sandy impact strength. This is a highly model-dependent result, for which the Galileo data provide only some new elements (see Chapman *et al.*, 1993). Compared with a self-consistent estimate for the collisional lifetime of Gaspra (500 My), this result means that Gaspra is relatively youthful. If Gaspra actually has the impact strength associated with ductile metal, which would be possible if it had a stony-iron composition, then it could be much older. A more controversial idea (Greenberg *et al.*, 1993; see also discussion by Chapman *et al.*, 1993) is that Gaspra behaves, ironically, as an exceptionally strong body because it is weakened by large impacts; these arguments, based on hydrocode models of asteroid collisions, would also lengthen Gaspra's lifetime and cratering age.

3. 243 Ida : Preliminary Observations

A single, highest resolution image of Ida, ~35 m/pixel, was the first Ida image to be returned, and the only one as of this writing. It presents an appearance similar to Gaspra in some respects, but very different in others. As had been expected from groundbased lightcurves, Ida is also very irregular. (Ida's shape in the single image is consistent with the second of two ambiguous pole solutions derived by Binzel *et al.*, 1993). Ida appears to be somewhat larger than was expected. It is not a contact binary, contrary to some pre-encounter suggestions, but whether it is a monolithic body, a rubble pile, or has some other shape characteristics is speculative until views from its other sides are returned. These will be very important for understanding whatever clues Ida might hold in store for us about how the Koronis family parent body was disrupted.

Ida, unlike Gaspra, is one of the most densely cratered objects in the solar system. The crater density is similar to that attained on surfaces that have been saturated by impacts and are in equilibrium. Theoretical studies of such equilibrium processes (applied to the lunar surface : see Chapman *et al.*, 1970) show that craters will follow a power law size distribution with differential index (slope) of -3, and that crater morphologies will range from fresh to degraded (degraded being the most common) with similar ratios between the morphological classes at all diameters. This is just what is seen on Ida (Fig. 2). The slope of the power-law for Ida's craters is about -3.3 (with error bars wide enough to include -3.0).

It is not always possible to unambiguously determine the underlying production function when the craters are in equilibrium. However, Ida's crater population could be produced by letting the same crater production function that is responsible for Gaspra's fresh craters impact on Ida for about 10 times as long. If this is true, Ida's surface may date back 2 billion years. If these asteroids have higher impact strength, then their corresponding ages are even older, and Ida could date back to the Late Heavy Bombardment.

Unlike Gaspra, Ida shows many suggestions of a deep regolith. There are boulders visible on Ida's surface (particularly near its ends), indications of downslope movement, and small-scale albedo variations (e.g. dark floored craters), in addition to the range of crater morphologies. There are some very large craters on Ida, so there are sources for the deeper regolith. Other geological features have been recognized on the high-resolution image of Ida, including lineaments that predominate on one part of the asteroid.

Preliminary reports by the NIMS experimenters suggest that Ida is much more spatially uniform in colors (spectral reflectance) than is Gaspra. This seems somewhat inconsistent with preliminary indications from groundbased observations, reported by Barucci *et al.* among others at ACM93, of large color differences on opposite sides of Ida. If the NIMS data hold up once the complete data set is returned in spring 1994, it could indicate either that Ida is compositionally homogeneous (perhaps it is an undifferentiated, chondritic body) or that it has been "painted grey" by global distribution of ejecta from the latest large impact.

Magnetometer observations during fly-bys of both Gaspra and Ida revealed anomalies that the experimenters have interpreted as probably reflecting interactions of these bodies with the solar wind. The original interpretations of the anomaly near Gaspra (Kivelson *et al.*, 1993) considered Gaspra to have remanent magnetization similar to that found in most meteorites (both differentiated and undifferentiated). Although such a net magnetic moment would be inconsistent with a randomized sandbank, it would not be inconsistent with a rubble pile having a size distribution (similar to that of the main-belt asteroid population) dominated volumetrically by the largest component. The magnetic anomaly observed near Ida has caused the investigators to begin rethinking the Gaspra anomaly (Kivelson, personal communication, 1993), so the implications of these unexpected data for either asteroid science or space physics remain to be developed.

Acknowledgements

I thank Galileo Project officials, the superb engineering team, and my scientific colleagues on Galileo for making these encounters possible and for sharing their results with me. In particular, I thank T. V. Johnson (the Project Scientist), M. Belton, J. Veverka, the other members of the SSI Team, J. Granahan, M. Kivelson, D. Davis, E. Ryan, W. Merline, R. Binzel, A. Harris, J. Bell, and P. Magnusson for discussions. P.S.I. is a division of Science Applications International Corp. This is P.S.I. Contribution No. 315.

References

Belton, M.J.S., Veverka, J., Thomas, P., Helfenstein, P., Simonelli, D., Chapman, C., Davies, M. E., Greeley, R., Greenberg, R., Head, J., Murchie, S., Klaasen, K., Johnson, T.V., McEwen, A., Morrison, D., Neukum, G., Fanale, F., Anger, C., Carr., M., and Pilcher, C. : 1992, "Galileo encounter with 951 Gaspra : first pictures of an asteroid." *Science*, **257**, 1647-1652.

Binzel, R.P., Slivan, S.M., Magnusson, P., Wisniewski, W. Z., Drummond, J., Lumme, K., Barucci, M.A., Dotto, E., Angeli, C., Lazzaro, D., Mottola, S., Gonano-Beurer, M., Michalowski, T., De Angelis, G., Tholen, D.J., Di Martino, M., Hoffmann, M., Geyer,

E. H., and Velichko, F. : 1993, "Asteroid 243 Ida : groundbased photometry and a pre-Galileo physical model." *Icarus*, **105**, 310-325.

Campo Bagatin, A., Cellino, A., Davis, D.R., Farinella, P., and Paolicchi, P. : 1993, "Wavy size distributions for collisional systems with a small-size cutoff." *Planetary and Space Sci.*, in press.

Carr, M.H., Kirk, R.L., McEwen, A., Veverka, J., Thomas, P., Head, J.W., and Murchie, S. : 1993, "The Geology of Gaspra." *Icarus*, in press.

Chapman, C.R., Mosher, J.A., and Simmons, G. : 1970, "Lunar cratering and erosion from Orbiter 5 photographs." *J. Geophys. Res.*, **75**, 1445-1466.

Chapman, C.R., Veverka, J., Belton, M.J.S., Neukum, G., and Morrison, D. : 1993, "Cratering on Gaspra." *Icarus*, in press.

Davies, M. E., Colvin, T. R., Belton, M.J.S., Veverka, J., and Thomas, P. C. : 1993, "The direction of the north pole and the control network of asteroid 951 Gaspra." *Icarus*, in press.

Dohnanyi, J.S. : 1971, "Fragmentation and distribution of asteroids." In *Physical Studies of Minor Planets* (T. Gehrels, Ed.) 263-295, NASA SP-267, U.S. Govt. Printing Office.

Granahan, J.C., Fanale, F.P., Robinson, M., Carlson, R.W., Kamp, L.W., Klaasen, K.P., Belton, M., Cook, D., Edwards, K., McEwen, A.S., Carcich, B.T., Helfenstein, P., Simonelli, D., Thomas, P., and Veverka, J. : 1993, "Galileo's multispectral synergistic view of 951 Gaspra." Abstract in *EOS, Transactions, Am. Geophys. Union*, **74** (No. 16/Suppl.), 197.

Greenberg, R., Nolan, M.C., Bottke, W.F. Jr., Kolvoord, R.A., and Veverka, J. : 1993, "Collisional history of Gaspra." *Icarus*, in press.

Helfenstein, P., Veverka, J., Thomas, P.C., Simonelli, D.P., Lee, P., Klaasen, K., Johnson, T.V., Breneman, H., Head, J.W., Murchie, S., Fanale, F., Robinson, M., Clark, B., Granahan, J., Garbeil, H., McEwen, A.S., Kirk, R.L., Davies, M., Neukum, G., Mottola, S., Wagner, R., Belton, M., Chapman, C., and Pilcher, C. : 1993, "Galileo Photometry of asteroid 951 Gaspra." *Icarus*, in press.

Housen, K.R. : 1992, "Crater ejecta velocities for impacts on rocky bodies." *Lunar and Planet. Sci. XXIII*, 555-556, Lunar and Planetary Institute, Houston.

Kivelson, M.G., Bargatze, L.F., Khurana, K.K., Southwood, D.J., Walker, R.J., and Coleman, P.J. Jr. : 1993, "Magnetic field signatures near Galileo's closest approach to Gaspra." *Science*, **261**, 331-334.

Magnusson, P., Barucci, M.A., Binzel, R.P., Blanco, C., Di Martino, M., Goldader, J. D., Gonano-Beuer, M., Harris, A. W., Michalowski, T., Mottola, S., Tholen, D.J., and Wisniewski, W. Z. : 1992, "Asteroid 951 Gaspra : pre-Galileo physical model." *Icarus*, **97**, 124-129.

Namiki, N. and Binzel, R.P. : 1991, "951 Gaspra : a pre-Galileo estimate of its surface evolution." *Geophys. Res. Lett.*, **18**, 1155-1158.

Rabinowitz, D. : 1993, "The size distribution of the Earth-approaching asteroids." *Astrophys. J.*, **407**, 412-427.

Thomas, P. C., Veverka, J., Simonelli, D., Helfenstein, P., Carcich, B., Belton, M.J.S., Davies, M.E., and Chapman, C. : 1993, "The shape of Gaspra." *Icarus*, in press.

Veverka, J., Belton, M., Klaasen, K., and Chapman, C. : 1993a, "Galileo's Encounter with 951 Gaspra : Overview." *Icarus*, in press.

Veverka, J., Thomas, P., Simonelli, D., Belton, M.J.S., Carr, M., Chapman, C., Davies, M.E., Greeley, R., Greenberg, R., Head, J., Klaasen, K., Johnson, T.V., Morrison, D., and Neukum, G. : 1993b, "Discovery of grooves on Gaspra." *Icarus*, in press.

DUST MEASUREMENTS IN THE OUTER SOLAR SYSTEM

EBERHARD GRÜN

Max-Planck-Institut für Kernphysik,
D 69117 Heidelberg, Germany
E-mail GRUEN@DUSTY.MPI-HD.MPG.DE

Abstract. In-situ measurements of micrometeoroids provide information on the spatial distribution of interplanetary dust and its dynamical properties. Pioneers 10 and 11, Galileo and Ulysses spaceprobes took measurements of interplanetary dust from 0.7 to 18 AU distance from the sun. Distinctly different populations of dust particles exist in the inner and outer solar system. In the inner solar system, out to about 3 AU, zodiacal dust particles are recognized by their scattered light, their thermal emission and by in-situ detection from spaceprobes. These particles orbit the sun on low inclination ($i \leq 30°$) and moderate eccentricity ($e \leq 0.6$) orbits. Their spatial density falls off with approximately the inverse of the solar distance. Dust particles on high inclination or even retrograde trajectories dominate the dust population outside about 3 AU. The dust detector onboard the Ulysses spaceprobe identified interstellar dust sweeping through the outer solar system on hyperbolic trajectories. Within about 2 AU from Jupiter Ulysses discovered periodic streams of dust particles originating from within the jovian system.

1. Introduction

The classical conception of interplanetary dust is derived from the interpretation of zodiacal light observations. Zodiacal light can be observed from non-polluted areas on the Earth with the naked eye. It appears as a diffuse glow in the west after twilight and in the east before dawn. It is symmetric with respect to the ecliptic and is wedge shaped with its widest and brightest parts near the Sun. For centuries it has been accepted wisdom that zodiacal light is caused by the reflection of sunlight from myriad interplanetary dust particles which are concentrated in the ecliptic plane. This concentration in the ecliptic plane is direct evidence for dust particles on low-inclination ($\leq 30°$) orbits (cf. Leinert and Grün, 1990) which contribute to the zodiacal brightness. The size distribution of interplanetary dust at 1 AU has been determined from lunar micro-craters and spacecraft data (Grün et al., 1985). From these observations it has been concluded that maximum scattering cross-section lies in the size range from 10 to 100 μm and hence it is this size range which contributes most to the zodiacal light.

It has been suggested by ground-based observations that the concentration of interplanetary dust particles is highest near the sun and decreases with heliocentric distance. Space probe measurements of the zodiacal light brightness from 0.3 AU (Leinert et al., 1981) to 3 AU (Toller and Weinberg, 1985) and beyond found that the intensity I of zodiacal light falls off with heliocentric distance r as $I \sim r^{-\gamma}$, with $\gamma = 2$ to 2.5. Beyond about 3 AU zodiacal light was no longer observable above the background light. Dumont and Levasseur-Regourd (1988) and Levasseur-Regourd et al. (1991) conclude from these and ground-based observations that there is a

A. Milani et al. (eds.), Asteroids, Comets, Meteors 1993, 367–380.

radial variation of the optical dust properties (scattering function, polarization and albedo) and hence the radial variation of the spatial dust density is approximately proportional to the inverse of the solar distance.

Besides a small increase in the visible brightness of zodiacal light in the anti-sunward direction (Gegenschein), which is caused by the scattering characteristics of small dielectric particles there is little evidence from observations in the visible wavelength range for dust outside the Earth orbit.

Because of the generally low albedo of interplanetary dust (Hanner, 1980) most of the energy from insolation is absorbed and re-emitted as thermal radiation at infrared wavelengths. Observations by e.g. the IRAS satellite (Hauser et al., 1984) show that maximum thermal emission from dust at the distance of the asteroid belt occurs in the 10 to 20 μm wavelength range. However, modelling of the interplanetary dust populations by Divine (1993) and by Staubach et al. (1993) show that both optical and thermal zodiacal light emissions can be explained to a large extent by a single "core" population of dust particles on low-inclination orbits which are concentrated towards the sun.

Recently Dubin and Soberman (1991) published a "Cosmoid"-hypothesis by which meteoroids flare up due to sublimation and disintegration into many smaller grains. This hypothesis was introduced in order to explain some (previously considered spurious) observations of the optical dust detector (AMD) on Pioneers 10 and 11. These ideas will not be discussed here because acceptance of them would invalidate the large body of experience gained from previous observations and analyses of interplanetary dust - the authors fail to substantiate these consequences.

In the following two chapters we will separate the discussion of in-situ dust measurements and interpretations inside and outside 3 AU because at about this distance there seems to be a significant transition of dust populations. In the fourth chapter the newly discovered dust emissions from the jovian system are discussed and in the fifth chapter evidence for interstellar dust in the solar system is presented.

2. Dust Inside 3 AU

Complementary to ground-based are in-situ observations by dust impact detectors on board interplanetary spaceprobes. In-situ measurements of interplanetary dust have been performed in the heliocentric distance range from 0.3 AU out to 18 AU (for a complete review see Leinert and Grün, 1990). We will start the discussion on the most recent results from the dust measurements on board the Galileo spacecraft (Grün et al., 1992a).

The Galileo spacecraft was launched in October 1989 on its trajectory to Jupiter. Before the Galileo spacecraft will reach Jupiter it has to perform a six year journey through the solar system. Figure 1 shows the Galileo trajectory through the inner solar system. A Venus swing-by took place in February 1990 and the spacecraft flew by the Earth in December 1990 for the first time. Until end of 1992 Galileo repeatedly traversed interplanetary space between 0.7 and 2.26 AU and had fly-bys of the asteroid Gaspra and the Earth for a second time.

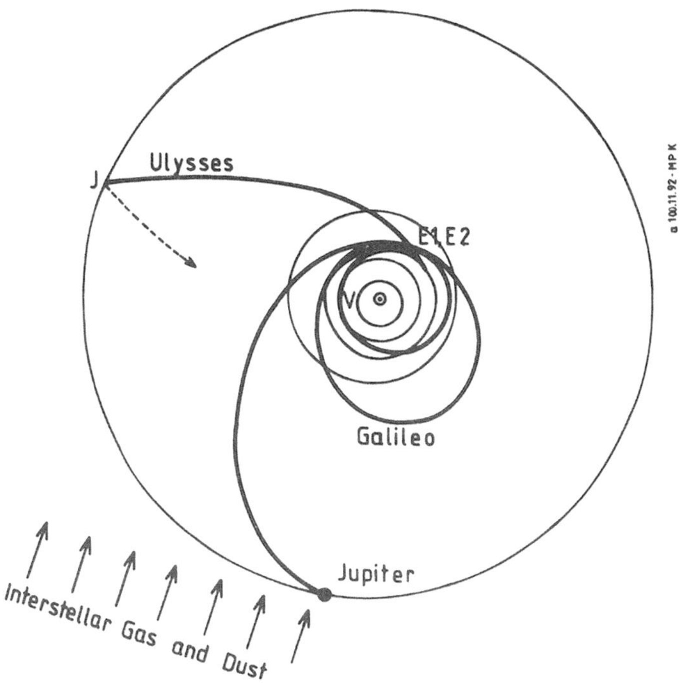

Fig. 1. Orbits of the Galileo and Ulysses spaceprobes through the solar system. The orbits of Mercury, Venus, Earth, Mars and Jupiter are indicated by thin lines. The dot at the end of the Galileo trajectory indicates that Galileo will become a satellite of Jupiter after swing-bys of Venus (V), and Earth (E1 and E2). The Ulysses trajectory was deflected by Jupiter out of the ecliptic plane onto a 80°-inclination orbit (dashed line). The inflow direction of interstellar gas and dust is shown for comparison.

The Galileo dust detector is a multi-coincidence impact ionization detector (Grün et al., 1992b). Masses and impact speeds of dust particles are determined from the measured amplitudes and rise-times of the positive and negative impact charge signals. The mass sensitivity threshold is $4 \cdot 10^{-15}$ g at 20 km/s and $6 \cdot 10^{-16}$ g at 40 km/s impact speed, as deduced from laboratory impact calibrations with carbon, silicate and iron dust particles (Göller and Grün, 1989). The accuracy of the speed determination is a factor of 2 and that of the mass determination is a factor of 10 in the calibrated range (Grün et al., 1992c).

The dust detector instrument is mounted on the spinning section of the spacecraft. During the initial phase of the mission the spin vector points away from the Sun. The viewing direction of the dust detector forms an angle of 55 degrees with respect to the spin vector. In December 1989 the dust detector was switched on for the first time, when Galileo was at a distance of 0.88 AU from the Sun. Figure 2 shows the measured dust flux during the initial Galileo mission phase. At the same heliocentric distance the impact rates were lower by a factor of ten when the spacecraft moved towards the Sun compared to the rates measured when it moved away from the Sun. After both Earth fly-bys the impact rate increased by about an order of magnitude. The general decrease of the flux with heliocentric distance is due to the decrease of the dust population with increasing distance from the sun. During the Gaspra fly-by no enhancement above the average count rate was observed.

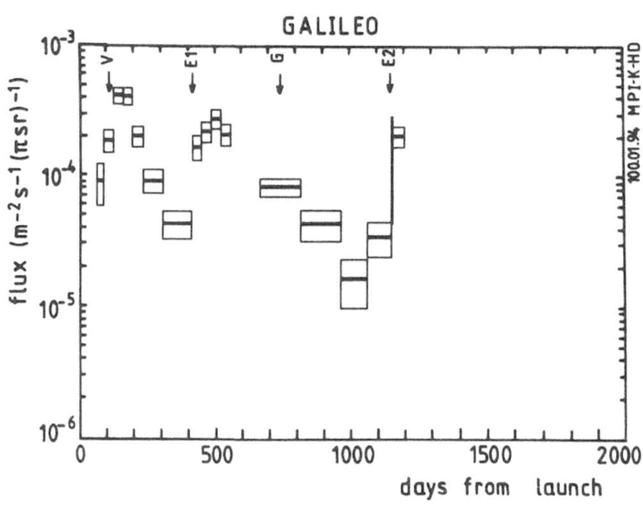

Fig. 2. Cumulative flux of dust particles recorded by the Galileo dust detector from launch (October 1989) until the end of 1992. The fluxes refer to a flat plate sensor (effective solid angle π sr) and to a positive impact charge of $\geq 10^{-13}$ C which corresponds to an effective mass threshold of $5 \cdot 10^{-14}$ g (for an assumed impact speed of 20 km/s). Each box represents the time interval over which the flux was averaged and the 1-σ uncertainty of this flux. Fly-bys of Venus (V), Earth (E1 and E2) and the asteroid Gaspra (G) are indicated.

The different fluxes observed by Galileo inbound and outbound at the same heliocentric distance R inside the asteroid belt are explained by the model of Divine (1993) as dust populations which show up differently under the different observation conditions along the Galileo trajectory (Fig. 3). In the Divine model a dust population is described by distributions of the particle mass, perihelion distance, eccentricity and inclination. Five distinct populations are required to match most interplanetary meteoroid data ranging from 10^{-18} to 1 g in mass and 0.1 to 20 AU in heliocentric distance. Inside about 3 AU the most important population to match both the Galileo data and the zodiacal light observations is the "core" population. The radial concentration of the core population falls off as $r^{-1.3}$ and the orbits have low inclinations ($i \leq 30°$) and moderate eccentricities (≤ 0.6). Among the other populations only the "inclined" population (small particles, with orbit inclinations of up to 60° and low eccentricities) contributes significantly to the Galileo flux at about Venus' distance. Outside about 3 AU distance the "halo" population becomes dominant.

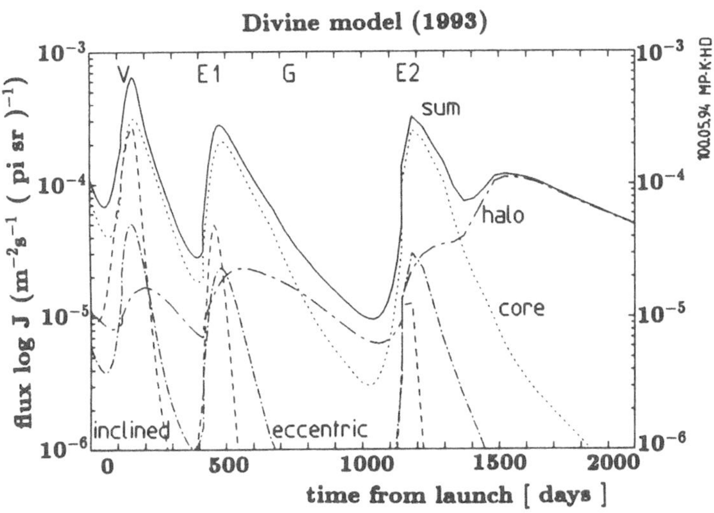

Fig. 3. Divine (1993) model of the Galileo flux (cf. Fig. 2) from launch until Jupiter arrival. Contributions to the flux from 5 different dust populations are shown separately as well as the sum of all. The model calculations take into account the orbit of Galileo, the detector geometry and sensitivity.

3. Dust Outside 3 AU : Early Measurements

The first spacecraft which passed through the asteroid belt and went beyond it were the Pioneer 10 and 11 probes (Fig. 4). Both spacecraft carried several detectors which measured different aspects of interplanetary dust. Only the meteoroid penetration detectors recorded significant concentrations of meteoroids outside 3 AU (Humes, 1980). These detectors are simple but highly reliable detectors which record the puncture of a pressurized cell by a meteoroid impact. Each detector consists of 234 individual detectors assembled in two independent data channels. On Pioneer 10 one channel malfunctioned early in the mission leaving only 108 pressurized cells available.

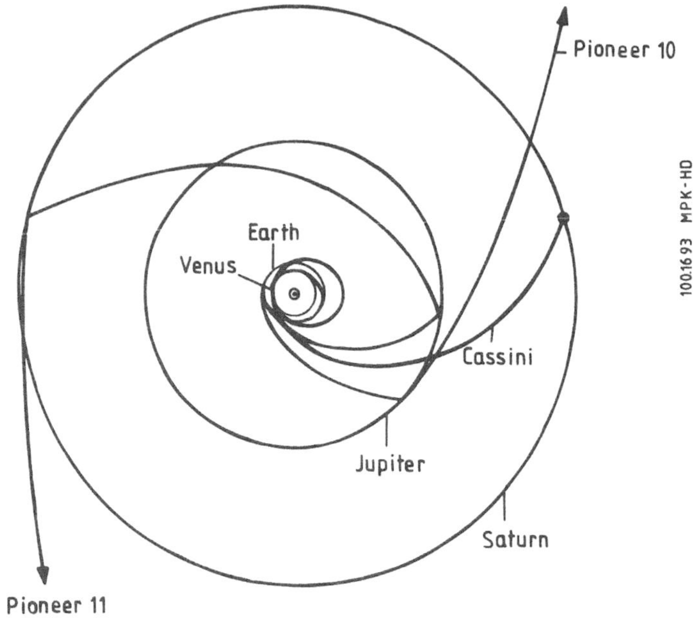

Fig. 4. Orbits of the Pioneer 10, 11 and Cassini spaceprobes projected onto the ecliptic plane. The orbits of some planets are shown as thin lines. The Pioneers 10 and 11 (launches in 1972 and 1973) had fly-bys of Jupiter (P10 and P11) and Saturn (P11) before they obtained solar system-escape-trajectories. After launch in 1997 the Cassini spaceprobe will fly-by Venus, Earth and Jupiter before becoming a satellite of Saturn.

From launch out to about 3 AU the impact flux decreased by about one order of magnitude and stayed constant for the remainder of the interplanetary tour (Fig. 5). No sign of the asteroid belt is recognized in the data. Out to 18 AU Pioneer 10 recorded 87 penetrations and Pioneer 11 recorded 87 penetrations past Saturn's fly-by. During fly-bys of the planets Jupiter and Saturn one to two orders of magnitudes increased impact fluxes were observed, mainly due to dust populations orbiting the planets. Pioneer 11 crossed three times the region between 3.7 and 4.9 AU at different impact configurations with respect to the meteoroid population (outbound, and after Jupiter fly-by inbound and outbound again).

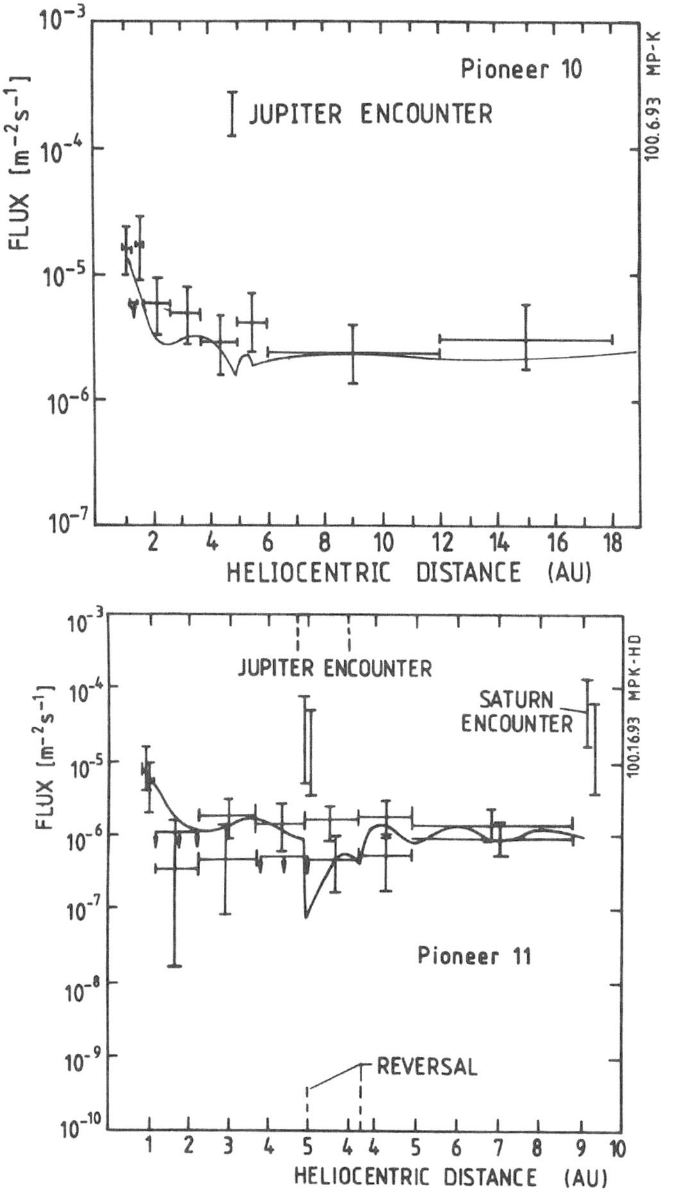

Fig. 5. Meteoroid fluxes recorded by the Pioneer 10 and 11 penetration detectors (after Humes, 1980). The threshold sensitivities of the Pioneer 10 and 11 instruments are $8 \cdot 10^{-10}$ and $6 \cdot 10^{-9}$ g, respectively, at an assumed impact speed of 20 km/s. The instruments were mounted on the back of the antenna dish which always pointed towards Earth. The solid line corresponds to the expected impact flux from a population of meteoroids on randomly inclined bound orbits of high eccentricities and with constant spatial density. Upper panel Pioneer 10 fluxes. Lower panel Pioneer 11 fluxes, two channels.

Extensive model calculations by Humes (1980) showed that the spatial density of $6 \cdot 10^{-9}$ to $8 \cdot 10^{-10}$ g meteoroids is essentially constant out to 18 AU. The multiple flux measurements with Pioneer 11 inside Jupiter's orbit allowed Humes to compare them with different inclination- and eccentricity-distributions of solar system dust. He found that the Pioneer 11 data obtained between 4 and 5 AU are best explained by meteoroids being in randomly inclined orbits of high eccentricity.

Recently Divine (1993) modelled the Pioneer 10 and 11 data by his "halo" population which has similar orbit characteristics as Humes' population except that the eccentricities needed only to reach moderate values (≤ 0.6). The impact rates recorded by Ulysses on its trajectory out to Jupiter are represented by the core and halo populations as well. However, additional information on these meteoroid populations is obtained from Ulysses data by the determination of impact direction, speed and particle mass.

4. Jupiter Dust Streams

The Ulysses and the Galileo dust detectors have a mass sensitivity 10^5 times higher than the Pioneer 10 and 11 penetration detectors. From 8 days prior to closest approach (CA) to Jupiter until 2 days after CA the Ulysses dust instrument sensitivity was reduced for reasons of instrument safety. The trajectory of Ulysses and the geometry of dust detection is explained in Fig. 6. The spacecraft spins around an axis which, along with the high gain antenna, points towards Earth. The dust detector onboard has a 140° conical field-of-view (FOV), and is mounted almost at right angle with respect to the Ulysses spin axis. The spin averaged sensitive area of the dust detector to a mono-directional stream of dust grains is ≤ 0.02 m^2; the maximum occurs when the center of the detector FOV passes through the stream during spacecraft rotation.

The impact rate observed by Ulysses (Fig. 7) of all particles (Baguhl et al., 1993) was low (about one impact per 2 days) for most of the time, although a statistically significant peak of big particles did occur at the time of Jupiter encounter (Grün et al., 1992d). For most of 1991 when Ulysses was inside 4 AU from the sun the impact rate of small particles was also low. However, within a few months of Jupiter fly-by, several bursts, in the impact rate were observed, which occurred with a remarkable periodicity of 28 ± 3 days around Jupiter fly-by (Grün et al., 1993a). No periodic dust phenomena in interplanetary space have been known before for such small grains.

The directions of sensor pointing at times when impacts were detected are shown in Fig. 8. Small particles ($m \leq 2.5 \cdot 10^{-14}$ g) appear mostly in well collimated streams of short duration. Most streams arrive from directions which include the direction to Jupiter (rotation angle 265° before CA and 53° post CA). Five characteristics of the streams, taken as a whole can be explained only if we assume a jovian origin; no interplanetary or interstellar source is capable of causing them :

1. Narrow, collimated streams must have a nearby source, otherwise they should be dispersed in space and time.

2. The streams are concentrated near Jupiter and the strongest stream was detected closest to it.

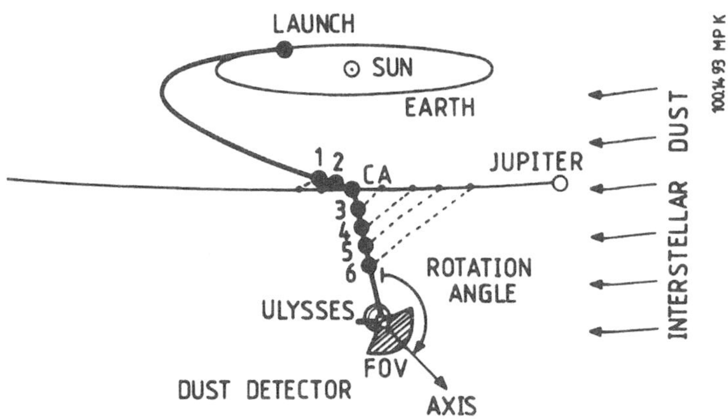

Fig. 6. Ulysses trajectory and geometry of dust detections. Oblique view from above the ecliptic plane also showing the sun and the orbits of Earth and Jupiter (in the foreground). Dots along the trajectory refer to positions of Ulysses at which major dust streams were detected - dotted lines point to Jupiter's positions. The rotation angle of the sensor axis at the time of a dust impact is measured from the ecliptic north direction. Arrows indicate the flow of interstellar dust (c.f. Grün et al., 1993a).

3. The streams are wider further away from Jupiter.

4. The streams before Jupiter CA approach Ulysses from directions almost opposite to the streams after CA; however, all streams radiate from close to the line-of- sight direction to Jupiter.

5. The observed periodicity strongly suggests that all streams derived from a single source and tends to rule out fortuitous cometary or asteroidal origins of individual streams.

Several mechanisms have been suggested which may explain the observed streams. Horanyi et al. (1993) show that small positively charged grains are accelerated and ejected by Jupiter's magnetosphere. Electromagnetic forces oppose and dominate jovian gravity. It is assumed that if Io is the source of the dust grains (Johnson et al., 1980, Morfill et al., 1980) its location in geographic and geomagnetic coordinates determines the exit direction of the escaping particles and provides a possible explanation of the observed periodicities.

An alternate source (Jupiter's gossamer ring) for the dust particles is discussed by Hamilton and Burns (1993). The periodicity of about 28 days is then caused by the modulation of the dust trajectories by the interplanetary magnetic field. Further analyses of Ulysses dust, magnetic field and solar wind data may give the clue to the correct mechanism. In addition, more data will become available when the Galileo spaceprobe reaches Jupiter in 1995.

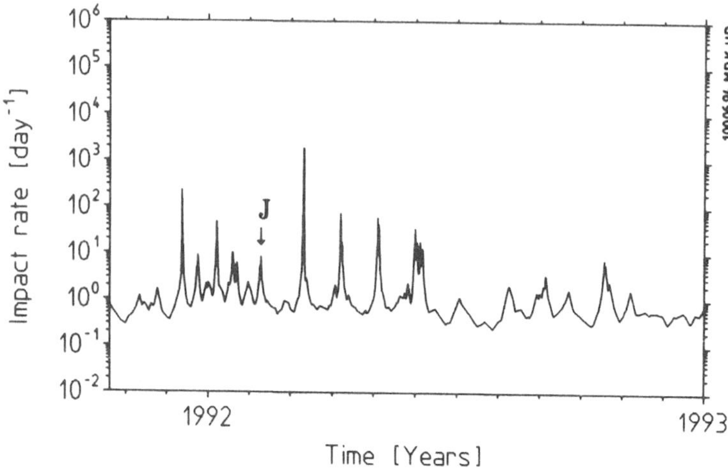

Fig. 7. Impact rate detected by the Ulysses dust detector from Oct. 1991 until end of 1992. Jupiter closest approach (J) occurred on Feb. 8, 1992. Running average continuously include 5 impacts.

5. Interstellar Dust

The Ulysses spaceprobe observed several aspects of interstellar material in the solar system. Besides ions of interstellar origin being picked-up by the solar wind (Geiss et al., 1993) the flow of interstellar helium through the solar system has been directly measured (Witte et al., 1993). The speed of the interstellar gas was 26 km/s outside the gravitational influence of the sun. It arrives from the direction : ecliptic longitude $l = 252°$ and latitude $b = 2.5°$. Grün et al. (1993a) identified a significant flux of interstellar grains at the distance of Jupiter. In a more recent paper (Grün et al., 1993b) a detailed analysis of the extended Ulysses data set (Baguhl et al., 1993) is given.

Dust particles emitted from the jovian system are identified by their smaller masses ($m \leq 2.5 \cdot 10^{-14}$ g) and their narrow bunching both in time and impact direction. The proof for the interstellar origin of some of the bigger particles outside the stream periods came from considerations of the impact direction and speed. Grün et al. (1993b) demonstrated that after the Jupiter flyby of Ulysses possible interstellar dust can be easily distinguished from interplanetary dust (generally on prograde orbits) by their retrograde trajectories. The authors identified those particles as of interstellar origin which had speeds above the maximum speed possible for particles on bound orbits.

The maximum speed which interplanetary dust can have on bound orbits varies with the impact direction from which dust is detected. Fig. 9 shows all dust particles

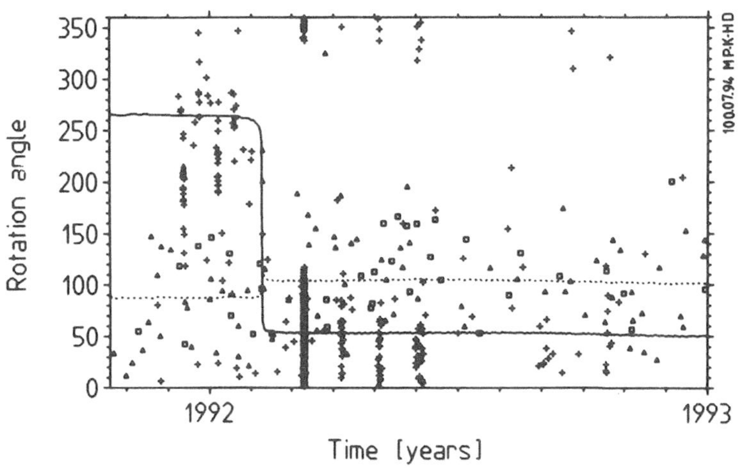

Fig. 8. Dust impact times versus spacecraft rotation angle observed by Ulysses. Only those impacts are shown (about 80% of all) for which correct masses and rotation angles were determined. The rotation angle is taken to be 0° when the dust sensor axis points closest to the north ecliptic pole. At rotation angle 90° the detector axis points parallel to the ecliptic plane in the direction of planetary motion. Impacts are marked according to their masses and impact speeds : squares : $m \geq 2.5 \cdot 10^{-14}$ g, $v \geq 26$ km/s, triangles : $m \geq 2.5 \cdot 10^{-14}$ g, $v < 26$ km/s, and crosses : $m < 2.5 \cdot 10^{-14}$ g, $v \geq 26$ km/s. The solid line indicates the direction to Jupiter. The arrival direction of interstellar dust (assumed to coincide in direction and speed with interstellar gas) is shown by the dashed line.

detected in the time interval from shortly after Jupiter flyby (day 43 of 1992) until end of 1992. During this time interval the radiant and impact speed of interstellar gas stayed almost constant. The radiant of interstellar gas and dust occurs at rotation angle 103°, that of Jovian dust on straight trajectories is at 50°.

Radiation pressure reduces the effect of solar gravity. The ratio $\beta = F_{rad}/F_{grav}$ (c.f. Burns et al. 1979) is independent of solar distance r and depends only on the size, density and the optical properties of a dust grain. The maximum heliocentric speed v_{max} of a particle on a bound orbit is

$$v_{max}(\beta) = \sqrt{2(1-\beta)\frac{\mu}{r}} \tag{1}$$

where $\mu = G\, M_{\odot}$, with gravitational constant G and solar mass $M_{\odot}$. The spacecraft speed ($v_{Uls} = 8.2$ km/s) has to be vectorially added in order to obtain the relative speed between spacecraft and dust particle. Fig.9 shows the limiting speeds between bound and unbound orbits (dashed lines) for different β-values. Speeds of particles on bound orbits are reduced for increasing β-values until at $\beta > 1$ no bound orbits exist. Assuming intermediate β-values for micron sized grains detected by Ulysses we conclude that about 80% of all big particles are on unbound orbits and hence may be of interstellar origin. Interstellar grains arriving at the solar system with a speed of 26 km/s will gain additional speed from the solar gravitational field (corresponding to the escape speed, 18 km/s at the distance of Jupiter), which is again reduced by radiation pressure. The resulting impact speed at Ulysses is $v_{is} \approx \sqrt{26^2 + (1-\beta)18^2 + 8.2^2}$, v_{is} varies from 32.6 to 27.2 km/s for β equal 0 and 1, respectively. In Fig. 9 we show the 1-σ uncertainty range (factor 2) for both cases.

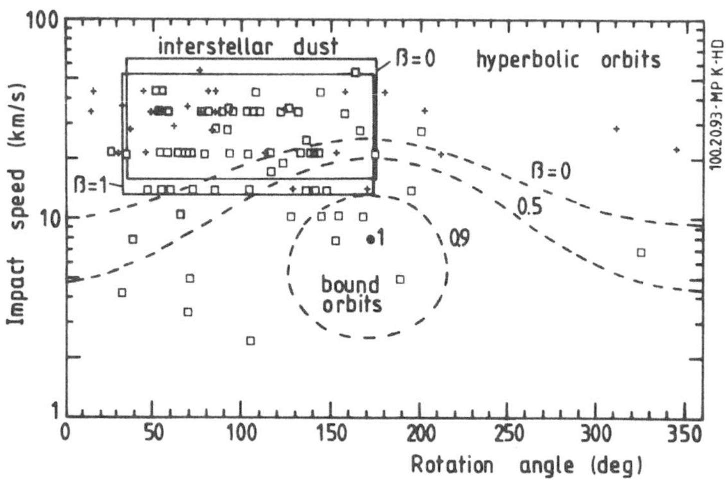

Fig. 9. Rotation angle vs impact speed of particles detected by Ulysses after Jupiter fly-by (impacts occurring during Jupiter streams are excluded). Impacts are marked according to their masses : squares : $m \geq 2.5 \cdot 10^{-14}$ g, and crosses : $m < 2.5 \cdot 10^{-14}$ g. The speed and rotation angle ranges for interstellar dust grains are shown by the boxes (both for β-values 0 and 1). The broken lines show the bound/unbound orbit limit for different β-values.

Grün et al. (1993b) identified dust grains with masses above $2.5 \cdot 10^{-14}$ g and impact speeds above 14 km/s as interstellar grains. 73 of these grains with an average mass of $3.2 \cdot 10^{-13}$ g have been recorded during the post Jupiter non-stream periods until end of 1992, which amounts to 278.4 days. Taking into account an effective sensitive area of 0.02 m^2 for the Ulysses detector an interstellar dust flux of $1.5 \cdot 10^{-4}$ m^{-2} s^{-1} and a mass flux of $5 \cdot 10^{-17}$ g m^{-2} s^{-1} is obtained.

The constant flux of 10^{-9} and 10^{-8} g particles out to 18 AU observed by Pioneers 10 and 11 could also be explained quite naturally by the assumption that the Pioneer detectors recorded impacts of 10 μm-sized interstellar grains. A consequence of this assumption, however, is that the flux of the big interstellar particles detected by the Pioneers had to arrive from the hemisphere opposite to the interstellar gas flow through the heliosphere.

The discovery of micron-sized and perhaps bigger interstellar grains will have some profound impact on our understanding of the interstellar medium : (1) because of the weak coupling of big grains to interstellar gas and fields they may provide information on their distant sources, (2) a significant mass fraction may hide in these optically invisible grains, the cosmochemical significance of which has to be explored (cf. Geiss et al., 1993) and (3) processes have to be found in which these grains can form in abundance. Chemical analyses of these particles will be attempted in the future by the Cosmic Dust Analyzer on the Cassini spacecraft.

Acknowledgements

A.C. Levasseur-Regourd provided helpful comments which improved the paper. The help by M. Baguhl and P. Staubach in preparing the manuscript is acknowledged. This work was supported by the Bundesminister für Forschung und Technologie.

References

Baguhl M., Grün E., Linkert D., Linkert G. and Siddique N. : 1993, "Identification of 'small' impacts in the Ulysses dust detector data", *Planet. Space Sci.*, in press

Burns J.A., Lamy P.L., Soter S. : 1979, "Radiation forces on small particles in the solar system", *Icarus*, **40**, 1-48

Divine N. : 1993, "Five populations of interplanetary meteoroids", *J. Geophys.Res.*, **E98**, 17029-17048

Dubin M. and Soberman R.K. : 1991, "Cosmoids : Solution to the Pioneer 10 and 11 meteoroid measurement enigma", *Planet. Space Sci.*, **39**, 1573-1590

Dumont R. and Levasseur-Regourd A.C. : 1988, "Properties of interplanetary dust from infrared and optical observations. I Temperature, global volume intensity, albedo and their heliocentric gradients", *Astron.Astrophys.*, **191**, 154-160

Geiss J., Gloeckler G., Mall U., von Steiger R., Galvin A.B. and Ogilvie K.W. : 1993, "Interstellar oxygen, nitrogen and neon in the heliosphere", *Astron. Astrophys.*, in press

Göller J.R. and Grün E. : 1989, "Calibration of the Ulysses/Galileo dust detectors with different projectile materials and at varying impact angles", *Planet. Space Sci.*, **37**, 1197-1202

Grün E., Zook H.A., Fechtig H. and Giese R.H. : 1985, "Collisional balance of the meteoritic complex", *Icarus*, **62**, 244-272

Grün E., Baguhl M., Fechtig H., Hanner M.S., Kissel J., Lindblad B.A,. Linkert D., Linkert G., Mann I., McDonnell J.A.M., Morfill G.E., Polanskey C., Riemann R., Schwehm G., Siddique N. and Zook H.A. : 1992a, "Galileo and Ulysses dust measurements : From Venus to Jupiter", *Geophys. Res. Lett.*, **19**, 1311-1314

Grün E., Fechtig H., Giese R.H., Kissel J., Linkert D., Maas D., McDonnell J.A.M., Morfill G.E., Schwehm G. and Zook H.A. : 1992b, "The Ulysses dust experiment", *Astron. Astrophys. Suppl.*, **92**, 411-424

Grün E., Fechtig H., Hanner M.S., Kissel J., Lindblad B.A., Linkert D., Morfill G.E. and Zook H.A. : 1992c, "The Galileo dust detector", *Space Sci. Rev.*, **60**, 317-340

Grün E., Zook H.A., Baguhl M., Fechtig H., Hanner M.S., Kissel J., Lindblad B.A,. Linkert D., Linkert G., Mann I., McDonnell J.A.M., Morfill G.E., Polanskey C., Riemann R., Schwehm G. and Siddique N. : 1992d, "Ulysses dust measurements near Jupiter", *Science*, **257**, 1550-1552

Grün E., Zook H.A., Baguhl M., Balogh A., Bame S.J. Fechtig H., Forsyth R., Hanner M.S., Horanyi M., Kissel J., Lindblad B.A,. Linkert D., Linkert G., Mann I., McDonnell J.A.M., Morfill G.E., Phillips J.L., Polanskey C., Schwehm G., Siddique N., Staubach P., Svestka J. and Taylor A. : 1993a, "Discovery of jovian dust streams and interstellar grains by the Ulysses spacecraft", ***Nature***, **362**, 428-430

Grün E., Gustafson B., Mann I., Baguhl M., Morfill G.E., Staubach P., Taylor A. and Zook H.A. : 1993b, "Interstellar dust in the heliosphere", ***Astron.Astrophys.***, in press

Hamilton D.P. and Burns J.A. : 1993, "Ejection of dust from Jupiter's gossamer ring", ***Nature***, **364**, 695-698

Hanner M.S : 1980, "On the albedo of interplanetary dust", ***Icarus***, **43**, 373-380

Hauser M.G., Gillett F.C., Low F.J., Gautier T.N., Beichman C.A., Neugebauer G., Aumann H.H., Band B., Boggess N., Emerson J.P., Houck J.R., Soifer B.T. and Walker R.G. : 1984, "IRAS observations of the diffuse infrared background", ***Astrophys. J.***, **278**, L15-L18

Horanyi M., Morfill G.E. and Grün E. : 1992, "Mechanism for the acceleration and ejection of dust grains from Jupiter's magnetosphere", ***Nature***, **363**, 144-146

Humes D.H. : 1980, "Results of Pioneer 10 and 11 meteoroid experiments : Interplanetary and near-Saturn", *J. **Geophys. Res.***, **85**, 5841-5852

Johnson T.V., Morfill G.E. and Grün E. : 1980, "Dust in Jupiter's magnetosphere : An Io source", Geophys. Res. Lett. 7, 305-308

Leinert C. and Grün E. : 1990, Interplanetary dust, in "Physics of the Inner Heliosphere", ***Eds. R. Schwenn and E. Marsch, Springer, Heidelberg***, 207-275

Leinert C., Richter I., Pitz E. and Planck B. : 1981, "The zodiacal light from 1.0 to 0.3 AU as observed by the Helios space probes", ***Astron. Astrophys.***, **103**, 177-188

Levasseur-Regourd A.C., Renard J.B. and Dumont R. : 1991, "The zodiacal cloud complex", in "Origin and Evolution of Interplanetary Dust", Eds., A.C. Levasseur-Regourd and H. Hasegawa, Kluwer Acad. Pub., Dordrecht, 131-138

Morfill G.E., Grün E. and Johnson T.V. : 1980, "Dust in Jupiter's magnetosphere : Origin of the ring", ***Planet. Space Sci.***, **28**, 1101-1110

Staubach P., Divine N. and Grün E. : 1993, "Temperatures of zodiacal dust", ***Planet. Space Sci.***, in press

Toller G.N. and Weinberg J.L. : 1985, "The change in near-ecliptic zodiacal light brightness with heliocentric distance", in "Properties and Interactions of Interplanetary Dust", Eds. R.H. Giese and P. Lamy, Reidel, Dordrecht, 21-25

Witte M., Rosenbauer H., Banaszkiewicz M. and Fahr H. : 1993, "The Ulysses neutral gas experiment : determination of the velocity and temperature of the interstellar neutral helium", ***Advances in Space Res.***, **13**, (6)121-(6)130

THE ROSETTA MISSION TO PRIMITIVE BODIES OF THE SOLAR SYSTEM

YVES LANGEVIN (ON BEHALF OF THE ROSETTA STUDY TEAM)

Institut d'Astrophysique Spatiale, Orsay, France
E-mail langevin@iaslab.span.cnes.fr

Abstract. The European Space Agency (ESA) has selected *Rosetta* as the next cornerstone mission, to be launched in 2003. The goal is to perfom one or more fly-bys to main belt asteroids, followed by a rendez-vous with an active comet. Advanced in situ analysis, both in the coma and on the surfaces of the nucleus, will be possible, as well as monitoring by remote sensing instruments of the nucleus and of the inner coma for a time span of more than one year, until perihelion. This paper outlines the scientific and technological choices done in the definition of the mission.

1. Background

A mission to small bodies was selected in 1983 as one of the four cornerstones of the "Horizon 2000" plan of the European Space Agency. During the first study phase, a comet nucleus sample return was studied, in the framework of an ambitious collaborative project with NASA. Following the cancellation of CRAF in early 1992, it was considered that the required NASA commitment would not be obtained for november 1993, at which time the order of implementation of the 3rd and 4th cornerstone would be selected between ROSETTA and FIRST, a submillimeter observatory mission. A new concept was th developed for ROSETTA, in which the spacecraft performs a rendez-vous with a comet after one/two asteroid fly-bys.

The highly successful close fly-bys of comet P/Halley, in 1986, by the GIOTTO and VEGA spacecraft (more distant flybys being performed by two Japanese probes) as well as the recent observations of cometary nuclei far from the Sun provided new insights on cometary nuclei. While cometary grains are expected to represent a relevant source of micrometeoroids, interplanetary flight dynamics results in undersampling of this source among the particles entering the atmosphere (see Dermott et al., this volume), and atmospheric entry result in the loss of major fractions of the volatile and organic components in the stratospheric dust particles which can be analyzed in the laboratory. There is therefore no representative sample of cometary material. Two approaches can be considered : collecting and returning to Earth a sample from a cometary nucleus, the initial strategy proposed for ROSETTA, or performing detailed in-situ investigations on or close to the nucleus, which is the selected mission scenario.

The situation is markedly different for asteroids, the other major class of primitive bodies. While recent chaotic transfer models suggest strong selection biases, the large and diverse collections of meteorites can be considered fairly representative of the main belt of asteroids, at least of its inner regions. There are however major problems in linking the types of meteorites to specific asteroids classes. The most

A. Milani et al. (eds.), Asteroids, Comets, Meteors 1993, 381–394.

striking example is that of ordinary chondrites, the most abundant meteorite type, which have as yet no unambiguous match among asteroid classes (see Farinella et al., this volume). The major effort should therefore be devoted to document the available samples, by observing at close range main belt asteroids in their diversity. The first steps were taken during the Galileo mission, which observed two relatively small S-type asteroids, Gaspra and Ida, on its way to Jupiter (see figure at page 361 in this volume). The more primitive C and D classes, as well as the highly evolved M and V classes have yet to be observed.

This science rationale led to the redefinition of ROSETTA as a rendez-vous with a comet focused on in-situ investigations of cometary material, one or two asteroid fly-bys being performed during the cruise to the comet. In this way it is possible to focus the mission on the major science goals of the sample return mission (nebular condensates, presolar grains), using remote sensing investigations and state of the art in-situ analyses both in the coma and at the very surface of the nucleus. On the other hand this redefinition makes the mission feasible within the technological and budget constraints of ESA. The nominal mission can be performed with European technology (Ariane 5 launcher, solar panels, ESA 15m ground stations + the German antenna at Weilheim). Scientific investigations of comet Schwassmann-Wachmann 3 (which has been selected for the baseline launch window) are performed during 16 months from the onset of activity, at nearly 4 AU from the sun, to perihelion. With larger receiving antennas, it would be possible to observe the nucleus since rendez-vous, 5 years after launch and 3 years before perihelion. With these characteristics, ROSETTA has been selected by the Science Planning Committee as the 3rd cornerstone of ESA's long term plan, with a launch in 2003.

2. Mission Scenario

The ROSETTA spacecraft is launched by an Ariane 5, then performs several gravity assists in the inner solar system to obtain the required energy to reach the aphelion of the target comet, at more than 5 UA (see Tab. I).

This strategy maximizes the mass budget for missions to the outer solar system, such as GALILEO and CASSINI. A number of interesting opportunities were found, some with very high mass margins (more than 1000 kg; see Table II). The selected nominal mission to Schwassmann–Wachmann 3 is the shortest (8 years from launch to end of mission), with a significant mass margin (more than 250 Kg) for the spacecraft.

During this cruise phase, which extends over 5 years, one or two asteroid fly-bys can be performed. The last leg of the orbit crosses the whole belt, making possible fly-bys in the mid and outer main belt of asteroids. The orbit matching manoeuvre is performed near aphelion, where the link budget is not high enough for comprehensive scientific investigations. The rendez-vous *per se* occurs 2 years later, when the heliocentric distance has been reduced to 3.25 AU. The following phases are implemented in the nominal mission scenario :

TABLE I
Rosetta Comet Rendez-vous Opportunities, Launch 2003–2004.

No.	Rendezvous with Comet	Asteroid Fly-bys at	Type	Launch Arrival Perihelion	Miss. Δv	AR 5 Perf.	S/C at Launch	Remarks
				date	km/s	kg	kg	
1	Schwassmann-Wachmann 3	Fennia 1985 QD1	VEE	2002/04/19 2009/12/31 2011/10/12	0.735	2708	1539	Early launch date
2	Wirtanen	Ministrobell Shipka	MEE	2003/01/22 2011/08/28 2013/10/21	1.490	2990	1948	
3	Schwassmann-Wachmann 3	Brita	ME	2003/07/18 2008/06/10 2011/10/12	1.924	2426	2232	98% tank filling
4	Finlay	1990 OK	MEE	2003/07/17 2013/09/05 2014/12/09	1.837	2597	2172	
5	Neujmin 2	1981 EG Virtus	VEE	2003/09/29 2012/05/21 2014/03/04	1.986	2499	2276	102% tank filling
6	duToit-Hartley		VEE	2003/10/20 2012/11/05 2013/08/16	1.884	3504	2204	One or two fly-bys above tank filling
7	Wirtanen	1982 DX3 1983 AD	VEE	2003/11/03 2012/10/23 2013/10/21	1.327	2698	1850	Rendezvous at 2.92 AU
8	Haneda-Campos	Isis	VEE	2003/11/21 2013/06/23 2016/11/08	1.465	2571	1932	Two fly-bys above tank filling
9	Schwassmann-Wachmann 3	1990 TJ	VE	2003/11/29 2008/10/28 2011/10/12	1 930	2352	2236	98% tank filling
10	Finlay	1982 BB Lunacharsky	VEE	2004/05/11 2013/12/01 2014/12/09	1.073	2651	1709	Maximum sun distance 5.8 AU
11	Brooks 2	Carr 1983 WM	VEE	2004/05/25 2011/11/23 2014/05/23	1.740	2413	2106	Minimum sun distance 0.6 AU

TABLE II

Comet opportunities, budgets and launch margins for the different mission opportunities

N°	Comet	Delta V Mission	Delta V Else	Propellant mass 400 N Engine	Propellant mass Attitude Thrusters Man.	Propellant mass Attitude Thrusters WOL	S/C Mass Dry	S/C Mass Launch	Ariane 5 capacity	Mission margin	S/Cgrowth capacity	Tank filling
		m/s	m/s	kg	kg	kg	kg	kg	kg	kg	kg	%
1	SW 3	735	300	418	7	8	1035	1528	2708	1180	972	38%
2	Wirtanen	1490	300	832	17	9	1035	1952	2990	1038	548	75%
3	SW 3	1924	300	1121	23	6	1035	2246	2426	180	254	100%
	Finlay	1826	300	1055	22	9	1035	2181	2590	409	319	95%
	Neujmin 2	1473	300	822	16	9	1035	1943	2463	520	557	74%
	du Toit Hartley	1677	300	952	19	9	1035	2075	3234	1159	425	85%
4	Wirtanen	1327	300	733	14	9	1035	1851	2698	847	649	66%
	Haneda - Campos	1559	300	878	18	11	1035	2001	2568	567	499	79%
5	SW 3	1930	300	1127	23	7	1035	2252	2352	100	248	101%
11	Finlay	1073	300	590	11	9	1035	1705	2651	946	795	53%
12	Brooks 2	1740	300	995	20	9	1035	2119	2413	294	381	89%

- Launch : July 2003
  Earth to Mars Cruise of 12 months
- Mars Swing-by : July 2004
  Mars to Earth Cruise of 21 months
- Earth Swing-by : May 2006
- Fly-by of Brita : January 2007
- Orbit matching manoeuvre with SW3 June 2008
- Comet Approach July 2010
  Science operations during 15 months
- Perihelion (End of nominal mission) October 2011

From March 2007 to July 2010, the distance from Earth to the spacecraft exceeds 3.25 UA, and the link budget is too low with the two 15 m ESA antennas at Villafranca (Spain) and Perth (Australia). During this phase, the 30 m antenna in

Weilheim (Germany) is used, in particular for the orbit matching manoeuvre. Science operations could be initiated at this stage (with some scheduling of resources for the payload) if additional 30 m antenna will be available in 2005 either in European networks or through a collaboration with DSN. Additional observations of the nucleus in its low activity phase could then be obtained.

The near comet operations are divided into 4 major phases :

- Detection and approach : after the orbit matching manoeuvre, the spacecraft will drift for 2 years at less than 500,000 Km from the nucleus of Schwassmann- Wachmann 3. A bias has been included towards the Sun, which guarantees appropriate illumination of the nucleus for detection with the imaging system. After recovering the nucleus, a series of low thrust manoeuvres will be initiated, to approach the comet from the sunward side, reducing by steps the relative distance and velocity. After 90 days, the spacecraft will be at less than 100 Km from the nucleus.

- global mapping and site selection : a global mapping phase of 20 days is scheduled at 5 to 10 comet radii from the nucleus. Such orbits are stable even for very small and irregular cometary nuclei. A resolution of 2000 pixels across the nucleus should provide an adequate data base for obtaining a detailed gravity field and rotation model of the nucleus and for selecting a few areas of interests (e.g. potentially active areas). It will then be possible to transfer to orbits with low pericenters (a few Km), so as to perform high resolution observations of the selected areas. After analyzing the results, the area most likely to present volatiles at or near the surface of the nucleus will be selected for implementing the surface science module

- surface science : a surface science module will be released from the spacecraft on an orbit impacting the nucleus at very low velocities (at most a few m/s). Several strategies can then be considered to minimize the rebound : crushable material absorbing most of the kinetic energy, an anchoring system, or a small thruster triggered by a contact sensor. The selected trajectory will have a negative total energy relative to the nucleus, so that in any event the surface science module will eventually come to rest on the surface of the nucleus. A minimum lifetime of 5 hours is required for operating the payload (see following sections). During this lifetime, the orbit of the main spacecraft will be selected so as to remain in sight of the surface science module. If contact is not directly obtained, a search pattern will be initiated.

- monitoring of the nucleus and of activity : during a time span of at least 10 months until perihelion, the spacecraft will perform in-situ analyses of dust and gas in the inner coma as well as remote sensing observations of the nucleus and jets. All orbit modifications will be commanded from the ground, so as to take into account possible hazards as well as targets of opportunities, such as new active areas and the associated dust and gas jets. Depending on the observed dust and gas fluxes, the distance to the comet will be modified so as to maintain the contamination of optical surfaces and solar panels within acceptable limits. Given the relatively low activity of ROSETTA target comets with respect to P/Halley, it should be possible to remain in very close orbits (a few Km to a few 10 Km) until perihelion.

If an improved ground segment is available in 2008 - 2011 (new 30 m European antennas or DSN support), the first phase can start immediately after the orbit matching manoeuvre. During the following 2 years, it would then be possible to obtain comprehensive observations of the bare nucleus, and to detect a possible activity very far from the sun (which has been reported for several other comets, including P/Halley).

3. Science Objectives

3. 1. THE COMETARY NUCLEUS

As indicated in the introduction, the primary science goal of ROSETTA is to obtain direct information on the chemical, mineralogical and isotopic composition of volatile and refractory cometary material, with implications on the formation of our solar system. In this context, observations of the cometary nucleus are important for two main reasons : they document the parent body in which the analyzed material formed and evolved; they provide detailed information on a surviving planetesimal from the outer parts of the solar nebula. Furthermore, such observations are essential for selecting candidate sites for the landing of a surface science package. The specific science objectives which can be addressed are the following :

- size, shape, and rotational state of the nucleus : they provide information on the history of the nucleus (cratering, sublimation), and represent key parameters for cometary activity as well as for mission planning. A resolution of 2000 pixels across the nucleus is adequate for these objectives

- mass, density, gravity field : together with the shape and rotational state, the mass and density provide constraints on the internal structure of the nucleus : can it be considered homogeneous, is there vertical layering, or is the nucleus formed from the accretion of diverse subnuclei? An important additional data would be the determination of higher moments of the gravity field, which requires orbits very close to the nucleus.

- Surface topography : the local topography, which together with rotation parameters define illumination conditions, is likely to be one of the parameters controlling activity. This issue is directly linked to the presence of ices at or close to the surface of the nucleus, and the relative role of local topography, local composition, and possible fracture zones between subnuclei needs to be evaluated. From the GIOTTO images (see Fig. 1), the size of active regions is likely to be a few 100 m, so that characterizing the topography of these areas requires stereo imaging with a resolution of at most a few meters.

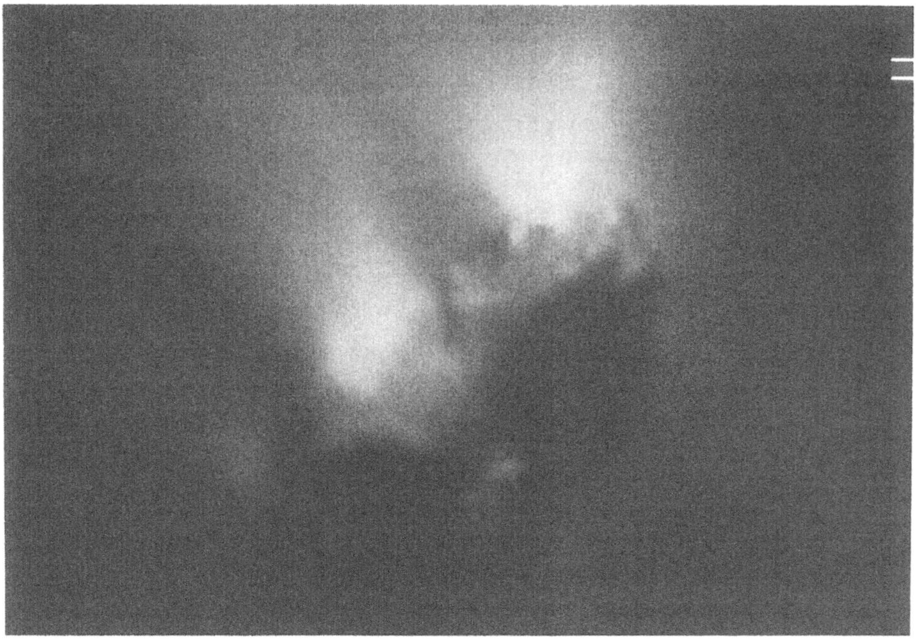

Fig. 1. Nucleus of comet P/Halley observed by GIOTTO (courtesy ESA).

- Spectral characteristics : ices, organic compounds and other major constituents of meteoritic material all present strong absorption bands in the visible and near infrared regions. In these spectral ranges, it is possible to determine whether ices are present at the very surface of the nucleus, and to obtain information on the mineralogical composition of active and non active areas. Given the typical size of active areas, a resolution of a few 10 m seems adequate.

- Thermal characteristics : the propagation of the solar heat wave is strongly dependent on the thermal conductivity of the surface layers and on the presence of heat sinks, such as pockets of ices, close to the surface. Observing thermal emission as the terminator crosses a surface element will therefore provide constraints on the subsurface structure. It would be highly desirable to obtain radiometric results at different wavelengths, so as to probe different depth scales.

3. 2. COMETARY MATERIAL

The characterization of cometary material is the primary science goal of the mission. It will be addressed by combining high quality investigations in the inner coma with ground truth obtained by the surface science package.

3.2.1 Surface science

Investigations of cometary material at the very surface of the nucleus are considered extremely important in combination with analyses in the inner coma. Obtaining ground truth information on physical, chemical and mineralogical characteristics for at least one site will provide unique insight on the conditions prevailing at the interface between cometary material in the nucleus and the cometary environment. It will also much improve the interpretation of remote sensing and inner coma investigations. The major science goals can be defined as follows :

- chemical composition of refractories and volatiles in bulk surface material : contrarily with coma investigations, which will provide analyzes on very small amounts of material, a surface package will provide information on the bulk composition (e.g. several tens of cm for gamma ray spectrometry). The characterization of the volatile and organic components is a particularly important goal, which can be addressed by evolved gas analyses. Obtaining the isotopic ratios of light elements represents a major technological challenge given the mass and power resources of a surface science package; this possibility should however be investigated given the high science priority.

- thermal and mechanical properties of surface material : with an accelerometer, a surface science package will also provide unique information on the mechanical properties of the cometary crust, which are at present totally unknown. Thermal probes will determine the temperature and conductivity of surface material, a key parameter for the thermal evolution of the nucleus during the diurnal and seasonal cycles. The potential of subsurface electromagnetic sounding and radio science using the link to the orbiter should also be evaluated.

Although a surface science package anywhere on the nucleus would provide a high science return, a site selection capability is desirable. If available, the preference should be given to active regions which are likely to present the largest concentrations of volatile material close to the surface.

3.2.2 Investigations in the inner coma

The ROSETTA mission scenario, and the relatively low activity of the target comets, make it possible to remain within a few km to a few 10 km of the nucleus until perihelion. These distances are far smaller than the few 1000 km at which extended gas sources have been observed for Halley. Gas species and even relatively volatile grains are likely to be very little disturbed during their travel from the cometary surface to the spacecraft. The major objectives of these investigations are the following :

- Chemical and isotopic composition of grains : it will be possible to analyze several ten thousand grains in the inner coma in terms of their chemistry and the isotopic ratios of light elements (H, C, N). Variations of several orders of magnitudes have been observed at a micrometer scale in candidate presolar grains from meteorites (see Fig. 2), which can be directly related to nucleosynthetic processes. Such large variations are well within the reach of a high resolution mass spectrometer. It will therefore be possible to characterize presolar components of cometary material, which are likely to be far more diverse than the highly refractory grains (SiC, diamonds) preserved in meteorites. The determination of the isotopic ratios for Oxygen, one of the most important nucleosynthetic tracers, represent a more difficult challenge since accuracies better than 1% are required over a wide dynamic range. It seems unlikely that the very high sensitivities required for noble gases or radiochronology will be available with space qualified instrumentation.

- Texture and elemental variations within dust grains : It will be possible to obtain high resolution images and elemental maps of dust grains by combining scanning electron microscopy with X-ray spectrometry. In combination with high resolution mass spectrometry, it will be possible to characterize presolar components as well as pristine condensates from the solar nebula. The texture and composition of cometary grains will also provide constraints on the link with interplanetary dust particles and stratospheric grains.

- Chemical and isotopic composition of gases in the inner coma : recent advances in the mass resolution of neutral and ion mass spectrometers will provide for gases the same level of information on the chemical and isotopic composition as that obtained on dust grains. Given the very short flight time, there should be a very close compositional relationship between the analyzed gases and the volatiles in the outer regions of the nucleus. This type of investigations is essential as the volatile component has been completely lost in stratospheric dust particles, the closest available equivalent to cometary material.

The long observation phase close to the nucleus (one year in the nominal mission) constitutes a tremendous asset for obtaining in the inner coma analyses of dust and gas which are representative of the outer layers of the nucleus. It will be possible to monitor the active regions during a large fraction of the activity cycle, hence to change the orbit so as to analyze material from each active region, and possibly from non active areas as well. The observation of compositional differences would be a strong argument in favor of the formation of the nucleus from the coalescence of subnuclei. The long observation time also leaves room for serendipity : if an outburst is observed, it may be possible to analyze material from a previously unexposed region. Finally, observing the evolution of active regions together with dust and gas activity will provide clues on the relationship between surface material and underlying material, as well as between material in the coma and surface material.

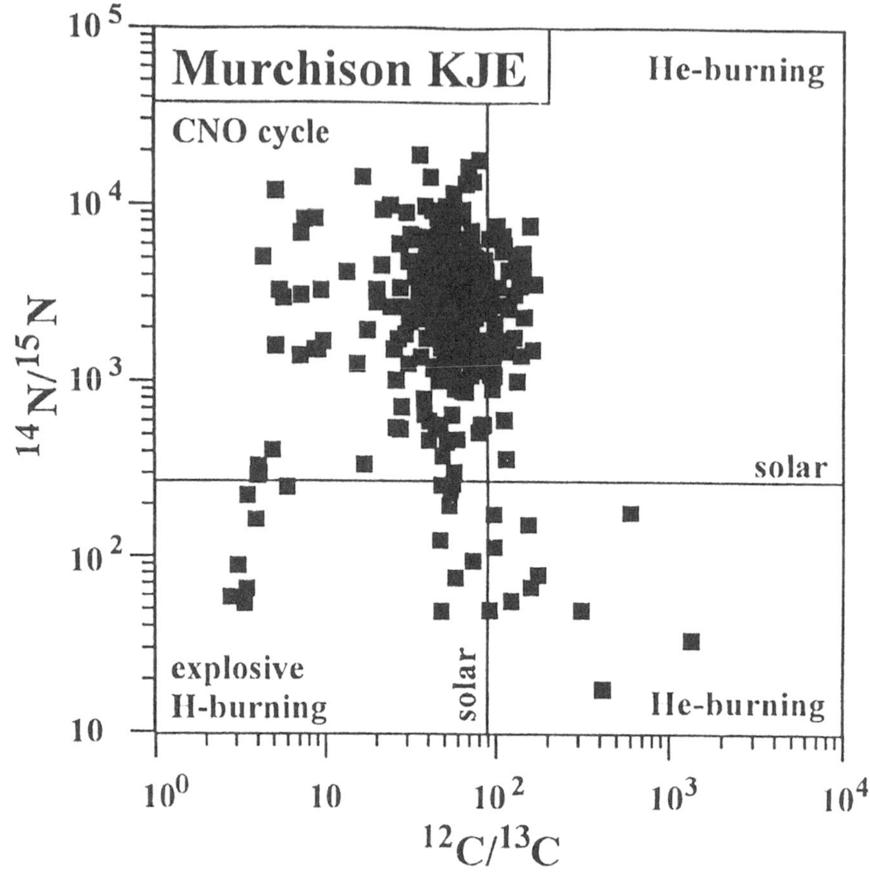

Fig. 2. Nitrogen and carbon-isotopic compositions of individual ~ 1 μm sized silicon carbide grains from the Murchison carbonaceous chondrite. The two lines labelled as "solar" indicate solar system ratios, dividing the plot into four quadrants for which the principal nucleosynthetic processes are indicated. Their anomalous isotopic composition indicate that the grains are of presolar origin. (Courtesy of Hoppe et al.)

4. Cometary Physics

While not dedicated mainly to the study of physical processes in the cometary environment, ROSETTA, with its more than one year of continuous observations, will provide high quality data on cometary processes. According to the mission scenario, the highest priority will be given to processes in the inner coma. The range of cometary processes is extremely wide, from solid state physics inside the nucleus to plasma physics in the outer coma. The major science goals can be very briefly summarized as follows :

- Temporal variations of cometary activity : as demonstrated by the Halley fly-bys and simultaneous ground observations, the relationship between the rotational state of the nucleus and the activity cycle is as yet not completely understood (see Rickman, this volume). ROSETTA will provide simultaneous observations of the evolution of active areas and the related activity. It will also be possible to relate outbursts (if any) to changes in surface configuration (e.g. ejection of part of the insulating mantle).
- Spatial anisotropies : detailed observations of dust and gas jets will be performed. Velocity vectors of dust grains impacting the spacecraft will be determined. It would be very useful to obtain remote sensing information on gas velocities using a high resolution sub-mm spectrometer.
- physico-chemical processes in the inner coma : observations of dust, gas and ionized species during more than one year in the inner coma will provide important information on parent molecules, the dust - gas interactions, and the processes through which parent molecules evolve into other neutral and ionized species.
- Interactions with the solar wind : At the beginning of the observation phase, it will be possible to investigate the interaction of the solar wind with the bare nucleus, then to observe this interaction during the growth phase of the coma. Excursions in the dust and plasma tail could be considered either in the nominal mission or during an extended mission, after perihelion.

5. Asteroid Science with ROSETTA

ROSETTA will perform at least one, possibly two fly-bys of asteroids in the main belt. These fly-bys will have similar characteristics to that already performed by GALILEO at Gaspra and Ida : a miss distance of a few 1000 km, and a relative velocity of 6 to 15 km/s. The main goal of asteroid fly-bys with ROSETTA is to complement the first survey of the diversity of asteroids initiated by GALILEO, as the two visited objects were relatively small S-type asteroids. On these grounds, the highest priority should be given to carbonaceous asteroids (C or D types), or very evolved asteroids (such as those of V or M types) which will not have been observed. These targets are also the most interesting in combination with the fly-bys of Gaspra and Ida to improve our understanding of the asteroid - meteorite relationship.

ROSETTA gives the possibility of observing two types of primitive bodies with the same remote sensing payload : the cometary nucleus which is the primary goal of the mission, and the asteroid(s) selected as fly-by target(s). It will therefore be possible to directly compare the two major classes of primitive solar system bodies. Indeed, the Halley fly-bys, then ground based observations, suggested similarities between cometary nuclei and the most primitive asteroids. On this basis, the priority for ROSETTA will be given among available targets to carbonaceous asteroids, the most likely to provide clues on the comet - asteroid relationship.

TABLE III

ROSETTA MODEL PAYLOAD
ORBITER SPACECRAFT

	Mass	Power
Remote Imaging System	20 kg	15 W
VIS and IR Mapping Spectrometer	23 kg	15 W
Neutral Gas and Ion Mass Spectrometer	15 kg	15 W
Cometary Matter Analyser	14 kg	20 W
Scanning Electron Microprobe	13 kg	22 W
Dust Flux Analyser	2.5 kg	1 W
Plasma Investigation	2.5 kg	3 W
	90 kg	

Mass Allocation: 96 kg
Additional Instruments: Microwave Spectrometer, γ-ray Spectrometer, Radar Mapper/Sounder, Radar Altimeter

ROSETTA MODEL PAYLOAD
NUCLEUS SURFACE INVESTIGATIONS

	Mass	Power
Gamma-Ray Spectrometer	0.9 kg	3 W
Alpha-Proton-X-Ray Fluorescence Spectrometer	1 kg	.5 W
Neutron Spectrometer	.4 kg	1 W
In-situ Imaging System	1.5 kg	2 W
Accelerometer	.2 kg	.5 W
Permittivity Probe	.5 kg	.2 W
Evolved Gas Analyser	3 kg	5 W
	7.5 kg	

Additional Instruments: Instrument Deployment Device 1.5 kg + 5 kg, Surface Radar (Sounder) 3 kg.

6. Model Scientific Payload

6. 1. THE ROSETTA ORBITER

The model payload which has been defined on the basis of the science objectives for the ROSETTA orbiter is divided in three groups (see Table III, top) :

- remote sensing instruments : an imaging system, with a navigation camera, a narrow angle camera (IFOV = 30 microradiants) and a wide angle camera (IFOV = 150 microradiants), and a visible and infrared imaging spectrometer (IFOV = 500 microradiants) extending from 0.3 μm to 5 μm (if possible higher in wavelengths) so as to cover major signatures of ices and minerals as well as reaching the thermal emission region.
- in-situ analyses instruments : a high resolution (a few 1000) spectrometer for dust particles; a high resolution (a few 1000) spectrometer for neutral and ionized species; a scanning electron microscope combined with X-ray analyzer for dust grains, with a sub-micrometer spatial resolution.
- environment monitoring instruments : a dust flux analyzer providing the mass, flux and velocity vector of incoming dust particles; a plasma package including electron density measurements, a key parameter for the local physico-chemical processes.

Additional instruments have been considered for the payload. In particular, a microwave instrument can provide important radiometric information on the temperature of the nucleus down to a depth of a few cm, much larger than that probed by the Visible - Near Infrared Imaging Spectrometer. It is possible to consider associated active microwave experiments, such as a high resolution microwave spectrometer for cometary gases, or an altimeter/sounder for obtaining information on the physical characteristics of the surface, and possibly the subsurface. The implementation of such instruments can be considered at various levels (e.g. additional electronic package associated with the telecommunications antenna, such as for CASSINI, or a dedicated small antenna). A gamma-ray spectrometer could also be considered, if the spacecraft can stay long periods of time within one radius of the nucleus.

6. 2. THE SURFACE SCIENCE PACKAGE

The total mass allocation for the surface science package is 45 kg, although there is a significant growth potential for all mission candidates. The proposed model payload (see Table III, bottom) includes an imaging system, to document the landing site. Two instruments dedicated to chemical analyses have been selected : a gamma ray spectrometer and an alpha-P-X-ray fluorescence spectrometer. Two instruments have been included for investigating organic and volatile material : an evolved gas analyzer, coupled with a gas chromatograph and/or mass spectrometer, and a neutron spectrometer (Hydrogen). An accelerometer will provide information on the mechanical characteristics of the surface material, while thermal and permittivity probes will determine major physical parameters.

Additional instruments which were considered included a small, mobile instrument deployment device and a sub- surface sounding experiment, if it can fit mass and power constraints. The potential of radio science using the link to the orbiter will also be evaluated.

Acknowledgments

The ROSETTA mission in its new scenario was defined by the ROSETTA Study Team : A. Bar-Nun, A. Barucci, E. Bussoletti, A. Coradini, M. Coradini, L. Colangeli, P. Eberhardt, E. Grün, M. Hechler, H.U. Keller, J. Kissel, J. Klinger, Y. Langevin, R. Laurance, J.A.M. McDonnell, A. Milani, G. Picardi, C. Pillinger, G. Schwehm, D. Stöffler, H. Wänke. The ROSETTA report has been published by ESA as SP SCI(93)7.

ASTEROID FAMILIES

VINCENZO ZAPPALÀ AND ALBERTO CELLINO

Osservatorio Astronomico di Torino
strada osservatorio 20, I-10025 Pino Torinese (TO), Italy
E–mail ZAPPALA@TO.ASTRO.IT

Abstract. In spite of their widely recognized importance in the framework of the modern understanding of the asteroidal population and its collisional evolution, asteroid dynamical families have long been a puzzling subject of research, due to the disagreement among the family lists published by different authors. In the present review, the definition and meaning of asteroid families are critically discussed, as well as the various problems which have to be faced by any suitable technique of family identification. In this respect, major improvements have been achieved during the last few years. The most recent family searches show an excellent agreement both in the number of reliably identified families, and in their members. Moreover, the overall performances of the most recent techniques of family identification have been tested by means of numerical simulations, with encouraging results. For these reasons, we believe that we are presently at the beginning of an era in which detailed physical studies of families can be attempted, and observational campaigns can be planned on the basis of solid evidence, like in the case of the recent spectacular results obtained for the family associated to the large asteroid 4 Vesta.

1. Introduction

Asteroid dynamical families have received a renewed interest recently, due to the recognition of their intrinsic importance in the framework of modern planetary research. Families are thought to be the outcomes of the catastrophic disruption of a number of sizable parent bodies. For this reason, they provide obvious opportunities for the studies of the mineralogical composition of asteroids, since many family members are made of material coming from the inner layers of their parent body, and put at the same time some essential constraints for any model of the collisional evolution of the asteroid belt. In addition to these well known facts, more recently it has been pointed out that families may be important also for the origin of meteorites, and more generally the interrelationships among minor bodies of the Solar System. As a matter of fact, simple estimates lead to conclude that in some cases the impacts responsible of the formation of a family within the asteroid main belt may also have injected a fraction of fragments into some of the most important resonances, like the 3/1 mean motion resonance with Jupiter at 2.5 AU, and the ν_6 secular resonance. As a consequence, through dynamical processes which have been recently extensively investigated (see Greenberg and Nolan, 1989, and Farinella *et al.*, in this volume), many fragments may have been injected into the inner regions of the solar system, where they may be discovered as Near-Earth-Asteroids, or even collide with terrestrial planets. An excellent example of these interrelationships among asteroid families, NEA asteroids and meteorites is given by the Vesta family, whose collisional origin has been recently confirmed by observations (Binzel and Xu, 1993; Burbine and Binzel, this volume). As a matter of fact, in the case

A. Milani et al. (eds.), Asteroids, Comets, Meteors 1993, 395–414.

of Vesta we may sketch a general scenario which accounts for all the observational constraints : the existence of the family, originated by an energetic cratering event; the existence of an hemispheric-scale albedo variegation on Vesta (clearly indicated by polarimetric and photometric data, see Cellino *et al.*, 1987, and Broglia and Manara, 1989), which could be related to the presence of the crater; the existence of a definite class of meteorites, the eucrites, and of a number of NEAs (Cruikshank *et al.*, 1991) which both exhibit a basaltic composition, and may plausibly be considered as originating from the fraction of high-speed fragments which were injected into the chaotic zones associated to the 3/1 and ν_6 resonances.

These considerations show how the study of asteroid families may be fruitful for our general understanding of many evolutionary processes active in the solar system, allowing to identify large-scale collisions as one of the main mechanisms which have shaped a number of apparently unrelated observational evidences which are often separately studied by different branches of planetary research. The recognition of the unifying role played by collisional evolution is actually one of the main achievements of modern planetary science.

However, the recognition of the importance of asteroid families should lead us to carefully take into account all the reasons which have until recently prevented us from establishing a well defined and coherent identification of those which are presently recognizable within the asteroid belt. In spite of the noticeable number of papers devoted by many authors to the problem of family identification – starting from the pioneering work of Hirayama at the beginning of the present century – it has not been possible for a long time to obtain an unambiguous list of families generally accepted by the majority of the scientists active in this field of research.

The present review is aimed at crytically re-analyzing the overall topic of the asteroid family identification and analysis. We start from the different meaning of the definition of families, depending upon the point of view (statistical or physical) adopted for this purpose. We review the problems encountered when an unambiguous identification of families is attempted, and we examine the current state of the art, showing that some important improvements have been obtained during the last few years by means of new methods, devised by different authors. Finally, we discuss what we can learn from a physical analysis of the presently known families, and we examine a list of open questions which still wait for a definitive answer. We can anticipate here that we hope to be able to convince the reader that the general understanding of asteroid families is presently much better than it used to be until a few years ago; in our opinion, we are at the beginning of an era in which the physical information hidden within families is going to be fully exploited. This is a consequence of the improvements in the techniques of identification of families and family memberships. The results start to be supported by some adequate observational evidence, and can be now used for testing the predictions of the presently available physical models of collisional events.

2. What is a Family ?

It is important to briefly discuss what is the meaning of the word "family". In fact, sometimes this word is used in slightly different ways by the specialists of dynamics, and by the experts of different branches of asteroid physics. From a dynamical point of view, families are groups of asteroids having orbital *proper elements* a', e', i' (proper semimajor axis, eccentricity and inclination, respectively) very similar. From this point of view, the main feature of families is to be *statistically significant*, that is clearly separable from random clusterings of objects in the space of proper elements. The immediate interpretation is that they are the outcome of the disruption of single parent bodies, since the laws of dynamics predict that fragments ejected with relatively low velocities, will conserve approximately the proper elements of their parent over long time scales. The general subject of asteroid proper elements is extensively covered by the Knežević and Milani chapter in this volume. For this reason, we do not enter here into details. We only recall that the Gauss equations connect the components of the post-breakup ejection velocities of the fragments from a catastrophic collision with the resulting differences in their proper elements $\delta a'$, $\delta e'$, $\delta i'$ (at least in the frame of the linear secular perturbation theory, see Brouwer, 1951; see also Zappalà *et al.*, 1984, and Chapman *et al.*, 1989).

From the point of view of the physics of asteroids, the meaning of families is just related to their collisional origin; in this sense, a set of fragments directly deriving from the disruption of a single parent body constitutes a family.

Even if the dynamical and physical definitions may appear essentially coincident, this is not always true in practice, and some misunderstandings have actually become apparent in the past. As a matter of fact, not always the two definitions of family lead to the same results : on one hand, there may be cases in which statistically significant groupings of objects do not share a common collisional origin; in other cases, "true" collisional families may be not identified on a purely statistical ground. Of course, mainly in the former case, a crucial role is played by the statistical criteria adopted for the identification. Actually, the procedures have to be carefully devised, in order to avoid to accept as significant, some groupings which could be more correctly rejected on the basis of improved criteria. On the other hand, some difficulties are intrinsic, and are a consequence of the actual structure of the asteroid belt. As an example, the Phocaea region (the populated zone located at semimajor axis values of 2.3–2.4 AU, and at high inclination values beyond 20°) is largely isolated by the rest of the belt. In such a case, since the only objects present there belong to this "populated island" in the space of proper elements, statistics might lead to accept that they form a family; this result, however, remains questionable, since there are reasons to believe that the Phocaea "island" is actually only a consequence of the depletion of the neighbouring regions due to the occurrence of important secular resonances. On the other hand, at the high inclination values of the Phocaea region, the proper elements used so far are not sufficiently precise to derive any definite conclusion. The situation should improve in the near future, due to the recent availability of a new large set of proper elements of good quality for high inclination objects (Lemaitre and Morbidelli, 1994). A preliminary analysis of these data seems to suggest that at least some objects in the Phocaea "island" may actually belong to a real family.

In order to avoid any misunderstanding due to an improper use of the word "family", Farinella *et al.* (1992) have proposed a new nomenclature. According to these authors, the term "family" itself should be used only in the cases in which the statistical and physical definitions are coincident. This means that "families"

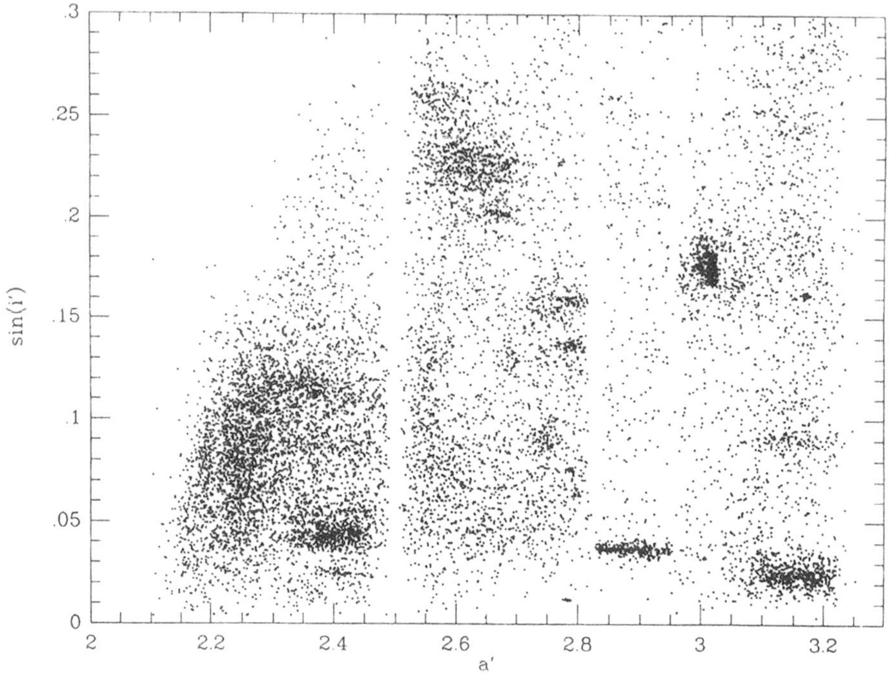

Fig. 1. $\sin i'$ vs. a' plot for the very large data set of asteroid proper elements (over 12,000 objects) provided by Milani and Knežević (1994)

should be groupings of objects for which both the statistical definition, based on the analysis of proper elements, and the physical definition based on observations confirming a common genetic relationship, coincide. With this definition of families, only a few of them can be presently definitely accepted, mainly due to the paucity of observational data on the surface properties (color indexes, albedos, taxonomic types) for most of the asteroidal population.

On the other hand, it is not easy to get satisfactory "proofs" of the existence of families in a purely physical sense. Even not taking into account the present lack of observational data, which will be hopefully overcome in the next years, some intrinsic problems remain : first, the outcome of an impact depends on the overall specific impact energy, i.e., the ratio between the kinetic energy of the projectile and the mass of the target : when very energetic impacts occur, the fragments may be ejected at high velocities, and spread over very large regions of the proper elements space. In these cases, what is formed is not properly a family, since no observable group of objects remains in the space of proper elements after the

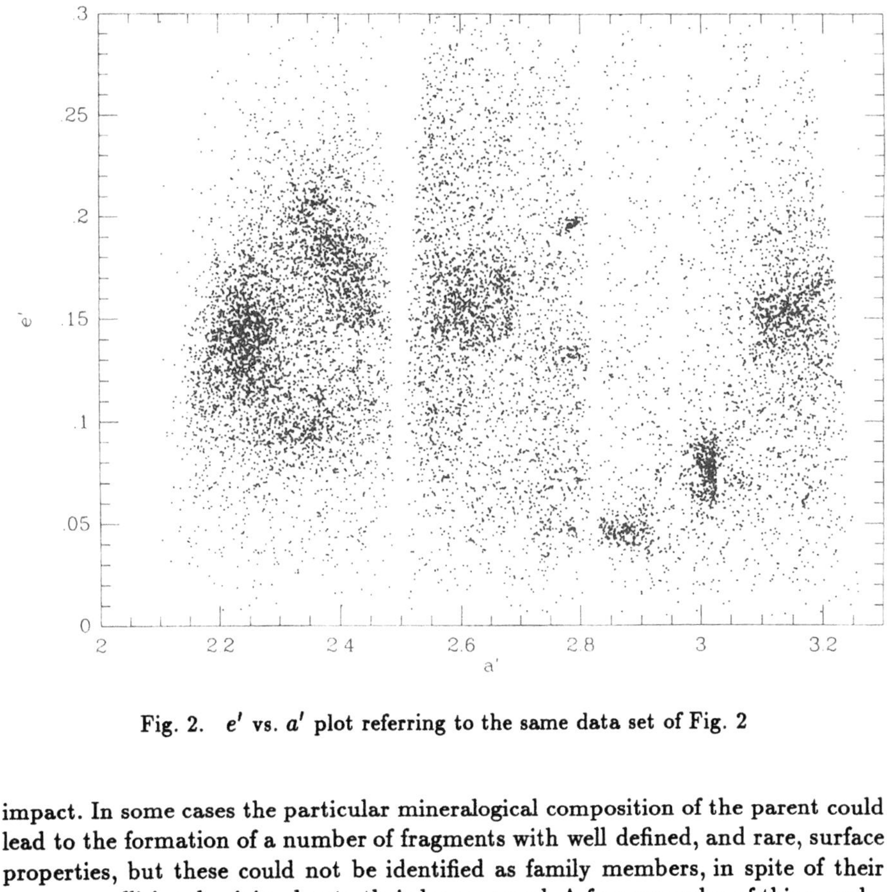

Fig. 2. e' vs. a' plot referring to the same data set of Fig. 2

impact. In some cases the particular mineralogical composition of the parent could lead to the formation of a number of fragments with well defined, and rare, surface properties, but these could not be identified as family members, in spite of their common collisional origin, due to their large spread. A few examples of this may be actually present in the asteroid belt (see, e.g., Burbine *et al.*, 1992), and this subject has also deep implications for the general problem of the origin of meteorites. In other cases, however, just the opposite may occur : some well defined groupings of asteroids in the proper element space may be plunged in a background of objects bearing the same general surface properties. In this case, a physical confirmation may be more difficult. One should also take into account that some basic physical properties of collisional outcomes, like their mass and velocity distributions, are not presently well known due to our unsatisfactory knowledge of the physics of catastrophic impacts. On the other hand, some recent theoretical improvements related to the elaboration of complex hydrocode models (Melosh *et al.*, 1992), or to the improvements of a semiempirical approach (Paolicchi *et al.*, 1993), allow to be confident about the possibility to compare observational data with reliable theoretical predictions in the near future. Moreover, from the experimental point of view, the increasing amount of data obtained by means of sophisticated laboratory techniques (see Fujiwara *et al.*, 1989 for a review) is encouraging. The next years will bring probably important results in the field of the physics of catastrophic collisions.

From the above discussion, it follows that the concept of family is based on concurring evidences coming from both statistics and physics : the first step in the process leading to the discovery of a family consists in identifying it as a statistically significant clustering of objects in the space of proper elements. The second step is to compare the physical properties of the supposed members with what is known about the outcomes of catastrophic impacts, and with the mineralogical properties of asteroidal bodies.

3. Family identification : problems and recent improvements

In the previous Section we have sketched a description of the ideal procedure which should be carried out for an unambiguous identification of asteroid families. In practice, however, the interplay of independent evidences coming from the statistical distribution of proper elements and from the observed physical properties of the supposed members has been rarely achievable so far.

This explains why for a long time, spanning over several tens of years, no definite agreement on the number of asteroid families, and on the memberships of these families, has been found by the authors who have published different family lists (see Valsecchi *et al.*, 1989, for an exhaustive review). This does not mean that all these searches were wrong : actually, many of them led to interesting findings over many respects (we can quote the important work carried out by Williams, 1979, 1989, as a good example of relevant contribution to the subject). The problem is that the results found by different searches contradicted each other.

Of course, this has entailed bad consequences for asteroidal science : as a matter of fact, a great deal of skepticism has arisen among many of those interested in the general process of collisional evolution of the asteroid belt, preventing them from seriously considering the constraints coming from the existence of the families. The skepticism was due both to the disagreement on the number of families identified by different authors (this number ranged from 15 to more than 100 on the basis of the papers published between 1978 and 1989) and to the observed cosmochemical inconsistency of some of the proposed families (Chapman *et al.*, 1989).

On the other hand, some important improvements have been recently achieved from the dynamical and statistical point of view. For this reason, we have now a number of well established groupings of objects for which we are waiting for a definitive confirmation coming from observations. These groups, according to Farinella *et al.* (1992) cannot yet be called "families", and other terms such as "clans" or "clusters" appear more appropriate. On the other hand, on the basis of the first observational tests, the situation is encouraging, and some spectacular confirmations, like in the case of the Vesta family (Binzel and Xu, 1993), allow us to be very confident for the future developments.

All this has been made possible by the simultaneous development of independent statistical procedures of identification, and by the availability of new very

large data bases of asteroid proper elements, computed by Milani and Knežević (1990, 1992) (versions from 4.2 to 5.7).
These data have been exploited both by searches based on the D-criterion of Southworth and Hawkins (1963) (Lindblad, 1992), and by investigations based on new methods, introduced by Zappalà et al. (1990, 1994), and by Bendjoya *et al.* (1991), and Bendjoya (1993). These procedures are completely independent, and are based on the hierarchical clustering techniques of multivariate data analysis, and on the application of a mathematical technique called "wavelet analysis", respectively. We do not recall here the technical details of these procedures, which are extensively explained in the quoted papers. What we want to note, is that for the first time two independent techniques have been applied to the same (large) data base of asteroid proper elements, and have given results in good agreement with each other. Moreover, both procedures are mostly based on automatic algorithms managed by computer, and are easily reproducible and fairly "objective". In particular, their performances have been tested by means of computer simulations, in which fictitious families have been created and plunged in random backgrounds of different densities. The results have shown that both techniques behave well even in particularly difficult conditions, and are reliable in cases bearing a resemblance to the properties of the actual asteroid belt (Bendjoya *et al.*, 1993).

These results are encouraging, taking into account that many searches for asteroid families carried out in the past by different authors, suffered from the (forced) adoption of small data bases on asteroid proper elements, and/or from the application of identification procedures mostly based on the (subjective) response of the human eye. Moreover, many of those searches did not include careful statistical methods for determining the reliability of the resulting families, or in some cases such criteria have been found to be questionable by other authors (such is the case of Williams' Poisson test, as will be noticed in the next Section).

Among other things, the most recent analyses have shown that, with large data sets of asteroid proper elements, some general, "philosophical" choices must be made, concerning the application of the identification methods. Apart from the obvious fact that groupings having no statistical significance have to be rejected, some problems can arise in some cases for the memberships of unambiguously identified groupings. In other words, the statistical criteria adopted can lead to different results due to the smooth transition that exists in the space of proper elements from the "core" of some families to the surrounding background. This implies that in several cases a statistically significant group is identified, but one has to make some choice about the criterion used to establish the identity of its members. One can be more conservative, and in this case only a fraction of the real members (the core of the cluster) will be accepted, but the resulting list will contain only a few chance interlopers, if any. The alternative is to be more liberal, and this implies to accept as family members also many objects lying in the outer part of the apparent cluster. In this case a larger number of "true" members will be recognized, but, as a drawback, one can expect also a significant amount of chance interlopers; the interlopers might account for a large fraction of the listed members in the worst cases (the "clans" according to the Farinella *et al.*, 1992, terminology).

The problem of the uncertain membership of asteroid families cannot be solved by means of any general algorithm, since it reflects a real physical fact, that is the progressive mixing of the family members ejected with larger ejection velocities with the background of objects present in the surrounding zones of the belt. The best approach is probably the following : one has first to identify the reliable family candidates, with their "sure" members, i.e., those belonging to the cores of these groupings. The membership of the objects lying in the "halo" of the family could then be established by means of physical observations, allowing to exclude from the halo those objects which are not plausible on the basis of the observed characterization of the core objects. Of course, this approach cannot work when the family members cannot be discriminated, with respect to the background, on the basis of any observable property.

4. The present situation

There are currently several published lists of families. Among these, we have decided to focus our attention only on the most recent ones. On the basis of this choice, three main lists of asteroid families, obtained by means of different techniques, will be discussed in this Section. These lists are those given by Zappalà *et al.*, 1994, Bendjoya, 1993, and Williams, 1989, 1992. Other independent searches have been recently performed by other authors, in particular Lindblad (1992), but the list of family memberships has not yet been published. The older family lists, published before 1989 (see Valsecchi et al., 1989) will not be analyzed here, even if we are aware that many of them have been historically important.

A comparison between the results obtained by Zappalà *et al.* (1994) by means of their most recent search based on the hierarchical clustering method, and by Bendjoya (1993), who used the most recent 3-D version of his wavelet analysis approach, is encouraging. Actually, Zappalà *et al.*, 1994, found 20 statistically significant and metric-independent family candidates, as well as 6 additional statistically significant candidates which were found to be metric-dependent, in the sense that their identification depends on the metric used to define the distance in the space of proper elements (see Zappalà *et al.*, 1990, and Milani *et al.*, 1992). In addition to these, 7 other groupings were found to be just below the threshold adopted for statistical reliability, and should be considered as dubious cases.

In the most recent application of his wavelet analysis method, Bendjoya (1993) found 26 statistically significant family candidates, plus 11 additional marginally significant groupings. The agreement between the resulting family lists can be considered good : taking into account just the groupings listed as statistically significant, those found in both the lists are 20; one of those listed by Zappalà *et al.* is listed as marginally significant in Bendjoya's list. Among the statistically significant groups listed by Bendjoya, 7 have no counterpart in the Zappalà *et al.* list. In most cases, however, they are small groupings; moreover, some of them are actually found by Zappalà *et al.* at the critical level of distance (Quasi Random Level QRL) adopted for the definition of reliable groupings, but they do not include the minimum required number of objects. The only one important discrepancy concerns a big group formed by 43 members, associated to the asteroid 79 Eurynome, found by Bendjoya, and which has no counterpart in the Zappalà *et al.* list. In three cases,

TABLE I

Comparison between the family lists found by Bendjoya (1993) (B) and Zappalà *et al.* (1994) (Z). The columns give, respectively, the identification Id(B) and number of members N_B of each grouping found by B; the corresponding identification, Id(Z), and number of members N_Z found by Z; the percentage of the common members with respect to N_B and N_Z; some remarks concerning a possible improvement of the agreement by relaxing the adopted criteria : for the criteria "2" and "3" of B see Bendjoya (1993); for the definition of the critical distance level QRL and its associated σ (Z) see the text.

Id(B)	N_B	Id(Z)	N_Z	%(B)	%(Z)	Remarks
8	368	8 + 244**	494	95	71	crit.2 (B) : 95%(B), 85%(Z)
4	79	4	64	71	88	
163	12	163	11	92	100	
272	21	1726 + 363	18	86	100	
170	52	170	50	90	94	
668	44	668	41	93	100	
145	26	145	28	100	93	
1272	37	1272	32	84	97	
15	156	15	180	96	83	crit.2 (B) : 95%(B), 93%(Z)
410	11	1403	8	73	100	$QRL + 1\sigma$ (Z) : 82%(B), 100%(Z)
490	12	490	11	92	100	
221	219	221	273	100	80	crit.2 (B) : 99%(B), 93%(Z)
158	161	158	207	100	78	crit.2 (B) : 100%(B), 90%(Z)
135	145	44	221	100	66	crit.2 (B) : 96%(B), 81%(Z)
213	17	808	10	59	100	$QRL + 1\sigma$ (Z) : 71%(B), 100%(Z)
1644*	11	1644	6	45	83	$QRL + 1\sigma$ (Z) : 82%(B), 90%(Z)
24	177	24	387	100	46	crit.2 (B) : 100%(B), 71%(Z)
						crit.3 (B) : 98%(B), 83%(Z)
10	29	10	20	66	95	$QRL + 2\sigma$ (Z) : 90%(B), 72%(Z)
569	13	569**	6	46	100	$QRL + 2\sigma$ (Z) : 62%(B), 100%(Z)
847	35	847 + 2354	16	40	88	$QRL + 2\sigma$ (Z) : 66%(B), 82%(Z)
79	43	–	–			in 44 at $QRL + 3\sigma$ (Z)
9	11	–	–			5 obj. at $QRL + 1\sigma$ (Z)
27	17	–	–			6 obj. at $QRL + 2\sigma$ (Z)
779	8	–	–			8 obj. at $QRL + 2\sigma$ (Z)
1639	5	–	–			4 obj. at QRL (Z)
262	21	–	–			6 obj. at QRL (Z)
869	9	–	–			3 obj. QRL (Z)
137	19	137	10	53	100	*tribe* for B
						crit.2 (B) : 63%(B), 100%(Z)

* : possible subgroup of 262 (B)

** : statistically reliable grouping with one metric only (see text) finally, one single group listed by Bendjoya includes two groups listed separately by Zappalà *et al.* As for the membership, the agreement is satisfactory, although, as explained in the previous Section, each statistically significant group should be separately studied in detail, beyond the "official" lists of members which constitute the output of the algorithms; actually, each candidate family has its own unique features, thus its structure should be analyzed separately. Even taking into account these *caveats*, however, the numbers and identities of family members published in the two searches are in a good agreement.

In Table I, the family candidates found by the two classifications are divided into three main categories : the first includes groupings for which the percentage of common members exceeds 70% of the number of members found by both classifications. The second category includes groupings for which the percentage of common members does not reach 70% of the number of members found by one of the two methods. In several cases, a better agreement is reached by relaxing slightly the criteria adopted for the definition of the membership, e.g. by using the so-called "criterion 2" (or "criterion 3") of Bendjoya, or accepting members above the nominal QRL of Zappalà *et al.* The third category includes the candidate families which are identified by one classification, but are not statistically reliable for the other (although they may be still recognizable, by relaxing the reliability criteria).

On the basis of these results, it appears that 20 good family candidates have been found in the asteroid belt. In addition to the three well known, "historical" families of Eos, Themis and Koronis, already found in the pioneering work of Hirayama at the beginning of the century (see references in Valsecchi *et al.*, 1989), a number of other interesting cases are apparent. Some of these were already found in the past by some of the authors who produced the older family lists, while others are new, like the interesting grouping associated to the large asteroid 10 Hygiea (see later).

Interestingly, these 20 family candidates include also, with one exception, the 14 main families listed by Lindblad (1992) in his recent analysis based on the D-criterion applied to the 4.2 version of the asteroid proper elements data base provided by Milani and Knežević. Actually, Lindblad (1992) did not publish a list of the members of his families; however, the names of the lowest-numbered members of each family, used to identify them, correspond to those found by Zappalà *et al.* and Bendjoya. The only exception is given by the group of Goberta, which was listed by Lindblad, but belongs probably to the Themis family found by the other authors. Due to the lack of additional information about Lindblad's results, it is not possible to perform here a more detailed comparison of the results. Lindblad (1992) noticed also that he found a relevant number of groupings formed by just a few members, but the statistical significance of such groupings is not clear.

Given the good agreement between the classifications of Bendjoya and Zappalà *et al.* (agreement that probably can also be extended to Lindblad's classification), a few comments should be made about the disagreement existing with the results of the extensive searches for asteroid families carried out over more than ten years by Williams (1979, 1989, 1992). These papers should be considered as milestones in the history of family searches, since Williams was able to compute accurate proper elements for a large data set of asteroids, and used this list for his comprehensive and systematic search work. He found a very high number of families, about 100, many of which are composed by a few members. Such a result has been criticized by other authors (including ourselves) who did not find some of these families plausible, due to the presence of assemblages of different taxonomic types not compatible with a common collisional origin. The high number of families found by Williams is in disagreement with the results of most of the other searches carried out recently, including, of course, Zappalà *et al.* (1994) and Bendjoya (1993).

The reasons for such a disagreement should be analyzed and understood correctly, in order to avoid possible misunderstandings and consequent criticism about the overall meaning and reliability of the existing family classifications.

The representation of families in terms of the "stalactite diagrams" introduced by Zappalà *et al.* (1990, 1994) appears here particularly useful. This kind of diagrams show the sizes of the groupings existing in a given region of the space of proper elements (corresponding to different zones of the asteroid belt) as a function of the level of the mutual distance D between the objects. The distance D is used for the identification of the groupings in the framework of the hierarchical clustering method; we recall here that in these diagrams, due to the relation between the components of the ejection velocities of fragments and the resulting final differences in the proper elements, the distance is expressed in m/s (see Zappalà *et al.*, 1990, and Milani *et al.*, 1992). From these diagrams, it is easy to see what kind of changes in the structure of a given grouping take place at different levels of D. A typical example is given in Fig. 3, which shows the different behavior exhibited by a typical "cluster" (very compact and stable at very different values of D) and a typical "clan", according to the nomenclature suggested by Farinella *et al.* (1992). The membership of the candidate families, identified by the hierarchical clustering method, is given by the objects belonging to the groupings found at a particular level of D, the so-called Quasi-Random Level (QRL). The QRL is determined on the basis of the stalactites which can be obtained in the different zones of the proper elements space by populations of fictitious objects distributed at random (on the scale of typical family sizes), but satisfying the constraint of reproducing the large-scale distributions of proper elements of the real asteroids.

The agreement (or disagreement) between different classifications can be investigated by analyzing the structure of the resulting groupings at different values of D, in order to find out whether the discrepancies between classifications obtained by means of different techniques may be due to the different criteria used for the acceptance/rejection of candidate families. In other words, we can check whether the discrepancies may be due to the fact that different authors "cut" a given stalactite at different levels.

We recall that in the case of the differences between the Zappalà et al. (1990) results and those obtained by Williams (1989, 1992) the disagreement cannot be due to the different sets of proper elements, as shown by Zappalà *et al.* (1992) : the problem must be due to the different statistical methods used for family identification.

While the methods of hierarchical clustering and of wavelet analysis are both based on automated procedures managed by computer, the first step of the method of Williams was a visual examination of stereo plots, leading to the identification of family candidates. A statistical test was performed on the chosen clusters only

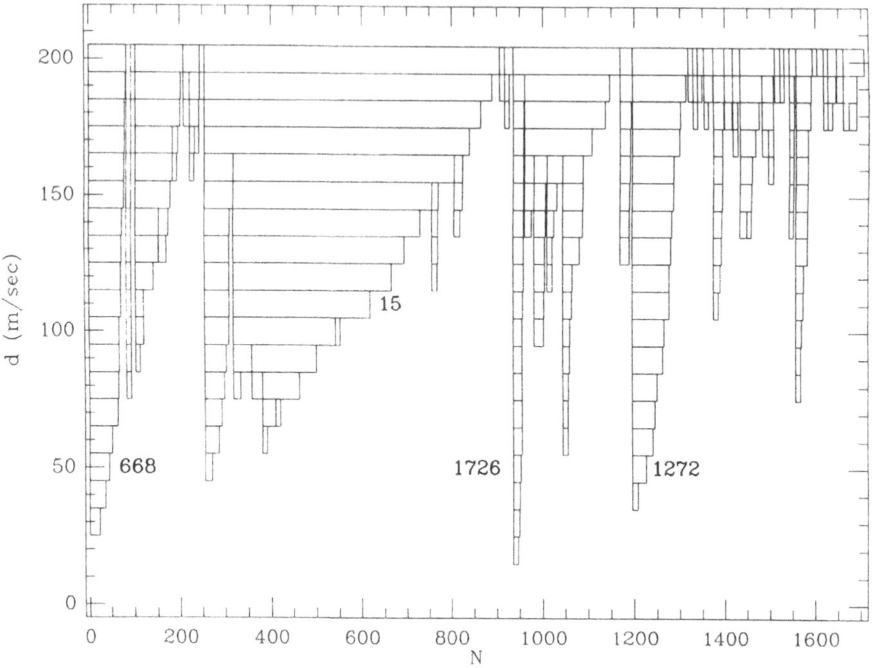

Fig. 3. A typical *stalactite diagram*, referring to the asteroids of the data set provided by Milani and Knežević (1994), in the zone of the belt delimited by $a' = 2.5$ and $a' = 2.825$. The size of the groupings found at each level of distance (expressed in m/s, see text) is shown in the abscissa. The critical Quasi Random Level (QRL) has not yet been determined for this data set, but should be around 140 – 150 m/sec. The differences between the behavior of a typical *clan*, the one associated to the asteroid 15 Eunomia, and some typical *clusters*, like those associated to the asteroids 668 Dora, 1726 Hoffmeister and 1272 Gefion, are evident (see the text)

a posteriori. He used Poisson statistics, based on the number of objects belonging to a cluster and the number expected by chance, derived from the background density and cluster volume. To be accepted as a family, the logarithm of the Poisson probability associated to a given identified clustering was required to be less than -3.5. 102 clusters passed this significance test.

It is our opinion that such a procedure is not completely safe, since Poisson statistics of density fluctuations would require a medium homogeneous on a large scale (and this is not the case in the present situation); moreover, the "boxes", in which a comparison is made between the number of objects really present and that expected on the basis of the average density, should have fixed sizes decided *a priori*. Choosing the boxes *a posteriori* on the basis of the volume of each observed surdensity, leads to maximize the derived probability of the groupings to be not due to chance.

However, we do not want here to discuss in detail the efficiency and/or the validity of this method, but only investigate the possible sources of the discrepancies between the results obtained in this way and those coming from the other techniques of family identification; in particular, we will refer to the hierarchical clustering method, since for this we have a direct experience of its mechanisms and we can more easily compare the results by using the stalactite diagrams.

To do this, we introduce a "quality code" q, describing the agreement - *family by family* - of the Williams (W) and Zappalà *et al.* (Z) classifications at different significance levels in the Z stalactite representation. If N_{int} is the number of members found to belong to a given family in both classifications (i.e., the intersection of a common Z and W family) and N_{wil} is the number of numbered objects found in the W family (we recall that Williams considered also a number of PLS objects which are not present in the data base used by Zappalà *et al.*) we assume to have a reasonable agreement when N_{int}/N_{wil} is larger than or equal to 1/3. Families behaving in this way at the nominal QRL of Z have been defined of quality $q = 1$. Since the QRL values were found by random simulations and are subject to a statistical variance σ, to families showing a satisfactory agreement ($N_{int}/N_{wil} \geq 1/3$) within 2 σ from QRL (i.e. cutting the Z stalactite at a more liberal significance level) we assigned a quality code $q = 2$, while $q = 3$ was assigned to families not reaching a satisfactory agreement within 3-σ from QRL. In other words, quality codes 2 and 3 mean that in order to find a satisfactory agreement between the Z and W classifications, we should accept in the framework of the Z method also groupings of very poor or no statistical significance.

On the other hand, we assume as a significance parameter for the W classification the logarithm of the Poisson probability (log P) obtained by the W statistical test. Then, we divide the W families in three categories : the high significance families having log P $<$ -10; the intermediate ones, having -10 $<$ log P $<$ -5; and the low significance families, having log P $>$ -5. At this point, we can make a comparison between these two sets of parameters. In particular, for each of the three categories of W families we check the number of families having quality codes 1, 2 or 3 (see Fig. 4). Note that the first category contains 27 families, the second 38 and the third 33 (for a few families of W we could not compute the q parameter, since they refer to high inclination asteroids, not present in the data set investigated by Z).

In general, we see that there is a good correlation between q and log P, in the sense that when log P increases the quality tends to be poor and viceversa. In particular, for log P $>$ -5 (smaller significance in the W Poisson test) we have 85% (23) of the families with $q = 2$ or 3 (poor or no intersection between the Z and W classifications), while for log P $<$ -10, 79% of the families (26) have quality code $q = 1$. Very interesting is the fact that these 26 families overlap almost perfectly to the set of high-reliability families found by Zappalà et al. and Bendjoya.

How to interpret this result in terms of the reliability of the various classifications? We have two possibilities : either to consider too optimistic the value of -3.5 (actually used by Williams) as a significance limit for log P in the W classification

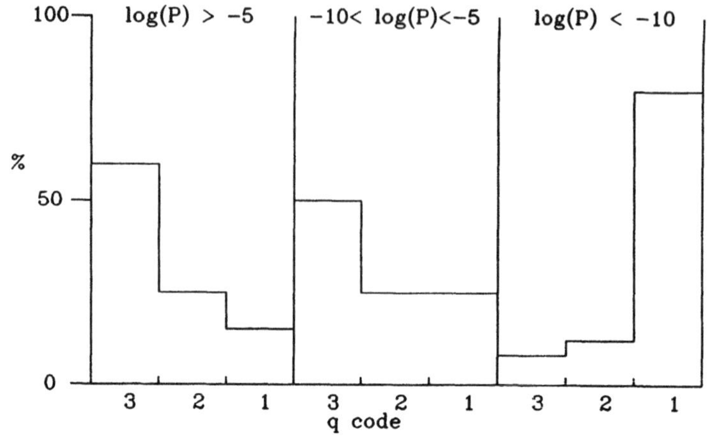

Fig. 4. A comparison between the results of the family classifications carried out by Williams (1989, 1992) and by Zappalà *et al.* (1994) shows that the agreement is very good (85%, corresponding to 26 family candidates with $q = 1$) for the groupings having the smallest values of $\log(P)$ listed by Williams. The agreement is not as good (majority of groupings having $q = 2$ or $q = 3$) for higher values of $\log(P)$ (see text).

(or also to consider not completely correct the *a posteriori* use of a Poisson test to groupings already selected by visual inspection); or to consider too pessimistic the nominal significance level (QRL) used by Zappalà *et al.* (1994). Obviously, as coauthors of the latter paper, we tend to be in favor of our own choice, which - we recall - has been tested by means of extensive numerical simulations (Bendjoya *et al.*, 1993). However, we do not want, after having tried to be as "objective" as possible with the identification procedure, to be now biased in excluding all the Williams families which do not agree with our results ! What is really important to stress is that the frustrating discrepancy between different classifications, which in some way hindered the research on families for several years in the past, can be interpreted now satisfactorily in most cases. In general, the present situation is characterized by a satisfactory agreement between the results obtained by means of the application of identification algorithms managed by computer. The remaining discrepancies concern mostly the statistical reliability of a significant number of groupings identified visually, due to differences in the optimistic/pessimistic criteria used to interpret the results.

Therefore, what we suggest is the following : to consider at present as highly reliable only the families found by the Zappalà *et al.* (1994) and by the Bendjoya (1993) classifications (which are also essentially coincident with the most significant candidates identified by Lindblad, 1992). The majority of the Williams' families should be put in a "waiting list" : we cannot exclude at all that some (if not most) of the smaller Williams' families could be real, even if they have not been found by other searches. But we think that at present their existence cannot be claimed on the basis of a solid statistical evidence. A strong increase in the number of asteroids having high-quality proper elements, in particular the addition of many smaller objects which have been so far excluded from the analyses, will certainly permit to discriminate the "good" candidates among the presently existing controversial cases.

5. Open problems and conclusions

The most recent family searches have indicated that two main types of plausible candidate families are apparent : the *clusters* have sharp limits, and their membership is particularly reliable, as it does not depend significantly on the criterion used to discriminate between family members and background objects in the space of proper elements. The *clans* are large associations, having very uncertain boundaries, and often characterized by the presence of several subgroupings. Their memberships cannot be defined easily without an adequate observational support. Five clans have been reliably identified so far; they are named after their lowest-numbered member : Flora, Nysa, Eunomia, Hygiea and Vesta. Clans are a challenging subject for future studies. It would be interesting to understand whether their structures are the consequences of some peculiar collisional events, or if they were generated by multicollisional phenomena. The observational data so far available do not allow to draw any definite conclusion, as remarked by Cellino and Zappalà (1993). Of course, in the cases of Vesta and Hygiea, due to the large sizes of their parent bodies (400-500 km diameter), the origin of the families is probably related to energetic, almost catastrophic, *cratering* events. It is hard to state whether the impact responsible for the formation of the Vesta family, which is also probably related to the hemispheric-scale albedo variegation detected by the observations (Cellino *et al.*, 1987; Cellino *et al.*, 1989), has been energetic enough to fracture the interior of this asteroid. Practically no data are available in the case of the Hygiea clan, located in the outer part of the main belt. Here, an observational effort should be carried out, since there are several analogies with the Vesta event.

The case of the Flora clan ($a' \sim 2.25$ AU) remains puzzling. Figures 1 and 2 show that this clan is located in a very crowded region of the inner belt. The accuracy of the proper elements in this zone is a little worse with respect to the other regions of the main belt, due to the presence of several resonances. For this reason, a reliable identification of this clan, and an assessment of its main features, is not easy. With a smaller data set of proper elements, with respect to the one which is presently available, and using also a less evolved version of the hierarchical clustering method, Zappalà *et al.* (1990) were not able to unambiguously identify this clan as a single statistically reliable grouping. On the other hand, an analysis of the asteroid size distribution in the main belt performed by Cellino *et al.* (1991) showed that the size distribution within the Flora region is very similar to that characterizing the major families (Eos, Themis and Koronis). Finally, Zappalà *et al.* (1994), as well as Bendjoya (1993) list now unambiguously the Flora group as one of the main clans present in the asteroid belt.

Another interesting case is the one of the Nysa clan. While Nysa itself (an E-type object) should be considered as a probable interloper, as well as the other large M asteroid Hertha, this clan is characterized by the presence of a large fraction of the presently known asteroids belonging to the rare F taxonomic type (see also Cellino

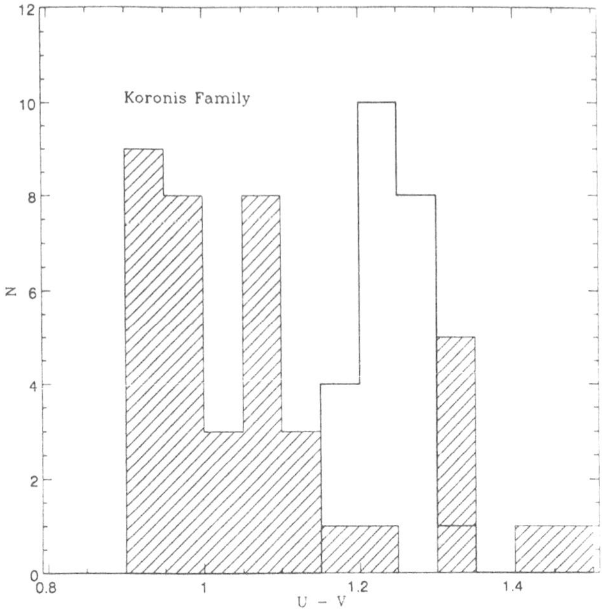

Fig. 5. U - V distribution for the members of the Koronis family (empty histogram), and for the background objects of the same zone of the belt (shaded histogram). The family appears well separated from the background.

and Zappalà, 1993). This seems very interesting, also taking into account that some F-type asteroids are present among the population of Near-Earth-Asteroids and that the clan lies close to the inner edge of the 3 : 1 resonance. Here, a specific observational campaign appears particularly useful and promising.

Figures 5–7 show some other facts which deserve further investigations. It is easy to see that both the Koronis and the Eos families show some peculiar distributions of the color-index U-V and the albedo p_V, with respect to the background objects of the zones in which they are located. The U-V distributions of these families are noticeably similar, being peaked at values of about 1.25. The Koronis members for which an evaluation of the taxonomic type is known, are classified as S, but they could constitute a definite subset of the S objects. As for the Eos family members, the available data indicate that they belong to the rare K taxonomic type. In addition to the U-V peculiarities, the Eos members have also albedo values well distinct with respect to the background objects in their zone of the belt (Fig. 7). This observational evidence should be kept in mind, since probably it is not due to chance, but reflects some important facts about the mineralogical differentiation of asteroid interiors, as well as, possibly, on the influence of collisional events on the observable optical properties of the asteroid surfaces.

Of course, the well established family candidates constitute a challenge for the

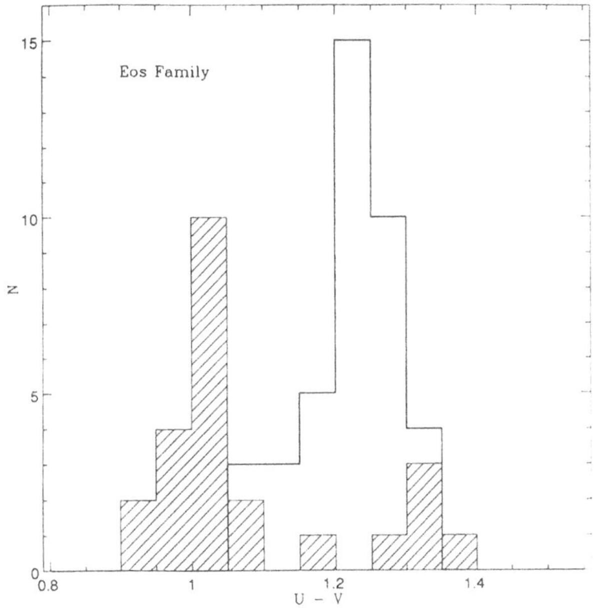

Fig. 6. The same as Fig. 5, but for the family of 221 Eos.

modern asteroidal science. On one hand, they deserve some dedicated observational campaigns, in order to derive the necessary information about the optical properties of the surfaces of the supposed members. This in turn would allow to discriminate in many cases between the "real" members and the chance interlopers. On the other hand, the families can provide the main observational evidence about the outcomes of catastrophic disruptions of bodies for which self-gravitation is not negligible. This would be very useful, if we take into account that recently a big effort has been made in order to improve the present physical understanding of collisional events (Melosh *et al.*, 1992, Asphaug and Melosh, 1993, Paolicchi *et al.*, 1989, 1993, 1994). Potentially, this can entail a better comprehension of the overall scenario of the collisional history of the asteroid belt. In particular, the abundance and size distribution of the presently identifiable families put some important constraints on any general model of the collisional evolution (see Davis *et al.*, 1989, for a review). Taking into account the case of the Vesta family, it is reasonable to expect that an analysis of the most important family candidates, especially in the case of the clans, could give some hints concerning the origin of the Near-Earth-Asteroids. In this respect, as stated above, the Nysa clan, with its large amount of F objects, looks particularly promising.

Of course, a number of problems have to be faced and should be solved : an example is given by the asymmetries in the ejection velocities of fragments from the catastrophic breakups which led to the formation of the present families. This

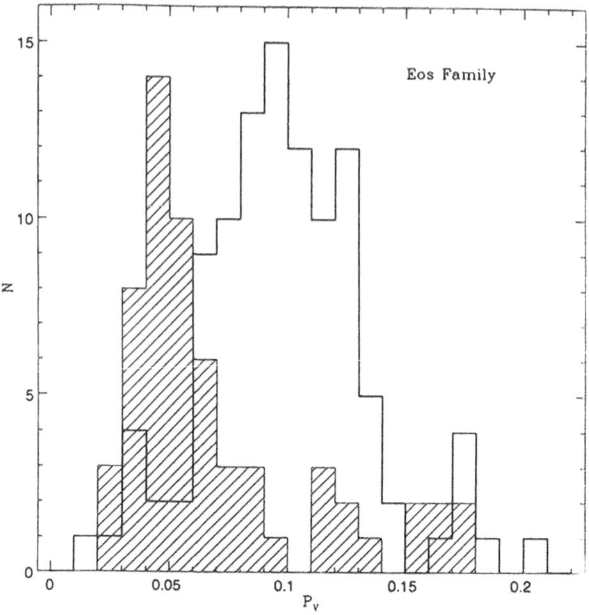

Fig. 7. Albedo p_V distribution for the members of the Eos family (empty histogram) and for the background objects in the same zone of the belt (shaded histogram).

problem has been known for many years (Brouwer 1951, Zappalà *et al.*, 1984, Chapman *et al.*, 1989). An analysis of the ejection velocity distributions resulting from the new, reliable candidate families seems now necessary.

Finally, we note that the continuous refinement of the techniques of asteroid proper elements computation, which has triggered the recent improvement of the situation concerning the identification of asteroid families, is not over. It is now possible to carry out a new extensive family search, based upon a data set of over 12,500 objects (Milani and Knežević, 1994; Knežević and Milani, this volume). Of course, this will be done in the near future. Such a job should give some answers about the structures of the candidate families identified so far, and about the possible statistical significance of some groupings which are presently marginal. Even the objects at high orbital inclination should now be accessible (Lemaitre and Morbidelli, 1994).

We can conclude this review by saying that the epoch of skepticism about asteroid families seems really over. We expect the near future to be very exciting, with large amounts of new data which will improve our understanding of the history of the asteroidal population, and its interrelationships with the other minor bodies of the solar system.

References

Asphaug, E. and Melosh, H.J. : 1993, "The Stickney impact of Phobos : a dynamical model." *Icarus*, **101**, 144–167.

Bendjoya, Ph. : 1993,"A classification of 6479 asteroids into families by means of the wavelet clustering method." *Astron. Astrophys. Suppl. Series*, **102**, 25–55.

Bendjoya, Ph., Slezak, E. and Froeschlé, Cl. : 1991, "The wavelet transform : a new tool for asteroid family determination." *Astron. Astrophys.*, **251**, 312-330.

Bendjoya, Ph., Cellino, A., Froeschlé, Cl. and Zappalà, V. : 1993, "Asteroid dynamical families : a reliability test for two identification methods." *Astron. Astrophys*, **272**, 651–670.

Binzel, R.P. and Xu, S. : 1993, "Chips off of asteroid 4 Vesta : evidence for the parent body of basaltic achondrite meteorites." *Science*, **260**, 186–190.

Broglia, P. and Manara, A. : 1989, "Rotational variations in the optical polarization of (4) Vesta." *Astron. Astrophys.*, **214**, 389–390.

Brouwer, D. : 1951, "Secular variations of the orbital elements of minor planets." *Astron. J.*, **56**, 9–32.

Burbine, T.H., Gaffey, M.J. and Bell, J.F. : 1992, "S-asteroids 387 Aquitania and 980 Anacostia : possible fragments of the breakup of a spinel-bearing parent body with CO3/CV3 affinities." *Meteoritics*, **27**, 424–434.

Cellino, A. and Zappalà, V. : 1993, "Asteroid clans : super families or multiple events ?", *Celest. Mech.*, **57**, 37–47.

Cellino, A., Zappalà, V., Di Martino, M., Farinella, P. and Paolicchi, P. : 1987, "Flattening, pole and albedo features of 4 Vesta from photometric data." *Icarus*, **70**, 546–565.

Cellino, A., Di Martino, M., Drummond, J., Farinella, P., Paolicchi, P. and Zappalà, V. : 1989, "Vesta's shape, density and albedo features." *Astron. Astrophys.*, **219**, 320–321.

Cellino, A., Zappalà, V. and Farinella, P. : 1991, "The size distribution of asteroids from IRAS data." *Mon. Not. Roy. Astr. Soc.*, **253**, 561–574.

Chapman, C.R., Paolicchi, P., Zappalà, V., Binzel, R.P. and Bell, J.F. : 1989, "Asteroid families : physical properties and evolution." In *Asteroids II*, (R.P. Binzel, T. Gehrels, M. S. Matthews, Eds.), 386–415, Univ. of Arizona Press.

Cruikshank, D.P., Tholen, D.J., Hartmann, W.K., Bell, J.F. and Brown, R.H. : 1991, "Three basaltic Earth-approaching asteroids and the source of the basaltic meteorites." *Icarus*, **89**, 1–13.

Davis, D.R., Weidenschilling, S.J., Farinella, P., Paolicchi, P. and Binzel, R.P. : 1989, "Asteroid collisional history : effects on sizes and spins." In *Asteroids II*, (R.P. Binzel, T. Gehrels, M. S. Matthews, Eds.), 805–826, University of Arizona Press.

Farinella, P., Davis, D.R., Cellino, A. and Zappalà, V. : 1992, "From clusters to families : a proposal for a new nomenclature." In *Asteroids Comets meteors 1991*, (A.W. Harris, E. Bowell, Eds.), 165–166, L.P.I., Houston.

Fujiwara, A., Cerroni, P., Davis, D.R., Ryan, E., Di Martino, M., Holsapple, K. and Housen, K. : 1989, "Experiments and scaling laws on catastrophic collisions." In *Asteroids II*, (R.P. Binzel, T. Gehrels, M. S. Matthews, Eds.), 240–265, University of Arizona Press.

Greenberg, R. and Nolan, M.C. : 1989, "Delivery of asteroids and meteorites to the inner solar system." In *Asteroids II*, (R.P. Binzel, T. Gehrels, M. S. Matthews, Eds.), 778–804, University of Arizona Press.

Lemaitre, A. and Morbidelli, A. : 1994, "Calculation of proper elements for high inclined asteroidal orbits." *Celest. Mech.*, in press.

Lindblad, B.A. : 1992, "A computer search for asteroid families." In *Asteroids Comets meteors 1991*, (A.W. Harris, E. Bowell, Eds.), 363–366, L.P.I., Houston.

Melosh, H.J., Ryan, E.V. and Asphaug, E. : 1992, "Dynamic fragmentation in impacts : hydrocode simulation of laboratory impacts." *J. Geophys. Res.*, **97**, 735–759.

Milani, A. and Knežević, Z. : 1990, "Secular perturbation theory and computation of asteroid proper elements." *Celest. Mech.*, **49**, 247-411.

Milani, A. and Knežević, Z. : 1992, "Asteroid proper elements and secular resonances." *Icarus*, **98**, 211–232.

Milani, A. and Knežević, Z. : 1994, "Asteroid proper elements and the dynamical structure of the asteroid main belt." *Icarus*, in press.

Milani, A., Farinella, P. and Knežević, Z. : 1992, "On the search for asteroid families." In *Interrelations between Physics and Dynamics for Minor Bodies in the Solar System*, (Cl. Froeschlé, D. Benest, Eds.), 85–132, Editions Frontières, Gif-sur-Yvette, France.

Paolicchi, P., Cellino, A., Farinella, P. and Zappalà, V. : 1989, "A semiempirical model of catastrophic breakup processes." *Icarus*, **77**, 187–212.

Paolicchi, P., Verlicchi, A. and Cellino, A. : 1993, "Catastrophic fragmentation and formation of families : preliminary results from a new numerical model." *Celest. Mech.*, **57**, 49–56.

Paolicchi, P., La Spina, A., Paolicchi, P. and Cellino, A. : 1994, "The interpretation of laboratory experiments in the framework of an improved semi-empirical model." *Planet. Space Sci.*, in press.

Southworth, R. and Hawkins, G. : 1963, "Statistics of meteor streams." *Smithson. Contrib. Astrophys.*, **7**, 261–285.

Valsecchi, G., Carusi, A., Knežević, Z., Kresák, Ľ. and Williams, J.G. : 1989, "Identification of asteroid dynamical families." In *Asteroids II*, (R.P. Binzel, T. Gehrels, M. S. Matthews, Eds.), 368–385, University of Arizona Press.

Williams, J.G. : 1979, "Proper elements and family memberships of the asteroids." In *Asteroids*, (T. Gehrels, Ed.), 1040-1063, University of Arizona Press.

Williams, J.G. : 1989, "Asteroid family identifications and proper elements." In *Asteroids II*, (R.P. Binzel, T. Gehrels, M. S. Matthews, Eds.), 1034-1072, University of Arizona Press.

Williams, J.G. : 1992, "Asteroid families – an initial search." *Icarus*, **96**, 251–280.

Zappalà, V., Farinella, P., Knežević, Z. and Paolicchi, P. : 1984, "Collisional origin of asteroid families : mass and velocity distributions." *Icarus*, **59**, 261–285.

Zappalà, V., Cellino, A., Farinella, P. and Knežević, Z. : 1990, "Asteroid families. I. Identification by hierarchical clustering and reliability assessment." *Astron. J.*, **100**, 2030-2046.

Zappalà, V., Cellino, A. and Farinella, P. : 1992, "A comparison between families obtained from different proper elements." In *Asteroids Comets meteors 1991*, (A.W. Harris, E. Bowell, Eds.), 675–678, L.P.I., Houston.

Zappalà, V., Cellino, A., Farinella, P. and Milani, A. : 1994, "Asteroid families. II. Extension to unnumbered multiopposition asteroids." *Astron. J.* **107**, 772–801.

COMET TAXONOMY AND EVOLUTION

DAVID G. SCHLEICHER
Lowell Observatory
1400 W. Mars Hill Rd.
Flagstaff, AZ 86001, USA
E-mail dschleicher@lowell.edu

Abstract. Measurements of physical and dynamical properties of comets have yielded considerable evidence for variations in the origin and evolution of these primordial objects. Given these findings, one might reasonably expect to observe differences in the chemical composition as well. This paper reviews our current state of knowledge of the similarities and differences in the composition of comets, and summarizes the recent findings by A'Hearn *et al.* (1994) for two taxonomic classes of comets and evidence of compositional variation due to origin and/or evolution.

1. Introduction

Comets are widely believed to be the most pristine remnants from the epoch of solar system formation and, as such, can provide important clues to the physical and chemical processes which took place at that time. For this reason, much of cometary research is aimed at obtaining an understanding of the origin, basic compositional and physical characteristics, and fate of cometary nuclei. However, the acquisition of the first direct samples from a nucleus still lies in the future, and so compositional studies have necessarily relied on measurements of cometary comae constituents. While the *in situ* measurements of Comet P/Halley by five spacecraft in 1986 (and subsequently of Comet P/Grigg-Skjellerup by Giotto) were extremely important in determining spatial densities within the coma and detecting species not observed from Earth, they were unable to address the issue of whether or not Halley's composition is "normal" and, therefore, typical of other comets.

Numerous papers in recent years have explored the dynamical evolution of cometary orbits and the evidence for two reservoirs of comets – the Oort cloud and the Kuiper belt – which are presumably the primary sources of dynamically new comets and Jupiter–family comets, respectively (cf. Duncan, Quinn, and Tremaine 1988; Mumma, Weissman, and Stern 1993). If this scenario is true, then compositional differences might be expected because the comets in these two groups would have formed at different locations and, thus, under different conditions within the proto–solar nebula. While not wholly immune from evolutionary processes while "stored" in either the Oort Cloud or Kuiper Belt (cf. Donn 1976; Stern 1990; Mumma, Weissman, and Sterns 1993), one would expect that any compositional differences would still be evident today. Thus far, quantitative studies of the chemical composition of comets – based on abundances and production rates of species within cometary comae – show only limited evidence for this type of trend with orbital dynamics.

Several spectroscopic studies prior to the 1980s reached the conclusion that the general composition of all comets is nearly the same, based on the recurrence of

A. Milani et al. (eds.), Asteroids, Comets, Meteors 1993, 415–428.

many atomic and molecular features in cometary spectra, and sometimes crude quantitative measures of relative abundances (cf. Arpigny 1976). More recent photometric and spectrophotometric studies of the UV, visible, and near–IR portions of the spectrum (see Figure 1 for a representative composite spectrum of Comet P/Tuttle) have reached the same basic conclusion, based on computed production rates of the observed species (cf. A'Hearn and Millis 1980; Weaver *et al.* 1981; Newburn and Spinrad 1984, 1989; Cochran 1990; Cochran *et al.* 1992; A'Hearn *et al.* 1994). The other general conclusion of these studies has been that the gas–to–dust ratio of comets is not constant, but instead varies by over two orders of magnitude, with no obvious trend with age.

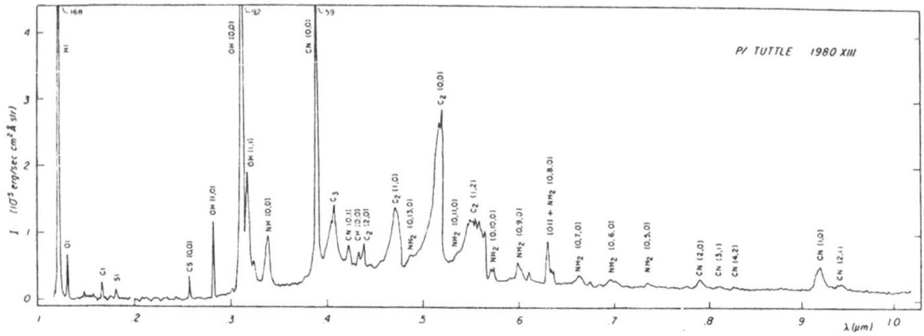

Fig. 1. Composite spectrum of Comet P/Tuttle (1980 XIII) is representative of the features observed in most cometary spectra. The UV portion of the spectrum was obtained with the IUE satellite (Weaver *et al.* 1981), while the remainder was obtained in two parts with the 1.54 m Catalina Station telescope of the University of Arizona. The composite was created by S. Larson and J. Johnson. Courtesy of S. Larson.

Radio observations, primarily of the OH molecule, have been obtained for many dozens of comets, but these have usually not been obtained concurrently with measurements of other species and, therefore, are not suitable for directly determining abundance ratios. In the last few years, observations of comets in the microwave region have become common, and have resulted in many new species being detected. The status of this new field and lists of the many species observed are given in several recent reviews (Krankowsky 1991; Mumma, Weissman, and Stern 1993; Crovisier 1992, 1994).

While it has been determined that the large majority of comets studied have the same basic composition, significant depletions or enhancements of certain species have also been measured in many comets. Evidence for abundance variations has been observed for several molecular species – in some cases apparently associated with the dust abundance – for which the number of comets measured is yet too small for meaningful statistical analyses to be performed. Examples include CO (Feldman 1983, 1986) and S_2 (Budzien and Feldman 1992) in the UV, H_2CO (formaldehyde) (Colom *et al.* 1992; Mumma, Weissman, and Stern 1993) and CH_3OH (methanol)

(Hoban *et al.* 1991; Mumma *et al.* 1993) in the IR and microwave, and N_2^+ (Lutz *et al.* 1993) in the visible. To investigate groupings and correlations of cometary composition, the remaining sections of this paper will necessarily concentrate on the findings based on the few larger sized data bases currently available.

2. Data Bases for Compositional Studies of Comets

To date, four long–term observational programs have obtained compositional measurements in a systematic manner for more than 15–20 comets. The earliest of these was begun in 1976 by A'Hearn and Millis (1980). Narrowband filter photometry has now been obtained by this group for a total of 88 comets. A'Hearn *et al.* (1994) report on summary results for the 85 comets observed through 1992, along with the outcome from taxonomic analyses performed on a subset of 41 comets for which multiple observations of OH, NH, CN, C_2, and C_3 were obtained. Their taxonomic results will be summarized in Section 3.

Two programs of visible spectrophotometry have been undertaken. The ongoing McDonald Observatory Faint Comet Survey by Cochran (1987) and colleagues has observed approximately 100 comets since its inception in 1979. Cochran, Green, and Barker (1989) analyzed 29 "low activity" comets, and Cochran (1990) briefly summarized results for the CN, C_2, and C_3 abundances of a larger sample of 48 comets. These studies concluded that only one comet – P/Giacobini–Zinner – stood out as having substantially different relative abundances; however, many of the comets have only "uninteresting" upper limits for C_2 and C_3. Giacobini–Zinner was, in fact, identified in several studies as having C_2 and C_3 depleted by more than a factor of 5 and NH and NH_2 depleted by lesser amounts (cf. Cochran and Barker 1987; Schleicher, Millis, and Birch 1987; Beaver *et al.* 1990). Recently, detailed analyses from the McDonald survey were extended to CH and NH_2 for 17 comets (Cochran *et al.* 1992), with none displaying unusual abundances.

The other program of visible spectrophotometry, begun in 1980 by Newburn and Spinrad, measures the emission of CN, C_2, C_3, CH, and O[^{1}D]. In their initial paper which included data for 17 comets, Newburn and Spinrad (1984) identified 2 comets – P/Borrelly and P/Swift-Gehrels – as being depleted in C_2 and C_3. In their more recent paper for 25 comets (including the original 17), Newburn and Spinrad (1989) note a clear trend of the C_2–to–CN ratio with heliocentric distance. However, Cochran *et al.* (1992) do not see a variation with distance, implying that any such trend may be an artifact of the scalelengths chosen for the reduction from column density to production rate. Newburn and Spinrad (1989) do see a clear correlation of CN abundance with dust, strongly implying a dust source for CN, in addition to the generally assumed parent gas source. This correlation is demonstrated in Figure 2. A similar effect is also seen by A'Hearn *et al.* (1994) for several individual comets which exhibit abundance ratio variations with heliocentric distance.

It was already noted in Section 1 that the gas–to–dust ratio varies enormously among comets. In their analysis of dust for 18 comets, Storrs, Cochran, and Barker (1992) show evidence that the heliocentric distance dependence of $Af\rho$, a measure of dust production, is less steep than is the dependence of the production rate of CN, a conclusion supported by A'Hearn *et al.* (1994). Storrs, Cochran, and Barker

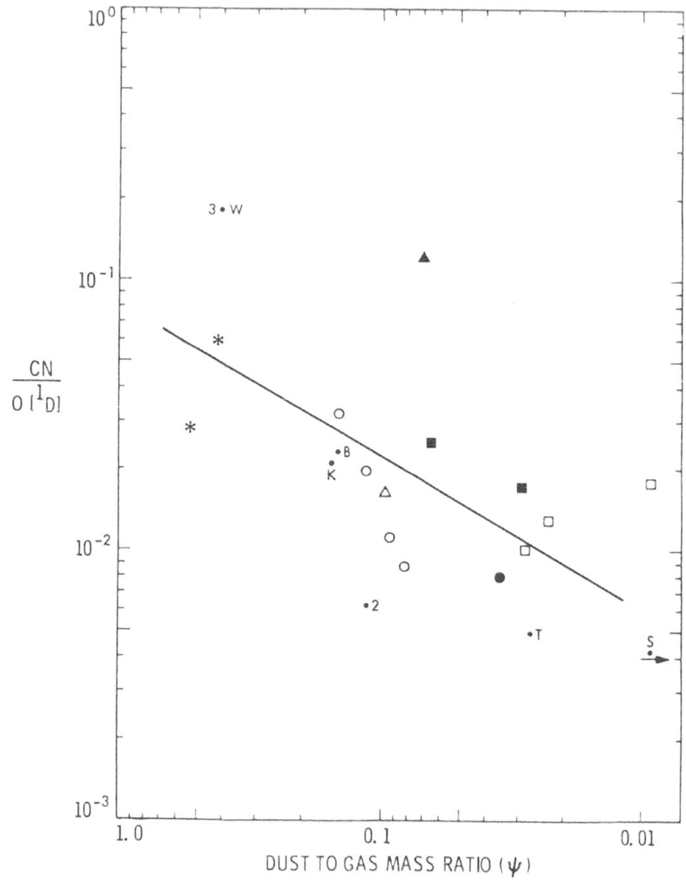

Fig. 2. The correlation of CN to dust is evident in this plot of the CN–to–O[^{1}D] ratio with respect to the dust–to–gas ratio. See the original paper for the correspondence between symbols and comets. From Newburn and Spinrad (1989).

also indicate that the gas–to–dust ratio shows a trend – with the ratio increasing with increased gas (CN) production – and attribute this effect to gas production through a porous mantle. While their assumption that CN is a good measure of the total gas production is true to first order, A'Hearn *et al.* show that the CN–to–OH ratio varies by as much as a factor of 4 among comets, implying that further work is required to better understand the relationship of gas with dust grains.

Finally, the ultraviolet program by Feldman, A'Hearn, Festou, Weaver and collaborators makes use of the International Ultraviolet Explorer (IUE) satellite. Approximately 50 comets have been observed since 1978. While the OH emission bands are ubiquitous in IUE spectra, other emission features are usually only observed in brighter comets. In more than a dozen comets for which CS was detected and measured, only P/Swift–Tuttle shows a significant variation from the mean CS–to–OH ratio, having a CS abundance that is high by more than a factor of two (Feldman, private communication).

3. A Comet Taxonomy

Until the last decade, the creation of a taxonomic classification system for comets based on chemical composition was stymied because too few species had been measured in too few comets. While the three more recent spectrophotometric studies discussed in Section 2 contain statistically useful numbers of comets, the results from these studies indicated that nearly all comets for which the usual molecular species were detected have about the same composition, and only a few comets clearly have significant anomalies. However, for many other comets only upper limits have been obtained for certain species due to the intrinsic faintness of the comets and/or the large amount of continuum in the spectrum.

An advantage of the narrowband filter photometry study by A'Hearn *et al.* (1994) is that relatively large photometer entrance apertures – usually between 30 and 150 arcsec — were employed in making the observations. The use of large apertures helps in several ways. For example, the effects of inaccurate scalelengths in the calculation of production rates are reduced because a much greater fraction of the comet is sampled. Moreover, this approach gives greater contrast of the emission bands to the underlying continuum resulting in many more actual detections of individual species. Another aspect of the narrowband filter photometry study by A'Hearn *et al.* is the simultaneous acquisition of OH measurements in the near–UV with measurements of the major emission features in the visible : CN, NH, C_2, and C_3. While both OH and O both are created by the dissociation of water, the bulk constituent of comets, deduction of the water production rate from OH is less dependent on imprecisely known parameters such as the branching ratio of H_2O, than are results based on measurements of $O[^1D]$. These aspects of the A'Hearn *et al.* program assisted in making taxonomic analyses viable, and it is primarily their results which are presented next.

To minimize the effects of spurious data and possible model dependencies in the computation of production rates, the A'Hearn *et al.* (1994) taxonomic analyses were restricted to a subset of the total data base of 85 comets. In particular, only data sets were included for which measurements of all 5 gas species (OH, CN, C_2, C_3, and NH) were made, and for which the comet's heliocentric distance was less than 3 AU. In addition, each included comet was required to have been observed on at least 2 nights and to have 3 or more individual data sets, as a verification of the quality of the data. These restrictions resulted in a subset consisting of 41 comets for which production rate ratios (i.e., abundance ratios) with respect to OH were determined. Log values of the mean ratios, weighted by the quality of the data, were computed for each comet, and the resulting 41 ratios for each minor species form the actual taxonomic database.

The most dramatic result for this database is shown in Figure 3, where the C_2–to–OH production rate ratio is plotted as a function of the CN–to–OH ratio. It is apparent that approximately 70% of the comets have a direct relationship of

C_2 with CN, while the remaining comets have lower C_2 abundances by varying amounts. This clear trend for the majority of comets indicates that both CN and C_2 vary together, but that they both vary by as much as a factor of 3–4 with respect to OH. Note that the line plotted in Figure 3 is not fitted to the data, but rather is a line with a slope of unity. This group is identified by A'Hearn *et al.* as having "typical" abundances. They also show that C_3 follows the same relationship as C_2, but with greater scatter, presumably due to poorer signal–to–noise ratio of the C_3 data. In contrast, no obvious relation is seen when the NH–to–OH ratio is plotted versus the CN–to–OH ratio. This indicates that for the "typical" class of comets the CN, C_2, and C_3 are all related to each other much more strongly than to either OH or NH. The remaining 30% of the comets form the second taxonomic group and exhibit significant depletions in the carbon–chain species (C_2 and C_3) – by as much as a factor of 20 from the mean value for the typical group – while CN and NH both have "typical" abundances with respect to OH.

As is evident from Figure 3, a transition region exists within which the appropriate classification is not obvious. Because of the direct relationship seen for C_2 with CN for the majority of comets, along with the results from a principle component analysis (discussed later), A'Hearn *et al.* use the C_2–to–CN ratio as the discriminator for classification. The resulting distributions of abundance ratios is shown in Figure 4, where histograms of each species with respect to both OH and CN are shown, and the "typical" comets are highlighted. Near–gaussian distributions are seen for the "typical" class in each case, while a long "tail" is seen for the carbon–chain species in the depleted class. Although the NH abundance is lower than average for several of the carbon–chain depleted comets (including previously mentioned Comet Giacobini–Zinner), it is not evident from these plots that additional classifications based on NH abundance are warranted. However, it it possible the groupings based on NH do exist within the apparently broad distribution, but are simply washed out due to the relatively poor S/N of the NH measurements and the resulting larger associated errors. Finally, it should be noted that some comets were observed to change their location on the diagram (Figure 3) as a function of heliocentric distance and/or time. For instance, several "typical" comets move along the diagonal line designating this class in the course of their respective apparitions. While this behavior can be partially explained as an artifact of the scalelengths used in the Haser model for calculating production rates, A'Hearn *et al.* indicate that the majority of the effect is real. This implies the existence of inhomogeneities within some individual nuclei which match the chemical variations measured among different comets within the "typical" class. That is to say, the same highly correlated variation in the abundances of the three carbon–bearing species with respect to water which is measured from one "typical" comet to another is also observed *within* some individual comets.

A'Hearn *et al.* also applied both the techniques of principle component analysis and cluster analysis to their taxonomic database. Because the database already consisted of four gas abundance ratios for each comet, the principle component analysis was not intended to reduce the number of variables to a smaller, physically meaningful parameter set. This contrasts to the situation in asteroid photometry where one can employ principle component analysis to reduce a large number of

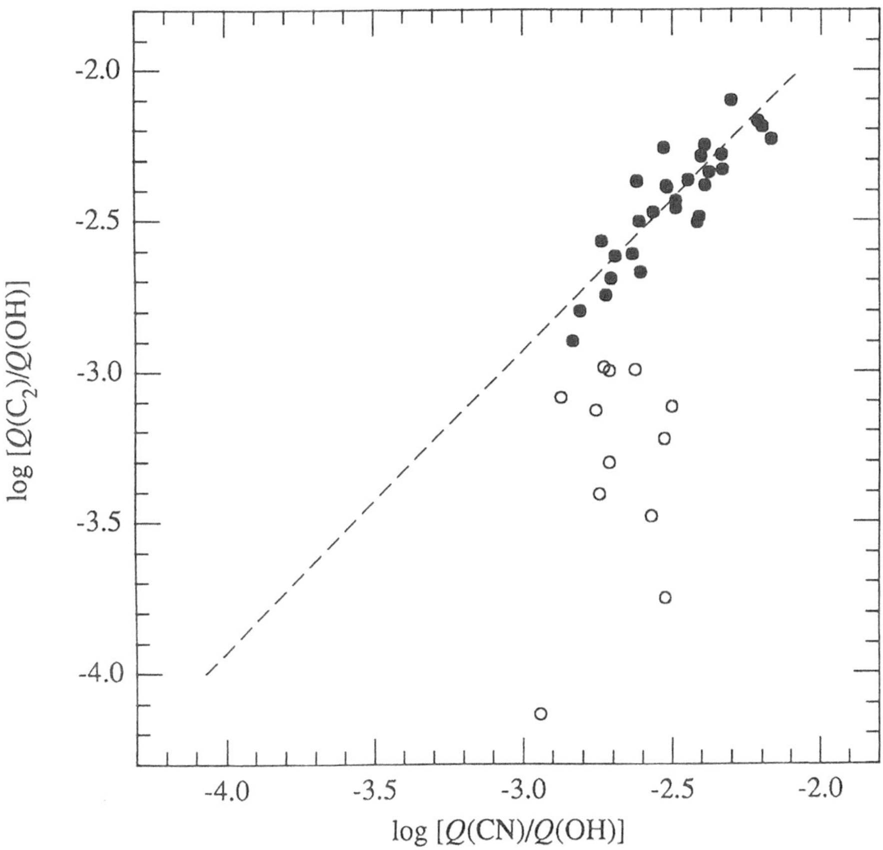

Fig. 3. Two taxonomic groups are identified based on the relative production rates of OH, CN, and C_2 for the 41 comets contained within the A'Hearn *et al.* (1994) restricted database. Within the "typical" group (filled symbols), CN and C_2 vary together with respect to OH (the line has a slope of unity). Based on A'Hearn *et al.* (1994).

filter magnitudes to a few diagnostic components (cf. Tholen 1984). In spite of this difference in situations, the analysis of the comet database results in the first principle component containing over 70% of the total variance, while the second principle component contains only about 15% of the variance. Not unexpectedly, the first principle component is primarily a near–equal combination of the C_2–to-OH and C_3–to–OH ratios, along with a smaller contribution from CN–to–OH. In other words, this axis goes from approximately the bottom and left–most data points shown in Figure 3 to the top-right points. The second principle component is nearly entirely composed of the NH–to–OH ratio, reflecting the generally broad distribution of this variable as seen in Figure 4. These results are therefore consistent with the basic taxonomic classifications of A'Hearn *et al.*

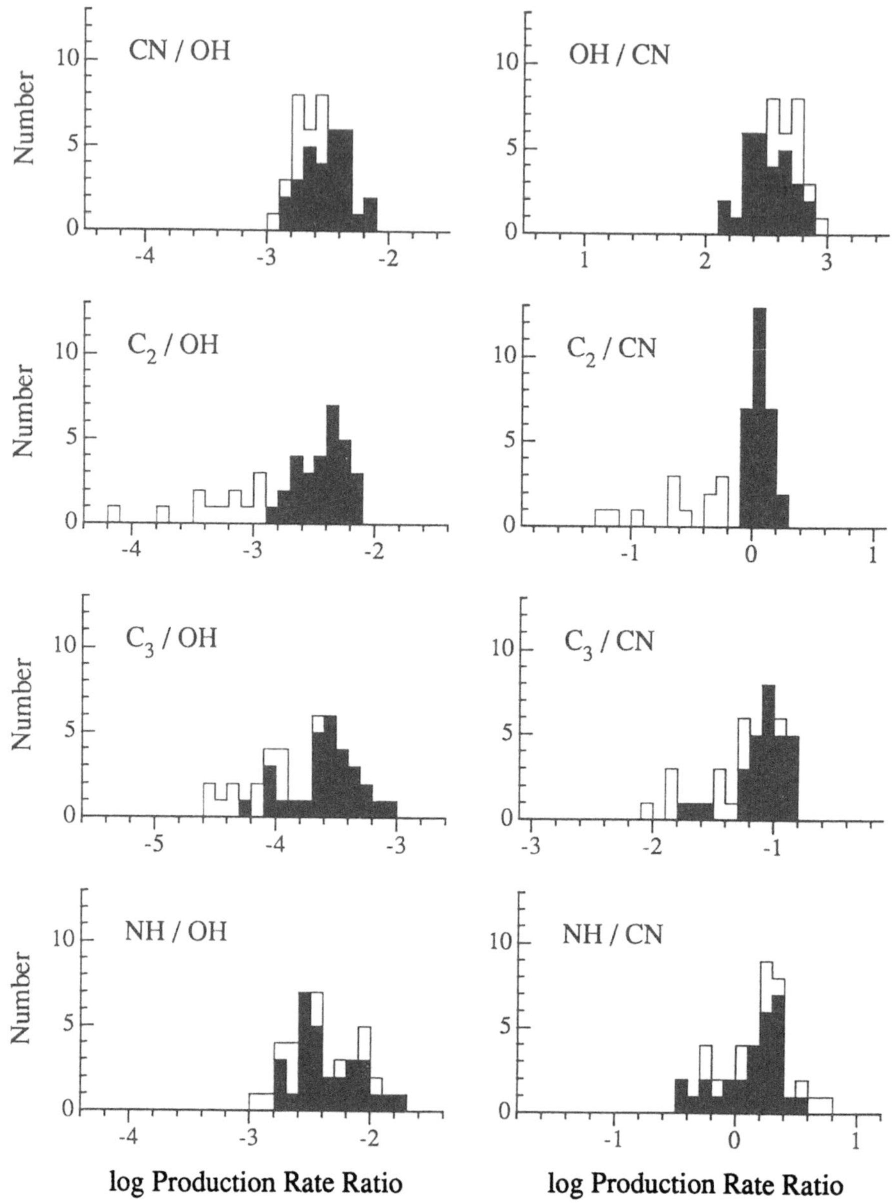

Fig. 4. Histograms showing the distribution of production rate ratios for the restricted database of 41 comets. Shaded regions are comets which are part of the "typical" group, and the histograms have been shifted horizontally to align the mean values of the "typical" group. From A'Hearn *et al.* (1994).

Using a complete linkage hierarchical clustering algorithm, A'Hearn *et al.* obtain two nearly equal-sized clusters or groups from their taxonomic database. These groups are basically the upper-half and lower-half of the points shown in Figure 3. The only significant difference between this cluster analysis grouping and the taxonomic groups identified previously is the placement of the comets at the lower-left end of the region defining the "typical" group in the diagram. As already noted, A'Hearn *et al.* consider the strong correlation of the abundances of CN, C_2, and C_3 for the majority of comets (i.e., comets along the diagonal line in Figure 3) as highly physically significant and, therefore, should override the cluster analysis result, since the clustering algorithm ignores this distinctive aspect of the data.

Comet Giacobini–Zinner was identified in Section 2 as a well-observed comet having depletions of several species. It is appropriate to consider Giacobini–Zinner as the prototype of A'Hearn *et al.'s* class of carbon-chain depleted comets, since it was identified as being unusual in its composition more than 60 years ago (Bobrovnikoff 1927). The most extreme example of this group from the A'Hearn *et al.* database is Comet P/Wolf-Harrington. Analysis of photometric and spectrophotometric data for this comet by Schleicher, Bus, and Osip (1993) indicate that both C_2 and C_3 are depleted by more than an order of magnitude with respect both to OH and to CN. This extreme example of a carbon-chain depleted comet is shown in Figure 5. Even more unusual abundances were reported for Comet Yanaka (1988r) by Fink (1992) based on spectrophotometric measurements. However, unlike comets in the carbon-chain depleted group identified by A'Hearn *et al.*, both CN and C_2 were undetected in the Yanaka spectrum, while both NH_2 and $O[^1D]$ were present (OH and NH were outside the spectral region and not observed). Based on a water production rate computed using the $O[^1D]$ measurement and upper limits for CN and C_2, Fink concludes that these species were depleted by a factor of at least 25 and 100, respectively, but that the NH_2 has normal abundance. It is interesting to note that comparable depletions of CN and C_2 with respect to water would place Yanaka near the lower-left corner of the A'Hearn *et al.* taxonomic diagram (Figure 3), close to where the diagonal line defining their "typical" group of comets extends.

4. Compositional Evidence for Comet Evolution

Much physical evidence exists indicating that comets evolve or age with time. For instance, dynamically new comets are observed to be preferentially brighter prior to perihelion than after, possibly due to the higher volatility of the original surface compared with material deeper within the nucleus (Whipple 1977). Well-established cases of short-period comets fading with successive apparitions and eventually disappearing are documented in the literature, while other comets have been mis-identified as asteroids due to their asteroidal appearance on some occasions (cf. Kresák 1991). However, changes in observing techniques, selection effects, sporadic variability, and changes induced by orbital perturbations must all be accounted for when interpreting a comet's behavior as evidence for evolution (Kresák, this volume; Rickman *et al.* 1991). Other expected evolutionary effects are often assumed, but data do not exist as yet to confirm them. For example,

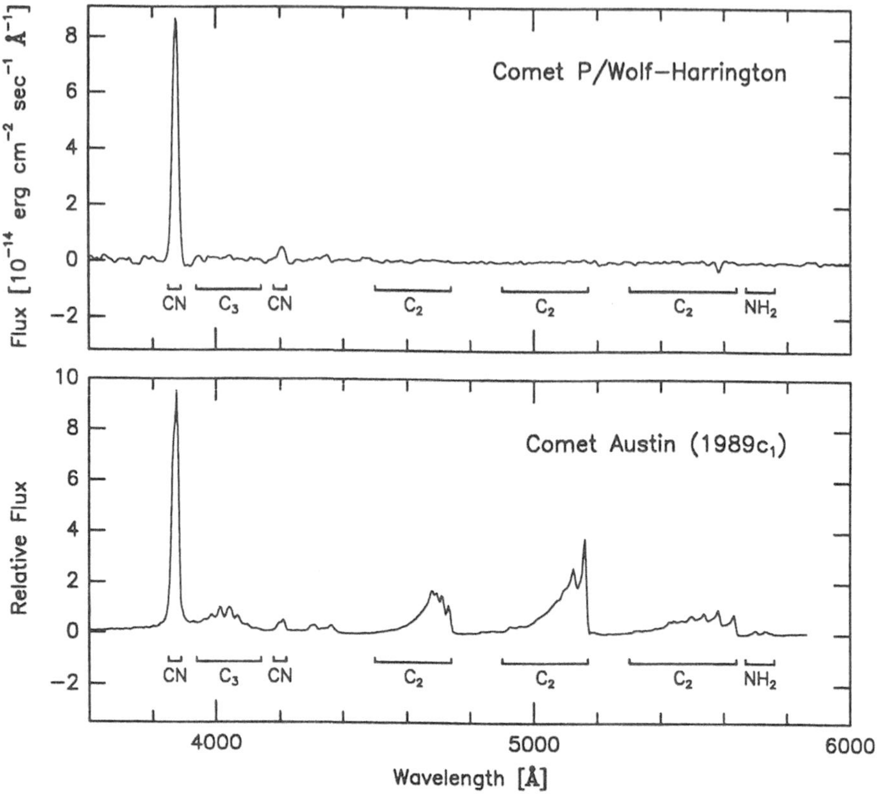

Fig. 5. Spectra of Comets P/Wolf–Harrington and Austin (1990 V). Note the apparent lack of C_2 and C_3 emission in Wolf–Harrington. From Schleicher, Bus, and Osip (1993).

dynamically new comets are often assumed to have a relatively fresh, bright surface as compared to periodic comets which have extremely low albedos. In fact, no albedo measurements have been made of the nucleus of a dynamically new comet, and a variety of processing of the surface is likely to have occurred over 4.5 billion years (cf. Mumma, Weissman, and Stern 1993). These and other lines of evidence for physical evolutionary processes have been reviewed recently by Kresák (1991; 1994) and Meech (1991).

The amount of data available for investigating evolution based on chemical composition is much more limited. Only a small number of comets have had compositional measurements made on multiple apparitions, and a similar small number of dynamically new comets have been observed in the larger database programs discussed in the previous two sections. However, several conclusions have been possible from this limited amount of data. Only 2 of the 11 periodic comets observed at multiple apparitions by A'Hearn *et al.* (1994) clearly show a change in their water production rate at a given heliocentric distance, and both of these comets had undergone significant orbital changes due to perturbations by Jupiter. Moreover, while the total production rates had changed in these comets, the relative abundances of the observed species remained constant. In contrast, variations in abundance ratios *are* observed to take place in many comets either as a function of heliocentric distance or as an asymmetry about perihelion, but these variations repeat from one apparition to the next. This behavior is interpreted by A'Hearn *et al.* as evidence for chemical inhomogeneities across the surfaces of some comet nuclei, coupled with changing orientation of the nucleus with respect to the sun. That is to say, variations of this type are believed to be driven by a seasonal effect.

Several compositional trends are observed as functions of orbital parameters, and presumably age or origin. For instance, dynamically new comets apparently preferentially exhibit an excess of pure water ice at large heliocentric distances (>3 AU) as they arrive near the sun for the first time (cf. A'Hearn *et al.* 1994). Dynamically new comets also apparently have higher overall production rates before perihelion rather than after, consistent with numerous brightness measurements (A'Hearn *et al.*). While these results strongly imply that the surface of a dynamically new comet is intrinsically different from the remainder of the material in the nucleus, many more dynamically new comets need to be observed over a large range of distances before these results can be considered conclusive.

A'Hearn *et al.* also find that while carbon–chain depleted comets constitute approximately 30% of their taxonomic database, most of these are Jupiter–family comets (i.e., having periods less than 20 years and/or aphelion distances less than 10 AU), resulting in about one–half of the observed Jupiter–family comets being classified as carbon depleted. In contrast, most but not all long–period comets are classified as "typical." Since only 1–3 comets from the taxonomic database are dynamically new (two comets have indeterminate values for $1/a_0$), the fact that none of these is depleted in the carbon–chain molecules is not statistically significant. A possible scenario consistent with these findings is that the temperature at which the comets formed in the proto–solar nebula not only controlled the basic abundances (cf. Yamamota 1985), but that the critical temperature occurred within the region from which Kuiper Belt comets originated. Therefore, Oort Cloud comets and as many as one–half of Kuiper Belt comets would be classified as "typical," while the remaining Kuiper Belt comets would be in the carbon–chain depleted class (A'Hearn *et al.* 1994). Alternatively, if about one–half of the Jupiter–family comets actually have an Oort Cloud origin – a fraction significantly greater than is generally assumed – then the A'Hearn *et al.* taxonomic classes would directly reflect a comet's origin.

A final evolutionary effect is noted for the dust–to–gas ratio observed in the A'Hearn *et al.* database. While there is no obvious trend with dynamical age, i.e. $1/a_0$, a clear trend with perihelion distance (q) *is* observed. This correlation is illustrated in Figure 6. A'Hearn *et al.* note that very little change is seen for the dust–to–gas ratio for a given comet as a function of heliocentric distance, consistent with the findings of Storrs, Cochran, and Barker (1992), ruling out a simple temperature or production rate cause of the correlation with q. Since this trend is observed in comets before perihelion as well as after, it also cannot be caused by a simple thermal lag mechanism. A'Hearn *et al.* postulate that the surface properties of the nucleus are controlled by the maximum temperature which it has previously attained, rather than the current temperature, and this in turn controls the release of dust particles observed in the visible. Incorporation of the many results from observations of dust in the IR should test this hypothesis.

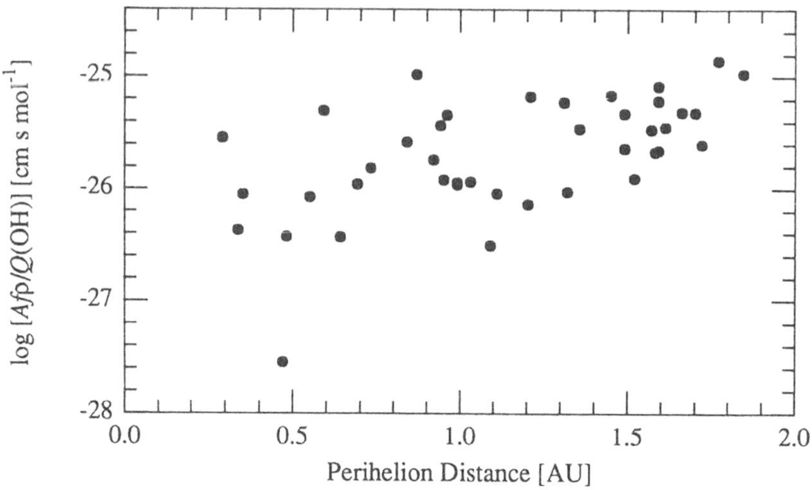

Fig. 6. **The dust-to-gas ratio for comets in the A'Hearn *et al.* restricted database as a function of perihelion distance. Note the clear trend, which may be caused by changes in the surface properties of the cometary nuclei which in turn are controlled by the maximum surface temperature a comet reaches each orbit. From A'Hearn *et al.* (1994).**

5. Summary

As evidenced by the many recent results reported on in this paper, the field of comet compositional studies has seen considerable progress in the past decade. Emission by molecular species in the UV and visible portions of the spectrum has now been measured in a sufficient number of comets that taxonomic classifications are not only possible, but have resulted in the discovery of 2 taxonomic groupings by A'Hearn *et al.* 1994. Moreover, a clear correlation is seen among the three carbon-bearing species within the larger, "typical" group, with their respective abundances varying together with respect to water. Evidence that the comets' site of origin is responsible for the basic grouping comes from the comets' orbital parameters, with approximately one-half of the sampled Jupiter-family comets being members of the carbon-chain depleted group, while very few members of this group are Halley-family or long-period comets. Correlations of CN production with dust abundance are observed in many comets, while the amount of dust measured in the visible is observed to vary inversely proportionally with a comet's perihelion distance. This latter effect implies that processing of the surface of a cometary nucleus, rather than bulk composition, controls the amount or type of dust grains observed in the cometary coma.

The number of comets observed in the visible will continue to increase in the future, presumably permitting additional taxonomic groupings to be determined. More exciting, however, is the prospect that, due to continued rapid improvements in instrumentation, taxonomic studies should be possible utilizing many additional cometary constituents which are only observed in the IR and microwave regions of the spectrum.

Acknowledgements

It is a pleasure to thank Mike A'Hearn and Bob Millis for innumerable discussions over many years on a wide range of cometary topics, including the contents of this review. I also thank David Osip for his many contributions, and the referees for their helpful comments. This work was supported by NASA Grant NAGW-1864.

References

A'Hearn, M. F., and Millis, R. L. : 1980, "Abundance Correlations Among Comets." *Astron. J.* **85**, 1528–1537.

A'Hearn, M. F., Millis, R. L., Schleicher, D. G., Osip, D. J., and Birch, P. V. : 1994, "Narrowband Photometry of Comets : A Database of 85 Comets." In preparation for submittal to *Icarus*.

Arpigny, C. : 1976, "Interpretation of Comet Spectra." In *The Study of Comets* (B. Donn *et al.*, Eds.) NASA SP-393, 797–838.

Beaver, J. E., Wagner, R. M., Schleicher, D. G., and Lutz, B. L. : 1990, "Anomalous Molecular Abundances and the Depletion of NH_2 in Comet P/Giacobini-Zinner." *Astrophys. J.* **360**, 696–701.

Bobrovnikoff, N. T. : 1927, "On the Spectra of Comets." *Astrophys. J.* **66**, 439–464.

Budzien, S. A., and Feldman, P. D. : 1992, "Upper Limits to the S_2 Abundance in Several Comets Observed with the International Ultraviolet Explorer." *Icarus* **99**, 143–152.

Cochran, A. L. : 1987, "Another Look at Abundance Correlations Among Comets." *Astron. J.* **93**, 231–238.

Cochran, A. L. : 1990, "Are All Comets Created Equal?" In *Asteroids, Comets, and Meteors III* (C. Lagerkvist *et al.*, Eds.), 281–284, Uppsala University.

Cochran, A. L., and Barker, E. S. : 1987, "Comet Giacobini-Zinner : A Normal Comet?" *Astron. J.* **93**, 239–243.

Cochran, A. L., Barker, E. S., Ramseyer, T. F., and Storrs, A. D. : 1992, "The McDonald Observatory Faint Comet Survey : Gas Production in 17 Comets." *Icarus* **98**, 151–162.

Cochran, A. L., Green, J. R., Barker, E. S. : 1989, "Are Low-Activity Comets Intrinsically Different from More Active Comets?" *Icarus* **79**, 125–144. Erratum : *Icarus* **80**, 446.

Colom, P., Crovisier, J., Bockellee-Morvan, D., Despois, D., and Paubert, G. : 1992, "Formaldehyde in Comets : I. Microwave Observations of P/Brorsen-Metcalf (1989 X), Austin (1990 V), and Levy (1990 XX)" *Astron. Astrophys.* **264**, 270–281.

Crovisier, J. : 1992, "The Infrared Spectrum of Comets." In *Infrared Astronomy with ISO,* (T. Encrenaz and M. Kessler, Eds.), 221–248, Nova Science Publishers.

Donn, B. : 1976, "The Nucleus : Panel Discussion." In *The Study of Comets* (B. Donn *et al.*, Eds.) NASA SP-393, 611–619.

Duncan, M., Quinn, T., and Tremaine, S. : 1988, "The Origin of Short-Period Comets." *Astrophys. J. Lett.* **328**, L69–L73.

Feldman, P. D. : 1983, "Ultraviolet Spectroscopy and the Composition of Cometary Ice." *Science* **219**, 347–354.

Feldman, P. D. : 1986, "Carbon Monoxide in Cometary Ice." In *Asteroids, Comets, and Meteors II* (C. Lagerkvist *et al.*, Eds.), 263–267, Uppsala University.

Fink, U. : 1992, "Comet Yanaka (1988r) : A New Class of Carbon–Poor Comet," *Science* **257**, 1926–1929.

Hoban, S., Mumma, M., Reuter, D. C., DiSanti, M., Joyce, R. R., and Storrs, A. : 1991, "A Tentative Identification of Methanol as the Progenitor of the 3.52–μm Emission Feature in Several Comets." *Icarus* **93**, 122–134.

Krankowsky, D. : 1991, "The Composition of Comets." In *Comets in the Post–Halley Era* (R. Newburn *et al.*, Eds.) vol. 2, 855–877, Kluwer Academic Press.

Kresák, L. : 1991, "Evidence for Physical Aging of Periodic Comets." In *Comets in the Post–Halley Era* (R. Newburn *et al.*, Eds.) vol. 1, 607–628, Kluwer Academic Press.

Lutz, B. L., Womack, M., and Wagner, R. M. : 1993, "Ion Abundances and Implications for Photochemistry in Comets Halley (1986 III) and Bradfield (1987 XXIX)." *Astrophys. J.* **407**, 402–411.

Meech, K. J. : 1991, "Physical Aging in Comets." In *Comets in the Post–Halley Era* (R. Newburn *et al.*, Eds.) vol. 1, 629–669, Kluwer Academic Press.

Mumma, M. J., Hoban, S., Reuter, D. C., and DiSanti, M. : 1993, "Methanol in Recent Comets : Evidence for Two Distinct Cometary Populations." In *Abstracts for IAU Symposium 160 : Asteroids, Comets, Meteors 1993*, LPI Contribution No. 810, 227.

Mumma, M. J., Weissman, P. R, Stern, S. A. : 1993, "Comets and the Origin of the Solar System : Reading the Rosetta Stone." In *Protostars and Planets III* (E. Levy and J. Lunine, Eds.), 1172–1252, University of Arizona Press.

Newburn, R. L., and Spinrad, H. : 1984, "Spectrophotometry of 17 Comets. I. The Emission Features." *Astron. J.* **89**, 289–309.

Newburn, R. L., and Spinrad, H. : 1989, "Spectrophotometry of 25 Comets : Post-Halley Updates for 17 Comets Plus New Observations for Eight Additional Comets." *Astron. J.* **97**, 552–569.

Rickman, H., Kamél, L., Froeschlé, C., and Festou, M. C. : 1991, "Nongravitional Effects and the Aging of Periodic Comets." *Astron. J.* **102**, 1446–1463.

Schleicher, D. G., Bus, S. J., and Osip, D. J. : 1993, "The Anomalous Molecular Abundances of Comet P/Wolf-Harrington." *Icarus* **104**, 157–166.

Schleicher, D. G., Millis, R. L., and Birch, P. V. : 1987, "Photometric Observations of Comet P/Giacobini–Zinner." *Astron. Astrophys.* **187**, 531–538.

Stern, S. A. : 1990, "ISM–Induced Erosion and Gas–Dynamical Drag in the Oort Cloud." *Icarus* **84**, 447–466.

Storrs, A. D., Cochran, A. L., and Barker, E. S. : 1992, "Spectrophotometry of the Continuum in 18 Comets." *Icarus* **98**, 163–178.

Tholen, D. J. : 1984, "Asteroid Taxonomy from Cluster Analysis of Photometry." Ph. D. Dissertation, University of Arizona.

Weaver, H. A., Feldman, P. D., Festou, M. C., A'Hearn, M. F., and Keller, H. U. : 1981, "IUE Observations of Faint Comets." *Icarus* **47**, 449–463.

Whipple, F. L. : 1977, "The Constitution of Cometary Nuclei." In *Comets, Asteroids, Meteorites* (A. Delsemme, ed.), 25–34, Univ. of Toledo Press.

Yamamoto, T. : 1985, "Formation History and Environment of Cometary Nuclei." In *Ices in the Solar System* (J. Klinger *et al.*, Eds.), 205–219, D. Reidel Publishing Co.

SMALL BODIES AROUND OTHER STARS

BRUNO SICARDY
Université Paris 6 et Observatoire de Paris, DAEC/EUROPA
F-92195 Meudon Cédex Principal, France
E–mail SICARDY@MESIOA.OBSPM.CIRCE.FR

Abstract. We briefly review recent advances in the observation and study of planetary bodies in extra-solar systems. We summarize in particular the main physical properties of the β-Pictoris dust disk, and the status of new disk observations. Theoretical implications of infalling discrete bodies are considered, in particular, the existence of possible perturbing planet(s) causing this influx. Such planets could spectacularly disturb circumstellar dust disks, thus revealing themselves in spite of their intrinsic faintness as mere point sources. Finally, we describe the recent possible discovery of at least two planets around a pulsar. This underlines the potential existence of planets in rather exotic circumstances.

1. Introduction

It may seem rather paradoxical to study small bodies around remote objects, when it is usually so difficult to observe them around our own Sun! However, we will see in this brief review that in some specific areas we can detect planets, dust, or even comets near other stars. It is not our aim to review the theoretical work on planetary formation, nor to give an exhaustive list of observational techniques to detect circumstellar material. These tasks would in effect take much more that the few pages allowed here (see Levy and Lunine, 1993). Rather, we present recent progress in this area, and give relevant references.

The regions surrounding young stellar objects are natural sites for searching for circumstellar material in general, and extra-solar planets in particular; they are important, since they represent cocoons in which planets like those in our solar system may form. The frequency of protoplanetary systems provides a fundamental clue to better understand the origin of our own solar system. We now live a quite special epoch, because of the increasing quality of observation techniques, and in particular because the infrared (IR) is revealing colder and colder material around stars.

A good example of "cold" material detection is given by the IR excess observed around the white dwarf G29-38 (Zuckerman and Becklin, 1987, Greenstein, 1988). Modelling the spectrum of the white dwarf, these authors show that the IR flux between 2 and 5 μm is actually a signature of non-stellar material (brown dwarf or dust) around the primary. IR images of the white dwarf GD 165 subsequently showed the presence of a low-temperature companion, also interpreted as a brown dwarf with mass between 0.06 and 0.08 solar masses (Becklin and Zuckerman, 1988). Data from the Infrared Astronomical Satellite (IRAS) have now shown that most of the nearby A, F, G stars exhibit an infrared excess suggesting the presence of circumstellar dust (Backman and Gillett, 1987, Auman, 1988). Direct imaging or spectroscopic observations from the ground show that between 25 and 50% of pre-main-sequence stars and T-Tauri stars have detectable circumstellar disks (see

A. Milani et al. (eds.), Asteroids, Comets, Meteors 1993, 429–442.

reviews by Beckwith *et al.*, 1990, Beckwith and Sargent, 1993, Basri and Bertout, 1993).

As an another example we cite the ground-based detection of an IR excess around the IRAS source HD 98800 (Zuckerman and Becklin, 1993*a*). This shows that HD 98800 is surrounded, within a few AU, by an amount of dust which is about six orders of magnitude larger than the zodiacal dust in the inner solar system. In another recent observation, the Hubble Space Telescope has revealed that nearly half of the stars in the Orion nebula have circumstellar disks (O'Dell *et al.*, 1993). Among these kinds of object, the β-Pictoris disk has been extensively observed, because of its larger intrinsic brightness (Smith and Terrile, 1984, 1987, Paresce and Burrows, 1987, Telesco *et al.*, 1988, Lecavelier des Etangs *et al.*, 1993, and see the review by Norman et Paresce, 1989, and Paresce, 1992).

As we now see, complementary observations can uncover several kinds of planetary material around these stars. First, the dust can be revealed either by direct imagery or through infrared excess measurements (Section 2 and 3). Second, more "exotic" detections are provided by spectroscopic evidence for infalling bodies, a process reviewed in more detail in Section 4. Also, we will see in Section 5 that planetary perturbations on a dust disk could betray the existence of planets around some of these stars. Finally, we describe in Section 6 what could be the first discovery of extra-solar planets around pulsars.

2. The β-Pictoris dust disk : physical properties and implications

The β-Pictoris star is believed to be a young A5 dwarf with an age less than $\sim 2 \times 10^8$ years, located at 17 parsecs from the Earth, with a mass of 1.5 $M_\odot$, and a luminosity of 6 $L_\odot$ (Norman and Paresce, 1989, Paresce, 1991). Coronographic images show a nearly edge-on disk with an extension of more than 1000 AU (Smith and Terrile, 1987), and an opening angle of 8° (Artymowicz *et al.*, 1989, and see figure 1). It is probable that this disk represents a unique example of an early planetary system. In a protoplanetary disk, the dust is either primordial (nebula condensates), or produced by collisions between larger planetesimals. The age of β-Pictoris, and the tenuous density of the gas, argue in favour of the second mechanism. According to Zuckerman and Becklin (1993*b*), this should be also the case for Vega and Fomalhaut. Then, the disk corresponds to the short phase of proto-planetary evolution, after the gas shell has been ejected and when the planetesimals are still accreting into a few larger bodies. The opening angle of the dust distribution also argues in favour of larger bodies stirring the disk and providing dust through collisions.

The images furthermore reveal an asymmetry in the structure of the two ansae of the disk at large distances, the northeast projection extending to more than 1100 AU, and the southwest projection extending to only $\sim$ 900 AU (Smith and Terrile, 1987). The inner part of the disk also exhibits an asymmetry in brightness, but reversed with respect to the outer regions (Vidal-Madjar *et al.*, 1992, Lecavelier des Etangs *et al.*, 1993). These latter observations show that the disk colour drops in the blue by a factor 4 from 75 AU to 30 AU, while the disk colour is neutral in V, R and Ic. This reddening could be explained by more and more dusty ice particles when going inward.

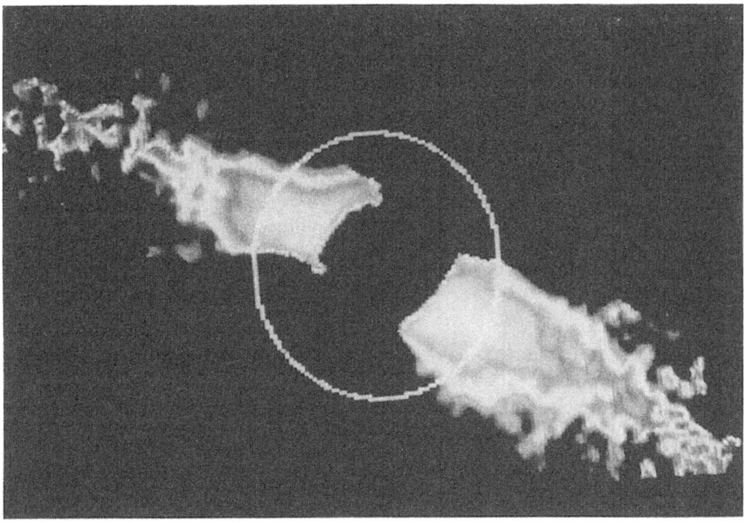

Fig. 1. The central part of the β-Pictoris disk, imaged in the V filter by an anti-blooming CCD camera (see Vidal-Madjar *et al.*, 1992 and Lecavelier des Etangs *et al.*, 1993, for detail). North is up and a 6-arcsec circle is drawn. The disk image has been radially flattened to strengthen the weakest parts. The disk is clearly detected down to 2.5 arcsec ($\sim$ 40 AU) from the star.

The combination of these data can be used to constrain the radial distribution of the dust around the star, together with the size distribution of the grains. Although there is no unique solution, some general structural features of the disk have been recognized. In one model (Artymowicz *et al.*, 1989), the grain radii are in the range 1-20 μm, with a high albedo > 0.5. The minimum grain radius is confirmed by independent observations (Norman and Paresce, 1989). The combination of IRAS and coronographic data requires that the inner part of the disk is largely cleared. However, while the existence of this clearing zone seems to be well established, its size is very model dependent, with a radius varying between 5 and 15 AU for μm-sized particles (Artymowicz *et al.*, 1989). According to these authors, the optical depth could be as large as 7×10^{-3} at the densest part of the disk and the estimated total mass of the dust disk is then one lunar mass, according to their model.

More refined models have been proposed by Backman *et al.* (1992), using visible data and IRAS fluxes, plus IR ground-based observations (10 and 20 μm). These models require a two-component disk. An outer ($r > 80$ AU) disk of icy particles, and an inner ($r < 80$ AU) disk would be made of refractory material, with a significant deficit of material with respect to an inward extrapolation of the outer component. The minimum grain size in the inner disk would be in the range $\sim 0.4 \div 3\mu$m, assuming a power law with an index -3.5 for the size distribution of the grains (i.e. the number of particles with radii between s and $s + ds$ is $dn \propto s^{-3\ 5} ds$).

The normal optical depth of their preferred model is $\sim 5 \times 10^{-4}$ around 20 AU. Finally, the inner disk should also have an inner limit between 1 and 30 AU, defining an innermost void.

Complementary observations include millimetre data as well as optical and IR data. The models derived from these observations indicate that large particles (at least 5 mm and possibly more) should be present around β-Pictoris (Chini *et al.*, 1991). These models require an inner cavity of 26 AU and a disk mass of $\sim$ 0.45 Earth masses ($\sim$ 35 lunar masses). A similar conclusion concerning the particle sizes is reached by Zuckerman and Becklin (1993*b*), for Vega, Fomalhaut and β-Pictoris. Also, these authors estimate that the total masses of these dust disks, contained in particles with radii $\sim$ 0.3 mm, lie in the range 0.1-10 lunar masses.

3. Other circumstellar disks : detections, images and structures

In spite of careful searches, the β-Pictoris system is so far the only firmly confirmed circumstellar disk imaged in the visible. Nevertheless, many disks are currently detected around other stars using a variety of methods. For instance, infrared images obtained by speckle or new adaptive optics technology are now revealing more and more circumstellar material (see for instance Koresko *et al.*, 1993, Ménard *et al.*, 1993, and the detection of a disk-like structure around the pre-main sequence binary system Z CMa, Malbet *et al.*, 1993). Also, among objects suspected to be surrounded by a dust shell, α-Piscis Austrini (Fomalhaut), τ_1-Eridani, ϵ-Eridani and α-Lyrae (Vega) have IR and millimetre excesses interpreted by circumstellar disks (Chini *et al.*, 1990, 1991). Structures are derived for these disks, with inner radii estimated to 40, 53, 7 and 40 AU respectively for Fomalhaut, τ_1-Eri, ϵ-Eri and Vega. Similarly, IR excesses, photometric variabilities, UV emissions, CO and polarization maps betray the existence of accretion disks around several pre-main-sequence stars (Beckwith and Sargent, 1993, and Basri and Bertout, 1993).

An interesting and original detection of a disk is also provided by the eclipsing component of ϵ-Aurigae, a spectroscopic binary which undergoes a partial eclipse of 2 years, every 27 years. The structure of the eclipse light curve reveals not only a disk, observed almost edge-on, around the companion of ϵ-Aurigae, but also ring structures inside this disk (Ferluga, 1990). The radii and width of these rings (of the order of AU) argue in favor of a thin dust disk with an inner clearing zone, and gaps reminiscent of the structure of Saturn's rings. Interestingly enough, the secondary itself could be a close binary, separated by less than 5 AU (Lissauer and Backman, 1984). Dynamical effects of such a binary on the disk could lead to resonant interactions with the disk (Ibid.).

From a recent observation, Stern, Festou and Weintraub (1993) report that a dust disk may also have been reconstructed at millimetre wavelengths, around the star Fomalhaut. A raster scan made at 1.3 mm from the IRAM station in Spain, reveals a disk-shaped object around the star. The aspect ratio of the disk is $\sim$ 2, and its greatest angular elongation from Fomalhaut is at least one arc minute, i.e. more than 500 AU from the star in linear distance, using a distance of $\sim$ 7 pc from the Earth. This experiment detects cold ($\sim$ 20 K) dust particles, with radii large compared to 1 μm. The total mass of the observed disk, inferred from this observation, would be typically 0.01 to 0.1 Earth masses, comparable in order of magnitude, with the mass of the β-Pictoris disk.

Since this observation has not been fully reconfirmed yet, careful independent experiments are now required to study this object and others, if any.

4. Throwing comets on to stars

Spectroscopic observations of accreting gas on to young stars is an active field of research with recent new results, for example the detection of infalling gas around HD 256 (Lagrange-Henri *et al.*, 1990), around the Herbig Be proto-planetary system HD 45677 (Grady *et al.*, 1993), or around 51 Ophiuchi (Grady and Silvis, 1993). For all these objects, modelling of these spectroscopic observations is in ongoing progress.

Again, the β-Pictoris system has proved to be one of the best documented objects. In particular, UV spectroscopic observations of this star show transient red-shifted absorption lines in the stellar spectrum. These lines are highly variable, on time scales ranging from some hours to some days, and show typical infalling velocities of $\sim$ 30-40 km sec^{-1}, which may sometimes be as high as 300-400 km sec^{-1}. There appears to be periods of "activity" of infalling material (several events per weeks in 1985-1986, or since late 1989), with quiescent intervals, like in 1987 (Ferlet *et al.*, 1987, Lagrange-Henri *et al.*, 1988, 1989, 1992, Norman and Paresce, 1989, Bogges *et al.*, 1991, Beust, 1991). These authors discuss the possible origins of these transient features, favoring infalling material from an extended disk, rather than from a nearby stellar envelope.

Coherent models have been built, in which sublimating comets (typically 10 km in size) can explain the intensity of the various absorption lines, as the comets cross the stellar disk along the line of sight (Beust, 1991 and see figure 2). This raises interesting issues, in particular the possibility of throwing small bodies on to a star. This question is now addressed from a more dynamical point of view.

As pointed out by Beust *et al.* (1990, 1991), Beust and Tagger (1993), and Beust and Lissauer (1994), the systematically red-shifted lines indicate that the comets, if any, are always thrown on to the star on almost parabolic orbits, at roughly the same angle with respect to the observer. This suggests that "showers" of comets are occurring (Ferlet *et al.*, 1993), although their origin remains unclear. A possibility is that a planet is presently perturbing a cloud of comets, through close encounters (Beust *et al.*, 1991). However, a problem associated with such a model is that it requires a high relative velocity during the encounters, i.e. either many comets on highly eccentric orbits ($e >\sim 0.6$), or one planet with an eccentric orbit.

A possibility is that the cometary orbital eccentricities are excited through resonant motion with a planet, and are then thrown on to the star during encounters with a second planet. As pointed out by A. Milani (1992, private communication), a more economical model has also been proposed by Bailey *et al.* (1992) to explain the occurrence of sungrazing comets, like the Kreutz family, in our own solar system. In this model, Jupiter secularly perturbs comets, originally on highly inclined orbits ($i \approx 90°$). Averaging the problem, and keeping only secular perturbations, they show that the energy of a comet, i.e. its reciprocal semi-major axis

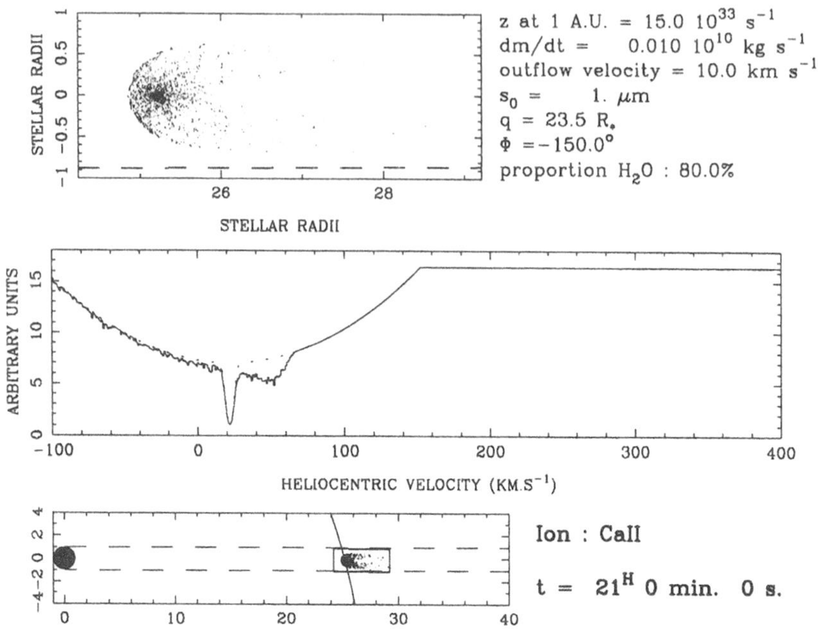

Fig. 2. Modelling the temporal variations the Ca II absorption line of β-Pictoris (from Beust *et al.*, 1990). The top panel shows the Ca II cloud around the nucleus of an infalling comet. The dashed line is the line of sight from the stellar edge to the observer (see the lower panel). The central panel is a synthetic spectrum of the Ca II absorption line, where the wavelength in abscissa has been translated in radial, heliocentric, velocity The lower panel shows a more general view of the comet orbit with respect to the star (black circle on the left). The box around the comet corresponds to what is shown in the upper panel.

a, is conserved, as well as its angular momentum perpendicular to Jupiter's orbit, i.e. $\sqrt{a(1-e^2)}.\cos(\imath)=$ constant. There is a third integral of motion (essentially the mean inverse distance of the comet to Jupiter, $< 1/\Delta >$), which makes this problem integrable. The integral curves may go to low inclination orbits ($\imath \approx 0$) for some appropriate initial conditions. The relation $\sqrt{1-e^2}.\cos(\imath)=$ constant thus requires that the eccentricity goes to unity as $\imath$ goes from 90° to 0°, so that the comet becomes a sungrazer. Bailey and colleagues then show that the probability that a comet enters the region of phase space eventually leading to $e \approx 1$ is rather large, so that sungrazing could be a much more common end-state than previously thought. In particular, the number of revolutions necessary to become sungrazer is of the order of 10^3, corresponding a rather short time scale (less than 10^5 years), i.e. at least ten time shorter than the dynamical ejection time scale (corresponding to a close encounter with Jupiter).

It would be interesting to see whether such a mechanism can apply to the case of β-Pictoris, or other stars. In particular, one should address the question of the total number of comets necessary to explain the frequency of events, and the mass of the hypothetical planet responsible for the showers. Another issue is to explain the existence of initially highly inclined cometary orbits. In any event, it seems important to include (hypothetical) planetary perturbations in order to fully describe the behaviour of small bodies around other stars. As we see in the following Section, this may be crucial for detecting planets, either through perturbations on the dust disk or through mutual perturbations between planets in the case of planets around pulsars.

More prospective work is in progress aimed at evaluating the effect of impacting bodies in stellar atmospheres. In particular, planetesimals could be detected while entering the deeper layers of young stars, explaining the flare activity of some of them (Andrews, 1991). More exotic cases have been studied, in which γ-ray bursts are caused by impacts of small bodies on to pulsars (Howard *et al.*, 1981, Harding and Leventhal, 1992). The question of planets around pulsars will be addressed in more detail in Section 6.

5. Dust-planet interaction : detecting a moderately small body

In this Section, we briefly address the question of planetary perturbations on a circumstellar dust disk. This problem may have important implications for understanding the behaviour of dust in our own solar system or in other systems (see the reviews by Dermott *et al.*, 1992 and Sicardy *et al.*, 1993), but also may serve to constrain the existence of planets (otherwise invisible) embedded in dust disks.

Dust disks immediately raise the question of time scales. Even without gas drag or interparticle collisions, μm-sized dust grains decay on to the star within a few million years through Poynting-Robertson (PR) drag (Sicardy *et al.*, 1993). A replenishing source is thus required to maintain such disks. In the course of this decay, planets may have an important influence in shaping the disk, especially by temporarily trapping particles at mean motion resonances. This may be the case both in our solar system (Jackson and Zook, 1989, 1992, Marzari *et al.*, 1991, Jayaraman and Dermott, 1993, Weidenschilling and Jackson, 1993), and in circumstellar disks (Scholl *et al.*, 1993, Lazzaro *et al.*, 1994, Roques *et al.*, 1994).

General conclusions on the behaviour of grains subjected to PR drag and perturbed by planets may be derived from the above works (see also Fig. 3) :

- Even though permanent trappings into resonances are *not* observed in any numerical experiments, the trapping time scale is comparable to the PR drag decay time scale, resulting in a possible accumulation of particles just outside the planet orbit.
- At typical planetary distances (e.g. ~ 20 AU) from a star like β-Pictoris, there is a critical planet mass of $\sim 10^{-5} M_\star$, i.e. about 5 Earth masses $M_\oplus$ (or 1/3 Uranian masses), above which trapping in mean motion resonances is very efficient for several millions of years (even though not permanently).
- Once they escape the resonances, the particles rapidly decay on to the star, due to their enhanced eccentricities in the resonance. This, combined with the long trapping time in resonances, leads to the creation an inner clearing zone by planets with mass larger than $\sim 5\ M_\oplus$.
- A moderate (10^{-2}) planet orbital eccentricity can create large scale azimuthal (arc-like) structures in the disk, see figure 3.

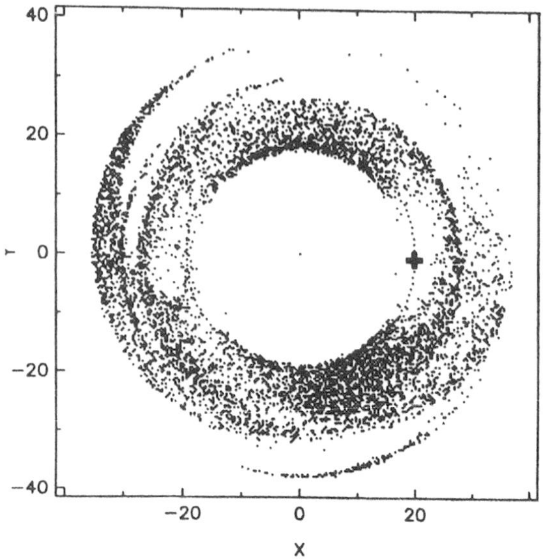

Fig. 3. Pole-on view of a simulated circumstellar disk around β-Pictoris (from Scholl *et al.*, 1993 and Roques *et al.*, 1994). The central star is represented by the dot, and the perturbing planet is at the black square on the right. Motion is counter-clockwise. Cartesian coordinates X and Y are in UA. The planet has a mass of 10^{-4} stellar masses, i.e. about 3 times the mass of Uranus, or half the mass of Saturn. Its orbital eccentricity is 0.01. The motion of 8192 particles is followed on a Connection Machine, taking into account the perturbation of the planet, and the effect of pressure of radiation and Poynting-Robertson drag. The ratio of the pressure of radiation to gravity is 0.3, corresponding to particles ~ 2 μm in radius around β-Pictoris. The particles, initially released between 32 and 33 AU, are shown here after ~ 0.6 Myears. Note the conspicuous arcs of material forced by the planet eccentricity. Note also the accumulation of matter ahead and behind the planet, as well as the void of particles around the planet.

Improving observational techniques, both from the ground and from space, could provide high-resolution images of circumstellar disks. It would be important to compare theoretical results against such data. In particular, Jovian or even Earth-like, planets are expected to be sufficiently massive to drive conspicuous structures in these disks (Paresce, 1992, Roques *et al.*, 1994). This would provide an efficient method for detecting extra-solar planets, otherwise too dim to be directly imaged.

6. Planets around pulsars

A planetary companion can be in principle detected through the motion of the star around the star-planet barycentre. Such detection is in practice quite difficult, although sub-milliarcsecond interferometry now yields very promising results (see e.g. Pan *et al.*, 1992).

The Doppler effect on spectral lines from a star directly provides its radial velocity with respect to the observer. Consequently, it may also reveal the presence of a low mass companion through motion of the star around the center of mass. This method has been used to detect a large Jovian planet, or a brown dwarf, orbiting in 84 days, at $\sim$ 0.4 AU around the star HD 114762 (Latham *et al.*, 1987).

Using the same approach, the detection of Earth-like planets around pulsars is a remarkable consequence of the amazing stability of the rotational period P of the pulsar. Typical values for the time derivative, $\dot{P}$, of millisecond pulsars are less than 10^{-15}, and as low as 1.21×10^{-19} for the pulsar PSR1257+12 around which a planetary system may have been detected (Wolszczan and Frail, 1992, and see below).

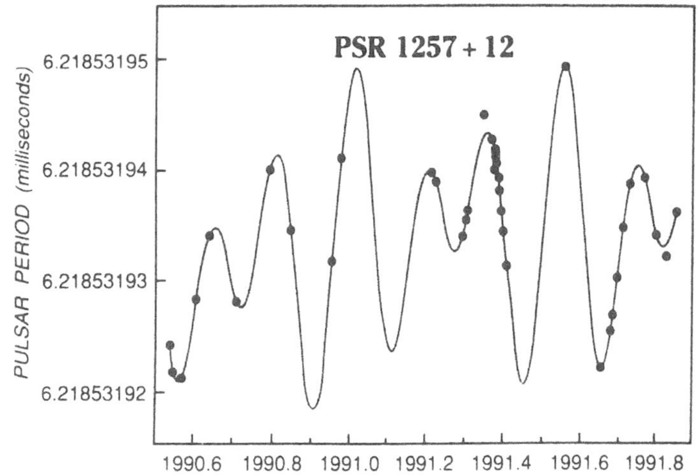

Fig. 4. The motion of the pulsar PSR 1257+12 around the center of mass of the system pulsar + planets causes a modulation of the apparent period of the radio pulses (in milliseconds) with time (in years). The dots represent the data, while the solid curve is the prediction of the model with two planets, as described in the text (Wolszczan and Frail, 1992, from *Sky & Telescope*, Fienberg, 1992).

This stability allows one to detect small modulations of the times of arrival (TOA) of radio pulses from the pulsar, forced by the periodic gravitational pull of small surrounding planets (see figure 4). More precisely, the motion of the pulsar around the barycentre of the system pulsar + companion(s) induces small, but detectable delays and advances of the TOA's, a simple version of the Doppler effect. The periodicity of these time residuals (with respect to a unperturbed pulsar) combined with the laws of celestial mechanics yields the "projected" mass of the companion, $m.\sin(i)$ and its "projected" semi-major axis $a.\sin(i)$, where i is the inclination of the orbital plane of the companion with respect to the plane of the sky. The factor $\sin(i)$ obviously arises because the TOA residuals are only sensitive to pulsar displacements along the line of sight.

A first report of TOA modulation was given by Bailes *et al.* (1991), who deduced the existence of a Uranus-like planet around the pulsar PSR1839-10 with an orbital period of 6 months for the planet. As later became apparent, however, the 6 month modulation was actually the imprint of the Earth motion. The latter was inaccurately removed from the observations, due to an offset of 7 arcmin on the actual position of the pulsar in the sky (Lyne and Bailes, 1992, Fienberg, 1992). An independent observation of a millisecond pulsar, PSR1257+12, led to the discovery of a *double* modulation, with respective periods of 98.2 and 66.6 days (Wolszczan and Frail, 1992). The projected masses of the putative planets causing these modulations are 2.8 and 3.4 $M_{\oplus}$, with projected semi-major axes of 0.47 and 0.36 AU, respectively. The orbital eccentricities are also detectable, with values of 0.020 $\pm$ 0.006 and 0.022 $\pm$ 0.007 for each planet.

An unexpected (and lucky) confirmation of this detection could be provided in the near future by celestial mechanics. The ratio between the two periods, $\sim$ 1.48, put the planets near a 3 to 2 mean-motion resonance. Then, mutual perturbations of the two bodies may build up over several years, yielding a slow, but large, modulation of the TOA's with respect to a model in which the planets would not interact at all. Such an effect would be undetectable were the planets not in resonant interaction. This slow modulation would provide irrefutable proof of the presence of the two planets and it would also yield the *absolute* value of the planet and pulsar masses, and thus the orbital inclination i, and the absolute values of the semi-major axes of both planets (Rasio *et al.*, 1992).

Malhotra *et al.* (1992), and Malhotra (1993), have shown that for values of $1/\sin(i)$ larger than about 10, the planets are actually in "exact" resonance. More precisely, the critical angle of resonance $3\lambda_1 - 2\lambda_2 - \tilde{\omega}_1$ (where λ is the mean longitude, $\tilde{\omega}$ is the longitude of periapse, and the indices refer respectively to each planet) then librates about 0 or 180°. The orbital eccentricities of both bodies also undergo excursions large enough to be easily detected on the TOA's modulations, over periods of the order of 10 years. However, as analyzed by Peale (1993), the present observations already rule out masses corresponding to values of $1/\sin(i)$ larger than about 4. This author also points out that the accuracy on the TOA's is such that an observational interval of $\sim$ 1000 days is necessary to detect the effect of mutual perturbations between the planets.

An interesting issue raised by these detections, if confirmed, is the possibility of accreting planets around catastrophically formed objects like pulsars (see the review by Fienberg, 1992). This was doubtful in the case of pulsar PSR1829-10, in view of its its estimated young age of $\sim 10^6$ years, given by its spin-down time $P/\dot{P}$. Nevertheless, some models were proposed in which a tiny fraction of the supernova remnant, which fell back around the pulsar, provided material for planetary formation, on a one-million years time scale or less (Lin *et al.*, 1991). Another possibility is that a former companion of the pulsar has been ablated by the latter (Krolik, 1991). The short time scale problem is avoided in the case of the millisecond pulsar PSR1257+12, whose spin-down age is of the order of 10^9 years.

This large time allows for more complex scenarios to be investigated. For instance, Tavani and Brookshaw (1992) propose that the original pulsar vaporizes a stellar companion. The ablated material than spirals outward, forming a disk from which planets accrete over several millions years.

New candidates are now on the list of pulsars with planets. For instance, a sub-Jovian planet, orbiting at about 7 AU from PSR 1620-26, could explain the anomalous spin period second derivative of the pulsar (Sigurdsson, 1993). According to this author, the presence of the planet around the pulsar could be the result of a capture during an encounter with a main sequence star, around which the planet previously revolved.

7. Conclusions

Improving observational techniques reveal complex circumstellar systems, where not only dust but also comets and probably planets, all interact. The increasing amount of data which is going to be collected on these objects in the near future will allow to fill the many gaps existing in our own solar system formation theory. Also, it is interesting to note that the discovery of planets, and in general, cold material, in exotic environments (pulsars, white dwarf, multiple systems, etc...) implies that planetary bodies may exist in situations greatly different from those previously expected. Ongoing observational efforts could thus show that small bodies are rather common in the universe, after all.

Acknowledgements

The author thanks Jean Schneider for providing many references on the various topics considered here, and Mark E. Bailey for a very careful reading of this paper. This work was partly supported by the Programme National de Planétologie.

References

Andrews, A.D. : 1991. "Investigation of micro-flaring and secular and quasi-periodic variations in dMe flare stars." ***Astron. Astrophys.***, **245**,219–231.

Artymowicz P., Burrows C. and Paresce, F. : 1989, "The structure of the Beta Pictoris circumstellar disk from combined *IRAS* and coronographic observations." ***Astrophys. J.***, **337**, 494–513.

Aumann, H : 1988, "Spectral class distribution of circumstellar material in main-sequence stars", ***Astron. J***, **96**, 1415–1419.

Backman D.E. and Gillett, F.C. : 1987, "Exploring the infrared : IRAS observations of the main sequence." In ***Cool stars, stellar systems and the Sun*** (Linsky, J.L. and Stencel, R.E. Eds.), 340–350, Springer-Verlag.

Backman, D.E., Gillet, F.C. and Witteborn, F.C. : 1992. "Infrared observations and thermal models of the β-Pictoris disk." ***Astrophys. J.***, **385**,670–679.

Bailes, M., Lyne, A.G. and Shemar, S.L. : 1991, "A planet orbiting the neutron star PSR 1829-10." ***Nature***,352,311–313.

Bailey, M.E., Chambers, J.E and Hahn, G. : 1992, "Origin of sungrazers : a frequent cometary end-state." ***Astron. Astrophys.***, **257**,315–322.

Basri, G. and Bertout, C. : 1993, "T-Tauri stars and their accretion disks." In ***Protostars and planets III*** (Levy, E.H. and Lunine, J.I. Eds.), 543–566,

Becklin, E.E. and Zuckerman, B. : 1988. "A low-temperature companion to a white dwarf star." *Nature*,**336**,656–658.

Beckwith, S.V.W., Sargent, A.I., Chini, R.S. and Günsten, R. : 1990. "A survey for circumstellar disks around young stars." *Astrophys. J.*, **99**, 924–945.

Beckwith, S.V.W. and Sargent, A.I. : 1993, "The occurrence and properties of disks around young stars." In *Protostars and planets III* (Levy, E.H. and Lunine, J.I. Eds.), 521–541, Univ. of Arizona Press.

Beust, H., Lagrange-Henri, A-M., Vidal-Madjar, A. and Ferlet, R. : 1990, "The β-Pictoris circum-stellar disk. X. Numerical simulations of infalling evaporating bodies." *Astron. Astrophys.*, **236**,202–216.

Beust, H : 1991, "Dynamique interne du disque protoplanétaire autour de l'étoile β-Pictoris." In *Thèse de Doctorat*, Univ. Paris 7.

Beust, H., Vidal-Madjar, A., Ferlet R. and Lagrange-Henri, A-M. : 1991, "The β-Pictoris circum-stellar disk. XII. Planetary perturbations in the disk and star-grazing bodies." *Astron. Astrophys.*, **247**,505–515.

Beust, H. and Tagger, M. : 1993, "A hydrodynamical model for infalling evaporating bodies in the β-Pictoris circumstellar disk." *Icarus*, **106**,42–58.

Beust, H. and Lissauer, J.J. : 1994. "The effect of stellar rotation on comets orbiting β-Pictoris." *Icarus*, *submitted.*

Boggess A., Bruhweiler F.C., Grady, C.A., Ebbets D., Kondo, Y., Trafton, L. M., Brandt, J. and Heap, S.R. : 1991, "First results from the Goddard high-resolution spectrograph : resolved velocity and density structure in the β-Pictoris circumstellar gas." *Astrophys. J.*, **377**, L49–L52.

Chini R., Krügel E. and Kreysa, E. : 1990, "Large dust particles around main sequence stars." *Astron. Astrophys.*, **227**, L5–L8.

Chini R., Krügel E., Shustov B., Tutukov A. and Kreysa, E. : 1991, "Dust disks around Vega-type stars." *Astron. Astrophys.*, **252**, 220–228.

Dermott, S.F., Gomes, R.F., Durda, D.D., Gustafson, B.Å.S, Jayaraman, S., Xu, Y.L. and Nicholson, P.D. : 1992. "Dynamics of the Zodiacal Cloud." In *Chaos, resonance and collective dynamical phenomena in the solar system*, IAU Symp. No. 152 (S. Ferraz-Mello Ed.),333–347, Kluwer Academic Publishers.

Ferlet, R., Hobbs, L.M. and Vidal-Madjar, A. : 1987. "The Beta Pictoris circumstellar disk V. Time variations of the Ca II-K line." *Astron. Astrophys.*, **185**,267–270.

Ferlet, R., Lagrange-Henri, A-M., Beust, H., Vitry, R., Zimmerman, J-P., Martin, M., Char, S., Belmahdi, M., Clavier, J-P., Coupiac, P., Foing, B., Sèvre, F. and Vidal-Madjar, A. : 1993. "The β-Pictoris protoplanetary system XIV. Simultaneous observations of the Ca II H and K lines; evidence for diffuse and broad absorption features." *Astron. Astrophys.*, **267**,137–144.

Ferluga, S. : 1990. "Espilon Aurigae I. Multi-ring structure of the eclipsing body." *Astron. Astrophys.*, **238**,270–278.

Fienberg, R.T. : 1992, "Pulsars, planets and pathos." *Sky and Telescope* **83**,493–495.

Grady, C.A. and Silvis, J.M. : 1993. "The circumstellar gas surrounding 51 Ophiuchus : a candidate proto-planetary system similar to β-Pictoris." *Astrophys. J.*, **402**,L61–L64.

Grady, C.A., Bjorkman, K.S., Shepherd, D., Shulte-Ladbeck, R.E., Pérez, M.R., de Winter, D. and Thé, P.S. : 1993. "Detection of accreting gas toward HD 45677 : a newly recognized, Herbig Be proto-planetary system." *Astrophys. J.*, **415**,L39–L42.

Greenstein, J.L. : 1988. "The companion of the white dwarf G29-38 as a brown dwarf." *Astron. J.*, **95**,1494–1504.

Harding, A.K. and Leventhal, M. : 1992. "Can accretion on to isolated neutron stars produce γ-ray bursts ?" *Nature*, **357**,388–389.

Howard, W.M., Wilson, J.R. and Barton, R.T. : 1981. "Radiation from an asteroid-neutron star collision." *Astrophys. J.*, **249**,302–307.

Jackson A.A. and Zook, H.A. : 1989. "A solar system dust ring with the Earth as its shepherd." *Nature*, **337**,629–631.

Jackson A.A. and Zook, H.A. : 1992. "Orbital evolution of dust particles from comets and asteroids." *Icarus*, **97**,70–84.

Jayaraman, S. and Dermott, S.F. : 1993. "Near-Earth resonance structure of the Zodiacal Cloud." In ***Asteroids, Comets, Meteors***, IAU Symp. No. 160, 145. [Abstract]

Koresko, C.D., Beckwith, S., Ghez, A.M., Matthews, K., Herbst, T.M. and Smith, D.A. : 1993. "Infrared images of Monoceros R2 IRS 3 : evidence for a circumstellar disk." *Astron. J.*, **105**,1481–1486.

Krolik, J.H. : 1991, "Creation by stellar ablation of the low-mass companion to pulsar 1829-10." *Nature*, **353**,829–831.

Lagrange-Henri, A-M., Vidal-Madjar, A. and Ferlet, R. : 1988. "The β-Pictoris circumstellar disk VI. Evidence for material falling on to the star." *Astron. Astrophys.*, **190**, 275–282.

Lagrange-Henri, A-M., Beust, H., Ferlet, R. and Vidal-Madjar, A. : 1989. "The β-Pictoris circumstellar disk VIII. Evidence for a clumpy structure of the infalling gas." *Astron. Astrophys.*, **215**,L5–L8.

Lagrange-Henri, A-M., Beust, H., Ferlet, R., Hobbs, L.M. and Vidal-Madjar, A. : 1990. "HR10 : a new β-Pictoris-like star ?". *Astron. Astrophys.*, **227**,L13–L16.

Lagrange-Henri, A-M., Gosset, E., Beust, H., Ferlet, R. and Vidal-Madjar, A. : 1992. "The β-Pictoris circumstellar disk VIII. Survey of the variable Ca II lines." *Astron. Astrophys.*, **246**,637–653.

Latham, D.W., Mazeh, T., Stefanik, R.P., Mayor, M. and G. Burki : 1987. "The unseen companion of HD 114762 : a probable brown dwarf." *Nature*, **339**,38–40.

Lazzaro D., Sicardy B., Roques, F. and Greenberg, R. : 1994. "Is there a planet around β-Pictoris? Perturbation of a planet on a circumstellar dust disk. II. Analytical model." *Icarus*, *in press*.

Lecavelier des Etang, A., Perrin, G., Ferlet, R., Vidal-Madjar, A., Colas, F., Buil, C., Sèvre, F., Arlot, J.-E., Beust, H., Lagrange-Henri, A-M., Lecacheux, J., Deleuil, M. and C. Gry : 1993, "Observations of the central part of the β-Pictoris disk with an anti-blooming CCD." *Astron. Astrophys.*, **274**,887–882.

Lin, D.N.C., Woosley, S.E. and P.H. Bodenheimer : 1991, "Formation of a planet orbiting pulsar 1829-10 from the debris of a supernova explosion." *Nature*, **353**,827–829.

Lissauer, J.J. and Backman, D.E. : 1984. "The Epsilon Aurigae secondary : a binary embedded within a disk ?". *Astrophys. J.*, **286**,L39–L41.

Levy, E.H. and Lunine, J.I., Eds., 1993, "Protostars and planets III." Univ. of Arizona Press.

Lyne, A.G. and M. Bailes : 1992, "No planet orbiting PSR 182-10." *Nature*, **355**,213.

Malbet, F., Rigaut, F., Bertout, C. and Léna, P. : 1993. "Detection of a 400 AU disk-like structure surrounding the young stellar object of Z CMa." *Astron. Astrophys.*, **271**,L9–L12.

Malhotra, R., Black, D., Eck, A. and A. Jackson : 1992, "Resonant orbital evolution in the putative planetary system of PSR 1257+12." *Nature*, **356**,583–585.

Malhotra, R. : 1993, "Orbital dynamics of PSR 1257+12 and its two planetary companions." In ***Planets around pulsars*** (J.A. Phillips, S.E. Thorsett, and S.R. Kulkarni, Eds.), workshop held at the California Institute of California, Pasadena, CA, April 30-May 1, 1992, *in press*.

Marzari, F., Weidenschilling, S.J., Fabris, M. and Vanzani, V. : 1991. "Temporary trapping of dust particles into orbital resonances with the Earth." *Lunar Planet. Sci. XXII*,861–862.[Abstract]

Ménard, F., Monin, J-L., Angelucci, F. and Rouan, D. : 1993. "Disks around pre-main-sequence binary systems : the case of Haro 6-10." *Astrophys. J.*, **414**,L117–L120.

Norman C. A. and Paresce, F. : 1989, "Circumstellar material around nearby stars : clues to the formation of planetary systems." In ***The Formation and Evolution of Planetary Systems*** (Weaver, H. and Danly, L. Eds.), 151–169, Cambridge Univ. Press.

O'Dell C. R., Zheng, W., Xi-Hai, H. : 1993. "Discovery of new objects in the Orion nebula on HST images : Shock, compact sources and protoplanetary disks." *Astrophys. J.*, **410**, 696–700.

Pan, X., Shao, M., Colavita, M.M., Armstrong, J.T., Mozurkewich, D., Vivekanand, M., Denison, C.S., Simon, R.S. and Johnston, K.J. : 1992. "Determination of the visual orbit of the spectroscopic α Andromedae with submilliarcsecond precision." *Astrophys. J.*, **384**,624–633.

Paresce, F. and Burrows, C. : 1987, "Broad-band imaging of the Beta Pictoris circumstellar disk." *Astrophys. J.*, **319**, L23–L25.

Paresce F. : 1991, "On the evolutionary status of β-Pictoris." *Astron. Astrophys.*, **247**, L25–L27.

Paresce, F. : 1992, "The search for extra-solar planetary systems." *Adv. Space res.*, **12**, (4)157–(4)167.

Peale, S.J. : 1993, "On the verification of the planetary system around PSR 1257+1." *Astron. J.*, **105**,1562–1570.

Rasio, F.A., Nicholson, P.D., Shapiro, S.L. and S.A. Teukolsky : 1992, "An observational test for the existence of a planetary system orbiting PSR 1257+12, *Nature*, **355**,325–326.

Roques, F., Scholl, H., Sicardy B. and Smith, B.A. : 1994. "Is there a planet around β-Pictoris? Perturbation of a planet on a circumstellar dust disk. II. The numerical model." *Icarus*, *in press*.

Scholl H., Roques, F. and Sicardy, B. : 1993. "Resonance trapping of circumstellar dust particles by an alleged planet." *Celest. Mech.*, **56**,381–393.

Sigurdsson, S. : 1993. "Genesis of a planet in Messier 4." *Astrophys. J.*, **415**,L43–L46.

Sicardy B., Beaugé, C., Ferraz-Mello, D., Lazzaro, D. and Roques, F. : 1993. "Capture of grains into resonances through Poynting-Robertson drag." *Celest. Mech.*, **57**,373–390.

Smith, B.A. and Terrile, R.J. : 1984, "Circumstellar disk around β-Pictoris" *Science*, **226**, 1421-1424.

Smith, B.A. and Terrile, R.J. : 1987, "The Beta Pictoris disk : recent optical observations." *Bull. Am. Astron. Soc.*, **19**, 829.

Stern, S.A., Festou, M.C. and Weintraub, D.A. : 1993, "α Piscis Austrini." *IAU Circ. 5732*

Stern, S.A., Festou, M.C. and Weintraub, D.A. : 1993, *Nature*, submitted.

Tavani, M. and L. Brookshaw : 1992, "The origin of planets orbiting millisecond pulsars." *Nature*, **356**,320–322.

Telesco, C.M., Becklin, E.E., Wolsrencroft, R.D. and R. Decher : 1988. "Resolution of the circumstellar disk of β-Pictoris at 10 and 20 μm." *Nature*, **335**,51–53.

Vidal-Madjar, A., Lecavelier des Etangs-Levallois, A., Perrin, G., Ferlet, R., Sèvre, F., Colas, F., Arlot, J.-E., Buil, C., Beust, H., Lagrange-Henri, A-M., Lecacheux : 1992, "Observations of the central part of the β-Pictoris disk with an anti-blooming CCD." *The Messenger*, **69**,45–48.

Weidenschilling S.J. and Jackson, A.A. : 1988. "Orbital resonances and Poynting-Robertson Drag." *Icarus*, **104**,244–254.

Whitmire D.P., Matese, J.J. and Tomley, L.J. : 1988, "A brown dwarf companion as an explanation of the asymmetry in the Beta Pictoris disk." *Astron. Astrophys.*, **203**, L13–L15.

Wolszczan, A. and D.A. Frail : 1992, "A planetary system around the millisecond pulsar PSR 1257+12." *Nature*, **355**,145–147.

Zuckerman, B. and Becklin, E.E. : 1987. "Excess infrared radiation from a white dwarf-an orbiting brown dwarf ?" *Nature*, **330**, 138–140.

Zuckerman, B. and Becklin, E.E. : 1993*a*. "Infrared observations of the remarkable main-sequence star HD 98800." *Astrophys. J.*, **406**, L25–L28.

Zuckerman, B. and Becklin, E.E. : 1993*b*. "Submillimeter studies of main-sequence stars." *Astrophys. J.*, **414**, 793–802.

FORMATION OF OUTER SOLAR SYSTEM BODIES : COMETS AND PLANETESIMALS

MARK E. BAILEY

School of Computing and Mathematical Sciences, Liverpool John Moores University, Byrom Street, Liverpool L3 3AF, U.K.

Abstract. Observations of massive, extended discs around both pre-main-sequence and main-sequence stellar systems indicate that protoplanetary discs larger than the observed planetary system are a common phenomenon, while the existence of large comets suggests that the total cometary mass is much greater than previous estimates. Both observations suggest that theories of the origin of the solar system are best approached from the perspective provided by theories of star formation, in particular that the protoplanetary disc may have extended up to $\sim 10^3$ AU. A model with a surface density distribution similar to a minimum-mass solar nebula, but extending further in radius, is derived by considering the gravitational collapse of a uniform, slowly rotating molecular cloud. The boundary of the planetary system is determined not by lack of mass, as in previous 'mass-limited' models (*i.e.* those with a sharp decrease in surface density Σ beyond the radius of the observed planetary system), but instead by the increasing collision time between the comets or planetesimals initially formed by gravitational instability beyond the planetary zone. Bodies formed beyond ~ 50 AU have sizes on the order of 10^2 km and represent a collisionally unevolved population; they are composed of relatively small, unaltered clumps of interstellar dust and ices with individual sizes estimated to range up to ~ 10 m. By contrast, bodies formed closer in, for example in the Uranus-Neptune zone, consist of larger agglomerations of dust and ices with individual sizes ranging up to ~ 1 km. Planetesimals formed by gravitational instability at smaller heliocentric distances r are typically much smaller than those formed further out, the masses m_p being proportional to $\Sigma^3 r^6$, but subsequent collisional aggregation in the planetary region is expected to produce bodies with sizes ranging up to 10^2 km or more. In both cases the first-formed solid objects may be identified with observed cometary nuclei; some accumulate to produce the outer planets, but the majority are ejected, either to interstellar space or into the Oort cloud. Observed comets represent a dynamically well-mixed group from various sources; they are expected to comprise a heterogeneous mix of both pristine and relatively altered material and to have a broad mass distribution ranging up to the size of the largest planetesimals.

1. Introduction

Comets and planetesimals play a key role in discussions of the origin of the solar system. They have sizes typically in the range 1–100 km, intermediate between interstellar grains and planets, and provide crucial clues to the evolution of interstellar dust and the processes leading to the formation of the sun and planetary system. Theories of the origin of the latter span a very wide range, manifested at one extreme by the standard planetesimal hypothesis, in which a ready-formed sun is assumed to be surrounded by a dense gas-and-dust disc with surface density proportional to $r^{-3/2}$ or r^{-2} and normalized to reproduce the observed planetary

A. Milani et al. (eds.), Asteroids, Comets, Meteors 1993, 443–459.

masses corrected to solar abundances. Such a model is here defined to be 'mass-limited' if it has an outer radius comparable to the size of the observed planetary system. Depending on the surface density within the planetary system such a model may or may not also be a 'minimum-mass' nebula. A mass-limited minimum-mass model of the protoplanetary disc typically has a total nebular mass $\lesssim 0.05\,M_{\odot}$.

At the other extreme are theories which approach the problem of solar system formation from the perspective provided by star formation, and which begin either with a collapsing protostellar or molecular cloud, or with accretion of gas and dust on to a pre-main-sequence star from an extended protoplanetary disc. These theories discuss both star and planet formation and usually consider a much greater mass of circumstellar material, ejecting most of it to interstellar space. However, although the initial conditions for such a scenario are generally more easily defined, for example from observations of molecular clouds or star-forming regions, it is unclear precisely which initial conditions are most likely to lead to the formation of a single solar-mass star surrounded by the observed nine planets. The existence of comets introduces further complications (e.g. Bailey et al. 1990), the major uncertainty of principle being the theoretical possibility that comets may only recently have been captured into the solar system (within the past $\sim 10^6$–10^8 yr), having originally been formed in molecular clouds or star-forming regions (e.g. Napier 1990) and captured as a result of gravitational or other perturbations.

These different ideas for the origin of the solar system and for the significance or otherwise of comets in the overall cosmogonical picture each have various points in their favour. The main problem, in view of the wide range of possibilities (e.g. Williams 1974; Woolfson 1993), is to identify the particular grains of truth contained in each theory and to strike an appropriate balance between them. Comets play a key role in these discussions, with the most detailed theoretical studies being devoted to the standard planetesimal picture. This hypothesis has been particularly successful in unifying concepts relating to planet, planetesimal and comet formation and formation of the Oort cloud, but it faces potentially severe difficulties in resolving questions such as the timescale for formation of the planets (particularly Uranus and Neptune) and the total mass of comets in the Oort cloud. These problems suggest that the standard model should be revised to include aspects of the star-formation approach, although it is possible that the outcome of such a course might then be to lose some of the advantages of the former approach by decoupling theories of comet formation from those of the origin of planets (e.g. Hills 1982) and leaving open the question whether recently discovered bodies orbiting beyond Neptune, the so-called Kuiper-belt candidates, are best thought of as giant comets, primordial planetesimals, or merely as stray asteroids ejected from the inner solar throughout its 4 billion year lifetime.

This review primarily focusses on the standard planetesimal theory for the formation of comets in the protoplanetary disc, but argues for an approach placing it firmly in the context of star-formation theories for the origin of the solar system and the interstellar-dust model for the origin of comets. So far as the standard model is concerned, the principal difference is that the revised model accommodates a much larger, more massive protoplanetary disc, with an outer radius R_d on the order of 10^3 AU and a total mass of solids in the range 10^2–$10^3\,M_{\oplus}$. Such a disc is not mass-limited, in the sense of having a sharp cut-off in surface density beyond the observed planetary system, but the surface density within the planetary region is nevertheless close (*i.e.* within a factor of a few) to that of previously considered minimum-mass nebulae.

Evidence for the existence of extended discs is provided by observations of pre-main-sequence stars (*e.g.* Sargent 1989; Strom *et al.* 1989a,b; 1993; Weintraub *et al.* 1989; Beckwith *et al.* 1990; Beckwith and Sargent 1993; Montmerle *et al.* 1993; *cf.* Pringle 1989) and main-sequence stars such as Beta Pictoris and Vega (*e.g.* Aumann *et al.* 1984; Aumann 1985; Telesco *et al.* 1988; Artymowicz *et al.* 1989; Piirola *et al.* 1992; Bachman and Paresce 1993; Sicardy 1994). These data strongly suggest that the boundary of the solar system is not determined solely by a rapid decrease in surface density of the primordial disc near the edge of the observed planetary system; rather, the reason there are no major planets beyond Neptune is principally a result of accretion dynamics. In particular, the observed boundary may be understood as a consequence of the rapid increase in the collision time of the first solid bodies to be formed by gravitational instability on the outskirts of the protoplanetary disc : the edge of the planetary system occurs where the planetesimal collision timescale first exceeds $\sim 10^8$–10^9 yr. In addition to providing a theoretically attractive explanation for why the edge of the *planetary* system lies where it is, such a model has the significant advantage of accommodating a large initial mass of planetesimals at great heliocentric distances. Comets formed throughout the protoplanetary disc may be ejected by gravitational or other perturbations to produce both the Oort cloud and the massive inner core necessary to replenish the dynamically unstable outer layers of the system. Here we briefly review the components of such a theory of comet formation : the origin of a massive, extended disc; interstellar dust evolution and grain growth; and the argument based on cometary masses that the Oort cloud cannot have originated from planetesimals produced in a conventional mass-limited minimum-mass model of the protoplanetary disc.

2. Massive disc

There have been many recent reviews of the planetesimal theory, including planet formation in both the inner and outer solar system (up to the distance of Neptune) and comet formation in the outer planetary region extending roughly from Jupiter through Neptune (*e.g.* Safronov 1969; Öpik 1973; Horedt 1979; Wetherill 1980, 1989, 1990; Safronov and Ruzmaikina 1985; Greenberg 1989; Weidenschilling *et al.* 1989; Bailey *et al.* 1990; Weidenschilling and Cuzzi 1993; Lissauer and Stewart 1993; Lissauer 1993). According to this picture, the observed planetary masses when corrected to solar abundances and distributed over rings of appropriate width and heliocentric distance r imply an initial disc surface density $\Sigma(r) \gtrsim r^{-\alpha}$, where estimates of α lie in the range 1.5–2 (*e.g.* Weidenschilling 1977; Nakano 1987; Tremaine 1990) and the outer radius is on the order of 50 AU.

However, observations of pre-main-sequence stars demonstrate the existence of more massive protoplanetary discs than those usually envisaged in the planetesimal hypothesis, showing that such structures are a common feature of young stellar systems. Here we argue that a disc with a similar density profile to that of the standard model, but extending much further in radius, could naturally arise as a consequence of the gravitational collapse of a protostellar cloud. This supports an approach from the direction of star formation whilst retaining the principal advantages of the planetesimal picture (*cf.* Cameron 1962, 1978, 1985; Cameron and Pine 1973; Biermann and Michel 1978; Lin and Papaloizou 1985).

Consider a cold, slowly rotating molecular cloud with outer radius R_c in the range 0.1–1 pc, temperature $T \simeq 10\,\mathrm{K}$ and total mass M_c in the range 1–2 $M_\odot$. Following Mestel (1963) and Mestel and Ray (1985), disc formation may be assumed to occur in two phases : formation of a 'projected disc' with radius R_0 similar to that R_c of the initial molecular cloud through gravitational collapse parallel to the rotation axis, followed by gradual contraction of the projected disc towards centrifugal equilibrium by a process that conserves both the mass and angular momentum of separate fluid elements within the projected disc. The original cloud is assumed to be threaded by a weak interstellar magnetic field, and to be uniformly rotating with a net rotational velocity $v_{\mathrm{rot}}(\varpi) = \beta(\varpi/R_c)v_{\mathrm{circ}}(R_c)$, where $v_{\mathrm{circ}}(r)$ denotes the circular velocity at radius r from the cloud centre and ϖ is the axial distance in the equatorial plane.

This leads to formation of a quasi-static disc, supported in the radial direction by magnetic stresses and in the direction perpendicular to the plane by thermal motions and possibly turbulence driven by residual infall of gas. Decay of the magnetic field by ambipolar diffusion (Mestel and Spitzer 1956; McKee *et al.* 1993) or magnetic reconnection results in slow contraction of the disc, culminating in a rotationally supported Keplerian configuration with radius $R_d \ll R_0$. Ignoring possible non-homologous collapse of the cloud core, expected to lead to the formation of a central star surrounded by a relatively small central accretion disc, and assuming that the original molecular cloud is uniform or has only a weak degree of central concentration (*e.g.* $\rho_c \simeq$ constant or $\rho_c \propto r^{-1}$), results in a Keplerian disc with surface density approximately proportional to $r^{-\alpha}$, with $\alpha \simeq 3/2$.

For example, if the cloud is uniform the projected disc has a surface density $\Sigma_{\mathrm{proj}}(\varpi) = 3M_c(1-x^2)^{1/2}/4\pi R_c^2$, where $x = \varpi/R_c$, and conservation of mass during evolution from an initial radius ϖ to a final radius r leads to a surface density $\Sigma(r)$ given by $\Sigma(r) = (\varpi/r)(d\varpi/dr)\Sigma_{\mathrm{proj}}(\varpi)$ The relation between r and ϖ evidently depends on the distribution of angular momentum in the projected disc, which in turn depends on that in the cloud. If the original cloud is uniformly rotating, for example $v_{\mathrm{rot}}(\varpi) = \beta\varpi(GM_c/R_c^3)^{1/2}$, then detailed conservation of angular momentum implies $rv_{\mathrm{circ}}(r) = \varpi v_{\mathrm{rot}}(\varpi)$, and hence $(\varpi/r)(d\varpi/dr) = R_c^{3/2}r^{-3/2}/4\beta$. The final Keplerian disc thus has $\Sigma(r)$ close to the standard form, *i.e.*

$$\Sigma(r) = \frac{1}{4\beta}R_c^{3/2}r^{-3/2}\Sigma_{\mathrm{proj}}(\varpi(r)) \tag{1}$$

where Σ_{proj} is a weak function of radius, $r = \beta^2\varpi^4/R_c^3$ and $R_d = \beta^2 R_c$.

This argument suggests that extended circumstellar discs will have surface densities that are generally close to the standard form $\Sigma(r) \propto r^{-3/2}$, although their radial extent will be primarily determined by the size and degree of rotation of the primordial molecular cloud. For values of β in the approximate range 0.1–0.5, the outer radius $R_d = \beta^2 R_c$ is much larger than the ~ 50 AU size of the observed planetary system. Normalizing the surface density to 500–1000 kg m^{-2} at $r_0 = 10$ AU, inferred from the observed planetary mass distribution, thus indicates a total disc mass between 10 AU and 1000 AU in the respective ranges 0.064–0.12 $M_\odot$ or 0.016–0.032 $M_\odot$, for values of $\alpha = 3/2$ or 2. If the effective surface density of solids (*i.e.* rock + ice) is close to 2% of the total mass density, *i.e.* $\Sigma_s(r) \simeq 0.02\Sigma$, the total mass of solids ~ 100–800 $M_\oplus$, depending on the assumed surface density normalization and power-law index α. An extended protoplanetary disc with radial extent $R_d \simeq 10^3$ AU could easily contain several 100 $M_\oplus$ of dust and ice, not only representing a reservoir from which to make comets but also the dominant sink of heavy elements (apart from the sun) in the solar system.

3. Interstellar dust

3. 1. ORIGIN AND PRESTELLAR EVOLUTION

Following the initial condensation of solid material in gaseous outflows associated with late stages of stellar evolution (red giants, novae, supernovae; *e.g.* Bode 1988; Gehrz 1989; Tielens 1991), newly formed grains spend typically 10^7–10^9 yr in the interstellar medium before finally being incorporated into bodies the size of planetesimals, comets or asteroids. During this phase of prestellar evolution an individual interstellar dust grain will probably undergo many transitions between different components of the interstellar medium, for example from the diffuse high-temperature phase to the warm neutral phase; into a cool H I cloud or molecular cloud, and back again. During these excursions the grain may grow by accretion of volatiles or by inelastic grain-grain collisions, and fragments may be lost as a result of disruptive collisions in shocks, through sputtering, or simply by thermal evaporation in a locally more intense radiation field (Seab and Shull 1986; Seab 1988; Jenkins 1989; Tielens 1989; McKee 1989).

The resulting interstellar grain aggregates, modified by thermal and radiative processing, have a complex structure and abundances and chemistry which reflect the history of each grain (*e.g.* Greenberg 1988, 1989; Greenberg and Hage 1990; Clayton and Liffman 1988; Clayton *et al.* 1989; Liffman 1990). A few aggregates are bound to be larger than the average, and in general a percentage of interstellar particles will be much larger than the $\sim$0.25 μm upper limit imposed by the canonical Mathis *et al.* (1977) law based on observations of interstellar extinction (*cf.* Bailey 1987a,b, 1988, 1991a). Several lines of argument demonstrate the existence of exceptionally large grains ($\sim$1–100 μm) in or around dense interstellar clouds (*e.g.* Lefèvre 1974; Jura 1980; Elmegreen 1981; Mathis and Wallenhorst 1981; Elsässer *et al.* 1982; Bhatt 1986), while it is also important to emphasize that observations of interstellar dust near Jupiter (Grün *et al.* 1993) and of apparently hyperbolic meteors on Earth (Baggaley *et al.* 1993; Taylor *et al.* 1994; *cf.* Brophy 1991; Fogg 1990; Napier 1990) also support the notion that interstellar dust includes a significant proportion of relatively large grains. Although hyperbolic meteoroids could be produced locally by sputtering or by erosion from the surfaces of much larger interstellar particles (*e.g.* comets), it seems reasonable to assume that whatever their immediate provenance such particles were originally produced by accretion as a result of grain-grain collisions in dense regions of the interstellar medium. They probably have a fluffy 'fractal' structure, in accordance with expectations based on models of random aggregation (e.g. Meakin *et al.* 1985; Donn and Hughes 1986; Meakin and Donn 1988; Donn 1990; Jewitt and Meech 1988; Brooks 1990, 1992). By the time a given parcel of interstellar gas has evolved to produce a gravitationally unstable cloud prior to forming a new solar system, it seems certain that it will include a significant number of large grains.

3. 2. PROTOSTELLAR COLLAPSE

A marginally unstable cloud of mass $M_c \simeq 1\,M_\odot$ destined to form a star has a radius (cf. Larson and Starrfield 1971) on the order of $R_c \simeq 0.41 G M_c / c_s^2 \simeq 10^4$ AU, where $c_s = (kT/\bar{m})^{1/2} \simeq 186\,\mathrm{m\ s^{-1}}$ is the isothermal sound speed in the cloud and the numerical value assumes $T = 10\,\mathrm{K}$ and a mean particle mass $\bar{m} = 4.0 \times 10^{-27}$ kg, appropriate for a molecular cloud of solar composition. Assuming the cloud is approximately uniform with a mean mass-density of solids $\rho_s = \zeta\rho_c \simeq 0.02\rho_c$, grains of mean radius $a \simeq 0.1\,\mu\mathrm{m}$ and bulk density ρ_g, the grain-grain collision time $t_{\mathrm{coll}} = 1/(n_g 4\pi a^2 v_{\mathrm{rel}})$ is on the order of

$$t_{\mathrm{coll}} = 4.4 \times 10^5 \left(\frac{\rho_g}{1\mathrm{g\ cm^{-3}}}\right)\left(\frac{a}{0.1\,\mu\mathrm{m}}\right)\left(\frac{1\,\mathrm{m\ s^{-1}}}{v_{\mathrm{rel}}}\right)\left(\frac{10\,\mathrm{K}}{T}\right)^3\left(\frac{M_c}{1\,M_\odot}\right)^2 \quad \mathrm{yr} \quad (2)$$

comparable to the collapse timescale $\sim(G\rho_c)^{-1/2}$ for the cloud. The larger particles in the system will thus grow during collapse of the cloud, and a substantial proportion of the smallest grains will accrete into systematically larger particles (Burke and Silk 1976; Arnold 1977; Kessel'man 1978, 1979). Insofar as the chemistry of a collapsing protostellar cloud differs in detail from that in a dense molecular cloud (Fegley and Prinn 1989; Van Dishoeck *et al.* 1993; Prinn 1993), so too any molecular ices condensing on or within these prestellar grain aggregates will reflect the conditions of their formation. Comets or planetesimals are thus expected to contain grains and interstellar condensates that reflect the 'interstellar' environment at the time of the grains' initial formation and growth.

3. 3. PROTOPLANETARY DISC

Following the Mestel-Ray prescription for the origin of a flattened Keplerian disc, a uniform, slowly rotating cloud with initial radius $R_c \approx 10^4$ AU and rotational velocity $v_{\mathrm{rot}}(\varpi)$ will evolve to produce a massive, extended protoplanetary disc with surface density $\Sigma(r) \propto\!\!\sim r^{-3/2}$ and outer radius R_d on the order of 10^2–10^3 AU, where we assume $\beta \approx 0.1$–0.5. Interstellar grains are expected to coalesce throughout the collapse phase, producing composite particles with a loosely bound fractal structure and a tendency for the volatile compounds and ices to reside on the surfaces of the growing grains and to fill the voids in the overall grain structure. Whereas grains in the inner regions of the disc are likely to be destroyed by collisions or by heating associated with formation of the pre-main-sequence star and its surrounding accretion disc (e.g. Morfill and Völk 1984; Tscharnuter and Boss 1993), those in the outer region are expected to retain their cosmic chemical memory (Clayton

1982; Sandford 1989) and to have a complex, hierarchical physical structure which may be sampled *in situ* next century by the European cometary mission Rosetta.

4. Grain growth

4. 1. GAS DISC

Following these prestellar phases of grain growth, the process of accumulation of grains in the protoplanetary disc may be divided into two phases : that in the presence of gas, lasting $\sim 10^6$–10^7 yr (Strom *et al.* 1993), and that in a gas-free environment following dispersal of the gas disc. Considering the first of these, if we take a grain of radius a, which we assume to be somewhat larger than the mean radius a_f of the background 'field' population, then its collision cross-section with respect to the background grains is $\sigma_c = \pi(a + a_f)^2 \simeq \pi a^2$. Turbulence in the gas, expected to occur, will drive significant relative velocities $v_{\rm rel}$ between grains of different sizes (Völk *et al.* 1980; Mizuno 1989; Weidenschilling and Cuzzi 1993), leading to grain growth at a rate given by

$$\frac{da}{dt} \simeq \frac{\rho_s v_{\rm rel}}{4\rho_g} \tag{3}$$

Assuming that the field particles have a Maxwellian velocity distribution with a one-dimensional velocity dispersion c_f and that the large grains have a significantly smaller velocity dispersion owing to their greater mass, then $v_{\rm rel} \simeq (8/\pi)^{1/2} c_f$ and hydrostatic equilibrium of the small grains perpendicular to the plane results in a Gaussian density distribution normal to the plane, given by $\rho_s(z) = \rho_s(0)\exp(-z^2/2h_f^2)$ where $h_f = c_f/\Omega = c_f(r^3/GM_\odot)^{1/2}$. The total surface density of solids in the disc is thus $\Sigma_s = \sqrt{2\pi}\rho_s(0)h_f$, so

$$\frac{da}{dt} = \frac{\Sigma_s \Omega}{2\pi\rho_g} \tag{4}$$

Thus the grain radius at time t is given by $a(t) = a_0 + \Sigma_s t/P\rho_g$, where $P(r) = 2\pi/\Omega$ is the orbital period of grains at heliocentric distance r.

We now consider a general power-law model of the form $\Sigma_s = k_s r^{-\alpha}$ for the surface density of solids, allowing α to be either 3/2 or 2 and choosing the normalization so that the dust surface density at 10 AU is 10 kg m^{-2}, roughly corresponding to that of a minimum-mass nebula within the planetary system. Ignoring the initial grain radius, equation (4) implies $a(t) = k_s(GM_\odot)^{1/2} r^{-\alpha-3/2}/2\pi\rho_g$, or

$$a(t) = \begin{cases} 0.32\left(\frac{100\,\mathrm{AU}}{r}\right)^3\left(\frac{t}{1\,\mathrm{Myr}}\right) \quad \mathrm{m} \quad \text{and} \quad \alpha = 3/2 \\ 0.10\left(\frac{100\,\mathrm{AU}}{r}\right)^{7/2}\left(\frac{t}{1\,\mathrm{Myr}}\right) \quad \mathrm{m} \quad \text{and} \quad \alpha = 2 \end{cases} \tag{5}$$

where we have assumed $\Omega = (GM_\odot/r^3)^{1/2}$. We conclude that substantial growth of grains occurs before dispersal of the gas disc, producing a population of large particles close to the equatorial plane. Equation (5) shows that after 3 Myr the dimensions of particles in the Uranus-Neptune zone ($r \simeq 25$ AU) are on the order of 50 m whereas those formed at a distance of 100 AU or more have radii $\lesssim 1$ m. Turbulent grain growth in the presence of gas thus leads to a strong gradient in mean particle size, producing bodies with sizes ranging from ~1–10 km in the Jupiter-Saturn zone to less than 1 m beyond 100 AU. This result jointly arises from the increase in orbital period at large heliocentric distances and the assumed decrease in the primordial disc surface density.

4. 2. DUST DISC

The phase of random accumulation in the presence of nebular gas ends with dispersal of the latter after $\sim 10^6$–10^7 yr. The clock is reset and the 'initial' conditions become those appropriate to a quiescent dust disc containing cometary building blocks of various sizes, comprising ice-covered interstellar dust aggregates and molecules reflecting the history of the dust in the presence of gas. Although grain growth by coagulation will continue, it seems likely that due to decreasing grain-grain relative velocities (no longer driven by turbulence) the process of random accretion will be overtaken by gravitational instability. As discussed by Goldreich and Ward (1973), the dust disc fragments into subdiscs with a characteristic length-scale that depends on the surface density and heliocentric distance. The first gravitationally unstable modes have a wavelength $\lambda_p = 4\pi^2 G\Sigma_s/\Omega^2$, while the most unstable mode has a wavelength about half this (Binney and Tremaine 1987). In this way, the mass of the first-formed planetesimals is approximately

$$m_p \simeq \pi \left(\frac{\lambda_p}{8}\right)^2 \Sigma_s = \frac{\pi^5 \Sigma_s^3 r^6}{4M_\odot^2} = \frac{\pi^5 k_s^3 r^{6-3\alpha}}{4M_\odot^2} \tag{6}$$

where we have assumed $\Omega = (GM_\odot/r^3)^{1/2}$. The first planetesimals in a disc with initial surface density $\Sigma_s \propto r^{-\alpha}$ thus have masses m_p proportional to $r^{6-3\alpha}$. Since the timescale for the instability to grow is on the order of $1/\Omega$, or about the rotation period at radius r, *i.e.* on the order of 3×10^4 yr at $r = 1000$ AU, the disc is expected to fragment rapidly and to form a large number of separate planetesimals with individual masses at radius r on the order of

$$m_p = \begin{cases} 6.9 \times 10^{18} \left(\frac{r}{100\,\mathrm{AU}}\right)^{3/2} \ \mathrm{kg} \quad \text{and} \quad \alpha = 3/2 \\ 2.2 \times 10^{17} \ \mathrm{kg} \quad \text{and} \quad \alpha = 2 \end{cases} \tag{7}$$

According to this theory, the first solid bodies to be formed by gravitational instability in the outer solar system may be identified with cometary nuclei or planetesimals. If they have a bulk density on the order of unity their diameters are on the order of 10^2 km, close to those of the 6 recently discovered Kuiper-belt candidates, namely 1992 QB$_1$ (220 km), 1993 FW (280 km), 1993 RO (180 km), 1993 RP (90 km), 1993 SB (180 km) and 1993 SC (280 km), adopting conventional low albedos ($\simeq 0.04$) for these outer solar system objects. This must be counted as a success of the theory. It is also worth noting that whereas the predicted planetesimal building blocks are ice-covered interstellar grain aggregates ranging in size from ~1 km (in the Saturn-Uranus zone) down to <1 m (beyond 100 AU), the predicted size of the resulting planetesimals or cometary nuclei is either constant or a slowly increasing function of heliocentric distance (*cf.* Tremaine 1990). The bodies formed further out are expected to be larger, and to comprise smaller building blocks, than those formed closer to the observed planetary system; observed comets should have the same range of properties.

4. 3. COLLISIONAL EVOLUTION

Detailed discussions of the collisional evolution of planetesimals formed by gravitational instability have been given, for example, by Nakagawa *et al.* (1983), Greenberg (1985, 1989), Wetherill (1990), Lissauer (1993), Lissauer and Stewart (1993) and Weidenschilling and Cuzzi (1993). In the planetary zones ranging from Jupiter to Neptune the planetesimals, identified as cometary or protocometary nuclei, are either accreted on to planets or dynamically ejected to interstellar space or to the inner and outer regions of the Oort cloud (Öpik 1973; Duncan *et al.* 1987). The process has been reviewed, for example, by Bailey *et al.* (1990), Fernández and Ip (1991) and Duncan and Quinn (1993a,b). However, as discussed by Greenberg *et al.* (1984) these theories are not without difficulties, and it remains uncertain whether the standard planetesimal theory can produce both comets and planets in the timescale available and explain, for example, the time-evolution of the inner solar system impactor flux. A further important question is the value of the placement efficiency, defined as the ratio of the initial cometary mass in the Oort cloud divided by the original cometary mass required in the protoplanetary disc. Current estimates (*e.g.* Bailey *et al.* 1990) suggest that this quantity may be quite small, *i.e.* $\lesssim 20\%$. This implies that the original cometary mass in the protoplanetary disc must have been at least 5 times the initial mass of the Oort cloud.

In the outer solar system, a discussion of the evolution of planetesimals produced by gravitational instability has been given by Yamamoto and Kozasa (1988). Their theory also provides an explanation of the recently discovered outer solar system bodies 1992 QB_1 and 1993 FW (Yamamoto *et al.* 1993). Here we briefly review these ideas, emphasizing again the good agreement between the predicted sizes of the initial planetesimals and those of the observed objects, and the fact that such a theory provides a natural explanation for the lack of planet-sized bodies beyond a heliocentric distance on the order of 30–50 AU.

Following previous authors (Safronov 1969; Yamamoto and Kozasa 1988), the newly formed planetesimals have relative velocities $v_{\rm rel}$ determined by the competition between acceleration during close encounters and collisional damping. A simple kinetic-theory approach contains the essential physics, and assuming that the particles have a Maxwellian velocity distribution with a one-dimensional velocity dispersion c_p, their mean relative velocity will be $\sqrt{2}\bar{v} = 4c_p/\sqrt{\pi}$. The density distribution of planetesimals normal to the disc is approximately Gaussian, with a scale height $h_p = c_p/\Omega$, and the central density is therefore $\rho_s \simeq \Sigma_s\Omega/\sqrt{2\pi}c_p$. Allowing for a factor $(1+2\theta)$ due to gravitational focussing, where $\theta = Gm_p/a_pv_{\rm rel}^2 = v_{\rm esc}^2/2v_{\rm rel}^2$, a_p is the radius of the first-formed planetesimals and $v_{\rm esc}$ is the escape velocity from their surface, and assuming a collision cross-section $\sigma_c = 4\pi a_p^2$ (since the planetesimals at each heliocentric distance are initially of comparable size), the expression for the mid-plane collision timescale $t_{\rm coll} = 1/[n_p\sigma_c v_{\rm rel}(1+2\theta)]$ becomes $t_{\rm coll} = 0.975(1+2\theta)^{-1}G^{-1/2}M_\odot^{-7/6}\rho_p^{2/3}r^{7/2}$, i.e.

$$t_{\rm coll} = 2.2\times10^{10}(1+2\theta)^{-1}\left(\frac{\rho_p}{1\,{\rm g\ cm^{-3}}}\right)^{2/3}\left(\frac{r}{100\,{\rm AU}}\right)^{7/2}\ {\rm yr} \qquad (8)$$

where $\rho_p \approx 1\,{\rm g\ cm^{-3}}$ is the bulk density of the newly formed planetesimals and the steady-state value of θ is close to 0.5. The significance of this expression (*cf.* Fig. 1 of Yamamoto and Kozasa 1988) is that once gravitational instability has occurred, the initial collision timescale is not only independent of the assumed surface density of the disc but is also a sharply increasing function of heliocentric distance. At $r \simeq 50$ AU the collision timescale is on the order of 10^9 yr, and effectively infinite beyond, suggesting that significant accumulation of planetesimals into planets can only occur within the region of the observed planetary system. If, as in the standard planetesimal theory, comets are interpreted as the result of the formation of planetesimals in the outer planetary region (*i.e.* in the Jupiter, Saturn, Uranus and Neptune accretion zones), bodies formed beyond ~50–100 AU are likely to represent a collisionally unevolved system, *i.e.* a population of pristine planetesimals with predominantly cometary characteristics.

5. Total cometary mass

Recent years have seen several lines of argument converge to the conclusion that typical cometary nuclei may be much larger than formerly believed and that the cumulative diameter distribution is approximately proportional to d^{-2} (Bailey *et al.* 1994). The mean cometary mass is dominated by the size of the largest bodies, implying a significant upwards revision of the total cometary mass in the solar system (*cf.* Bailey and Stagg 1988; Bailey 1990, 1991b). For example, we may consider a spherically symmetrical model of the Oort cloud with an inner edge corresponding to a semi-major axis $a_0 = 4000$ AU, a value $a_t = 3.3\times10^4$ AU for the semi-major axis above which new comets are directly injected into the inner solar system by the Galactic tide, and a power-law energy index $\gamma = 0$ (leading to a total number of comets roughly proportional to a_0^{-1}). The total number of comets brighter than visual absolute magnitude $H_{10} = 7$ is then on the order of 5.7×10^{11} for the above parameters, of which 5.2×10^{10} have semi-major axes greater than a_t and the remainder comprise the cloud's dense inner core. Following Donnison (1986) and Hughes (1987), the cumulative luminosity distribution of observed long-period comets can be written in the form $\log[N(\leq H_{10})] = 0.3H_{10} + {\rm constant}$, a result which may be converted into a cumulative mass distribution if the mass-magnitude relation is known, and hence into a cumulative diameter distribution given the mean density of the cometary nucleus. Adopting three recent determinations for these quantities (Bailey *et al.* 1992; Weissman 1990; Hughes 1987) and a density for the cometary nucleus on the order of $1\,{\rm g\ cm^{-3}}$, the corresponding expressions for the mean cometary mass per object brighter than $H_{10} = 7$ as a function of the maximum diameter $d_{\rm max}$ in the size distribution can be calculated. The respective results are $\bar{m}/1\,{\rm kg} = 4.6\times10^{16}(d_{\rm max}/300\,{\rm km})^{1.125}$, $4.0\times10^{15}(d_{\rm max}/300\,{\rm km})^{0.75}$ and

$3.5 \times 10^{16}(d_{\max}/300\,\text{km})^{1.50}$, suggesting that despite large uncertainty in the mean cometary mass and the precise form of the cumulative diameter distribution the present cometary mass must be at least $380\,M_{\oplus}$ if the upper limit on the diameter is taken to be as large as 300 km. The original mass must have been substantially more than this owing to the combined placement and survival probability being $\lesssim 20\%$. It is clear that the total cometary mass in the solar system cannot be accommodated within a standard mass-limited minimum-mass planetesimal picture (*cf.* Mendis and Marconi 1986), the evidence from cometary masses thereby providing an important additional reason for modifying the the theory to encompass a more massive, extended disc. However, we note for completeness that an extended disc model does not provide the unique solution to this problem; other ideas for the origin of comets, also rooted in the star-formation approach to the origin of the solar system, should perhaps be given equal weight (*e.g.* Hills 1981, 1982; Cameron 1988; Marochnik and Mukhin 1988; Marochnik *et al.* 1988, 1989).

6. Discussion

Theorists investigating the origin of the solar system have tended to follow one of two paths. Some have approached the problem from the point of view of star formation, starting with the gravitational collapse of a dense molecular cloud to form the sun surrounded by a massive, extended accretion disc in which the planets and lesser bodies are eventually formed as a result of coagulation of ice-covered interstellar dust grains and gravitational instabilities in an extended dust disc. Others have followed the standard planetesimal hypothesis, a theory founded not on star formation but on planet formation. This assumes a ready-formed sun initially surrounded by a relatively low-mass disc of gas and dust with surface density proportional to $r^{-3/2}$ and an outer boundary on the order of 30–50 AU.

Here, we have presented the case for adopting a modified planetesimal theory, involving aggregation of ice-covered interstellar dust grains during distinct phases of prestellar evolution culminating in the formation and evolution of a relatively massive, extended protoplanetary disc. The theory has been developed particularly by Yamamoto and colleagues, and appears to be consistent with a wide range of observational data, not least the argument that cometary ices have much in common with those expected to occur in cool interstellar clouds (*e.g.* Knacke 1989; Bar-Nun and Kleinfeld 1989; Lunine 1989; Engel *et al.* 1990; Yamamota 1991; Mumma *et al.* 1993). Several aspects of the extended disc model clearly need further study, for example the mechanisms by which comets are formed at large heliocentric distances and ejected to produce the Oort cloud (and, of course, *its* long-term evolution and structure), and the detailed differences in size, physical structure and chemical composition expected for comets formed in different parts of the disc, and whether such differences are detectable by remote observations. But despite these uncertainties, the outer solar system bodies — both comets and planetesimals — provide an important link between the sun and other stellar systems. The problem of their origin, properly understood, offers the exciting prospect of placing theories of the origin of the solar system at the heart of current developments in observational and theoretical astrophysics; its solution will provide key insights into the formation not only of our solar system but also that of other stellar and planetary systems in the Galaxy.

Acknowledgements

It is a pleasure to thank the organizers for their invitation to present this review and for financial support enabling my attendance at the meeting.

References

Arnold, J.R. : 1977. "Condensation and agglomeration of grains." In ***Comets, Asteroids, Meteorites : Interrelations, Evolution and Origins*** (A.H. Delsemme, Ed.), IAU Coll. No. 39, 519–524, University of Toledo Press.

Artymowicz, P., Burrows, C. and Paresce, F. : 1989. "The structure of the Beta Pictoris circumstellar disk from combined IRAS and coronographic observations." ***Astrophys. J.***, **337**, 494–513.

Aumann, H.H. : 1985. "IRAS observations of matter around nearby stars." ***Publ. Astron. Soc. Pac.***, 97, 885–891.

Aumann, H.H., Gillett, F.C., Beichman, C.A., De Jong, T., Houck, J.R., Low, F.J., Neugebauer, G., Walker, R.G. and Wesselius, P.R. : 1984. "Discovery of a shell around Alpha Lyrae." ***Astrophys. J. Lett.***, **278**, L23–L27.

Backman, D.A. and Paresce, F. : 1993. "Main-sequence stars with circumstellar solid material : the Vega phenomenon." In ***Protostars and Planets III*** (E.H. Levy, J.I. Lunine, Eds.), 1253–1304, University of Arizona Press, Tucson.

Baggaley, W.J., Taylor, A.D. and Steel, D.I. : 1993. "The influx of meteoroids with hyperbolic heliocentric orbits." In ***Meteoroids and their Parent Bodies*** (J. Štohl, I.P. Williams, Eds.), 53–56. Slovak Acad. Sci., Bratislava, Slovakia.

Bailey, M.E. : 1987a. "Giant grains around protostars." ***Quart. J. Roy. Astron. Soc.***, **28**, 242–247.

Bailey, M.E. : 1987b. "The formation of comets in wind-driven shells around protostars." ***Icarus***, **69**, 70–82.

Bailey, M.E. : 1988. "Comets in star-forming regions." In ***Dust in the Universe*** (M.E. Bailey, D.A. Williams, Eds.), 113–120, Cambridge University Press.

Bailey, M.E. : 1990. "Cometary masses." In ***Baryonic Dark Matter*** (D. Lynden-Bell, G. Gilmore, Eds.), 7–35. Kluwer, Dordrecht.

Bailey, M.E. : 1991a. "Comets and molecular clouds : the sink and the source." In ***Molecular Clouds*** (R.A. James, T.J. Millar, Eds.), 273–289, Cambridge University Press.

Bailey, M.E. : 1991b. "Comet craters versus asteroid craters." ***Adv. Space Res.***, **11**, No. 6, (6)43–(6)60.

Bailey, M.E., Clube, S.V.M. and Napier, W.M. : 1990. "The Origin of Comets." Pergamon Press, Oxford.

Bailey, M.E., Chambers, J.E. and Hahn, G. : 1992. "Detection of comet nuclei at large heliocentric distances." ***Mon. Not. Roy. Astron. Soc.***, **254**, 581–588.

Bailey, M.E., Clube, S.V.M., Hahn, G., Napier, W.M. and Valsecchi, G.B. : 1994. "Hazards due to giant comets : climate and short-term catastrophism." In ***Hazards due to Comets and Asteroids*** (T. Gehrels, Ed.), University of Arizona press, in press.

Bailey, M.E. and Stagg, C.R. : 1988. "Cratering constraints on the inner Oort cloud : steady-state models." ***Mon. Not. Roy. Astron. Soc.***, **235**, 1–32.

Bar-Nun, A. and Kleinfeld, I. : 1989. "On the temperature and gas composition in the region of comet formation." ***Icarus***, **80**, 243–253.

Beckwith, S.V.W. and Sargent, A.I. : 1993. "The occurrence and properties of disks around young stars." In ***Protostars and Planets III*** (E.H. Levy, J.I. Lunine, Eds.), 521–541, University of Arizona Press, Tucson.

Beckwith, S.V.W., Sargent, A.I., Chini, R.S. and Güsten, R. : 1990. "A survey for circumstellar disks around young stellar objects." ***Astron. J.***, **99**, 924–945.

Bhatt, H.C. : 1986. "Large dust grains in the dark cloud B5." ***Mon. Not. Roy. Astron. Soc.***, **222**, 383–391.

Biermann, L. and Michel, K.W. : 1978. "The origin of cometary nuclei in the presolar nebula." ***Moon and Planets***, **18**, 447–464.

Binney, J. and Tremaine, S. : 1987. "Galactic Dynamics." Princeton University Press, Princeton.

Bode, M.F. : 1988. "Obervations and modelling of circumstellar dust." In ***Dust in the Universe*** (M.E. Bailey, D.A. Williams, Eds.), 73–102, Cambridge University Press.

Brooks, A.M. : 1990. "Aggregation of grains in the protosolar nebula." In ***Asteroids Comets Meteors III*** (C.-I. Lagerkvist, H. Rickman, B.A. Lindblad, M. Lindgren, Eds.), 25–28, Uppsala Univ. Reprocentralen.

Brooks, A.M. : 1992. "Random Aggregation and the Formation of Comets." Thesis, University of Manchester.

Brophy, T.G. : 1991. "Does debris from the formation of other planetary systems impact Earth?" ***Icarus***, **94**, 250–254.

Burke, J.R. and Silk, J. : 1976. "The dynamical interaction of a newly formed protostar with infalling matter : the origin of interstellar grains." ***Astrophys. J.***, **210**, 341–364.

Cameron, A.G.W. : 1962. "The formation of the sun and planets." ***Icarus***, **1**, 13–69.

Cameron, A.G.W. : 1978. "Physics of the primitive solar accretion disk." ***Moon and Planets***, **18**, 5–40.

Cameron, A.G.W. : 1985. "Formation and evolution of the primitive solar nebula." In ***Protostars and Planets II*** (D.C. Black, M.S. Matthews, Eds.), 1073–1099, University of Arizona Press, Tucson.

Cameron, A.G.W. : 1988. "Origin of the solar system." ***Annu. Rev. Astron. Astrophys.***, **26**, 441–472.

Cameron, A.G.W. and Pine, M.R. : 1973. 'Numerical models of the primitive solar nebula." ***Icarus***, **18**, 377–406.

Clayton, D.D. : 1982. "Cosmic chemical memory : a new astronomy." ***Quart. J. Roy. Astron. Soc.***, **23**, 174–212.

Clayton, D.D. and Liffman, K. : 1988. "Monte-Carlo histories of refractory interstellar dust." In ***Dust in the Universe*** (M.E. Bailey, D.A. Williams, Eds.), 343–354, Cambridge University Press.

Clayton, D.D., Scowen, P. and Liffman, K. : 1989. "Age structure of refractory interstellar dust and isotopic consequences." ***Astrophys. J.***, **346**, 531–538.

Donn, B.D. : 1990. "The formation and structure of fluffy cometary nuclei from random accumulation of grains." ***Astron. Astrophys.***, **235**, 441–446.

Donn, B.D. and Hughes, D.W. : 1986. "A fractal model of a cometary nucleus formed by random aggregation." In ***20th ESLAB Symposium on the Exploration of Halley's Comet*** (B. Battrick, E.J. Rolfe, R. Reinhard, Eds.), ESA SP-250 Vol. III, 523-524, ESA Publications, Noordwijk.

Donnison, J.R. : 1986. "The distribution of cometary magnitudes." ***Astron. Astrophys.***, **167**, 359-363.

Duncan, M. and Quinn, T. : 1993a. " The long-term dynamical evolution of the solar system." ***Ann. Rev. Astron. Astrophys.***, **31**, 265–295.

Duncan, M. and Quinn, T. : 1993b. "The long-term dynamical evolution and stability of the solar system." In ***Protostars and Planets III*** (E.H. Levy, J.I. Lunine, Eds.), 1371–1394, University of Arizona Press, Tucson.

Duncan, M., Quinn, T. and Tremaine, S. : 1987. "The formation and extent of the solar system comet cloud." ***Astron. J.***, **94**, 1330–1338.

Elmegreen, B.G. : 1981. "Grain formation behind shocks and the origin of isotopically anomalous meteoritic inclusions." ***Astrophys. J.***, **251**, 820–833.

Elsässer, H., Birkle, K., Eiroa, C. and Lenzen, R. : 1982. "On the infrared sources 1 and 2 in NGC 7538." ***Astron. Astrophys.***, **108**, 274–278.

Engel, S., Lunine, J.I. and Lewis, J.S. : 1990. "Solar nebula origin for volatile gases in Halley's comet." ***Icarus***, **85**, 380–393.

Fegley, B. and Prinn, R.G. : 1989. "Solar nebula chemistry : implications for volatiles in the solar system." In ***The Formation and Evolution of Planetary Systems*** (H.A. Weaver, L. Danly, Eds.), 171–211, Cambridge University Press.

Fernández, J.A. and Ip, W.-H. : 1991. "Structure and evolutionary aspects of cometary orbits." In ***Comets in the Post-Halley Era*** (R.L. Newburn Jr., M. Neugebauer, J. Rahe, Eds.) IAU Coll. No. 116, Vol. 1, 487–535, Kluwer, Dordrecht.

Fogg, M.J. : 1990. "Interstellar planets." ***Comments Astrophys.***, **14**, 357–375.

Gehrz, R.D. : 1989. "Sources of stardust in the Galaxy." In ***Interstellar Dust*** (L.J. Allamandola, A.G.G.M. Tielens, Eds.), IAU Symp. No. 135, 445–453. Kluwer, Dordrecht.

Goldreich, P. and Ward, W.R. : 1973. "The formation of planetesimals." ***Astrophys. J.***, **183**, 1051–1061.

Greenberg, J.M. : 1988. "The interstellar dust model of comets : post Halley." In ***Dust in the Universe*** (M.E. Bailey, D.A. Williams, Eds.), 121–143, Cambridge University Press.

Greenberg, J.M. : 1989. "The core-mantle model of interstellar grains and the cosmic dust connection." In ***Interstellar Dust*** (L.J. Allamandola, A.G.G.M. Tielens, Eds.), IAU Symp. No. 135, 345–355. Kluwer, Dordrecht.

Greenberg, J.M. and Hage, J.I. : 1990. "From interstellar dust to comets : a unification of observational constraints." ***Astrophys. J.***, **361**, 260–274.

Greenberg, R. : 1985. "The origin of comets among the accreting outer planets." In ***Dynamics of Comets : Their Origin and Evolution*** (A. Carusi, G.B. Valsecchi), IAU Coll. No. 83, 3–10, Reidel, Dordrecht.

Greenberg, R. : 1989. "Planetary accretion." In ***Origin and Evolution of Planetary and Satellite Atmospheres*** (S.K. Atreya, J.B. Pollack, M.S. Matthews, Eds.), 137–164, University of Arizona Press, Tucson.

Greenberg, R., Weidenschilling, S.J., Chapman, C.R. and Davis, D.R. : 1984. "From icy planetesimals to outer planets and comets." ***Icarus***, **59**, 87–113.

Grün, E., Zook, H.A., Baguhl, M., Balogh, A., Bame, S.J., Fechtig, H., Forsyth, R., Hanner, M.S., Horanyi, M., Kissel, J., Lindblad, B.-A., Linkert, D., Linkert, G., Mann, I., McDonnell, J.A.M., Morfill, G.E., Phillips, J.L., Polanskey, C., Schwehm, G., Siddique, N., Staubach, P., Svestka, J. and Taylor, A. : 1993. "Discovery of jovian dust streams and interstellar grains by the Ulysses spacecraft." ***Nature***, **362**, 428–430.

Hills, J.G. : 1981. "Comet showers and the steady-state infall of comets from the Oort cloud." ***Astron. J.***, **86**, 1730–1740.

Hills, J.G. : 1982. "The formation of comets by radiation pressure in the outer protosun." ***Astron. J.***, **87**, 906–910.

Horedt, G.P. : 1979. "Cosmogony of the solar system." ***Moon and Planets***, **21**, 63–121.

Hughes, D.W. : 1987. "The history of Halley's comet." ***Phil. Trans. R. Soc. Lond.***, **A323**, 349–367.

Jenkins, E.B. : 1989. "Insights on dust grain formation and destruction provided by gas-phase element abundances." In ***Interstellar Dust*** (L.J. Allamandola, A.G.G.M. Tielens, Eds.), IAU Symp. No. 135, 23–36. Kluwer, Dordrecht.

Jewitt, D.C. and Meech, K.J. : 1988. "Optical properties of cometary nuclei and a preliminary comparison with asteroids." ***Astrophys. J.***, **328**, 974–986.

Jura, M. : 1980. "Origin of large interstellar grains toward ρ Ophiuchi." ***Astrophys. J.***, **235**, 63–65.

Kessel'man, V.S. : 1978. "The effect of turbulence on the coagulation of interstellar dust." ***Sov. Astron.***, **22**, 276–278.

Kessel'man, V.S. : 1979. "Influence of turbulence on the growth of solid particles in contracting interstellar clouds." ***Sov. Astron.***, **23**, 475–479.

Knacke, R.F. : 1989. "Comet dust : connections with interstellar dust." In ***Interstellar Dust*** (L.J. Allamandola, A.G.G.M. Tielens, Eds.), IAU Symp. No. 135, 415–428. Kluwer, Dordrecht.

Larson, R.B. and Starrfield, S. : 1971. "On the formation of massive stars and the upper limit of stellar masses." ***Astron. Astrophys.***, **13**, 190–197.

Lefèvre, J. : 1974. "Coagulation of interstellar grains in the Rho Ophiuchi dark cloud." ***Astron. Astrophys.***, **37**, 17–19.

Liffman, K. : 1990. "The effect of catastrophic collisional fragmentation and diffuse medium accretion on a computational interstellar dust system." ***Astrophys. J.***, **355**, 518–535.

Lin, D.N.C. and Papaloizou, J. : 1985. "On the dynamical evolution of the solar system." In ***Protostars and Planets*** *II* (D.C. Black, M.S. Matthews, Eds.), 981–1072, University of Arizona Press, Tucson.

Lissauer, J. : 1993. "Planet formation." ***Annu. Rev. Astron. Astrophys.***, **31**, 129–174.

Lissauer, J. and Stewart, G.R. : 1993. "Growth of planets from planetesimals." In ***Protostars and Planets*** *III* (E.H. Levy, J.I. Lunine, Eds.), 1061–1088, University of Arizona Press, Tucson.

Lunine, J.I. : 1989. "Primitive bodies : molecular abundances in Comet Halley as probes of cometary formation environments." In ***The Formation and Evolution of Planetary Systems*** (H.A. Weaver, L. Danly, Eds.), 213–242, Cambridge University Press.

Marochnik, L.S. and Mukhin, L.M. : 1988. "The Halley missions : does the solar system contain copious unseen mass?" ***Sov. Astron. Lett.***, **14**, 241–242.

Marochnik, L.S., Mukhin, L.M. and Sagdeev, R.Z. : 1988. "Estimates of mass and angular momentum in the Oort cloud." ***Science***, **242**, 547–550.

Marochnik, L.S., Mukhin, L.M. and Sagdeev, R.Z. : 1989. "The distribution of mass and angular momentum in the solar system." ***Sov. Sci. Rev. E. Astrophys. Space Phys.***, **8**, 1–55.

Mathis, J.S., Rumpl, W. and Nordseick, K.H. : 1977. "The size distribution of interstellar grains." ***Astrophys. J.***, **217**, 425–433.

Mathis, J.S. and Wallenhorst, S.G. : 1981. "The size distribution of interstellar particles. III. Peculiar extinctions and normal infrared extinction." ***Astrophys. J.***, **244**, 483–492.

McKee, C.F. : 1989. "Dust destruction in the interstellar medium." In ***Interstellar Dust*** (L.J. Allamandola, A.G.G.M. Tielens, Eds.), IAU Symp. No. 135, 431–443. Kluwer, Dordrecht.

McKee, C.F., Zweibel, E.G., Goodman, A.G. and Heiles, C. : 1993. "Magnetic fields in star-forming regions : theory." In ***Protostars and Planets*** *III* (E.H. Levy, J.I. Lunine, Eds.), 327–366. University of Arizona Press, Tucson.

Meakin, P. and Donn, B. : 1988. "Aerodynamic properties of fractal grains : implications for the primordial solar nebula." ***Astrophys. J. Lett.***, **329**, L39–L41.

Meakin, P., Vicsek, T., and Family, F. : 1985. "Dynamic cluster-size distribution in cluster-cluster aggregation : effects of cluster diffusivity." ***Phys. Rev. B***, **31**, 564–569.

Mendis, D.A. and Marconi, L. : 1986. "A note on the total mass of comets in the solar system." ***Earth, Moon, Planets***, **36**, 187–191.

Mestel, L. : 1963. "On the Galactic law of rotation." ***Mon. Not. Roy. Astron. Soc.***, **126**, 553–575.

Mestel, L. and Ray, T.P. : 1985. "Disc-like magneto-gravitational equilibria." ***Mon. Not. Roy. Astron. Soc.***, **212**, 275–302.

Mestel, L. and Spitzer, L. : 1956. "Star formation in magnetic dust clouds." ***Mon. Not. Roy. Astron. Soc.***, **116**, 503–514.

Mizuno, H. : 1989. "Grain growth in the turbulent accretion disk solar nebula." ***Icarus***, **80**, 189–201.

Montmerle, T., Feigelson, E.D., Bouvier, J. and André, P. : 1993. "Magnetic fields, activity and circumstellar material around young stellar objects." In ***Protostars and Planets*** *III* (E.H. Levy, J.I. Lunine, Eds.), 689–717, University of Arizona Press, Tucson.

Morfill, G.E. and Völk, H.J. : 1984. "Transport of dust and vapor and chemical fractionation in the early protosolar cloud." ***Astrophys. J.***, **287**, 371–395.

Mumma, M.J., Weissman, P.R. and Stern, S.A. : 1993. "Comets and the origin of the solar system : reading the Rosetta stone." In ***Protostars and Planets III*** (E.H. Levy, J.I. Lunine, Eds.), 1177–1252, University of Arizona Press, Tucson.

Nakagawa, Y., Hayashi, C. Nakazawa, K. : 1983. "Accumulation of planetesimals in the solar nebula." ***Icarus*** **54**, 361–376.

Nakano, T. : 1987. "Formation of planets around stars of various masses – I. Formulation and a star of one solar mass." ***Mon. Not. Roy. Astron. Soc.***, **224**, 107–130.

Napier, W.M. : 1990. "Interstellar planetesimals." In ***Dusty Objects in the Universe*** (E. Bussoletti, A.A. Vittone, Eds.), 103–110. Kluwer, Dordrecht.

Öpik, E.J. : 1973. "Comets and the formation of planets." ***Astrophys. Space Sci.***, **21**, 307–398.

Piirola, V., Scaltriti, F. and Coyne, G.V. : 1992. "Circumstellar disks deduced from subarcsecond polarization observations of two young stars." ***Nature***, **359**, 399–401.

Pringle, J.E. : 1989. "The Egg Nebula : a protostellar disc remanant around an evolved star." ***Mon. Not. Roy. Astron. Soc.***, **238**, 37P–40P.

Prinn, R.G. : 1993 "Chemistry and evolution of gaseous circumstellar disks." In ***Protostars and Planets III*** (E.H. Levy, J.I. Lunine, Eds.), 1005–1028, University of Arizona Press, Tucson.

Rouan, D. and Léger, A. : 1984. "Large grains in Orion are indicated by IR polarization and flux data." ***Astron. Astrophys.***, **132**, L1–L4.

Safronov, V.S. : 1969. "Evolution of the Protoplanetary Cloud and Formation of the Earth." Israel programme for scientific translations, Jerusalem 1972.

Safronov, V.S. and Ruzmaikina, T.V. : 1985. "Formation of the solar nebula and the planets." In ***Protostars and Planets II*** (D.C. Black, M.S. Matthews, Eds.), 959–980, University of Arizona Press, Tucson.

Sandford, S.A. : 1989. "Interstellar dust in collected interplanetary dust particles." In ***Interstellar Dust*** (L.J. Allamandola, A.G.G.M. Tielens, Eds.), IAU Symp. No. 135, 403–413. Kluwer, Dordrecht.

Sargent, A.I. : 1989. "Molecular disks and their link to planetary systems." In ***The Formation and Evolution of Planetary Systems*** (H.A. Weaver, L. Danly, Eds.), 111–129, Cambridge University Press.

Seab, C.G. : 1988. "Grain destruction and growth." In ***Dust in the Universe*** (M.E. Bailey, D.A. Williams, Eds.), 303–326, Cambridge University Press.

Seab, C.G. and Shull, J.M. : 1986. "Shock processing of interstellar grains." In ***Interrelations among Circumstellar, Interstellar, and Interplanetary Dust*** (J.A. Nuth III, R.E. Stencel, Eds.), NASA CP-2403, 37–53, NASA, Washington DC.

Sicardy, B. : 1994. "Small bodies around other stars." This book.

Strom, K.M., Strom, S.E., Edwards, S., Cabrit, S. and Skrutskie, M.F. : 1989a. "Circumstellar material associated with solar-type pre-main-sequence stars : a possible constraint on the timescale for planet building." ***Astron. J.***, **97**, 1451–1470.

Strom, S.E., Edwards, S. and Strom, K.M. : 1989b. "Constraints on the properties and environment of primitive stellar nebulae from the astrophysical record provided by young stellar objects." In ***The Formation and Evolution of Planetary Systems*** (H.A. Weaver, L. Danly, Eds.), 91–109, Cambridge University Press.

Strom, S.E., Edwards, S. and Skrutski, M.F. : 1993. "Evolutionary timescales for circumstellar disks associated with intermediate- and solar-type stars." In ***Protostars and Planets III*** (E.H. Levy, J.I. Lunine, Eds.), 837–866, University of Arizona Press, Tucson.

Taylor, A.D., Baggaley, W.J., Bennett, R.G.T. and Steel, D.J. : 1994. "Radar measurements of very high velocity meteors with AMOR." ***Planetary and Space Science***, in press.

Telesco, C.M., Becklin, E.E., Wolstencroft, R.D. and Decher, R. : 1988. "Resolution of the circumstellar disk of β Pictoris at 10 and 20 μm." ***Nature***, **335**, 51–53.

Tielens, A.G.G.M. : 1989. "Dust in dense clouds." In ***Interstellar Dust*** (L.J. Allamandola, A.G.G.M. Tielens, Eds.), IAU Symp. No. 135, 239–262. Kluwer, Dordrecht.

Tielens, A.G.G.M. : 1991. "Characteristics of interstellar and circumstellar dust." In ***Origin and Evolution of Interplanetary Dust*** (A.C. Levasseur-Regourd, H. Hasegawa, Eds.), IAU Coll. No. 126, 405–412, Kluwer, Dordrecht.

Tremaine, S. : 1990. "Dark matter in the solar system." In ***Baryonic Dark Matter*** (D. Lynden-Bell, G. Gilmore, Eds.), 37–65. Kluwer, Dordrecht.

Tscharnuter, W.M. and Boss, A.P. : 1993. "Formation of the protosolar nebula." In ***Protostars and Planets III*** (E.H. Levy, J.I. Lunine, Eds.), 921–938, University of Arizona Press, Tucson.

Van Dishoeck, E.F., Blake, G.A., Draine, B.T. and Lunine, J.I. : 1993. "The chemical evolution of protostellar and protoplanetary matter." In ***Protostars and Planets III*** (E.H. Levy, J.I. Lunine, Eds.), 163–241, University of Arizona Press, Tucson.

Völk, H.J., Jones, F.C., Morfill, G.E. and Röser, S. : 1980. "Collisions between grains in a turbulent gas." ***Astron. Astrophys.***, **85**, 316–325.

Weidenschilling, S.J. : 1977. "The distribution of mass in the planetary system and solar nebula." ***Astrophys. Space Sci.***, **51**, 151–158.

Weidenschilling, S.J. : 1987. "Accumulation of solid bodies in the solar nebula." ***Gerlands. Beitr. Geophysik, Leipzig***, **96**, 21–33.

Weidenschilling, S.J., Donn, B. and Meakin, P. : 1989. "The physics of planetesimal formation." In ***The Formation and Evolution of Planetary Systems*** (H.A. Weaver, L. Danly, Eds.), 131–150, Cambridge University Press.

Weidenschilling, S.J. and Cuzzi, J.N. : 1993. "Growth of planets from planetesimals." In ***Protostars and Planets III*** (E.H. Levy, J.I. Lunine, Eds.), 1031–1060, University of Arizona Press, Tucson.

Weintraub, D.A., Sandell, G. and Duncan, W.D. : 1989. "Submillimeter measurements of T Tauri and FU Orionis stars." ***Astrophys. J. Lett.***, **340**, L69–L72.

Weissman, P.R. : 1990. "The cometary impactor flux at the Earth." In ***Global Catastrophes in Earth History*** (V.L. Sharpton, P.D. Ward, Eds.), Geol. Soc. Amer. Spec. Pap. 247, 171–180.

Wetherill, G.W. : 1980. "Formation of the terrestrial planets." ***Ann. Rev. Astron. Astrophys.***, **18**, 77–113.

Wetherill, G.W. : 1989. "The formation of the solar system : consensus, alternatives, and missing factors." In ***The Formation and Evolution of Planetary Systems*** (H.A. Weaver, L. Danly, Eds.), 1–30, Cambridge University Press.

Wetherill, G.W. : 1990. "Formation of the Earth." ***Annu. Rev. Earth Planet. Sci.***, **18**, 205–256.

Williams, I.P. : 1974. "Planetary formation." In ***Exploration of the Planetary System*** (A. Woszczyk, C. Iwaniszewska, Eds.), IAU Symp. No. 65, 3–12, Reidel, Dordrecht.

Woolfson, M.M. : 1990. "The solar system — its origin and evolution." ***Quart. J. Roy. Astron. Soc.***, **34**, 1–20.

Yamamoto, T. : 1991. "Chemical theories on the origin of comets." In ***Comets in the Post-Halley Era*** (R.L. Newburn Jr., M. Neugebauer, J. Rahe, Eds.) IAU Coll. No. 116, Vol. 1, 361–376, Kluwer, Dordrecht.

Yamamoto, T. and Kozasa, T. : 1988. "The cometary nucleus as an aggregate of planetesimals." ***Icarus***, **75**, 540–551.

Yamamoto, T., Mizutani, H. and Kadota, A. : 1993. "Are 1992 QB_1 and 1993 FW remnant planetesimals?" ***Preprint***, submitted to Nature.

THE ACM93 SMALL BODIES DATA BASES

The papers in this section present data bases on the solar system small bodies. Th idea was not to create a single centralized data base, not even a single provider of th data. Centralized data bases run the risk of being too rigid, and therefore of becomin very quickly obsolete. On the contrary in a such a rapidly moving field of research th guarantee that the data bases will be continously updated is more important than th guarantee of stability of the data. Therefore we are here only providing the informatio on the existence of data bases which are maintained and upgraded by the people wh created them. The only requirement we have enforced is that the data have to b available to everybody, either by a self service system (such as **anonymous ftp** on th **internet** computer network), or upon request (to be forwarded by either E–Mail o ordinary mail). This commitment to availability has been accepted by all the provider of the data bases listed in the following papers; moreover, the service of giving acces to the data cannot be a commercial entreprise, although in some case the cost of th magnetic media/paper copies/electronic subscription has to be payed by the user.

The following eight papers describe, for each data base, the data sets currentl available, the source of the data, the access method, the address of the contact person(s) as well as providing other useful information, such as references, availability of publi domain software and/or user interfaces. Some papers also include comments on th quality and reliability of the data, and predictions on their future upgrades; however more complete information has to be looked for in the data bases themselves (e.g. i **README** and other documentation files and in online help) and/or asked to the contac person. This is necessary not only to save space, but also because the time scales fo upgrade of the data bases are often comparable to the time scale of the publication o this book. The readers will also notice that we have given up the publication of th printout of the data: the size of the data bases is now much bigger than the informatio which can be printed in a volume such as this one; the printed versions are generall available upon request to the contact persons.

A. Milani et al. (eds.), Asteroids, Comets, Meteors 1993, 461.

IRAS MINOR PLANET SURVEY

EDWARD F. TEDESCO
Mission Research Corporation, One Tara Boulevard, Nashua, NH 03062 USA
E-mail etedesco@dirac.plh.af.mil

Abstract. This chapter describes the contents and organization of the IRAS Minor Planet Survey Data Base and tells how to obtain hard-copy and machine-readable versions of the data.

The primary purpose of the Infrared Astronomical Satellite (IRAS) was to survey the sky in four wavelength bands centered near 12, 25, 60 and 100 μm. The satellite was launched in January 1983 and obtained observations until November 1983. In this period it surveyed approximately 96 percent of the sky. The IRAS mission, data processing and data products are described in a series of catalogs. These catalogs all deal with sources which are fixed on the sky. The *IRAS Asteroid and Comet Survey, 1986* provided observations for moving sources, the asteroids and comets.

This chapter summarizes the creation of the second IRAS catalog for moving sources, the *IRAS Minor Planet Survey* (IMPS). It describes the contents and organization of this catalog and tells how to obtain hard-copy and machine-readable versions of the data.

1. The IRAS Minor Planet Survey

In contrast with the *IRAS Asteroid and Comet Survey (1986)*, IMPS processed only IRAS survey observations of asteroids; comets were not processed. Low Resolution Spectrometer, Serendipitous, and Additional Observations data were not processed. IMPS did process all asteroids with reasonably-well-known orbits as of December 1990. In particular, IMPS updates the processing of asteroids numbered 1 through 3318 and extends this processing to asteroid number 4679 plus 2,632 asteroids with preliminary (two or more opposition) orbits.

Note that, as with the *IRAS Asteroid and Comet Survey (1986)*, the IMPS catalogs and databases are fundamentally different from those catalogs and databases produced for fixed sources. Asteroids move and their apparent emission levels can vary by large amounts. Consequently different methods and criteria were used for processing potential asteroid sightings. The IMPS documentation describes these differences.

2. The IRAS Minor Planet Survey Catalog

The *IRAS Minor Planet Survey (1992)* consists of a 460 page hard-copy document. This document presents a history of the IRAS asteroid task, identifies the ground-based asteroid data used in associating asteroids with IRAS sources and in deducing albedos and diameters from their IRAS-measured infrared fluxes, and describes how asteroid sightings were identified and processed. It presents analyses of the asteroid associations and accepted sightings, an overview of the results obtained. The descriptive part concludes with a summary chapter which discusses

A. Milani et al. (eds.), Asteroids, Comets, Meteors 1993, 463–466

the completeness and reliability of the survey, the statistical adjustments made to the derived results, and details the major differences between this catalog and the *IRAS Asteroid and Comet Survey (1986)*.

Following the descriptive section the document describes the IMPS data products and presents all of the *IRAS Minor Planet Survey* catalogs, i.e., all of the printed data products. This section contains technical details about the data, including data formats. It presents a general description of the final data products, gives the formats of the data bases and catalogs, presents a sub-set of the ground-based data used (the corresponding data base contains all the ground-based data used), and presents albedos and diameters derived for all asteroids with multiple sightings and, separately, for those asteroids with only a single IRAS sighting at one wavelength. Additional products present data on detection statistics, rejected sightings, and missed-predictions. Also included are a table of acronyms and glossary of terms, and the IRAS flux look-up table (used to convert an IRAS flux into a visual geometric albedo).

3. Final Data Products

The printed data products are referred to as "catalogs" and the machine-readable products as "data bases". All of the catalogs appear in the hard-copy document; the machine-readable data bases are available via ftp from the National Space Science Data Center (NSSDC) at the Goddard Space Flight Center under the name The *IRAS Minor Planet Survey Catalog and Data Base, 1992*.

The machine-readable data (Tedesco, *et al.*, 1992) may be obtained from the National Space Science Data Center by sending an electronic mail message to :

REQUEST@NSSDCA.GSFC.NASA.GOV (Internet) or to :

NSSDCA: : REQUEST (NSI-DECnet).

Arrangements to obtain them may also be made via telephone at (301) 286-6695 or by writing to : NSSDC Coordinated Request and User Support Office, NASA/Goddard Space Flight Center, Code 633.4, Greenbelt, MD 20771 U.S.A.

The hard-copy document [*IRAS Minor Planet Survey (1992)*] is only available from Dr. Stephan Price, Phillips Laboratory, Geophysical Directorate, Backgrounds Branch/GPOB, Hanscom Air Force Base, MA 01731-3010 (Internet : price@dirac.plh.af.mil). Furthermore, the only supported version of the machine-readable data base is that at the NSSDC.

4. Catalogs and Data Bases

Table 1 presents a concise summary of the available data products. With the exception of a portion of the *IMPS Ground-Based Data Catalog* (FP 107) and the entire *IMPS Sightings Data Base* (FP 108) all products exist as both a "Catalog", *i.e.*, as tables appearing in the hard-copy documentation, and as a "Data Base", *i.e.*, as ASCII data file(s).

TABLE I
IMPS DATA PRODUCTS

Final Product No. (Size in Kb)	Final Product Name	Remarks
100 (NA)	*The IRAS Minor Planet Survey*	Hard-copy documentation.
101 (20)	IMPS Final Products Format Catalog	This catalog gives the formats of all the machine-readable data products.
102 (239)	IMPS Albedos and Diameters Catalog	A distilled summary of the averaged results for the 1,796 numbered asteroids and 88 unnumbered asteroids with at least two accepted observations.
103 (15)	IMPS Singleton Catalog	Same as Final Product 102 but for the 94 numbered and 26 unnumbered asteroids which have only a single accepted sighting in a single band.
104 (193)	IMPS Statistics Catalog	A summary of the number of times each asteroid was sighted, the number of times it was predicted to be scanned, and possible reasons for any failure to be detected.
105 (100)	IMPS Reject Catalog	A summary of the number of rejected sightings for each asteroid and possible reasons for their rejection.
106 (320)	IMPS Missed-Predictions Catalog	A summary of asteroids which were scanned by the IRAS focal plane array but which did not generate any associations.
107 (2,385)	IMPS Ground-Based Data Catalog	A listing of ground-based data used in IMPS data processing.
108 (4,538)	IMPS Sightings Data Base	A listing of 7,937 accepted sightings associated with 1,890 numbered asteroids and 273 accepted sightings associated with 114 unnumbered asteroids.

For the most part, the formats of the Catalog and Data Base versions of each data product are the same. In some cases, however, there are differences introduced to improve the readability of the Catalog version. These differences are noted in the appropriate places.

The hard-copy documentation, together with the machine-readable files of the final data products, constitute *The IRAS Minor Planet Survey Catalog and Database, 1992*. This supplants the 1986 version *IRAS Asteroid and Comet Database and Catalog*. A cross reference between final data products from the *IRAS Asteroid and Comet Database and Catalog, 1986* and *The IRAS Minor Planet Survey Catalog and Database, 1992* is given in the hard-copy documentation.

5. Acknowledgements

Creation of the IMPS catalogs and data bases was made possible through grants from the Phillips Laboratory, Geophysical Directorate under funding provided by the Strategic Defense Initiative Organization.

References

Infrared Astronomical Satellite Asteroid and Comet Survey : Preprint Version No. 1, 1986, edited by Matson, D.L., JPL Internal Document No. D- 3698. (Jet Propulsion Laboratory, Pasadena)

IRAS Minor Planet Survey, 1992, edited by Tedesco, E.F. (Phillips Laboratory Technical Report No. PL-TR-92-2049. Hanscom Air Force Base, MA.)

Tedesco, E.F., Veeder, G.J., Fowler, J.W., and Chillemi, J.R (1992). The IRAS Minor Planet Survey Data Base, National Space Science Data Center, Greenbelt, MD.

A COMPOSITE CATALOGUE OF ASTEROID PROPER ELEMENTS

ANDREA MILANI
Dipartimento di Matematica, Via Buonarroti 2, I-56127 Pisa, Italia
E–mail milani@dm.unipi.it

EDWARD BOWELL
Lowell Observatory, 1400 W. Mars Hill Road, Flagstaff AZ 86001, USA
E–mail elgb@lowell.edu

ZORAN KNEŽEVIĆ
Astronomska opservatorija, Volgina 7, 11050 Beograd, Yugoslavia

ANNE LEMAITRE
Dép. Mathématique, FUNDP, Rempart de la Vierge 8, B–5000 Namur, Belgium
E–mail alemaitr@math.fundp.ac.be

ALESSANDRO MORBIDELLI
Observatoire de la Côte d'Azur, BP 229, 06304 Nice Cedex 04, France
E–mail morby@rameau.obs-nice.fr

KARRI MUINONEN
Observatorio, PL 14, 00014 Helsingin yliopisto, Suomi–Finland
E–mail karri@ceres.helsinki.fi

1. Contents of the catalogue and sources

We have assembled the asteroid proper elements computed by different authors, using different methods, and for different regions of the asteroid belt. Asteroids on planet crossing orbits are not yet included. The use of very different algorithms is dictated by the dynamics of the different regions. For the asteroids of the main belt, having semimajor axes between 2.1 and 3.8 AU, and with proper eccentricity and sine of inclination less than 0.3, proper elements are computed by a fully analytical iterative theory developed by Milani and Knežević (1990,1992,1994; hereafter M&K). For high inclination and/or eccentricity main–belt asteroids, having $1.8 < a < 3.8\,AU$ and either e or $\sin I$ larger than 0.24, proper elements are computed by a semianalytical theory developed by Lemaitre and Morbidelli (1994; hereafter L&M). For Trojans, proper elements are computed by a synthetic theory (that is, from the output of a numerical integration for a few Myr) by Milani (1993). For the Hilda asteroids in the 3 : 2 resonance, proper elements have been computed by a synthetic theory by Schubart (1982, 1991). For a discussion of the state of the art in the computation of proper elements, see Knežević and Milani, this volume; for Trojans, see Milani, this volume; for some other cases (e.g. proper

A. Milani et al. (eds.), Asteroids, Comets, Meteors 1993, 467–470.

elements for resonant asteroids), see Froeschlé and Morbidelli, this volume. When and if new proper elements will be available for other asteroids, they will be added to the public domain file, together with updates and upgrades of the existing catalogues.

Users of this composite catalogue need to be aware that different authors use different definitions of the proper elements. For the main belt, mean semimajor axes are computed by analytical removal of short periodic perturbations with an algorithm developed by Knežević (1988), Milani and Knežević (1990, 1994) (the output of this intermediate computation is a catalogue of mean elements, also available); this mean semimajor axis is used as proper a by both M&K and L&M. For Trojans and Hildas, the mean semimajor axis is not useful; the meaningful proper element is the amplitude of the oscillation of the semimajor axis around the mean value; as an alternative, the amplitude of the libration of the critical argument can be used. For proper eccentricities and sine of inclination, the values computed by M&K can be interpreted as the average values over a time much longer than the time scale of the secular perturbations; in contrast, L&M give two values for both proper e and $\sin I$, which can be interpreted as maximum and minimum value (over one cycle of the argument of perihelion). The arithmetic mean of the two values given by L&M is close, at least for moderate inclinations, to the M&K value. For Trojans and Hildas, proper e and $\sin I$ have a geometric interpretation similar to that of the M&K elements, although the computational procedure is completely different.

In most cases, the use of proper elements defined in different ways (e.g. to identify asteroid families) should not give rise to problems, because different definitions are used in widely separated regions. However, the M&K and the L&M elements do have an overlap region (for eccentricity and/or sine of inclination between 0.24 and 0.3). In this region the two methods have comparable accuracy; the overlap is necessary for asteroid family studies, to avoid loss of information on families extending across this range; the users are warned that to take a difference between M&K and L&M asteroids might result in a significant loss of accuracy (with respect to differences between elements computed by means of the same algorithm).

2. Accuracy and reliability

Proper elements have been computed for essentially all the asteroids for which the available osculating elements are accurate enough (this does not apply to the Hildas, for which the catalogue is not entirely up to date). We have included all the numbered asteroids, together with the asteroids observed at more than one opposition. For the orbits computed on the basis of observations performed at one opposition only, we have used the covariance matrices of the orbital elements computed by Muinonen and Bowell (1993; see also Muinonen et al., 1994) to assess their precision. If we assume that the main purpose of the proper elements is to study asteroid families, then for each metric used to compute the distance between two asteroids in the proper elements space there is a corresponding covariance, which can be computed as a function of the full covariance matrix of the elements. If the metric is the simplest possible one (for a discussion of the metric problem, see Zappalà et al. 1990, Milani et al. 1992) :

$$d = \sqrt{\frac{1}{4}(\frac{\delta d}{a})^2 + 2(\delta e)^2 + 2(\delta \sin I)^2}$$

then the covariance to be used is :

$$\iota_0^2 = e^2(\Sigma_{\omega\omega} + \Sigma_{\Omega\Omega} + 2\Sigma_{\omega\Omega}) + \Sigma_{ee} + \Sigma_{\Omega\Omega}\sin^2 i + \Sigma_{ii}\cos^2 i + \Sigma_{aa}\frac{1}{4a^2}.$$

where the $\Sigma_{..}$ are the Bowell and Muinonen covariances. We have therefore computed proper elements for the single opposition orbits with $\iota_0 \leq 0.01$; orbits determined in a poorer way (e.g., by observations over a very short arc) would not be useful to find families, since the error in the proper elements cannot be smaller than the error in the osculating ones. For single–opposition orbits used in this catalogue, the quality of the osculating elements is indicated by a Quality Code for Osculating elements (QCO), on a decibel scale in which $QCO = 0$ corresponding to $\iota_0 = 10^{-4}$ and $QCO = 20$ corresponding to $\iota_0 = 10^{-2}$.

As a result, the current version of the osculating elements file used as input catalogue for the computation of mean elements contains 14127 orbits. Mean elements have been computed for 13456 main belt asteroids. The proper elements M&K catalogue contains 12573 asteroids; L&M contains 2511 asteroids, the Trojans are 184 and the Hildas 57. The proper elements files currently amount to about $1150\,KB$; the mean elements, the input catalogue of osculating elements, the software and the auxiliary files amount to about $4\,MB$.

The accuracy and reliability of the computation of the mean and proper elements are indicated by other quality codes. A Quality Code for Mean elements (QCM) indicates the cases in which the removal of short periodic perturbations results in degraded accuracy because of some nearby low–order mean–motion resonance. For the M&K proper elements, a Resonance FLag (RFL) indicates the cases in which the results might be degraded by some secular resonance; the RFL encodes the small divisor responsible for the divergence of the iterative theory. For the L&M proper elements, the flag $IC = 0$ means that the process is convergent, $IC \neq 0$ gives by its value an indication about the non convergence reason (see the documentation for more details); the flag $IL = 0$ indicates a perihelion circulation case and $IL = \pm 1$ a libration case. For the Trojans, a separate file contains estimates of the stability for all the proper elements and frequencies, as obtained by the running box method.

3. Public–domain software

Together with the data files, the distribution files contain also the software used to produce some of the data. In particular, for the M&K mean and proper elements, all the software is already public domain and the users should be able to compute mean and proper elements by themselves. Although transportability cannot always be guaranteed, the computer programs are written in standard *FORTRAN 77*; they are reasonably robust and have been tested on several different CPUs. For the *UNIX* environment, a `makefile` is provided, and compilation should be straightforward. Some macros for compilation in a *DOS* environment are available upon request.

To test the stability of the proper elements, at least for low–eccentricity orbits, the numerical integration software package `orbit8v` is also provided. Some essential documentation is available, both as `README` files and as TEX sources; however, the users should remember that this is not commercial software, and documentation might not be always complete.

4. Access

The catalogues of proper elements, the auxiliary files and the public domain software are available by `anonymous ftp` to the server of the Department of Mathematics, University of Pisa, at the `internet` address `gauss.dm.unipi.it` (absolute address 131.114.6.55). The subdirectory is `pub/propel`.

The complete distribution file (with a name like `versNNN.tar.Z`) has been packed and compressed (with the standard *UNIX* utilities `tar` and `compress`); the size is currently about 3 *MB*. The users of different operating systems should `get` the uncompressed files they need one by one, by looking in the subdirectories `propels, meanels, oscels, sources, doc`. In case of any problem, please contact the relevant author by E-mail.

References

Knežević, Z. 1988. "Asteroid mean orbital elements". *Bull. Astron. Obs. Belgrade* **139**, 1–6.

Lemaitre, A., and Morbidelli, A. 1994. "Calculation of proper elements for high inclined asteroidal orbits". *Celest. Mech.*, in press.

Milani, A. 1993. "The Trojan asteroid belt : proper elements, stability, chaos and families". *Celest. Mech.*, **57**, 59–94.

Milani, A., and Knežević, Z. 1990. "Secular perturbation theory and computation of asteroid proper elements". *Celestial Mechanics* **49**, 347–411.

Milani, A., and Knežević, Z. 1992. "Asteroid proper elements and secular resonances". *Icarus*, **98**, 211–232.

Milani, A., and Knežević, Z. 1994. "Asteroid proper elements and the dynamical structure of the asteroid main belt". *Icarus*, in press.

Milani, A., Farinella, P., and Knežević, Z. 1992. "On the search for asteroid families". In *Interrelations between physics and dynamics for minor bodies in the solar system* (D. Benest, C. Froeschléi, Eds.), 85–132, Editions Frontières, Gif–sur–Yvette.

Muinonen, K., and Bowell, E. 1993. "Asteroid orbit determination using Bayesian probabilities". *Icarus* **104**, 255–279.

Muinonen, K., Bowell, E. and Wasserman, L.H. 1994. "Orbital uncertainities of single–opposition asteroids." *Planet. Space Sci.*, in press.

Schubart, J. 1982. "Three characteristic parameters of orbits of Hilda type asteroids". *Astron. Astrophys.* **114**, 200–204.

Schubart, J. 1991. "Additional results on orbits of Hilda type asteroids". *Astron. Astrophys.* **241**, 297–302.

Zappalà, V., Cellino, A., Farinella, P., and Knežević, Z. 1990. "Asteroid families I : identification by hierarchical clustering and reliability assessment". *Astron. J.* **100**, 2030–2046.

THE UPPSALA ASTEROID DATA BASE

P. MAGNUSSON, C.-I. LAGERKVIST, M. DAHLGREN, A. ERIKSON

Astronomiska observatoriet, Box 515, S-751 20 Uppsala, Sweden
E-mail Per.Magnusson@astro.uu.se.

M.A. BARUCCI

Observatoire de Paris-Meudon, F-92195 Meudon, France

I. BELSKAYA

Astronomical Observatory, Sumskaya Str. 35, Kharkov 310022, Ukraine

M.T. CAPRIA

Istituto di Astrofisica Spaziale, CNR, Viale dell'Università 11, I-00185 Rome, Italy

Abstract. A collection of data and results of asteroid photometry and related observations has been made available by anonymous ftp. It contains all published asteroid lightcurve photometry, a table of reliable rotation periods, and a collection of all asteroid spin vector determinations. Here we describe the database in some detail.

1. Introduction

Data sets made in the course of extensive asteroid research at Uppsala Astronomical Observatory and in collaboration with other institutes are made available to the asteroid community. They are divided into three parts described in the following sections. Access is freely available by anonymous ftp.

2. Asteroid Photometric Catalogue Data Base

The Asteroid Photometric Catalogue was first published in printed form by Lagerkvist *et al.* (1987). Two subsequent updates are now available (Lagerkvist *et al.*, 1989, 1992) and a third one is in preparation. These volumes contain illustrations of all published asteroid lightcurves, replotted in standard formats. In addition, extensive reference lists, aspect data, and details of the observations are included.

We now make available the digital data used in the production of these printed volumes. Software for easy digestion of the data is also supplied.

A. Milani et al. (eds.), Asteroids, Comets, Meteors 1993, 471–476.

2. 1. CONTENTS OF THE PHOTOMETRIC PART

The database currently contains all asteroid lightcurve photometry published up to and including the year 1992.

This part of the database consists of :

- Documentation;
- A table of references to all original publications; these currently number in eccess of 300;
- A table with recomputed aspect data for all observing nights.
- The lightcurves with detailed information (more than 4000 lightcurves);
- Fortran subroutines for reading and converting the digital lightcurves.

We discuss the last three items in more detail in the following subsections.

2. 2. ASPECT DATA

In order to establish aspect data of uniform accuracy and reliability we recompute all the observing geometries instead of relying on the aspect data published by the observers. This is achieved using the integrator Radau (Everhart 1985) applied to a 9-planet Solar System. The actual aspect values are computed with the Jet Propulsion Laboratory software EPHJPL. In addition to the five standard aspect parameters appearing in the printed volumes full high-accuracy information on the relative 3-body geometry Sun-asteroid-Earth for individual data points in the lightcurves can be extracted from the database.

2. 3. LIGHTCURVES

The data on the individual lightcurves contain a general header containing the following information :

- Asteroid number and name;
- Reference to the original publication;
- Acquisition (usually original data from the observer or a digitization from the publication)*;
- Aspect data (see discussion above);
- Observatory name and geographic coordinates*;
- Telescope aperture and name*;
- Detector type*;
- Specification of the photometric system (*e.g.*, Johnson UBV);
- Applied corrections, *e.g.* light time correction and magnitude reduction;

- Zero points and units for the time and magnitude data;
- Composite formation (in order to minimize the model-dependence of the data, we try to avoid composite lightcurves as discussed in Sec. 2.6).

The items marked with an asterisk (*) are applicable to current insertions into the database — old data may still lack some of this information.

This header information is followed by a table containing for each integration typically the time and the measured magnitude with an error estimate, if available. The general philosophy is to keep this tabulation as close as possible to that supplied by the observer (digitally, or in a diagram) and to use the header as a specification of how to interpret this table. The advantage is that the database contents can easily be checked against the original data. The draw-back is that the resulting data have a complicated inhomogenous format, which is difficult to read automatically without a lot of programming effort. The software described in the next section is the remedy to this problem.

2. 4. SOFTWARE AND FORMATS

Fortran software running on Sun and Vax machines are supplied with the data. In particular, subroutines for reading the data in the "human-friendly", but complicated, format described in the previous subsection (called the *atlas* format) is available. Users may convert these subroutines to fit as "front-ends" to their own application software. Alternatively, these subroutines are part of a self-contained program (also supplied) which reads data given in the atlas format and converts it into a more "machine-friendly" format called *homo*. The resulting homogeneous data files are easy to use as input to user-made software.

2. 5. SIZE

The current size of the photometric part of the database is about 5 MByte (1-2 MByte will be added in the third update). It can be compressed to less than 2 MByte. The size of the data for individual asteroids range from less than 1 kByte up to 150 kByte. The corresponding storage requirements for the non-resident homogeneous format *homo* (see Sec. ??) is about twice those given above.

2. 6. UPDATING AND OBSERVER INPUT

The photometric part of the database is updated every second year. Continuous updating is under consideration.

We depend very much on collaboration from observers submitting a copy of their lightcurves in digital form. This ensures that the data will be available for future researchers in an easily accessible form at full accuracy. We strongly encourage observers to send their observations to us as soon as it has been accepted for publication somewhere. From experience, we know that later attempts to recreate the data from multiple versions on backup tapes *etc.* is time consuming and error prone. On request, we can keep data in a confidential queue and not include it in the public domain database until they appear in the literature.

In order to minimize the model-dependency of the database, we prefer to receive observations which have not been composited or transformed to standard solar phase angles. We therefore urge observers who publish data as composites or transformed beyond standard instrumental and atmospheric corrections to store one copy for the database containing observed magnitudes and times of observation.

3. Asteroid Rotation Periods

The rotation periods and lightcurve amplitudes in this part have been determined from the photometric part. For asteroids with reliable spin vectors the periods are taken from the spin vector part. The contents are :

- The rotation periods.
- The lightcurve amplitudes and the corresponding ecliptic longitudes.
- A reliability code.

For further information please contact Claes-Ingvar Lagerkvist at Internet address `Classe@astro.uu.se`

4. Asteroid Spin Vector Determinations

All published asteroid spin vector determinations are collected. Supplementary information on shape models and albedo variegation is also included, but only when part of a spin vector determination.

This part of the database consists of :

- A documentation file describing the database in detail;
- A file containing more than 100 references to the original papers containing the spin vector determinations;
- The main tabulation of results contains information on the types of data used, all spin vector directions (including symmetric solutions, basis for rejection of solutions, and constraints on the spin axis), sidereal rotation periods, shape modeling, and information on albedo variegation.

An earlier version of the collection was published by Magnusson (1989). The current version has 50% more determinations. In addition, the format has been improved to more clearly separate solutions that differ in sense of rotation.

For an introduction to the various techniques commonly used for asteroid spin vector determinations see Magnusson *et al.* (1989).

4.1. SYNTHESIS OF INDEPENDENT RESULTS

For some asteroids a large number of independent solutions have been published. This is of interest to researchers studying the spin vector determination problem. Other readers may want a quick answer to the question : Is the spin vector for asteroid NN known and reliable enough for use in applications? For their benefit

Ia *synthesis* of our knowledge of the spin vector is included for some asteroids. We estimate that these synthesis results have a high reliability and an accuracy in the spin vector direction of 10° or less. They were obtained by taking averages of the most recent independent results, with weights based on the method used and the amount and type of the input data. This procedure is necessarily somewhat subjective, and can't replace a careful evaluation of the original results.

4. 2. SIZE AND FORMATS

The main tabulation is available as TEX files, a PostScript file, and in ascii (fo-otnotes are missing in the ascii version). The supplementary reference list and documentation is available in ascii only.

The TEX files and the ascii version each take up less than 100 kBytes. The PostScript file currently requires about 150 kBytes (14 pages).

4. 3. UPDATING

This part is updated continuously. Comments, suggestions for improvements and references to new asteroid spin vector determinations will be appreciated (contact the first author).

5. Access to the Data Base

Access to the database is available by anonymous ftp. If you are connected to the Internet you type :

`ftp ftp.astro.uu.se`

When prompted for the username, give `anonymous`. Identify yourself when prompted for a password. Proceed to the `pub/Asteroids` directory. The three parts described above reside in the sub-directories `Photometry`, `RotationPeriods`, `SpinVectors`. Each of these directories contains a file `README` giving further instructions.

Readers without access to ftp may request by email from the first author the contents in `RotationPeriods` and `SpinVectors`. Due to its size, we can only send small sub-sets of the photometric database. In special cases, the entire database can be supplied on diskettes. The printed volumes of the Asteroid Photometric Catalogue showing all lightcurves in graphical form (Lagerkvist *et al.* (1987, 1989, 1992) can be ordered from C.-I. Lagerkvist.

Acknowledgements

We are indebted to many observers who kindly submitted their observations in machine-readable form. Their efforts have made our job much easier and improved the quality of the database. Here we pay special tribute to contributors of digital data to the third update : R.P. Binzel, M.W. Buie, E. Dotto, T.F. Flower, P. Gammelgaard, R. Gil Hutton, M. Gonano-Beurer, A.W. Harris, M. Hoffmann, L.K. Kristensen, D.F. Lupishko, S. Mottola, W. Neely, G. Tancredi, D.J. Tholen, F.P. Velichko, C.J. Wetter, and W.Z. Wisniewski.

This project was supported by the Swedish Natural Science Research Council (NFR), and the Swedish Delegation for Space Research (DFR).

References

Everhart, E., 1985, In ***Dynamics of Comets : Their Origin and Evolution, Proceedings of the IAU COLL 83*** (A. Carusi, G.B. Valsecchi, Eds), 185, Reidel, Dordrecht.

Lagerkvist, C.-I., Barucci, M.A., Capria, M.T., Fulchignoni, M., Guerriero, M., Perozzi, E., Zappalà, V. : 1987, ***Asteroid Photometric Catalogue*** Consiglio Nazionale Delle Ricerche, Rome.

Lagerkvist, C.-I., Barucci, M.A., Capria, M.T., Fulchignoni, M., Magnusson, P., Zappalà, V. : 1989, ***Asteroid Photometric Catalogue, First Update*** Consiglio Nazionale Delle Ricerche, Rome.

Lagerkvist, C.-I., Barucci, M.A., Capria, M.T., Dahlgren, M., Erikson, A., Fulchignoni, M., Magnusson, P. : 1992, ***Asteroid Photometric Catalogue, Second Update*** Uppsala Astronomical Observatory.

Magnusson, P. : 1989, "Pole determinations of asteroids". In ***Asteroids II*** (R.P. Binzel, T.Gehrels, M.S. Matthews, Eds.), 1180–1190, Univ. of Arizona Press.

Magnusson, P., Barucci, M.A., Drummond, J.D., Lumme, K., Ostro, S.J., Surdej, J., Taylor, R.C., Zappalà, V. : 1989, "Determination of pole orientations and shapes of asteroids". In ***Asteroids II*** (R.P. Binzel, T.Gehrels, M.S. Matthews, Eds.), 66–97, Univ. of Arizona Press.

A PUBLIC–DOMAIN ASTEROID ORBIT DATABASE

EDWARD BOWELL
Lowell Observatory
1400 W. Mars Hill Road, Flagstaff AZ 86001, USA
E–mail elgb@lowell .edu

KARRI MUINONEN
Observatory
P.O. Box 14, FIN–00014, University of Helsinki, Finland
E–mail karri@ceres.helsinki.fi

LAWRENCE H. WASSERMAN
Lowell Observatory
1400 W. Mars Hill Road, Flagstaff AZ 86001, USA
E–mail lhw@lowell.edu

Abstract. We are constructing a database of J2000.0 osculating elements for all asteroids that have reasonably calculable orbits, and to make it publicly accessible via anonymous ftp. The database will be a very dynamic one; it will change on a quasi–daily basis as new observations and orbits become available. At present, an incomplete version of the database is online. In this paper, we present and discuss sample records of asteroid orbital data as they might appear in more mature form.

1. PURPOSE OF THE DATABASE

Conventionally, asteroid orbits resulting from four or more observations are calculated by means of least–squares differential correction of a preliminary orbit. Planetary perturbations are taken into account when : (1) observations from more than one apparition are available; (2) observations from a single apparition span a long arc; (3) a single–apparition asteroid has been observed near Earth. These are some of the precepts adopted by the Minor Planet Center in preparing the *Minor Planet Circulars* (*MPC*s), the prime source of information on asteroid orbits. We largely agree with that approach.

Our recent work on asteroid orbital errors (Muinonen and Bowell 1993a, 1993b; Muinonen *et al.* 1994) has made it clear to us that the orbits in our public–domain database could usefully contain information, in addition to that published in the *MPC*s, such as error parameters and more complete documentation of perturbation schemes. Accordingly, we plan to construct an asteroid orbital element database that will contain much more information than is customary.

2. CONTENT OF THE DATABASE

Our principal file of osculating elements will be called `astorb.dat`. In Table I we present partial sample records for two asteroids, one the very well–observed

A. Milani et al. (eds.), Asteroids, Comets, Meteors 1993, 477–481.

TABLE I
Sample Orbital Database Records for Two Asteroids

(263) Dresda	1991 RT_2	number–name/designation
331.37020559	34.82453961	M, deg (1)
161.45751209	82.40590120	ω, deg (1)
216.82169458	229.55741997	Ω, deg (1)
1.30831575	4.30946682	i, deg (1)
0.0793944410	0.1958489932	e (1)
2.8860918975	2.2386595195	a, AU (1)
19940217.0	19910921.0	epoch of osculation, TDT (yyyymmdd.d) (1)
87.98 91	15 30	arc, years or days; number of observations (2)
19050731 19930724	19910909 19911009	dates of first, last observations (2)
1.03 1.01	0.68 0.57	rms (O–C) residuals in α, δ, arcsec (2)
E. Bowell	E. Bowell	Orbit computer/ *MPC* or *EMP* reference (3)
10.40 0.15	16.0 0.15	H, mag, G (4)
		covariance matrix, rad (5)
0 0 0 0 0 0 0 0 0 0	0 0 0 0 0 0 0 0 0 0	integer–coded information (6)
M–N,1,2,4		perturbation scheme (7)
DE200		planetary ephemeris and masses(8)
2.0D9 9.2D9 7.3D9		asteroid inverse solar masses (8)
19880704 19931124	19910122 19930207	dates of record entry and last amendment (9)

(multi–apparition) main–belt asteroid (263) Dresda, the other the single–apparition asteroid 1991 RT_2. We comment on the items in each asteroid's record :

(1) J2000.0 elements are given to higher precision than is customary, even though their accuracy may be much less. This allows accurate representation of the observations and computation of residuals. Normally, the epoch of osculation for integrated orbits is the nearest 100–day Julian Date (e.g., 1993 November 11, 1994 February 17) to the current date; for 2–body orbits it is the 20–day Julian Date nearest the midtime of the orbital arc. Customized files in which, for example, the epoch of osculation is not the aforementioned, will be available on request.

(2) For a multi–apparition orbit (meaning that the asteroid passed through solar conjunction in longitude at least once between the first and last observations), the orbital arc is given in years. The number of observations is that used in the orbit computation.

(3) Orbit computers' names are given, except where reference is made to the *Minor Planet Circular* or *Ephemerides of Minor Planets* in which the orbit was published.

(4) Absolute magnitudes H and slope parameters G for numbered asteroids are

those adopted by IAU Commission 20. G defaults to 0.15 unless photometric observations suggest otherwise.

(5) Orbital element covariance matrices (not shown) contain twenty–one nonredundant elements of a 36–element matrix in M, ω, Ω, $\imath$, e, and a. They are only calculable for sufficiently well–observed asteroids. In general, at least five good-quality observations are required to yield a meaningful covariance matrix. For poor–quality (single–apparition) orbits, we will include other error parameters (to be decided) based on the results given in Muinonen *et al.* (1994). With the publication of new observations, it will usually be possible to update the covariance matrix without recomputing an orbit.

(6) We invite suggestions for the content of these important integer identifiers (there could be more than the nine integers indicated). Uses we have in mind at present (and in no particular order) are :

(a) Orbit based on uncertain linkage(s).

(b) Orbit includes radar observations (number given).

(c) Orbit includes the effects of general relativity.

(d) The asteroid is a planet–crosser, according to Shoemaker *et al.*'s (1979) definition. The parameter could also indicate crossing with respect to other planets' orbits and/or the presence of one or more protective resonances.

(e) Family membership.

(f) Observed in one or more of the major surveys (P–L, T–1, T–2, T–3, UCAS).

(g) Orbit status flag, indicating, for example, an orbit in need of updating or one that does not result from all the available accurate observations.

(h) Orbit contains fixed elements. In the case of poor or indeterminate orbits, it is customary to fix one or more elements (typically e, a, q, or $\imath$) at "reasonable" values.

(i) Observations needed. This is our version of the Minor Planet Center's "critical list". It will be based on criteria deriving from the error analysis.

(7) The rather unwieldy string for (263) Dresda is meant to indicate that all the planets except Pluto are included in the calculation of perturbations, that the Earth and Moon are treated as separate bodies, and that the effects of asteroids (1) Ceres, (2) Pallas, and (4) Vesta—but no other asteroid—are included. Since there are a limited number of perturbation schemes used in practice, all this information (and that of other schemes) could be condensed to a single parameter. (It is to be hoped that orbit computers will move toward adopting a standardized method of orbit computation.) However, we do plan to allow for the gravitational effects of additional selected asteroids, whenever close asteroid–asteroid encounters are identified. No perturbation scheme is given for the orbit of 1991 RT_2 because that orbit is a 2–body one. In general, it is our view that the selection of orbits for which perturbed elements can usefully be calculated should be based on a threshold value (to be decided) of one or more error metrics (see below and Muinonen and Bowell 1993a). A rule of thumb, applicable to main–belt asteroids, is that some single–lunation orbits, most two–lunation orbits, and almost all longer–arc orbits should be integrated. Such orbits often lead to the identification of images on search or archive plates, whereas 2–body

TABLE II
Orbital Error Parameters for 1991 RT_2

Least–Squares Orbit (J2000.0)		
Epoch	=	1991 9 21.0 TDT = 2448520.50 JDT
Perihelion passage T	=	1991 5 25.65133072 ±0.18370150 d
Mean anomaly M	=	34.82453961 ±0.03582387 deg
Argument of perihelion ω	=	82.40590120 ±0.05129419 deg
Longitude of ascending node Ω	=	229.55741997 ±0.03034063 deg
Inclination i	=	4.30946682 ±0.00464656 deg
Eccentricity e	=	0.1958489932 ±0.0002624519
Semimajor axis a	=	2.2386595195 ±0.0007856229 AU
Mean motion n	=	0.2942537489 ±0.0001548957 deg/d
Period P	=	3.3495793586 ±0.0017632244 y
Energy	=	−0.2169652124 ±0.0000761406 (deg AU/d)^2
Angular momentum	=	1.4461224662 ±0.0002712855 deg AU^2/d
Semiminor axis b	=	2.1953057876 ±0.0007887974 AU
Perihelion distance q	=	1.8002203066 ±0.0008975046 AU
Aphelion distance Q	=	2.6770987325 ±0.0010662278AU
Heliocentric Equatorial Rectangular Coordinates (J2000.0)		
Position X	=	1.9098854236 ±0.0009740748 AU
Position Y	=	0.0395827014 ±0.0000831233 AU
Position Z	=	0.1318170525 ±0.0001578007 AU
Velocity $\dot{X}$	=	2.0712419890 ±0.0057775864 km/s
Velocity $\dot{Y}$	=	21.4207118167 ±0.0062698051 km/s
Velocity $\dot{Z}$	=	8.1956900467 ±0.0044504393 km/s

orbits may not. For near–Earth or other planet–approaching asteroids, we prefer to calculate perturbed orbits for all but the shortest arcs.

(8) We are currently using the DE200 planetary ephemeris supplied by the Jet Propulsion Laboratory. Planetary masses are defined as part of this and other such ephemerides. However, the mass of Ceres that we have adopted is slightly different from that used in computing the DE200 ephemeris.

(9) Two dates indicate when a record was first entered into the database and when the record was last amended, respectively. The latter will point the user to an auxiliary file (`astorb.log`) in which changes to the main database, including the identity of deleted records, may be traced. Thus, for example, it will be a simple matter to identify all changes during the last month.

It is possible to recover uncertainties in the orbital elements using the methods given in detail in Muinonen and Bowell (1993a). In Table II, we list some of these uncertainties. The overall quality of the orbit can be assessed using various metrics (not listed). One particular application is assessing the suitability of an orbit for proper–element calculation using the so–called incomplete *sleak* metric (Muinonen *et al.* 1994) The covariance matrix itself may be used to compute the 1–σ sky–plane uncertainty as a function of time. Also, by performing statistical tests on the distributions of the sky–plane (O–C) residuals in right ascension and declination, it is usually possible to decide which observations should be omitted to provide an improved orbit. We hope to make software available to users of the orbit database so they can perform the aforementioned calculations and tests themselves.

At present, we are only able to make publicly available a subset of the asteroid orbital data (minus error analyses), although we have already constructed a pilot version of the complete database, containing 22,735 orbits (records) of which 11,147 are integrated and the remainder 2–body. A number of tasks remain to be completed before we start on the full database. As of January 1994, we have developed software for orbit computation and differential correction (including general relativity), and subroutines for Bayesian error analysis. We will shortly combine the aforementioned software into a single mainline, and then develop a method of automatic orbit determination suitable for the analysis of the majority of the orbits to be calculated. Adding the treatment of radar observations and assessing the effects of close asteroid–asteroid encounters (other than those involving Ceres, Pallas, and Vesta) will follow. We anticipate constructing most of the database during 1994. Interested readers are invited to use `anonymous ftp` at the `internet` address `lowell.edu` (absolute address `192.103.11.2`). The password is your usual login name; then type `cd pub` and use `ls -lt astorb.*` to look for files `astorb.dat`, `astorb.log`, and `astorb.txt`, the last of which contains information on the current content of the database.

We invite potential users of the osculating element database to discuss its proposed content and to make suggestions for improvement.

Acknowledgements

Research funding was provided by NASA grants NAGW-1470 and NAGW–3397 (EB, KM, and LHW), and by NASA grant NAGW-1912 and the Lowell Observatory endowment (EB and LHW). We thank Hannu Karttunen for guiding us through the almost impenetrable thickets of TEX, and Andrea Milani for helpful comments.

References

Muinonen, K., and Bowell, E. : 1993a, "Asteroid orbit determination using Bayesian probabilities." *Icarus*, **104**, 255–279.

Muinonen, K., and Bowell, E. : 1993b, "Collision probability for Earth-crossing asteroids on stochastic orbits." *Bull. Amer. Astron. Soc.*, **25**, 1116.

Muinonen, K., Bowell, E., and Wasserman, L.H. : 1994, "Orbital uncertainties of single–apparition asteroids." *Planet. Space Sci.*, in press.

Shoemaker, E.M., Williams, J.G., Helin, E.F., and Wolfe, R.F. : 1979, "Earth–crossing asteroids : Orbital classes, collision rates with Earth, and origin." In *Asteroids*, (T. Gehrels, Ed.), 253–282, Univ. of Arizona Press.

AN OVERVIEW OF DATASETS ON SMALL BODIES AVAILABLE THROUGH THE PLANETARY DATA SYSTEM AND SOARD

DONALD R. DAVIS
Planetary Science Institute,
620 N. 6th Avenue, Tucson, Arizona 85705
E-mail drd@psi.edu

MICHAEL A'HEARN
University of Maryland, College Park, Maryland 20742
E-mail ma@astro.umd.edu

EDWIN GRAYZECK
University of Maryland, College Park, Maryland 20742
E-mail grayzeck@stars.dnet.nasa.gov

MARK V. SYKES
University of Arizona, Tucson, Arizona 87521
E-mail grayzeck@stars.dnet.nasa.gov

E.M. ALVAREZ DEL CASTILLO
National Optical Astronomy Observatories,
950 North Cherry, Tucson, Arizona 85726
E-mail ealvarez@noao.edu

DAVID THOLEN
Institute for Astronomy,
2680 Woodlawn Drive, Honolulu, Hawaii 96822
E-mail tholen@hale.ifa.hawaii.edu

KEVIN GARLOW
Steward Observatory,
University of Arizona, Tucson, Arizona 85721
E-mail garlow@as.arizona.edu

Abstract. In this chapter we present an overview of NASA's Planetary Data System project with emphasis on the Small Bodies Node (SBN) and describe a separate relational database project, the Steward Observatory Asteroid Relational Database (SOARD). We summarize the datasets that are currently available through SBN and SOARD as well as those planned to be ingested in the future. Procedures for accessing data from the SBN and SOARD via Internet are included.

A. Milani et al. (eds.), Asteroids, Comets, Meteors 1993, 483–488.

1. Planetary Data System Overview

The mission of the Planetary Data System (PDS) is to provide the scientific community electronic access to planetary spacecraft data and groundbased data whose science content has been reviewed and verified, to prepare datasets for long-term curation, and to provide knowledgeable scientific expertise to the user community in order to enhance and maximize the efficiency of data identification and access. Eventually, a user should be able to login to the PDS, identify, examine, extract or order a subset of the available data, request special products, and be provided with scientific and technical support. PDS also distributes datasets and provides user support so far as resources permit.

PDS comprises the following elements (Table 1) : The Central Node, seven Discipline Nodes, and a Support Node, the Data Distribution Laboratory (DDL). One institution serves as the principal site for each node, and one scientist at each Discipline Node serves as that node manager.

Table 1

NODE NAME	LOCATION
Central Node	JPL
Discipline Nodes	
Geosciences	Washington University
Atmospheres	University of Colorado
Plasma Interactions	UCLA
Small Bodies	University of Maryland
Imaging	USGS - Flagstaff
Planetary Rings	NASA Ames
Navigation and Ancillary Information Facility	JPL
Support Node: Data Distribution Lab	JPL

The roles of the various nodes are as follows :

CENTRAL NODE

• Responsible for maintaining a high-level catalog of all datasets in the PDS, and interface software allowing the user to browse through this catalog and be pointed to the appropriate Discipline Node (eventually being seamlessly connected to that Discipline Node).

• The Central Node does not maintain any datasets, but may process orders for datasets made through them by passing those orders along to NSSDC or the Discipline Node. The CN is responsible, in consultation with the Discipline Nodes, for the development of standards for dataset content and format. They are charged with ensuring that those standards are maintained in the data ingestion process across all Discipline Nodes.

• Responsible for maintaining standards and formats of distributed data sets on CDROMs across all Discipline Nodes.

DISCIPLINE NODES (DN)

• Responsible for the identification and ingestion of datasets pertaining to their subject area.
• Responsible for verification of dataset scientific content and quality.
• Responsible for interfacing with science teams of ongoing missions to design and archive data products consistent with PDS standards and then accomplish the ingestion of that data, thereby giving the science community widespread access to the data.
• Responsible for maintaining a copy of the above datasets which are to be made available online as practical with available equipment.
• Responsible for the development of their own online interface software allowing the users to examine in some detail the datasets for which the Discipline Node is responsible.
• Provide scientific expertise and human interaction to the user helping them to navigate and understand the available data. Respond to user requests insofar as resources allow for special processing of datasets suited for their research needs and data subsets automatically, without human intervention on the part of the Node.
• Responsible, in coordination and collaboration with the CN and DDL, for the production and large-scale distribution of mission datasets (usually undertaken by the missions themselves).

DATA DISTRIBUTION LABORATORY (DDL)

• Responsible for incorporating into the Discipline Nodes advanced technology for data archiving coming from work funded independently of PDS.
• Support DN scientists in the production and distribution of datasets using CDROMs.

2. PDS Small Bodies Node Overview

The Small Bodies Node (SBN) of the Planetary Data System (PDS) provides expertise on comets, asteroids, and interplanetary dust. The main node, located at the University of Maryland, College Park, acts as the primary contact point for the community. It processes requests for its holdings, acquires and archives new datasets, coordinates peer reviews of the archives, and provides software support for browsing the data archives and working with the archived dataset. The main node also provides scientific expertise on the cometary datasets. Two other subnodes, the Interplanetary Dust node at the University of Arizona, and the Asteroids node at the University of Hawaii and at the Planetary Science Institute in Tucson, provide scientific expertise in their respective fields to the user community as well as assisting in the collecting and archiving tasks of the SBN. An additional subnode at Konkoly Observatory, Budapest, Hungary, handles distribution of SBN holdings to the European scientific community and assists in collecting and archiving data from selected space missions in Europe. The current status of SBN datasets is summarized in Table 2. A more detailed listing of the current holdings (e.g., file size, etc.) and planned additions to the SBN may be obtained from E. Grayzeck.

Current projects at the SBN include development of a Windows-based interface into the node data holdings which will provide catalog searching and some data browsing capabilities, integration of the NAIF software into the node interface to provide ephemeris calculations for comets and asteroids, and inclusion of the Steward Observatory Asteroid Relational Database (SOARD) in the SBN node interface.

Table 2

	Reviewed, Ingested and Available	Being Revised Following Review	Collected But Not Reviewed
Asteroid Data	Pole positions 24-color Asteroid name file	Eight-color Taxonomy Magnitude and slope parameters (note that these include many ancillary files like the filter transmission curves for the ECAS filters)	Cunningham bibliography Beginnings of a lightcurve database Zappalà's families File of proper elements Ostro's radar results IRAS diameters
Comet Data	IHW Earth-based - 34 datasets		IHW Space Based - Giotto IHW Space based - Vega IHW Space based - Suisea/Sakigaka/Ice
Dust Data	IRAS Filters Scan history Med. Res. ZOHF Low Resolution ZOHF		Super Sky Flux

3. An Interactive Database for Asteroid Data : SOARD

Two of us (MS and EAdC) saw the need to have a tool for answering scientific questions using large datasets. Supported, in part, by the Space Engineering Resource Center (SERC) at the University of Arizona, they developed SOARD, a menu-driven database management system. This program uses relational database software (dBase IV) in a fashion transparent to the user to provide interactive and manipulative tools for accessing a variety of datasets. The initial development phase of SOARD is now complete and future improvements to it are supported jointly by the Asteroids subnode of the SBN and SERC. The datasets currently available in SOARD and the principal contributing scientists for the dataset are listed in Table 3.

Currently, SOARD has more asteroid datasets than does SBN; however, the SOARD datasets have not been peer-reviewed. It is our plan to replace the unreviewed datasets with the peer-reviewed versions for PDS/SBN as they become available. Thus, over time, the datasets within the two projects will become identical. We also plan to make SOARD accessible via the SBN interface, once that interface has been developed.

Table 3

Orbital Elements:	Osculating Orbital Elements[1] (E. Bowell, MPC) Proper Orbital Elements (Milani, Williams)
Classification Schemes:	Taxonomic Classifications (Tholen, Barucci, Tedesco, CMZ) Family Designations (Williams, Zappalà)
Physical Parameter Datasets:	Magnitudes, Albedos, and Diameters (Tholen, IRAS) Photometric Parameters (MPC) Pole Orientations (Magnusson) Radar Albedos and Cross-sections (Ostro)
Observational Data:	Individual IRAS Observations (IRAS.FPND) Polarimetry data (IRAS.FPND) Lightcurve data (IRAS.FPND) UBV Observations (IRAS.FPND) 8-Color Spectroscopy (D. Tholen) 24-Color Spectroscopy (C. Chapman and M. Gaffey) 52-Color Spectroscopy (J. Bell) 3 Micron Spectroscopy (T. Jones)
Miscellaneous Data:	Asteroid Bibliography (C. Cunningham) Preliminary Designations and Discovery Circumstances (MPC)

[1]This file contains orbits for more than 21,000 asteroids with an osculation date of 1993/04/23. This file will be updated on approximately a quarterly basis; see the chapter by E. Bowell for additional information on osculating orbits.

SOARD allows users to : (1) browse through datasets to view parameters as well as the actual data, (2) define and extract subsets of database parameters for subsets of asteroids, (3) output files (tailored for plotting) of multi-band photometry or spectroscopy for individual asteroids, and (4) search for bibliographic references satisfying users' search criteria. After using SOARD to create the desired dataset, the user creates ASCII output to be "ftp'ed" back to the home computer. In addition, users may download the original datasets in ASCII format without using SOARD to extract subsets of data.

We are interested in collecting other datasets relating to comets, meteors and meteorites that could be incorporated into SOARD. If you have datasets to contribute, please contact Dr. Donald R. Davis or Dr. Mark V. Sykes at the appropriate address listed below.

4. Accessing the Datasets and Future Additions

Information for accessing the data sources described above via telnet is given in Table 4. Also listed in this Table are the scientific and system support personnel for each data source.

We anticipate further development in both SBN and SOARD capabilities, both through acquisition of additional datasets and by updates of existing ones. Many of the datasets are continuously updated and we plan to periodically update the various available datasets through both the SBN and SOARD. Information about the current version of a dataset will be supplied on-line.

Table 4

				Address		
Data Source	**Internet Address**	**User Name**	**Password**	**E-mail**	**Mail**	**Telephone**
Small Bodies Node	pdssbn.astro.umd.edu (128.8.251.13)	pdsuser	None required but user is asked to register with name/e-mail address			
M. A'Hearn				ma@astro.umd.edu	Astronomy Department Bldg. 224, Room 206 University of Maryland College Park MD 20742	301/405-6076
E. Grayzeck				grayzeck@stars.dnet.nasa.gov	Astronomy Department University of Maryland College Park MD 20742	301/405-1539
A. Raugh				raugh@astro.umd.edu	Astronomy Department University of Maryland College Park MD 20742	301/405-6855
SOARD	SPIDER.AML.ARIZONA.EDU (128.196.148.2)	guest	visitor			
SOARD Programmer (Kevin Garlow)				SOARD@SPIDER.AML.ARIZONA.EDU or GARLOW@aAS.ARIZONA.EDU		602/621-2054
M. Sykes				msykes@as.arizona.edu	Steward Observatory University of Arizona Tucson AZ 85721	602/621-5351
D. Davis				drd@psi.edu	Planetary Science Institute 620 N. 6th Avenue Tucson AZ 85705	602/622-6300
Konkoly Observatory				pds@astro.rmki.kfki.hu	H-1525 Budapest P. O. Box 67 Hungary	36-1-1755866

MPC AND ICQ DATABASES

B. G. MARSDEN, D. W. E. GREEN AND G. V. WILLIAMS
Harvard-Smithsonian Center for Astrophysics,
60 Garden Street, Cambridge, MA 02138, U.S.A.
E–mail marsden@cfa.harvard.edu

Abstract. Databases of astrometric and orbital data for minor planets and comets and of photometric data for comets are maintained in connection with the publication of the *Minor Planet Circulars* and the *International Comet Quarterly*. Aspects of these publications and the content and availability of these databases are discussed.

1. Astrometric observations of minor planets

Astrometric observations of minor planets received by the Minor Planet Center (MPC) are generally published in the monthly batches of *Minor Planet Circulars*. Most of the observations made during the twentieth century have been photographic, but CCD data are rapidly increasing in number. In addition to a designation for each object, each observation contains the UTC of observation (usually mid-exposure, although sometimes the beginning and/or end of a trail will be used), generally to 0.00001 day; the FK_5/J2000.0 right ascension and declination, generally to 0.01 second of time and 0.1 second of arc, respectively; and a three-figure code that defines the location from which the observation was made. Most of the positions are topocentric, although some have been reduced to the geocenter. Discovery observations of new objects are denoted with an asterisk, some observations include a magnitude (assumed Johnson B unless otherwise stated), and there may also be an observational note (e.g., bad seeing, time uncertain, etc.) or a code number that identifies an observation as made in the course of a particular program at the observatory location.

In the printed *Minor Planet Circulars* the observations are listed by observatory code and sublisted by designation (unnumbered objects first) and date. This format permits the convenient incorporation of more details concerning the observing programs. The observations are also made available in machine-readable form in the "Extended Computer Service" (ECS) operated by the MPC and the Central Bureau for Astronomical Telegrams (CBAT). Subscribers log in to one of the MPC/CBAT computers and give an address to which the data should be e-mailed. Options produce the files of observations of numbered and unnumbered minor planets from the current batch of *Circulars*. It should be noted that the distinction between numbered and unnumbered is that existing *before* the batch of *Circulars* was prepared, and minor planets given new permanent numbers in the batch will still therefore be unnumbered. The observations are filed in a format that is somewhat different from what is used on the pages of the *Circulars*. The ECS eliminates the notes shown in the printed *Circulars* but adds one indicating whether the observation has simply been transformed to J2000.0, or whether it was obtained directly in the FK_5/J2000.0 system using a CCD, micrometer or meridian

A. Milani et al. (eds.), Asteroids, Comets, Meteors 1993, 489–494.

circle (photography being the default). The ECS also allows one to access the files from the previous batch of *Circulars*.

The MPC also merges the monthly files of observations into its full system of J2000.0 observational files. It is at this point that observations of newly-numbered minor planets (each of which may involve several different provisional designations) are transferred from the files of observations of unnumbered to those of numbered minor planets. The files are also amended according to the initial pages of the *Minor Planet Circulars*, as observations are corrected, deleted or redesignated.

Soon after the MPC was established at the Cincinnati Observatory in 1947 it was decided that files should be maintained of all observations of minor planets back to the beginning of 1939, although earlier data were available in some cases, notably the extensive series of photographic data extending back to the 1890s obtained at the Heidelberg Observatory. It was not practical to carry out much checking, and most of the observations, even current data, were simply added to the file as published or as provided by the observers (the only modification of significance being a possible transformation of positions to equinox 1950.0). Frequently the observations were only approximate.

On the MPC's transfer to the Smithsonian Astrophysical Observatory in 1978 it was decided that current data should be subjected to some checking before publication and filing, that approximate positions would not be accepted, and that there should be at least some record of each provisional designation (even if this did involve approximate positions) back to the introduction of the modern designation system in 1925. Effort was also expended on checking the observations in that portion of the file inherited from Cincinnati, and many observations that had been attributed to low-numbered minor planets were (with suitable documentation in the *Minor Planet Circulars*) corrected, redesignated, reidentified and—in some cases—simply deleted. More and more pre-1939 photographic observations have been added to the files, and during the past year or two attention has also been given to the incorporation of older micrometric observations, re-reductions being automatically performed using positions of the reference stars updated from the AGK_1 (for example) to the *PPM* catalogue. Some meridian observations, notably those obtained in recent years with the Carlsberg Automatic Meridian Circle on La Palma and published elsewhere, are incorporated in the MPC files without republication in the printed *Minor Planet Circulars*.

The MPC makes complete copies of its files of astrometric observations, which as of October 1993 number 752,896, available on magnetic tape from time to time, although it has not been practical to update these exportable files more frequently than once every two or three years. Some users are able to make their own updates using the computer files issued monthly in the ECS, but this is not particularly easy to do. The ECS does therefore include an additional feature that allows subscribers to obtain from the latest MPC files all the available observations of objects specifically of interest to them.

2. Astrometric observations of comets

Much of the preceding section applies also to astrometric observations of comets. One difference is that publication in the *Minor Planet Circulars* is by object and date. The observations made at the various locations are combined, with an initial list mentioning each program concisely. If magnitudes are provided, they are specified as either "total" (T = m_1) or "nuclear" (N = m_2), without consideration of the waveband to which the observations refer. Although quite detailed information concerning a comet's physical appearance has sometimes been included in the *Circulars* until very recently, it has been decided that this information is more appropriately published in the *International Comet Quarterly* (*ICQ*), and the notes published in the *Minor Planet Circulars*, like those for minor planets, tend to relate more to the "mechanics" of the observations.

Discoveries of new comets and recoveries of returning periodic comets are first announced in the CBAT's *IAU Circulars*. When comets are discovered visually by amateur astronomers the initial positions are necessarily very rough, and these positions are not carried over to the *Minor Planet Circulars* and MPC files. The *IAU Circulars* also include some accurate positions, generally up to the point that an initial orbit (usually a parabola) can be computed with some reliability, and these positions are repeated in the *Minor Planet Circulars*.

The MPC files of cometary observations are complete only back to 1964, although many earlier observations are also included. The total number of observations is 46,046. The ECS allows the files of cometary observations to be accessed, and the data are included on the occasional magnetic tapes that are issued.

3. Orbital elements of minor planets

Numerous sets of orbital elements of minor planets appear in the monthly batches of *Minor Planet Circulars*. These sets can include redeterminations for already-numbered minor planets, determinations for minor planets that are being newly numbered, identifications and linkages of observations at different oppositions, and results from observations at a single opposition. Except for the last group, the orbits are generally least-squares Newtonian representations and allow for the perturbations by the eight planets Mercury-Neptune and Ceres—although the effects of Pluto, Pallas and Vesta, the Earth and Moon separately, and in some cases relativity, may also be considered. In order that users can judge better on their reliability, the results for newly numbered minor planets and new multiple-opposition identifications are shown in full, with the (O–C) residuals listed for all the available observations. In the single-opposition cases perturbations are generally ignored, unless there is a close approach to the Earth or the length of the observed arc is more than 75 days. In some cases the computations are contrained by holding the orbital eccentricity fixed at some particular value.

Perturbed orbital elements are given in the FK_5/J2000.0 system for standard epochs of osculation such that the Julian date is divisible by 200 with remainder 0.5 day; the angular elements are given to a precision of 0.00001 deg, and there is general compatibility for the other elements. Unperturbed orbital elements take for the epoch an instant when the Julian date is divisible by 20 with remainder 0.5 day, the angular elements being given to a precision of 0.01 deg.

Orbital data for individual minor planets (numbered, multiple-opposition unnumbered and recent single-opposition) can be obtained from the MPC/CBAT standard Computer Service (CS); there are also procedures therein for computing an unperturbed ephemeris and for listing the positions of all the numbered minor planets in a particular area of the sky at some time. The ECS provides orbital data from the current and immediately-previous *Minor Planet Circulars*, much as it provides observations. The current files contain orbital elements for the 5,700 numbered minor planets (one of which, 719 Albert, is lost) and for 16,347 unnumbered minor planets (not all of which will be unique). Still other options show all the provisional designations involved in the new numberings or the new identifications and double designations involving unnumbered objects.

Low-precision orbital data and identification information for all the unnumbered minor planets are issued from time to time (most recently early in 1992) in printed and diskette form as the *Catalogue of Orbits of Unnumbered Minor Planets*. The recent *Catalogue of High-precision Orbits of Unnumbered Minor Planets* (also in printed and diskette form) applies only to the cases where perturbations have been applied and to the standard epoch 1993 Aug. 1.0. This catalogue also provides some information about opposition dates and ephemerides; it is intended to be a companion to the annual *Ephemerides of Minor Planets* and STAMP software issued by the Institute of Theoretical Astronomy, St. Petersburg, for the numbered minor planets and will also be issued annually (with epochs updated to the last "200-day date" in each year) if there is sufficient demand.

The monthly schedule of the *Minor Planet Circulars* is not frequent enough for attending to newly-discovered Earth-approaching objects. These objects have traditionally therefore been covered on the *IAU Circulars*, which are themselves available in printed form and in the CS, delivery by e-mail also being possible. The increasing frequency of discoveries, and the hitherto largely unmet need for updating the data beyond an initial orbit determination, have overtaxed the *IAU Circulars*. Accordingly, a series of *Minor Planet Electronic Circulars* (*MPEC*s) has been inaugurated to provide these data more adequately. The *MPEC*s are available in the CS and by e-mail but are *not* issued in printed form, on the grounds that all the important material in them (astrometric observations, final orbits) will appear in the next batch of *Minor Planet Circulars*.

4. Orbital elements of comets

Much of the preceding section applies also to orbital elements of comets. When perturbations are applied, which again usually requires observed arcs of 75 days or more, the osculation epoch is taken to be the "40-day Julian date" closest to perihelion passage. In unperturbed cases no epoch is specified, because of the convention of specifying the time of perihelion passage; likewise, the semimajor axis (and mean motion) is replaced by the perihelion distance. Predicted orbital elements are given around May of year n for the comets expected at perihelion in the year $n + 3$. Elements (and ephemerides) are also included in the annual *ICQ*

Handbook if there is a reasonable chance that observations can be made in that year, but this *Handbook* is not available in computer form.

Cometary orbits in the CS include those of the long-period comets of the two or three preceding years and the reasonably predictable short-period comets, even when they are far from perihelion. Preliminary orbits are given with the astrometric observations in the *IAU Circulars* and thence in the CS. Cometary information of moderate urgency can also be given in the *Minor Planet Electronic Circulars.*

The complete *Catalogue of Cometary Orbits* has been issued eight times since 1972, and there seems to be interest in now having it updated annually. The 1993 edition, which contains 1,392 orbits for 855 individual comets (174 of them of short period), is available in printed form, by diskette and by e-mail. The diskette and the e-mail editions only contain the general catalogue and the "original" and "future" orbits of long-period comets in well-determined cases. The printed edition also supplies references and much other information, including the nongravitational parameters that should be taken into account when orbits are updated to other epochs.

5. Photometric and other physical data on comets

The *ICQ* was established in 1978 as a clearinghouse and archive for photometric observations of comets, filling a need for such an international center that had not previously existed. The establishment of the *ICQ* archive has meant that many data previously published in the *IAU Circulars* and the *Minor Planet Circulars* are now usually published only in the *ICQ*, with minor exceptions. The *ICQ* is published in January, April, July and October and contains general news and articles pertaining to the study of comets, although the central purpose of the *ICQ* is the publication of tabulated data and descriptive textual information pertaining to the physical (photometric) observation of comets.

All of the tabulated data that are published in each issue of the *ICQ* are sorted into the general *ICQ* archive. A tabulated observation is an 80-character line that contains a coded identification for each comet, the date (given to 0.01 day in UT), a 5-character code for the observer, information concerning the instrument used, whether the observation was visual or non-visual (i.e., photographic, CCD, etc.), an estimate of the comet's total magnitude (along with coded information regarding the method used for making the estimate, the source from which comparison-star magnitudes were obtained for the estimate, and whether or not a correction was made for atmospheric extinction), estimates of the comet's coma diameter and tail length (with the position angle of the tail), and an estimate of the "degree of condensation". Of course, not all of the above information is provided with each observation, though observers are encouraged to report as many of these data as possible. For publication, it is required that observers contributing magnitude estimates provide the method employed, the source for comparison-star magnitudes, and complete instrumentation details; data made prior to 1982 (or where the given comet is poorly observed) are exempt from this requirement. The requirement was established in an effort to standardize the way in which magnitude estimates are made, thereby improving the quality of the archive, and the scatter has been greatly reduced during the past decade as a result of this procedure.

As of October 1993, there were 59,869 observations in the archive, 94-percent of them containing some sort of magnitude information. "Negative" magnitude estimates (in which limiting magnitudes are estimated in situations where the comet was not detected) are present for 1,558 of these observations. Among the data containing magnitude estimates, there are 32,807 observations of 199 long-period comets and 23,656 of 125 short-period comets. The observations of long-period comets currently include comets from 1932 through 1993, while those of short-period comets begin with Halley's Comet in 1909. Halley's Comet has 6,779 observations with magnitude information in the *ICQ* archive, leading all comets in this regard. Current efforts are aimed first at publishing all recent observations, and second at collecting older data from various groups worldwide (many of which have never been previously published). A third effort is being made at extracting older data from the literature, going back prior to 1900, in which such published data enter the *ICQ* archive directly without republication in the *ICQ*.

About a dozen "observation coordinators" have been set up throughout the world to collect observations within certain countries or regions, and these coordinators usually enter data in machine-readable form for relay to the *ICQ* editor.

The data in the *ICQ* archive are available in their entirety on 9-track magnetic tape to interested researchers.

6. Availability of data

Since the databases discussed in this paper have been prepared and maintained in connection with publications for which a subscription charge is necessary, the databases are themselves made available only by subscription. Those interested in subscribing to the data should contact one of the authors.

THE UPPSALA COMET MAGNITUDE DATABASE

LARS KAMÉL

Astronomiska observatoriet, Box 515, S-751 20 Uppsala, Sweden
E–mail lars@laban.uu.se.

Abstract. The second part of the Comet Light Curve Catalogue/Atlas (CLICC/A) has been made available by anonymous ftp. This database contains magnitude estimates of periodic comets, reduced to a geocentric distance of 1 AU and corrected for observational effects. The database is described in some detail.

1. Introduction

CLICC/A is a compilation of magnitude estimates and measurements of periodic comets. The period covered is 1832-1989 for comet P/Encke and 1899-1989 for the other comets. It consists of three parts, two catalogue parts and the atlas. Although only the second part is available by anonymous ftp, all three parts are briefly described in the following sections; more information can be found in the introductions to the catalogue (Kamél, 1991) and the atlas (Kamél, 1992).

2. The three parts of CLICC/A

2. 1. THE COMPILED OBSERVATIONS.

The first part is a catalogue of the compiled observations, including time of observation, observer(s), location, reference(s), telescope and codes describing type of observation, magnitude and telescope. It has been published as a book (Kamél, 1991) and the data are also available on magnetic tape and floppy disks.

2. 2. THE CORRECTED MAGNITUDES.

The second part is a catalogue of corrected magnitudes, that can be used to obtain light curves of the comets. Corrected means that the magnitudes have been reduced to a geocentric distance of 1 AU and corrected for observational effects, e.g. the delta effect and telescope aperture. These data are available only in digital form. There is one data file per apparition, and the names of the files are based on, but not necessarily the same as, the designations of the apparitions (the exact names depend on the computer system for which the files are intended). The files are text files and their names is the only thing that differs between different computer systems.

2. 3. THE ATLAS.

The data in the second catalogue part has been used to obtain the light curves that form the atlas. This is available only on paper and has been published by Kamél (1992).

A. Milani et al. (eds.), Asteroids, Comets, Meteors 1993, 495–496.

3. The availability of the data.

Access to the data in the second catalogue part is available by anonymous ftp. If you are connected to Internet you type :

`ftp ftp.astro.uu.se`

When prompted for the username, give `anonymous`. Identify yourself when prompted for a password. Proceed to the `pub/Comets/CLICC` directory.

Here you find the MS-DOS version (The reason for this choice is that the MS-DOS version can be used on any other computer system, while the opposite is not true) of the second part of CLICC/A. This means that all the data files have suffixes *.red*. A file name is the designation of the apparition, but with roman numbers converted to arabic, e.g. the data for the apparition 1987 XXXIII (P/Borrelly) are found in the file *198733.red*. There is also a file called *README* and one file called *appar.all*, containing designations (also those not included in CLICC/A) of apparitions up to 1989 and a file called *intro*, containing a digital version of the introduction to the catalogue. The data files are available as compressed ZIP- and tar-archives in the files *clicc2.zip* and *clicc2.tar.Z*, respectively. Note that these archives should be transferred as binary files.

Readers without access to ftp may order single data files via E-mail from the author. Because of the size of the data set, only small parts of it will be supplied in this way. Those interested in extensive data sets, or the data in the first catalogue part, may order them on floppy disks or magnetic tape from the author.

References

Kamél, L. : 1991, *The Comet Light Curve Catalogue/Atlas. Part I The compiled observations.* Uppsala Universitet.
Kamél, L. : 1992, *Astron. Astrophys. Suppl. Ser.* **92**, 85-149.

METEOROID ORBITS AVAILABLE FROM THE IAU METEOR DATA CENTER

B.A. LINDBLAD

Lunds Observatorium, Box 43, S–22100 Lund, Sweden
E–mail : linasu@gemini.ldc.lu.se

D.I. STEEL

Anglo-Australian Observatory, Private Bag, Coonabarabran, NSW 2357; and Department of Physics and Mathematical Physics, University of Adelaide, G.P.O. Box 498, Adelaide, SA 5001, Australia
E–mail : DIS@AAOCBN3.AAO.GOV.AU

Abstract. Since it was founded early in the 1980's, the IAU Meteor Data Center (IAU MDC) has accumulated a large number of the meteoroid orbits measured world-wide so as to make these freely available to all interested researchers. The total number of orbits available is about 68,000, of which about 6,000 were determined using optical techniques (photographic or TV), the bulk having been detected using decameter radars. The observation sites range from various locations in the U.S.A., Canada, Australia, and in the former Soviet Union and Czechoslovakia; radar orbits from the Soviet Equatorial Expedition to Somalia are also archived. About 39,000 of the 62,000 radar orbits are derived from the Harvard Radar Meteor Project. Most of these programs were carried out during the 1960's and 1970's, but still represent our best knowledge of the orbital distribution of interplanetary particles in the size range from 100 μm to 1 meter. A new survey currently in progress in New Zealand has so far rendered over 350,000 orbits, and it is anticipated that these will soon become available through the IAU MDC. Presently the 68,000 orbits archived in the IAU MDC are only available on magnetic recording media, but it is planned that they will shortly be made accessible via anonymous ftp.

1. The IAU Meteor Data Center

Starting in the 1930's it became possible to determine the orbits of individual particles observed as meteors using photographic techniques, so that at least the nighttime meteor showers could be investigated and definitively linked to specific comets. The development of suitable (decameter) radars in the 1940's meant that by the 1950's the daytime showers were also amenable to study, and the smaller particles (fainter meteors) detectable with such radars also led to a much enhanced detection rate. During the 1950's the various researchers at the Jodrell Bank station of the University of Manchester (England) developed a set-up which could render meteoroid orbits by determining the atmospheric speed from Fresnel oscillations in the echo amplitude, and using the time of closest approach to each of a series of receiver stations separated by several kilometers to deduce the meteor radiant. With the time of detection known the heliocentric orbit could then be found. The Jodrell Bank team determined over 2,500 orbits in this way, comprising the largest repository of small particle orbits available in the late 1950's for investigations of the

A. Milani et al. (eds.), Asteroids, Comets, Meteors 1993, 497–501.

origin and evolution of such bodies. Some hundreds of orbits of larger meteoroids had also been measured using photographic methods, in particular by the team at Harvard led by F.L.Whipple.

These pioneering Jodrell Bank orbits have unfortunately been lost over the years, and so are no longer available to researchers wanting to investigate their orbital distribution, nor indeed historians of astronomy. With the realization that such epoch-making data are of importance, and should not be lost since they are still of scientific value, the first author of this paper established the IAU Meteor Data Center with a view to ensuring that all subsequent meteoroid orbit data be archived for use by new researchers. Five immediate justifications for such a data center spring to mind : (i) Such meteor research programs are ephemeral, so that unless a central respository exists, as a program terminates and the researchers disperse the data are often mislaid or deliberately discarded; (ii) Although several programs were carried out during the years 1960–75, since that time there has been very little activity in this area, making the old data the *only* available data; (iii) Some millions of dollars and roubles were spent in obtaining the data initially, so that it would not be sensible to let them be lost for the sake of a few thousand units of the same currencies, which is all that was required to assure their security; (iv) New techniques (and new minds) have made alternative types of analysis possible, in particular with the huge improvement in the computing power which has become available; (v) Newly-discovered comets and asteroids present themselves as possible parent objects for previously-observed meteoroids, the original orbits being needed for testing of such genetic relationships.

In fact the decision to collect all meteoroid orbits together was made only just in time. For example, well over 50% of the orbits now archived come from the Harvard Radar Meteor Project observation campaigns of 1961–65 and 1968–69, and the only source for these available was two computer tapes which had been stored in Z. Sekanina's attic in Pasadena, California, for some years. Thankfully those tapes were successfully read by C.D. Murray and S.F. Dermott, and the data saved. Similarly the orbits from the Adelaide radar surveys of 1960–61 and 1968–69, the only such data collected from the southern hemisphere until very recently, were rescued from racks of dusty and decaying cards in an old data store which was cleared out soon thereafter.

Despite some close escapes most of the known orbits have been collected together, with Lindblad basing the IAU MDC at the Lund Observatory in Sweden (Lindblad, 1991a). Its foundation results from a proposal to the IAU at the General Assembly of 1976 from Commission 22, confirmed at the GA in 1982.

2. Meteoroid orbits available

The meteoroid orbits available from the IAU MDC in machine-readable form are listed in Table 1. These have been divided here into two separate groups, according to whether the meteors were observed optically (*i.e.* nighttime meteors only) or with radars (both night- and day-time observations). The survey names are given here only for the purposes of easy identification. A complete description of the observing techniques, sites, references to original publications on each, *etc.*, is available from the IAU MDC (see below).

Table 1 : Meteoroid orbits archived at the IAU Meteor Data Center.

Survey	Operation	Orbits
Optical Surveys		
Harvard P Photo[1]	1936–59	1,245
Harvard G Photo[2]	1952–54	2,529
Prairie Network[3]	1963–75	336
MORP[4]	1971–84	218
Ondřejov[5]	1947–89	285
Soviet Photo[6]	1940–83	1,111
TV meteors[7]	1982–84	531
Japanese small camera[8]	1964–89	325
Radar Surveys		
Adelaide[9]	1960–61	2,092
Harvard[10]	1961–65	19,327
Obninsk[11]	1967–68	9,358
Adelaide[12]	1968–69	1,667
Harvard[13]	1968–69	19,818
Mogadisho[14]	1968–70	5,328
Kharkov[15]	1975	5,317

[1] Harvard Precisely-reduced photographic meteors, observed from Massachusetts and New Mexico.

[2] As above, except for with a less-accurate Graphical reduction. Some of the meteors in this catalogue are in fact also included in the precise orbits.

[3] Prairie Network fireballs, U.S.A.

[4] Meteorite Orbit and Recovery Program fireballs, Canada.

[5] Fireballs observed from Ondřejov Observatory in the Czech Republic (formerly part of Czechoslovakia), one station of the present European Network of fireball observatories. Other fireball orbits are also available directly from Ondřejov; see text.

[6] These precisely-reduced photographic orbits were observed from Dushanbe, Odessa and Kiev, in the former Soviet Union.

[7] Television meteors (generally fainter than observed in the photographic surveys) recorded from Ontario, Canada.

[8] Japanese 35 mm camera orbits (Koseki *et al.*, 1990; Lindblad, 1991b).

[9] Radar meteors observed near Adelaide, Australia.

[10] Harvard Radar Meteor Project (RMP); observatory site near Havana, Illinois. Very faint radar meteors.

[11] Obninsk, south of Moscow, Russia.

[12] Second Adelaide survey, fainter limiting magnitude.

[13] Harvard RMP Synoptic Year study.

[14] Observatory built near Mogadisho, Somalia, by Soviet Equatorial Expedition; all catalogued orbits have southern radiants.

[15] Kharkov radar survey of faint meteors. Another 150,000 orbits have yet to be supplied to the IAU MDC.

The optical meteors cover a wide range of initial particle sizes, from fireballs having masses of a kilogram or more down to faint TV-detected meteors having masses of order 0.1 mg (*i.e.* sizes from a decimeter down to a millimeter). A description of all orbits determined photographically has been given by Lindblad (1991a).

Turning to the radar orbits, the masses of the larger particles detected (corresponding to limiting radar magnitudes of $+6 \div 8$, as for the Adelaide, Obninsk and Mogadisho surveys) are similar to those of the TV meteors. However, the Kharkov and Harvard surveys detected meteors as faint as magnitude $+12 \div 13$ (corresponding to meteoroids of masses down to $\sim 1\ \mu$g, or sizes of $\sim 100\ \mu$m).

It has generally been found in orbit programs that distinct streams/showers provide 30% or more of the total number of orbits, with a reducing fraction in streams as fainter meteors (smaller meteoroids) are accessed in surveys using more powerful radars. The personal experience of the second author in manipulating these data has confirmed this to be the case; he finds that streams/showers are more preponderant amongst the Adelaide and Obninsk orbits, compared to the other radar surveys. For typical orbital distribution plots, see Steel (1991).

3. Other meteoroid orbits

Apart from the surveys listed in Table 1, there are some other photographic orbits available from the IAU MDC, although not all have yet been added to the checked and formatted repository. These include 25 orbits from the program at New Mexico State University in the 1970's. The 325 orbits determined by Japanese amateur groups using 35 mm cameras (Koseki *et al.*, 1990; Lindblad, 1991a,b) are a recent addition. In fact there are a number of amateur groups elsewhere who are now producing excellent sets of orbits, usually for specific streams such as the Perseids and the Geminids; for references see Lindblad (1991a).

There are also some hundreds of fireball orbits measured from the Ondřejov Observatory (251 65 Ondřejov, Czech Republic) which are available directly from that institution; the appropriate contact person is P.Spurný.

After development stretching through the 1980's, a new powerful meteoroid orbit radar began operations in New Zealand in 1990. This has so far rendered over 350,000 individual orbits to radar magnitude $+12 \div 13$ (Baggaley *et al.*, 1993, 1994). Although observing campaigns to investigate background/sporadic counts have been run, much of the data has been collected during the epochs of specific showers; in particular the η Aquarids and the δ Aquarids. The present antenna pattern results in most detected meteors having radiants between declinations of $+5^\circ$ and -40°, but this will soon be reversed so as to give coverage to the south equatorial pole, since deep southern radiants are very poorly represented in the data presently available. It is anticipated that these orbits will be made available through the IAU MDC in the future.

4. How to obtain the data

The meteor orbits listed in Table 1 are available directly from B.A.Lindblad at the address given at the head of this paper, in the form of a nine-track 1600 bpi tape. Email contact may be made through linasu@gemini.ldc.lu.se. Geophysical data are also available for most of the photographic orbits (*e.g.* start and end heights of the meteor trains, magnitudes, *etc.*).

If some other data format (*e.g.* exabyte, 3.5- or 5.25-inch disks) is required, the interested party should approach D.I.Steel; email contact is preferable, through either dis@aaocbn3.aao.gov.au or dsteel@physics.adelaide.edu.au.

It is planned that these data will be available through anonymous ftp at some stage during 1994. Even if this is the case, it will likely be highly desirable to any researcher to have access to the paper documentation on the data sets, and this documentation is obtainable from B.A.Lindblad.

Acknowledgements

Various scientists have kindly made these data available to the general astronomical community, either through allowing their release or else searching out the data from where they were stored and supplying them to the IAU MDC in a machine-readable format. Financial support for the center has come from the International Astronomical Union, the Swedish Natural Science Research Council, and the Kungliga Fysiografiska Sällskapet. D.I.S. checked and formatted the radar orbits at the IAU MDC in 1987 when he was supported by the European Space Agency; his subsequent work on the data has been supported by the Australian Research Council.

References

Baggaley, W.J., R.G.T.Bennett, D.I.Steel & A.D.Taylor (1994), "The Advanced Meteor Orbit Radar Facility : AMOR." ***Ql. J. Roy. Astron. Soc.***, **35**, in press.

Baggaley, W.J., Taylor, A.D. and Steel, D.I. : 1992, "The Southern Hemisphere Meteor Orbit Radar Facility : AMOR." In ***Meteoroids and their parent bodies*** (J. Štohl and I.P. Williams, Eds.), 245–248, Astron. Inst., Slovak Acad. Sci., Bratislava, Slovakia.

Koseki, M., Sekiguchi, T. and Ohtsuka, K. : 1990, "Photographic meteor observations in Japan.". In ***Asteroids, Comets, Meteors III*** (C.-I. Lagerkvist, H. Rickman, B.A. Lindblad and M. Lindgren, Eds.), 547–550, University of Uppsala, Sweden.

Lindblad, B.A. : 1991a, "The IAU Meteor Data Center in Lund." In ***IAU Colloq. 126 : Origin and Evolution of Interplanetary Dust*** (A.C. Levasseur-Regourd and H. Hasegawa, Eds.), 311–314, Kluwer, Dordrecht, Holland; reprinted in ***IAU Information Bulletin***, **70**, 10–13 (1993).

Lindblad, B.A. : 1991b, "A study of meteor orbits obtained in Japan." In ***IAU Colloq. 126 : Origin and Evolution of Interplanetary Dust*** (A.C. Levasseur-Regourd and H. Hasegawa, Eds.), 299–302, Kluwer, Dordrecht, Holland.

Steel, D. : 1991, "The orbital distribution and origin of meteoroids." In ***IAU Colloq. 126 : Origin and Evolution of Interplanetary Dust*** (A.C. Levasseur-Regourd and H. Hasegawa, Eds.), 291–298, Kluwer, Dordrecht, Holland.

INDEX OF COLLABORATING AUTHORS

GPSR Compliance

The European Union's (EU) General Product Safety Regulation (GPSR) is a set of rules that requires consumer products to be safe and our obligations to ensure this.

If you have any concerns about our products, you can contact us on ProductSafety@springernature.com

In case Publisher is established outside the EU, the EU authorized representative is:

Springer Nature Customer Service Center GmbH
Europaplatz 3
69115 Heidelberg, Germany

Batch number: 10620878

Printed by Printforce, the Netherlands